Map-Scale Structures

Contact between rock units; exactly located (solid), approximately located (dashed), concealed or indefinite (dotted)

High-angle fault: U–upthrown side, D–downthrown side; exactly located (solid), approximately located (dashed), concealed or indefinite (dotted)

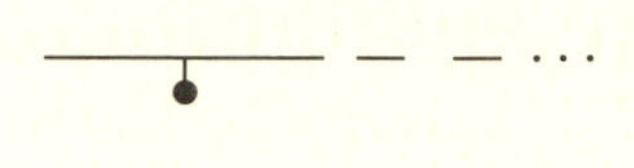

Normal fault: ball and stick on on downthrown side; exactly located (solid), approximately located (dashed), concealed or indefinite (dotted)

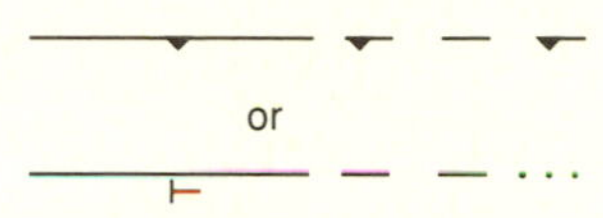

Thrust fault, teeth or T on hanging wall; exactly located (solid), approximately located (dashed), concealed or indefinite (dotted)

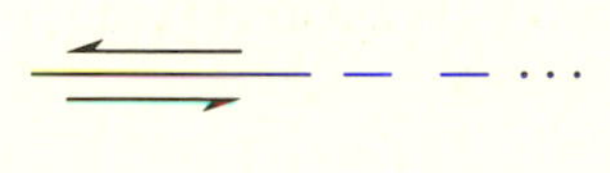

Sinistral (left-lateral) strike-slip fault; exactly located (solid), approximately located (dashed), concealed or indefinite (dotted)

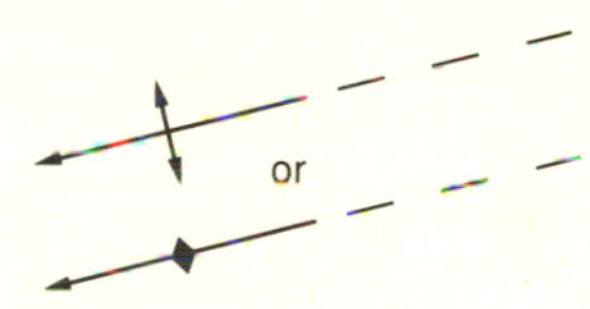

SW-trending anticline (or antiform); arrow on SW end indicates plunge; exactly located (solid), approximately located (dashed)

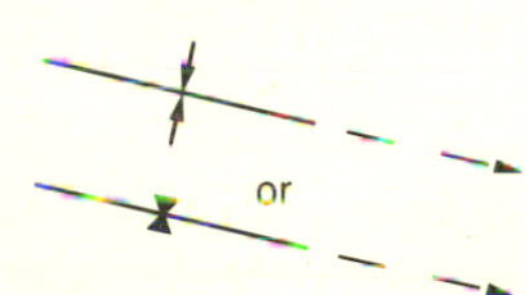

SE-trending syncline (or synform); arrow on SE end indicates plunge; exactly located (solid), approximately located (dashed)

NE-trending overturned anticline (or antiform); arrow on NE end indicates direction of plunge

NE-trending overturned anticline (or antiform); triangle on axis indicates dip direction of axial surface; arrow indicates plunge direction of axis

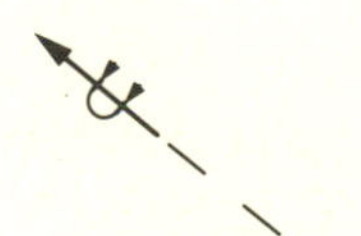

NW-trending overturned syncline (or synform); arrow on NW end indicates plunge

Structural Geology

Principles, Concepts, and Problems

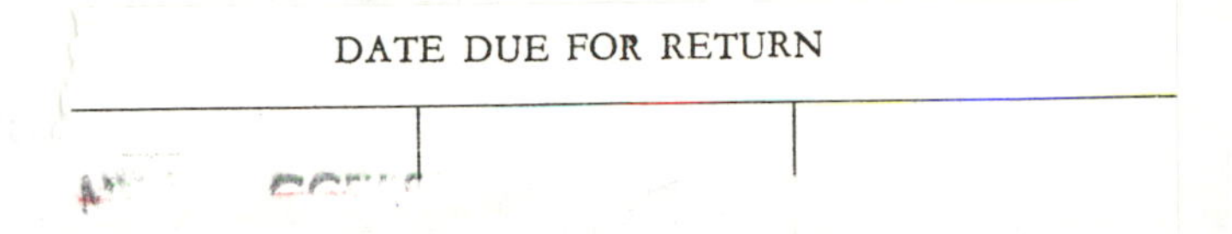

Robert D. Hatcher, Jr.

University of Tennessee–Knoxville

and

Environmental Sciences Division

Oak Ridge National Laboratory

Merrill Publishing Company

A Bell & Howell Information Company

Columbus Toronto London Melbourne

Cover Photo: NASA shuttle photo, taken by astronaut with hand-held camera. Location: 27.5°N 67°E, Kirthar Range, Pakistan. Slide provided by Professor William R. Muehlberger, University of Texas–Austin.

Published by Merrill Publishing Company
A Bell & Howell Information Company
Columbus, Ohio 43216

This book was set in Century Schoolbook

Executive Editor: Steve Helba
Production Coordinator: Linda Bayma
Art Coordinator: Lorraine Woost
Cover Designer: Brian Deep
Photo Editor: Terry Tietz

Library of Congress Catalog Card Number: 89–62573
International Standard Book Number: 0–675–20626–X
Printed in the United States of America
1 2 3 4 5 6 7 8 9—93 92 91 90

This book is dedicated to
My parents
My wife and daughters
and
My teachers, particularly those who taught me to think freely, clearly and critically, and to be a careful observer in the field: Robert H. Barnes, Douglas W. Rankin, Richard G. Stearns, and George D. Swingle.

PREFACE

STRUCTURAL GEOLOGY HAS PROBABLY CHANGED MORE IN THE last 15 years than in the previous 50 because we know more about the crust and the processes that deform it and because of the application of techniques developed both in structural geology and in the allied disciplines of geophysics and materials science. Also, more of the Earth's crust—and other planets—has been studied today by structural geologists than ever before. We have recently benefited, through the work of John Ramsay and a number of other geologists, from development and application of the techniques of fabric analysis that were invented in the early part of this century. We now routinely measure strain in deformed rocks, something done by a relative few a decade or so ago. We also today measure structures, such as kinematic indicators, that were not measured in the 1960s to provide greater insight into the deformation and motion of rock masses. Moreover, most structural geologists today are familiar with seismic reflection imaging of the crust because of the resolution of structure by this technique, and many also routinely use paleomagnetism and other geophysical methods. Modern computer technology has made possible rapid development and widespread use of seismic techniques in the petroleum industry and more recent application in academic research. Few junior-senior level texts that were published in the 1970s and early 1980s considered all of these aspects of structural geology.

Structural geology has also become more quantitative during the 1970s and 1980s. A goal of this book is to bridge the gap between older, less quantitative, texts and newer, but more advanced, mathematical structural geology books. An attempt is made herein to present concepts and discuss processes—many of which have a rigorous mathematical basis—with minimal use of advanced mathematics: mathematics used is mostly algebra and trigonometry. Use of higher mathematics is restricted to two or three places—where it is needed, and where understanding may be enhanced by incorporating it.

TO THE USER OF THE BOOK

Structural Geology: Principles, Concepts, and Problems is intended to provide a one-semester junior-senior level course in modern structural geology and to present balanced coverage of the entire subject. Each chapter is designed to contain more material than can be covered during the allotted class time and may thus serve partly as a resource for material presented from a different point of view than the instructor's.

The twenty chapters are arranged in five parts. The introductory part is intended to provide background and review largely nonstructural, but closely related, topics. The "Rock Mechanics" part investigates the concepts of stress, strain, deformation mechanisms, and methods of strain measurement that form the basis for study of all the structures discussed in the remainder of the book. Several reviewers suggested that I introduce strain before stress, for good reason: structural geology is largely a study of strain—the structures we see clearly manifest strain features, with only rare vestiges of the stresses that produced them. But, because more structures and rock-mechanics experiments provide evidence suggesting stress produces strain, and therefore indicate a logical progression, I chose to introduce stress first. The chapters dealing with strain, and related topics (Chapters 4 through 7), may be taught independently of Chapter 3 on stress.

The part "Fractures and Faults" follows "Rock Mechanics" and stands largely independent of the later parts "Folds and Folding," and "Fabrics, Structural Analysis, and Geophysics." Consequently, any one can be taught independent of the others in the order preferred by the instructor. My own preference is to consider faults first, hence the placement of this part.

The last chapter—"Geophysical Methods"—is intended to introduce geophysical techniques useful in structural geology, not provide a course in geophysics. The techniques introduced here, principally seismic reflection and refraction, Earth magnetism, and gravity, have become important tools for understanding crustal structure because of the large amount of data available today. Several examples of structures imaged by seismic reflection and other geophysical techniques are scattered throughout the book, so I believe incorporation of this chapter is essential in an elementary structural geology text. This chapter also helps form a transition to study of large structures and tectonics.

Organization of individual chapters is intended to aid the learning process. Most terms and concepts are accompanied by illustrations. Problems are worked and examples are presented in most chapters to bring students closer to the subject. Words or phrases in ***boldface italic*** should, in my opinion, be understood and learned; those in *light italic* are terms that some instructors may not consider important enough to learn; other instructors may want to emphasize all italicized terms, bold and light. Most such terms are in the Glossary. The Glossary also contains a few terms that are not italicized in the text. Questions at the ends of chapters have a dual purpose: *review,* with answers readily accessible in the chapter, and *challenge,* to require additional thought about concepts presented in the chapter to reason about structural problems. Essays are intended to provide interesting nutshell-size sidelights to kindle further thought, provide different viewpoints, illustrate applications, or bring in controversial ideas. The interpretations in the Essays frequently represent my own, whereas attempts were made to balance all points of view in the body of the text.

Appendices One and Two are intended to provide background information on fabric diagrams and structural measurements. Both will probably be covered in lab, but, because fabric diagrams and structural measurements are used throughout the text, some discussion is essential here. Appendix Three consists of a data set collected from an outstanding exposure of complexly deformed rocks at Woodall Shoals in the southern Blue Ridge. These data were plotted in a series of fabric diagrams and maps in the Essay for Chapter 19, and the structure there discussed further in the Chapter 16 Essay. Additional problems can be designed around this data set.

As a teacher, I want students to enjoy their structural geology course. The learning process need not be difficult or painful, but should be challenging, and can be approached as a game; the rules of the game should be spelled out by the instructor at the beginning of the course. The degree to which an instructor becomes involved in the course from the beginning largely determines the quality of the course and how much the students derive from it. The text chosen helps determine the level at which the course is taught and the kinds of material to be covered. The balanced coverage in this text is intended to help with the involvement of both students and instructors in structural geology. I hope this book will kindle the interests of some of the students who use it so they will choose to become structural geologists.

ACKNOWLEDGMENTS

This text has been extensively reviewed at several stages of development at either the request of the publisher, or colleague reviews solicited by myself. Reviewers of the text at different stages include John Anderson (Kent State University), Tom Anderson (University of Pittsburgh), Ed Beutner (Franklin and Marshall College), Enid Bitner (Auburn University), Wally Bothner (University of New Hampshire), Donald Davidson (Northern Illinois

University), Ian Duncan (Southern Methodist University), Jeremy Dunning (Indiana University), Terry Engelder (Pennsylvania State University), Eric Erslev (Colorado State University), Peter Geiser (University of Connecticut), Gary Girty (San Diego State University), Art Goldstein (Colgate University), Brann Johnson (Texas A&M University), Roy Kligfield (University of Colorado), Elizabeth Miller (Stanford University), Charles Onasch (Bowling Green State University), John Palmquist (Lawrence University), Donal Ragan (Arizona State University), Arthur Reymer (North Carolina State University), Jim Sears (University of Montana), Bruce Smith (Bowling Green State University), James Snook (Eastern Washington University), Rolfe Stanley (University of Vermont), John Tabor (University of Tennessee), Tom Tharp (Purdue University), Bill Travers (Cornell University), Daniel Tucker (University of Southwestern Louisiana), Brian Wernicke (Harvard University), Ian Williams (University of Wisconsin, River Falls), and Nick Woodward (University of Tennessee). Art Goldstein and Charlie Onasch did yeoman service by reviewing the entire text twice and exerted a very positive influence on the final version. Graduate students John Costello, Tim Davis, Teunis Heyn, Robert Hooper, Janet Hopson, Peter Lemiszki, and Beth McClellan reviewed several chapters, provided thin sections and hand samples of structures, and helped ferret out needed references. John Hanchar (Vanderbilt University) helped with photography of shatter-cone specimens. The efforts of all are very much appreciated.

Several colleagues and friends who provided key information, hand samples, interesting quotes, guidance to localities exposing spectacular structures, and needed conversation at critical times include Dick Campbell, Avery Drake, Bill Dunne, Steve Edelman, Hu Gabrielse, David Gee, André Michard, Bill Muehlberger, Bob Neuman, Alain Pique, Donald Ramsay, Brian Sturt, John Tabor, Bill Thomas, Rudolf Trümpy, Don Turcotte, George Viele, Rick Williams, Dave Wiltschko, and Nick Woodward. They might think their contributions small, but they improved the book, so these contributions cannot be small. Others who generously provided photos and diagrams that were used are acknowledged in figure captions.

Lynne E. Gaskin, Nancy L. Phalen, Karen M. Keener, and Nancy L. Meadows provided secretarial help by assisting with preparation and revision of chapters. Nancy Meadows helped additionally by editing some of the final chapters, obtaining permissions, and preparing the references cited, glossary, and indexes. Donald G. McClanahan, graphics specialist and artist, was also very helpful in skillfully preparing many of the diagrams and photographs for the book. This text would not have been completed without their assistance.

Free-lance editors Wendell Cochran and Jean Simmons Brown made my writing more clear and understandable. I learned a great deal from them about writing, yet remain culpable for all errors of scientific content or writing style. I appreciate the efforts of Merrill editorial, art, and production staff, principally Linda Bayma, Lorraine Woost, and Bruce Johnson, who paid close attention to detail and maintained high quality standards. I am grateful to the Merrill College Division Editors for their patience, their confidence in the project, and for shepherding it to completion.

Finally, I thank my wife Diana and daughters Melinda and Laura for their support while this and several other large projects were underway at the same time.

Bob Hatcher

CONTENTS

PART 1 INTRODUCTION 1

CHAPTER 1 INTRODUCTION AND REVIEW 2

Structures and Structural Geology 2
Fundamental Concepts 6
Plate Tectonics 10
Geochronology 12
Equilibrium 18
Geologic Cycles 21

CHAPTER 2 NONTECTONIC STRUCTURES 24

Primary Sedimentary Structures 26
Sedimentary Facies 31
Dewatering Structures 31
Unconformities 32
Primary Igneous Structures 36
Gravity-Related Features 38
Impact Structures 43

PART 2 ROCK MECHANICS 49

CHAPTER 3 STRESS 50

Definitions 50
Stress on a Plane 53
Stress at a Point 55
Mohr Construction 58
Mohr's Hypothesis 61
Stress Ellipsoid 61

CHAPTER 4 STRAIN 67

Definitions 68
Kinds of Strain 68
Strain Ellipsoid 71
Mohr Circles for Strain 72
Simple and Pure Shear 73

CHAPTER 5
MEASUREMENT OF STRAIN IN ROCKS 80

Kinds of Strain 80
Strain Markers 82
Flinn Diagram 85
Strain Measurement Techniques 87
Determination of Pressure-Solution Strain 99

CHAPTER 6
MECHANICAL BEHAVIOR OF ROCK MATERIALS 100

Definitions 100
Elastic (Hookean) Behavior 102
Permanent Deformation–Ductility 104
Controlling Factors 107
Behavior of Crustal Rocks 108
Strain Partitioning 112

CHAPTER 7
MICROSTRUCTURES AND DEFORMATION MECHANISMS 116

Lattice Defects and Dislocations 116
Deformation Mechanisms 121
Unrecovered Strain, Recovery, and Recrystallization 130
Laboratory Models of Deformation Processes 135

PART 3
FRACTURES AND FAULTS 141

CHAPTER 8
JOINTS AND SHEAR FRACTURES 142

Fracture Analysis 144
Joints and Fracture Mechanics 149
Joint in Plutons 157
Nontectonic and Quasitectonic Fractures 157

CHAPTER 9
FAULT CLASSIFICATION AND TERMINOLOGY 161

Anatomy of Faults 162
Andersonian Classification 164
Criteria for Faulting 165

CHAPTER 10
FAULT MECHANICS 176

Formation of Fractures: Griffith Theory 176
Anderson's Fundamental Assumptions 177
Anderson's Fault Types 179
Role of Fluids 182
Frictional Sliding Mechanisms 183
Movement Mechanisms 184
Brittle versus Ductile Faults 185
Shear Zones 190
Shear-Sense Indicators 191
Shear-Zone Kinematics 198

CHAPTER 11
THRUST FAULTS 201

Nature of Thrust Faults 202
Detachment within a Sedimentary Sequence 208
Small-Scale Features of Thrust Sheets 213
Propagation and Termination of Thrusts 213
Features Produced by Erosion 219
Crystalline Thrusts 221
Thrust Mechanics 224
Gravity versus Compression 230
Mechanics of Crystalline Thrusts 231
The Room Problem 231
Discussion 236

CHAPTER 12
STRIKE-SLIP FAULTS 240

Properties and Geometry 240
Environments of Strike-Slip Faulting 242
Nature of Fault Zones 243
Mechanics of Strike-Slip Faulting 246
Fault Geometry and Other Fault Types 247
Terminations of Strike-Slip Faults 247
Transforms 249

CHAPTER 13
NORMAL FAULTS 256

Properties and Geometry 256
Environments and Mechanics 259

PART 4
FOLDS AND FOLDING 277

CHAPTER 14
FOLD GEOMETRY AND CLASSIFICATIONS 278

Descriptive Anatomy of Simple Folds 279
Map-Scale Parallel Folds and Similar Folds 288
Fold Classifications 292

CHAPTER 15
MECHANICS OF FOLDING 315

Fold Mechanisms and Accompanying Phenomena 316
Deformation Mechanisms and Strain 338
Noncylindrical and Sheath Folds 339
Lineations and Fold Mechanisms 346
Parallel Folds and Similar Folds 347

CHAPTER 16
COMPLEX FOLDS 351

Occurrence and Recognition 352
Fold Interference Patterns 353
Recognition of Multiple Fold Phases 355
Noncylindrical and Sheath Folds 355
Formation of Complex Folds 357
Mechanical Implications of Complex Folding 357

PART 5 FABRICS, STRUCTURAL ANALYSIS, AND GEOPHYSICS 365

CHAPTER 17
CLEAVAGE AND FOLIATIONS 366

Definitions 366
Cleavage-Bedding Relationships 373
Cleavage Refraction 373
Early Ideas of the Origin of Slaty Cleavage 375
Mechanics of Slaty-Cleavage Formation 379
Progressive Development in Fine-Grained Sediment 383
Strain and Slaty Cleavage 387
Crenulation Cleavage 389
Cleavage Fans and Transecting Cleavages 390
Transposition 392

CHAPTER 18
LINEAR STRUCTURES 399

Definitions 400
Lineations as Shear-Sense Indicators 405
Folds and Lineations 405
Deformed Lineations 407
Interpretation of Linear Structures 410

CHAPTER 19
STRUCTURAL ANALYSIS 413

Definitions 414
Resolving Structures in Multiply Deformed Rocks 414
Mesoscopic Analysis 417
Symmetry of Fabrics 418
Structural-Analysis Procedures 422

CHAPTER 20
GEOPHYSICAL TECHNIQUES 434

Potential-Field Methods 434

Seismic Reflection 454

Seismic Refraction 458

Electrical Method 462

APPENDIX ONE
FABRIC DIAGRAMS 465

APPENDIX TWO
STRUCTURAL MEASUREMENTS AND OBSERVATIONS 475

APPENDIX THREE
WOODALL SHOALS FABRIC DATA 482

GLOSSARY 487

REFERENCES 507

AUTHOR INDEX 519

SUBJECT INDEX 523

PART ONE

Introduction

OUTLINE

1 Introduction and Review
2 Nontectonic Structures

THIS SECTION IS INTENDED TO REVIEW AND REINTRODUCE MANY concepts and principles that form the basis for understanding and deciphering geologic structures that first appeared in other geology courses. It is intended to form a bridge between the ideas of the introductory physical and historical geology courses, stratigraphy, and modern elementary structural geology. The fundamental geologic laws, concepts of scale, time relationships, cycles, plate-tectonics processes, and equilibrium all are involved when an attempt is made to decipher the structure of an area or a single structure, no matter how large or small. Primary sedimentary structures, discussed in Chapter 2, are an important frequently used tool. Distinguishing nontectonic from tectonic structures is a critical aspect of accurately deciphering individual structures or the structure of an area. Structures in salt and ice formed near or at the surface also are useful analogs for similar tectonic structures formed at great depth. Structures formed by the impact of large objects on the Earth are also considered here.

1

Introduction and Review

A stone, when it is examined, will be found a mountain in miniature. The fineness of Nature's work is so great, that, into a single block, a foot or two in diameter, she can compress as many changes of form and structure, on a small scale, as she needs for her mountains on a large one; and, taking moss for forests, and grains of crystal for craggs, the surface of a stone, in by far the plurality of instances, is more interesting than the surface of an ordinary hill; more fascinating in form and incomparably richer in colour—the last quality being most noble in stones of good birth (that is to say, fallen from the crystalline mountain ranges).

JOHN RUSKIN, *Modern Painters*

STRUCTURES AND STRUCTURAL GEOLOGY

FOR CENTURIES WE HAVE BEEN FASCINATED BY THE SHAPES OF continents and ocean basins, the linearity of mountain chains, the distribution of volcanoes, and the motion along large faults producing earthquakes. We know today that most such features are produced by the present and past configurations of tectonic plates on the Earth (Figure 1–1). We are constantly reminded of the dynamic character of the Earth by earthquake and volcanic activity that indicate the awesome forces that drive the plates. The lives of most of the Earth's population are influenced every day by tectonic activity; unfortunately, many are threatened by the potential for earthquakes and volcanic eruptions. Aside from the imminent danger and practical need to comprehend the processes that produce earthquakes and volcanoes, most scientists feel a basic scientific urge to understand these processes. Structural geologists are also concerned with why parts of the Earth have been bent into smoothly curved shapes, but others have been broken. We are also concerned with structures on all scales and how they are related. Ruskin's statement about rocks illustrates how closely geologic processes are interwoven, whether parts of entire mountain ranges or in structures we can observe in a hand specimen.

Tectonic structures are produced in rocks in response to stresses generated, for the most part, by plate motion within the Earth and include different kinds of faults and folds, along with other structures. They make up the framework of the Earth. The kinds of structures that form in a rock mass are determined by (1) temperatures and pressures that prevail in the environment surrounding the mass, (2) composition, (3) layering, (4) contrast in properties with direction between and within individual layers (***anisotropy***) or the lack of contrast (***isotropy***), and (5) amount and character of fluids within the rock mass. How rapidly the mass is deformed and the orientations of stresses applied to it also influence the kinds of structures produced. These factors determine whether deformation will be continuous (***ductile deformation***) or discontinuous (***brittle deformation***), producing a great variety of structures both within and at plate boundaries in the rocks of the Earth and in other planets (Figure 1–2). Continuous deformation produces certain kinds of folds, ductile faults, cleavages, and foliations; discontinuous deformation produces other kinds of folds, brittle faults, and joints. Structures may also form as products of nontectonic processes, such as gravitational forces and extraterrestrial

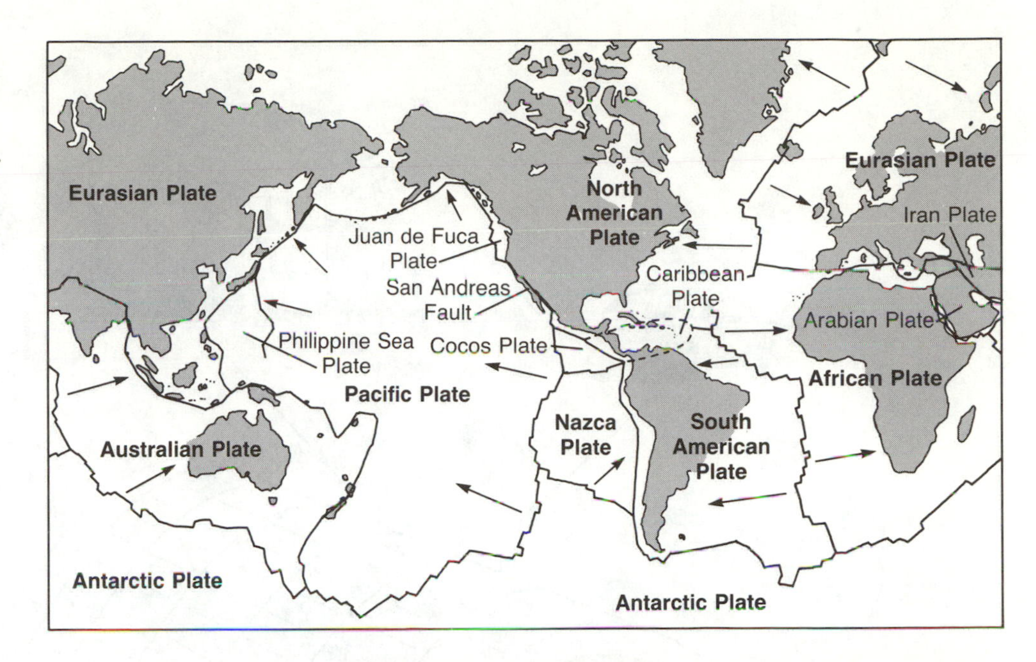

FIGURE 1–1
Distribution of tectonic plates on the Earth's surface. (U.S. Geological Survey.) Arrows indicate principal directions of plate motion. Solid lines indicate plate boundaries.

impacts. It is important that we distinguish between tectonic and nontectonic structures (Chapter 2) because some nontectonic structures closely resemble—even mimic—structures formed by tectonic processes. Our main purposes here are to describe the various structures that have been observed in rocks and to understand how they form.

Structural geology deals with the origin, geometry, and kinematics of formation of structures. Structural geology is similar to architecture in that both require an ability to visualize objects in three dimensions as they change through time (Figure 1–3). There is a close parallel between the shapes of structures and the physical conditions that formed them. In particular, the contrast in shape and type between structures that form near the Earth's surface and those that form at great depth under the weight of overlying rocks and at high temperature indicates profound differences in physical conditions. An appreciation of the geometry of structures provides greater insight into the origins of structures.

(a)

(b)

FIGURE 1–2
Continuous (ductile) and discontinuous (brittle) structures in rocks. (a) Folds produced by ductile flow in Precambrian gneiss near Central City, Colorado. Quarter indicates scale. (b) Brittle fracturing at the same locality as (a) produced several sets of joints. Older ductile structures are outlined by color banding. Scale indicated by small trees 3 to 5 m tall. (RDH photos.)

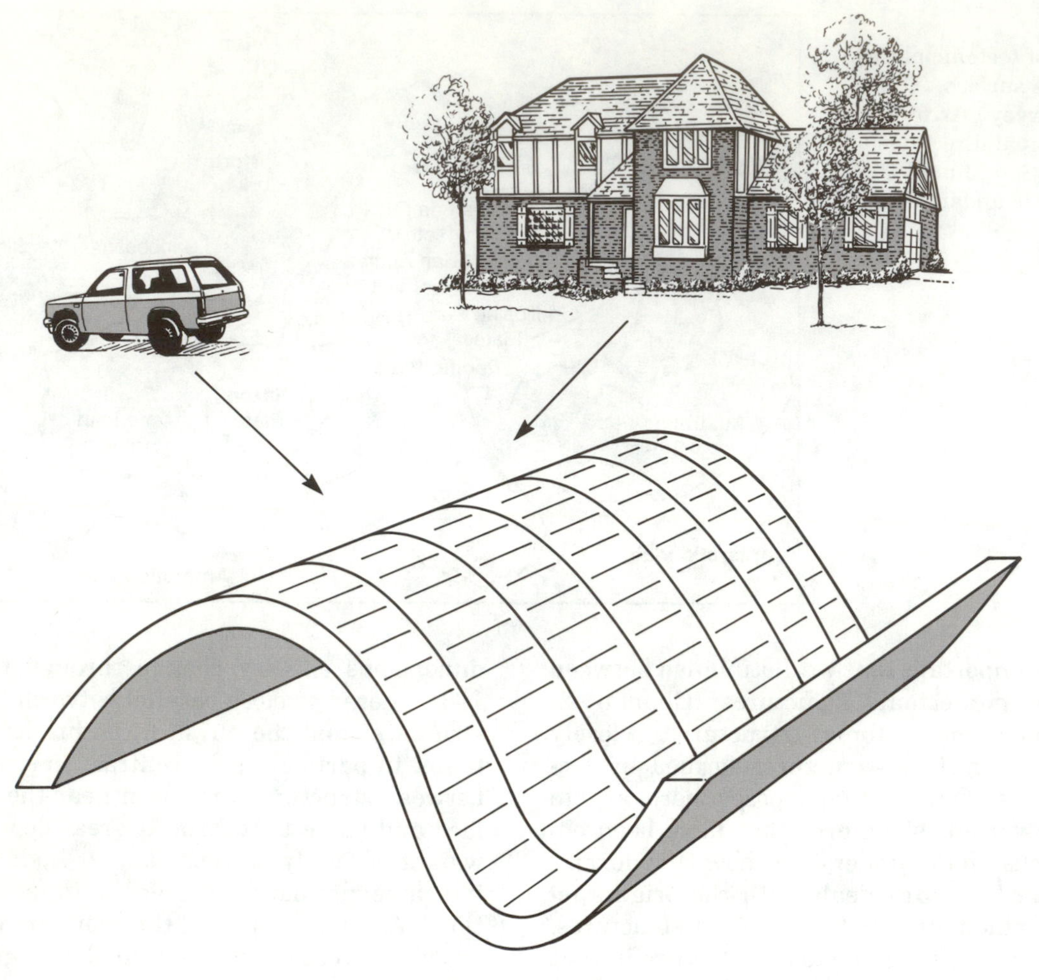

FIGURE 1–3
Visualizing objects in three dimensions. Start with a familiar object, like your house, car, or a familiar building, then begin to think about geologic structures.

This discussion of structural geology may be your first encounter with visualizing things in three dimensions. You might begin by attempting to think about familiar objects in three dimensions, such as your car, house, or room, then move on to work with less familiar tectonic structures. Keep in mind that most of us had difficulty at the beginning, but that we learned to visualize objects in three dimensions through practice. Because structural geology is also a traditional field discipline, we must develop an ability to recognize subtle features in rocks. In that sense, structural geology is partly an observational science. Also, a better understanding of physical and chemical principles, along with the ability to use computers and mathematics, is needed today to bridge the gaps between laboratory, theoretical, and field studies.

The link between laboratory and field studies is both essential and supportive, for structural geology is divisible into subdisciplines, most of which overlap. For example, laboratory studies determining fluid pressure that facilitates movement on faults are supported by field observations of evidence that fluid was present when a fault was active.

Rock mechanics is the application of physics to the study of rock materials. It deals with rock properties and the relationships between forces and the resulting structures, as well as with the study of structures produced in the laboratory in an attempt to duplicate natural structures (Figure 1–4). In the laboratory we can simulate the higher temperatures and pressures thought to exist at great depths. Alternatively, very weak materials, such as salt, gelatin, clay, putty, and paraffin, which behave like rocks at higher temperatures, may be used to produce experimental structures at room temperature. A severe handicap of laboratory experiments in rock mechanics is that they cannot be run over geological time—thousands to millions of years. They must be run at temperatures and pressures far above those normally occurring in nature so that deformation rates will occur rapidly enough to be observed in a

FIGURE 1–4
Experimental structures made in a centrifuge from viscous materials of different densities and fluid properties. (From Hans Ramberg, *Tectonophysics* v. 19, Fig. 15. © 1973, Elsevier Science Publishers B.V., Amsterdam.)

reasonable time. Artificial or natural materials that simulate the behavior of rock materials must be scaled to approximate natural processes.

The study of ***field relations*** is an exceptionally important aspect of structural geology because they provide constraints for formulating models of regional deformation. In structural geology, we try to understand how small structures form and how they are related to larger structures and, ultimately, to crustal stresses. A geologist undertaking structural studies will (1) measure orientations of small structures to provide information about the shapes and relative positions of larger structures in the field, (2) study the sequence of development and superposition of different kinds of structures in an area to determine the sequence of conditions of deformation, and (3) try to apply rock-mechanics data to relate structures to stress that was present in the Earth at the time of deformation. Quite often a structural geologist will also try to relate structures in one area to those elsewhere that may have been formed by similar mechanisms.

Tectonics and ***regional structural geology*** involve larger features. Studies of mountain ranges, parts of entire continents, trenches and island arcs, oceanic ridges, and entire continents and ocean basins, and the relationships to stresses and tectonic plates, are involved in these subdisciplines. ***Plate tectonics*** deals specifically with plate generation and motion and their interactions.

Separation of tectonics from regional structural geology is difficult. Regional structural geology is more concerned with continental structures and uses more data from detailed studies of small structures to reconstruct the deformational history and tectonics of a large region. Moreover, geophysical data and information derived from other disciplines of geology must be integrated with structural data for use in regional structural geology and tectonics. Use of geophysical data in structural geology is becoming more common because technology is making available more data of higher quality, particularly gravity, magnetic (including paleomagnetic), and seismic-reflection data.

It is easy to see that the many subdivisions of structural geology are related to other disciplines in geology as well as to the other sciences. Direct applications are made from physics to the origin of geologic structures. Isotopic data are frequently useful in working out the absolute time of formation of structures, and geochemical data may help determine mobility of elements or isotopes during deformation. The chemical composition of highly deformed rocks may indicate the nature of the original material (*protolith*) and the environment of formation before deformation—a key factor in tectonics.

Structural geology can be applied to other fields. It is readily applied to engineering problems that involve bridges, dams, and power plants, where large excavations are necessary, as well as highways, where excavations extend for long distances. Studies of geologic structures beneath buildings, dams, and highway cuts are of great importance because of the potential for renewed motion along faults and other fractures, as well as concern for the stability of slopes and geologic materials. Many power plants (both nuclear and conventional), large buildings, airports, and dams are under construction in different parts of the world. Siting these structures within active fault zones is not desirable, but sometimes it is impossible to build them in tectonically quiet areas. Therefore geologists and engineers must work together from the design stage through construction to work out the timing of movement of all structures that might affect engineering works (Figure 1–5) and to minimize both cost and hazards.

Structural geology has long had a close working relationship with petroleum and mining geology. The geometric techniques of understanding and projecting fault surfaces, geologic contacts, and struc-

FIGURE 1–5
Complexly deformed metamorphic rocks exposed in the excavation for construction of Mica Dam in southern British Columbia. (RDH photo.)

tures to depth have been used to great advantage by exploration and mining geologists and others. Similar applications of structural geology have been used for many decades in petroleum geology. The principles of tectonics have been applied to understanding larger trends and regional processes that control the concentration of mineral deposits and hydrocarbons.

The concept of ***scale*** is also of great importance in structural geology. Structures—such as geologic contacts and some foliations, faults, and folds—are commonly observed in the field in both hand specimens and at outcrop (or ***mesoscopic***) scale. Small structures that require magnification to be observed, such as many foliations and linear structures, are called ***microscopic.*** Mountainside to map-scale structures of all kinds are called ***macroscopic*** structures. We must be constantly aware of the relationships between structures at all scales (Figure 1–6a). Scales and geometric perspectives of geologic cross sections must be maintained between the map from which the section is constructed and the section itself (Figure 1–6b).

Structures that occur as single features, like a fault or an isolated fold, are termed ***nonpenetrative structures.*** These are not present on all scales. Other structures, like slaty cleavage, foliations, and some folds, occur on any scale that we may choose for observation and are termed **penetrative.** Some structures, such as joints, may be penetrative on one scale but not on another. Joints are penetrative only on the macroscopic scale, but not on the mesoscopic or microscopic scale.

FUNDAMENTAL CONCEPTS

The fundamental—almost simple—relationships to be discussed here provide us with the most powerful tools available to begin investigating the subject of structural geology. Without understanding them, we would be so severely handicapped that no technologically advanced equipment, like sophisticated computers, seismological equipment, or other analytical tools, could help solve our structural problems.

Probably the most important foundation doctrine in geology is uniformitarianism. It was first stated by James Hutton, an eighteenth-century Scottish farmer and scientist. Because his writing style was obscure, his ideas did not become widely known until they were rewritten by John Playfair in the early nineteenth century. The ***doctrine of uniformitarianism*** states that *processes occurring today upon and within the Earth have probably gone on similarly in the past and will continue in the future;* that is, *the present is the key to the past.* Hutton's conclusions were based on his observations along the coast of Scotland (Figure 1–7), where he could see waves wearing down rocks and producing pebbles that were further reduced to sand. He observed that sand bars and beaches were constantly being created and destroyed by storms, then slowly rebuilt again. Hutton also recognized that the sand in sandstone is the same as that moving about on a modern beach. He concluded that layers of sandstone turned on end were originally deposited horizontally and that an immensity of time must have

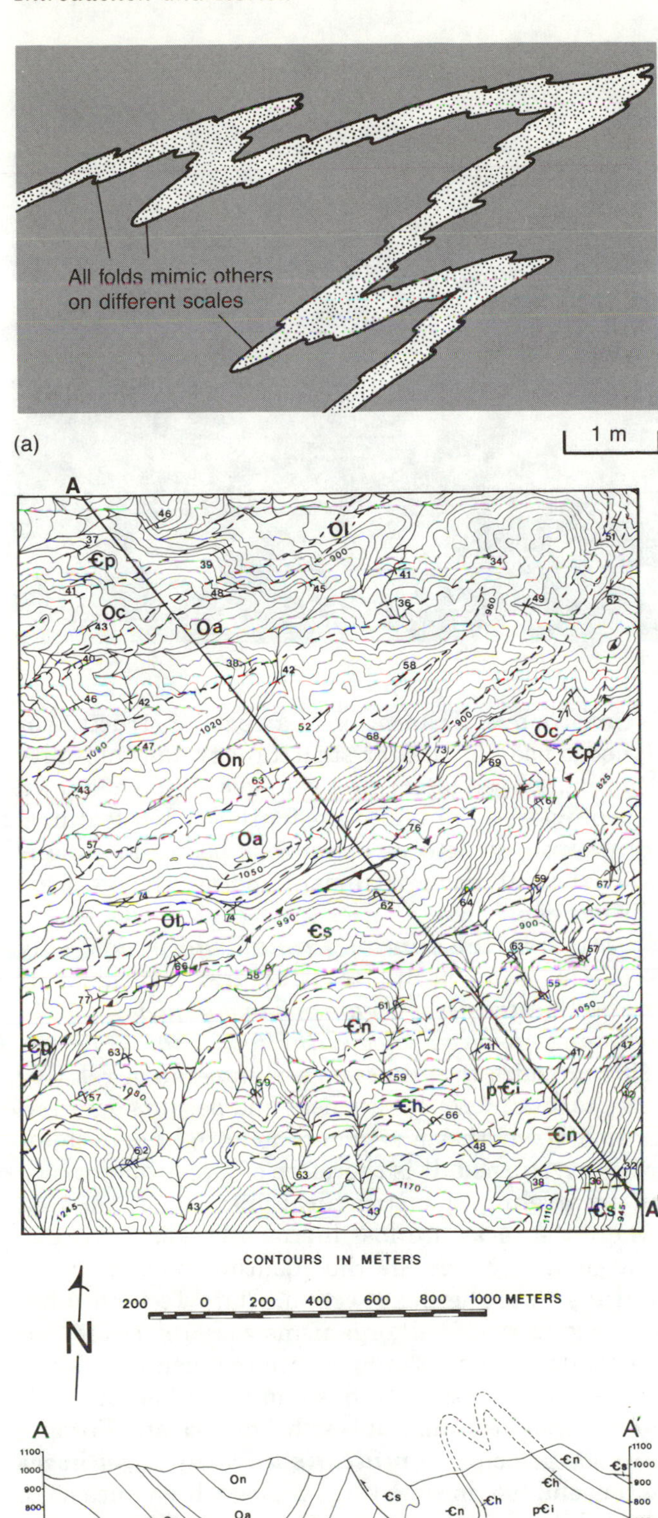

FIGURE 1–6
(a) Relationships between small and large folds in the same structure. (b) Geologic map and cross section. Note that constructing an accurate cross section requires close attention to scale and position of geologic contacts on the topographic surface.

elapsed since the sand grains were formed, became consolidated into rock layers, and were turned on end by crustal forces. His observations mark the beginning of modern geology—for the first time it was recognized that a huge amount of time is both available and necessary to carry out geologic processes. Before Hutton and long afterward, the prevailing notion was that unknown catastrophic events were responsible for geologic processes and features. Uniformitarianism immediately led to conflict with religious dogma, resulting in numerous debates between scientists and theologians during the nineteenth century, particularly with the rise of other theories such as evolution.

Today we recognize that most geologic processes require immense amounts of time. We also have realized that, although the movement across a large fault may total many tens of kilometers, a large part of the motion may have occurred as relatively small displacements (or by nearly instantaneous movements producing earthquakes), not by continuous slippage through time. Study of active faults indicates that some segments move by continuous creep, but other segments undergo instantaneous catastrophic movement—earthquakes. Therefore Hutton was probably correct when he recognized the immensity of time involved in geologic processes, and so uniformitarianism is the best means of thinking about geologic processes through time. *These long-term effects may, however, represent the sum of many instantaneous and even catastrophic events randomly distributed over the continuum of geologic time.*

The doctrine of uniformitarianism may break down if certain aspects of Precambrian geology are considered. The iron formations in the Lake Superior region probably resulted from a different atmospheric composition, and some geologists say that the smaller continental nuclei that existed during the Archean may show that the rules of plate tectonics may not have held during that early part of Earth history. Contrasts in the nature of the crust formed during Archean and Proterozoic time may further indicate fundamental differences in processes operating before and after about three billion years ago. Archean processes resulted in a crust dominated by greenstones intruded by large granitic batholiths, but Proterozoic crust involves appearance of the first platform sediments and cratonic basins, as well as the apparent addition of new crust around old nuclei, a process that continues today.

The ***law of superposition*** is another cornerstone of geologic thought. It states that *within a layered sequence, commonly sedimentary rocks, the oldest rocks will occur at the base of the sequence and*

FIGURE 1–7
Rocks along the coast of Scotland, like this scene near Loch Eribol in the footwall of the Moine thrust, enabled James Hutton to formulate the doctrine of uniformitarianism. (RDH photo.)

successively younger rocks will occur toward the top, unless the sequence has been inverted through tectonic activity. Geology could not function as a science and the understanding of many processes would be greatly impaired without this law and the doctrine of uniformitarianism. The first statement of the law of superposition was made during the seventeenth century by Nicholas Steno (Niels Stensen), a Danish physician with interests in geology. The law is of great importance in structural geology because it is necessary to determine whether the stacking order in a sequence has been disrupted tectonically. The sequence may have been tilted, completely overturned, or repeated by folding or faulting (Figure 1–8a). Superposition is therefore an inviolate second principle in the study of geologic structure.

The ***law of original horizontality*** is another fundamental geologic law stating that *bedding planes within sediments or sedimentary rocks form in a horizontal to nearly horizontal orientation at the time of deposition.* This law is fundamental in structural geology because bedding is the common initial reference frame (Figure 1–8b).

Another law that goes hand in hand with working out the structural history of an area is the ***law of cross-cutting relationships,*** applied as either the ***law of igneous cross-cutting relationships*** or the ***law of structural relationships*** (Figure 1–8c). Both state virtually the same thing, that *an igneous body or a structure—that is, a fold or fault—must be younger than the rocks it cuts through.* In other words, the rocks that contain a structure or that form the host for an igneous body must have been there before the structure or igneous body formed. These laws provide a basis for placing structures in a time context. Truncation of an earlier structure or igneous body by a later structure, an unconformity (to be discussed later), or a younger pluton of known age provides a minimum age for the earlier features. Bracketing structures or igneous bodies in time is an essential part of understanding the geologic history of an area.

At first, the ***law of faunal succession*** may seem far from useful to structural geologists. It states that *the fossil organisms in a sequence should be more advanced toward the top of the sequence.* This provides the basis for assignments of relative ages to fossiliferous sequences and permits determination of whether a sequence is upright or tectonically overturned. It is therefore of major importance in unraveling the structural history of an area where the rocks are fossiliferous. Fossils have also played a key role in the identification of exotic terranes (see the next section, Plate Tectonics) because groups of fossil organisms were restricted to a particular region. Examples are the fusulinid fauna of the late Paleozoic Tethys and the different Cambrian trilobite faunas of North America and Europe.

The principle of ***multiple working hypotheses*** is a useful tool in structural geology. It enables us to formulate more than one possible explanation of the same data, to evaluate each, and to select the most likely hypothesis. Suppose you are working in a field area that lacks a critical exposure needed to correctly interpret a particular contact of an igneous body with the overlying sedimentary section (Figure 1–9). You have hypothesized that the contact can be (1) an intrusive contact, (2) a fault, or (3) an unconformity. Each hypothesis may be equally valid. You begin by sorting through your previous observations

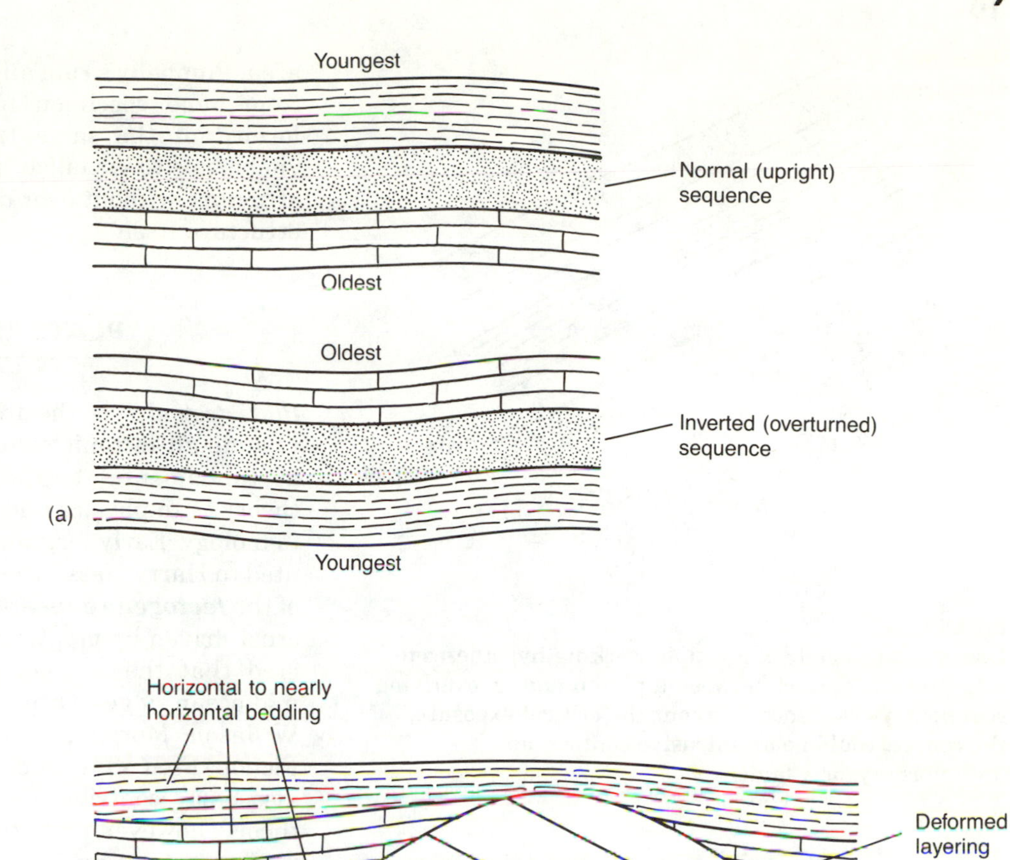

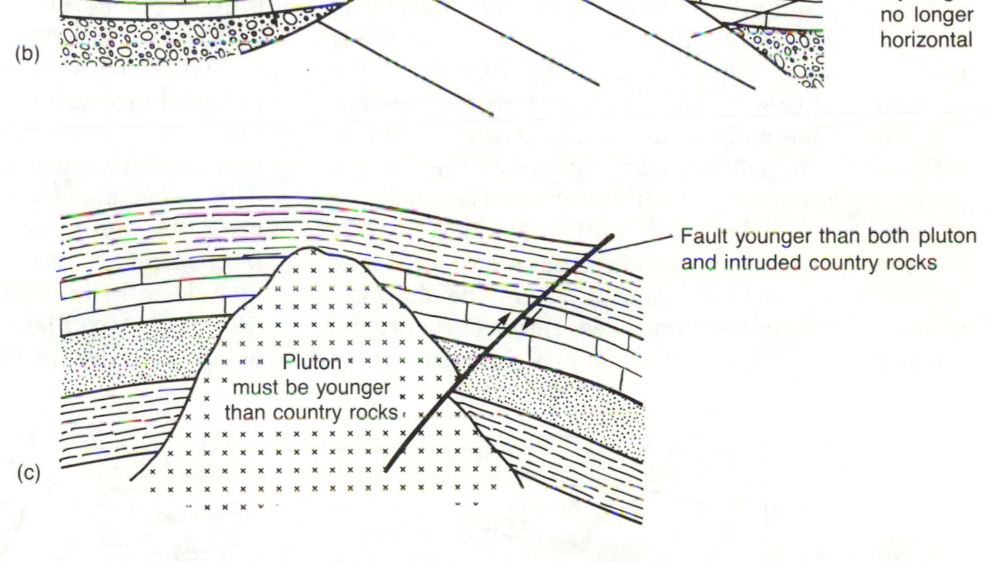

FIGURE 1–8
Laws of superposition (a), original horizontality (b), and cross-cutting relationships (c).

and ask, with regard to (1), have you observed baking or other alteration of the sedimentary rocks near the contact? With regard to (2), have you observed crushed rocks or other evidence of faulting near the contact? With regard to (3), have you observed clasts of the igneous rocks incorporated into the base of the overlying sedimentary sequence? One or two exposures of baked rocks along the contact eliminates the second and third possibilities, but before discovery of the critical pieces of data all working hypotheses were equally valid.

There is also much value in the ***outrageous hypothesis*** (W. M. Davis, 1926; Wise, 1963) as an alternative working hypothesis because it provides a focus for critical pieces of data on a solution to the problem at hand. An outrageous hypothesis appears to be an impossible solution to the problem from the moment it is formulated. On consideration of the data, it may gain the position of a credible alternative working hypothesis or it may be quickly abandoned as other likely working hypotheses are formulated.

Another relationship of great importance in structural geology is ***Pumpelly's rule.*** It was first applied to geology by Raphael Pumpelly in the late nineteenth century and states that *small structures*

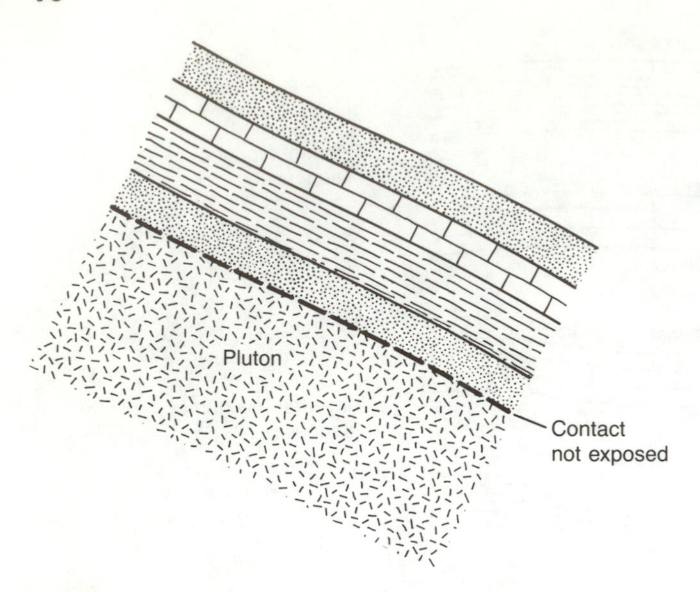

FIGURE 1–9
Use of the principle of multiple working hypotheses to interpret the contact between a pluton and an overlying sedimentary sequence. Without the critical exposure, the contact could be an intrusive contact, an unconformity, or a fault.

are a key to and mimic the styles and orientations of larger structures of the same generation within a particular area. (The quotation by Ruskin at the beginning of this chapter is an earlier statement of the rule.) Pumpelly's rule holds if all structures referred to were formed at the same time by the same stresses in rocks of similar properties and were deformed similarly at all scales. Pumpelly's rule provides a basis for presuming that small and large structures of the same generation may be shown to be related within the same area. Because we are not always able to observe structures on all scales in an area, Pumpelly's rule allows us to assume similarity from hand specimen to map scale of structures formed at the same time (Figure 1–6). It thus enables us to visualize the configuration of a larger structure without ever directly observing the entire structure itself.

PLATE TECTONICS

Plate tectonics is the framework within which we assume all tectonic structures form. The concept is as fundamental to the earth sciences as atomic theory is to physics and chemistry and evolution is to biology. Early formulation of the theory is attributed to Harry Hess, who during the 1930s conceived of the ***tectogene concept*** of the subsiding crumpling crust driven by mantle convection. He later discovered that the seafloor is spreading apart at the mid-ocean ridges. Others, including Robert S. Dietz, W. Jason Morgan, Dan P. McKenzie, Xavier Le Pichon, Fred Vine, and Drummond Matthews, were also early contributors to different aspects of the theory; however, plate tectonics was first stated as a unified theory by Bryan Isacks, Jack Oliver, and Lynn Sykes (1968).

The Earth's surface is divisible into seven major plates that contain all the continents and oceans, and several smaller ones, which are being formed as new oceanic crust at the oceanic ridges and are being consumed by subduction in the trenches (Figure 1–1). Motion of plates is thought to affect the ***lithosphere,*** which is about 100 km thick, and includes all the crust and part of the upper mantle (Figure 1–10). Plate motion may be characterized using an Eulerian theorem that describes the mo-

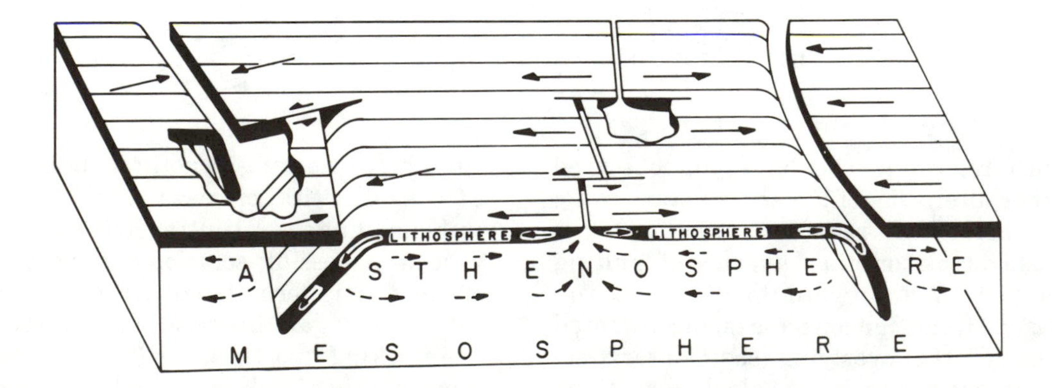

FIGURE 1–10
Generation of lithospheric plates at spreading centers and destruction at subduction zones. Differences in rate of motion are compensated by transforms at ridges and other boundaries. Arrows indicate direction of motion. (From B. Isacks, J. Oliver, and L. Sykes, Seismology and the new global tectonics: *Journal of Geophysical Research,* v. 78, p. 5855–5899, fig. 2, © 1968 by the American Geophysical Union.)

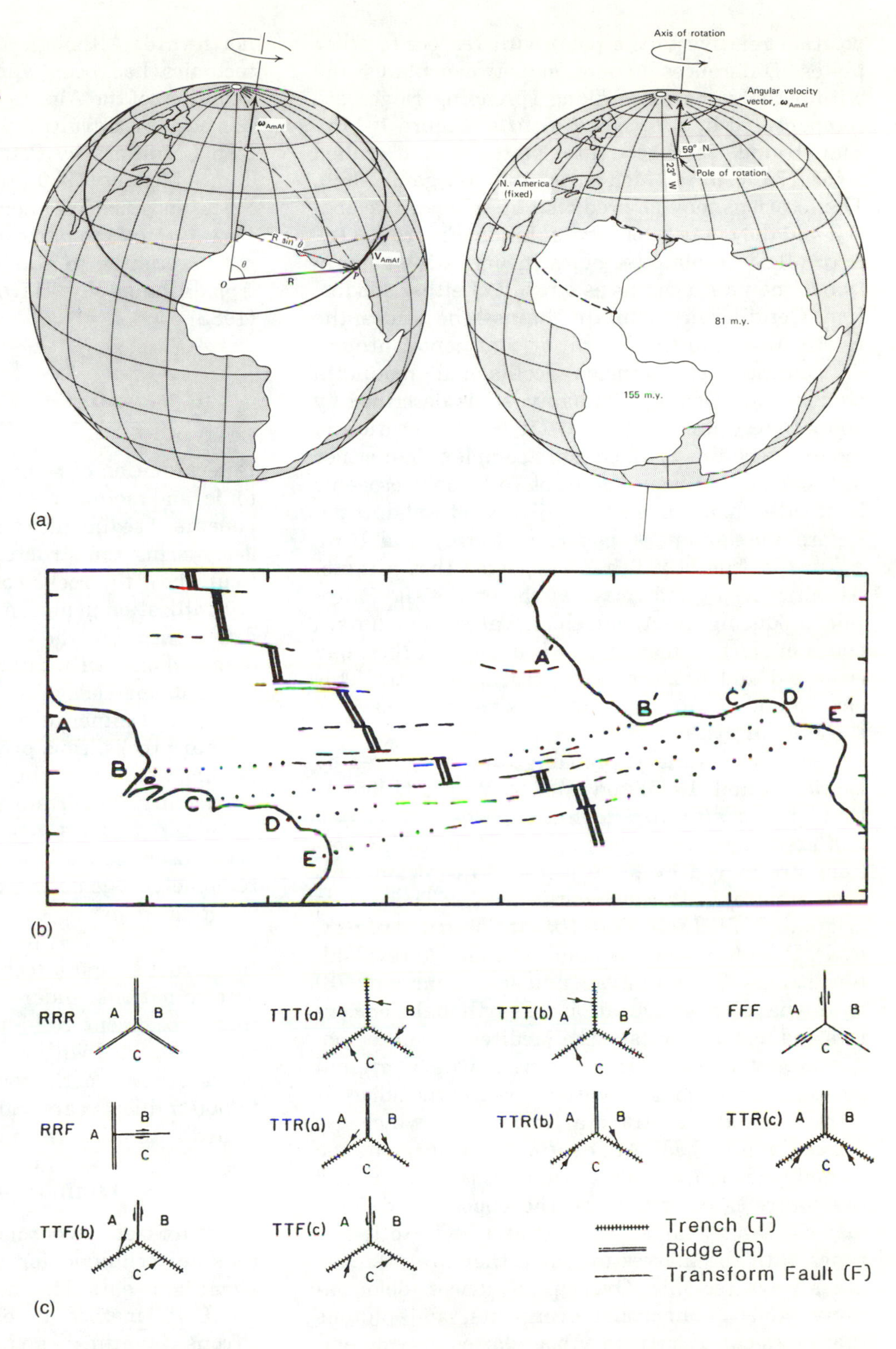

FIGURE 1–11
(a) Angular displacement of Africa predicted from the laws of spherical geometry; ω is angular velocity, *R* is the radius vector, *V* is particle velocity for a reference point in Africa. (From B. E. Hobbs, W. D. Means, and P. F. Williams, *An Outline of Structural Geology,* © 1976, John Wiley & Sons, Inc.)
(b) Transforms at plate boundaries and within plates function to compensate differences in relative motion. (From J. T. Wilson, reprinted by permission from *Nature,* v. 207, p. 343–347 © 1966, Macmillan Magazines Ltd.)
(c) Several kinds of triple junctions. (From D. P. McKenzie and W. J. Morgan, reprinted by permission from *Nature,* v. 224, p. 125–133 © 1969 Macmillan Magazines Ltd.)

tion of plates on a sphere in which displacement on the surface increases vectorially away from the spreading (rotation) axis. Angular displacement of a plate involves rotation about a line passing through the center of the Earth (Figure 1–11a), and rotation of a plate about the spreading axis may be expressed by its angular velocity (ω) on the sphere. Velocity increases away from the pole of the spreading axis; angular velocity remains constant. Consequently, along a spreading center (at a mid-ocean ridge), the spreading rate should increase to a maximum velocity 90° from the pole (at the equator). This variation in spreading rate has been demonstrated along the Mid-Atlantic Ridge (Le Pichon, 1968; Morgan, 1968). Each plate would have an angular velocity on the sphere (determined by the absolute motion and position relative to the pole) with respect to other plates. Differences in motion between plates and within the same plate along spreading ridges are compensated by ***transform faults*** (Figure 1–11b). Plate boundaries where three plates meet are called ***triple junctions*** (McKenzie and Morgan, 1969). The locations may be predicted by spherical geometry, and they may be of several kinds (Figure 1–11c). A corollary to plate-tectonics theory is the generation of mountain chains as a result of either subduction (Cordilleran mountain chains—the Andes, the North American Cordillera) or continent-continent or continent-arc collision (collisional mountain chains—the Alps, the Himalayas), as described by John Dewey and Jack Bird (1970). Generation of mountain chains is much more complex than either of those mechanisms, for most collisional orogenic belts also had an earlier history of subduction. Before the landmark paper by Dewey and Bird, J. Tuzo Wilson (1966) had suggested that a proto-Atlantic Ocean had closed at the end of the Paleozoic, producing the Appalachian-Variscan mountain chain of North America and Europe, and then had reopened and produced the present Atlantic. This closing and opening of oceans has become known as the ***Wilson cycle.***

Another corollary, probably first recognized by Emile Argand (1925) and later by Wilson (1968), is that of ***accretionary tectonics,*** whereby *suspect* and *exotic terranes* of less than continental proportions are moved by plate motion to collision with each other or with continental masses. These were originally called ***microcontinents,*** or ***microplates,*** by D. P. McKenzie (1970) and by John Dewey, Walter Pittman, William Ryan, and Jean Bonnin (1973) on the basis of configurations of earthquake epicenters and tectonic units in the Mediterranean region. A ***suspect terrane*** is a mass in which original position is questionable with respect to the adjacent terrane or stable continental land mass to which it is presently attached. An ***exotic terrane*** bears no resemblance to the mass to which it is attached and the source is commonly on the opposite side of a major ocean. Boundaries of suspect and exotic terranes with the masses to which they are attached are always tectonic. Overlap sequences, deformational and metamorphic overprints, and plutons that cross-cut accretionary boundaries provide evidence of "docking" and frequently the time of docking of terranes. Warren B. Hamilton's compilation (1979) of the geology of the Indonesian region demonstrates that a complex of volcanic arcs, continental fragments, oceanic crust, and large continental blocks (such as Australia) are all in the initial stages of being accreted to Asia as Australia moves northward. Although the concept of accretionary tectonics has been applied to the plate-tectonics evolution of the Alps by Dewey and others (1973), it has been most fruitfully applied to the North American Cordillera by Peter Coney, David Jones, and James Monger (1980), then by Monger, Raymond A. Price, and Dirk Templeman-Kluitt (1982) and David Howell (1985) to parts or all of the same chain, and more recently to the precollision history of the Appalachians by Harold Williams and Hatcher (1983).

GEOCHRONOLOGY

Determination of radiometric ages of volcanic, plutonic, and metamorphic rocks, and the detrital components of sedimentary rocks, is an important aid in deciphering the structural history of an area, particularly if the rocks contain no fossils. The age of crystallization in plutonic and volcanic rocks is often determined, but the exact age of metamorphism is obtained only with difficulty.

Radiometric ages can sometimes be determined for detrital minerals in sedimentary and metasedimentary rocks, thus providing an age of the source terrane from which the sediments were derived; it is particularly important in attempting to work out the nature and origin of accreted terranes. Next is a brief discussion of the principal methods used in radiometric age dating of rocks.

Refinement of sampling and analytical techniques in recent years has permitted reduction in the errors in most techniques for radiometric age determinations under optimum conditions to less than two percent. Rubidium-strontium ages are commonly reported with a possible error of ± 2 million years if the age of the rock body is on the order of 300 to 500 million years. Similar results have been obtained with other techniques.

Uranium-Lead Method

Uranium-lead geochronology is probably the most reliable technique for rocks wherein ages exceed 10 million years. The most common way of applying the U-Pb method to date rocks is by analyzing zircons, apatites, and sphenes separated from crushed rock samples. These minerals contain small quantities of the radioactive elements uranium and thorium. Neither parent nor daughter elements are easily released from the zircon lattice during deformation or metamorphism, and, consequently, it is possible to determine the ages of many zircons by analysis of several uranium and thorium isotopes and the daughter products by using a mass spectrometer. These elements decay through a series of intermediate radioactive elements to stable daughter products:

$$^{238}U \rightarrow {}^{206}Pb \text{ (half-life} = 4.5 \times 10^9 \text{ y)} \quad \textbf{(1–1)}$$

$$^{235}U \rightarrow {}^{207}Pb \text{ (half-life} = 0.7 \times 10^9 \text{ y)} \quad \textbf{(1–2)}$$

$$^{232}Th \rightarrow {}^{208}Pb \text{ (half-life} = 1.4 \times 10^9 \text{ y)} \quad \textbf{(1–3)}$$

Ratios of ^{206}Pb to ^{238}U, ^{207}Pb to ^{235}U, and ^{208}Pb to ^{232}Th are used to determine the age of the zircons by plotting them on a ***concordia curve*** (Figure 1–12). The curve shows the variation in abundances of these elements through time. Zircons that plot on the curve are said to be *concordant;* those that do not are *discordant* and must have acted in part as open systems. A chord may be drawn through the points along which discordant zircons plot, yielding upper and lower intercepts on the concordia curve. The lower intercept yields a lesser age and indicates *episodic lead loss,* commonly due to thermal resetting of the zircons. The upper intercept represents the original age of the zircons.

The U-Pb method provides the most accurate ages for determining the time of crystallization of zircons in igneous and volcanic rocks. Sometimes metamorphic ages also may be obtained by this technique. Lower intercepts of discordant zircons are frequently metamorphic ages. Upper intercepts may either represent the original age of the body or show that the magma assimilated zircons from another rock mass. The latter are known as ***inherited zircons.*** Reset and inherited zircons (either detrital or assimilated) from an earlier tectonic cycle provide a major source of error in determining absolute ages of different rock materials.

The U-Pb method is particularly good because the temperature at which the radiogenic isotopes are locked into the zircon lattice is relatively high. This is called the ***blocking temperature.*** Consequently, zircons are less subject to resetting by lower-temperature events; however, other factors, such as circulating water and radiation damage to the lattice by radioactive constituents, may also reopen the system and cause loss of lead.

A technique invented by Thomas Krogh (1973) involves selective dissolution or abrasion of the outer layers of zoned zircon crystals to determine the composition and age of the outer layer. The remaining cores of the crystals may then be dissolved and analyzed. An age greater than that of the younger rim may be obtained. Zoned zircons are very common. It is important in zircon geochronology not only to separate zoned from unzoned zircons, but also to separate euhedral from anhedral or rounded

FIGURE 1–12
Concordia curve showing several hypothetical discordant zircon analyses with upper and lower intercepts at 1.05 billion years and 350 million years. Small elongate hexagons indicate margin of analytical error. Note that points that plot above the curve indicate uranium loss; points that plot below the curve indicate lead loss (or uranium gain).

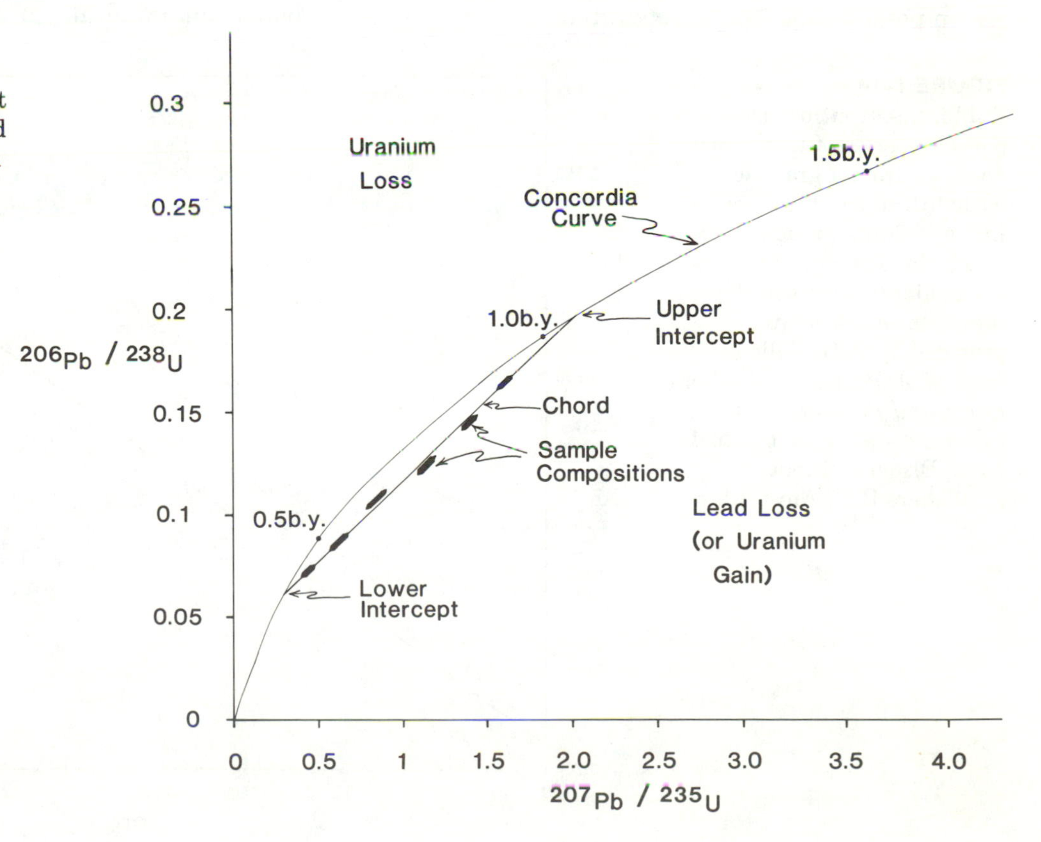

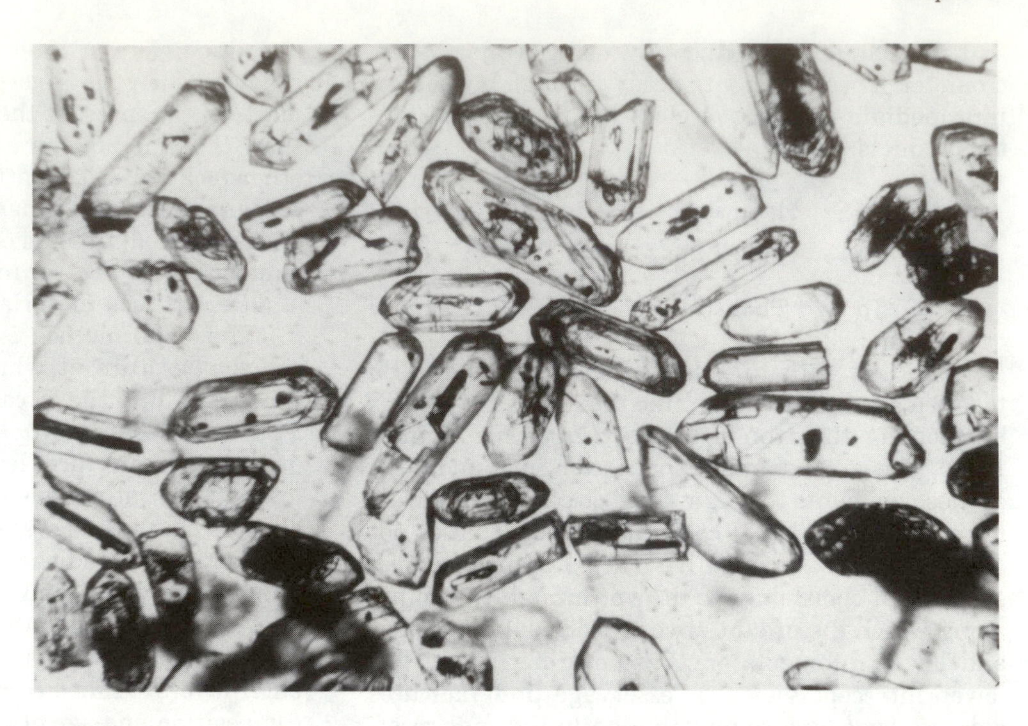

FIGURE 1–13
Zircon sample from the High Shoals Granite, North Carolina, illustrating various morphological types—euhedral, rounded, zoned. The largest zircons are approximately 1 mm long. This sample yielded a concordant age of 317 Ma. (From J. W. Horton, J. F. Sutter, T. W. Stern, and D. J. Milton, 1987, *American Journal of Science,* v. 287.)

crystals. Other variations in size, shape, or internal characteristics—such as found in clear or colored zircons—may indicate mixed populations that will yield different ages or, if not separated, discordant ages (Figure 1–13). This complexity creates analytical problems, but it may also provide important keys to the tectonic history of an area if the different zircon populations can be separated.

Rubidium-Strontium Method

Rubidium-strontium age determination is also most applicable in rocks of ages that exceed 100 million years. The age is determined by analyzing ^{87}Rb, ^{87}Sr, and ^{86}Sr in a mass spectrometer. Ages of potassium minerals, primarily micas and feldspars,

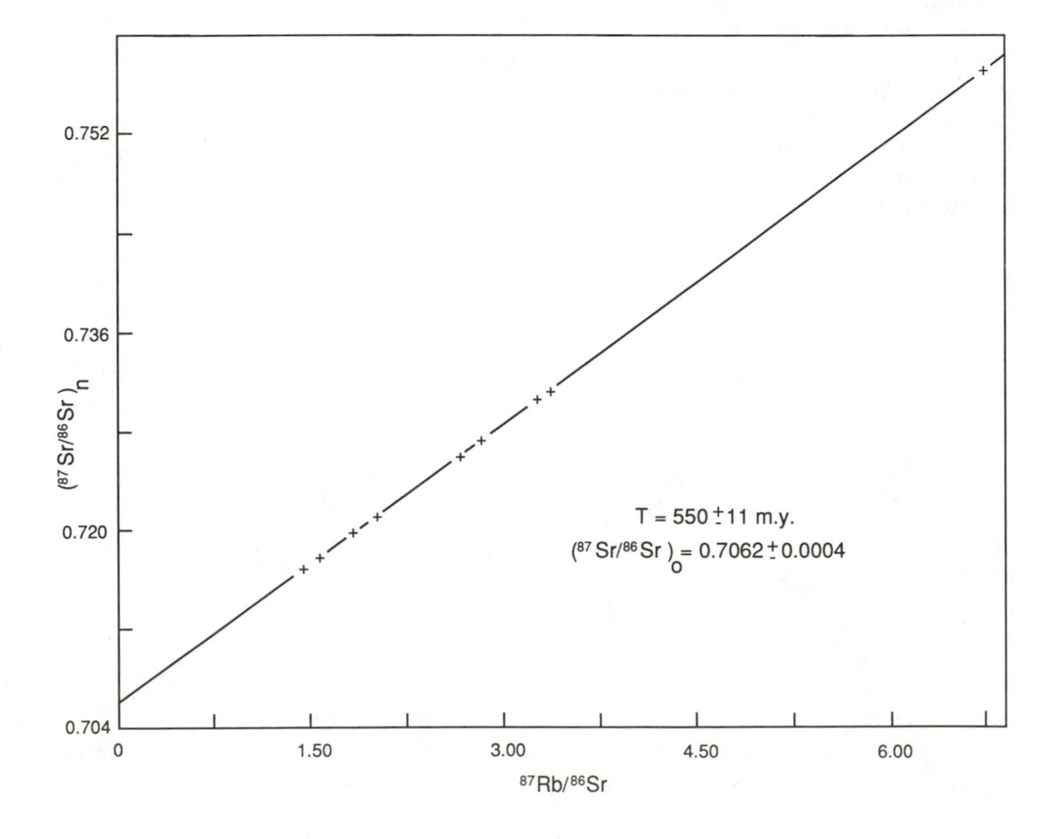

FIGURE 1–14
Rubidium-strontium isochron plot of a series of whole-rock analyses from a granitic batholith in south-central Libya yielding an age of 550 ± 11 Ma. The age of the rock is calculated from the slope of the isochron. (Isochron provided by P. D. Fullagar from W. J. Pegram and others, *Earth and Planetary Science Letters,* v. 30, fig. 3, p. 126 © 1976, Elsevier Science Publishers B.V., Amsterdam.)

may be determined because the parent element Rb most commonly occurs in potassium minerals. Rb-Sr mineral ages are less reliable than ***whole-rock ages,*** where the Rb and Sr isotopes are determined in the entire rock. The decay scheme for this process is

$$^{87}Rb \rightarrow {}^{87}Sr \text{ (half-life} = 48.8 \times 10^9 \text{ y)} \quad (1\text{–}4)$$

The extremely long half-life makes this method best suited to dating very old rocks. The age of the rock or mineral is obtained by plotting $^{87}Sr/^{86}Sr$ versus $^{87}Rb/^{86}Sr$ (Figure 1–14) and statistically fitting the points to the best straight line. The slope of the line gives the age of the rock or mineral. Points should plot very close to or on the line. If they do, the line is called an ***isochron*** and represents the age of the rock. If the points do not plot close to the statistically determined line, the line is called a ***mixing line,*** or *scatterchron,* and the age determined may represent an age between the real age of the rock or mineral and a more recent thermal or other event (such as hydrothermal activity) that has reopened the Rb-Sr system. The blocking temperature for the Rb-Sr whole-rock system is generally lower than that for U-Pb in zircons and for minerals is about the same as for the K-Ar method (discussed in the next section). The principal advantage of the Rb-Sr system over K-Ar is that no daughter product that must be determined is gaseous, and so each has a better chance of being preserved in the material, particularly if whole-rock analyses are made.

The isochron may be projected to the point where it intercepts the $^{87}Rb/^{86}Sr$ axis (Figure 1–14). This yields a value for $^{87}Sr/^{86}Sr$ that represents the value at the time the rock formed and is called the ***$^{87}Sr/^{86}Sr$ initial ratio.*** Studies of isotopic compositions have shown that middle to upper continental crustal rocks have an initial ratio >0.706, but oceanic crust, Rb-depleted lower continental crust, and mantle rocks have initial ratios <0.706. This tool is very useful in determining the sources of magmas for plutons and volcanic rocks. It has also been used for reconstructing ancient continental margins by simply plotting the distributions of $^{87}Sr/^{86}Sr$ initial ratios on a map (Figure 1–15). If a clear-cut division

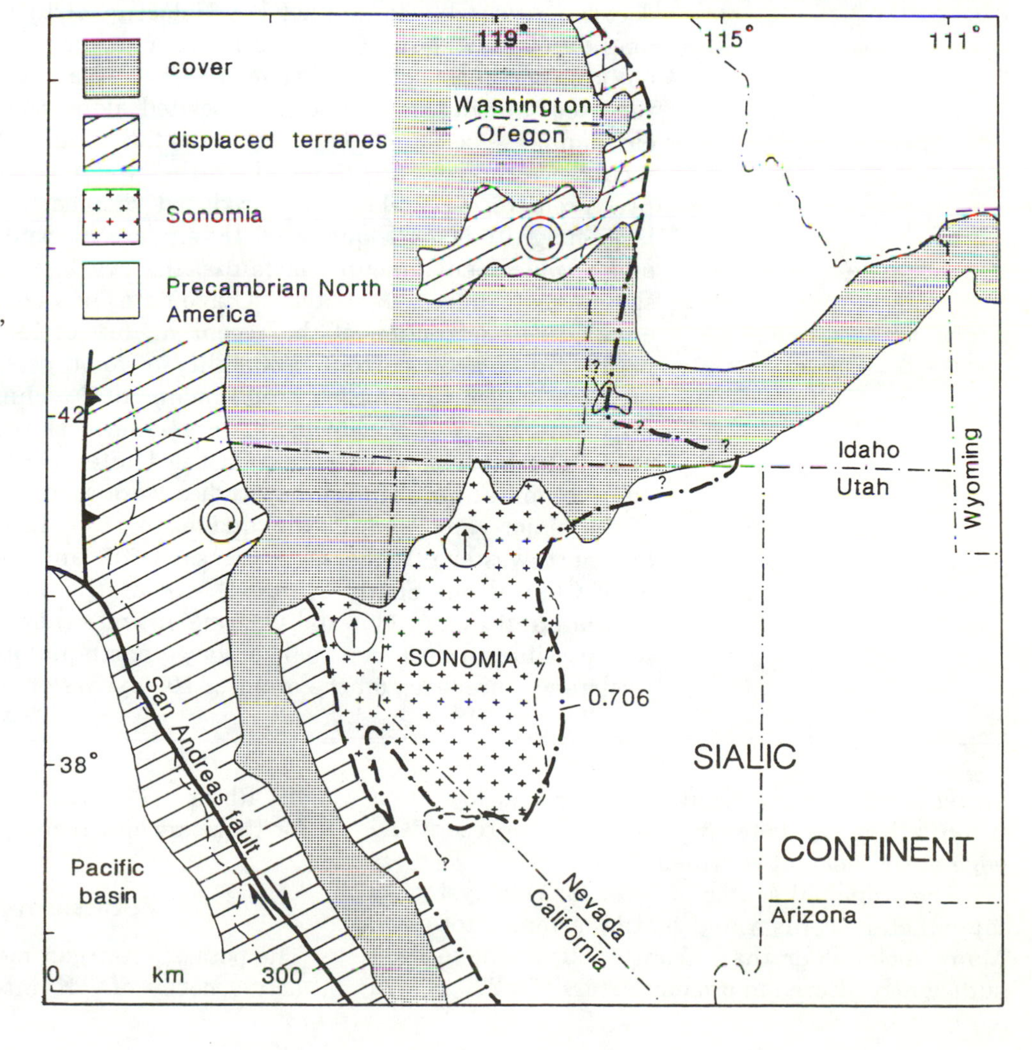

FIGURE 1–15
Map of the western United States showing the 0.706 line. Plutons west of the line have initial ratios <0.706; those east of the line have ratios >0.706. Arrows indicate rotation sense determined for different blocks. S.A.F.—San Andreas fault. (Modified from R. C. Speed, AAPG *Memoir* 34, fig. 2. © 1982. Reprinted by permission of American Association of Petroleum Geologists.)

ESSAY

Capability of Tectonic Structures

The Code of Federal Regulations (Appendix A, Part 100) specifies that documentation of the antiquity of geologic structures in the foundations of critical buildings (such as dams and nuclear power plants) must be provided to show they have not experienced motion in the last 500,000 years. As a result, faults discovered near or within excavations must be carefully studied to show whether they have moved during this time and whether they are capable of moving during the projected useful life of the buildings planned. To some it has been a very costly and difficult regulation, for it requires a level of study and documentation never achieved before. But to many geologists it has provided a body of new and useful information on the structural history of areas where these projects have been undertaken. Faults have been exposed that would not have been known otherwise, and details of movement history have been brought to light. The techniques used in resolving these details have ranged from the applications of the classical geologic laws of cross-cutting relationships and superposition to modern isotopic and strain-analysis techniques.

In tectonically active areas like the West Coast of the United States, documentation of antiquity of structures involves careful study of overprinting structures and age determinations using recent fossils or ^{14}C dating of organic material preserved in the fault zones or in undeformed sediments that truncate these zones. One such study was carried out by Kerry E. Sieh (1984) of the California Institute of Technology as part of his doctoral research. He studied the Holocene history of an excavation 50 m long by 50 m wide by 5 m deep along a segment of the San Andreas fault, 55 km northeast of Los Angeles. The excavation revealed evidence of repeated faulting in sediments deposited along the fault, indicating twelve earthquakes occurred between A.D. 260 and 1857, with an average recurrence interval of 145 years. Of the twelve earthquakes he documented, five prehistoric earthquakes produced displacements comparable to the large-magnitude earthquake of 1857. Careful study of displacement sense and displacement amounts of faults, the overlap (superposition) of younger sediments (Figure 1E–1), and ^{14}C age dates of wood fragments in sediments enabled reconstruction of the recent history of faulting here and led to a prediction of a 60 percent probability of another large earthquake in this area by the year 2000. Sieh used fundamental techniques to work out a very important interval of geologic history—the most recent prehistoric past—in one of the most populous areas in the United States, pointing out a major environmental hazard. Similar studies were carried out earlier in the region of the major New Madrid earthquakes of 1811–12 in southeastern Missouri and northwestern Tennessee by Russ (1979) and more recently in the area near Charleston, South Carolina, site of the large 1886 earthquake, by Talwani and Cox (1985) and Obermeier and others (1985). In both areas, it was possible to document earlier large earthquakes, but with much longer recurrence intervals, on the order of 1000 years or more.

appears between the values greater and less than 0.706, a line can be drawn that may represent the edge of the ancient continental crust.

The principal source of error for this system is later metamorphism and hydrothermal alteration. Many rocks older than about 1500 Ma have been sufficiently altered to produce errors of >200 m.y. in the Rb-Sr ages. Thus zircon ages are more reliable for these ancient rocks.

Potassium-Argon Method

The potassium-argon method is based on the radioactive decay of ^{40}K into ^{40}Ar by a process called

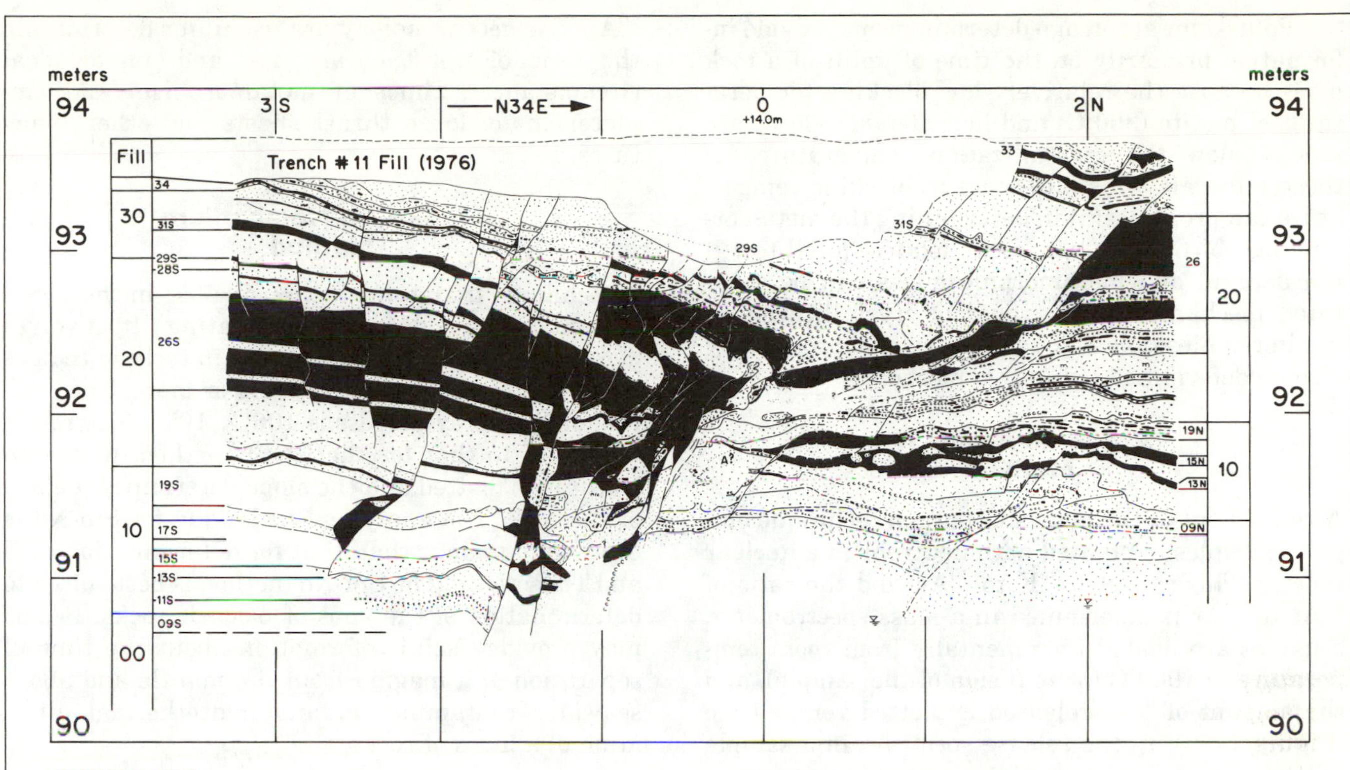

FIGURE 1E–1
Section in part of the trench along Pallett Creek on the San Andreas fault studied by Sieh (1984). It shows several overlapping sequences of sediment deposited during the Holocene and disrupted by recurrent movement along the fault. Most disruption events that produced cross-cutting faults were overlapped by deposition of younger sediments. (From K. E. Sieh, *Journal of Geophysical Research*, v. 89, pl. 46, © 1984 by the American Geophysical Union.)

References Cited

Obermeier, S. F., Gohn, G. S., Weems, R. S., and Gelinas, R. L., 1985, Geologic evidence for recurrent moderate to large earthquakes near Charleston, South Carolina: Science, v. 227, p. 408–410.

Russ, D. P., 1979, Late Holocene faulting and earthquake recurrence in the Reelfoot Lake area, northwestern Tennessee: Geological Society of America Bulletin, v. 90, p. 1013–1018.

Sieh, K. E., 1984, Lateral offsets and revised dates of large prehistoric earthquakes at Pallett Creek, southern California: Journal of Geophysical Research, v. 89, p. 7641–7670.

Talwani, Pradeep, and Cox, John, 1985, Paleoseismic evidence for recurrence of earthquakes near Charleston, South Carolina: Science, v. 229, p. 379–381.

branching decay. Most of the ^{40}K actually decays into ^{40}Ca, with a much smaller proportion decaying into ^{40}Ar, but it is no problem because of the abundance of potassium in minerals, including the micas, feldspars, and amphiboles. The decay process is

$$^{40}K \rightarrow {}^{40}Ar \text{ (half-life} = 1.2 \times 10^9 \text{ y)} \qquad \textbf{(1–5)}$$

The technique is generally useful for determining the ages of minerals a million years old or more. The K-Ar method suffers from having a gaseous daughter product that must be trapped in the crystal lattice of the potassium-bearing mineral. Biotite, muscovite, and hornblende lattices retain argon better than do other minerals.

Potassium-argon age determinations provide information primarily on the time of uplift of a rock mass because the relatively low blocking temperatures of biotite (300°C) and hornblende (500°C) are much below the crystallization temperatures of these minerals. This difference in blocking temperature can prove useful in determining the metamorphic age of fine-grained rocks (slates, phyllites) if the detrital and metamorphic micas can be separated mechanically. The ages of crystallization of unaltered Mesozoic and Tertiary volcanic rocks may also be determined using this method.

Argon-40/Argon-39 Method

A refinement of the potassium-argon technique employs samples irradiated with neutrons in a nuclear reactor. This converts ^{39}K to ^{39}Ar, and the ratio of ^{40}Ar to ^{39}Ar is determined in a mass spectrometer. Samples are heated incrementally from room temperature to 1000°C (or to fusion of the sample), and the amount of ^{39}Ar released is plotted versus time (Figure 1–16). In the release spectrum in a sample with no excess argon, a plateau results for similar apparent ages for gas released with increasing temperature. An age thus determined from the plateau is assumed to be the cooling age of the rock mass for the mineral being determined. An irregular *release spectrum* permits identification of samples containing excess argon and samples that have had an uneven history of opening and closing of the system to release or absorb argon. Studies using K-Ar and $^{40}Ar/^{39}Ar$ geochronology are useful for determining the time of uplift of an area and, under ideal circumstances, times of metamorphism and emplacement of large thrust sheets and other structures.

Samarium-Neodymium Method

A technique that gained extensive use in the 1980s is samarium-neodymium age dating. It involves determination of ^{147}Sm and ^{143}Nd in rocks with ages of several hundred million years or more. The half-life for the decay process is 106×10^9 y. The result is similar to that for the Rb-Sr method, where an isochron is plotted and the slope determines the age of the rock. The analytical technique for Sm-Nd is even more tedious than that for determination of U and Th isotopes. The Sm-Nd method is best suited to determination of the ages of basaltic rocks. It also may provide useful information about the time of separation of a magma from the mantle and about seawater contamination and hydrothermal alteration of a mass of rocks.

EQUILIBRIUM

The Earth is a dynamic system. Energy from radioactive decay, the Earth's gravity field, and a smaller component from our sun and moon drive processes within the Earth. Heat converted to work drives processes within the Earth that move plates, deform

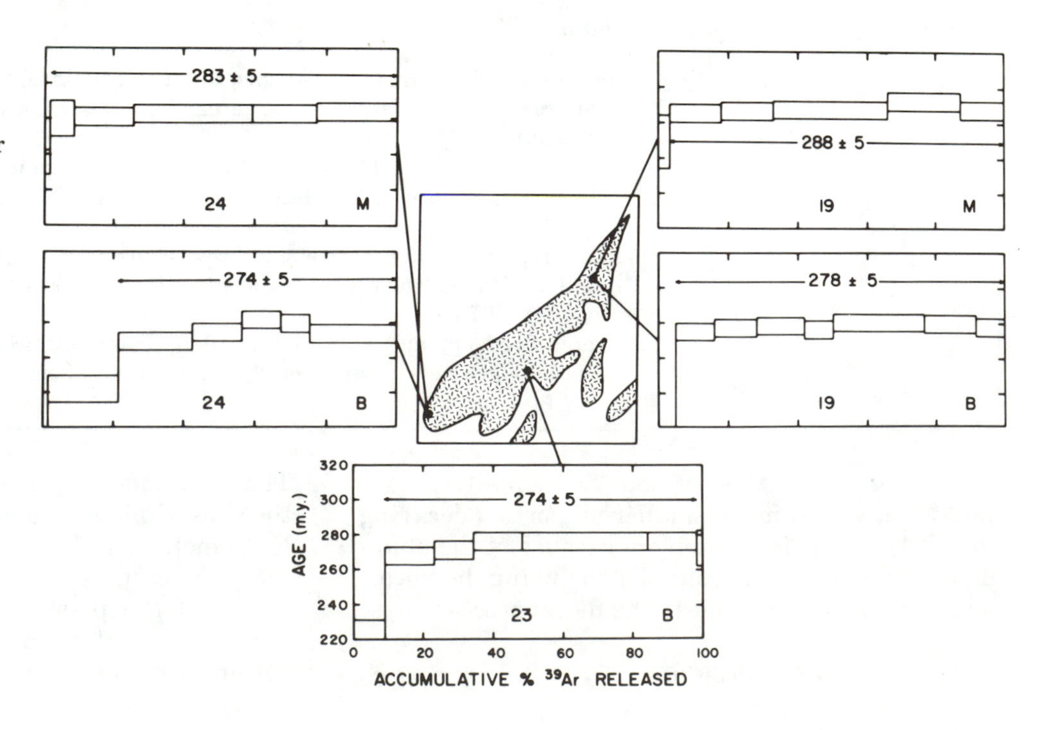

FIGURE 1–16
Incremental argon-release spectra from a New England pluton. (From R. D. Dallmeyer and Otto Van Breeman, *Contributions to Mineralogy and Petrology,* v. 78, fig. 7. © 1981, Springer-Verlag, Heidelberg.)

rocks of the lithosphere, and produce melts and metamorphism. Generally, some excess energy results and must be dissipated to restore a state of rest, or ***equilibrium,*** to the part of the lithosphere where the excess occurs. The balance may be restored by volcanic eruption, breaking the crust along a fault, or some other process whereby heat may be converted into mechanical energy. The second law of thermodynamics predicts that a certain amount of energy is never available to do work and will be lost in any energy-consuming process, as long as the process is not 100 percent efficient. This quantity, called *entropy,* increases with time as more energy is expended. From a structural point of view, an increase in entropy is reflected in an increase in deformation relative to the undeformed state.

All processes in nature move toward a state of equilibrium. If heat is added to a system, the system will readjust to once again establish a state of equilibrium at the new temperature. The readjustment may be in the form of recrystallization, chemical reaction, change in deformation style from brittle to ductile (or vice versa), or some other process. Similar readjustments take place in response to changes in pressure. Striking a rock with a hammer produces elastic rebound if it is not struck hard enough to exceed the elastic strength of the rock (Chapter 6). If we strike the rock hard enough to break it, permanent deformation in the form of a fracture is produced and any excess energy remaining is dissipated as a very small temperature increase in the vicinity of the fracture.

A large-scale attempt to restore equilibrium occurred in northern Europe and North America after melting of the Pleistocene ice sheets. When the ice sheets formed and loaded the continents with additional mass, the more rigid lithosphere sank to a lower level in the less rigid asthenosphere to

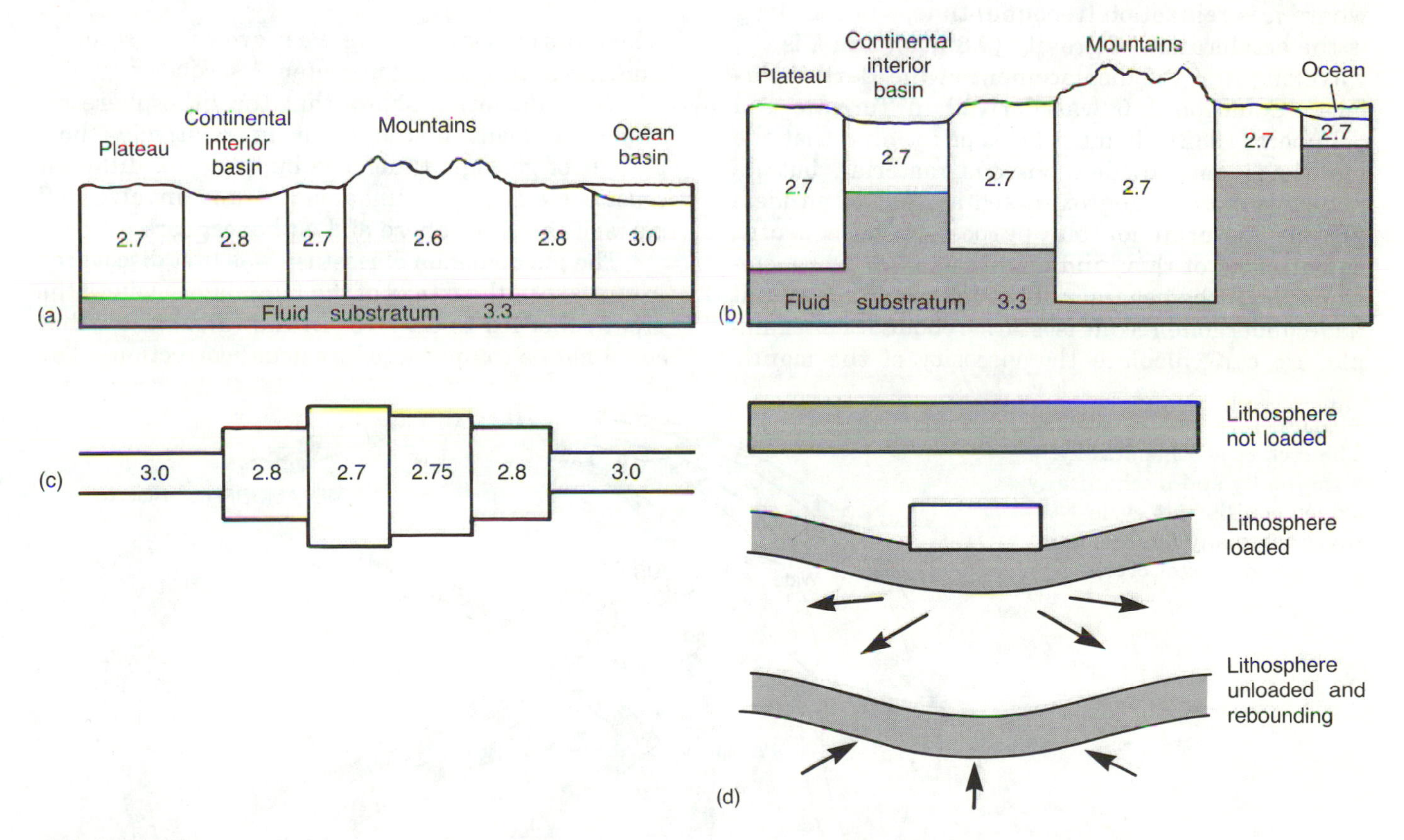

FIGURE 1–17
Isostatic equilibrium between crustal blocks of different densities and thicknesses, as well as between the continents and oceans; (a) and (b) are the early models of Pratt and Airy based separately on different density (in units of g/cm^3) and different size of blocks. We realize today that both density and size of crustal blocks affect the isostatic equilibrium of the blocks (c), and are involved in isostatic compensation; (d) shows the effect of loading and unloading of a mass on the lithosphere. Arrows indicate directions of compensating flow in the asthenosphere during and after loading.

attain a new equilibrium state. As the ice melted, the lithosphere was again forced out of equilibrium and accordingly began rebounding to restore a new equilibrium state. This process is continuing today. The greatest rebound occurs where the ice was thickest. This readjustment process, involving a state of equilibrium between blocks within the continents and between continents and the adjacent oceans, is called ***isostatic equilibrium*** (Figure 1–17). It is possible to calculate the viscosity of the mantle from the rate of isostatic rebound of the continents where information on the rate of uplift can be obtained. A good example of this is in the determination of rate of uplift of raised beaches from ^{14}C age determinations of wood fragments found in successive beach levels. The viscosity (μ) of the mantle beneath the uplifted beaches may be calculated by

$$\mu = \tau_r \rho g \lambda \, (4\pi)^{-1} \qquad \textbf{(1–6)}$$

where τ_r is relaxation (rebound) time, ρ is density, g is the acceleration of gravity (9.8 m/s), and λ is the wavelength of the displacement of the Earth's surface. (Equation 1–6 was derived in Turcotte and Schubert [1982].) It must be kept in mind that the mantle is not an ideal viscous material, but its behavior may be approximated as that of an ideal viscous material for our purposes. Consequently, calculations of this kind enable us to draw conclusions about the behavior of the mantle in areas that have undergone recent isostatic rebound. For example, we can calculate the viscosity of the mantle beneath the central Canadian shield by determining the uplift rate of beach terraces along the shore of James Bay in northeastern Ontario and by using estimated dimensions of the Keewatin ice sheet that covered this area during the Pleistocene. If the oldest beaches in that area are now 180 m above sea level and it is assumed that 20 m more of uplift will occur, we can obtain the rate of uplift from the time of retreat of the glacial ice from this region, which was about 8000 years ago. Best estimates of the width of the Keewatin ice sheet give it dimensions of about 9000 km, which is the wavelength of the displacement of the surface in equation 1–6. The density of the mantle is assumed to be 3300 kg/m^3. Calculating μ from equation 1–6,

$$\begin{aligned} \mu &= [(8000\ \text{y})(365\ \text{d/y})(24\ \text{h/d})(3600\ \text{s/h})] \qquad \textbf{(1–7)} \\ &\quad \times 3300\ \text{kg/m}^3 \times 9.8\ \text{m/s}^2 \times (9 \times 10^6\ \text{m}) \\ &\quad \times (4\pi)^{-1} \\ &= 5.84 \times 10^{21}\ \text{Pa s} \end{aligned}$$

(The units of viscosity here, Pa s, are pascal seconds. One pascal is 1 kilogram meter^{-1} second^{-2}.)

The calculation shows that the lithosphere responds to loads placed on it in relatively short periods of geologic time. The buoyancy of different crustal elements is fundamental and involves all parts of the lithosphere and asthenosphere.

The phenomenon of isostasy was first discovered in surveys in the flanks of the Himalayas, where the great relief led to an error in the calculations that could not be compensated by usual corrections. The

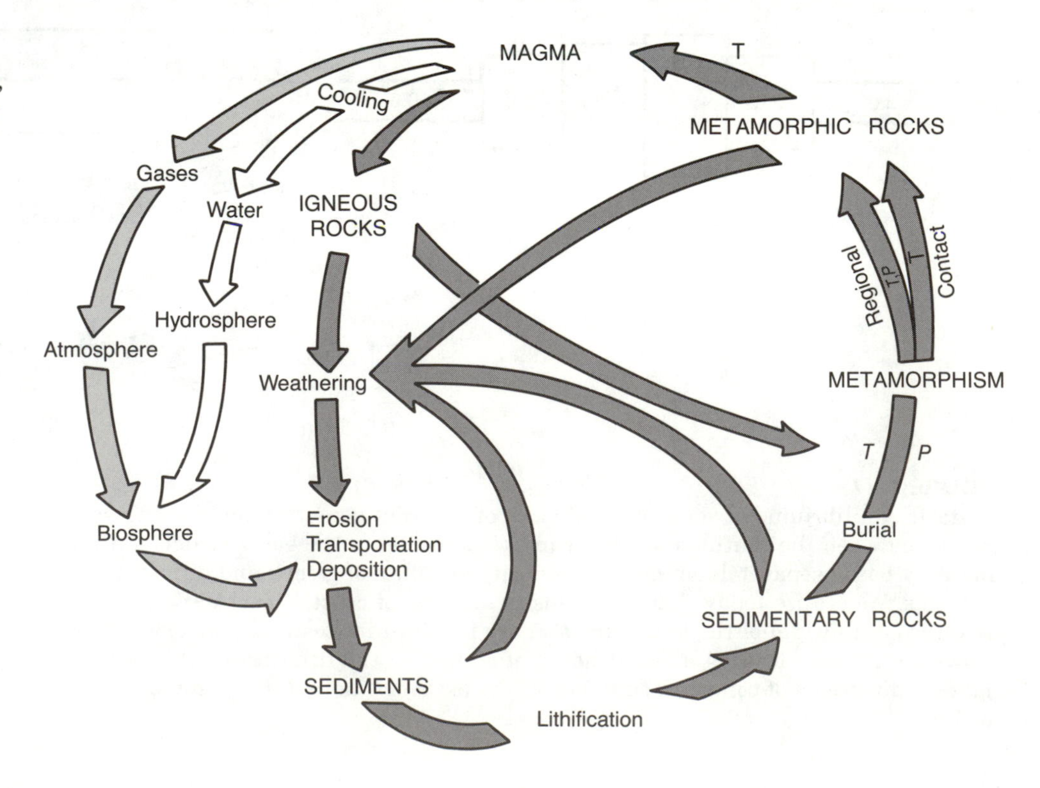

FIGURE 1–18
The rock or geochemical cycle, a thermally and mechanically driven equilibrium cycle involving many intermediate stages and shorter cycles.

early isostatic models based on either volume or density failed to satisfy the need for correction. Later models incorporating both volume and density changes (Figure 1–17), along with flexural bending of the crust, best corrected the errors in the surveys and demonstrated the fundamental nature of the principle of isostatic adjustment.

The emplacement of large thrust sheets (Chapter 11) having areas of hundreds of square kilometers and thicknesses of 5 to 10 km would thicken the lithosphere in the immediate area where the thrust sheet was emplaced and would require profound adjustments in the asthenosphere beneath to accommodate the increase in the lithospheric thickness above. Similarly, crustal extension processes, like those affecting the Basin and Range Province in the western United States (Chapter 14), serve to unload and thin the lithosphere, thus decreasing the amount of lighter material above the asthenosphere. Profound adjustments also occur here in both the lithosphere and asthenosphere. Because the amount of light material and total mass of the lithosphere is decreased, a significant amount of uplift does not occur after extension.

GEOLOGIC CYCLES

Most geologic processes are driven by cyclic changes of energy fluxes, often over millions of years. The ***rock*** or ***geochemical cycle*** is probably the most familiar of these geologic cycles (Figure 1–18). Each stage in the cycle, from the crystallization of magma

FIGURE 1–19
The Wilson cycle of opening and closing of an ocean basin. The cycle may be complicated by the formation and movement of suspect terranes, partial closing of small oceans, and the lack of continent-continent collision to terminate the cycle.

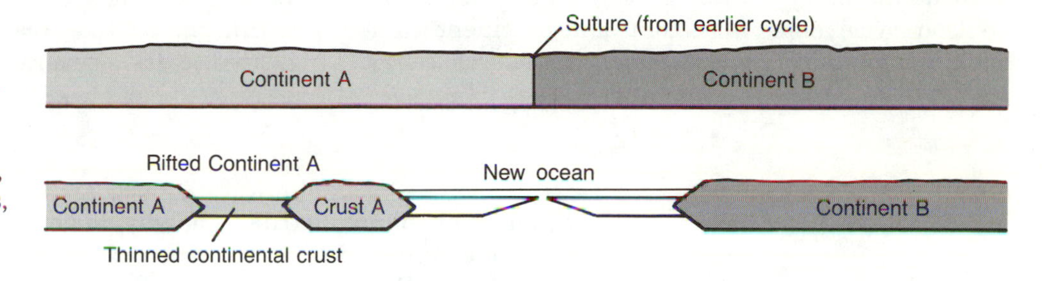

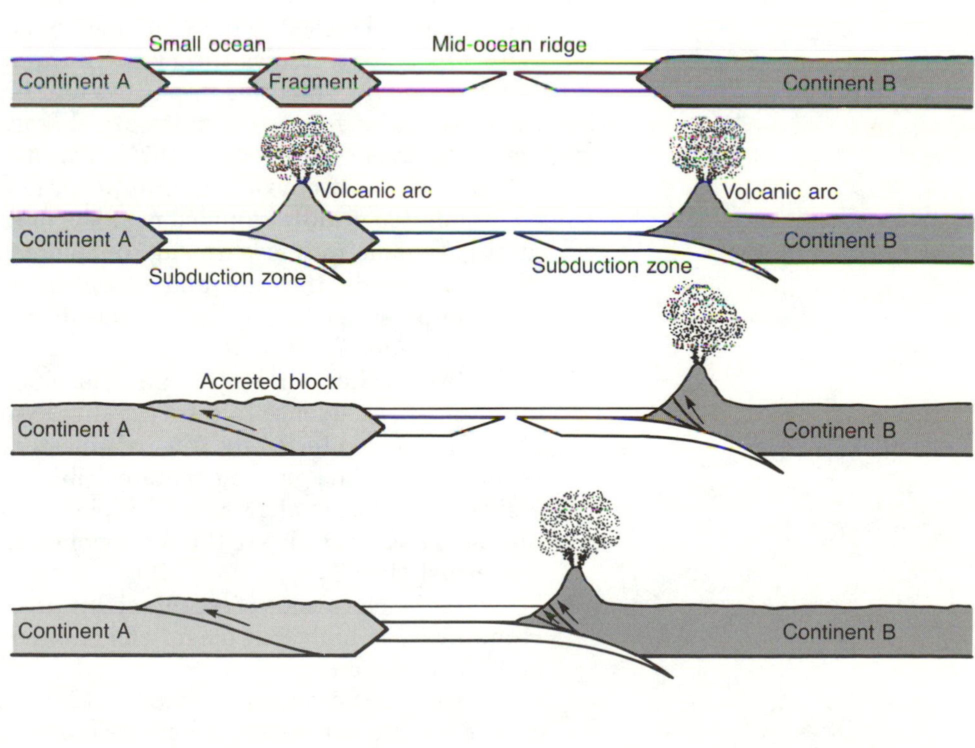

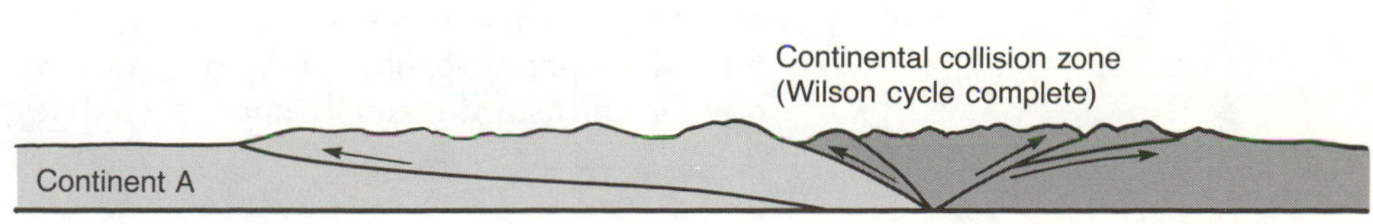

to conversion of sedimentary or igneous rocks into metamorphic rocks, is in some way driven by thermal processes and, to a lesser degree, by changes in pressure. Inputs of heat or mechanical energy at particular places short-circuit the cycle. Chemical changes accompany deformation in several stages in the geochemical cycle. All stages tend to restore equilibrium to the system.

The ***tectonic,*** or *Wilson,* ***cycle*** (defined in the section on plate tectonics) involves plate motion, beginning with the opening of an ocean basin and producing a trailing plate margin like the present-day East Coast of the United States (Figure 1–19). The phase is terminated by formation of a subduction zone along the trailing margin, subducting oceanic crust, generating heat and pressure, and forming either a volcanic island arc offshore or a continental magmatic arc on the old continent. The Wilson cycle ends with continent-continent collision, closing the old ocean. Stages in the cycle reflect response to changing physical conditions in an attempt to restore a state of equilibrium to all or part of the plate system. Mountain building may be considered a direct consequence of a partial or completed Wilson cycle. The world's highest mountains formed in response to energy expenditure in a collision zone between India and Asia during closing of part of the ancient Tethys Ocean. The exceptional height of the Himalayas is due to an extreme isostatic imbalance in the crust because of the great thickness of continental crust there. Erosion is rapidly reducing the elevations in this chain to levels that will be closer to equilibrium, but rapid uplift has thus far outstripped the erosion rate.

All processes operating upon or within the Earth act to achieve and maintain equilibrium. Energy is constantly being dissipated to keep the Earth in a dynamic state. Work is done to melt rocks within the Earth to restore equilibrium, and energy is used to drive several cyclic processes.

Questions

1. In terms of energy distribution, why do geologic processes tend to reach a state of equilibrium?
2. How are the Wilson cycle and the doctrine of uniformitarianism related?
3. How can the laws of superposition, faunal succession, and cross-cutting relationships be used to date terrane boundaries?
4. Why was plate-tectonics theory not formulated in the nineteenth century like the unifying theories in physics and biology?
5. Zircons from a foliated granite are discordant, plotting on a chord that yields an upper intercept of 1870 m.y. and a lower intercept of 480 m.y. What is the most reasonable interpretation of the two numbers?
6. An unfoliated granite sampled near the foliated pluton in Question 5 yields an Rb-Sr whole-rock isochron age of 365 Ma and an $^{87}Sr/^{86}Sr$ initial ratio of 0.712. What do the age and initial ratio tell you about the time of metamorphism of the rocks and the initial ratio about the source of the granitic magma?
7. $^{40}Ar/^{39}Ar$ studies on biotite and hornblende from a metasedimentary-metavolcanic sequence in a thrust sheet yield plateau ages of 320 Ma and 380 Ma. Slate interlayers from a limestone in the footwall of the thrust sheet contain fine-grained metamorphic muscovite that, once separated, yields a conventional K-Ar and a plateau age of 460 Ma. What do these numbers mean relative to the timing of metamorphism and emplacement of the thrust sheet?
8. Why are the elevations of recent mountain chains, like the Andes, Alps, and Himalayas, so high, but elevations of older chains, like the Appalachians and Urals, so low?
9. Using the mantle viscosity of 5.84×10^{21} Pa s in equation 1–7, calculate the effects on a lithosphere (100 km thick, $\rho = 2900$ kg/m^3) of the emplacement of a large thrust sheet 300 km long, 100 km wide, and 10 km thick and having a density, ρ, of 2700 kg/m^3. Assume the thrust sheet was emplaced during an instantaneously short period of geologic time.

Further Reading

Adams, F. D., 1954, Birth and development of the geological sciences: New York, Dover Publications, 506 p.

Provides an interesting summary of the evolution of geological science from classical times through the beginnings of modern geology with Hutton, Lyell, Darwin, and others in the nineteenth century.

Cloud, P. E., 1970, Adventures in Earth history: San Francisco, W. H. Freeman and Company, 992 p.

A compendium of classic papers on the foundations of ideas on the origin of the Earth, the atmosphere and life, the geologic record, and geologic processes.

Dott, R. H., and Batten, R. L., 1981, Evolution of the Earth: New York, McGraw-Hill Book Company, 113 p.

This is a historical geology book but is an excellent compendium of the fundamental principles upon which many observations in structural geology are based. It also contains a wealth of information on the tectonic evolution of North America.

Faure, Gunter, 1986, Principles of isotope geology, 2nd ed.: New York, John Wiley & Sons, 464 p.

An outline of the techniques of radiometric age dating with problems and discussions of the shortcomings of each.

Glen, William, 1982, The road to Jaramillo: Critical years of the revolution in Earth science: Stanford University Press, 459 p.

Outlines the history of the development of plate-tectonics theory, particularly through use of paleomagnetic measurements.

Hamilton, W. B., 1979, Tectonics of the Indonesian region: U.S. Geological Survey Professional Paper 1078, 345 p.

A synthesis of the geology of the Indonesian region, containing numerous maps showing the elements of a dispersed group of terranes ranging from Precambrian basement to recent volcanic-arc materials in the initial stages of being swept back into the Asian continent as Australia moves northward.

Howell, D. G., 1985, Terranes: Scientific American, v. 253, no. 5, p. 116–125.

Summarizes the distribution of microplates, or terranes, in and around the Pacific basin, presenting the background of plate tectonics and accretion concepts.

Wilson, J. T., 1966, Did the Atlantic close and reopen? Nature, v. 211, p. 676–681.

This short paper set the stage for the concept of the Wilson cycle.

2

Nontectonic Structures

> It has been said that stratigraphy is the basis of all geology (Weller, 1947). This is true of the metamorphic rocks as well as the sedimentary rocks. Cognizant of the importance of petrography, physical chemistry, and structural geology, the investigator of metamorphic rocks should nevertheless give more adequate treatment to the fascinating problems in stratigraphy, sedimentation and paleogeography that await solutions in the interesting rocks to which, perhaps foolishly, he has dedicated his scientific career.
>
> MARLAND P. BILLINGS, 1950, Geological Society of America *Bulletin*

SEDIMENTARY STRUCTURES HAVE A NONTECTONIC ORIGIN AND may help determine the facing (younging) direction in a sequence of rocks to ascertain if the sequence is upright or overturned. Sedimentary and volcanic structures originate in the environment where the rocks form. Except for fossils, primary structures are probably the best tools for working out the structural geometry and history in deformed rocks. In the 1950s, Robert Shackleton used primary sedimentary structures to determine that a large part of the rocks in the southern Highlands of Scotland are upside down and concluded that they make up the inverted

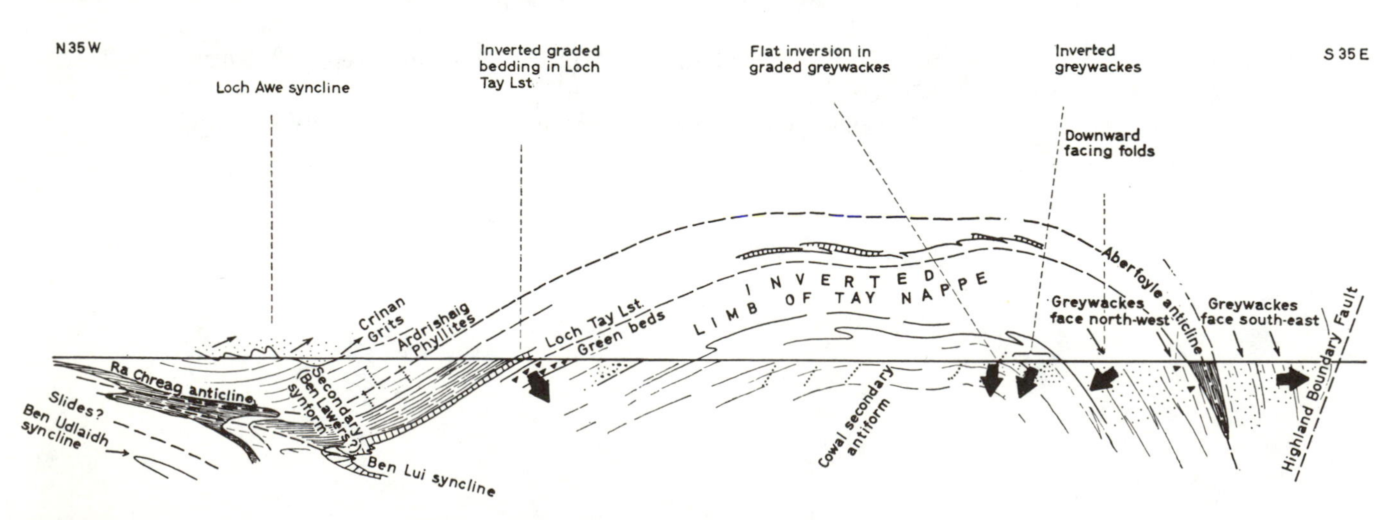

FIGURE 2–1
Geologic cross section of the Tay nappe in the Scottish Highlands showing relationships between overturned structures and facing directions of sedimentary sequences determined from sedimentary structures. Arrows indicate facing directions. (Reproduced by permission of the Geological Society from Downward facing structures of the Highland Border, R. M. Shackleton, in *Geological Society of London Quarterly Journal,* v. 113, 1958.)

limb of a large overturned fold (Figure 2–1). He later published a paper outlining the significance of downward-facing structures (Shackleton, 1958).

In studies of deformed rocks, we often need to know whether the observed structures have a tectonic or nontectonic origin (Figure 2–2). Many structures formed in a primary depositional environment may mimic structures in rocks formed in response to tectonic deformation. Thus it is important to make the distinction. Moreover, many sedimentary structures that form at or near the surface provide useful models to compare with rocks that form at elevated temperatures and pressures. For example, structures formed by ductile flow (Chapter 6) in water-

FIGURE 2–2
(a) Folds formed by slumping of nearly plastic, water-saturated sediment during freezing (from Byron Stone, 1976, *Journal of Sedimentary Petrology,* v. 46) and (b) the Great Red Spot in the atmosphere of Jupiter (*Voyager I* photograph, National Aeronautics and Space Administration) exhibit almost the same characteristics as tectonic folds (c) that form at high temperatures and pressures deep in the Earth. The example in (c) is sillimanite-grade gneiss in the Thor-Odin dome in the Shuswap complex in southern British Columbia. (RDH photo.)

saturated silts, glacial ice, and salt (halite) deposits at surface pressure and temperature (P-T) conditions are of the same styles as those formed at much higher P-T conditions in rocks deformed at depths of 15 to 20 km.

Distinguishing tectonic from nontectonic structures often requires observation of particular critical features. If the structure in question has overprinted or cross-cuts an existing, obviously tectonic structure (such as a fault), then it probably had a tectonic origin. If, on the other hand, tectonic structures overprint the structure in question but cannot be shown to overprint earlier tectonic structures, it may be difficult to determine if the origin was tectonic or nontectonic. Careful study of the orientation and sequence of development of minerals in all structures present is sometimes necessary before reaching a conclusion. Orientation of small-scale tectonic structures should be consistent over a large area and related to larger mappable structures (Pumpelly's rule), but the orientations of primary (nontectonic) structures will be related to features (for example, currents) in depositional environments, not to large tectonic structures.

PRIMARY SEDIMENTARY STRUCTURES

Bedding

The most common characteristic and most diagnostic feature of sedimentary rocks is ***bedding,*** and, because it forms with horizontal orientation, it is the first-order reference surface for most structural measurements. ***Bedding planes*** are mechanical zones of weakness that form in the primary sedimentary environment (Figure 2–3). They exist for several reasons, most commonly because of slight compositional or textural differences at the interface between adjacent beds. The differences in composition and texture reflect subtle changes in sedimentary environment in a depositional basin. As an example, sediment composed of particles of different size, shape, or (in many instances) composition may be deposited during sediment influx. Most of the sediment previously deposited may be of another particle size or composition (such as mud or clay) and separate the dominant sand layers. The textural difference produces a physically distinct boundary—a *bedding plane*—between the material deposited during the first influx of sediment and the material deposited later.

A second mechanism involves compaction of sediment already deposited. When the next influx of sediment occurs, it is compacted less than that deposited earlier. Consequently, a discontinuity may form between newly deposited material and material deposited previously. This discontinuity is also a bedding plane (Figure 2–3).

Graded beds form where sediment of widely different particle sizes is deposited rapidly in the same depositional environment (Figure 2–4). The largest particles settle to the bottom first, followed by successively smaller particles. In some graded beds, particle sizes range from pebbles or boulders at the base to clays at the top. After lithification, the bottom of the bed may be conglomerate, the middle part sandstone, and the top of the bed shale, with a gradual transition between units. Graded beds form in both sedimentary and pyroclastic deposits and are important tools for determining the *younging direction* (*facing direction,* or the top of a sequence). Channels, shown in Figure 2–4, and other primary features may be associated with graded beds. *Reversed graded bedding* is relatively rare but may form as a primary structure during deposition of pumice fragments in water. Larger pumice fragments will be at the tops of beds and smaller particles beneath because pumice fragments commonly float and settle more slowly than denser particles. Debris flows may also produce reversed graded bedding. The term has also been applied to describe metamorphosed normal graded beds where the fine-grained top of the bed is recrystallized to very coarse mica that is larger than the granular material at the original bottom of the bed.

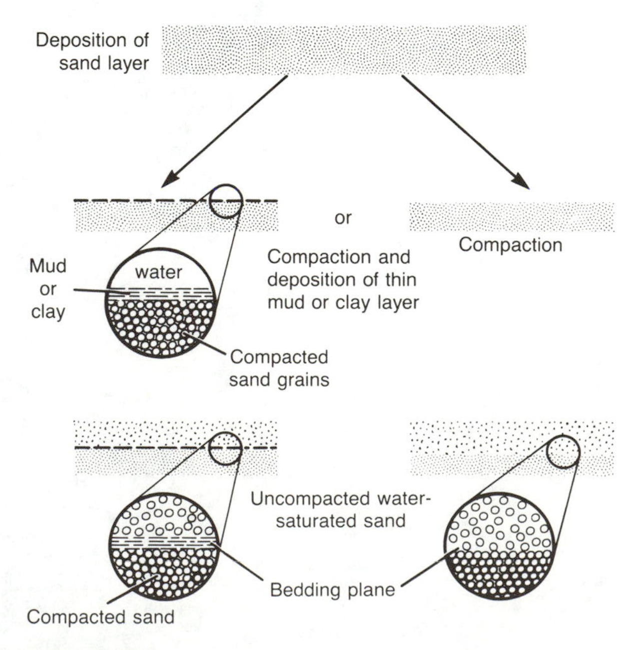

FIGURE 2–3
Formation of bedding planes in sediments.

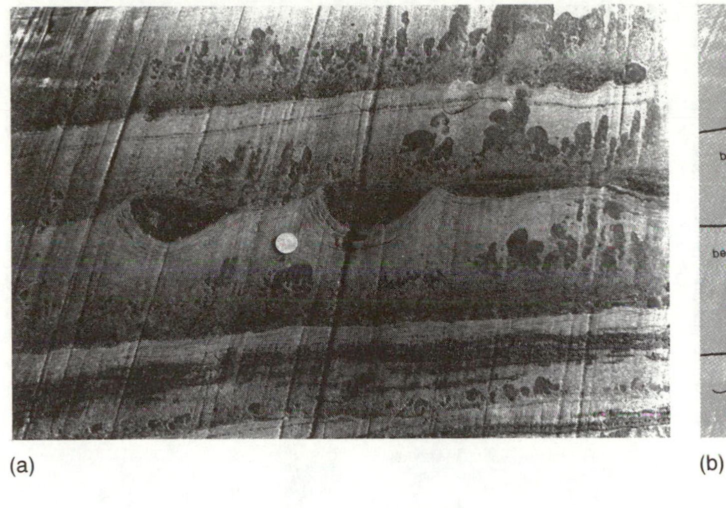

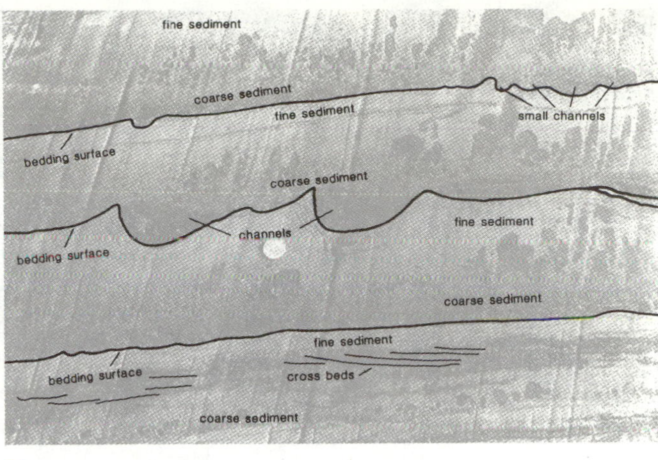

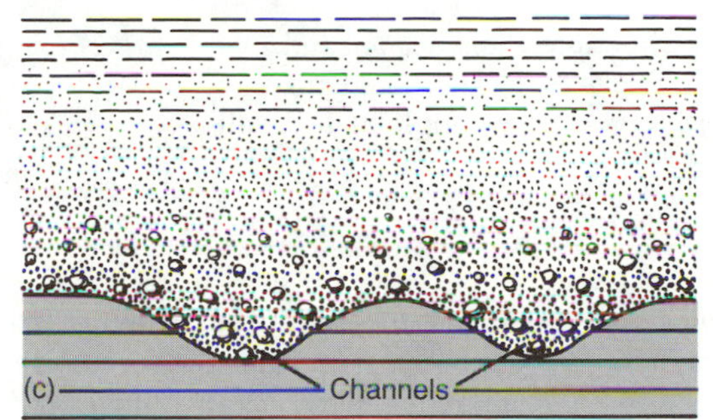

FIGURE 2–4
(a) Graded beds in Middle Silurian Ekeberg Graywacke exposed on a glacially scoured surface near Jämtängen, Sweden. Note there are three cycles of graded beds and fining-upward sequences, and cross beds appear in the lowest bed. (RDH photo.) (b) Sketch of photo in (a). (c) General relationships of graded beds and associated textures.

Cross bedding, or cross stratification, is common in sandstones. It forms where sediment grains are transported by water or wind and produce moving mounds, or dunes, with inclined laminations within an individual bed. Two common types of cross beds are *normal* (trough-festoon) and *torrential* or *planar* (Figure 2–5). Normal cross beds may be used for determining facing direction; torrential cross beds, even though inclined, provide the same perspective whether overturned or upright and so cannot be used to determine facing direction. A third type, called *ripple cross laminations* and *hummocky laminations*, are small-scale cross beds that form under low-velocity conditions and may be useful for determining tops (Reineck and Singh, 1975; Harms and others, 1982). Those useful in determining tops

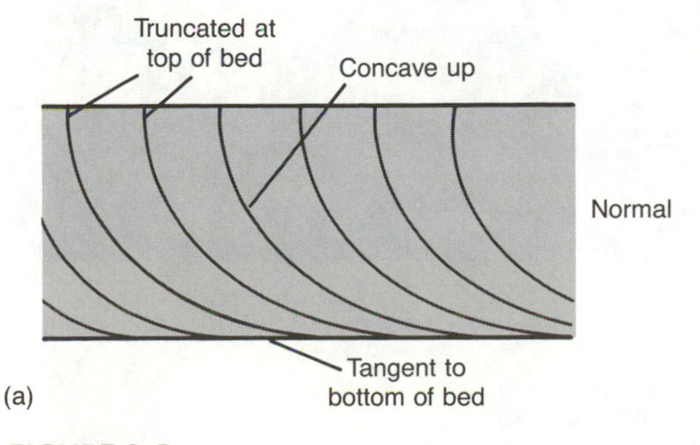

Current direction
Planar
(torrential)
(b)

FIGURE 2–5
Normal (a) and torrential, or planar (b), cross beds.

(a)

(b)

FIGURE 2–6
(a) Normal cross beds allow determination of facing direction; Pleistocene sand, Elmore County, Idaho. (H. E. Malde, U.S. Geological Survey.) (b) Deformed cross beds defined by thin dark laminae that intersect bedding at a low angle at Eidsvoll quarry, near Oppdal, southern Norway. (RDH photo.)

(normal) are generally concave upward, tangent to the bottom of the primary bedding plane, and generally truncated at the top (Figure 2–6).

Mud Cracks

Fine sediment exposed to the atmosphere forms extensional shrinkage cracks called ***mud*** or ***desiccation cracks*** (Figure 2–7). In plan view, they are polygonal, bounded by cracks in the surface of the sediment that taper downward and terminate. Because of its fine texture, the surface layer may separate from that below and curl up, providing another indicator of tops. Mud cracks occur in the geologic record, but, because of the fragility in the subaerial environment where they form, not as frequently as cross beds or graded beds. Where preserved, they are quite useful in determining the facing direction and may also be used as indicators of natural strain (deformation) in rocks.

(a)

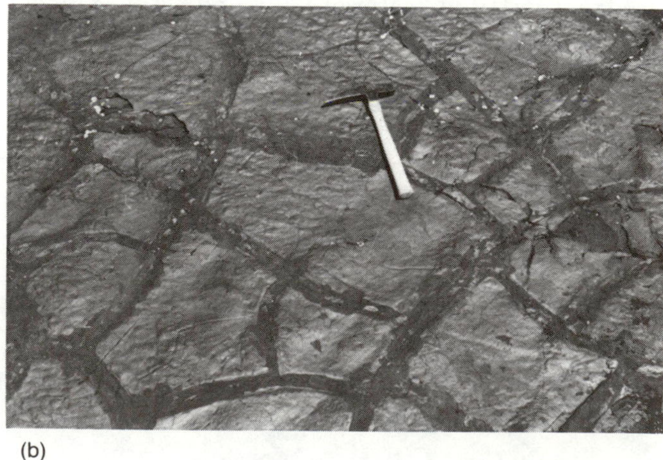
(b)

FIGURE 2–7
(a) Mud cracks in a dried-up shallow body of water near Salton Station, Riverside County, California. (G. K. Gilbert, U.S. Geological Survey.) (b) Ancient mud cracks in siltstone, Isle Royale National Park, Michigan. (N. K. Huber, U.S. Geological Survey.)

Ripple Marks

Ripple marks may form where sediment finer than about 0.6 mm is moved by a current or where the bottom sediment surface is otherwise disturbed by water moving above a threshold velocity. The setting is very common along beaches and streams as well as in deeper water where bottom currents or surface waves interact with bottom sediment. Two kinds of ripple marks may be distinguished: ***current*** (translational) and ***oscillatory***. Current ripples form where a prevailing direction of transport and deposition of sediment occurs; they are commonly asymmetric, and their steep sides face downstream, in the direction of transport (Figure 2–8). They present the same shape regardless of whether they are upright or overturned. Oscillatory ripples are symmetrical and generally consist of alternating high and low crests. They form where currents oscillate. The crests are symmetrical, with sharp peaks and rounded troughs (Figure 2–9). Oscillatory ripples may therefore be used for determining tops in sediments. Unfortunately, current ripples are more common than oscillatory ripples and generally cannot be used for top determination.

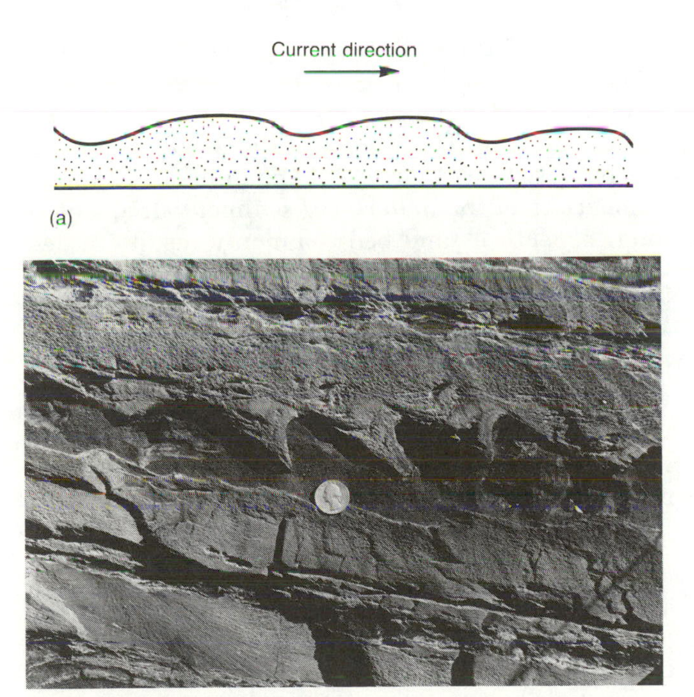

FIGURE 2–8
(a) Current (translation) ripple marks. (b) Slightly deformed current ripples in silty carbonate and siltstone, Maddens Branch, Ocoee Gorge, Polk County, Tennessee. Bedding plane below (where the quarter is located) is minimally distorted, whereas the rippled top has been flattened into a series of asymmetric foldlike shapes. (RDH photo.)

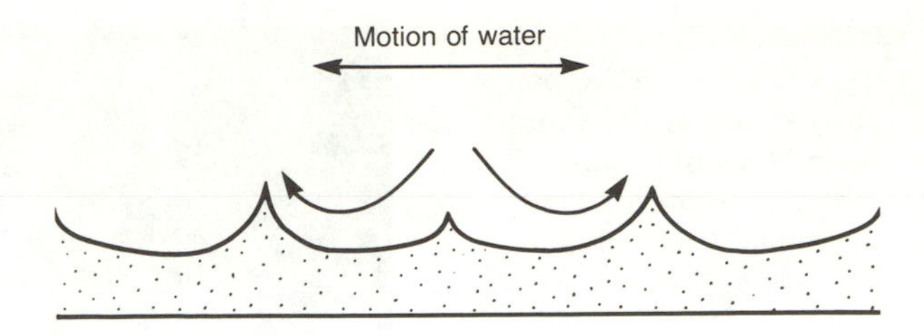

FIGURE 2–9
Oscillatory ripple marks.

Rain Imprints

Rain imprints are sometimes preserved in the sedimentary record where rain falls on fine sediment that is intermittently exposed to the atmosphere (Figure 2–10). If the sediment is covered quickly by another layer of sediment, the imprints may be preserved and can be used to determine

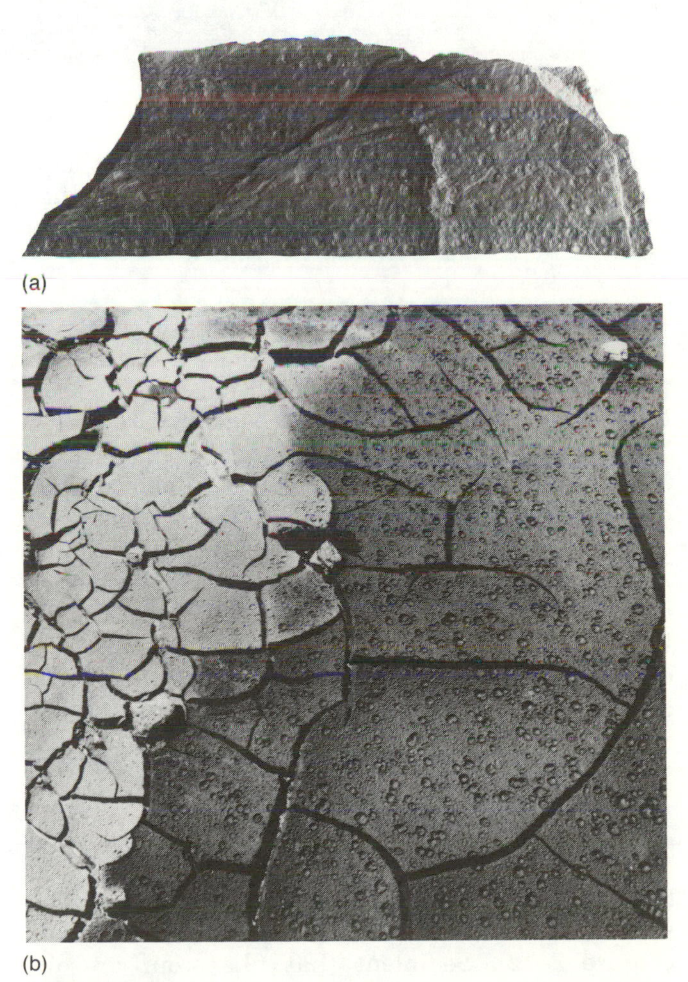

FIGURE 2–10
(a) Ancient rain imprints in siltstone, Isle Royale National Park, Michigan. (b) Rain imprints on Recent mud cracks, Isle Royale National Park, Michigan. (Both photos by N. K. Huber, U.S. Geological Survey.)

FIGURE 2–11
Track left by an arthropod as it crawled across the Cambrian seafloor; Pyramid Shale Member, Titanothere Canyon, Inyo County, California. Tracks are approximately 1 cm wide. (A. R. Palmer, U.S. Geological Survey.)

facing direction. Because they are relatively delicate, they are not preserved as often as many other primary structures.

Tracks and Trails

Tracks and ***trails*** left by organisms are frequently preserved and under certain conditions may be used to determine facing direction of beds (Figure 2–11). The weight of an organism produces a track or trail, and if the impression is filled with sediment during deposition of the overlying bed the top may be determined. In some instances, partly filled burrows may be useful for determining tops if the fillings can be properly recognized and interpreted. The widely known cylindrical *Skolithos* tubes of the Cambrian of North America and the Scottish Highlands have also been used as strain indicators (Chapter 5).

Sole Marks, Scour Marks, Flute Casts

Marks formed as an object has moved across a bedding surface or as currents scour a bedding surface are ***sole*** and ***scour marks*** and ***flute casts*** (Figure 2–12). Sediment that fills scour or sole marks by subsequent deposition may allow the marks to be preserved and thus indicate tops. Flute casts consist of scoop-shaped structures formed when currents scour and erode a surface.

Load Casts

Load casts form as a result of gravitational instability at the interface between a layer of water-saturated sand and underlying mud after deposition and dewatering. The weight of the newly deposited overlying sediment forces out the interstitial water. Compaction of the underlying sediment (frequently mud) beneath a sand bed commonly results in depressions in the bed below (Figure 2–13) as water is expelled from the mud during compaction. Load casts may be used with relative ease to determine facing direction. Load casts are common in deep-water turbidites and in any accumulation of sediment where there is a contrast in grain size, significant dewatering, and opportunity for differential compaction.

Fossils

The remains of organisms that are preserved in the geologic record, ***fossils,*** provide very useful indicators of relative age. Fossils may also be used to determine tops of sequences by studying the relative positions in a sequence where organisms of different ages are found. Sometimes partly filled cavities occur inside molds of organisms; other organisms, such as trees or bottom-living attached animals, may be preserved in the position in which they lived (Figure 2–14). If so, they may be used to determine the top of a sequence.

(a)

(b)

FIGURE 2–12
(a) Sole marks on siltstone beds, Borden Formation, Bullitt County, Kentucky. (R. C. Kepferle, U.S. Geological Survey.) (b) Flute casts on the sole of a turbidite bed near Lenhartsville, Berks County, Pennsylvania. (G. G. Lash, U.S. Geological Survey.)

Reduction Spots

Reduction spots are sedimentary structures produced by a small grain or fragment of organic matter or other material that is chemically different from the surrounding mass of sediment. The chemical difference may produce a nearly spherical area of reduction expressed as a color change in the immediate vicinity of the grain in the otherwise oxidized sediment (Figure 2–15). These features cannot be used to determine the facing direction in a sequence, but they serve as an important indicator of strain if the rock mass was deformed ductilely. Deformed reduction spots are common in slate and set important constraints on the origin of slaty cleavage (Chapters 5 and 17). It is important to know when the reduction spots formed in order to study their role in the strain history of a rock mass, a subject to be discussed in greater detail in Chapter 5.

SEDIMENTARY FACIES

It was recognized early in the twentieth century that sedimentary (and volcanic) rock units undergo lateral changes in the kind of sediment as the environment of deposition changes. Such lateral differences of sediment (later, rock) type are called *sedimentary facies* (Figure 2–16). Before the principle was recognized, numerous errors were made in interpreting the ages and nature of different rock types deposited in a particular area. It was assumed that each rock type was a separate unit, especially if there were no fossils to provide information on the ages of the rocks. The possibility of one rock type grading into another within the same unit was not considered. Careful studies of relations between interlayered and gradational sedimentary rocks and associated fossils showed that, although rock types may change laterally, different rock types, or *facies,* of the same rock unit may be deposited at the same time and should not be considered different formations solely because of lateral differences in rock type.

DEWATERING STRUCTURES

Interesting analogs have been observed in soft sediment where deformation at hand-specimen or larger scale may approximate that occurring in hard rock at deeper levels in the Earth. Ductile, brittle, and fluid behavior have all been observed in soft sediment. An interesting phenomenon in soft sediment is dewatering: water trapped between individual grains in sediment may remain under pressure as long as the sediment is in the stable confined mass. If an abrupt change occurs in the plumbing within the mass of sediment and the water escapes suddenly, disruption of bedding and other primary structures results (Figure 2–17). Earthquakes have been known to cause dewatering in Recent sediments and have been suggested as the cause of

FIGURE 2–13
Load casts in underlying mudstone, Sunnyside No. 1 coal, Carbon County, Utah. (J. O. Maberry, U.S. Geological Survey.)

disruption of layering in partly consolidated sands (Talwani and Cox, 1985).

UNCONFORMITIES

A break in the geologic record, where some part of geologic history is missing, is called an ***unconformity.*** Unconformities are produced by nondeposition or erosion (or both). Either results in a loss of strata for a part of the geologic record.

Three fundamental types of unconformities are disconformities, angular unconformities, and nonconformities. ***Disconformities*** (Figure 2–18a) are produced by deposition of a sequence of rock units, followed by uplift or a drop in sea level. Erosion (or nondeposition) then occurs without tilting or deformation of the sequence; then comes subsidence and

FIGURE 2–14
Heads of several Middle Cambrian trilobites (*Paradoxides*) on a bedding surface in volcanic mudstone from near Batesburg, South Carolina. (Donald T. Secor, Jr., University of South Carolina.)

FIGURE 2–15
Deformed reduction spots in Eocambrian Metawee Slate from near Rutland, Vermont. The broken surface on top of the specimen exposes light-colored semicircular reduction spots, whereas the sawed surface in front displays thin vertical sections through the deformed reduction spots. The sawed front of the specimen is 15 cm high. (RDH photo.)

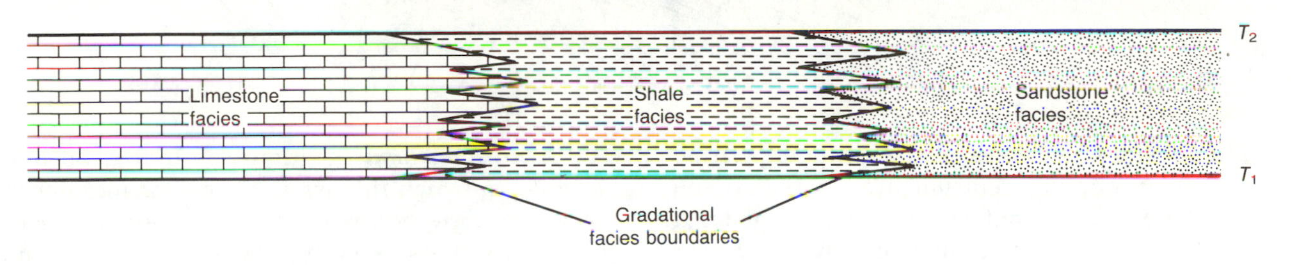

FIGURE 2–16
Facies changes within the same rock unit in a sequence of sedimentary rocks. T_1 and T_2 refer to earlier and later time boundaries.

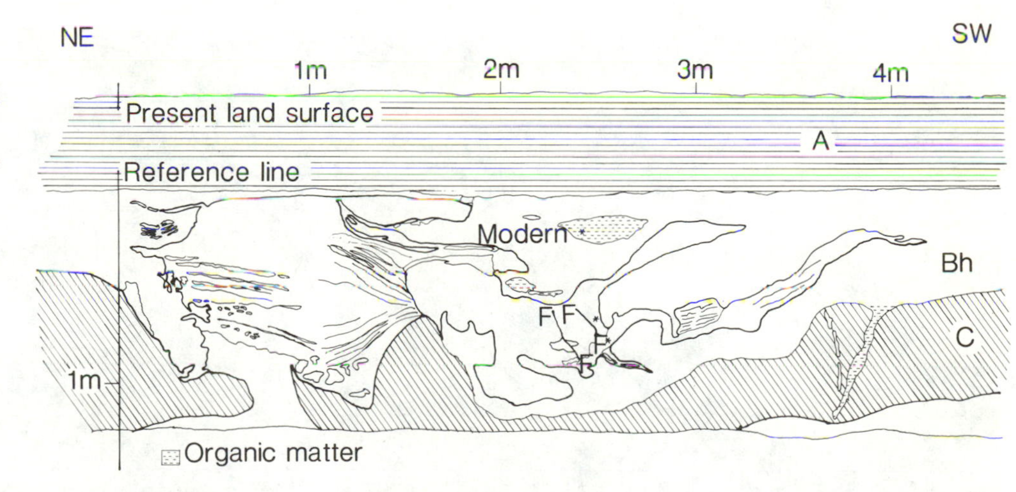

FIGURE 2–17
Dewatering structures in Holocene sediments, attributed to a large prehistoric earthquake. F marks faults; A, Bh, and C are layers of sediment. (From Pradeep Talwani and John Cox, *Science,* v. 229, fig. 2, p. 380. Copyright 1985 by the AAAS.)

renewed deposition. Bedding in the rocks above and below the unconformity remains parallel. Disconformities are recognized where rock units are missing either because they were eroded or because they were never deposited. There may be considerable topographic relief along the unconformity. In places where there is little relief on an unconformity, and bedding remains parallel on both sides, the term *paraconformity* may be used.

Angular unconformities (Figure 2–18b) are produced where a sequence has been deposited and later tilted as a result of the tectonic processes of faulting or folding. Erosion commonly accompanies or follows the episode of folding or faulting. When

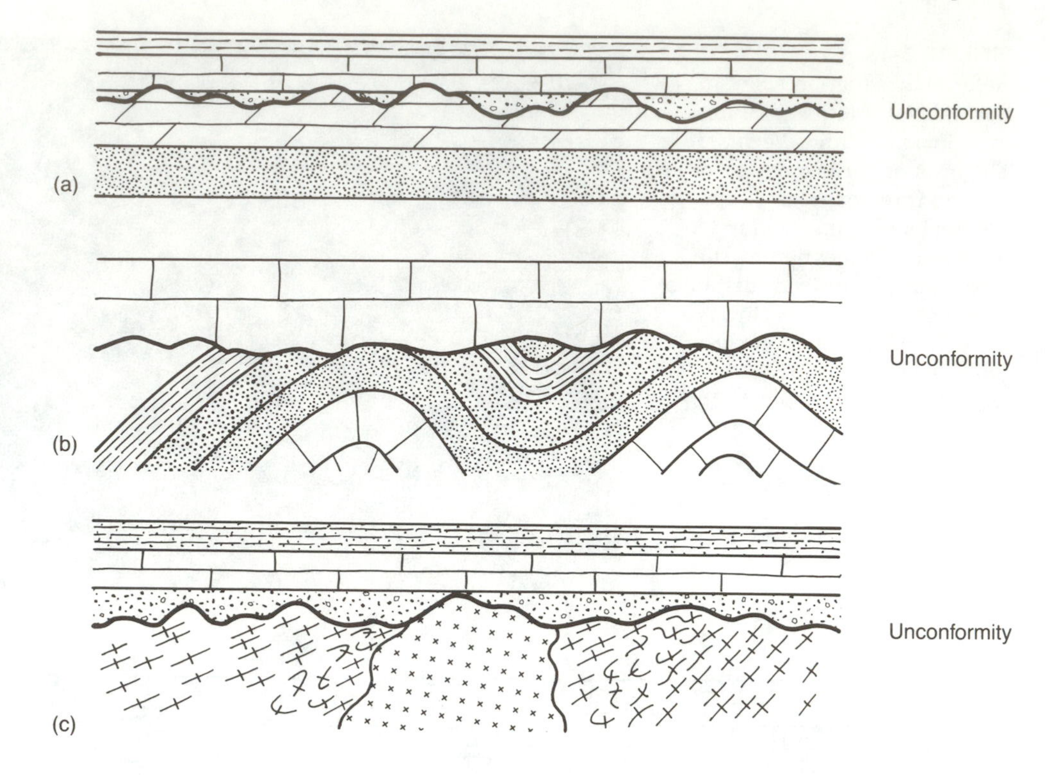

FIGURE 2–18
Types of unconformities.
(a) Disconformity.
(b) Angular unconformity.
(c) Nonconformity.

the area subsides, or sea level rises, and deposition is renewed, an angular relationship exists between the rocks below the unconformity and those above it (Figure 2–19). The angular unconformity at Siccar Point in Scotland is probably the best known in the world, because there James Hutton first observed evidence that led him to the principle of uniformitarianism.

A ***nonconformity*** (Figure 2–18c) is an unconformity in which intrusive igneous or metamorphic rocks (or both) occur below the erosion surface and sedimentary rocks occur above. No distinction is made between an angular relationship and topographic relief on the surface. In fact, both may exist along the erosion surface. The characteristics of igneous or metamorphic rocks below the unconfor-

(a)

(b)

FIGURE 2–19
(a) Angular unconformity, southeastern Alaska. (D. J. Miller, U.S. Geological Survey.) (b) Tilted angular unconformity at Kingston, New York, between Ordovician Austin Glen Graywacke and Silurian Binnewater Sandstone. (RDH photo.)

FIGURE 2–20
Basal conglomerate in Late Proterozoic Torridonian Sandstone, Assynt District, Scotland. Pebbles (light-colored) accumulated in the basal sand on the erosion surface on the darker Lewisian Gneiss that had been intruded earlier by light-colored felsic dikes, forming a nonconformity. (RDH photo.)

mity and sedimentary rocks above distinguish between this and other types of unconformities. This feature alone implies that a large time interval passed between crystallization of the igneous rocks, or recrystallization of the metamorphic rocks, at great depths in the Earth and deposition of the sediment atop the eroded crystalline rock mass. The characteristics of the rocks above help identify the boundary as an unconformity rather than an igneous contact because there is no contact metamorphic aureole. A contact metamorphic zone in the sedimentary rocks would be expected if the contact is intrusive. A basal conglomerate composed of clasts of the underlying crystalline rocks may also occur in the sedimentary sequence at the contact (Figure 2–20).

Angular unconformities, nonconformities, and possibly even disconformities may all be represented at an unconformable surface if the rocks below the unconformity are metamorphic. Layering or foliation in the metamorphic rocks is commonly truncated at the unconformity, producing both an angular unconformity and a nonconformity. Appreciable relief on the unconformity surface suggests that the unconformity is also a disconformity. The classic unconformity at the base of the Precambrian sedimentary sequence in the Grand Canyon has all those characteristics (Figure 2–21).

Another clue that a geologic boundary is an unconformity may be obtained from examination of the boundary: pebbles, boulders, and other debris accumulated from erosion may remain as part of the base of the unit deposited on the erosion surface (Figure 2–20). Not all unconformities have residual debris, either because no material was deposited at the site of observation or because all debris was removed from the area. Careful study of the rocks

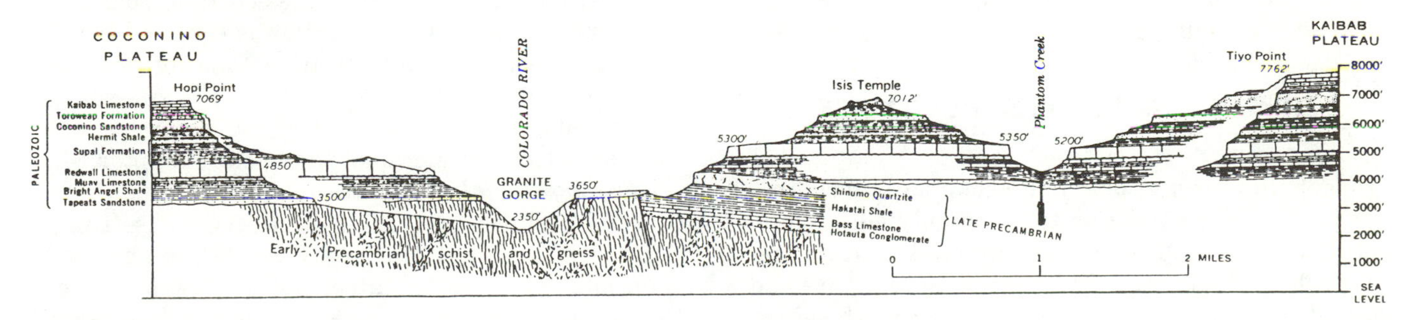

FIGURE 2–21
Cross section through the Grand Canyon showing the unconformity at the Precambrian-Paleozoic boundary. Note that the boundary is both an angular unconformity and a nonconformity and has considerable relief. (From F. E. Matthes, 1962, The Grand Canyon of the Colorado River: U.S. Geological Survey Bright Angel Quadrangle, scale 1/62,500.)

ESSAY

Deciphering a Major Structure in the Southern Highlands of Scotland

The study by Robert M. Shackleton cited at the beginning of Chapter 2 provides an important key to later work in the southern Highlands of Scotland (Figures 2–1 and 2E–1) and illustrates the power of using primary sedimentary features to work out the structure of complexly deformed rocks.

To appreciate what Shackleton (1958) accomplished, it is important to know something about the deformed state of the rocks in this region. The sequence he studied consists of late Precambrian to early Paleozoic clastic sedimentary rocks that rarely contain fossils. At least one early Paleozoic Caledonian regional metamorphic event recrystallized the sequence to greenschist and amphibolite facies assemblages. During the thermal event, the rocks were ductilely deformed and subjected to several events of very tight folding, each of which was overprinted onto the earlier episodes (see Chapter 16 for further discussion of complex folding). Ductile deformation also produced a strong foliation in most of the rocks.

Shackleton used the fairly abundant graded bedding and cross bedding that survived the rigors of Caledonian multiple deformation and metamorphism to determine that most of the rocks of the area (covering several hundred square kilometers) in the southern Highlands are downward-facing and therefore overturned. His conclusion must have been initially astounding, but it was reaffirmed by his many determinations of facing direction based on observation of primary sedimentary structures. This kind of observation should be made routinely in the course of the detailed structural study of any complexly deformed area. Other structural or stratigraphic criteria may help determine if a sequence is upright or overturned, but primary structures may be the *only* criterion available to enable this important assessment of regional facing directions.

Reference Cited

Shackleton, R. M., 1958, Downward facing structures of the Highland Border: Quarterly Journal of the Geological Society of London, v. 72, p. 361–392.

immediately beneath an unconformity may prove that the rocks along the old surface underwent a period of prolonged weathering without accumulation of erosional debris. Ancient soils (paleosols) have been recognized beneath many unconformities.

PRIMARY IGNEOUS STRUCTURES

Igneous plutons and lava flows commonly form in a variety of shapes ranging from roughly equidimensional to tabular. Less commonly, they contain features that result from flow within the magma and resemble sedimentary structures. Cross bedding, graded bedding, and other kinds of layering have been observed in igneous bodies. The most common structures in igneous bodies include primary (or flow) foliation, compositional banding, and vesicles. A foliation resulting from flow of crystallizing magma is difficult to distinguish from a tectonic foliation. Careful study of the structures in the enclosing rocks and texture of the igneous body may be necessary to document cross-cutting relationships or to show that the foliation is confined to the igneous body and is not a tectonic foliation present in the country rock (Figure 2–22). Flow processes in magma also may produce folds that resemble tectonic structures in metamorphic rocks (Chapter 14).

Compositional banding in an igneous body (Figure 2–23a) may result from crystal settling, differentiation, and fractional crystallization and from multiple parallel intrusions or flow processes that flatten xenoliths. Careful study of a layered pluton may reveal a differentiation or fractional-crystallization sequence that made the bottom of a pluton more mafic and the top more felsic. If layering can be shown to have been produced by differentiation and gravitational settling, the upward

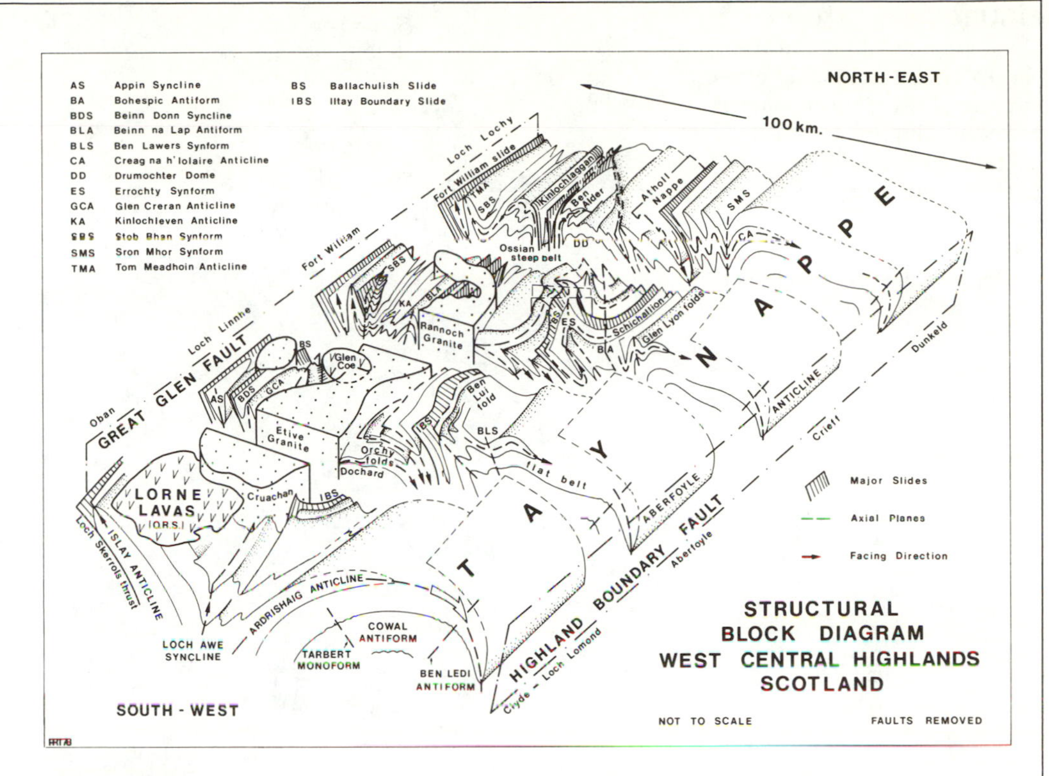

FIGURE 2E–1
Block diagram showing the major structures of the western Central Highlands of Scotland. Note the size of the Tay nappe relative to the other structures in this region. (Reproduced by permission of the Geological Society from The Caledonides in the British Isles—Reviewed Thomas, P. R., in Harris, A. L., Holland, C. H., and Leake, B. E., eds., *The Caledonides in the British Isles—Reviewed*, Geological Society of London Special Publication 8.)

change from mafic to felsic composition may be used to determine the top direction. Graded bedding or cross bedding may exist in rare instances, such as in the Skaergaard intrusion in Greenland (Wager and Deer, 1939). These features may be used in igneous bodies to determine facing directions in the same way as in a sedimentary sequence, except in a few places where they occur on the near-vertical walls of plutons.

Vesicles are cavities left by gas bubbles that form in magma as the pressure decreases (Figure 2–23b). Vesicles occur in lava flows and less commonly in plutons where magma has moved from deep to shallow levels in the crust. The bubbles move upward, being less dense, and make their way toward the top of a shallow pluton or lava flow, producing a cavity-filled zone that may be preserved and used as a facing criterion. Vesicles may become filled with secondary minerals and as such are called *amygdules* (Figure 2–23c). Their presence obviously would not affect determination of the facing direction in a flow or sill, but special care must be taken to not mistake the amygdules for phenocrysts. Amygdules and vesicles may occur in layers, thus revealing the horizontal plane. They may also be useful strain indicators (Chapter 5).

Contact metamorphic zones associated with sills and flows may be useful in determination of tops and in distinguishing sills from flows. Ideally, a sill (Figure 2–24a) could produce contact metamorphic zones on both the top and bottom contacts with the country rock. Therefore, the use of this feature for determination of tops is limited. A lava flow (Figure 2–24b), on the other hand, should metamorphose only the material below it, and there would be no metamorphism in a subsequently deposited overlying sequence. Thus, if a concordant tabular igneous body occurs in a sequence of sediments and a contact

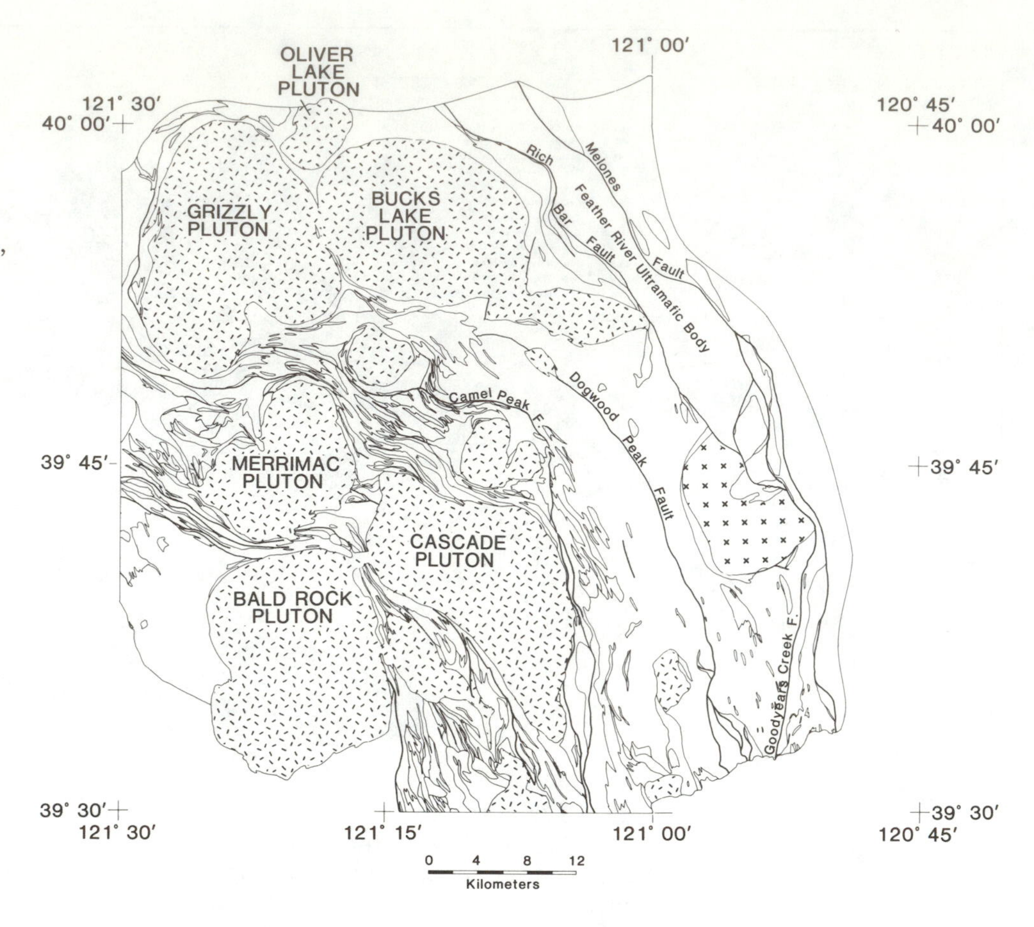

FIGURE 2–22
Geologic map of part of the Sierra Nevada showing cross cutting of layering in country rocks by plutons. Granitic plutons shown by a random pattern, gabbro by an x pattern. (After Anna Hietanen, 1981, U.S. Geological Survey Professional Paper 1226–B.)

metamorphic aureole occurs only below it, the body must be a lava flow and is useful for determination of tops.

Pillow structures form where lavas are extruded beneath or flow into water. Because of their shape, they can be used to determine facing direction (Figure 2–25). The vesicular glassy curved tops and V-shaped nonvesicular bases provide a criterion for determining the top of a flow. Borradaile and Poulsen (1981) argued that the criterion is not useful in strongly deformed pillows, but others have used it successfully where the pillows are slightly to moderately deformed, particularly in areas where other criteria are scarce or absent. M. E. Wilson (1941, 1962) and Allard (1976) have used pillow structures successfully to determine tops in the Noranda, Rouyn, and Chibougamau areas of Québec.

Granophyric quartz toward the top of gabbroic sills has been used successfully as a top criterion in several places where differentiation has occurred. The texture occurs in the upper part of the Palisades sill in New Jersey and New York (Walker, 1940; Poldervaart and Walker, 1962).

GRAVITY-RELATED FEATURES

Landslides and Submarine Flows

Landslides occur both above and below sea level, and the results in either case frequently resemble tectonic structures. Slides may be triggered by earthquakes or other tectonic activity, either deep in the crust or on the surface many kilometers away from the landslide.

The most obvious factor contributing to landslides is a slope, but not all sloping surfaces produce landslides. In addition to slopes (some of less than one degree), parallelism of bedding, foliation, or fractures at the surface may provide conditions favoring landslides. Weak material at or near the surface may also produce favorable conditions for a landslide, but not actually cause one.

Landslide-prone conditions exist in many places. A variety of things trigger landslides: earthquakes, overloading of slopes, high precipitation, streams undercutting and oversteepening a slope, and, recently, human activities. Poorly consolidated

(a)

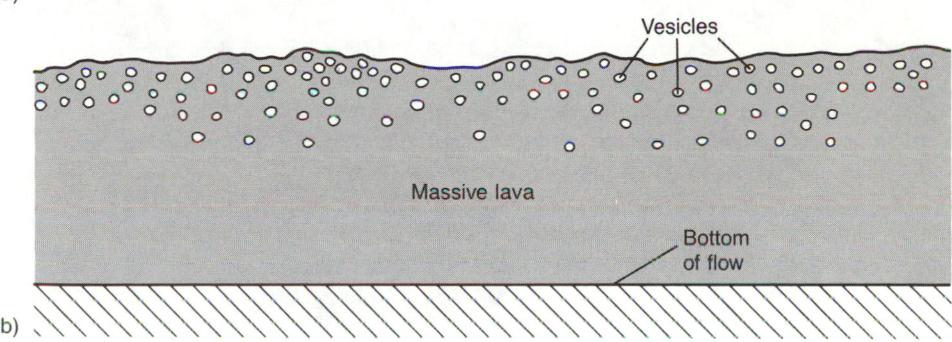

(b)

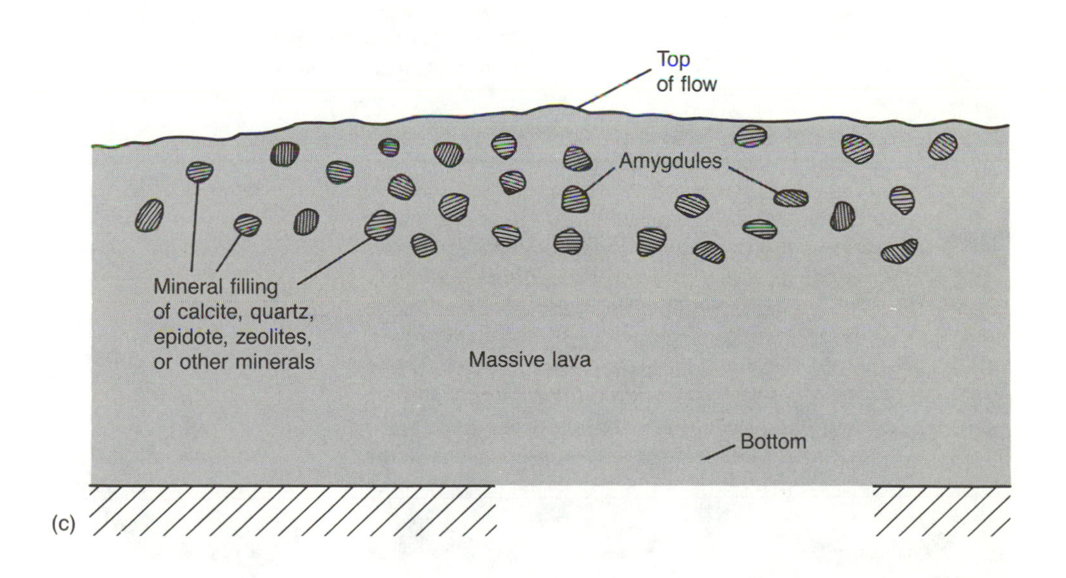

(c)

FIGURE 2–23
(a) Compositional banding in a layered gabbro south of Koppervik, island of Karmøy, southern Norway. Dark layers are rich in mafic minerals; light layers are rich in plagioclase. (RDH photo.) (b) vesicles, and (c), amygdules, in volcanic rocks.

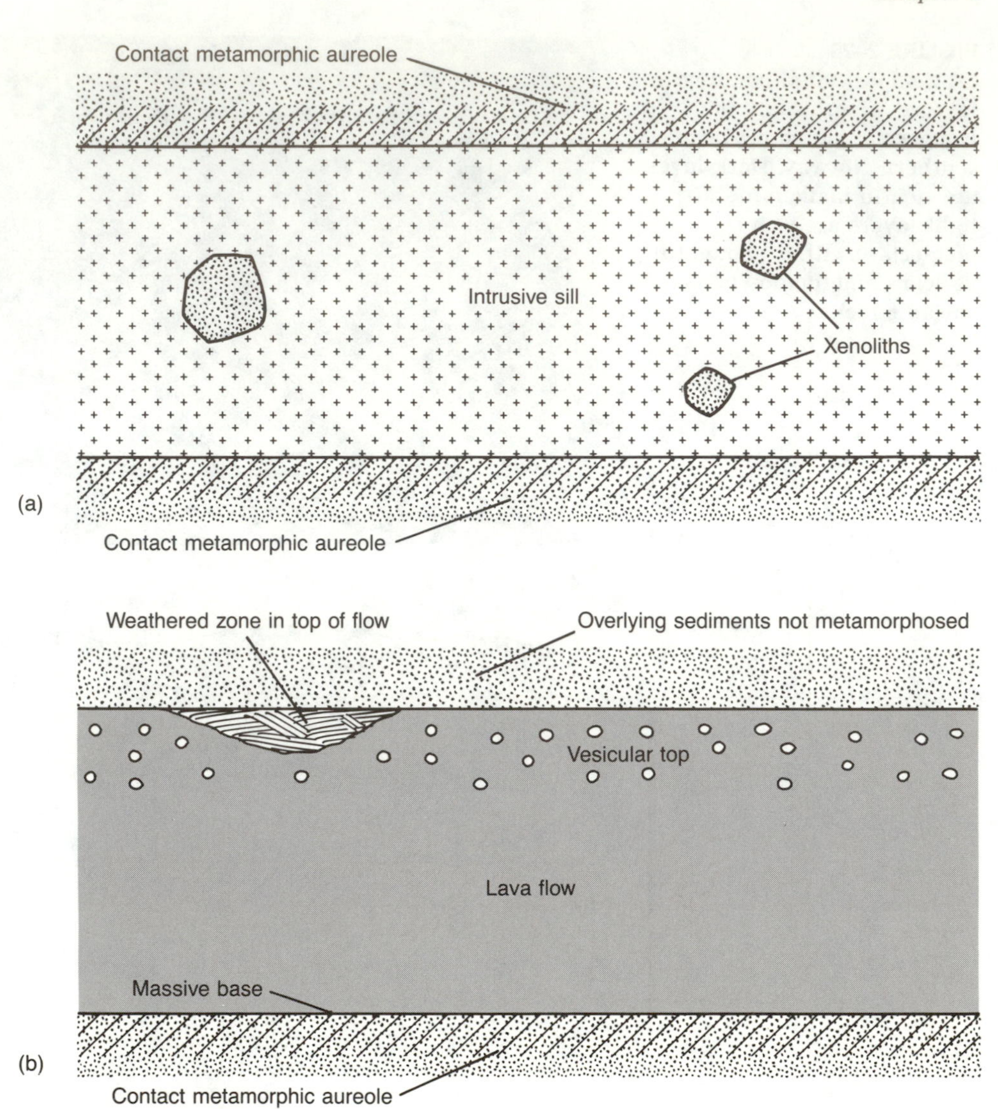

FIGURE 2–24
Distinguishing sills (a) from flows (b) by using contact metamorphic aureoles.

(a)

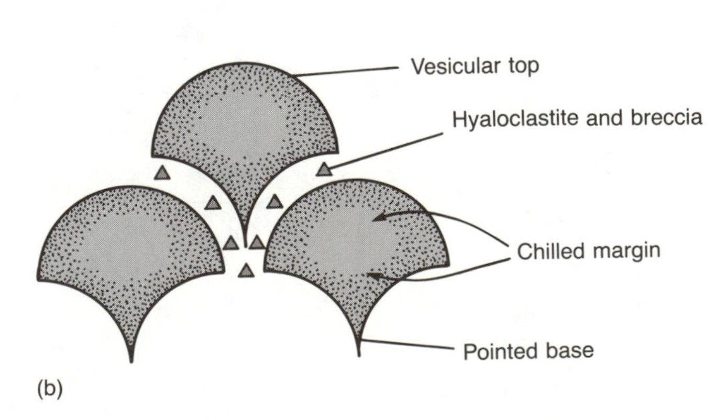

(b)

FIGURE 2–25
Pillow structures (a) in Perchas Lava Member, Barrio Gato, Puerto Rico. (H. G. Berryhill, Jr., U.S. Geological Survey.) Cross section (b) shows structure.

FIGURE 2–26
Olistostrome (wildflysch) near Albany, New York, composed of blocks of Ordovician Austin Glen Graywacke enclosed in Ordovician Normanskill Shale and derived from an approaching thrust sheet. (RDH photo.)

material—commonly of fine silt or sand size—may spontaneously liquefy when loaded and produce a landslide. Also, the material may suddenly liquefy when subjected to earthquake shock.

Landsliding may occur along the toe of an advancing thrust sheet or any other escarpment consisting of poorly consolidated materials, particularly in the submarine environment. Blocks may fall from the escarpment and be deposited in finer sediment. Masses of material such as this are known as *olistostromes* (bedded) and *diamictites* (no obvious bedding). Masses containing rounded clasts may be called pebbly mudstones. Tillites may be included as diamictites. Olistostromes may consist of exotic blocks (olistoliths), of any size ranging from a few centimeters in diameter up to several kilometers, contained within a fine-grained matrix (Abbate and others, 1970). Olistostromes are commonly interlayered with nonchaotic sediments and overall are lenticular (Figure 2–26).

Mélanges consist of mixtures of weak and strong rock materials—such as fragments of sandstone or basalt in a clay matrix—of either tectonic or nontectonic origin (Figure 2–27). E. Greenly (1919) first used the term mélange for terranes in Anglesey (part of Wales) that are characterized by fragments of stronger rock embedded in a sheared matrix of weaker materials (from Raymond, 1975). Nowadays we believe that tectonic mélanges commonly form in accretionary prisms and consist of coherent masses of different size of variously deformed materials contained in a pervasively sheared matrix (Hsü, 1968). Other mixtures of coarse and fine sediment form as olistostromes from material breaking off an escarpment and may have either tectonic or nontectonic origin. Here the matrix may not be sheared if the mass is preserved relatively undeformed, but the blocks in the mass would be both chaotic and of diverse lithology (Figure 2–28). The term *wildflysch* has been applied to a heterogeneous accumulation of blocks and smaller particles generally deposited in a deep-water environment. Such an accumulation also may result from tectonic or nontectonic processes.

Turbidites are deposits produced by rapid (turbulent) flow of a sediment-laden turbidity current down a slope on the seafloor or in a large lake. They generally consist of an unsorted mass of sediment that cascades to lower levels, spreads out on the sea-

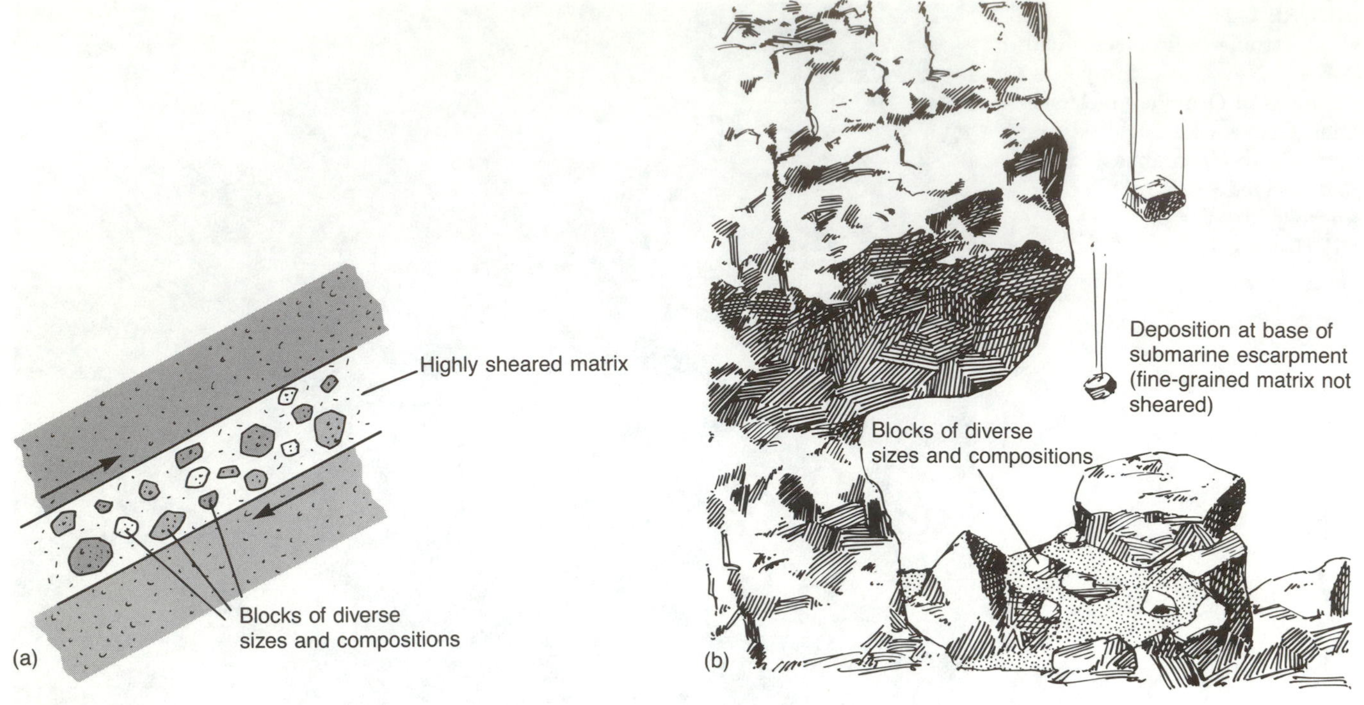

FIGURE 2–27
Formation of mélanges by tectonic (a) and nontectonic (b) mechanisms.

FIGURE 2–28
Mélange terrane near Tarnilat in northern Morocco. The high hills in the distance are held up by mélange blocks and the lower hills and lowlands are underlain by shale. Rocks in the foreground are part of an underlying unit. (RDH photo.)

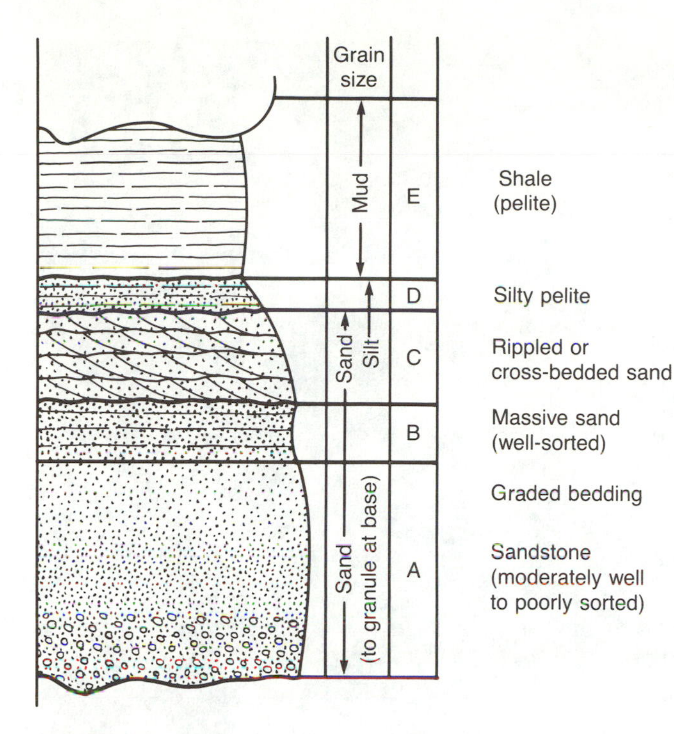

FIGURE 2–29
Bouma sequence. (From Bouma, A. H., *Geologie en Mijnbouw*, v. 21, Fig. 8. © 1959, Elsevier Science Publishers B. V., Amsterdam.)

floor, and settles as graded beds. Arnold H. Bouma (1962) recognized that these deposits may develop as graded, channeled, massive, and cross-bedded sequences that grade upward one into another (Figure 2–29). An intact or partial *Bouma sequence* is another facing criterion.

Salt Structures

Evaporite deposits occur in sedimentary sequences at shallow crustal levels in many parts of the world. Layers of anhydrite and gypsum undergo ductile deformation (Chapter 5) more readily than the more common sedimentary rock types like sandstone, dolostone, limestone, and shale. Rock salt, composed mainly of halite, will flow more readily than any other common rock type mentioned so far. Tens or hundreds of meters of rock salt from the evaporation of seawater occur in the United States in the Texas-Louisiana Gulf Coast, West Texas, Kansas, Michigan, and New York and in West Germany, Spain, Jordan, Iran, and elsewhere. Salt will flow under the influence of its own weight, low density, and the force of gravity in contrast with the greater density of the overlying sediments. A great variety of structures is produced, ranging from salt glaciers on the surface (Figure 2–30a) to salt pillows, intrusive stocks, and domes at depth (Figure 2–30b). The internal structure of these salt features provides abundant evidence of ductile flow, with folds, foliations, and other structures resembling those formed by high pressure and temperature in metamorphic rocks (Figure 2–31; Muehlberger, 1968).

Salt or other material—frequently water-saturated "mud lumps"—that move upward and gravitationally intrude the overlying sediments are called ***diapirs.*** They are common in the Mississippi delta and serve as important hydrocarbon traps there and elsewhere (Figure 2–30b). Salt and mud diapirs provide useful analogs for deep crustal diapirs of lower-density rocks (such as granitic gneisses) that move upward through denser rocks and form gneiss domes.

IMPACT STRUCTURES

The surfaces of our moon and the planets with thinner atmospheres than the Earth reveal a history of impacts spanning billions of years. There is no reason to indicate Earth should not have experienced a similar history, but evidence of this extensive history is lacking because of the dynamic surficial and tectonic processes that have changed the surface of the Earth through time. Considering the evidence for impacts of large objects on the other terrestrial planets throughout the history of the Solar System, large objects should have impacted the Earth over intervals of every few million years. Layered mafic intrusive bodies, like the Sudbury irruptive, have been interpreted by a number of geologists as being related to large impacts.

Structures generally having a circular or elliptical outline have been identified in different parts of the world but are not obviously related to tectonic processes. Unfortunately, many of these structures are deeply eroded, having formed during the Paleozoic or Mesozoic or even earlier. Others, like Meteor Crater in Arizona, have an origin more obviously extraterrestrial. Some of the older structures, including the Wells Creek structure in Tennessee, Jephtha Knob in Kentucky, and Serpent Mound in Ohio, were at one time considered *cryptovolcanic* structures, related to explosive volcanic activity at depth that disrupted the surface rocks but did not produce surface evidence of volcanic activity; hence the prefix *crypto-* (hidden).

Direct evidence of meteorite impact has been found at Meteor Crater in the form of meteorite fragments and the silica polymorphs coesite and

(a)

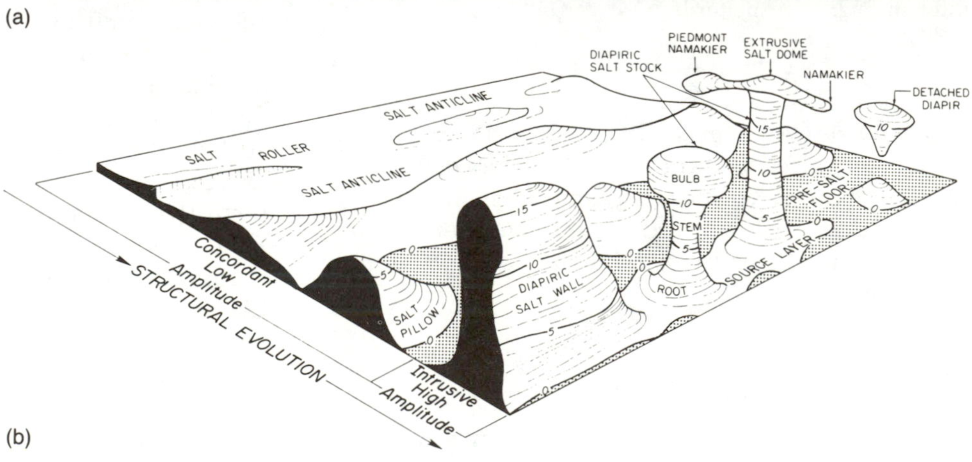

(b)

FIGURE 2–30
(a) Oblique aerial view of an eroded salt dome in Iran showing folded internal structure and boundaries of the dome. Layering in salt and country rocks is truncated at the contact. (Augusto Gansser, Eidg. Technisches Hochschule, Zurich.) (b) Shapes of different types of large salt structures. (From M. P. A. Jackson and C. J. Talbot, 1986, Geological Society of America *Bulletin,* v. 97.)

stishovite, which form at high pressure and low temperature. The other structures are more difficult to assess, but the cryptovolcanic interpretation has been questioned because almost none show any vestige of the effects of volcanic activity at depth. One of the best-exposed impact craters is the Wells Creek structure. It was first investigated by Bucher (1936), who cited it as an outstanding example of a cryptovolcanic structure (Figure 2–32), but it was discovered that Wells Creek and similar structures contain *shatter cones* (Figure 2–33), thought to be produced by brittle deformation like that observed where an explosive charge at the bottom of a drill hole produces a cone-shaped fracture propagating down and away from the explosion. Occasionally, shatter cones are obvious in highway cuts, radiating downward from the base of an exhumed drill hole. Where shatter cones can be seen in place, their apices always point upward (like those in an exploded drill hole), suggesting impact from above.

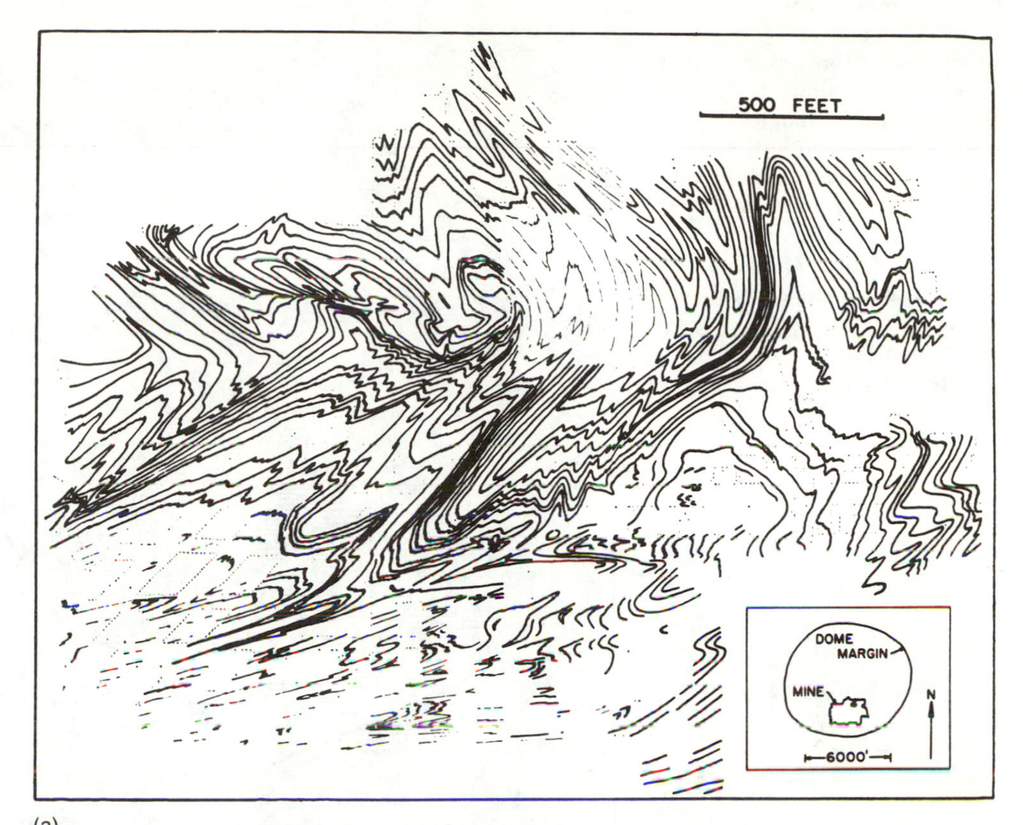

(a)

(b)

FIGURE 2–31
(a) Map of structures in salt in the Grand Saline salt mine, Grand Saline salt dome, Texas. (From W. R. Muehlberger, 1968, Geological Society of America *Special Paper* 88.)
(b) Ductilely deformed strongly foliated and folded salt inside the Grand Saline salt dome. Each dark mark in the salt is a shot hole spaced about two meters from the adjacent hole. (W. R. Muehlberger, University of Texas, Austin.)

Careful measurement of the orientation of many shatter cones at Wells Creek indicates a common orientation (Figure 2–34), suggesting impact from above by a body with a trajectory inclined about 10° from the vertical. Estimates based on experiments and measurements of meteors traveling through the atmosphere suggest that the object that produced the Wells Creek structure was about 300 m in diameter, weighed 1.8 to 9 $\times$ 10^{10} kg, and was traveling at a velocity of about 40 to 50 km/s at the time of impact (C. W. Wilson and Stearns, 1968).

Studies of this and similar structures in North America, Europe, and on other continents led to the conclusion that they formed by meteorite impact. Some very large circular structures, like the eastern part of Hudson Bay, have even been suggested as impact structures (Dietz, 1960). The importance of large-body impacts on our moon and on planets with thinner atmospheres than the Earth is widely known and led to a controversy during the 1980s about the role of impacts in the extinctions of dinosaurs and other organisms (Silver and Schultz, 1982).

FIGURE 2–32
Geologic map of the Wells Creek structure in west-central Tennessee. (From W. H. Bucher, 1936, 16th International Geological Congress, v. 2.) Note the circular character of the outcrop pattern produced by doming and radial and concentric faults. Oldest units are exposed in the central part of the structure. s_1 and s_2 are depressions; a_1 and a_2 are uplifted areas.

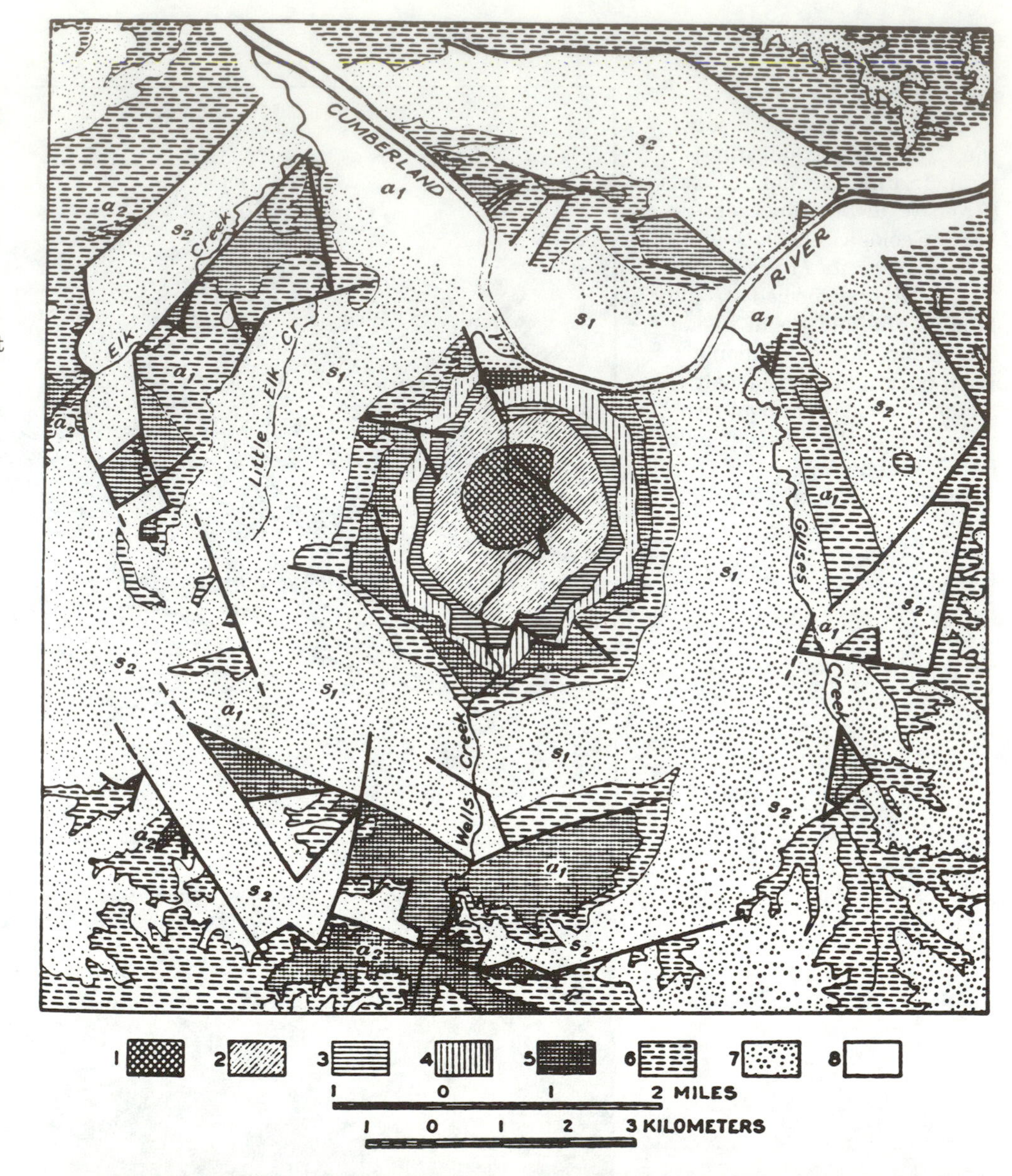

FIGURE 2–33
Shatter cones from the Well Creek structure in Tennessee. (Specimen courtesy of Richard G. Stearns, Vanderbilt University.)

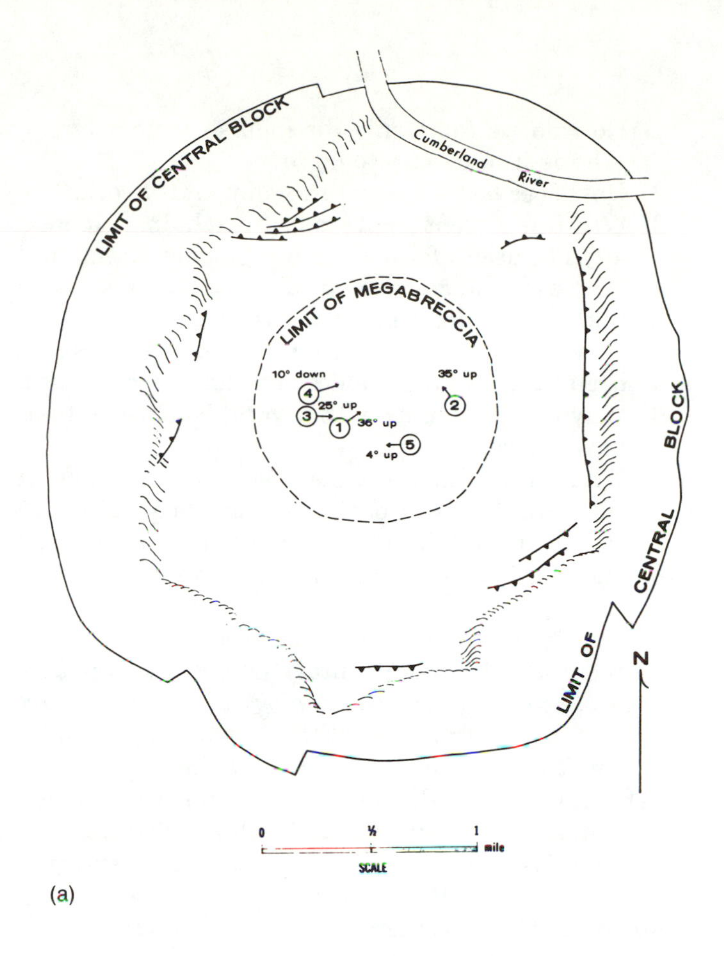

(a)

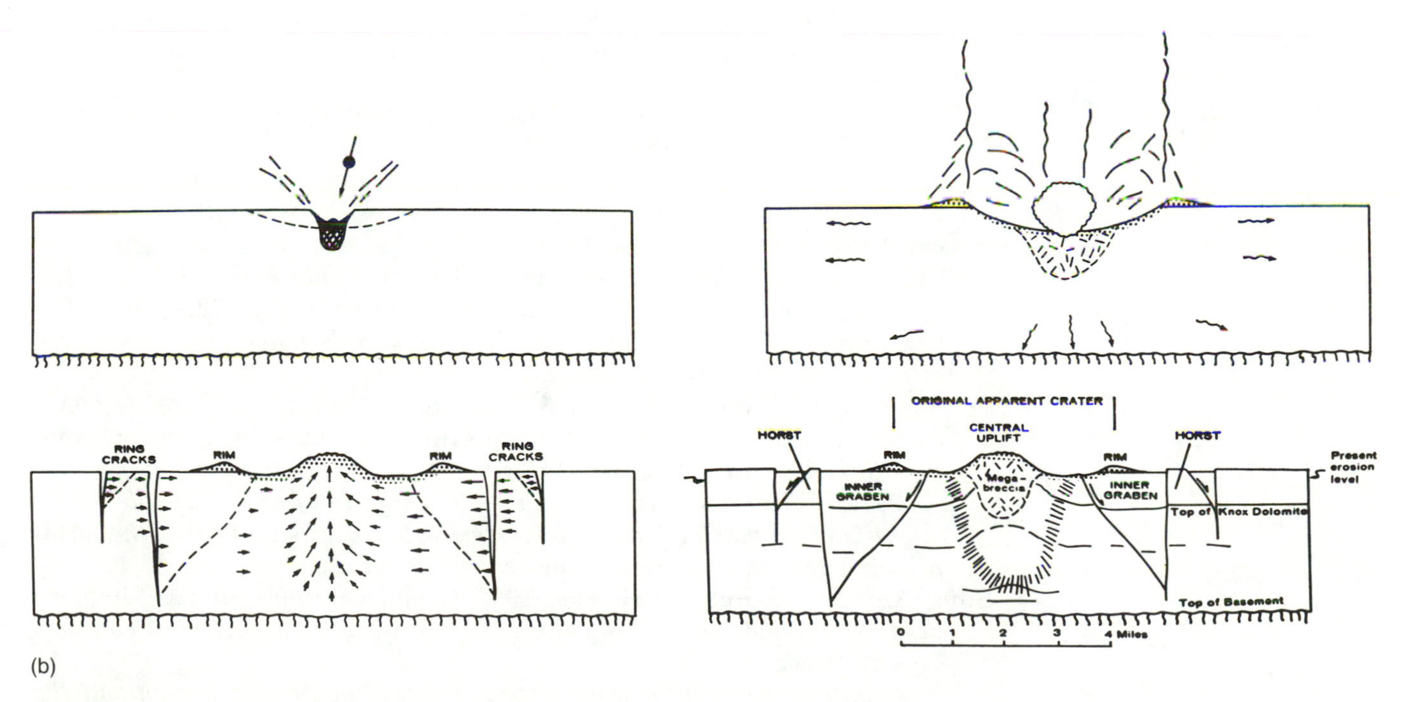

(b)

FIGURE 2–34
Reconstruction of the average orientations (numbered localities with arrows indicating directions and inclinations) of shatter cones at Wells Creek structure. Map (a) indicates an upward but not vertical orientation for the body at impact. Cross section (b) shows the sequential history of impact. (From C. W. Wilson, Jr., and R. G. Stearns, 1968, Tennessee Division of Geology *Bulletin* 68.)

Questions

1. How do we commonly distinguish tectonic from nontectonic structures? Why is it necessary to do so?
2. How does bedding form in sedimentary rocks?
3. Why are current ripple marks (excluding ripple cross laminations) generally not useful for determining facing directions?
4. Why are mud cracks and rain imprints less useful for top determinations than are graded beds and cross beds?
5. How can you tell the difference between a sill and a lava flow that had been buried in a sequence of sedimentary rocks?
6. Why do some areas with several conditions that favor landsliding never have slides?
7. What is the difference between a mélange and an olistostrome?
8. Why are the lower parts of Bouma sequences most frequently preserved?
9. How can you tell an igneous intrusive contact from a nonconformity?
10. How can a nonconformity also be an angular unconformity?

Further Reading

Abbate, Ernesto, Bortolotti, Valerio, and Passerini, Pietro, 1970, Olistostromes and olistoliths: Sedimentary Geology, v. 4, p. 521–557.
Development of the terminology and processes related to the formation of olistostromes and olistoliths. Examples primarily from the Alpine orogen are discussed clearly and summarized.

Blatt, Harvey, Middleton, Gerard, and Murray, Raymond, 1980, Origin of sedimentary rocks, 2d ed.: Englewood Cliffs, N.J., Prentice-Hall, 782 p.
Contains an excellent discussion of sedimentary structures (in Chapter 5), as well as details of sedimentary processes.

Harms, J. C., Southland, J. B., and Walker, R. G., 1982, Structures and sequences in clastic rocks: Society of Economic Paleontologists and Mineralogists Short Course 9, 249 p.
An excellent discussion of recognition and interpretation of the structures that form in clastic sediments. May be difficult to find except in larger libraries.

Reineck, H.-E., and Singh, I. B., 1975, Depositional sedimentary environments: Berlin, Springer-Verlag, 439 p.
Contains outstanding illustrations and photographs of sedimentary structures and textures and discussions of modern depositional sedimentary environments.

Shackleton, R. M., 1958, Downward facing structures of the Highland Border: Quarterly Journal of the Geological Society of London, v. 72, p. 361–392.
The author employed sedimentary structures to decipher a major tectonic structure in the southern Highlands of Scotland.

Shanmugam, G., Damuth, J. E., and Moiola, R. J., 1985, Is the turbidite facies association scheme valid for interpreting ancient submarine environments? Geology, v. 13, p. 234–237.
Raises fundamental questions related to use of the facies principle in the analysis of submarine fans. It discusses earlier works that apply the facies principle to both ancient and modern environments.

Silver, L. T., and Schultz, P. H., eds., 1982, Geological implications of impacts of large asteroids and comets on the Earth: Geological Society of America Special Paper 190, 528 p.
The controversy about extraterrestrial impacts has raged throughout the 1980s after Luis and Walter Alvarez and others reported that a rare element, iridium, is concentrated in clays at the Cretaceous-Tertiary boundary. This publication is a compendium of papers from a meeting held to discuss various aspects of the problem, ranging from probability of impacts, through the geologic record of impacts, to other possible extinctions and the Cretaceous-Tertiary boundary problem.

PART TWO

Rock Mechanics

OUTLINE

3 Stress
4 Strain
5 Measurement of Strain in Rocks
6 Mechanical Behavior of Rock Materials
7 Microstructures and Deformation Mechanisms

THIS SECTION EXAMINES THE CONCEPTS OF STRESS AND STRAIN, ways to measure strain, models for behavior of rocks during deformation, and microstructures and deformation mechanisms that can be directly observed in naturally deformed rocks. We can observe and measure strain in rocks—as deformed objects and geologic structures—but can only crudely estimate the magnitudes and directions of the stresses we think produced them. Stress is discussed before strain so that continuity can be maintained with the related chapters that follow. Chapter 7 attempts to relate lattice-scale processes and those we can observe under the microscope to textures and structures in rocks we can see without magnification. These processes ultimately are related to structures discussed in later chapters.

3

Stress

In all quantitative studies on the relationship between the original forces and the resulting deformations, an intermediate field of investigation enters, the condition of stress in the earth's crust. The original forces—whether primary or secondary, whether internal body or external boundary forces—set up a state of stress in the earth's crust which, in turn, produces the observed distortions.

W. HAFNER, Geological Society of America *Bulletin*

WE ARE ALL FAMILIAR WITH DAILY EFFECTS OF FORCES AND stresses. The most pervasive force affecting us is gravity: it holds the atmosphere, the oceans, and us to the Earth, and keeps the Earth and other planets from fragmenting. ***Forces*** change the velocity or direction of motion of a body. We use forces to open and close doors, ride bicycles and exercise machines, and turn on lamps and in other everyday activities. In the Earth, forces act on rock bodies and drive deformation, but force is not as meaningful as stress in the study of rock deformation because we must consider the area affected by the force.

Stress is force applied to an area, or force per unit area. We commonly think of tectonic structures as products of stress (Figure 3–1). Yet we mostly observe the effects of stress in tectonic structures without being able to measure that stress. A tectonic structure is a manifestation of strain (distortion and rotation) that is assumed to result from stress applied to the mass of rock. Stress originates in processes that generate, move, and consume lithospheric plates, with the aid of gravity, which is a component of tectonic deformation and also is primarily responsible for deformation in salt and impact structures. Thus, the nature of stress is important to understanding tectonic structures. A principal goal of structural geology is to reconstruct the orientations and magnitudes of stresses that produced the structures we encounter in the field. Unfortunately, these stresses are usually no longer present. Direct measurement of the present-day stress field does not provide information about the nature of most structures formed during ancient periods of Earth history, but may provide us with a basis for estimating the magnitudes of ancient stresses. Experimental attempts to duplicate natural structures in rocks also help us estimate the orientations and magnitudes of stresses.

DEFINITIONS

A ***scalar*** possesses only magnitude; a ***vector*** possesses both magnitude and direction (Figure 3–2). A scalar is a number like the price of oil, the score in a baseball game, or the thickness of a rock unit. A vector is a number with an indication of direction. For example, if we say a car is traveling northwest at 100 km per hour, we have defined a vector; if we say only that it is traveling 100 km per hour—with no indication of direction—we have defined a scalar. In this book, vectors in equations are indicated by bold type.

A ***tensor*** is a mathematical way of representing a physical quantity by referring it to an appropriate coordinate system. A tensor has the same value in any coordinate system, but the magnitude of its

FIGURE 3–1
The fault and accompanying folds exposed in this cliff in Alaska, like most others, are assumed to be a result of stress in the crust. (T. H. Moffitt, U.S. Geological Survey.)

components depends on the choice of coordinate system. Tensors have different ranks (or orders): a *zero-rank tensor* is a scalar and has only one component; an example is the temperature at a particular time. A *first-rank tensor* consists entirely of vectors and has three components. A *second-rank tensor* consists of vectors and nine components; an example is the northeastward motion of the Gulf Stream at a surface-water velocity of 150 cm/s. The number of components in a tensor may be determined from 3^n, where n is the rank.

A ***force*** (Figure 3–3a) is a vector that produces a change in the velocity, direction, or acceleration of a body. The body may already be in motion, or it may be stationary. Newton's *second law of motion* states that

$$\mathbf{F} = m\mathbf{a} \qquad (3\text{–}1)$$

where **F** is force (a vector), m is mass (a scalar), and **a** is acceleration (also a vector). *Body forces* act equally on all parts of a body. An example is the effect of gravity on a mass. Body forces must be

FIGURE 3–2
Scalars and vectors.

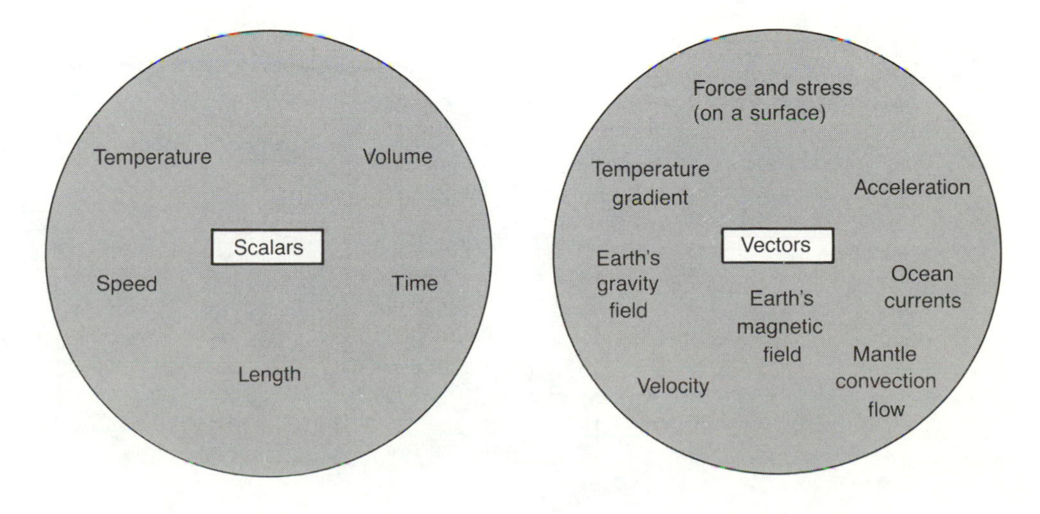

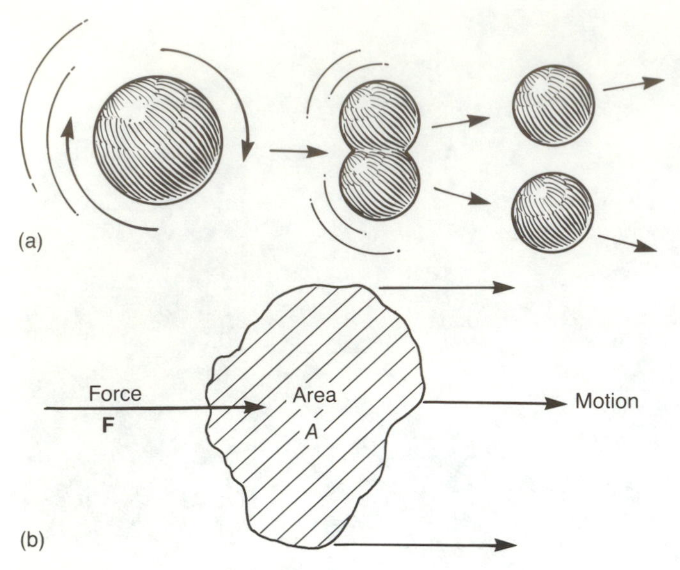

FIGURE 3–3
Forces (a) and stresses (b).

considered if the behavior of fluid or ductile material is involved. *Surface forces* act on the surface of a mass and include forces acting along a fault or a major plate boundary. The magnitude of a body force is proportional to the amount of mass present, but the magnitude of a surface force is proportional to the surface area affected. Moreover, a surface force can be resolved into mutually perpendicular components: one normal to the surface, and one or two parallel to the surface. This relationship will reappear when we discuss normal and shear stress.

Force may be converted to stress by dividing by the area affected by the force; that is, stress is force per unit area (Figure 3–3b). Stress acting on a surface is also a vector regardless of the area of the surface. Stresses may be ***tensional*** (pulling apart) or ***compressional*** (pushing together).

A stress may be resolved into two components; ***normal stress, σ,*** acts perpendicular to a reference surface; ***shear stress, τ,*** acts parallel to a surface. A *hydrostatic state of stress* occurs where normal stress is the same in all directions and no shear stress exists; it may be thought of as the confining stress of a mass submerged in water at a known depth. In the

FIGURE 3–4
Stress on a body 10 km under water (a) or 10 km deep in the crust (b), assuming the stresses are distributed homogeneously.

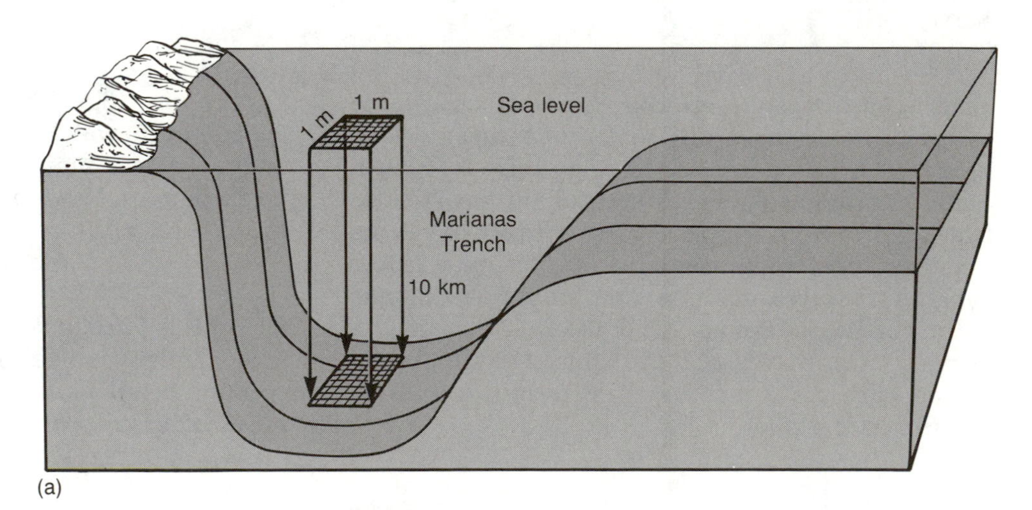

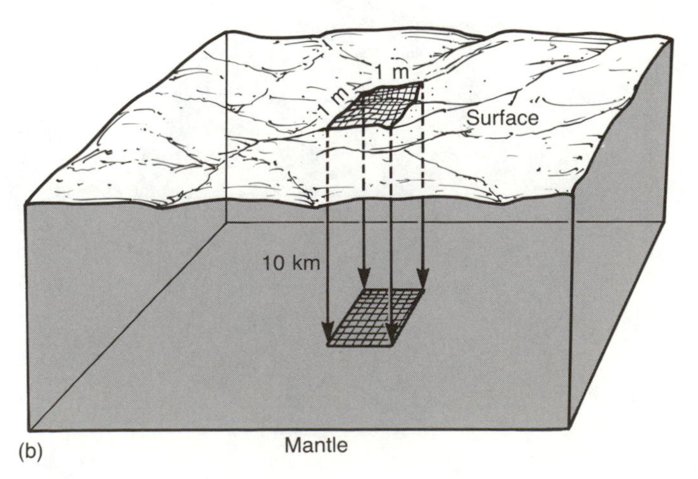

case of a body buried in the Earth, the weight of the column of rock (instead of water) per unit area above it is called the *lithostat* (Figure 3–4). Stress in the Earth, however, is not hydrostatic: shear stress does exist in solids.

For example, the force on a body submerged at a depth of 10 km near the bottom of the Mariana Trench would equal the weight of the mass of water on it, expressed as

$$\begin{aligned}\mathbf{F} &= \text{weight of water}\\ &= \text{height} \times \text{density of seawater}\\ &\quad \times \text{area} \times \text{acceleration of gravity} \\ &= 10^4\ \text{m} \times 1030\ \text{kg/m}^3 \times 1\ \text{m}^2 \times 9.8\ \text{m/s}^2\\ &= 1.009 \times 10^8\ \text{kg m/s}^2\end{aligned} \quad (3\text{–}2)$$

The hydrostatic stress is obtained by dividing by the area—that is, taking the water depth times the density of seawater times the acceleration of gravity, g, or

$$\begin{aligned}&10^4\ \text{m} \times 1030\ \text{kg/m}^3 \times 9.8\ \text{m/s}^2\\ &= 101 \times 10^6\ \text{kg/m/s}^2 = 101\ \text{MPa}\end{aligned} \quad (3\text{–}3)$$

(1 Pa = 1 pascal = 1 kg/m/s^2; 1 MPa = 1 megapascal, the standard unit of stress in the Earth; also equal to 10 bars or 9.8 atmospheres of pressure.) If the same body were buried 10 km deep in the continental crust, the lithostat could be calculated by substituting the average density of the column of rock above the body, assumed to be 2750 kg/m^3. Substituting for seawater density in equation 3–3, the lithostatic value of stress is

$$\begin{aligned}&10^4\ \text{m} \times 2750\ \text{kg/m}^3 \times 10\ \text{m/s}^2 = 275\\ &\times 10^6\ \text{kg/m/s}^2 = 275\ \text{MPa}\end{aligned} \quad (3\text{–}4)$$

Differential stress is the difference between the maximum (commonly designated $\boldsymbol{\sigma}_1$) and minimum (commonly $\boldsymbol{\sigma}_3$) principal normal stresses ($\boldsymbol{\sigma}_1 - \boldsymbol{\sigma}_3$; see equation 3–6 and related text). *Mean stress* is $(\boldsymbol{\sigma}_1 + \boldsymbol{\sigma}_2 + \boldsymbol{\sigma}_3)/3$. *Deviatoric stress* is the nonhydrostatic component of stress, expressed as actual stress with mean stress substracted from the normal stress components. If the differential stress exceeds the strength of the rock, permanent deformation (*strain*) results. The *strength* of a material is the stress required to rupture the material or cause other permanent deformation.

STRESS ON A PLANE

Now that we have defined the terms used in describing and measuring stress, we should consider ways of using them. If we define a plane P in a mass of rock subjected to a vertical stress $\boldsymbol{\sigma}_i$ (Figure 3–5), the stress across a small part of the plane can be written as

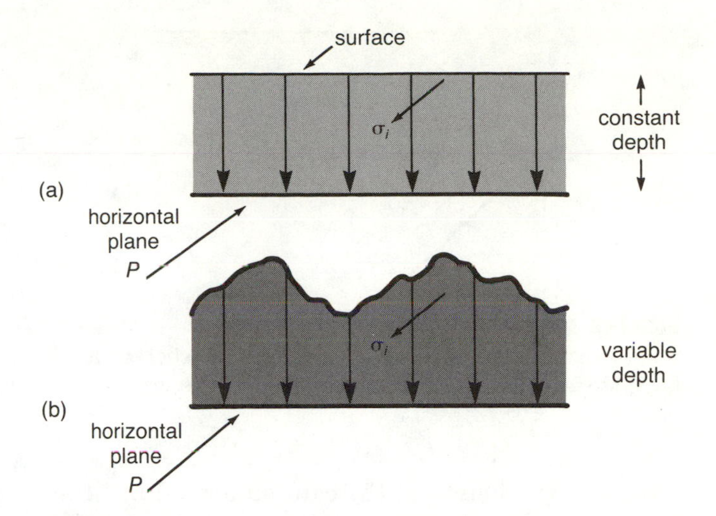

FIGURE 3–5
Stress vectors acting on a horizontal plane near the surface beneath smooth topography (a) and irregular topography (b).

$$\boldsymbol{\sigma}_i = \frac{\Delta\mathbf{F}}{\Delta A} \quad (3\text{–}5)$$

or, if we assume this segment of the plane is infinitesimally small (equivalent to the statement of stress at a point, discussed in the next section),

$$\boldsymbol{\sigma}_i = \lim_{\Delta A \to 0} \frac{\Delta\mathbf{F}}{\Delta A} \quad (3\text{–}6)$$

or

$$\boldsymbol{\sigma}_i = \frac{d\mathbf{F}}{dA} \quad (3\text{–}7)$$

Equation 3–6 shows that stress on a plane, $\boldsymbol{\sigma}_i$, is a vector quantity—it is the product of a vector (force) and a scalar (1/area)—and that each plane has a unique stress vector. Stress on any plane we choose in a mass of rock, particularly near the surface, is likely to vary from place to place on the plane, either because the amount of overburden varies or because the plane is inclined to the surface (Figure 3–6). At depths of several kilometers in the Earth, stress on a horizontal plane is related to the density and height of the column of rock above it (= ρgh = density × gravity × height). Determining the stress on an inclined plane is more difficult. The area of the plane, the density of the column of

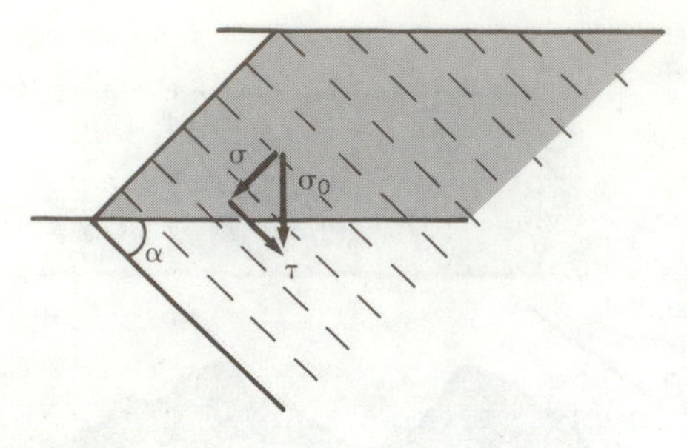

FIGURE 3–6
Stress vectors acting on an inclined plane, such as a fault surface, several kilometers deep in the crust.

rock, and the height of the column are all involved, but the angle between the plane and the principal stress directions must also be known. Because the plane is inclined, the height of the column of overburden varies along the plane. Within the plane, the stress $\boldsymbol{\sigma}_i$ may be resolved into components of shear stress ($\boldsymbol{\tau}$) and normal stress ($\boldsymbol{\sigma}$).

If we assume that our area of 1 m^2 in Figure 3–7 is a plane inclined 45° to the horizontal, we should be able to calculate the vertical stress on the plane (Figure 3–7a). Because the plane is now inclined above the 1 m^2 area, the area of the plane is greater—1.41 m^2—and so the vertical force, using $\mathbf{F} = m\mathbf{a}$ = volume × acceleration of gravity × density, becomes

$$\mathbf{F} = 10^4 \text{ m}^3 \times 9.8 \text{ m/s}^2 \times 2750 \text{ kg/m}^3 = 2.7 \times 10^8 \text{ kg m/s}^2 \quad \textbf{(3–8)}$$

Conversion of force to stress involves dividing by the area of 1.41 m^2, which yields a stress on the inclined plane of 191 MPa. Note that the area has increased, and so the stress on the inclined surface has decreased.

Because any plane will have a unique stress vector resolvable into normal and shear components, the normal and shear components of stress (Figure 3–7b) can also be calculated from

$$\boldsymbol{\sigma} = \boldsymbol{\sigma}_i \cos 45° = 191 \text{ MPa} \times 0.707 = 135 \text{ MPa} \quad \textbf{(3–9)}$$

and

$$\boldsymbol{\tau} = \sigma_i \sin 45° = 191 \text{ MPa} \times 0.707 = 135 \text{ MPa} \quad \textbf{(3–10)}$$

where $\boldsymbol{\sigma}_i$ is the vertical stress component. Note that we are considering only vertical stress here; much greater complexity would be introduced if we considered horizontal stresses that result from expansion due to vertical loading.

If a plane P exists with a vertical stress $\boldsymbol{\sigma}_i$ acting on it, any section of the plane may be assigned a thickness (perpendicular to the page) of one unit, and MN is a segment of the plane across which the normal stress is directed (Figure 3–8), we can derive several expressions for normal and shear stresses on the plane. If a triangular segment ABC represents a small prismatic element with two sides, AC and BC, perpendicular to each other, and AC makes the angle α with AB (which lies in the larger element MN), BC and AC are also assumed to be perpendicular to the greatest and least principal (normal) stress components—$\boldsymbol{\sigma}_1$ and $\boldsymbol{\sigma}_3$. If dA represents the area of the hypotenuse AB (one unit thick) of the prismatic element, and $\boldsymbol{\sigma}$ and $\boldsymbol{\tau}$ are normal and shear stresses acting on this surface, we can derive several useful relationships from the prismatic element.

Assume that the prism ABC is infinitesimally small, so that the weight is negligible compared to the forces acting on it. We can also assume that the prism is in a state of mechanical equilibrium: all forces acting to move the prism up are countered by equal and opposite forces acting to move the prism down. The same is true for forces acting to move the prism in either direction horizontally. (These are statements of Newton's *third law of motion.*) Equilibrium is maintained under the influence of surface

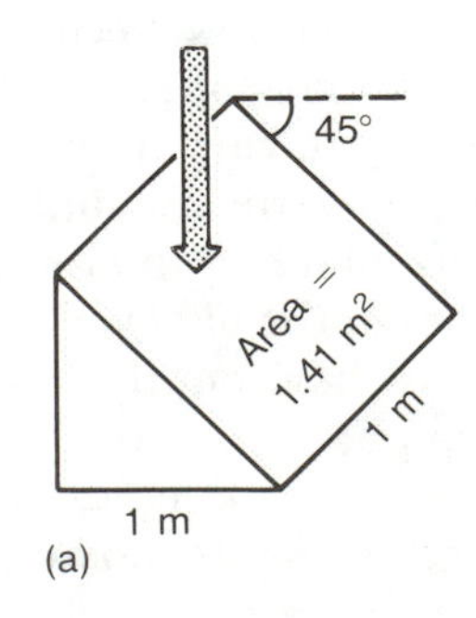

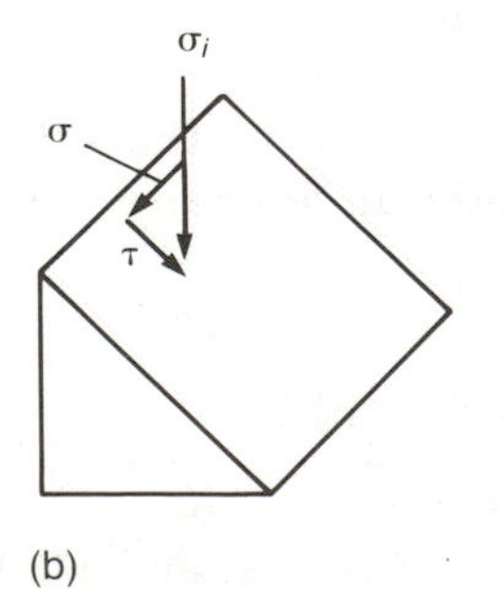

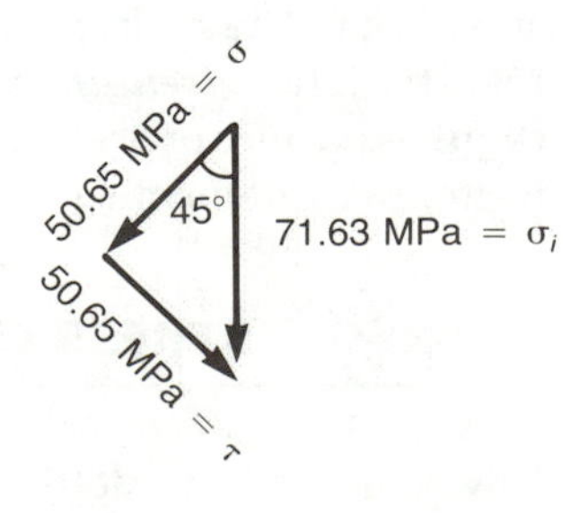

FIGURE 3–7
(a) Vertical stress acting on an inclined plane with an area of 1.41 m^2. (b) Resolution of normal and shear stress components of the vertical stress in (a) and calculation of the magnitude of stresses.

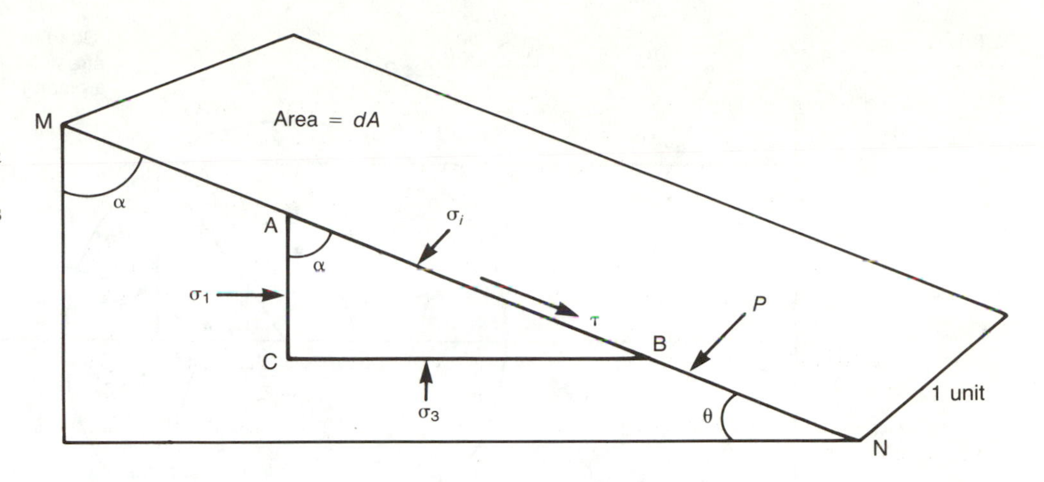

FIGURE 3–8
Relationships between stress σ_i and resolved principal normal stresses σ_1 and σ_3, shearing stress τ, the angles α and θ, and the plane *P* with area *dA* on which the stress is applied.

forces caused by stresses σ_1, σ_3, σ, and τ. Maintaining a state of equilibrium in our hypothetical prism makes possible an equation that expresses the equilibrium condition of horizontal forces (forces acting to move the prism to the left are equal to the forces acting to move the prism to the right, parallel to plane MN):

$$F = \sigma_3 \cos\alpha\,(\sin\alpha\, dA) - \sigma_1 \sin\alpha\,(\cos\alpha\, dA) \tag{3–11}$$

Dividing through by *dA* and collecting terms,

$$\tau = (\sigma_1 - \sigma_3)(\sin\alpha \cos\alpha) \tag{3–12}$$

A similar equation may be written to express equilibrium of forces acting normal to plane MN (forces acting to move the prism up equal forces acting to move the prism down, normal to MN):

$$F = \sigma_1 \cos\alpha\,(\cos\alpha\, dA) + \sigma_3 \sin\alpha\,(\sin\alpha\, dA) \tag{3–13}$$

Dividing through by *dA* and carrying out indicated multiplications,

$$\sigma = \sigma_1 \cos^2\alpha + \sigma_3 \sin^2\alpha \tag{3–14}$$

Because $\sin\alpha \cos\alpha = (\sin 2\alpha)/2$, equation 3–12 becomes

$$\tau = \frac{\sigma_1 - \sigma_3}{2} \sin 2\alpha \tag{3–15}$$

Also, because $\cos^2\alpha = 1/2(1 - \cos 2\alpha)$, and $\sin^2\alpha = 1/2(1 - \cos 2\alpha)$, equation 3–14 becomes

$$\sigma = \frac{\sigma_1 + \sigma_3}{2} + \frac{\sigma_1 - \sigma_3}{2} \cos 2\alpha \tag{3–16}$$

Equations 3–15 and 3–16 are equations for a circle—the ***Mohr circle*** (Figure 3–9). The radius of the circle is $(\sigma_1 - \sigma_3)/2$, and its center is at $(\sigma_1 + \sigma_3)/2$. Given α, σ_1, and σ_3, values of σ and τ may be calculated for any plane in $\sigma\tau$ space.

Despite the rather complicated mathematical nature of stress, the Mohr circle allows us to determine the unique stress vector for a plane with any orientation. While we can do this here in two dimensions, some very important relationships will be produced from this type of analysis. DePaor (1986) represented normal and shear stress with a graphic method employing an "orthonet." His technique provides a way to obtain values for σ and τ without trigonometry.

The angle α is measured between σ_3 and the plane in the ***Mohr construction.*** The relationship may also be stated in terms of the angle θ, which is the complement of α, or the angle between σ_1 and plane *P* in Figure 3–9. These angles will reappear in Mohr circles as 2α and 2θ.

STRESS AT A POINT

Having considered stress on a plane, we can now consider stress in three dimensions. Principal normal stresses, designated σ_1, σ_2, and σ_3 ($\sigma_1 > \sigma_2 > \sigma_3$), may be thought of as oriented parallel to coordinate axes *x*, *y*, and *z*, so that σ_1 corresponds to σ_x, σ_2 to σ_y, and σ_3 to σ_z (Figure 3–10). These normal stresses may also be considered parallel to the edges of a cube. If the cube is reduced to infinitesimal size, a stress, σ_0, applied to the cube may be considered as being applied at a point O (Figure 3–11a). We may also assume, using Newton's third law of motion, that the stresses on all sides of the cube balance and cause no rotation or translation, regardless of the size of the cube and

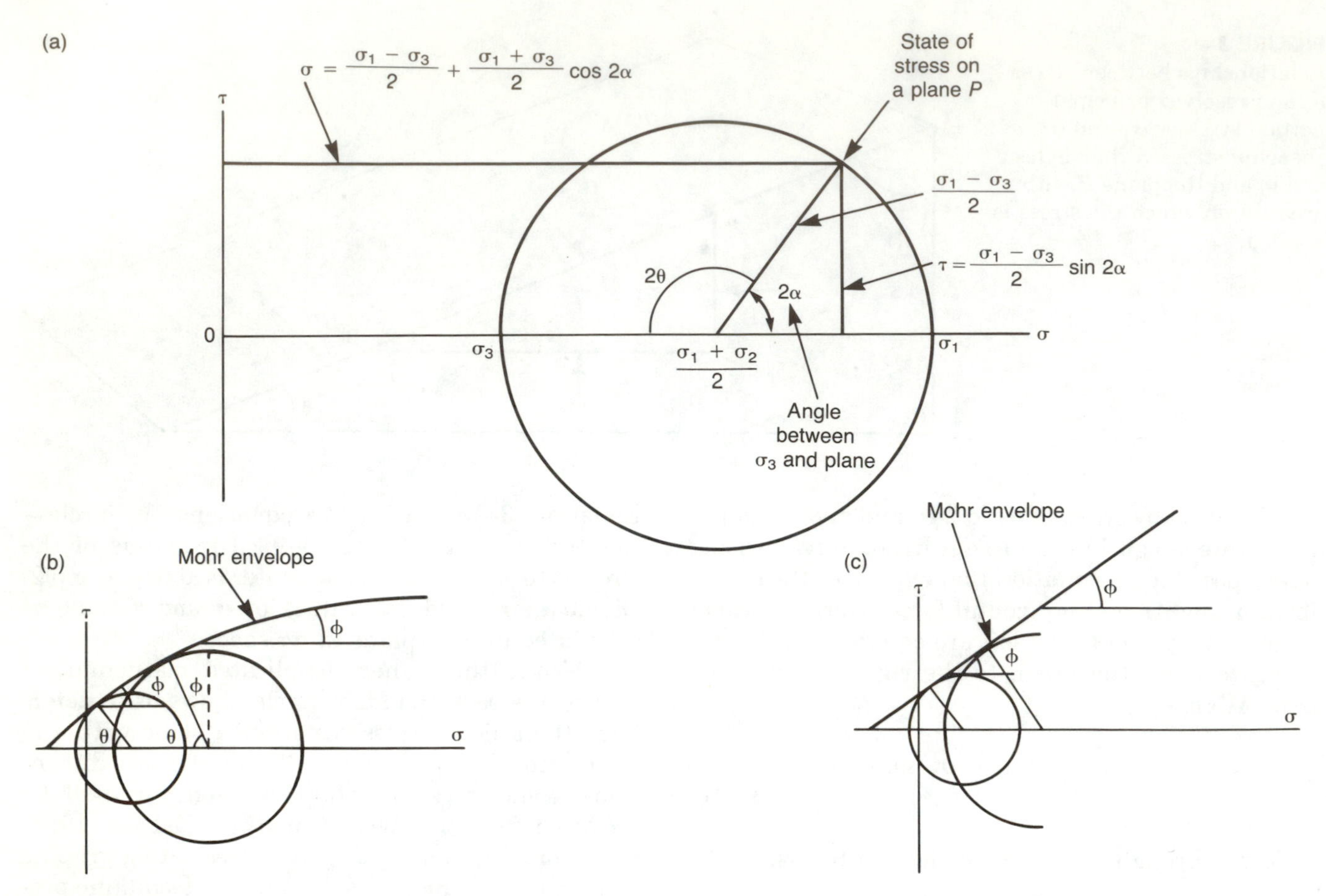

FIGURE 3–9
(a) Mohr circles plotted on σ and τ axes, the envelopes, and the relationships between 2α and 2θ as defined by Hubbert (1951). Note the differences in (b) and (c). With a curved envelope (b), ϕ is variable; with a straight envelope (c), ϕ is constant.

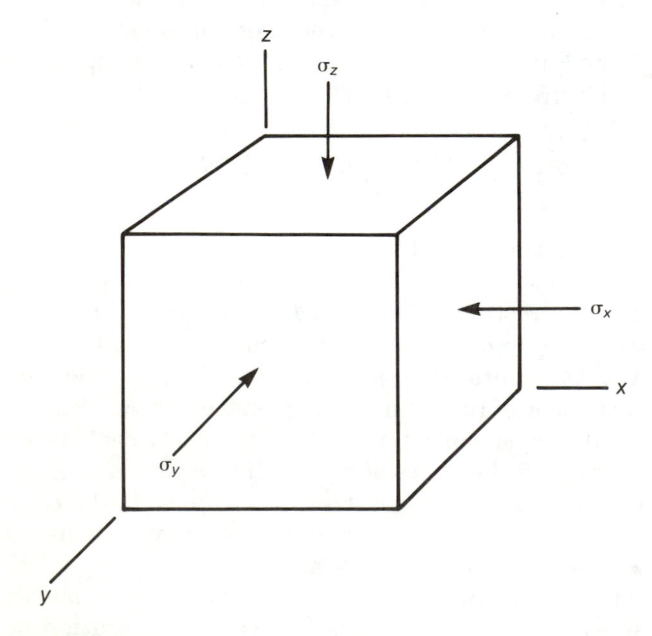

FIGURE 3–10
Principal normal stresses oriented parallel to the edges of a cube in *x-y-z* space. Note the position of the cube relative to the positive ends of the axes.

regardless of whether the stresses are normal or shearing stresses. The stress σ_0 has an orientation and defines a second-order tensor (not a vector here) that may be expressed in both normal and shear components, σ and τ. In three-dimensional space, nine components (Figure 3–11b) are required to describe the stress σ_0 at point O. They include the three components of normal stress, σ_{xx}, σ_{yy}, and σ_{zz}, oriented parallel to the three coordinate axes, and six components of shear stress lying within the six faces of the cube defined as planes *xy, yz,* and *xz,* with parallel planes *yx, zy,* and *zx.* The components of shear stress are then expressed as τ_{xy}, τ_{yz}, τ_{xz}, τ_{yx}, τ_{zy}, and τ_{zx}. (Each subscript represents the association of a shear stress with a particular face on the cube.) Using this set of resolved stresses, we can write simple equations for stress components on each face of the cube. Stresses on the *xy* face (normal to the *z* axis) include σ_{zz}, normal to the face, and shearing stresses τ_{zx} and τ_{zy}, parallel to the face. Stresses on the *zy* face (normal to the *x* axis) are σ_{xx}, normal to the face, and shearing stresses τ_{xy} and τ_{xz}, parallel to the face. Finally, stresses on the *zx* face (normal to the y axis) include the normal stress σ_{yy} and shear stresses τ_{yx} and τ_{yz}.

All nine components of the second-rank tensor for stress at a point, σ_{ij}, where i and j take the values of x, y, and z, may be arranged in matrix form as

$$\begin{pmatrix} \sigma_{xx} & \tau_{xy} & \tau_{xz} \\ \tau_{yx} & \sigma_{yy} & \tau_{yz} \\ \tau_{zx} & \tau_{zy} & \sigma_{zz} \end{pmatrix} \tag{3–17}$$

This matrix is called a ***stress tensor.*** The rows (horizontal) from top to bottom represent components of stress on faces of the cube (Figure 3–11b) normal to x, y, and z. The columns (vertical) from left to right represent the directions of stress components parallel to the x, y, and z axes. The left column represents stresses acting parallel to the x axis, the middle column the direction parallel to the y axis, and the right column the direction parallel to the z axis. If we designate the elements of the tensors i and j—again with i and j taking values of x, y, and z—we can state that the tensor in equation 3–16 is symmetric because $\tau_{ij} = \tau_{ji}$ (that is, elements i and j may be reversed without changing the solution). Symmetry is a requirement of the condition of force balance, or equilibrium, assumed at the beginning of this discussion. If the x-y-z coordinate system is oriented parallel to the principal normal stress axes (parallel to one of the edges of the cube), no shear stress will be present on the faces (all values of $\tau = 0$). The same thing could be accomplished by rotating the cube in the stress field until only

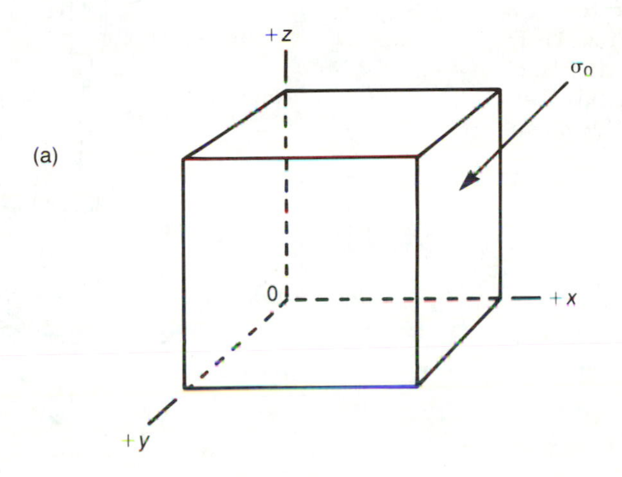

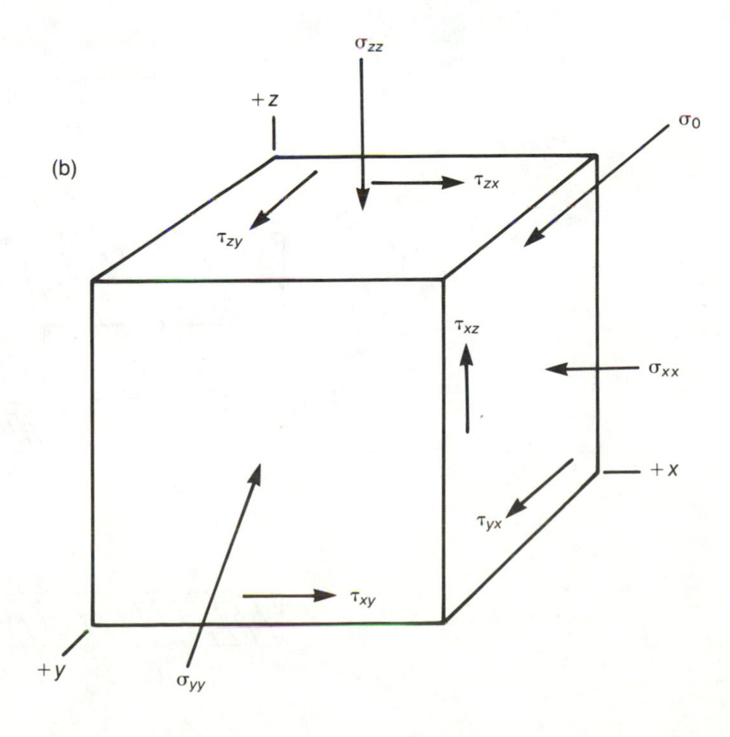

FIGURE 3–11
(a) Randomly oriented stress σ_0 applied to an infinitesimally small (point size) reference cube in x-y-z space. (b) Enlarged reference cube shows resolution of nine shearing and normal stress components.

normal stresses are present on the faces. The stress tensor for zero shear stresses may be written as

$$\begin{pmatrix} \sigma_{xx} & 0 & 0 \\ 0 & \sigma_{yy} & 0 \\ 0 & 0 & \sigma_{zz} \end{pmatrix} \tag{3–18}$$

The three normal stresses are called the ***principal stresses.*** They act parallel to the x, y, and z axes where shear stresses are zero.

MOHR CONSTRUCTION

Now we can use experimental results to calculate values of material properties. Plotting σ versus τ on two coordinate axes results in the Mohr construction (Figure 3–9a). Construction of the Mohr circle from values of σ_1 and σ_3 from laboratory measurements in a triaxial load apparatus (Figure 3–12) is remarkably simple. A cylindrical sample of rock, concrete, or other material is placed in the test cylinder. The

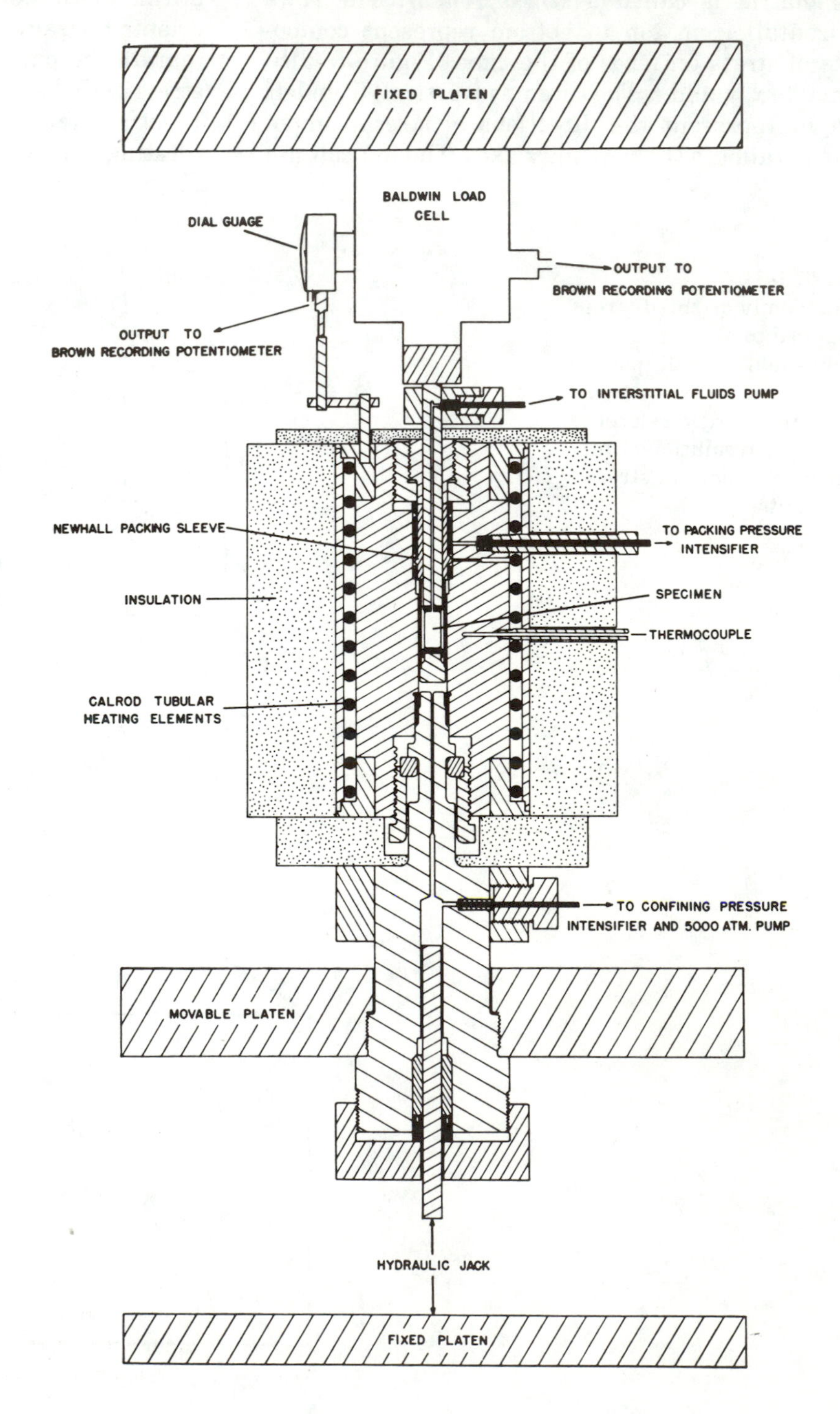

FIGURE 3–12
Triaxial test apparatus. Axial load (σ_1) is applied with a hydraulic jack. The axial load is measured directly by the load cell. The dial gauge provides a direct measure of strain as percent shortening of the rock cylinder. (From D. T. Griggs, F. J. Turner, and H. C. Heard, 1960, Geological Society of America *Memoir* 79.)

specimen is jacketed in aluminum or copper foil, or a thermal plastic, to protect it and the equipment and permit easy monitoring of the experiment. The apparatus is also jacketed so that a fluid (such as water or oil) can be used to vary the confining pressure externally. The confining pressure is equal to both σ_2 and σ_3 because pressure is being applied equally around the cylinder. The load on the ends of the cylinder is usually increased until the material ruptures. This load is the axial load, σ_1. If the rock is loaded to the rupture point, a new cylinder of the same material is required for each run at increasing confining pressure σ_3. Table 3–1 shows values of σ_1 and σ_3 (in kilograms per square centimeter) derived from successive runs to failure on cylindrical specimens of limestone in an axial load apparatus. Plotting the data on the σ, or horizontal, axis permits construction of a Mohr diagram for the limestone (Figure 3–13a). Plots on the σ axis determine the diameter of the circle for each run. The *Mohr envelope* can be constructed with ease, as the tangent to the circles. The Mohr envelope also separates *unfailed* (undeformed) from *failed* (deformed) regions in the diagram, with the unfailed region lying within the envelope (Figure 3–13b).

A Mohr circle that does not intersect the envelope indicates that the sample experienced no permanent deformation—that it did not rupture with the values of σ_1 and σ_3 employed in that run. As a result, the Mohr envelope can be constructed using only circles obtained when the sample actually ruptured.

Curvature of the Mohr envelope is frequently related to inherent properties of the material (Figure 3–14). Brittle materials tend to produce relatively straight Mohr envelopes with steep slopes. With increasing confining stress (σ_3), even a brittle material behaves megascopically in a ductile manner. If the materials are ductile, they experience some permanent nonrecoverable strain before rupture (Chapter 6). With an increase in ductility, the Mohr envelopes become flattened, resulting in an overall curvature. In general, the greater the curvature, the greater the amount of ductile strain before rupture, but Griffith materials (Chapter 10), which are actually brittle, produce curved envelopes. Their behavior is related to the propagation of fractures through glass.

TABLE 3–1

σ_3	σ_1
0	750
250	1750
500	2400
1050	3550

The angle 2θ, termed the ***conjugate shear angle,*** is related to the coefficient of internal friction μ_i (or $\tan \phi$, as will be defined in equation 3–24), and θ is the angle between σ_1 and the fracture that forms. The relationship is

$$90 - \phi = 2\theta \tag{3–19}$$

(Here θ is defined as in Figures 3–9 and 3–14.) As the radii of the Mohr circles increase and approach a limit (with increasing stress difference and confining pressure), the slope of the envelope flattens as a function of increased ductility, and 2θ approaches a maximum value of 90°. Therefore the maximum shear angle, θ, is 45°. This angle is most commonly attained in ductile materials.

Several relationships within Mohr diagrams are summarized in Figure 3–15. Note that the τ axis separates compressional from tensional normal stress fields (Figure 3–15a). Materials with no tensile strength, such as dry sand, have an envelope that terminates at the origin (Figure 3–15b). Most geologic materials have compressive strengths much greater than their tensile strengths (Figure

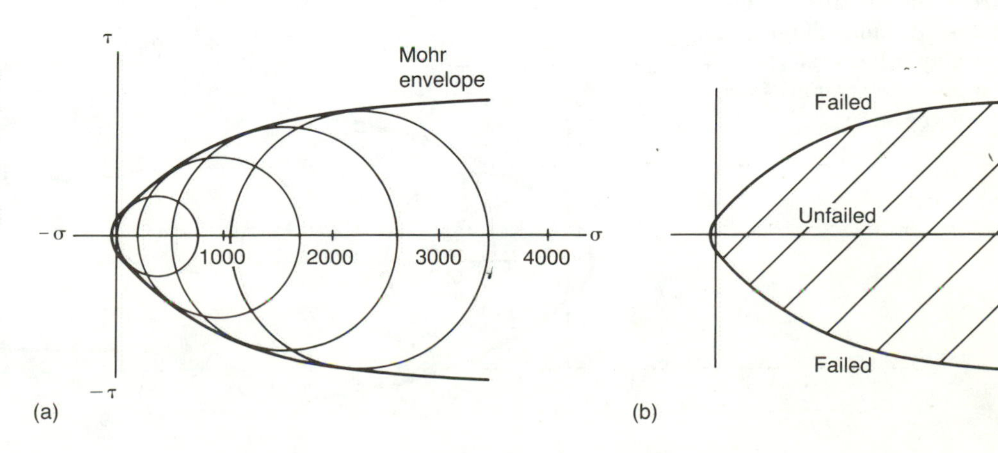

FIGURE 3–13
(a) Mohr diagram plotted from the values obtained from the experimental deformation of a limestone at different confining pressures. (b) Mohr diagram showing unfailed and failed regions.

FIGURE 3–14
Mohr diagrams for brittle (a) and ductile (b) materials. Mohr diagrams for the Oil Creek Sandstone and the Blaine Anhydrite. (From M. K. Hubbert and D. G. Willis, 1957, *American Institute for Mining, Metallurgical, and Petroleum Engineers Transactions,* v. 210.)

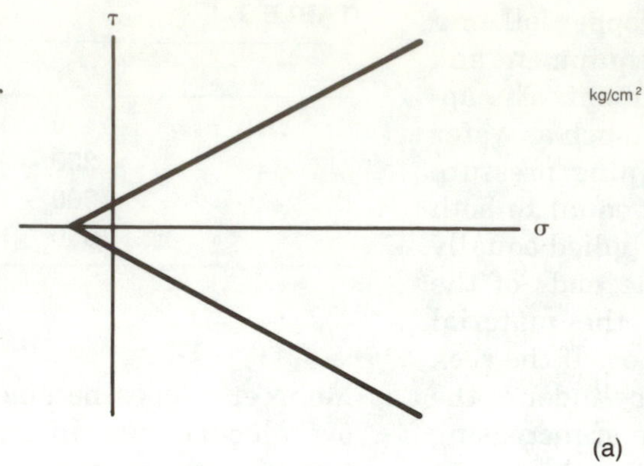

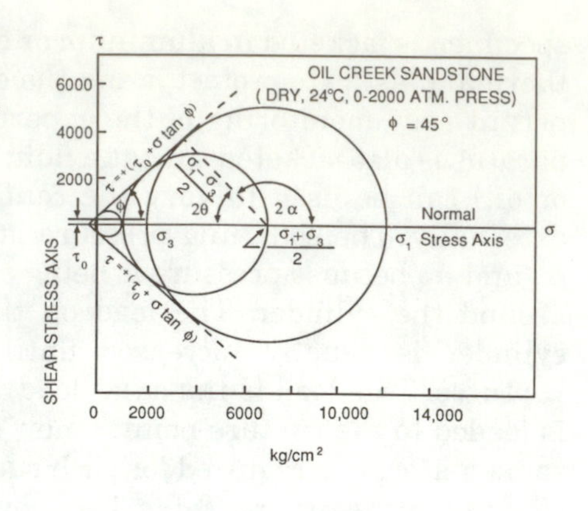

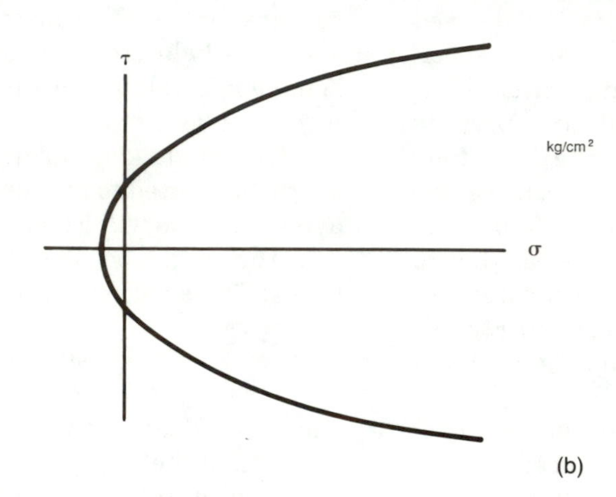

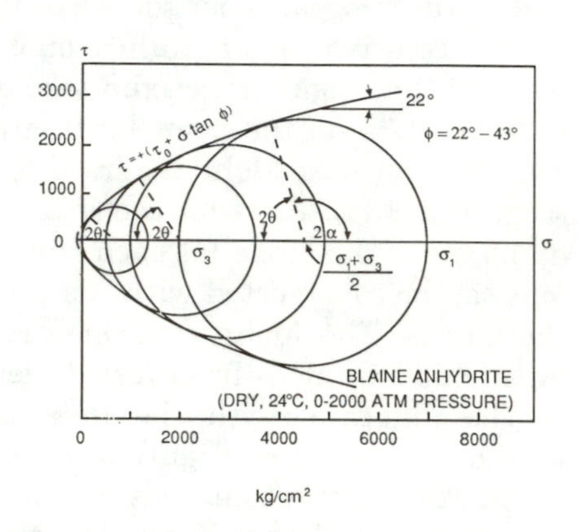

FIGURE 3–15
(a) Axes for Mohr diagrams showing extensional ($-\sigma$) and compressional ($+\sigma$) fields. (b) Mohr diagram for quartz sand, a material with no tensile strength. (c) Mohr diagram for an average rock material that exhibits tensile strength at low stress and ductility at higher stress. (d) Mohr diagram for wet clay, where tensile and compressive strengths are about the same.

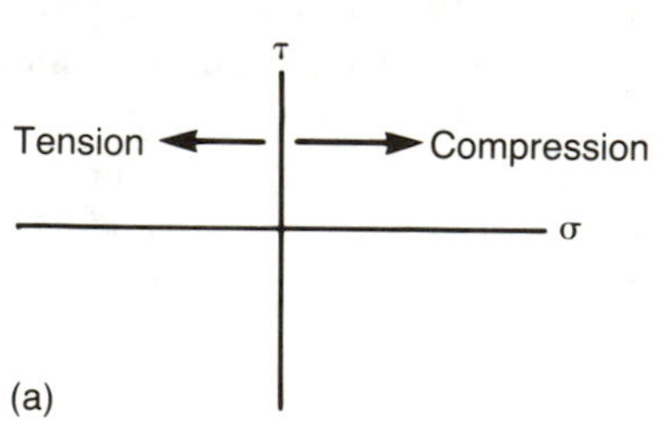

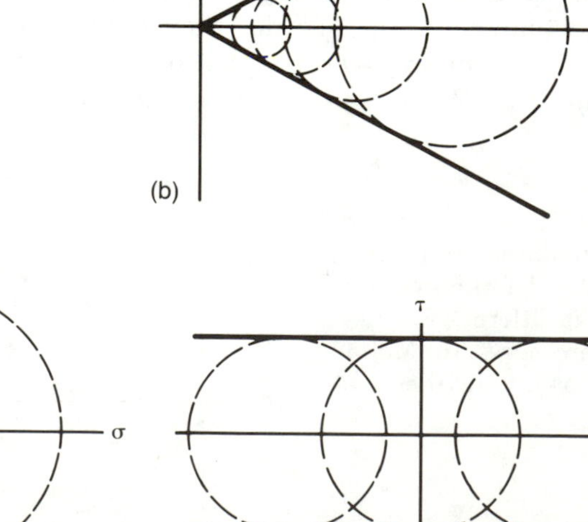

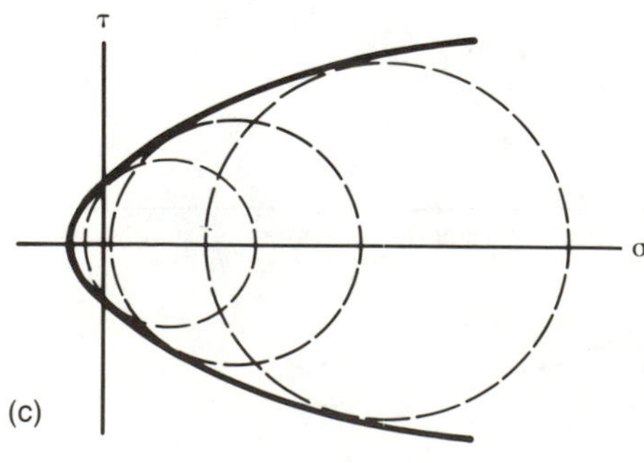

τ
σ
(d)

3–15c). Very few materials have tensile and compressive strengths even roughly the same, but wet clay (Figure 3–15d) is one that does. Note that these comparisons are also between cohesive (rocks, clays) and noncohesive (dry sand) materials.

MOHR'S HYPOTHESIS

Otto Mohr, the German engineer who in 1882 developed the Mohr construction, suggested in 1900 that shear strength is a function of normal stress,

$$|\boldsymbol{\tau}_s| = f(\boldsymbol{\sigma}) \qquad (3\text{–}20)$$

where $\boldsymbol{\tau}_s$ is *shear strength*—a material's resistance to shear stress. Mohr's hypothesis, and the derivation of the equations for resolved shear and normal stresses on a plane (equations 3–15 and 3–16), are both incorporated in the Mohr construction (Figure 3–9).

A relationship similar to *Mohr's hypothesis* may be derived from Amontons' first law,

$$\mathbf{F} = \mu_s \mathbf{W} \qquad (3\text{–}21)$$

where **F** is the shear force along a contact surface, μ_s is the coefficient of sliding friction along the surface, and **W** is the force perpendicular to the surface (Jaeger and Cook, 1976). The coefficient of sliding friction is a measure of the resistance of a material to sliding along a surface. If we divide both sides of equation 3–21 by area (A), it becomes

$$\boldsymbol{\tau}_s = \mu_s \boldsymbol{\sigma} \qquad (3\text{–}22)$$

Another relationship, which predates Mohr's hypothesis by more than a century and bears directly on the Mohr construction, is the ***Coulomb criterion of failure,*** proposed in 1773 by a French physicist, Charles A. Coulomb. It states that the absolute value of shear strength, $\boldsymbol{\tau}_s$, is the sum of the inherent shear strength, $\mathbf{S}_0$, and the coefficient of internal friction, μ_i, or static friction, multiplied by the normal stress $\boldsymbol{\sigma}$

$$|\boldsymbol{\tau}_s| = \mathbf{S}_0 + \mu_i \boldsymbol{\sigma} \qquad (3\text{–}23)$$

Equation 3–23 predicts that fracturing will occur when the shear stress on a plane (such as a fault plane) reaches a critical value. The Coulomb equation is an equation of a line and defines the Mohr envelope (Figure 3–9). It is commonly expressed today as

$$|\boldsymbol{\tau}_s| = \boldsymbol{\tau}_0 + \boldsymbol{\sigma} \tan \phi \qquad (3\text{–}24)$$

Here $\boldsymbol{\tau}_0$ is the inherent shear strength of the material at zero normal stress, and $\mu_i = \tan \phi$; ϕ is the angle of internal friction, and $\tan \phi$ is the coefficient of internal friction.

STRESS ELLIPSOID

The Mohr construction represents stress characteristics in two dimensions. A graphic means of showing the relationships between the principal stresses is the ***stress ellipsoid*** (Figure 3–16). It is a triaxial ellipsoid in which the greatest, intermediate, and least principal axes are $\boldsymbol{\sigma}_1$, $\boldsymbol{\sigma}_2$, and $\boldsymbol{\sigma}_3$. Principal planes in the stress ellipsoid contain the principal axes. The $\boldsymbol{\sigma}_1$-$\boldsymbol{\sigma}_2$, $\boldsymbol{\sigma}_2$-$\boldsymbol{\sigma}_3$, and $\boldsymbol{\sigma}_1$-$\boldsymbol{\sigma}_3$ planes are principal planes. The $\boldsymbol{\sigma}_1$-$\boldsymbol{\sigma}_3$ plane defines the maximum stress difference. Two other planes, parallel to circular sections, are shear planes. Stress vectors oriented between the principal stresses terminate on the surface of the ellipsoid. Circular sections through the ellipsoid define the planes of maximum shear stress.

We end our survey of stress and its mechanical basis with this short discussion of the stress ellipsoid. Now we turn in the next several chapters to a discussion of strain—the presumed effect of stress.

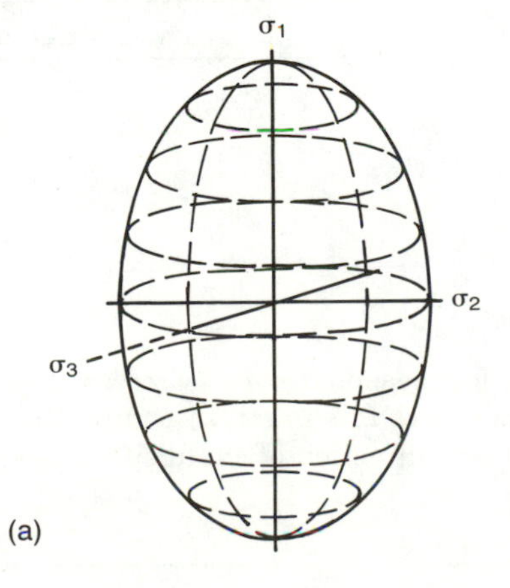

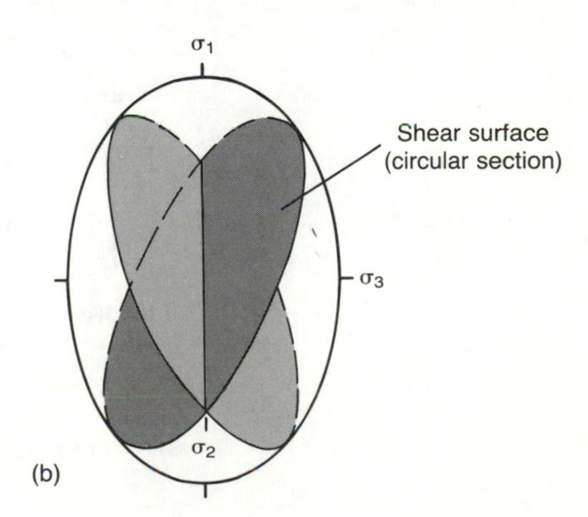

FIGURE 3–16
(a) The stress ellipsoid, a triaxial ellipsoid in which axes are the principal normal stresses $\boldsymbol{\sigma}_1$, $\boldsymbol{\sigma}_2$, and $\boldsymbol{\sigma}_3$. (b) The circular sections are planes of maximum shear stress.

ESSAY

Measuring Present-Day Stress in the Earth

Present-day stress in the Earth can be measured by placing a strain gauge in rock and recording the *in situ* elastic strain. Such measurements are most frequently carried out by drilling a hole into bedrock, inserting the instruments, and recording the amounts and orientations of elastic strain. Two techniques are commonly used for measuring *in situ* stress: *overcoring* and *hydraulic fracturing*.

Overcoring involves drilling a hole 3 to 4 cm in diameter, then drilling a larger core (12 to 15 cm) with the same center as the smaller hole (Figure 3E–1; Hooker and Bickel, 1974). Before coring the larger hole, an instrument called a dilatometer is inserted into the smaller hole to permit measurement of the expansion (relaxation) of the rock mass as the larger hole is cored. Changes in shape of the small hole are recorded as it is overcored. The orientation of the small hole is known, so changes in its shape can be measured as it relaxes from a circle to an ellipse. This provides a measure of the orientation of the present-day stress ellipsoid. If several measurements can be made at the same locality of overcores in different orientations—as is possible in a mine, tunnel, or quarry—a very good measure of the orientation of the stress ellipsoid may be obtained. Attempts have also been made to measure the magnitude of the principal stresses using this technique, but the accuracy has been severely questioned when compared with values obtained in other ways. Consequently, this method is not now much used if hydraulic fracturing is feasible.

Hydraulic fracturing involves drilling a vertical hole, sealing off a part of the hole with packers, and increasing the hydraulic pressure in the sealed-off part of the hole until the rocks fracture (Figure 3E–2; Zoback and Haimson, 1982). Both the amount of pressure needed to produce fractures and also the orientations of hydraulic fractures are measured. That provides a measure of both the amount of compressive stress present at the site and also the orientation of maximum and minimum principal stresses. It is possible to make several measurements in the same hole if it is drilled to a minimum

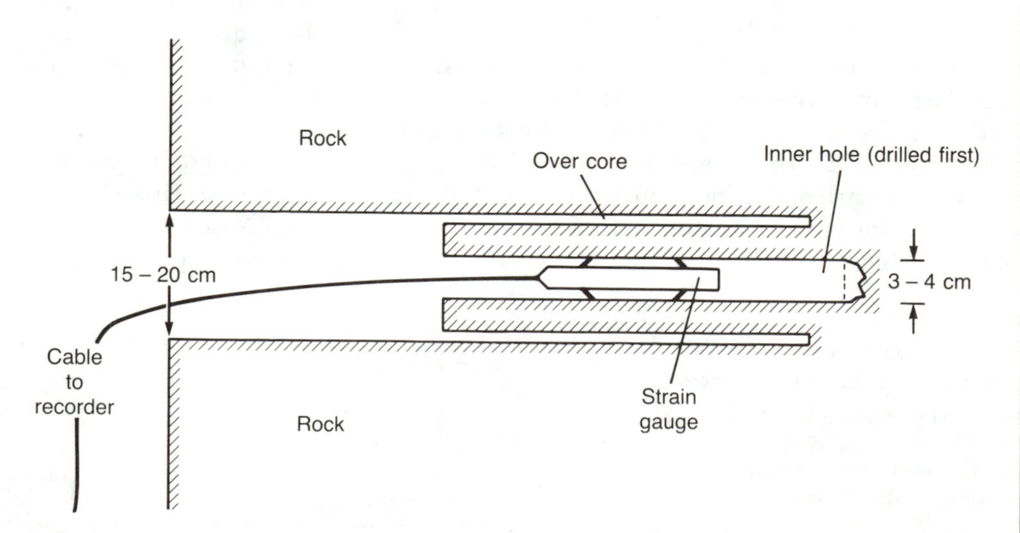

FIGURE 3E–1
Overcoring technique for measuring *in situ* stress. The strain gauge rests inside the small borehole and measures *in situ* stress—while the overcore is being drilled—by recording the amount and direction of change from a circular to an elliptical borehole.

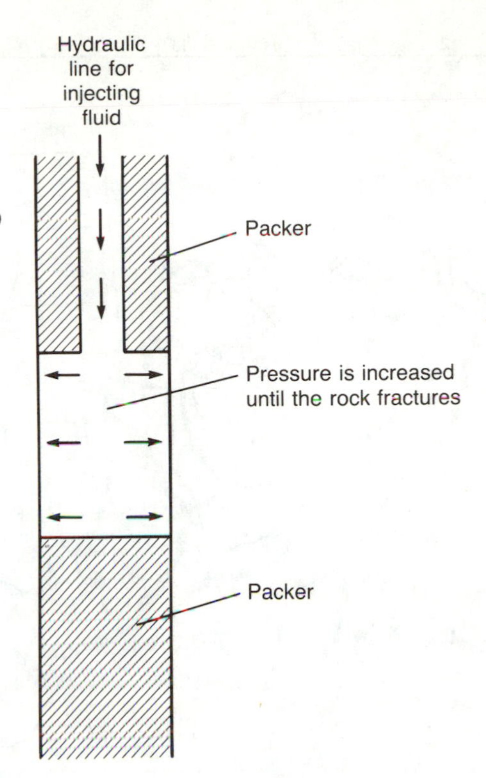

FIGURE 3E–2
Hydraulic fracturing method for measuring *in situ* stress. (Modified from R. O. Kehle, *Journal of Geophysical Research,* v. 69, © 1964 by the American Geophysical Union.)

depth of 300 to 400 m. The hydraulic fracturing method requires that the hole drilled be nearly vertical, for it is assumed that one of the principal normal stresses is vertical and that the hydraulic fracture in a vertical borehole propagates perpendicular to $\boldsymbol{\sigma}_3$ (Hubbert and Willis, 1957). Zoback and Zoback (1980) showed that field data support the relationship between hydraulic-fracture propagation and $\boldsymbol{\sigma}_3$.

Numerous measurements of *in situ* stress have been made in the United States. Coupled with other indicators of stress orientation, they permit identification of areas or domains of common maximum principal stress orientation (Figure 3E–3). Knowledge of these orientations bears directly on our understanding of the causes of seismic activity (Zoback, Tsukahara, and Hickman, 1980; Zoback and Zoback, 1980).

In situ stress in a tectonically active area has been measured by Zoback, Tsukahara, and Hickman (1980) in a series of wells about 240 m deep drilled as far as 34 km from the San Andreas fault. They also made several measurements in a 1-km well about 4 km from the fault. Their goal was to evaluate variation in the magnitude of the principal stresses with depth and to calculate the magnitude of shear stress along the fault. They used hydraulic fracturing to measure stress and obtained the magnitudes of $\boldsymbol{\sigma}_1$ and $\boldsymbol{\sigma}_3$ by using an equation derived by Hubbert and Willis (1957):

$$P_b = 3\boldsymbol{\sigma}_3 - \boldsymbol{\sigma}_1 - P_p + T \qquad \textbf{(3E–1)}$$

where P_b is breakdown (fracture formation) pressure, P_p is pore pressure, and T is the tensile strength of the rock mass. P_b is measured directly as the pressure necessary to hydraulically break the rocks. P_p is calculated from the pressure of a column of water (using ρgh) at the depth at which hydraulic

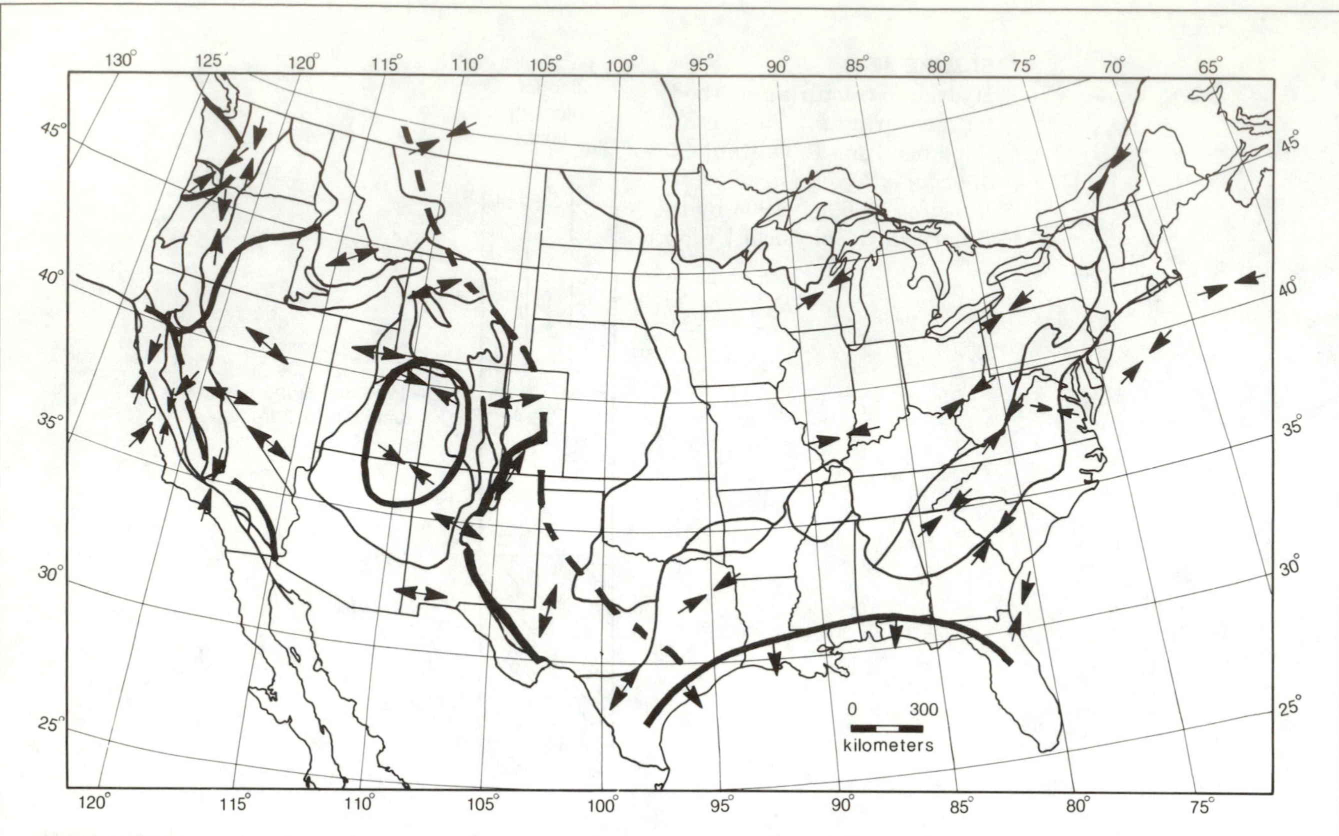

FIGURE 3E–3
Domains of common orientation of present-day maximum principal stress in the United States. Arrowheads indicate whether stress is extensional or compressional. (From M. L. Zoback and M. D. Zoback, 1989, in press, Geological Society of America *Memoir.*)

fracturing was carried out, assuming the pores in the rocks are interconnected and communicate with the surface. The tensile strength of the rock mass can be estimated (as here) or measured for core samples obtained from the well. The magnitude of σ_3 is measured directly, and its azimuth is obtained from orientations of hydraulic fractures in the well. The value of σ_1 is then calculated from equation 3E–1.

Results of measurements in the two areas studied (Table 3E–1) indicate that the maximum principal horizontal stress, σ_1, is oriented about 45° from the strike of the San Andreas fault. Shear stress, determined by plotting the maximum principal stresses on a Mohr diagram, increases from about 2.5 MPa (25 bars) at depths of 150 to 300 m to about 8.0 MPa (80 bars) at 750 to 850 m.

Plotting principal stress orientations on a map enables us to relate the present-day stress in the Earth to plate motion (Figure 3E–3). The nearly

TABLE 3E–1
Hydrofracture data along the San Andreas fault

Well	Depth (m)	Distance to San Andreas Fault (km)	Hydrofracturing Data: Breakdown Pressure (bars)	Hydrofracturing Data: Fracture Opening Pressure (bars)	Hydrofracturing Data: Pore Pressure (bars)	Principal Stress (bars): σ_3	Principal Stress (bars): σ_1	Principal Stress (bars): σ_2 Vertical	Tensile Strength (bars)	τ_{max} (bars)	Direction of Maximum Compression
1	167	4	200	69	17	73	133	45	131	30	N 4° W
1	196	4	209	74	20	77	138	53	135	31	N 1° E
2	238	4	109	63	34	74	125	91	46	26	N 43° W
2	561	4	163	130	56	150	264	152	33	57	N 20°W
2	787	4	192	124	78	183	346	213	68	82	N 19° W
3	80	2	144	24	8	23	38	18	120	8	N 20°W
3	185	2	250	73	19	56	73	43	177	6	N 23° W
4	167	4	139	47	17	51	89	45	92	19	N 83° E
4	230	4	164	85	23	83	140	62	79	29	N 14° W

From Zoback, Tsukahara, and Hickman (1980), *Journal of Geophysical Research,* v. 85. Used by permission.

uniform N 70° E orientation of stress fields in the eastern United States is thought to be related to "ridge-push" from the Mid-Atlantic Ridge, and the markedly different orientations in western states are related to interaction between the Pacific and North American plates (Figure 1–1; Richardson, Soloman, and Sleep, 1979).

References Cited

Hooker, V. E., and Bickel, L. D., 1974, Overcoring equipment and techniques used in rock stress determination: U.S. Bureau of Mines Information Circular 8618, 32 p.

Hubbert, M. K., and Willis, D. G., 1957, Mechanics of hydraulic fracturing: Journal of Petroleum Technology, v. 9, p. 153–168.

Richardson, R. M., Soloman, S. C., and Sleep, N.H., 1979, Tectonic stress in the plates: Reviews of Geophysics and Space Physics, v. 17, p. 981–1019.

Zoback, M. D., and Haimson, B. C., 1982, Status of the hydraulic fracturing method for *in situ* stress measurements, *in* Society of Mining Engineers, Proceedings 23rd Symposium on Rock Mechanics, New York, American Institute of Mining and Metallurgical Engineers, p. 143–156.

Zoback, M. D., Tsukahara, H., and Hickman, S., 1980, Stress measurements at depth in the vicinity of the San Andreas fault: Implications for the magnitude of shear stress at depth: Journal of Geophysical Research, v. 85, p. 6157–6173.

Zoback, M. L., and Zoback, M. D., 1980, State of stress in the conterminous United States: Journal of Geophysical Research, v. 85, p. 6113–6156.

Zoback, M. L., and Zoback, M. D., 1989, Tectonic stress field of the continental United States, *in* Pakiser, L., and Mooney, W., Geophysical framework of the continental United States: Geological Society of America Memoir, in press.

Questions

1. Why do we commonly deal with stress, rather than force, in the earth sciences?
2. What does the shape of the Mohr envelope tell you about the strength of the material being deformed?
3. Why does the Mohr construction work using only values of σ?
4. What is actually happening in materials that produce straight Mohr envelopes?
5. Calculate the lithostatic value of stress on a fault plane inclined at 30° and located at a depth of 7 km in oceanic crust (ρ = 3000 kg/m^3).
6. Does the limestone represented by the Mohr diagram in Figure 3–13 exhibit brittle or ductile behavior (units of kg/m^3)? Would values of σ_3 of 1000 kg/m^3 and σ_1 of 2500 kg/m^3 produce failure in a cylinder of this limestone? Explain.
7. Hydrofracture measurement of principal stresses along the San Andreas fault at a depth of >600 m in a drill hole by Zoback, Tsukahara, and Hickman (1980) yielded a value for σ_3 of 141 bars and for σ_1 of 258 bars. Determine the value of shear stress at that point.

Further Reading

Hubbert, M. K., 1951, Mechanical basis for certain familiar geologic structures: Geological Society of America Bulletin, v. 62, p. 355–372.
This is a classic study relating experimentally produced structures to both theory and real tectonic structures as observed in the field. It also presents a clear, simple, and concise derivation of the Mohr circle equations.

DePaor, D. G., 1986, A graphical approach to quantitative structural geology: Journal of Geological Education, v. 34, p. 231–236.
A new graphical method is described for use in structural geology, determining normal and shear stress without trigonometry.

Means, W. D., 1976, Stress and strain: New York, Springer-Verlag, 339 p.
A clearly written introduction to the concepts of stress and strain through the principles of continuum mechanics. Mathematical concepts and derivations are presented understandably, using problems (with solutions) related to geologic situations. Emphasizes elastic and viscous strain.

4

Strain

Stress conditions within the earth's crust change during the progress of time, and these changes often lead to the permanent deformation of crustal rocks. One of the prime aims of the structural geologist is to determine the nature and amount of these displacements.

JOHN G. RAMSAY, *Folding and Fracturing of Rocks*

STUDY OF STRUCTURAL GEOLOGY IS LARGELY A STUDY OF deformation. We assume deformation is produced as a direct result of stress, but the most direct evidence to support this assumption is derived from our ability to measure elastic strain and detect earthquakes. Indirect evidence of the connection between stress and nonelastic strain is derived from observing plastically folded layers of rock exposed by erosion on the present surface of the Earth. These layers were folded at great depths where crustal rocks were undergoing recrystallization. A more direct analogy may exist between similar structures—brittle near the surface and plastic at depth—that may be observed in glaciers. It is possible to measure rates of plastic deformation deep inside glaciers, providing a more direct indication of the relationships between stress and nonelastic strain in a material—glacial ice—that will flow under surface conditions.

Our ability to observe tectonic structures is largely restricted to a two-dimensional observation platform—the surface of the Earth—with the third dimension limited to a very few kilometers of vertical exposure in young mountain chains. Most structures we can observe were formed long ago and several kilometers deep in the crust, and, to understand how they form, we must resort to experiments that model conditions in the crust. Fortunately, many tectonic structures we can observe on our two-dimensional platform do not parallel the Earth's erosion surface, but are inclined to it obliquely (Figure 4–1). In such cases, sometimes we can observe an entire structure that formed over a range of depths in the crust and later was tilted and exposed by erosion.

One way we gain a greater understanding of tectonic structures and the conditions that formed them is by comparing their present *deformed* (strained) *state* to the original *undeformed* (unstrained) *state* for the same rock mass. This is frequently a difficult undertaking because for comparison we need an undeformed model of each structure. Sometimes parts of structures or objects within them help us solve the problem. For example, deformed fossils are easily compared with their undeformed counterparts (Figure 4–2). An ideal study of strain would consist of a comparison of deformed and undeformed states, but the complexity of tectonic structures frequently frustrates our efforts. Recent efforts to describe strain quantitatively have provided an important and useful aid in our understanding of strain in rocks and also have helped us to construct models that help restore structures to an undeformed state. In this first discussion of the kind and amount of strain affecting a mass of rocks, we will consider only the initial and final states of

FIGURE 4–1
Gently plunging folds in Middle Proterozoic gneiss near Central City, Colorado. These rocks were deformed at elevated temperature and pressure that produced ductile conditions. (RDH photo.)

deformation of the mass—not the *deformation path* or the stages of *progressive deformation* affecting the mass in intermediate stages.

DEFINITIONS

Strain is the change in angles and length of lines as a presumed response to force applied to an area. Strain may also be thought of as *distortion* of the shape, length, or volume of a mass (Figure 4–3a). In this more restrictive definition (Figure 4–3b), *displacement* or *rotation* of a mass without distortion is not considered strain.

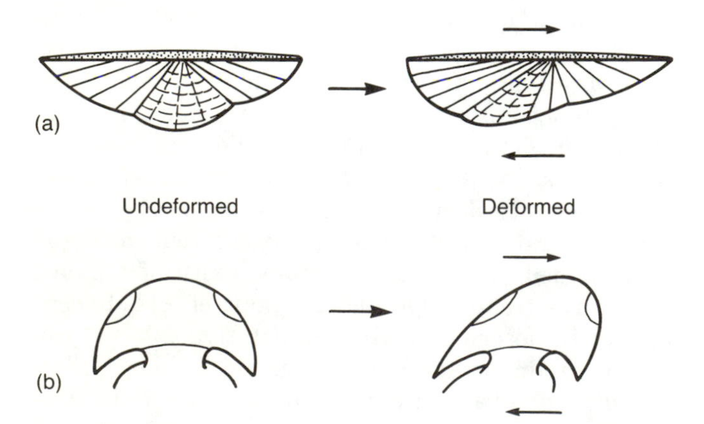

FIGURE 4–2
Undeformed and deformed fossils. Shear direction is indicated on deformed specimens. (a) Brachiopod. (b) Trilobite cephalon.

In *homogeneous strain,* lines that are straight and parallel before deformation remain straight and parallel after deformation (Figure 4–4). *Inhomogeneous strain* is the opposite: straight or parallel lines before deformation do not remain straight (or parallel) after deformation. Lines may also be broken during inhomogeneous deformation. Another factor in identifying the kind of strain affecting a rock mass is scale: homogeneous strain affecting a rock mass on a scale of several kilometers may on a scale of centimeters be resolved into inhomogeneous strain.

Strain may occur in infinitesimally small amounts as ***infinitesimal strain,*** or ***finite strain,*** which consists of a comparison of the difference in the present state of strain with some previous, less deformed state. Strain that occurs in events separated from other deformational events in the strain history of the body is ***incremental strain.*** These separable events may be infinitesimal or finite. Most natural strain we observe in the field is finite strain and may or may not be separable into events of incremental strain. Moreover, incremental finite strain is *progressive strain.*

KINDS OF STRAIN

Strain may occur in several ways to effect deformation of a rock body. It may be recognized as change in line length, angles between lines, and volumetric shape. If some fundamental dimensions are known,

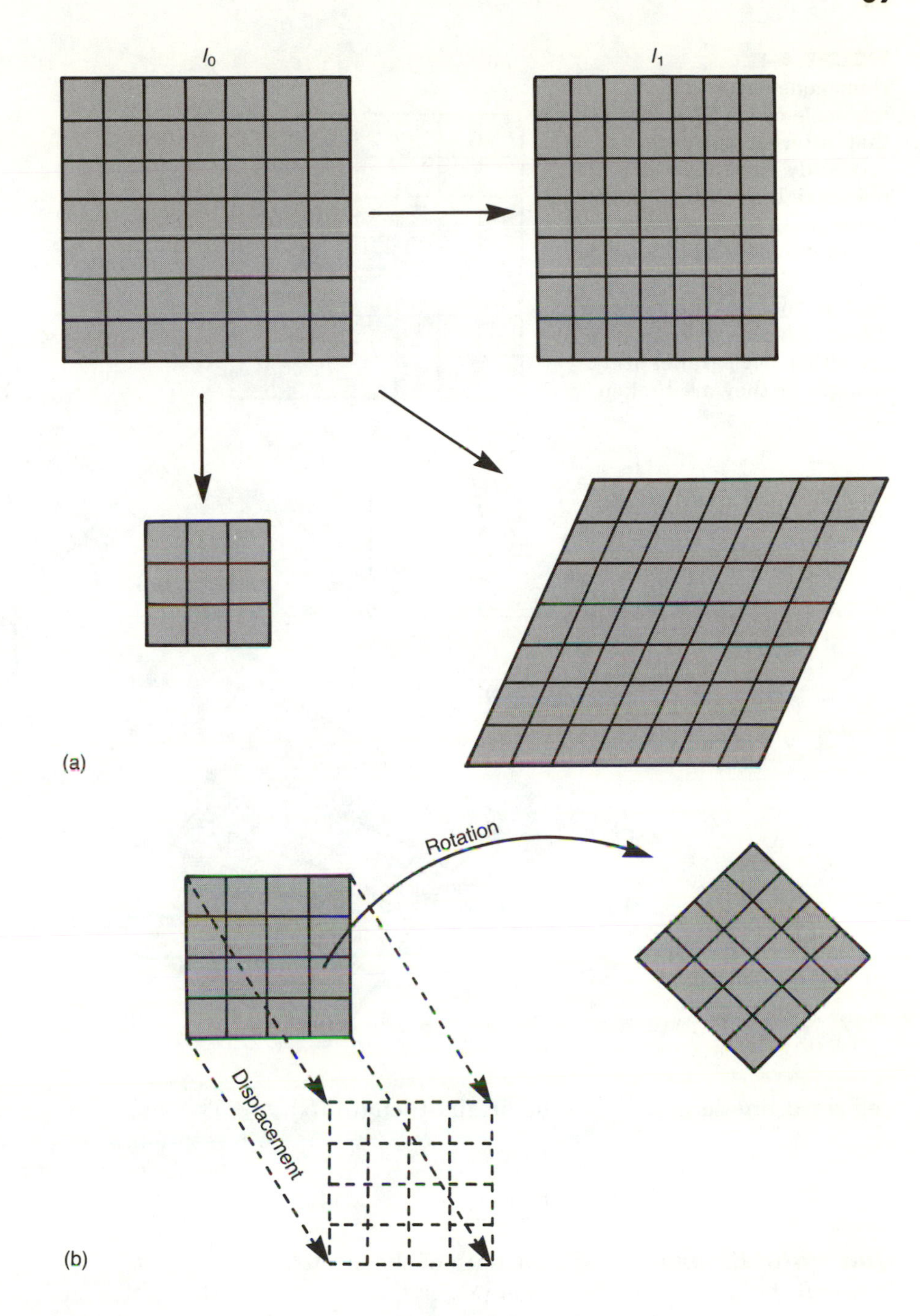

FIGURE 4–3
(a) Strain as distortion of initially parallel and perpendicular lines involving a change of length, shape, or volume of a mass.
(b) Displacement or rotation of a mass without accompanying distortion.

several kinds of strain may be calculated for the same body.

Linear Strain

Elongation. The *elongation* (ϵ) of a reference line in a rock mass (Figure 4–5) is the ratio of the length of the line in the deformed mass (l_1) minus the original length (l_0) to the original length; it is written mathematically for a finite strain as

$$\epsilon = (l_1 - l_0)/l_0 = \Delta l/l_0 \tag{4–1}$$

For an infinitesimal strain,

$$\epsilon = dl/l \tag{4–2}$$

If $\epsilon > 1$, or positive, an elongation or extension is indicated; if $\epsilon < 1$, or negative, shortening is indicated. (Note that engineers use a negative sign to indicate extension and a positive sign for compression.)

Stretch. The *stretch* (S)—or engineer's stretch—of a reference line in a rock mass is the ratio of the

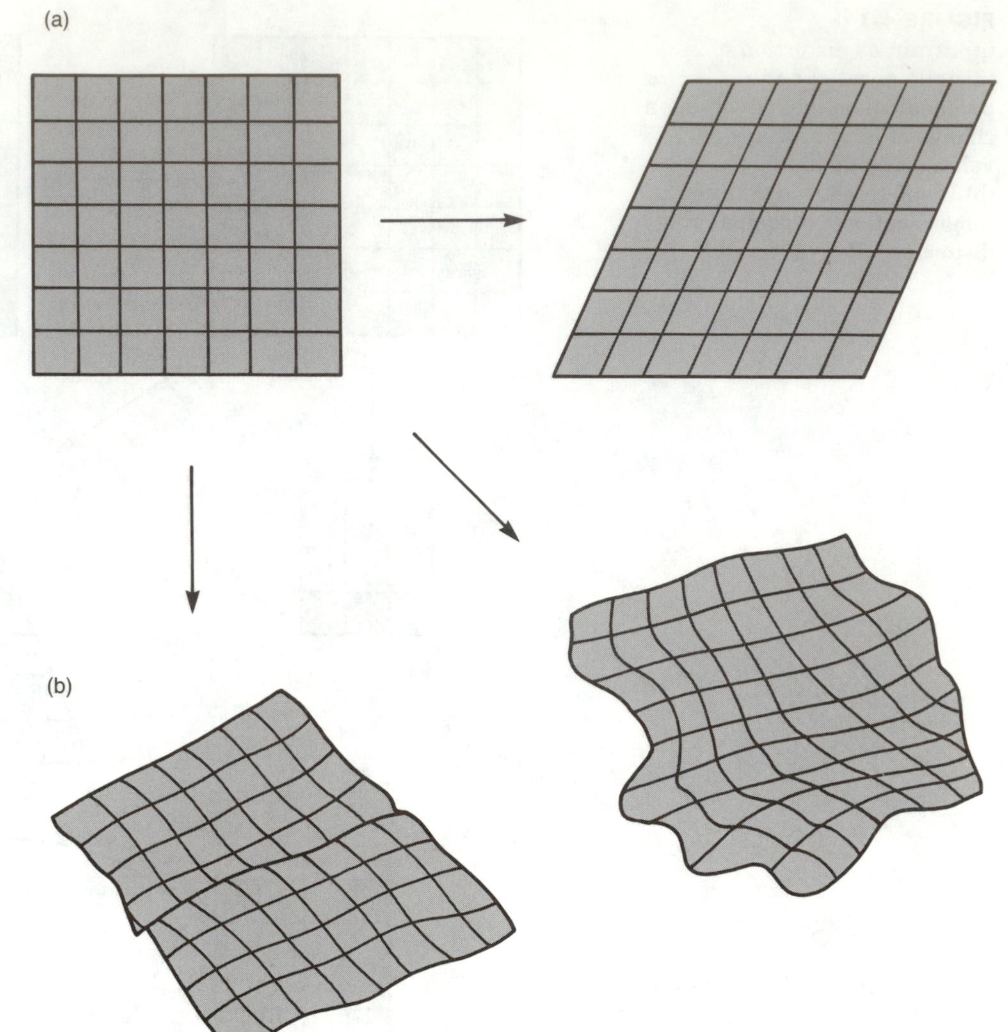

FIGURE 4–4 Homogeneous (a) and inhomogeneous (b) strain. Note that, after deformation, originally parallel or straight reference lines within the body undergoing homogeneous strain remain straight or parallel. Similar reference lines in the body undergoing inhomogeneous deformation are either not parallel or straight or they are broken.

deformed line length (l_1) to the original length (l_0), or

$$S = l_1/l_0 = (1 + \epsilon) \quad \textbf{(4–3)}$$

Quadratic Elongation. The square of the stretch is called the *quadratic elongation* (λ), or

$$\lambda = (l_1/l_0)^2 = (1 + \epsilon)^2 \quad \textbf{(4–4)}$$

All the strains involving lines may easily be related to each other; so, if one is known, the others can be calculated (Figure 4–5). They have no units—they are only ratios.

Shear Strain

The *shear strain* (γ) involves movement of parts of a rock body past one another so that angles between parts of the rock mass are distorted (Figure 4–6). The angle ψ, or *angular shear,* is the angle of rotation of reference lines within the mass. Shear strain is related to angular shear by

$$\gamma = \tan \psi \quad \textbf{(4–5)}$$

The history of differently oriented lines may be traced during progressive deformation for a shear strain applied to a reference cube in a rock mass (Figure 4–7). Lines in particular orientations undergo only extension; others undergo shortening and then extension.

Volume Changes

Changes in volume of a material, or *dilation,* may occur by at least three different mechanisms (Figure 4–8): closing voids between grains, producing a negative volume change; dissolving away part of the

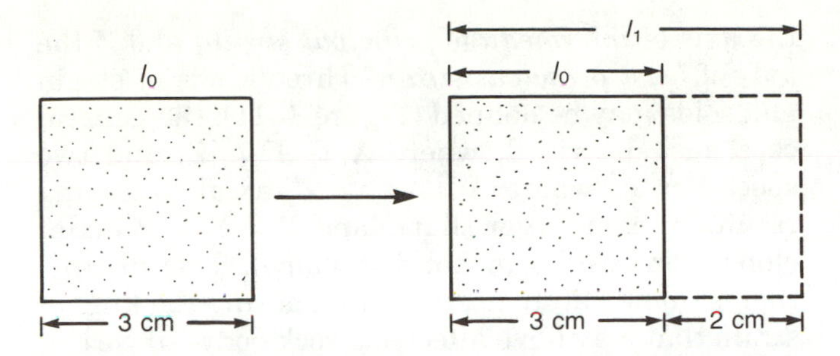

FIGURE 4–5
Dimensions of linear strains. The elongation strain here is 0.67, calculated from equation 4–1. The stretch is 1.67 (from equation 4–3), and quadratic elongation is 2.78 (equation 4–4).

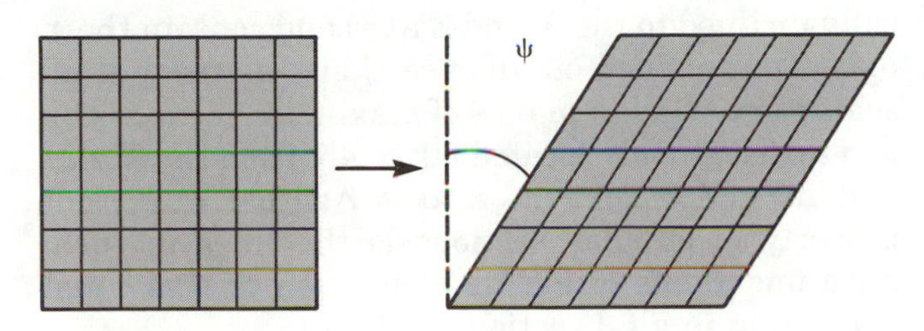

FIGURE 4–6
Shear strain. The rotation angle, ψ, is 39°, so the shear strain γ is 0.81 (from equation 4–5).

rock mass by pressure solution, resulting in negative volume change; and fracturing the mass, producing positive volume change (Figure 4–9). A finite volumetric or dilational strain (Δ) is the ratio of the volume change to the original volume of the mass, expressed as

$$\Delta = (V_1 - V_0)/V_0 = \delta V/V \qquad (4\text{–}6)$$

If the strain is infinitesimal,

$$\Delta = dV/V \qquad (4\text{–}7)$$

STRAIN ELLIPSOID

The ***strain ellipsoid*** (Figure 4–10) is a useful tool for relating strain to particular orientations and ultimately to stress. It consists of a triaxial ellipsoid, the equation for which is

$$\frac{x^2}{X^2} + \frac{y^2}{Y^2} + \frac{z^2}{Z^2} = 1 \qquad (4\text{–}8)$$

Properties of the ellipsoid include three mutually perpendicular axes x, y, and z, with principal directions X, Y, and Z, in which the length of $X > Y > Z$. Therefore X is the axis of *greatest principal strain*, Y

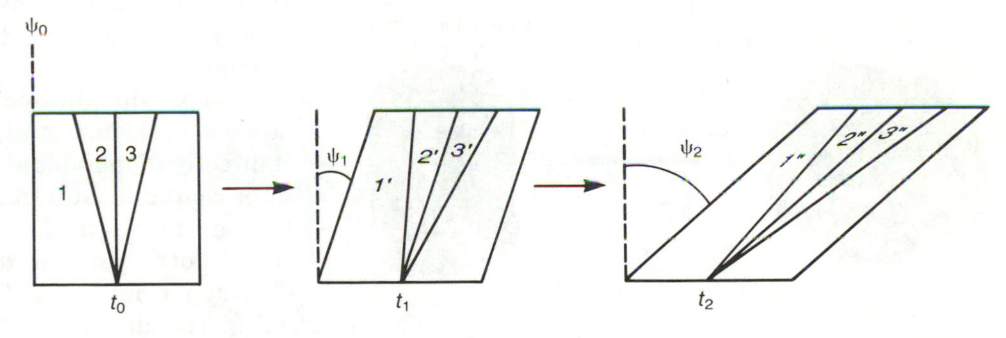

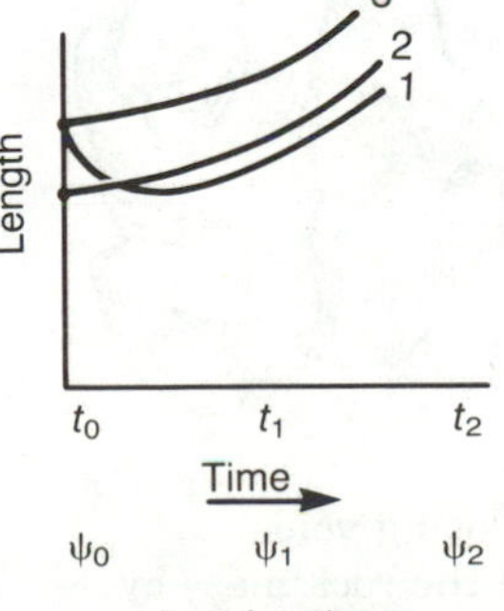

FIGURE 4–7
History of progressive deformation by application of a shear strain ψ in which reference line 1 undergoes initial shortening, then extension from t_0 to t_1 during shear strain, then extension from t_1 to t_2. Reference lines 2 and 3 with a different orientation experience extension only from t_0 to t_2. This kind of information is useful in attempts to decipher the progressive strain history in different parts of the same body, as well as the deformation of particular features as a function of orientation relative to strain.

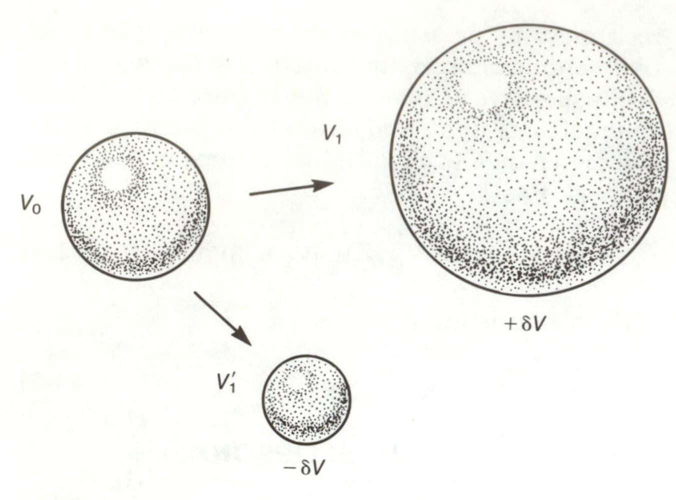

FIGURE 4–8
Positive and negative dilation, illustrated by soap bubbles that have expanded or contracted. If the radius of the bubble V_0 is 1.5 cm and that of the expanded bubble $+\delta V$ is 2.75, the dilation Δ is 6.16 (from equation 4–6). If the radius of the bubble $-\delta V$ is 1.1 cm, the dilation is 0.39 (from equation 4–6).

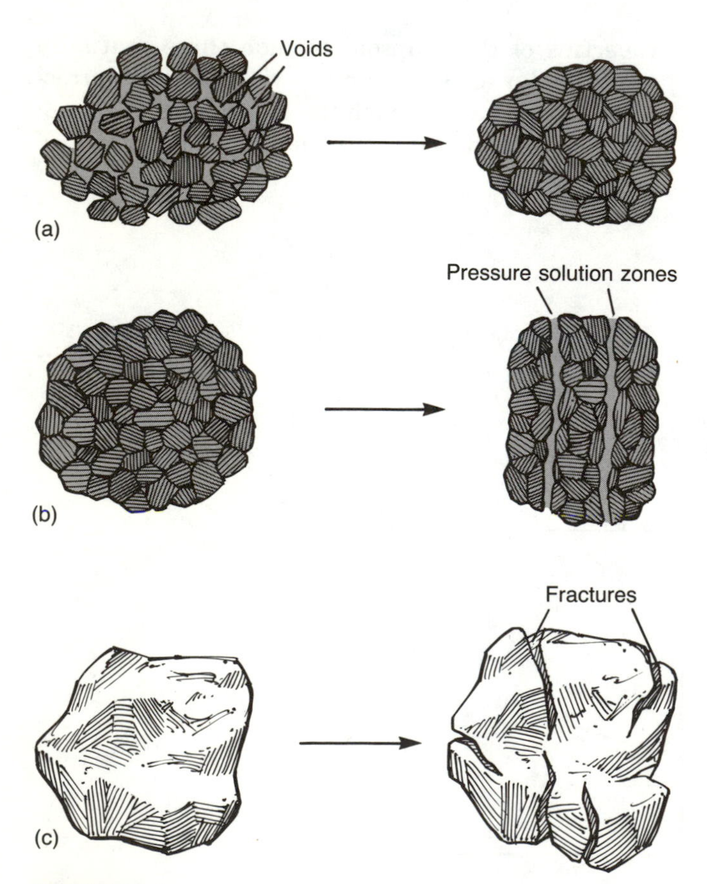

FIGURE 4–9
Three kinds of volume changes. (a) Closing voids between grains. (b) Dissolving part of the rock mass by pressure solution. (c) Fracturing the rock body; (a) and (b) produce negative volume changes, whereas an increase in volume occurs in (c).

the axis of *intermediate principal strain,* and Z the axis of *least principal strain.* Three kinds of strain ellipsoids may be defined (Figure 4–11): the general case, just described, where $X > Y > Z$, and two special cases where $X = Y > Z$ (axial flattening results in hamburger shape) and $X > Y = Z$ (axial elongation results in hot-dog shape). This distinction is immediately useful in diagnosing the kind of strain that may have affected a rock body—if we can find a natural strain marker (Chapter 7) such as deformed oöids.

Sections through the general strain ellipsoid that are parallel to either of the principal strain axes—that is, the XY, ZY, and XZ sections—are ellipses (Figure 4–12). In the other special cases, XY and YZ sections are circular. In addition, any strain ellipsoid has circular sections that are symmetrically inclined to the X and Z axes and contain the Y axis. The inclined circular sections in strain ellipsoids correspond to planes of maximum shear strain and are commonly located 60° to 45° from the X axis and 30° to 45° from the Z axis. Another important property of circular sections is the lines in them have undergone either no elongation or else equal elongation in all directions.

In the simplest case, the relationship of the strain ellipsoid to the stress ellipsoid is one of an inverse correspondence of axes. X, the axis of greatest principal strain, corresponds to σ_3, the axis of least stress. Likewise, the Z axis corresponds to σ_1; the two intermediate axes, Y and σ_2, are coincident. This relationship holds as long as the strain is coaxial.

The strain ellipsoid is particularly useful in the study of geologic bodies, for we can frequently relate it directly to individual structures. As we will see in later chapters, fault planes (Chapter 10) correspond to shear planes in the strain ellipsoid, joints (Chapter 8) form parallel to the XZ plane, and slaty cleavage (Chapter 17) forms parallel to the XY and YZ (perpendicular to X) planes of the strain ellipsoid. Also, many fold axial surfaces (Chapters 14 and 15) form parallel to the XY plane.

MOHR CIRCLES FOR STRAIN

It is possible to construct Mohr circles for strain that are closely analogous to those used for showing the relationships between shear and normal stresses. The Mohr construction for finite strain involves plotting values of reciprocal quadratic elongation, λ' $(= 1/\lambda)$, on the horizontal axis and modified shear strain, γ' $(= \gamma/\lambda)$, on the vertical axis (Figure 4–13). These variables are plotted,

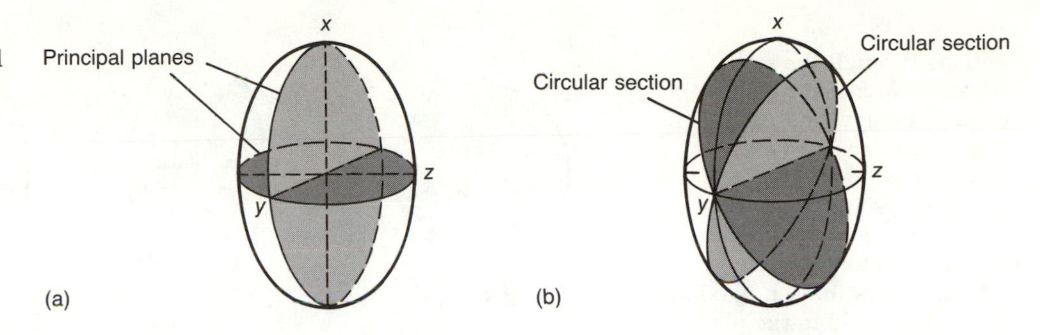

FIGURE 4–10
Strain ellipsoid with principal axes *X*, *Y*, and *Z* and circular sections.

rather than λ and γ, for mathematical convenience. A more detailed account of the mathematical basis of the Mohr circle for finite strain may be found in Ramsay and Huber (1983). Values of λ′ and γ′ may be expressed in the form of equations of a circle (analogous to equations 3–15 and 3–16):

$$\lambda' = \frac{\lambda'_1 + \lambda'_2}{2} - \frac{\lambda'_2 - \lambda'_1}{2} \cos 2\phi' \qquad \textbf{(4–9)}$$

and

$$\gamma' = \frac{\lambda'_2 - \lambda'_1}{2} \sin 2\phi' \qquad \textbf{(4–10)}$$

(Ramsay and Huber, 1983). The angle ϕ' appears again and is analogous to the coefficient of internal friction ϕ. The angular strain ψ may be found for any value of ϕ' from

$$\psi = \tan^{-1} \gamma = \gamma'/\lambda' \qquad \textbf{(4–11)}$$

We can determine λ and γ values for each ϕ' if we know values of the elongations ϵ_1 and ϵ_2. It should therefore also be possible to determine a Mohr strain circle from values of elongation determined from deformed objects in the field. (See Chapter 5 for more details of techniques for measuring strain; also Ramsay and Huber, 1983.) In contrast to the Mohr circle construction for stress, which requires values of present-day stress to obtain estimates of shear stress and fracture angles, the Mohr construction for finite strain provides an approximation of the fracture angle in rocks deformed millions of years ago by stress long since dissipated. With such tools we can construct models of former strain conditions in the Earth.

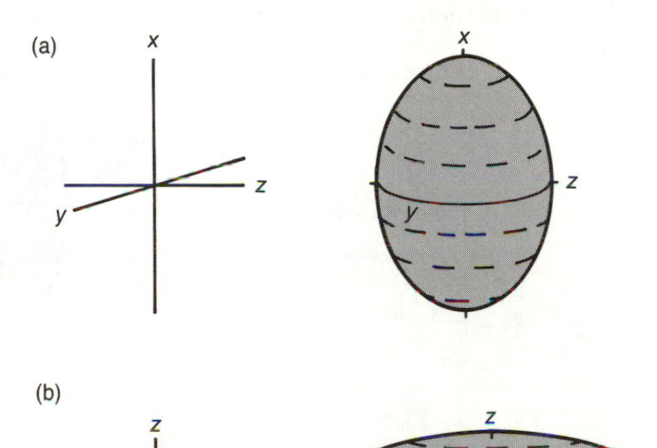

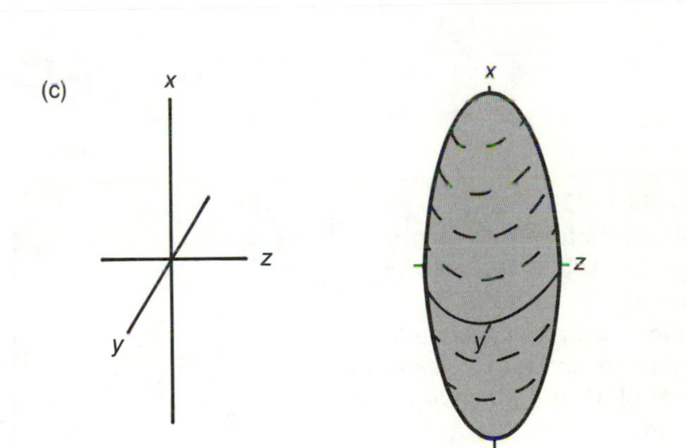

FIGURE 4–11
Three types of strain ellipsoids. (a) Triaxial ellipsoid, with $X > Y > Z$. (b) Oblate biaxial spheroid, with $X = Y > Z$ (axial shortening producing hamburger shape). (c) Prolate biaxial spheroid, with $X > Y = Z$ (axial elongation producing hot-dog shape).

SIMPLE AND PURE SHEAR

Homogeneous deformation may be considered either rotational or irrotational. Rotational homogeneous strain is most easily compared to the motion observed when the cards in a deck are riffled so that each moves the same distance past the next (Figure 4–14). The relative motion of adjacent cards is parallel, with a plane of displacement between

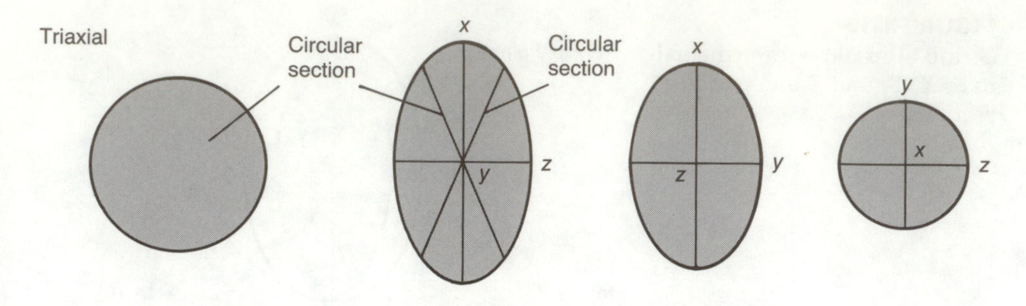

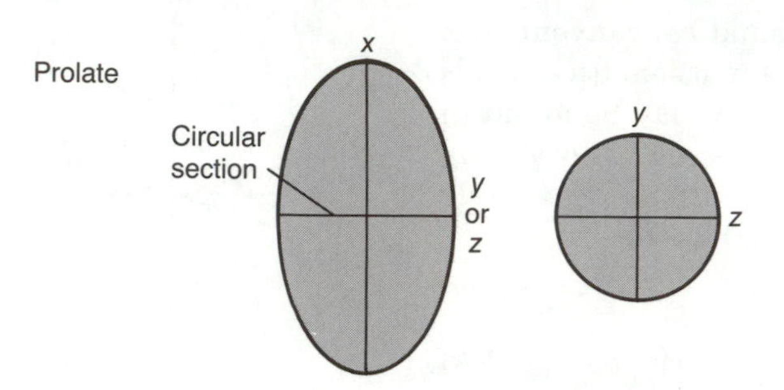

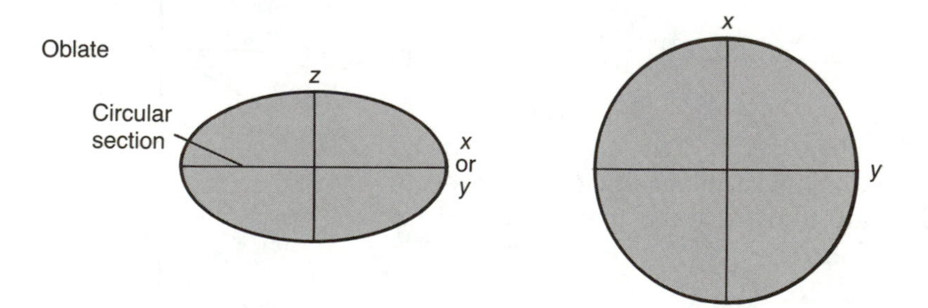

FIGURE 4–12
Sections through the strain ellipsoid. Note that all sections, except the two special circular sections, are ellipses, but any section through the biaxial ellipsoids (Figure 4–11) normal to the unique axis is circular. A set of circular sections occurs in each biaxial ellipsoid inclined to the plane of the two axes of equal length (the equatorial plane).

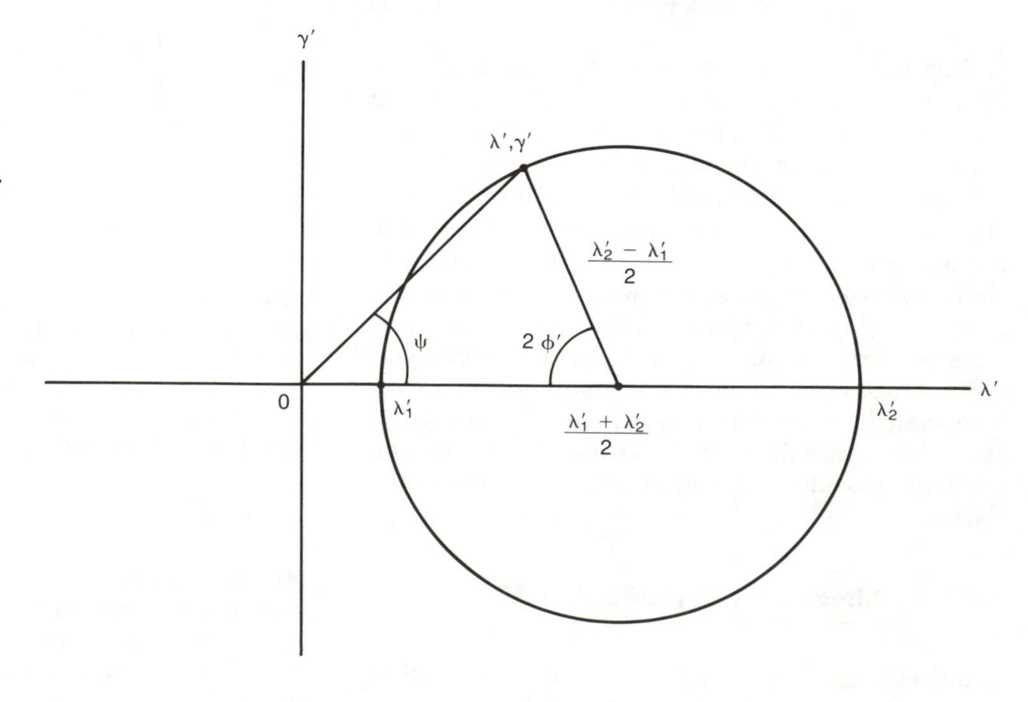

FIGURE 4–13
Mohr circle for finite strain. Note that $(\lambda_1' + \lambda_2')/2$ is the center of the circle. A Mohr envelope may be constructed tangent to several circles after values of λ_1' and λ_2' have been plotted.

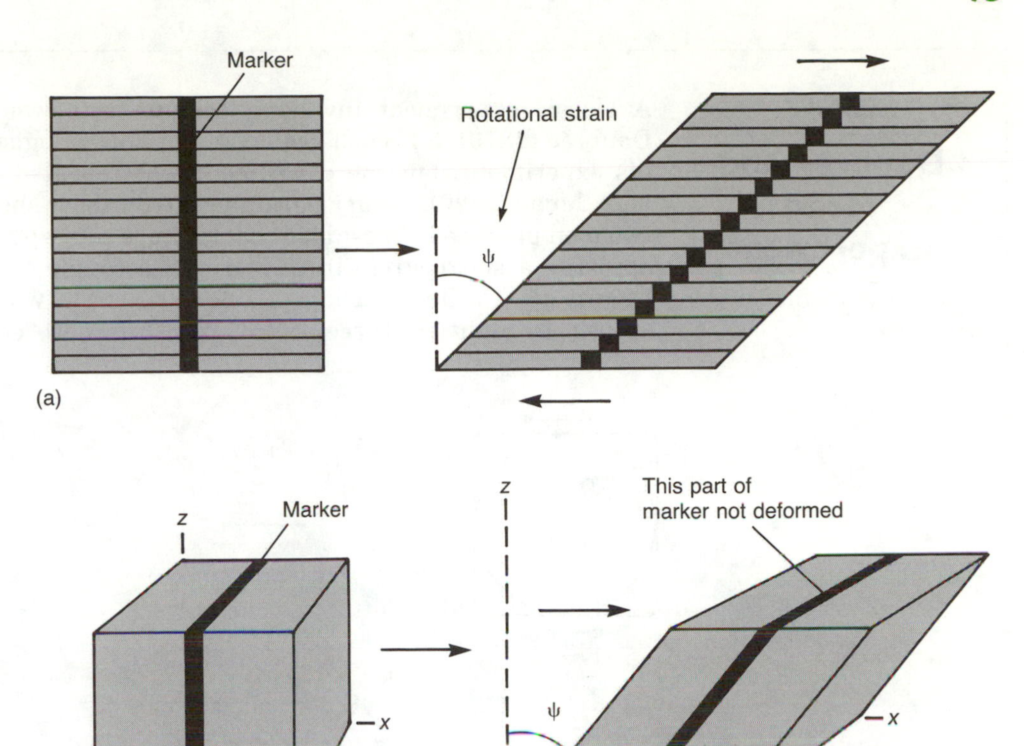

FIGURE 4–14
(a) Card-deck model of simple shear. (b) Deformation of a cube by homogeneous simple shear.

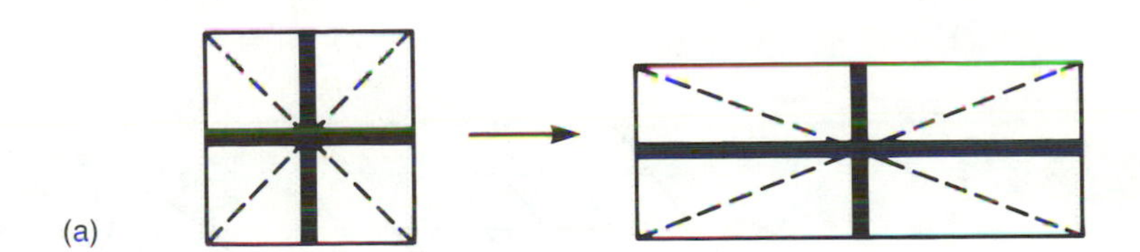

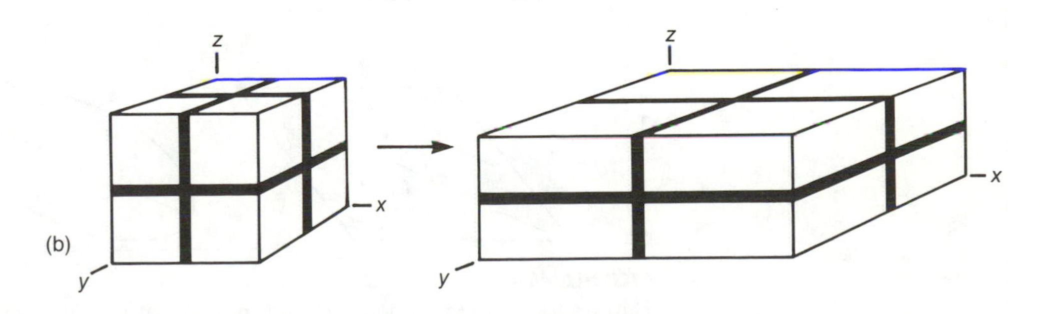

FIGURE 4–15
(a) Deformation by homogeneous pure shear. Note that the angles between the principal axes (heavy solid lines) remain unchanged, but the angles between other lines (dashed) do change. (b) Pure shear deformation of a cube.

ESSAY

Daubrée and Mead Experiment

A classic experiment involving fracture sets was made first by Auguste Daubrée (1879), a French geologist and mineralogist. He used plate glass for his experiment, but the experiment was repeated several decades later by W. J. Mead (1920), using a paraffin-coated rubber sheet in a square frame. The two men produced almost identical fracture sets within the glass or paraffin by applying a shear stress in the form of a couple (oppositely directed shears; Figure 4E–1). On shearing, a reference circle was deformed into a strain ellipse. In addition, three sets of fractures were easily recognized. One set

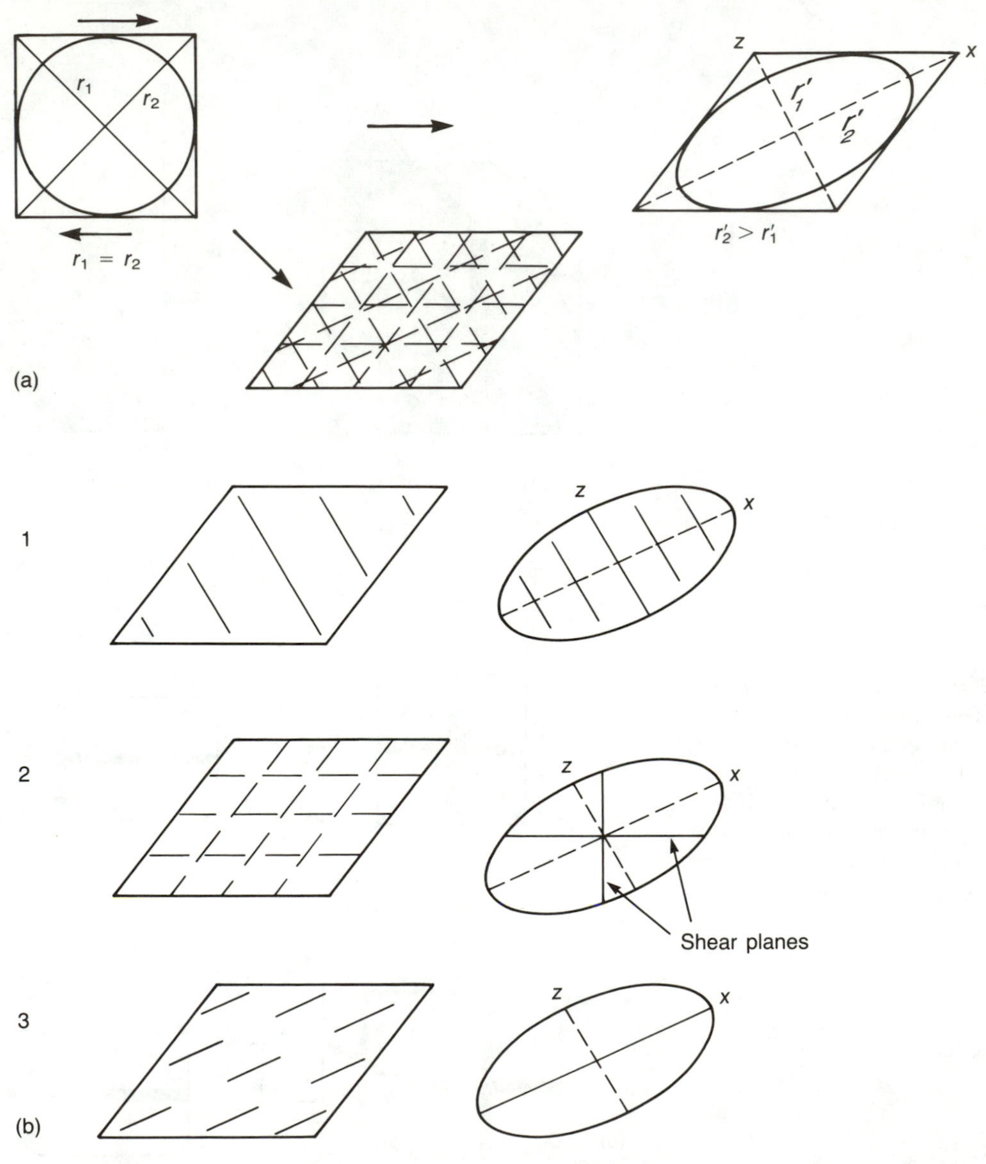

FIGURE 4E–1
Daubrée-Mead experiment (a) and interpretations of fracture sets in terms of the strain ellipse. (a) Four fracture sets in glass or paraffin. (b) Separating fracture sets. 1. Set parallel to *Z*, therefore, are extension fractures (joints). 2. Shear planes produced by simple shear. 3. Set parallel to *X*. May be thrust faults (?) or fractures produced by pure shear.

(a)

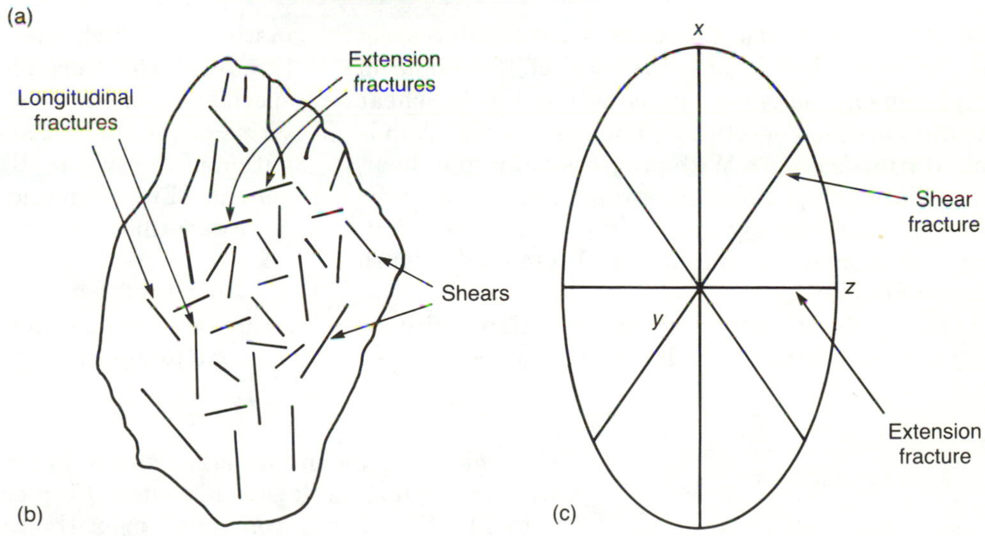

FIGURE 4E–2
(a) Fragment of safety glass from a wrecked motor car (unfortunately Hatcher's) illustrating a fracture pattern almost identical with that produced by Daubrée and Mead. Deformation of the window glass probably occurred by homogeneous simple shear. (b) Sketch of the fracture pattern in (a). (c) Strain ellipse with principal and shear planes oriented parallel to most of the fractures in the glass.

paralleled the long axis of the strain ellipse (X), another paralleled the short axis (Z), and a third set was actually a conjugate (paired) set involving two fracture directions parallel to the edges of the glass plate or square frame.

Considering the work by Daubrée and Mead, it is easy to see that a fracture set parallel to the short axis of the strain ellipse results from extension. Extension fractures are known to form parallel to the XZ and YZ planes of the strain ellipsoid and normal to X. We can also see that these fractures formed normal to X and parallel to Z in the essentially two-dimensional strain model.

The conjugate set of fractures parallel to the edges of the glass plate are readily explained as shear fractures. If we consider that the stresses were applied parallel to the edges of the frame in Figure 4E–1, they must be considered shears, for they are oblique to the orientation of the normal stresses.

The fracture set parallel to the X direction is harder to explain. The set formed normal to σ_1 and Z—assuming that the stress field did not rotate during deformation—and formed parallel to a newly oriented shear plane. If so, such faults could be considered thrust faults (Chapters 9, 10, and 11), especially because they frequently form inclined to the surface of the glass or paraffin—*not* normal to σ_1. Otherwise they are called "compression fractures." Any homogeneous material of approximately the same shape strained in similar fashion would yield roughly the same fracture pattern (Figure 4E–2).

References Cited

Daubrée, A., 1879, Études synthétiques de géologie expérimentale: Paris, p. 306–314.
Mead, W. J., 1920, Notes on the mechanics of geologic structures: Journal of Geology, v. 28, p. 505–523.

them. The net displacement of one card with respect to the one below it is a function of the angle of displacement, as shown by equation 4–5 for shear strain. This kind of rotational homogeneous strain is called ***simple shear.*** With progressive simple shear, the relationship between stress and strain axes changes continuously and so the inverse relationship between stress and strain ellipsoids described earlier does not hold.

Pure shear results from distortion—with or without translation—involving homogeneous deformation, in which the principal axes do not rotate (Figure 4–15). Pure shear may be thought of as a special case of simple shear and can be considered in terms of the ideal cases of axial flattening or elongation of the strain ellipsoid. Thus, by adding components of rotation and translation, pure shear may be transformed into simple shear.

Now that we can identify different kinds of strain, we can turn in the next chapter to the means we have to measure strain.

Questions

1. The angle between the foliation and a shear surface was measured in a westward-dipping fault zone in eastern Connecticut. The angle ranges from 20° to 30°. What can be concluded about the range in strain in the rocks in this fault zone, using information provided in Chapter 4?
2. We have several ways to measure strain that involve change in line length. Why?
3. Why is the 45° maximum shear angle rarely attained in natural materials?
4. Construct a Mohr circle for finite strain in a deformed pebble, given the values $\epsilon_1 = 0.6$ and $\epsilon_2 = 1.2$ for elongation strain. To solve the problem, plot the values corresponding to λ and γ on the Mohr circle at intervals of 20° $2\phi'$, corresponding to values of ϕ of 10°, 15°, etc.

5. Describe a real situation involving either natural or artificial materials where strain begins as pure shear (flattening) and then is transformed into simple shear.

Further Reading

Hubbert, M. K., 1951, Mechanical basis for certain familiar geologic structures: Geological Society of America Bulletin, v. 62, p. 355–372.
This is a classic attempt to relate experimentally produced structures to both theory and real tectonic structures as we observe them in the field.

Means, W. D., 1976, Stress and strain: New York, Springer-Verlag, 339 p.
A well-written introduction to the concepts of stress and strain through the principles of continuum mechanics. Mathematical concepts and derivations are presented understandably, making problems (with solutions) related to geologic situations useful.

Ramsay, J. G., and Huber, M. I., 1983, The techniques of modern structural geology: Volume 1: Strain analysis: London, Academic Press, 307 p.
A detailed, well-illustrated treatment of strain and strain analysis using mathematics expected of junior or senior geology majors. Many concepts are presented by means of solved problems, using the illustrations in the book.

5

Measurement of Strain in Rocks

In 1966 we discovered that in addition to the relationship between lineation, oölite extension, and cleavage, the striation on slickensided surfaces, and growths of fibrous minerals in fractures were also related. If the *ac* [*XZ*] plane is normal to the fold axes that same surface contains the lineation *a* [*X*], the long axes of deformed oöids and the maxima for striations and mineral growths. This suggests a similar and uniform deformation plan for all elements, stratigraphically from the Precambrian at least to the Silurian. . . . To determine these facts it was necessary to sample the information and treat it statistically in graphs, charts, diagrams, and maps . . .

ERNST CLOOS, *Microtectonics along the Western Edge of the Blue Ridge, Maryland and Virginia*

DISTORTED OBJECTS IN ROCKS REPRESENT DEFORMATION AND strain of objects of known original shape—an observation first reported by J. Phillips and D. Sharpe in the mid-nineteenth century. Then Henry C. Sorby (1856) first linked the distorted objects and the deformation (strain) ellipsoid.

The first to measure strain, using fossils, were Albert Heim in 1878 and A. Wettstein in 1886 (Wood, 1973). Not until the twentieth century was major use made of this early work; Ernst Cloos published the first of his widely known works on the distorted oöids in the South Mountain fold in the central Appalachians of Maryland. All those studies paved the way for modern quantitative work on distortion of originally spherical and nonspherical objects, growth of fibers, and other strain markers. Together these methods permit us to measure strain in a rock mass.

Today we can quantitatively assess the state of strain of a rock mass using relatively simple techniques, but suitable ***strain markers*** must be present. Strain markers are features where original shapes or distributions are known well enough in the rock so that the current shapes or distributions reflect the strain. With enough strain markers, we can determine the distribution of strain as well as the amount of strain produced by different deformation mechanisms.

Most structural studies investigate the effects of stress on rocks as manifested in the kinds of structures that can be observed (Figure 5–1). General observations of the direction of overturning (*vergence*) of folds (Chapter 14), orientations of fibers and slickenlines on fault or bedding surfaces, and the sense of offset on faults—all enable us to make qualitative and sometimes quantitative measurements of strain and displacement. Many other things can be used to measure the strain in rocks more precisely. Now we will look at some strain-measurement techniques and apply them in several examples.

KINDS OF STRAIN

Strain is change in shape or volume of a reference object between the initial undeformed state and the final deformed state. Strain may take the form of volume change (dilation), or changes in shape (distortion), or both (Figure 5–2). As previously defined, brittle strain occurs as a result of rupture. Brittle deformation results in loss of cohesion and development of fractures separating masses of undeformed material. In contrast, ductile strain—except for certain types of spaced cleavages (Chapter 17)—involves the entire mass, so that deformation is continuous from an initial unstrained state to a final strained state. Throughout a period of strain accumulation, any number of events may affect the state of strain in a rock body. The amount of strain in each

FIGURE 5–1
A rock mass subjected to homogeneous finite strain may be useful in determining the amount of strain—if it contains a useful strain marker. Undeformed brachiopods from the Devonian Hamilton Group, New York. (E. B. Hardin, U.S. Geological Survey.)

event may vary considerably, but total strain in the rock under study is the result of a ***progressive deformation.***

These terms, discussed in Chapter 4, are redefined here to stress their importance in measuring strain: ***Finite strain*** relates the instantaneous shape of a rock mass at any one time to initial undeformed shape. Most deformation we observe in a rock mass represents a state of finite strain—the final or a cumulative state in a series of deformational events. ***Incremental strain*** involves separate steps that occur in small distortion or dilation events through progressive deformation. Finite strain is the sum of incremental strains. ***Infinitesimal strain*** is restricted to very small strain relative to an initial condition.

Observation of fractures and folds permits us to make qualitative or semiquantitative estimates of the amount of strain in a rock mass, but the actual amount of finite or incremental strain in the rock mass is difficult to determine by measuring fracture orientations or by determining the sense of rotation of a fold. Sometimes we can determine the amount of strain if we assume the rocks have been folded by a relatively simple mechanism, but most fold mechanisms involve strains that are difficult to reconstruct to an undeformed state. Sometimes microscopic techniques, such as the *calcite strain gage* in slightly deformed rocks (Groshong and others, 1984) or other method involving microscopic features of individual grains, help in determining finite-strain geometry.

It is useful to note whether the geometry of strain is ***homogeneous*** or ***inhomogeneous.*** Homogeneous strain represents strain occurring in a uniform manner so that straight and parallel reference lines in a material being deformed remain straight and parallel (Figure 4–4). Strain within one part of a rock body is the same as in another part. In contrast, inhomogeneous strain involves deformation in which strain is nonuniform throughout the body. The end result is that lines originally parallel are parallel no longer, angular relationships have changed in a nonuniform manner, and straight lines are no longer straight.

Scale-dependence of the kind of strain affecting a mass of rocks also exists. Strain may seem homogeneous on a scale of kilometers but inhomogeneous on the outcrop or hand-specimen scale. For example,

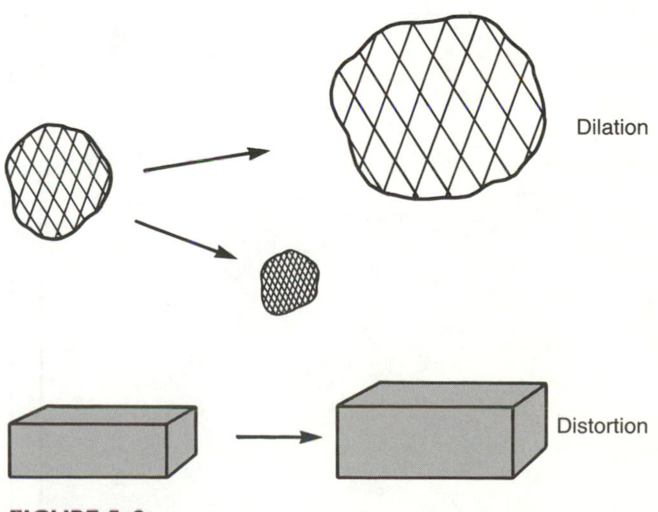

FIGURE 5–2
Kinds of strain.

cleavage may seem to occur in all parts of a rock body on the map scale, but at the mesoscopic scale where measurements are made in the field, cleavage is restricted to shales and does not occur in sandstones or limestones in a sequence.

STRAIN MARKERS

Any deformed feature in a rock mass in which original shape can be quantitatively inferred from the present deformed shape may be used as a ***strain marker.*** Two characteristics are needed if a strain marker is to be used as a finite-strain indicator. First, the precursor of the feature should be identifiable and original shape known or determinable. Second (for most strain measurement techniques), the marker should have the same mechanical properties as the rest of the rock mass. But very few strain markers meet the latter requirement. The best strain markers were originally spherical. Some that were originally spherical can serve as finite-strain ellipsoids in deformed rocks; they include oöids, reduction spots, and vesicles. Most natural strain markers, however, were not originally spherical.

Reduction spots are small, mostly spherical features in fine-grained sediments where the red oxidized sediment has been chemically reduced to a greenish material (Figure 5–3a). Reducing conditions are produced locally by a grain or area of

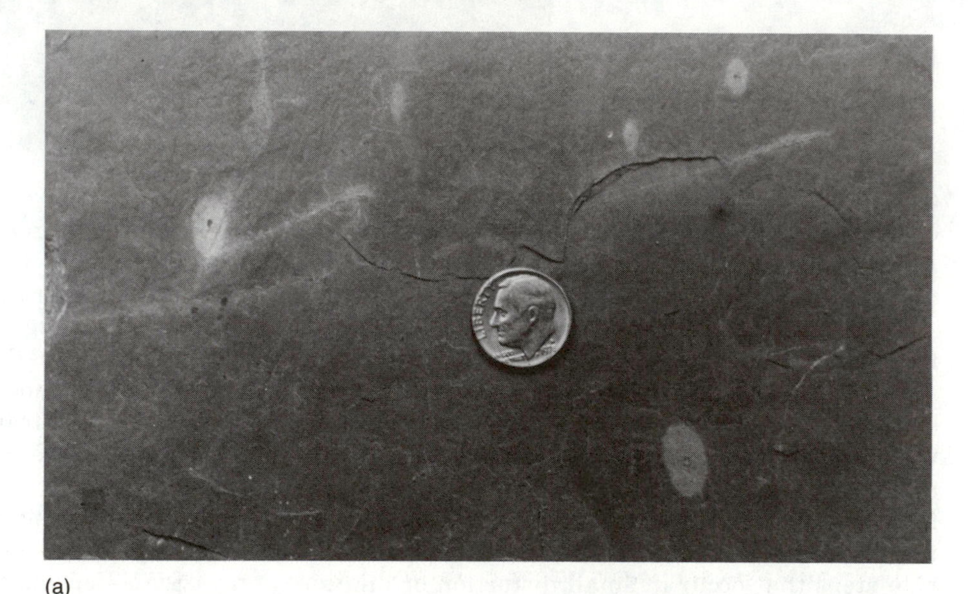

(a)

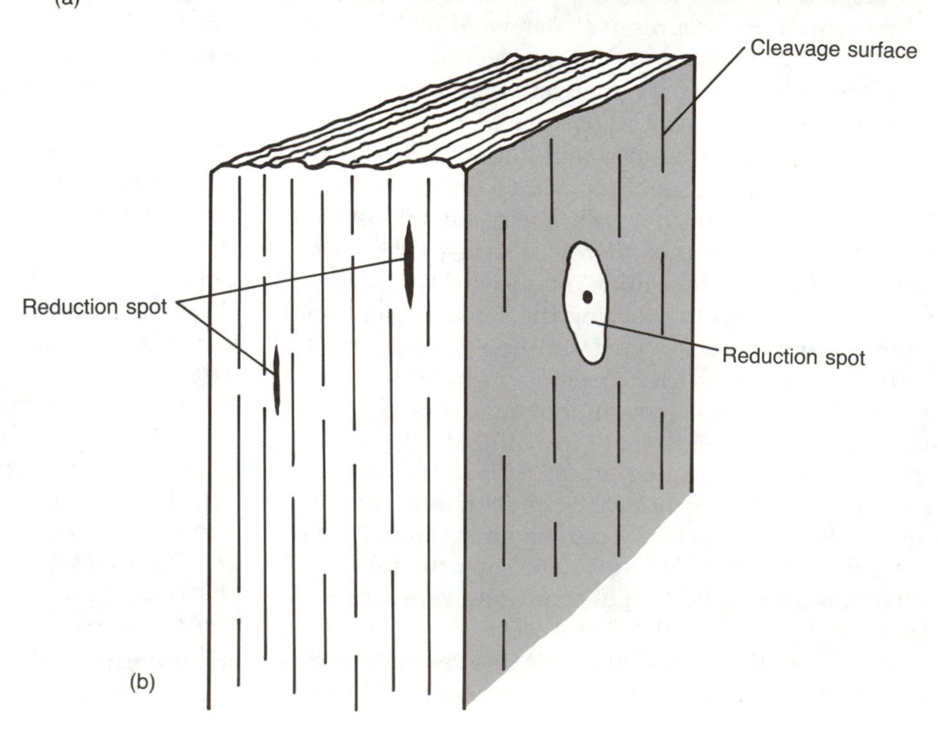

(b)

FIGURE 5–3
(a) Reduction spots in Metawee Slate near Rutland, Vermont. (Also see Figure 2–15; RDH photo.) (b) Relationship between reduction spots and slaty cleavage.

FIGURE 5–4
Pebbles in the Late Proterozoic Bygdin conglomerate at Bygdin, southern Norway, locally exhibit very large amounts of strain (locally greater than 700 percent elongation according to Hossack, 1968). Note that sections of pebbles in the vertical face to the left of the knife appear much less deformed than those on the sloping face to the right. Knife is 10 cm long. (RDH photo.)

different composition. Ideally, these are perfect markers for determining finite strain because they probably have the same mechanical properties as the enclosing rock mass. The only difference is the color. Most reduction spots occur in shales and mudstones, and so their use for finite-strain measurement associated with slaty cleavage planes can indicate the amount of strain in slates. Interesting studies of finite strain have been made by John G. Ramsay and Dennis S. Wood (1972), who measured reduction spots in the Cambrian slates of Wales and Vermont. They concluded that elongations greater than 100 percent occurred by flattening in some samples in the *XY* plane of the strain ellipsoid and parallel to the slaty cleavage (Figure 5–3b). Use of reduction spots is complicated because not all were originally spherical, and their time of formation relative to deformation may not always be clearly known.

Pebbles are among the most frequently used strain indicators. Originally, most were elliptical, but, if the amount of strain is very large, the originally spherical shape may be assumed and the resulting measurement error will be reduced. Where the amount of strain is relatively small and the pebbles were not originally spherical, however, the assumption may lead to a large error in determination of finite strain (Figure 5–4). Also, mechanical properties of the pebbles and matrix may differ and introduce significant errors.

Oöids and ***pisolites*** are good indicators of finite strain. They most commonly form in carbonate rocks and ironstone and are nearly spherical to slightly ellipsoidal before deformation. A classic study of oöid deformation was made by Ernst Cloos (1971) in the oölitic limestones in the South Mountain area of Maryland, Pennsylvania, and northern Virginia (Figure 5–5). After several decades and more than 10,000 measurements of strain in oriented limestone samples, Cloos resolved details of the strain accumulated during penetrative deformation of the Blue Ridge anticlinorium.

Fossils have been used to considerable advantage in determining finite strain in rocks (Figure 5–6). Generally, we know the original shape of an organism from specimens found in undeformed rocks, and we can readily determine if a fossil has been deformed; few organisms are originally spherical, however, and reconstruction of a deformed fossil to its undeformed

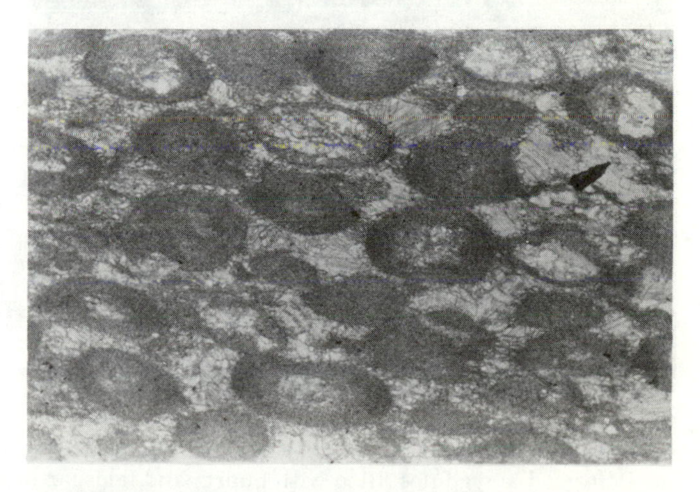

FIGURE 5–5
Deformed oöids in the Conococheague Limestone in the Great Valley, near Hagerstown, Maryland. (Charles M. Onasch, Bowling Green State University.)

FIGURE 5–6
Deformed trilobite (*Angelina*) from the Lower Ordovician of North Wales. (W. Stuart McKerrow, Oxford University.)

shape may be geometrically difficult. Trilobites, belemnites, graptolites, and brachiopods have been used as strain markers. Because we know their original shapes, many other fossils may also be used to determine shear strain (Chapter 4).

Vesicles (gas bubbles) in volcanic rocks may be used as finite-strain indicators—provided they were not appreciably deformed during initial outpouring of lava. Vesicles that become filled with mineral material (*amygdules*) may be used with equal ease (Figure 5–7).

In submarine lava flows (commonly basaltic), ***pillows*** may be used to determine finite strain. Generally, these form as roughly equidimensional masses with a glassy or vesicular top and a coarser-grained and more massive base (Figure 5–8). In many areas they have been severely deformed, and occasionally they have been used as finite-strain markers. The usefulness of pillows in accurate determination of finite strain is limited, however, for their original shapes are not uniform. Where strains are large, pillows may enable us to estimate finite strain.

(a)

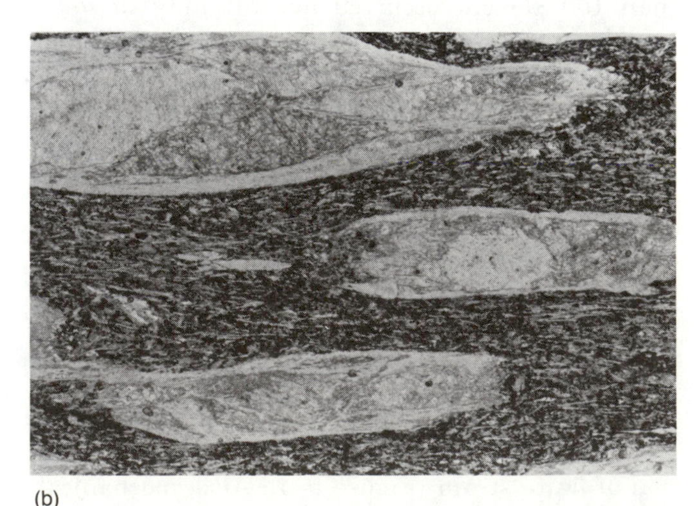

(b)

FIGURE 5–7
(a) Deformed amygdules filled with quartz and feldspar in amphibolite near Berner, Georgia. (Specimen courtesy of Robert J. Hooper, University of South Florida.) (b) Deformed amygdules filled with calcite in Eocambrian Sams Creek Formation near Union Bridge, Maryland. (Charles M. Onasch, Bowling Green State University.)

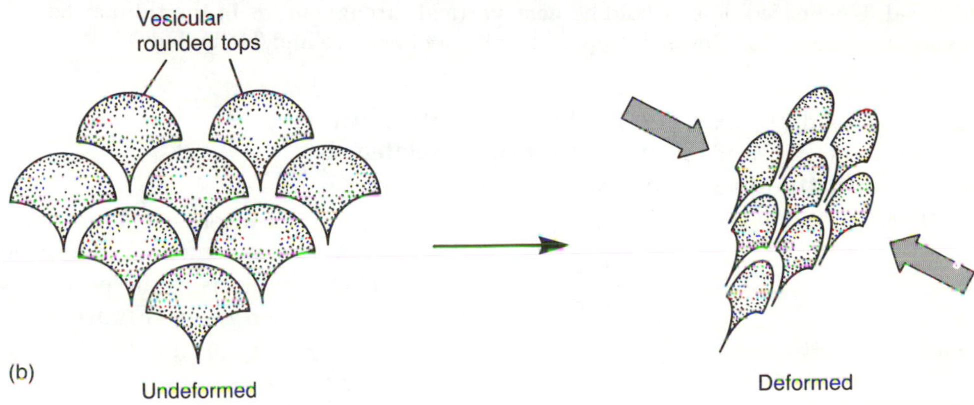

FIGURE 5–8
(a) Deformed pillow in the Chibougamou Lake area, Québec. (G. O. Allard, University of Georgia.) This pillow was compressed from the bottom and top of the photo. Compare with undeformed pillows in Figure 2–25. (b) Comparison of pillows before and after deformation.

Indirect indicators of activity of organisms, such as ***burrows,*** may be used in determining finite strain. Most useful are almost-cylindrical burrows oriented normal to the surface of the sea bottom in clean sand environments. For example, the Cambrian Pipe Rock Sandstone in Scotland contains abundant *Skolithos* tubes. Deformation of the Pipe Rock Sandstone has been studied extensively (Figure 5–9); they were initially cylindrical and circular sections were deformed into ellipses. Thus they may be used to determine finite strain in the plane of bedding (Coward and Kim, 1981; Fischer and Coward, 1982). They may also be used to determine shear strain parallel to bedding—if the shear strain is penetrative—because the *Skolithos* tubes are characteristically oriented normal to bedding. *Skolithos* tubes are also commonly found in clean sandstones in Europe, North America, Asia, and Australia.

FLINN DIAGRAM

One of the most useful means of displaying constant-volume finite strain was invented by Derek Flinn (1962), who assumed that ellipsoidal objects like reduction spots were initially spherical. Thus the principal planes (XY, YZ) of the strain ellipsoid may be determined from principal axes and planes of the deformed objects.

Flinn showed that ratios of the principal axes would be easy to determine and calculate and defined a value called k as

$$k = \frac{a - 1}{b - 1} = \frac{R_{xy} - 1}{R_{yz} - 1} = \tan q \qquad \textbf{(5–1)}$$

where a is the length of the major axis divided by the length of the intermediate axis and b is the length of

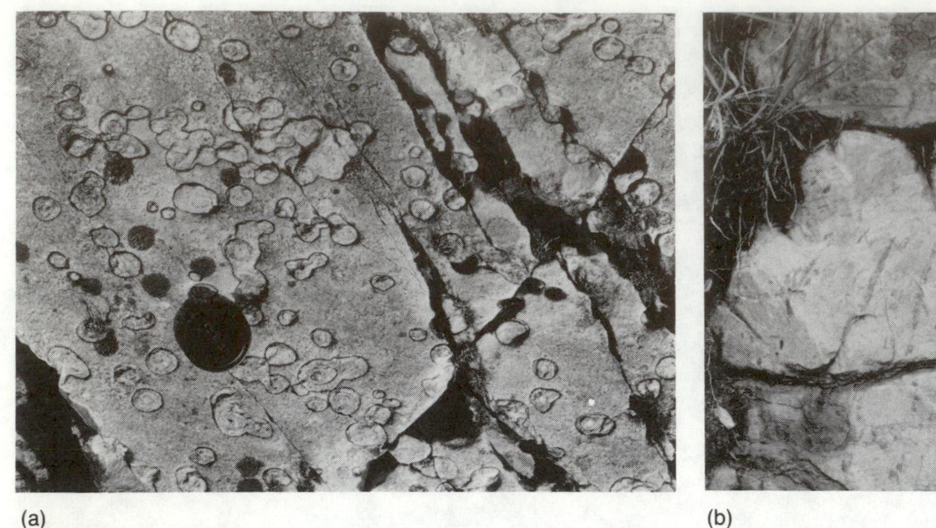

(a) (b)

FIGURE 5–9
Deformed *Skolithos* tubes (pipes) in the Lower Cambrian Pipe Rock Sandstone, Northern Highlands, Scotland. (a) Plan view on a bedding surface showing uniformly oriented elliptical sections of pipes. (b) Vertical section through a bed showing curving—deformed—pipes with greater shear deformation near the top of the bed. Unsheared pipes would be near vertical throughout, as in the thinner bed beneath. (Michael P. Coward, Imperial College of Science and Technology.)

the intermediate axis divided by the length of the minor axis. R_{xy} and R_{yz} refer to strains related to changes in the lengths in axes in the XY and YZ planes expressed as

$$R_{xy} = \frac{1 + \epsilon_1}{1 + \epsilon_2} \quad (5\text{–}2)$$

and

$$R_{yz} = \frac{1 + \epsilon_2}{1 + \epsilon_3} \quad (5\text{–}3)$$

where ϵ is elongation strain (equation 4–1). The angle q is measured between the abscissa and the line connecting the point p with the origin in the ***Flinn diagram*** (Figure 5–10). An ellipsoid plotting at point p in the diagram has a k value of tan q. Two-

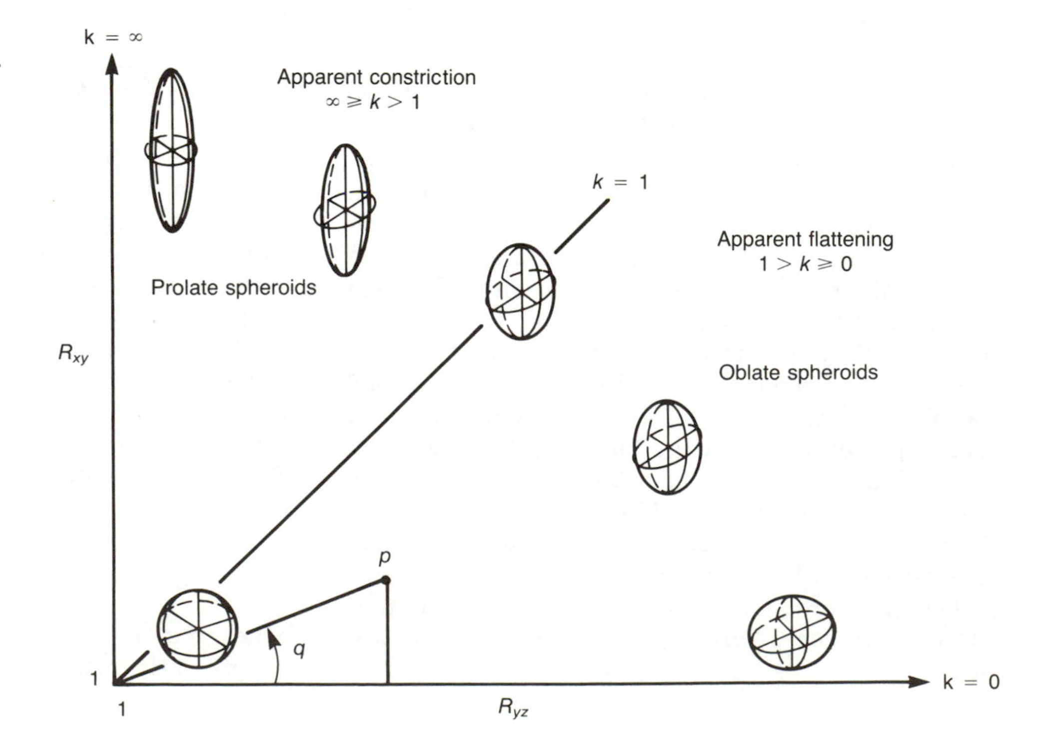

FIGURE 5–10
Flinn diagram. The k value for any ellipsoid, such as at point p, may be obtained from tan q. Note the difference in shape of $k = 1$ ellipsoids that have low total strain and plot close to (1,1) or have high total strain and plot far from (1,1).

dimensional plots of R_{xy} versus R_{yz} yield a valuable relationship portraying three-dimensional strain. Along the horizontal axis, k equals 0; on the vertical axis, k goes to infinity. A line with a slope of 1 yields $k = 1$. As k approaches infinity, the shape of a deformed object begins to resemble a hot dog (prolate spheroid) and involves *axial elongation* (or constriction). Values near the horizontal axis indicate *axial flattening,* which produces a hamburger shape (oblate spheroid). *General strain* occurs in the region away from the coordinate axes because a deformed object is shaped like a triaxial ellipsoid, with axes of unequal length. A value of $k = 1$ implies *plane strain* (where the Y axis has the same length as the diameter of the initial sphere) or simple shear. If plane strain can be proved, the Flinn diagram may also be used as an indicator of volume change.

STRAIN MEASUREMENT TECHNIQUES

Wellman's Method

A simple geometric technique for determining both the orientation and shape of the strain ellipse was invented by H. G. Wellman (1962). His method is based on the angular distortion of reference lines originally aligned 90° to each other. The technique requires at least ten strain-marker objects, and all must be arranged in the same plane. Accuracy of the method depends on the occurrence of enough fossils (or other strain markers with bilateral symmetry) in a small planar area. A distinct disadvantage is that a concentration of fossils must occur on a bedding surface; rarely do we find even two or three deformed fossils like brachiopods in a small area, and ten are very difficult to find frequently enough to characterize the strain of an area. Consequently, several techniques described in later sections have greater utility, even if they are not as easy to apply.

The most convenient and commonly used strain markers that meet the requirement of original 90° reference lines are brachiopods lying on a bedding surface (Figure 5–11). Either a photograph or an accurate tracing of the surface containing the fossils is needed.

Using brachiopods as the strain marker, an arbitrary reference line AB, at least 10 cm long, is drawn in any orientation, not necessarily parallel to the hinge lines of any of the brachiopods, and the ends labeled. Lines are also drawn along the hinges of the brachiopods on the photograph or tracing and parallel to the symmetry line (commonly marked by a sinus or fold) of each fossil.

A pair of lines is then drawn for each fossil parallel to both the hinge line and symmetry line passing through points A and B. If there is no strain in the bedding plane containing the fossils, the result is a rectangle. If the fossils are deformed parallel to the plane, a parallelogram results (Figure 5–12). This procedure is repeated for the hinge and symmetry lines of each brachiopod in the plane. The corners of the parallelograms, once connected, outline an ellipse. The lengths and orientations of the axes of the ellipse indicate the amount of strain in the plane of bedding. Obviously, the more fossils

FIGURE 5–11
Deformed brachiopods on a bedding plane from the Ordovician Davidsville Formation, Gander Lake, central Newfoundland. (Robert B. Neuman, U.S. National Museum, and R. Frank Blackwood, Newfoundland Department of Mines; U.S. National Museum Specimen.)

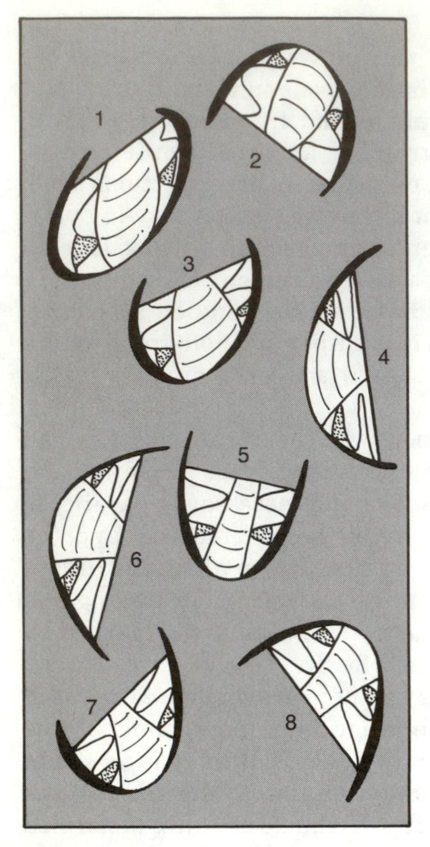

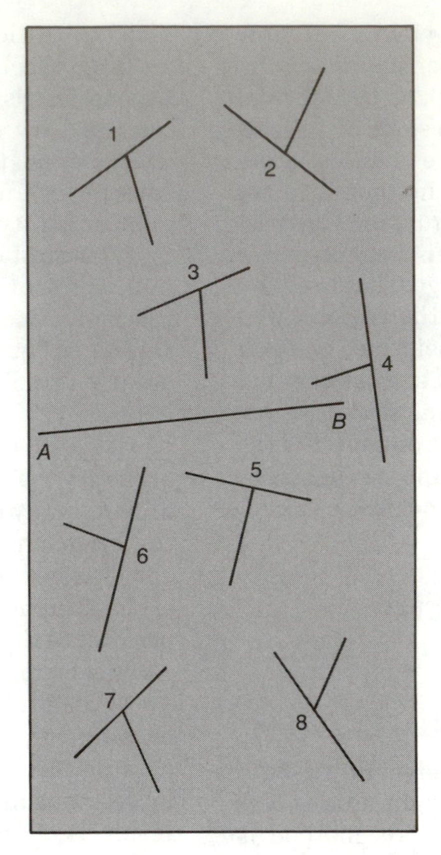

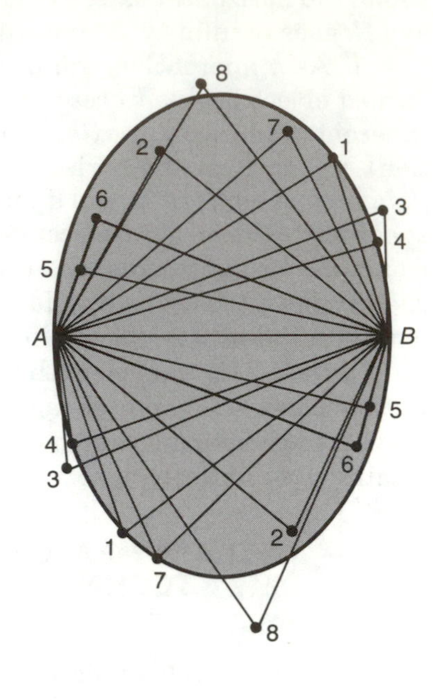

FIGURE 5–12
Determination of strain in fossils using Wellman's method.

available for construction of lines, the better defined the resulting strain ellipse.

Determination of Finite Strain of Initially Elliptical Markers

Strain markers with an initial elliptical shape, such as pebbles or sand grains, present another variable for determining finite strain. Three variables are involved: initial ellipticity, initial orientation, and orientation and magnitude of the principal strain axes. Pebbles, oöids, sand grains, and other initially elliptical markers may be randomly oriented during deposition. If so, the average initial shape can be considered spherical and the initial ellipticity is less important. If such markers were deposited by a current, as in a stream channel, the long axes may tend to parallel the current. Pebbles in a channel may thus be very strongly oriented and even imbricated downstream. A difficulty might arise in distinguishing a preferred orientation due to strain from a primary depositional orientation. Primary orientation does occur, but strain orientation is generally more consistent. The difficulty may be resolved by noting the relative parallelism of the present orientation of the objects to a known tectonic feature such as cleavage or foliation. Because most rocks commonly studied are marine, primary orientation is not so much of a problem, but it does occur as a primary bedding-plane orientation. Primary ellipticity is the rule.

The next few sections summarize the R_f/ϕ and several rapid methods for determining finite strain of initially elliptical objects. A more detailed and comprehensive discussion of these methods may be found in Ramsay and Huber (1983).

***R_f/ϕ* Method.** John Ramsay (1967) introduced the ***R_f/ϕ* method,** which was further developed by Dunnet and others in several papers and is comprehensively and clearly presented in the book by Ramsay and Huber (1983). Deformation of initially elliptical objects having an initial ellipticity R_i by a homogeneous strain results in objects that remain elliptical. The shape of the final ellipse (R_f) is determined by the initial shape and orientation of the starting ellipse (R_i) along with the shape or ellipticity (R_s) and orientation of the strain ellipse (Figure 5–13). The angle made with an initial reference orientation is designated as ϕ. The initial angles will change

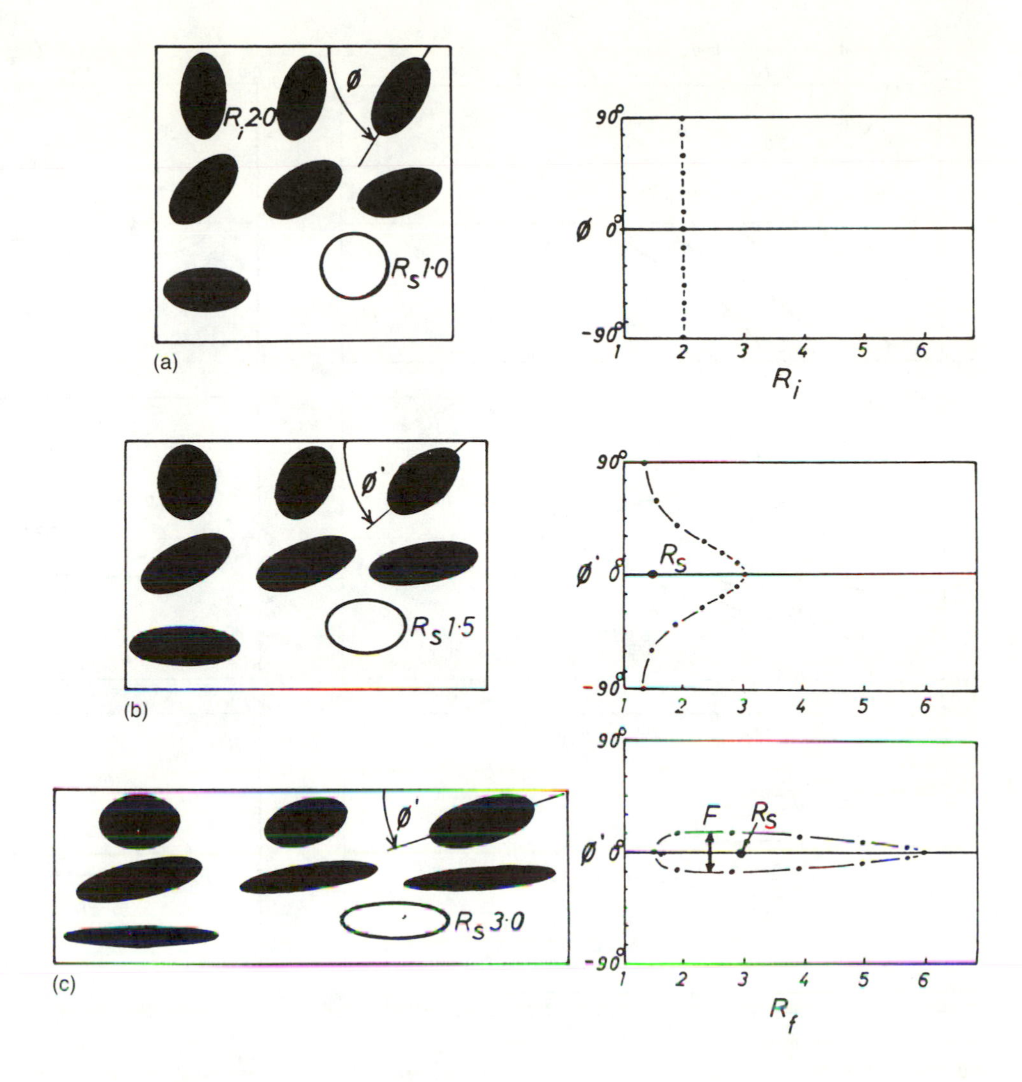

FIGURE 5–13
Relationships between initial ellipticity (R_i) and orientation (ϕ), final orientation (ϕ') and ellipticity (R_f) in an array of elliptical markers.
(a) Undeformed.
(b and c) Deformed, with vertical imposed flattening R_s = 1.5 and R_s = 2.0. F is the fluctuation. (From J. G. Ramsay and M. I. Huber, 1983, *The techniques of modern structural geology,* Academic Press, Volume 1.)

during deformation to ϕ', an angle more nearly parallel to that of the long axis of the strain ellipse. The only orientations that will not change will be those of ellipses that are oriented either parallel (ϕ = 0°) or normal (ϕ = 90°) to the long axis of the strain ellipse (using a reference line parallel to a principal axis). As a result, those parallel become *more* elliptical and those normal become *less* elliptical. Ellipticities of others with values of ϕ between 0° and 90° will lie between the two extremes, but most will have greater ellipticity. As the amount of deformation increases, R_f increases and deviation from the mean value of ϕ' decreases. Plots of R_f versus ϕ' become more and more closed, and this deviation, termed the *fluctuation* (F), decreases so that curves become more closed and onion-shaped; the fluctuation is less than 90° (ranging from less than +45° to greater than −45°) for small to moderate values of R_i (1.5 to 3.0) and moderate to large values of R_s of about 3.0 (Figure 5–14).

The ultimate objective of this analysis is determination of the relative contributions of initial ellipticity and R_s. An exercise in curve-fitting may determine the value of R_i that best fits the resultant R_f/ϕ' curve; this can be done manually or with a computer and generally results in a curve relating initial elliptical shapes to high values of R_i. Curves are always symmetrical with respect to the long axis of the strain ellipse—unless there was an original preferred orientation of pebbles before deformation. Also, on each curve, points derived from a group of ellipses wherein initial orientation was random will cluster toward high values of R_f.

Ramsay and Huber (1983) described two situations that can arise: maximum R_i greater than strain R_s, and maximum R_i less than R_s (Figure 5–13). In the first case, the result will be data with a fluctuation of 180°, but points will concentrate about the maximum R_f value (Figure 5–14). The direction of maximum concentration of data will also indicate

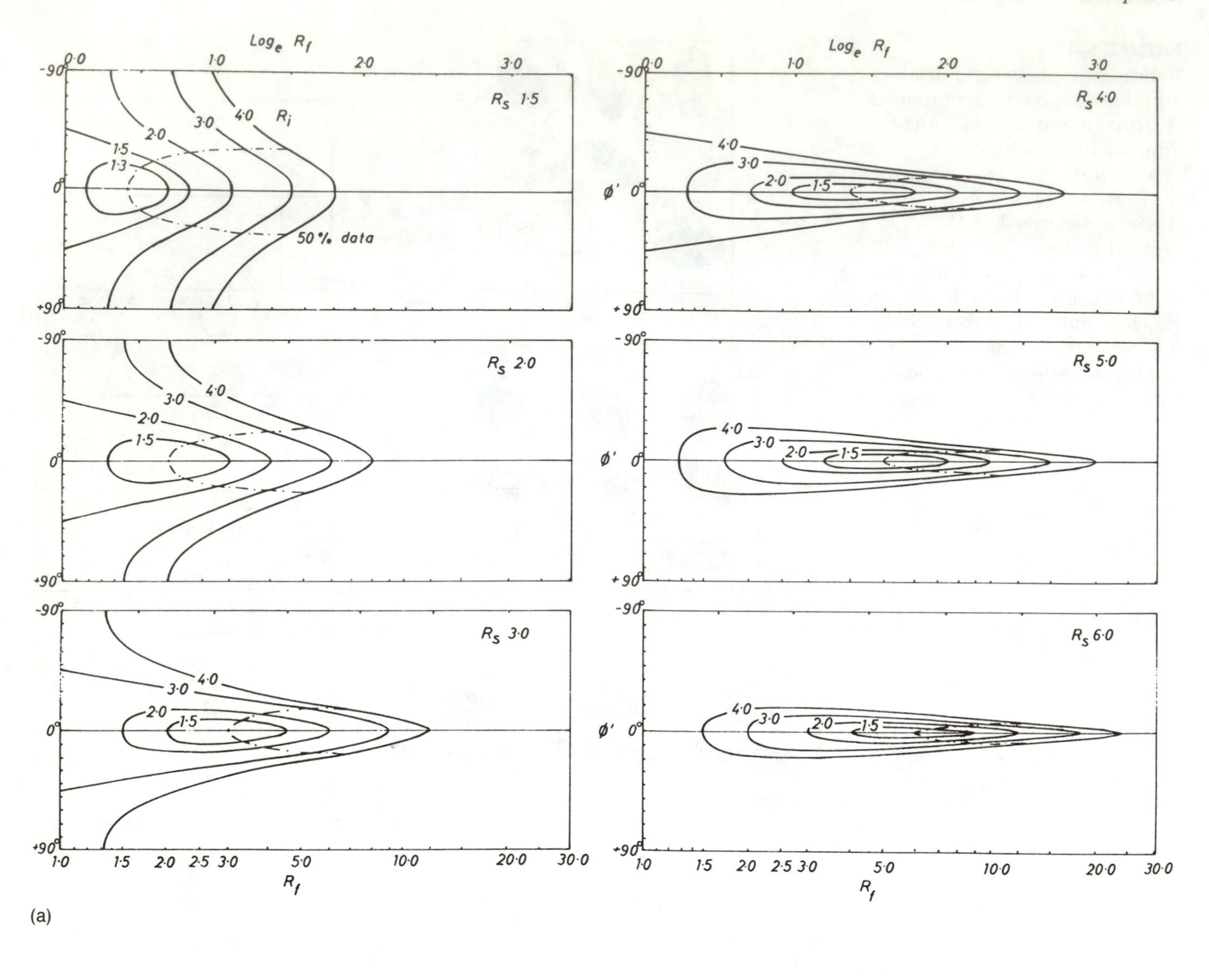

(a)

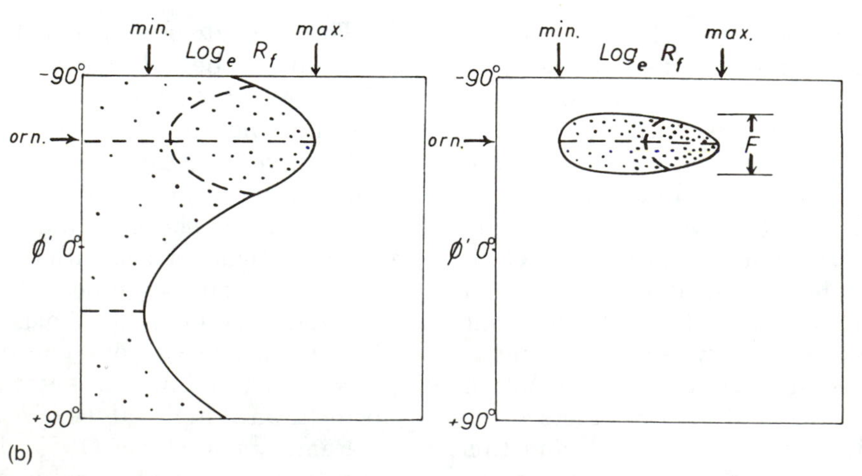

(b)

FIGURE 5–14

R_f/ϕ plots showing different relationships between R_i and R_s and the resulting curves. (a) R_f/ϕ reference curves for different values of initial ellipticity and the strain ellipse, R_s. (b) Features of R_f/ϕ plots used for calculating strain. Note the symmetry of the fluctuation (F) about the orientation of the strain ellipse. In the left diagram, $R_i > R_s$; in the right, $R_s > R_i$. (From J. G. Ramsay and M. I. Huber, 1983, *The techniques of modern structural geology,* Academic Press, Volume 1.)

the orientation of the long axis of the strain ellipse, and data should be symmetrically distributed on either side of the maximum. If not, the initial distribution of ellipses was not random. The relationships are given by

$$R_{f\,max} = R_s R_{i\,max} \tag{5–4}$$

and

$$R_{f\,min} = R_{i\,max}/R_s \tag{5–5}$$

Solving for R_s in both equations and substituting for R_s from equation 5–5 in equation 5–4,

$$(R_{f\,max} R_{f\,min})^{1/2} = R_{i\,max} \tag{5–6}$$

Substituting for $R_{i\,max}$ from equation 5–5 in equation 5–4 and solving for R_s yields

$$(R_{f\,max}/R_{f\,min})^{1/2} = R_s \tag{5–7}$$

The second and opposite case, where R_i maximum is less than R_s, produces clustered data with closed envelope curves. Data should cluster toward $R_{f\,max}$, be symmetrical about the long-axis orientation of the strain ellipse, and show minimal fluctuation (less than 90°). Relationships between $R_{f\,max}$ and $R_{f\,min}$ are

$$R_{f\,max} = R_s R_{i\,max} \tag{5–8}$$

and

$$R_{f\,min} = R_s/R_{i\,max} \tag{5–9}$$

Solving equation 5–9 for $R_{i\,max}$, substituting in equation 5–8, and solving for R_s yields

$$R_s = (R_{f\,max} R_{f\,min})^{1/2} \tag{5–10}$$

Solving equation 5–9 for R_s, substituting in equation 5–8, and solving for $R_{i\,max}$ yields

$$R_{i\,max} = (R_{f\,max}/R_{f\,min})^{1/2} \tag{5–11}$$

Equations 5–6 and 5–7, and 5–10 and 5–11, enable calculations of the contributions of initial strain and tectonic strain, respectively, in a randomly oriented set of initially elliptical objects. If any preferred orientation of ellipses exists, an asymmetric distribution of data points will appear in the R_f/ϕ plot, and it will be difficult to obtain useful values of R_i and R_s.

Another problem common in using this technique is the difficulty in deciding on the best-fitting envelope curve. Distribution of points often makes curve-fitting a tedious and even subjective part of the analysis, even with randomly oriented particles.

The R_f/ϕ technique is both rigorous and time consuming without a computer and digitizer. Techniques discussed next are faster (but less accurate) and allow differentiation of R_i from R_s, and tectonic strain; their reliability increases with increasing tectonic strain.

Center-to-Center Method. The ***center-to-center method,*** also originally devised by John Ramsay (1967), is based on the principle that the distance and angular relationships between an aggregate of initially randomly oriented and closely packed objects (sand grains, pebbles, and oöids) should help determine the orientation of the strain ellipse in the deformed aggregate (Figure 5–4). The method involves measurement of the distances and angles between a reference grain and the nearest neighbors (Figure 5–15a). Measurements of distances and angles should be discarded for grains for which connecting lines to the marker grain cross other grains:

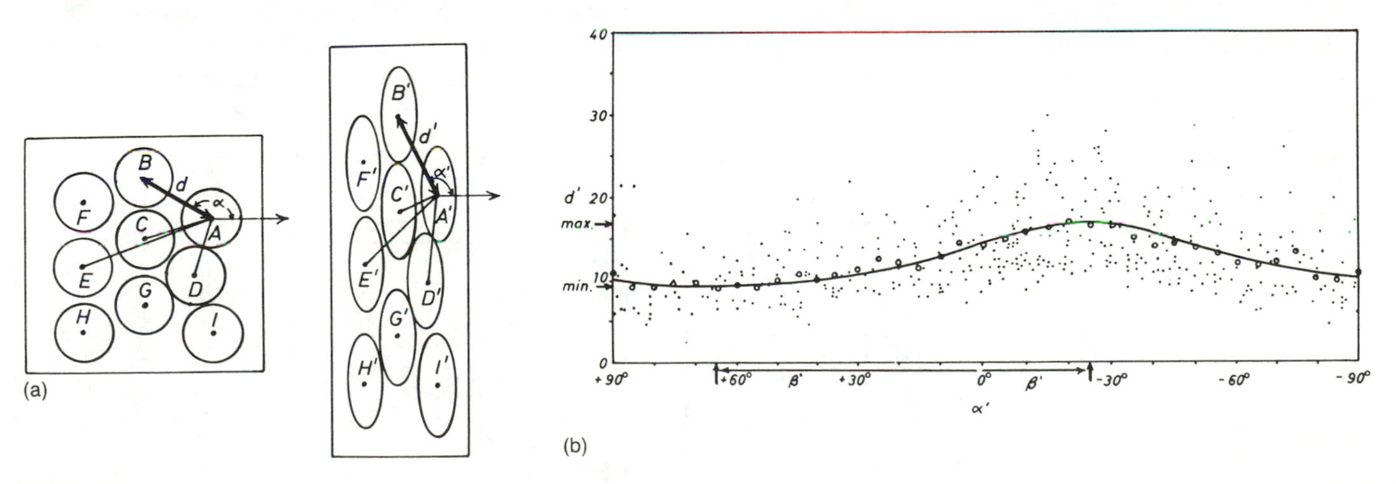

FIGURE 5–15
(a) Construction of tie lines in a deformed rock for determination of finite strain by the center-to-center method. (b) Plot of d' versus α'. The circles are averages of d' over 36 10-degree sectors. (From J. G. Ramsay and M. I. Huber, 1983, *The techniques of modern structural geology,* Academic Press, Volume 1.)

the technique is based on measurement of immediately adjacent grains only. A serious problem is that properly identifying nearest neighbors may be impossible.

Application of the technique is rather simple, but it is slow without a computer and digitizer. A transparent overlay is placed on an oriented photograph (at convenient magnification) of the objects to be measured. The overlay could also be placed onto a sawed slab of the rock, if the objects to be measured (pebbles, pisolites, amygdules) are large enough. A reference line is drawn on the overlay. The center of each object is then marked with a dot, and tie lines are drawn connecting each point with its nearest neighbors. Distances (d') and angles ($+\alpha'$ or $-\alpha'$) between objects are then measured between the tie lines and the reference marker (azimuth) line.

A plot may be made of d' versus values of α' ranging from $+90°$ to $-90°$ (Figure 5–15b). The principal difficulty lies in constructing the best-fit curve and then determining the maximum and minimum values and the symmetry axes of the curve. Ramsay and Huber (1983) suggested determining the arithmetic means of the α' values at certain intervals (say every 10°) and using them to plot the best-fit curve. Maximum and minimum values of the curve will not represent exact values for tectonic strain, but a value for R (ellipticity of the strain ellipse) may be obtained from

$$R = \frac{d'_{max}}{d'_{min}} \quad \textbf{(5–12)}$$

Values of d'_{max} and d'_{min} are obtained directly from the d'-α' plot. The position of d'_{max} on the axis also represents orientation of the long axis of the strain ellipse.

The Fry Method. A technique invented by Norman Fry (1979) is a simpler version of the center-to-center method and works on the same principle. Angular relationships and angles between particles are modified according to the nature and amount of accumulated strain. The result of the Fry technique is a diagram containing a set of points with a circular to elliptical blank area of relative shape and orientation proportional to the shape and orientation of the strain ellipse. A circular area indicates there is no strain in the rock.

To use the ***Fry method*** graphically, an overlay is made and a central reference point identified (Figure 5–16). A second overlay is made and the center of each particle is marked with a numbered point. (Centers may be marked directly on a photograph if other prints are available.) The central reference mark on the first overlay is placed over a point in the center of the photograph, and dots corresponding to all the numbered centers of the particles are marked on the overlay. The central reference mark is then moved (maintaining a constant orientation) to another numbered point in the array of particle centers, and the process of placing dots on the overlay is continued. The procedure is repeated until all the numbered centers have been used; a vacant field usually emerges. If no vacant field emerges, or if the field is circular, the rock is undeformed in the orientation used—or the distribution of particles was not random. An elliptical vacant area allows determination of the orientation and approximates the shape of the strain ellipse. Undeformed oöids and conglomerates—either of which may have a primary ellipticity—will yield mostly elliptical vacant fields.

Discussion. The R_f/ϕ, center-to-center, and Fry methods all yield estimates of strain in elliptical objects, but each has advantages and disadvantages. The R_f/ϕ method is probably the most accurate, but is very slow unless the parameters can be digitized

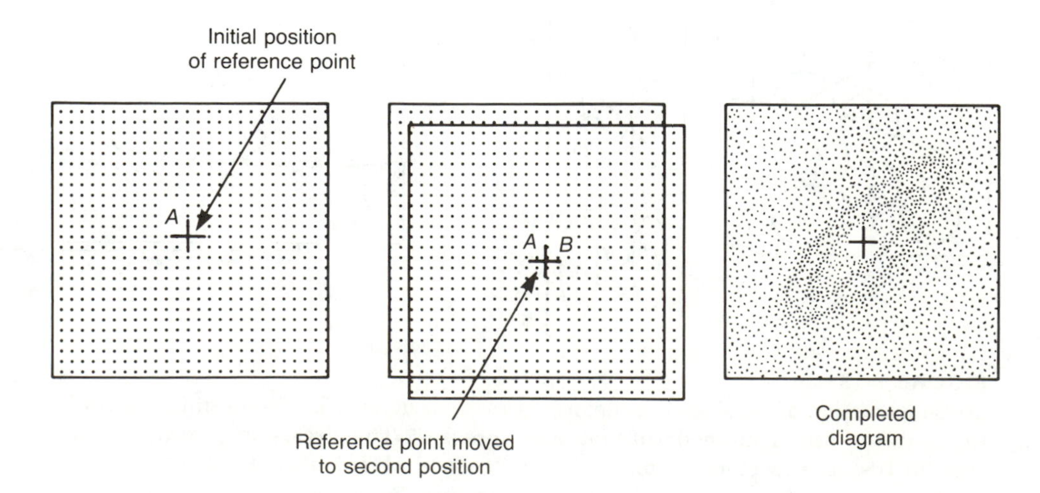

FIGURE 5–16
Stepwise plotting of a Fry-method diagram.

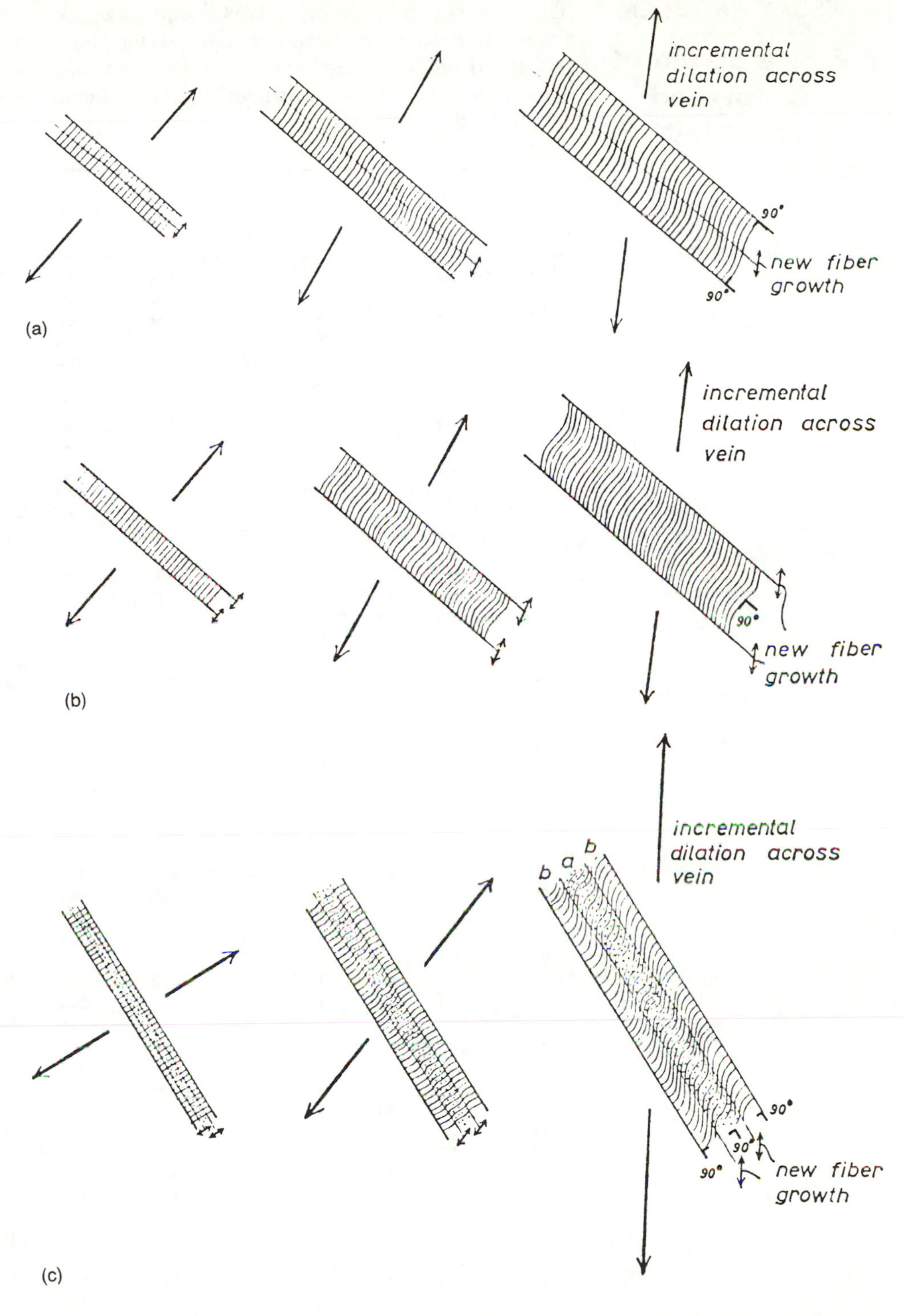

FIGURE 5–17
Several types of tectonic extension veins showing multiple stages of opening and filling. (a) Type 1—syntaxial crystal growth. New crystals in the vein form in optical continuity with wall-rock crystals. (b) Type 2—Antitaxial crystal growth. (c) Type 3—Composite new crystals b are in optical continuity with wall-rock crystals. (From D. W. Durney and J. G. Ramsay, *in* Kees De Jong and Robert Scholten, eds., *Gravity and tectonics,* copyright © 1973, John Wiley & Sons. Reprinted by permission of John Wiley & Sons, Inc.

for processing by a computer program. The center-to-center method has a severe disadvantage: the user may not be able to determine nearest neighbors accurately. The Fry method is probably the most rapid and can be used in conjunction with appropriate computer programs. But all the methods are designed to determine two-dimensional homogeneous strain, if determination is made only on one surface. For complete determination of three-dimensional strain in a rock body, the technique should use three perpendicular sections, preferably oriented parallel to the three principal planes of the strain ellipse. Programs to run on desktop computers will transform two-dimensional strain to three dimensions and make three-dimensional analyses. McEachran and Marshak (1986) have pointed out that graphics capabilities of some small computers, particularly the Apple Macintosh®, permit writing

ESSAY PROBLEM

Finite Strain of Deformed Pebbles

The data set in Table 5E–1 was compiled by B. K. Morgan as part of a senior research project at Clemson University. For his study, Morgan measured deformed pebbles weathered from the Draytonville metaconglomerate in the Kings Mountain belt in the southern Appalachian Piedmont of South Carolina (Figure 5E–1).

TABLE 5E-1
Finite-strain data for Draytonville metaconglomerate

						Extension (%)					
Sample #	X (cm)	Y (cm)	Z (cm)	Volume (cm^3)	Undeformed Diameter (cm)	X	Y	Z	$a = Z/Y$	$b = Y/X$	k
1	3.2	4.7	11.9	85	5.46	−41%	−14%	118%	2.5	1.5	3
2	3.0	5.1	10.4	60	4.86	−38%	5%	114%	2.0	1.7	1.4
3	2.4	4.2	8.4	30	3.86	−38%	9%	118%	2.0	1.75	1.3
4	2.2	3.8	10.3	35	4.06	−46%	−6%	154%	2.7	1.7	2.4
5	2.4	4.1	8.1	35	4.06	−41%	1%	100%	2.0	1.7	1.4
6	2.0	3.5	6.8	20	3.37	−41%	4%	102%	1.9	1.75	1.2
7	3.1	5.4	12.4	105	5.86	−47%	−8%	112%	2.3	1.7	1.7
8	3.4	4.0	9.2	60	4.86	−30%	−18%	89%	2.3	1.2	6.5
9	2.4	3.4	7.0	20	3.37	−29%	0.9%	108%	2.1	1.4	2.75
10	2.2	3.0	5.9	19	3.31	−34%	−9%	78%	2.0	1.4	2.5
11	1.9	3.3	8.9	25	3.63	−48%	−9%	145%	2.7	1.7	2.4
12	2.2	2.8	7.5	19	3.31	−34%	−15%	127%	2.7	1.3	5.7
13	2.5	4.5	8.2	39	4.2	−40%	7%	95%	1.8	1.8	1.0
14	2.7	4.6	8.7	48	4.5	−40%	2%	93%	1.9	1.7	1.3
15	2.1	3.1	8.6	20	3.37	−38%	−8%	155%	2.8	1.5	3.6
16	1.2	2.2	3.1	3	1.79	−33%	23%	73%	1.4	1.8	0.5
17	1.9	3.0	7.0	19	3.31	−43%	−9%	111%	2.3	1.6	2.2
18	1.9	2.2	4.2	8	2.48	−23%	−11%	69%	1.9	1.2	4.5
19	2.3	2.4	7.3	20	3.37	−32%	−29%	117%	3.0	1.0	---
20	2.0	3.3	7.8	25	3.63	−45%	−9%	115%	2.4	1.65	2.2
21	4.0	6.6	16.1	190	7.13	−44%	−7%	126%	2.4	1.45	2.2
22	2.1	3.1	7.1	20	3.37	−38%	−8%	111%	2.3	1.5	2.6
23	1.8	3.4	6.7	19	3.31	−46%	3%	102%	2.0	1.9	1.1
24	2.2	4.1	5.5	21	3.42	−36%	20%	61%	1.3	1.9	0.33
25	1.4	2.0	3.8	4	1.97	−29%	2%	93%	1.9	1.4	2.2
26	2.2	4.3	6.4	20	3.37	−35%	28%	90%	1.5	2.0	0.5
27	1.6	3.7	9.3	24	3.58	−55%	3%	160%	2.5	2.3	1.2
28	1.6	3.0	5.4	10	2.67	−40%	12%	102%	1.8	1.9	0.9
29	2.6	4.6	9.2	50	4.57	−43%	0.7%	101%	2.0	1.8	1.3
30	1.6	3.0	5.6	15	3.06	−48%	−2%	83%	1.9	1.9	1.0
31	2.7	4.1	9.7	40	4.24	−36%	−3%	129%	2.4	1.5	2.8
32	2.0	3.1	7.8	20	3.37	−41%	−8%	131%	2.5	1.55	2.7
33	2.8	5.3	8.1	58	4.8	−42%	10%	69%	1.5	1.9	0.6
34	1.9	3.5	6.9	25	3.63	−48%	−4%	90%	2.0	1.8	1.3
35	1.4	1.6	5.3	5	2.12	−34%	−25%	150%	3.3	1.1	23.0
36	1.8	2.9	5.6	10	2.67	−33%	9%	110%	1.9	1.6	1.5
37	3.0	6.1	9.4	80	5.35	−44%	14%	76%	1.5	2.0	1.5
38	2.0	2.5	5.4	10	2.67	−25%	−6%	102%	2.2	1.3	4.0
39	2.6	3.4	6.5	22	3.48	−25%	−2%	87%	1.9	1.3	3.0
40	3.0	4.1	8.1	45	4.4	−32%	−7%	84%	2.0	1.4	2.5
41	2.2	2.9	5.9	19	3.31	−34%	−12%	78%	2.0	1.3	3.3
42	2.2	3.7	5.7	20	3.37	−35%	10%	69%	1.5	1.7	0.7
43	1.9	3.2	4.7	15	3.06	−38%	5%	54%	1.5	1.7	0.7
44	1.3	2.4	5.4	8	2.48	−48%	−3%	118%	2.3	1.8	1.6
45	1.8	3.5	4.6	12	2.84	−37%	23%	62%	1.3	1.9	0.3
46	1.6	4.0	4.7	15	3.06	−48%	31%	54%	1.2	2.5	0.1
47	1.3	2.7	3.5	3	1.79	−27%	51%	96%	1.3	2.1	0.3
48	1.6	4.7	8.6	20	3.37	−54%	39%	155%	1.8	2.9	0.4
49	3.0	7.0	10.4	90	5.56	−46%	26%	87%	1.5	2.3	0.4
50	3.0	4.3	9.0	63	4.94	−39%	−13%	82%	2.1	1.4	2.8
51	1.5	3.0	6.8	15	3.06	−51%	−2%	122%	2.3	2.0	1.3
52	1.8	2.7	8.8	22	3.48	−48%	−22%	153%	3.3	1.5	4.6
53	1.4	2.7	5.4	10	2.67	−48%	1%	102%	2.0	1.9	1.1
54	1.8	3.2	7.5	15	3.06	−41%	5%	102%	2.3	1.8	1.6
55	2.2	3.9	13.6	50	4.57	−52%	−15%	198%	3.5	1.8	3.1
56	1.8	3.1	6.5	19	3.31	−46%	−6%	103%	2.7	1.3	5.7
57	1.6	2.0	5.3	8	2.48	−35%	−19%	114%	2.7	1.3	5.7
58	3.6	5.7	11.3	120	6.12	−41%	−7%	85%	2.0	1.6	1.7
59	2.0	2.2	4.4	8	2.48	−19%	−11%	77%	2.0	1.1	10.1

The Draytonville metaconglomerate is a relatively clean conglomerate consisting of pebbles ranging up to 15 cm long in a matrix of predominantly recrystallized quartz grains. Most of the pebbles consist of recrystallized vein quartz, although a few (<2 percent) may consist of other rock types. Both the pebbles and matrix have been totally recrystallized. Microscopic grains within

TABLE 5E-1 *(continued)*

Sample #	X (cm)	Y (cm)	Z (cm)	Volume (cm^3)	Undeformed Diameter (cm)	Extension (%)			$a = Z/Y$	$b = Y/X$	k
						X	Y	Z			
60	2.1	3.8	6.5	20	3.37	−38%	−13%	93%	1.7	1.8	0.9
61	1.3	1.8	4.3	4	1.97	−34%	−9%	118%	2.4	1.4	3.5
62	2.2	3.5	5.7	15	3.06	−28%	8%	86%	1.7	1.5	1.4
63	3.6	6.8	12.1	98	5.72	−37%	19%	112%	1.8	1.9	0.9
64	2.3	3.0	6.2	19	3.31	−31%	−9%	87%	2.1	1.3	3.7
65	1.7	3.2	4.5	15	3.06	−44%	5%	47%	1.4	1.9	0.4
66	1.0	2.1	4.5	8	2.48	−60%	−15%	81%	2.1	2.1	1.0
67	2.1	2.8	4.4	10	2.67	−21%	5%	65%	1.6	1.3	2.0
68	1.4	2.7	5.9	10	2.67	−48%	1%	121%	2.2	1.9	1.3
69	1.9	2.9	5.1	18	3.25	−42%	−11%	57%	1.8	1.5	1.6
70	2.8	4.2	7.9	36	4.10	−32%	2%	93%	1.9	1.5	1.8
71	2.1	2.4	5.1	10	2.67	−21%	−10%	91%	2.1	1.1	11.0
72	2.5	4.1	9.1	40	4.24	−41%	−3%	115%	2.2	1.6	2.0
73	1.5	2.4	5.9	10	2.67	−44%	−10%	121%	2.5	1.6	2.5
74	1.2	2.5	4.7	6	2.25	−47%	11%	109%	1.9	2.1	0.8
75	1.0	1.7	3.2	3	1.79	−44%	−5%	79%	1.9	1.7	1.3
76	1.6	2.9	5.4	8	2.48	−36%	17%	118%	1.9	1.8	1.1
77	2.2	4.2	9.6	35	4.06	−46%	3%	137%	2.3	1.9	1.4
78	1.5	2.9	6.9	15	3.06	−51%	−5%	126%	2.4	1.9	1.6
79	1.7	2.4	6.8	15	3.06	−44%	−22%	122%	2.8	1.4	4.5
80	2.3	4.3	8.3	40	4.24	−46%	1%	96%	1.9	1.9	1.0
81	2.0	2.6	5.3	12	2.84	−30%	−8%	87%	2.0	1.3	3.3
82	1.1	2.5	4.3	8	2.48	−56%	0.8%	73%	1.7	2.3	0.5
83	2.4	3.7	7.4	30	3.86	−38%	−4%	92%	2.0	1.5	2.0
84	2.9	4.3	8.5	55	4.72	−39%	−9%	80%	2.0	1.5	2.0
85	2.4	2.9	5.8	20	3.37	−29%	−14%	72%	2.0	1.2	5.0
86	1.6	2.7	3.9	12	2.84	−44%	−5%	37%	1.4	1.7	0.6
87	1.5	1.6	3.5	3	1.79	−16%	−11%	96%	2.2	1.1	12.0
88	0.8	1.6	4.1	2	1.56	−49%	3%	163%	2.6	2.0	1.6
89	2.7	5.3	9.8	80	5.35	−50%	−0.9%	83%	1.8	2.0	0.8
90	2.5	4.4	7.9	50	4.57	−45%	−4%	73%	1.8	1.8	1.0
91	3.6	6.6	10.2	110	5.94	−39%	11%	82%	1.5	1.8	0.6
92	1.4	3.1	6.2	15	3.06	−54%	1%	103%	2.0	2.2	0.8
93	2.1	3.9	9.4	35	4.06	−48%	−4%	132%	2.4	1.9	1.6
94	1.5	3.9	8.1	20	3.37	−55%	16%	140%	2.1	2.6	0.7
95	2.9	6.1	12.6	95	5.66	−40%	8%	123%	2.1	2.1	1.0
96	1.6	1.8	3.6	3	1.79	−11%	6%	101%	2.0	1.1	10.0
97	1.3	2.3	4.1	4	1.97	−30%	17%	108%	1.8	1.8	1.0
98	2.9	3.4	5.9	20	3.37	−14%	0.9%	75%	1.7	1.2	3.5
99	2.4	3.2	5.9	19	3.31	27%	−3%	78%	1.8	1.3	2.7
100	3.0	3.2	10.1	40	4.24	−29%	−25%	138%	3.2	1.1	22.0
101	2.6	2.9	8.5	23	3.53	−26%	−18%	141%	2.9	1.1	19.0
102	3.7	5.0	15.9	110	5.94	−38%	−16%	168%	3.2	1.4	5.5
103	3.4	5.0	12.5	110	5.94	−43%	−16%	110%	2.5	1.5	3.0
104	2.7	4.6	13.0	63	4.94	−45%	−7%	163%	2.8	1.7	2.6
105	2.6	4.0	11.5	60	4.86	−47%	−18%	137%	2.9	1.5	3.8
106	1.8	3.5	4.9	8	2.48	−27%	41%	98%	1.4	1.9	0.4
107	1.0	1.9	3.4	3	1.78	−44%	7%	91%	1.8	1.9	0.9
108	3.2	4.6	9.4	58	4.80	−33%	−4%	91%	2.0	1.4	2.5
109	2.2	4.1	8.6	38	4.18	−47%	−2%	106%	2.1	1.9	1.2
110	3.3	5.6	13.4	120	6.12	−46%	−8%	119%	2.4	1.7	2.0
111	0.8	1.8	3.3	2	1.56	−49%	15%	112%	1.8	2.3	0.6
112	0.6	1.3	2.9	2	1.56	−62%	−17%	86%	2.2	2.2	1.0
113	1.0	1.4	2.9	2	1.56	−36%	−10%	86%	2.1	1.4	2.8
114	1.1	1.9	3.4	3	1.78	−38%	7%	91%	1.8	1.7	1.1
115	2.0	2.4	5.1	10	2.68	−25%	−10%	90%	2.1	2.2	0.9
116	1.8	2.5	4.0	9	2.58	−30%	−3%	55%	1.6	1.4	1.5
117	1.4	2.7	5.1	10	2.68	−48%	7%	90%	1.9	1.9	1.0

the conglomerate show no sign of the deformation evident in the shapes of the pebbles today. Also, few of the pebbles bear evidence of deformation by pressure solution. Long axes of pebbles in the metaconglomerate are oriented parallel to upright folds that plunge gently northeast, parallel to a major synclinal structure (Figures 5E–1, 5E–2).

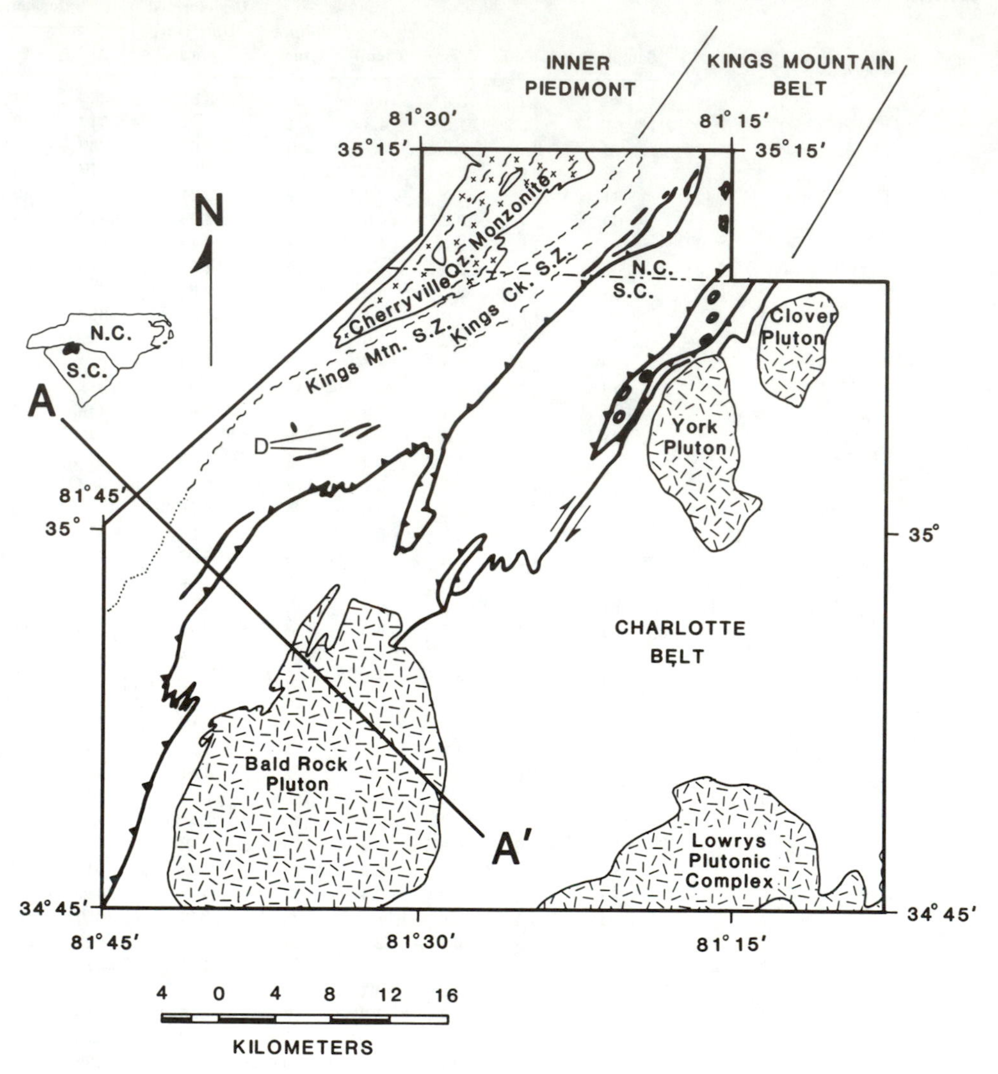

FIGURE 5E–1
Geologic map of the Kings Mountain belt in part of the Carolinas showing the location of the study area in South Carolina. D—Draytonville metaconglomerate localities studied. Metaconglomerate bodies are patterned black. Barbed line is probably a suture between a Paleozoic exotic terrane and North America. Dotted lines are traditional geologic subdivision boundaries labeled along the northeastern edge of the map. Wavy lines indicate shear zones (S.Z.).

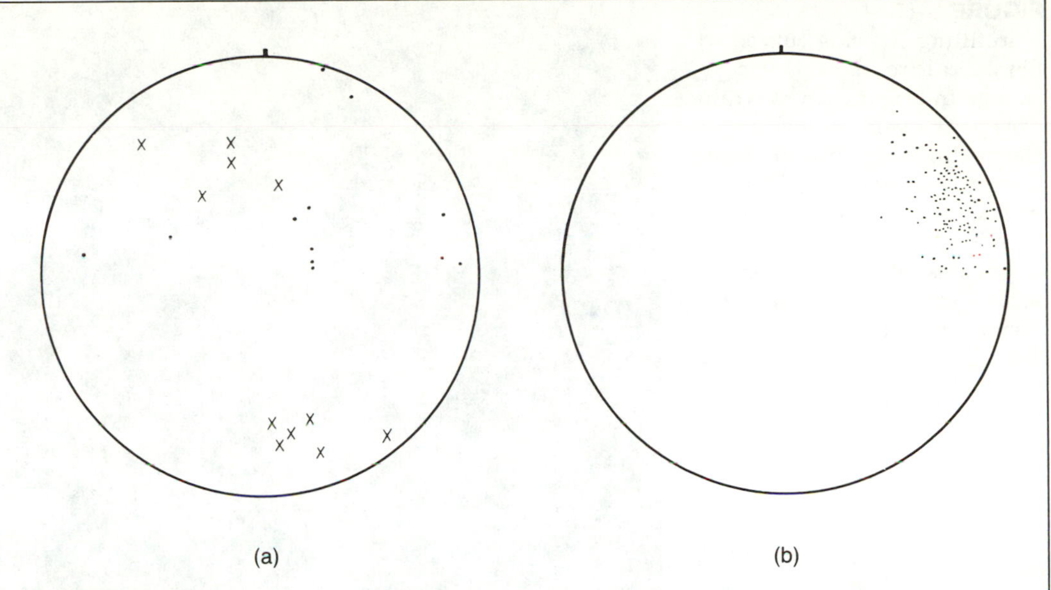

FIGURE 5E–2
Fabric diagram of long axes of 143 deformed pebbles in the Draytonville metaconglomerate, measured in place. (a) ×—pole to the dominant foliation; dot—pole to a younger foliation. (b) Orientations of 143 long axes of deformed pebbles.

Morgan collected more than 100 intact pebbles weathered from the Draytonville metaconglomerate and measured the major axes with a caliper to determine total strain involved in the deformation. He also measured orientations of the long axes of 143 pebbles in place to determine the finite-strain ellipsoid. Because the long axes of the pebbles all cluster about N 55° E, 15° NE (Figure 5E–2), it is unlikely that any of the pebbles had appreciable original ellipticity unless original orientation and that of the axis of major compression were the same. Observations at localities where deformation is minimal confirmed that the pebbles were mostly originally spherical.

Using any technique you think is appropriate and data from Table 5E–1 and Figure 5E–2, determine the orientation of the finite-strain ellipsoid and comment on the magnitude of strain affecting the rock mass. What additional sources of error may be present?

One approach to solving this problem is to plot a Flinn diagram of the data and estimate the magnitude of strain. The center-to-center and Fry methods require observation and measurement of objects in their original positions relative to each other, so we cannot use them here. In compiling Table 5E–1, Morgan used an immersion technique to determine the volume of each pebble and back-calculated the original spherical dimensions. What sources of error could be important in this method of determining finite strain?

FIGURE 5–18
Vein filling showing curved fibrous calcite, indicating a change in orientation of strain axes as the vein opened. Oölitic limestone bed in Upper Cambrian Nolichucky Shale near Oak Ridge, Tennessee. Vein is approximately 2 mm thick; crossed polars. (Thin section courtesy of Peter J. Lemiszki, University of Tennessee.)

simple programs that simulate strain and permit calculations to be made. Programs also exist for calculation of two-dimensional strain using the Fry, center-to-center, and R_f/ϕ methods.

Measurement of Progressive Strain and Deformation from Fibers

Growth of fibrous quartz, calcite, and other minerals in veins (Figure 5–17), pressure shadows, and movement surfaces make possible direct measurement of incremental strain. Veins that contain several layers of fibers (Figure 5–18) indicate separate events of opening of the vein (crack-seal) and reveal part of the incremental strain history of the rock mass. Orientations of fibers also indicate the direction of opening of the vein, and curved fibers indicate rotation of the strain field or shear during formation.

Study of the multiple-opening history of extension veins yields details of progressive deformation.

FIGURE 5–19
Pressure-solution cleavage in Martinsburg Slate near the Delaware Water Gap, New Jersey, showing quartz grains partly dissolved. Note the truncation of layers by pressure solution in the partly dissolved detrital muscovite grain in the lower center of the photo. Width of the long dimension of the photo is approximately 5 mm; plane light. (Thin section courtesy of Timothy L. Davis, University of Tennessee.)

Most veins are filled with massive, randomly oriented crystals, but those nucleating fibrous growth during multiple events of progressive deformation provide a useful aid in studying the incremental strain history of an area.

DETERMINATION OF PRESSURE-SOLUTION STRAIN

Under ideal conditions, pressure-solution strain may account for 20 to 30 percent of the total strain in a rock body. This strain is commonly observed as stylolites, not all of which need be tectonic, and pressure-solution cleavage. Direct effects are best observed in thin sections cut normal to surfaces formed by pressure solution (Figure 5–19).

Estimates of the bulk strain in rock may be obtained by several means: construct a model of partly dissolved clasts across several pressure-solution surfaces and calculate the relationship to total surface area in the rock, unfold folded veins produced as a product of pressure-solution shortening, and determine the ratios of insolubles (such as clays and heavy minerals) in dissolved zones to those in the bulk undissolved rock mass.

Now that we have surveyed the different methods of measuring strain in rock, we can go on to Chapter 6, where we will consider the various models and examples of the mechanical behavior of rock materials.

Questions

1. How are finite and incremental strain related?
2. What are the desirable characteristics of a good strain marker?
3. What primary depositional processes can add to the difficulties of determination of finite strain of initially elliptical markers?
4. What characteristic of Flinn's k enables us to determine the kind of strain that distorted a marker?
5. Why does Wellman's method work?
6. With the strain markers in Figure 5–4, determine
   (a) R_s, R_i, and ϕ', using the R_f/ϕ technique,
   (b) R and α using the center-to-center technique,
   (c) shape and orientation of the strain ellipse, using the Fry method.
7. Why does the Fry method work?
8. Why are fossils better than many other strain markers for quantitative determination of shear strain?

Further Reading

Cloos, Ernst, 1971, Microtectonics along the western edge of the Blue Ridge, Maryland and Virginia: Baltimore, Johns Hopkins University Press, 234 p.
Strain in a very large area in the central Appalachian Blue Ridge was documented by measuring dimensions of more than 10,000 oöids in a study lasting several decades.

Hossack, J. R., 1968, Pebble deformation and thrusting in the Bygdin area (S. Norway): Tectonophysics, v. 5, p. 315–339.
Determination of strain in pebbles in the Bygdin conglomerate and the relationships to thrust faulting are demonstrated. Some pebbles here exhibit very large elongation strains and the rock mass containing them has been called a "walking stick" conglomerate because some pebbles are up to 1 m long and only 1 to 2 cm in diameter.

Ramsay, J. G., and Huber, M. I., 1983, The techniques of modern structural geology, Volume 1: Strain analysis: New York, Academic Press, 307 p.
Detailed summary of strain-analysis techniques. The mathematics is not too rigorous for undergraduate geology students.

6

Mechanical Behavior of Rock Materials

The fact that rock does show marked differences in behavior in natural deformation provides the geologist with an unusual opportunity to learn more about the conditions of deformation—if only he can relate these differences in some systematic manner to environmental and rock factors.

FRED A. DONATH, *American Scientist*

THE EFFECTS OF THE MECHANICAL BEHAVIOR OF MATERIALS ARE visible all around us, in deformed rocks and in the technology of everyday life. In any motor car, the fenders and engine block display contrasting physical properties produced by different physical conditions involved in their manufacture. Much the same steel is used in both, but for fenders it is rolled into sheets and stamped (deformed) into shape; for the block it is cast roughly in molten form and cooled under conditions that yield the needed strength and temperature resistance, and then it is machined into shape.

Plastics and ceramics are quite different from metals, but they, too, are pertinent here. What we now call plastics are mostly artificial polymers made from hydrocarbons. They are manufactured at temperatures high enough to soften—make them plastic—but not melt them. Ceramic materials in common use today—bricks, semiconductors, glasses, fine china, and ceramic internal-combustion engines—all exhibit elastic properties at room temperature but are made from different geologic raw materials at high temperatures, where they behave ductilely or as viscous melts.

Much of the theory we use in structural geology to explain the behavior of rock materials was originally developed to explain the behavior of metals under different conditions. Ceramic materials consist of compounds with properties more like those of minerals than those of metals, and so structural geologists are now benefiting from theoretical developments in the emerging field of materials science as well as from metallurgy.

Physical models that simulate the mechanical behavior of rocks, soils, concrete, steel, and ceramics have been used for many years because they enable us to describe and measure the response of these to stress (Figure 6–1). The three ideal end-member behavior models—elastic, viscous, and plastic (to be discussed shortly)—also show us more clearly how to write mathematical expressions for ideal behavior. Some rock materials approach ideal behavior under particular conditions; others exhibit more complex types of behavior under any conditions. Most rock materials never exhibit ideal behavior, but frequently respond to stress in ways that approach one of the ideal types. Our discussions of stress and strain in Chapters 3 and 4 set the stage for us to look at the behavior of rock materials under various conditions. Now, examining the ideal behavior types will enable us to understand some of the bulk properties of rocks.

DEFINITIONS

Properties of ***homogeneous*** materials are the same throughout any sample of any size; those of ***inhomogeneous*** materials vary spatially with location either in a hand specimen or in a region, demon-

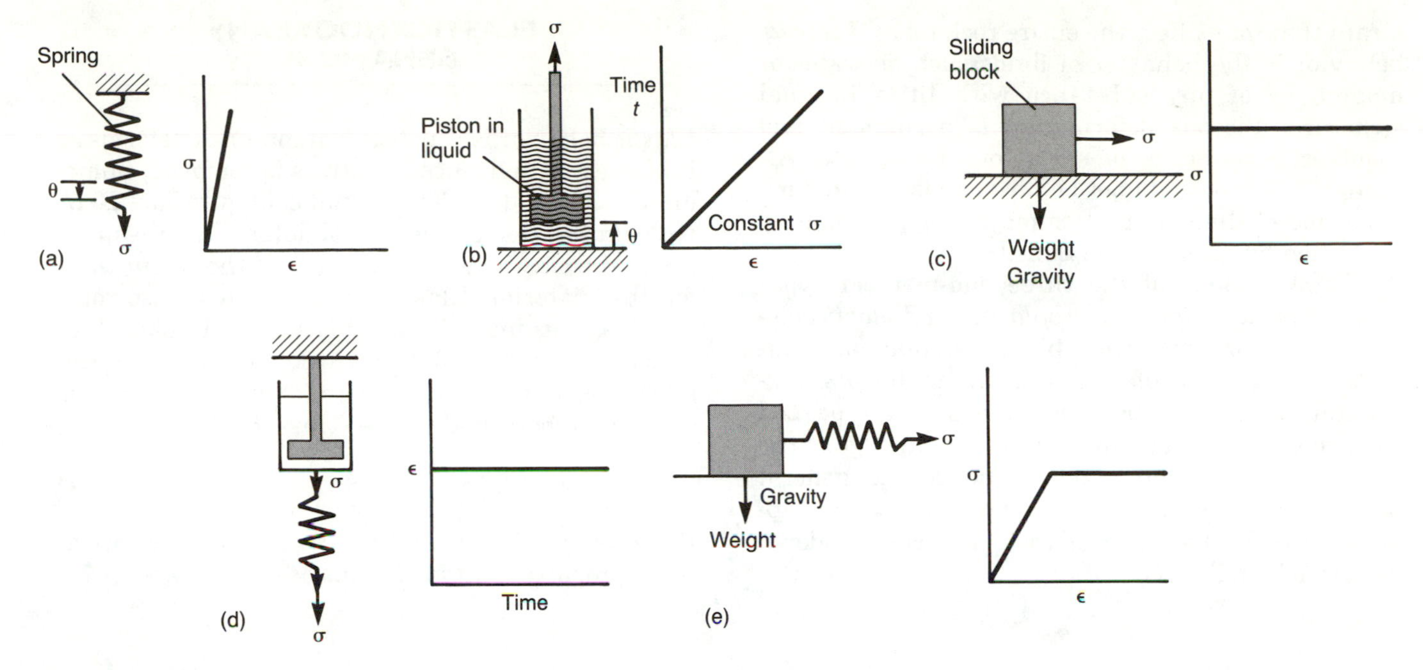

FIGURE 6–1
Ideal mechanical models and stress-strain or strain-time curves for material behavior. Think about the mechanical models in terms of how a spring, a piston in a liquid, and a sliding block behave independently or in combination. Strain is plotted versus time for viscous and viscoelastic materials because strain in viscous materials (liquids) is time-dependent. Stress exists in a fluid only if it is in motion.

strating scale-dependence of rock-material behavior. ***Isotropic*** materials have the same properties in all directions, contrasting with ***anisotropic*** materials, where properties vary with direction. Some rocks, like shale and schist, behave as strongly anisotropic materials on the scale of a hand specimen, but may behave isotropically on a scale of a rock mass of tens to hundreds of meters across. Other rock masses, such as plutons, may be homogeneous and isotropic in response to stress.

The crust of the eastern United States behaves homogeneously with respect to the present-day stress field (Figure 3E–3); but earthquakes do occur in the East, and so there must be inhomogeneities that interrupt the stress field, concentrate stress, and rupture the crust. Layered rocks are strongly anisotropic to stress, but the degree of expression of the anisotropy depends on the direction in which the stress is applied. So, rupturing of the crust producing eastern earthquakes probably occurs as a result of localized stress concentrations at suitably oriented anisotropies.

We commonly think of strain in terms of three end-member behavior types: *elastic, plastic,* and *viscous* (Figure 6–2). Strain that is recovered instantaneously on removal of applied stress is called ***elastic.*** The object deformed elastically returns to the original undeformed shape once the strain is removed. No other behavior mode exhibits this reversible property. ***Plastic*** behavior involves permanent strain, generally irreversible, occurring without loss of cohesion as a result of rearranging chemical bonds in the crystal lattices in minerals by one (or more) of the dislocation mechanisms described in Chapter 7; obviously it is a pervasive

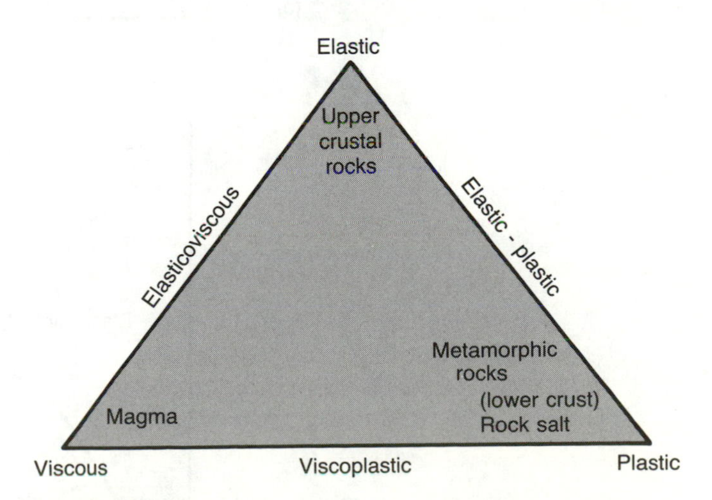

FIGURE 6–2
Material behavior-modes triangle. Most rock materials do not exhibit ideal behavior, but their behavior plots close to corners and edges—not in the middle of the diagram.

strain that may affect the entire rock mass. ***Viscous*** behavior is the behavior of fluids such as water or magma or of any substance with little internal structure. Viscous deformation is permanent and involves dependence of stress on strain rate. Although most rocks do not behave as viscous materials, some of their properties may be approximately explained by assuming viscous behavior.

Combinations of the three end-member types produce *elastic-plastic, viscoplastic,* and *elasticoviscous* behavior. Other combinations may approximate the behavior of most geologic materials. Fortunately, plots of the behavior of most rock types fall close to one of the end members or to an edge of the triangle—not in the middle—and so the general behavior models discussed in the next sections approximate the behavior of rock materials under a variety of conditions.

ELASTIC (HOOKEAN) BEHAVIOR

Elastic behavior is illustrated if material deforms as it is stressed and, once the stress is removed, immediately rebounds to its original configuration. ***Brittle behavior,*** as defined in Chapter 1, implies failure in the elastic range—actually at the *elastic limit* for the material. Elastic behavior in an isotropic homogeneous material is described by Hooke's law, which is named for Robert Hooke (1635–1703), English physicist; in it the strain (ϵ) in the material is directly proportional to the stress (σ):

$$\sigma \propto \epsilon \tag{6–1}$$

Equation 6–1 is converted to an equality by inserting a proportionality constant as

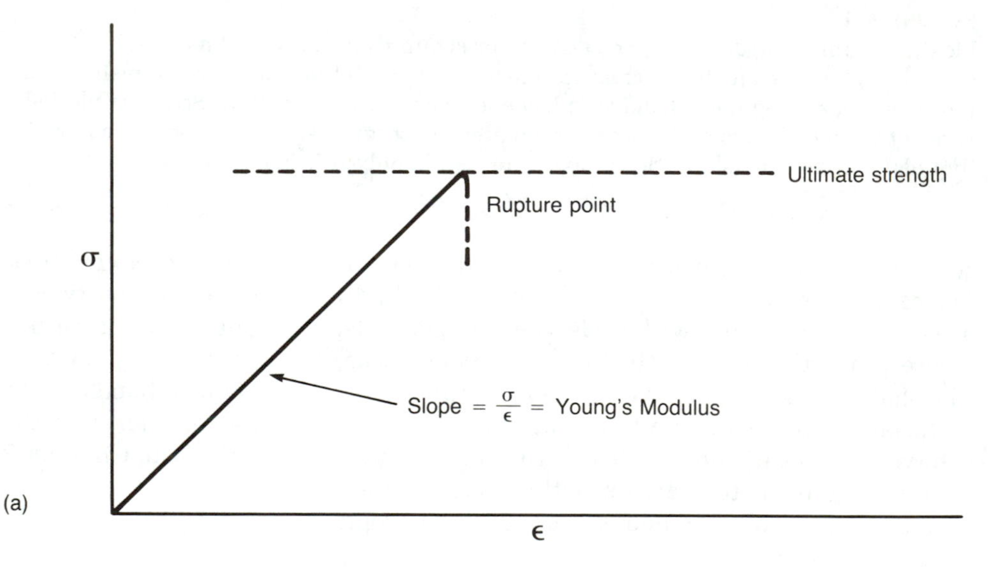

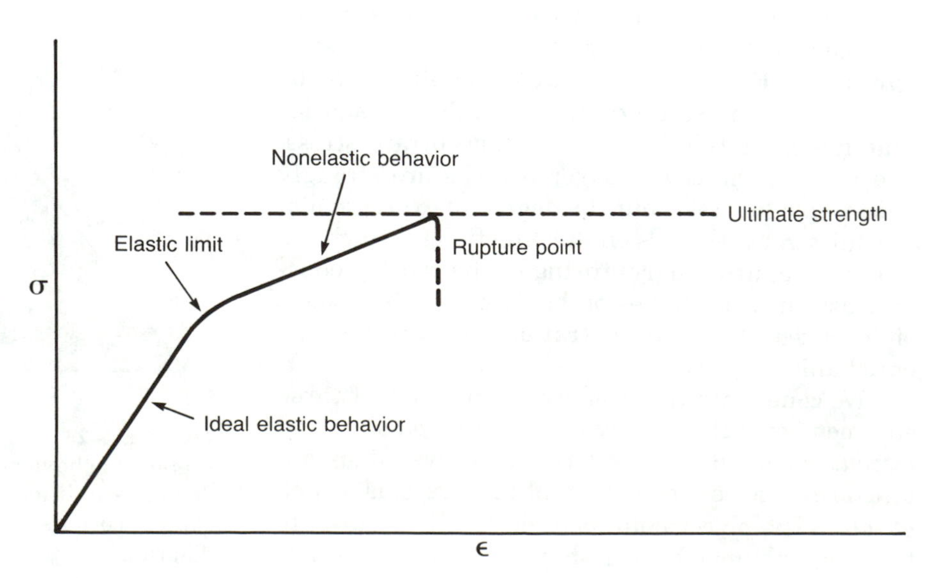

FIGURE 6–3
Stress-strain diagrams for an ideal elastic material (a) and a real elastic material (b). The real elastic material exhibits ideal behavior initially, but begins to deviate and becomes nonelastic at higher strains.

$$\sigma = E\epsilon \quad (6\text{–}2)$$

E is ***Young's modulus,*** an experimentally determined elastic constant in which

$$E = \frac{\sigma}{\epsilon} = \frac{\mathbf{F}/A}{\Delta l/l_0} \quad (6\text{–}3)$$

In equation 6–3, **F** is force, A is area, Δl is the change in length of a reference line, and l_0 is the initial length of the line. Note that the strain in equations 6–1, 6–2, and 6–3 is elongation (equation 4–1). For elastic behavior under shear stress,

$$\tau = G\gamma \quad (6\text{–}4)$$

where τ is shear stress, γ is shear strain, and G is another experimentally determined elastic constant, ***rigidity,*** or ***shear, modulus,*** defined as the ratio of shear stress to shear strain.

A linear relationship exists in ideal elastic behavior between the application of stress and the resulting strain. When plotted on a stress-strain diagram (Figure 6–3), the slope of the line is determined by the elastic constant for the material used in the experiment. The stress-strain curve for an elastic material begins at the origin and terminates at the ***yield point*** where Hooke's law no longer holds (nonelastic behavior begins) or the material ***ruptures.*** Permanent strain occurs in an ideal elastic material as rupture. A change in slope of the stress-strain curve also indicates that direct proportionality between stress and strain no longer exists. The highest point on the curve is the ***ultimate strength*** for the material. The ***elastic limit*** is the point on the curve beyond which the material begins to undergo permanent deformation or ruptures (or both). Below the elastic limit, an elastic material rebounds instantaneously to the original shape when the stress is removed.

In a more realistic stress-strain curve for elastic materials (Figure 6–3b), the slope decreases (but does not necessarily become negative) before reaching the ultimate strength of the material. This point where the decrease in slope occurs is both the yield point and the elastic limit. Above the elastic limit, the material exhibits nonelastic behavior.

Another elastic constant, frequently determined in elastic materials, is ***Poisson's ratio***—a measure of compressibility named in honor of Simeon-Denis Poisson (1781–1840), a French mathematician. It is commonly determined from

$$\nu = -\epsilon_1/\epsilon_3 \quad (6\text{–}5)$$

where ν is Poisson's ratio and ϵ_1 and ϵ_3 are linear elastic strains derived from changes of length measured in uniaxial compression experiments (Figure 6–4). An axial test cylinder is shortened by ϵ_3, a

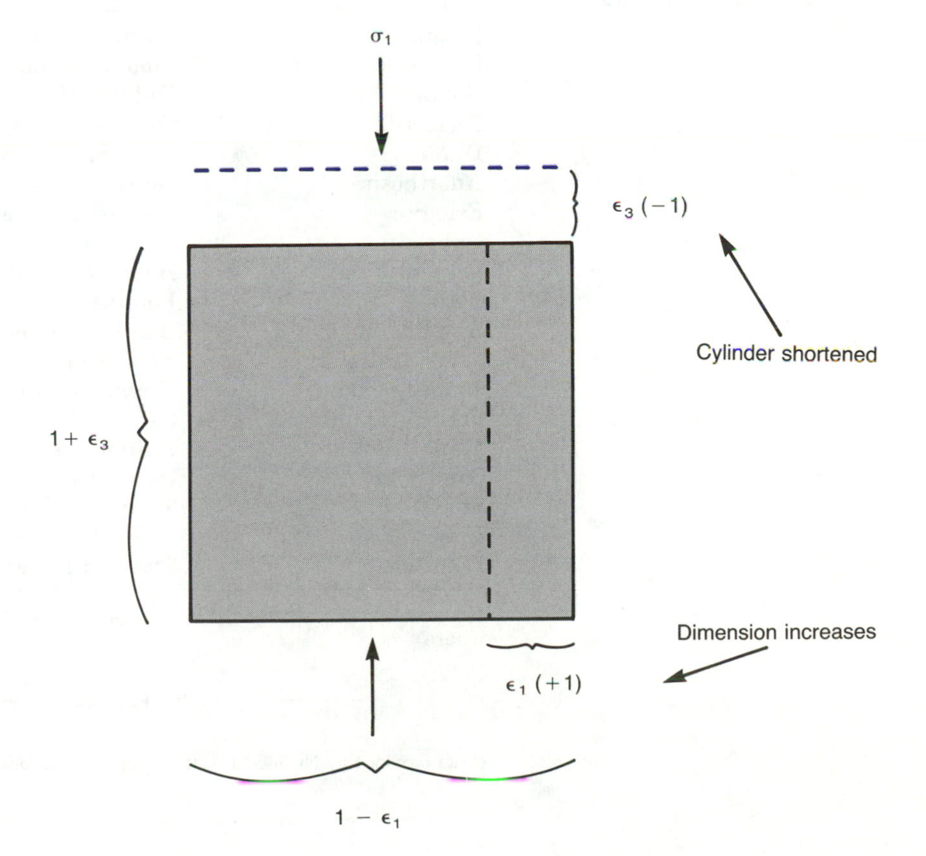

FIGURE 6–4
Poisson's ratio compressibility diagram showing the derivation of ϵ_1 and ϵ_3. The rock cylinder is compressed parallel to σ_1 during the experiment and expands normal to σ_1. Poisson's ratio is the ratio of shortening to extension.

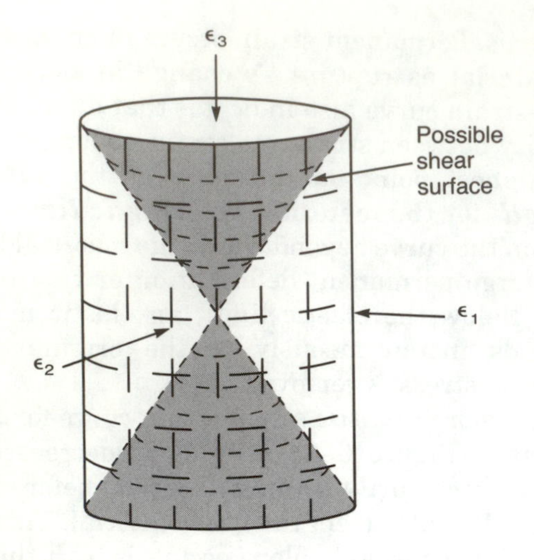

FIGURE 6–5
Axial test cylinder of rock showing orientation of strain axes and that $\epsilon_1 = \epsilon_2$. The conical areas inside the cylinder are locations of potential shear surfaces.

principal strain, but the lateral dimension is simultaneously increased by another principal strain ϵ_1, equal to ϵ_2 in the test cylinder (Figure 6–5). Values of ν are positive because one of the strains (ϵ_3) is negative and so cancels the negative sign in equation 6–5. For incompressible materials, $\nu = 0.5$, indicating that the material maintains a constant volume at any stress. Most rock materials have Poisson's ratios of less than 0.5, indicating that they are not ideal elastic materials. In contrast, soft biological tissues and rubber have Poisson's ratios of about 0.5; the ratio for cork is near zero; metallic lead, 0.45; aluminum, 0.33; most steels, 0.27; and polymers, 0.1 to 0.4 (Lakes, 1987). Table 6–1 contains representative values of Poisson's ratio for a variety of common rocks.

PERMANENT DEFORMATION—DUCTILITY

Permanent strain in the form of viscous or plastic deformation—ductility—may occur in a material beyond the elastic limit. Viscous behavior occurs in fluids, whereas plastic deformation occurs in solids below their melting points. Rutter (1986) maintained that ***ductility*** should reflect the capacity for large amounts of nonlocalized homogeneous strain. He would include all flow processes in the usage of the term ductile and restrict the use of the term

TABLE 6–1
Poisson's ratios for a variety of rock types.

Rock Type	Locality	Poisson's Ratio
Granite	Westerly, Rhode Island	0.25
Diorite	Mount Rainier, Washington	0.28
Gabbro	Duluth, Minnesota	0.30
Peridotite	Cypress Island, Washington	0.27
Dunite	Twin Sisters, Washington	0.28
Anorthosite	Adirondack Mountains, New York	0.31
Eclogite	Healdsburg, California	0.26
Quartzite	Baraboo, Wisconsin	0.10
Serpentinite	Burro Mountain, California	0.35
Slate	Poultney, Vermont	0.30
Amphibolite	Bantam, Connecticut	0.26
Quartz mica schist	Thomaston, Connecticut	0.31
Tonalite gneiss	Torrington, Connecticut	0.27
Felsic granulite	Saranac Lake, New York	0.26
Mafic granulite	Adirondack Mountains, New York	0.29
Sandstone	Berea, Ohio	0.26
Shale	Thorn Hill, Tennessee	0.26
Limestone	Thorn Hill, Tennessee	0.32
Dolostone	Thorn Hill, Tennessee	0.29

Poisson's ratios calculated from compressional (V_p) and shear (V_s) wave velocity measurements, where

$$\nu = \frac{1}{2}\left[1 - \frac{1}{(V_p/V_s)^2 - 1}\right] \text{ at 200 MPa confining pressure}$$

(Data provided by Nicholas I. Christensen, Purdue University.)

"plastic" to rocks that have been deformed by diffusion processes involving plastic deformation of crystal lattices. Scale-dependence is important when considering deformation that affects a rock mass because the kind of deformation occurring in a few cubic centimeters of rock may not be the same as that occurring in several cubic kilometers of rock, even if the larger mass includes the smaller. Consequently, the term "ductile" will be used to exclude those processes that have a brittle component. Ductile deformation is permanent and affects all parts of a rock mass at all scales.

Viscous Behavior

Viscous behavior is fluid behavior. Natural fluids, including water, magma, gases, and the molten part of the Earth's core, are all viscous materials. We can also argue that the mantle is a viscous material, using the example in Chapter 1 for calculating mantle viscosity from glacial rebound rates.

Stationary fluids will not transmit shearing stresses, and are called *perfect fluids*. Most moving fluids experience shear stresses on planes of differential motion including the boundaries of the fluid. Fluids in which there is a proportional relationship between stress (σ) and shear-strain rate are ***Newtonian fluids,*** expressed as

$$\sigma \propto \frac{d\gamma}{dt} = \dot{\gamma} \tag{6–6}$$

or

$$\sigma = \eta\dot{\gamma} \tag{6–7}$$

where γ is shear strain and t is time, so that $d\gamma/dt$, or $\dot{\gamma}$, is the shear-strain rate, and η is the viscosity of the fluid and the proportionality constant. (The SI unit for viscosity is the *poise,* and 10 poise equals 1 Pa s.) The reciprocal of viscosity, $1/\eta$, is called the *fluidity,* which is the ability of a fluid to move, rather than the resistance to motion measured by the viscosity. Stress-strain curves for viscous materials are not as useful as strain-time curves (Figure 6–6) because of the dependence of viscosity upon strain rate. Stress exists in fluid only when it is in motion—hence the time dependence.

Rocks at high temperatures near their melting points behave in a nearly viscous fashion, but their behavior is better described as plastic (to be discussed) than viscous. Useful models of crustal dynamics can, however, be formulated assuming viscous behavior.

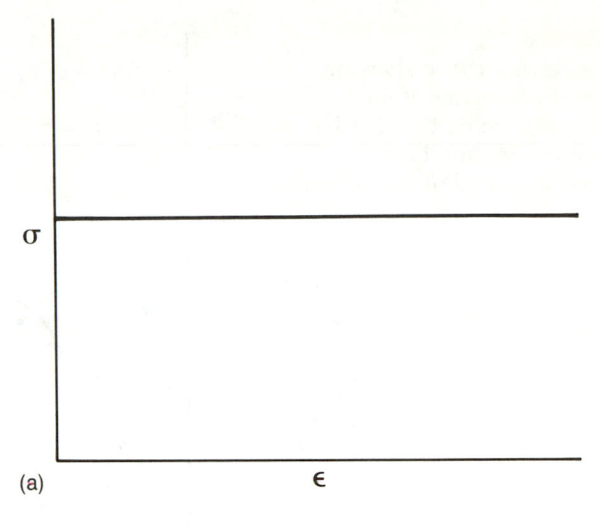

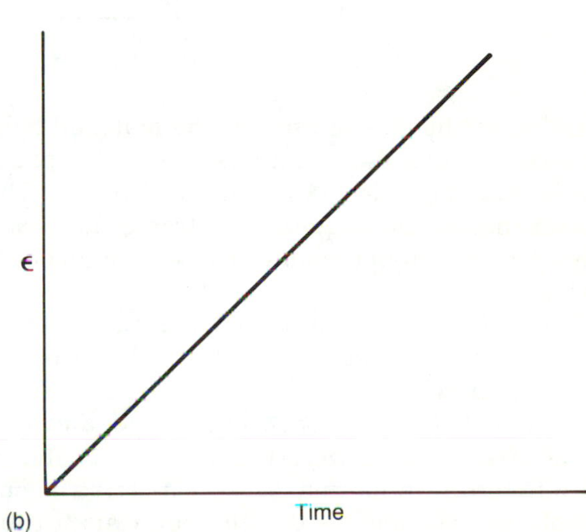

FIGURE 6–6
Stress-strain (a) and strain-time (b) diagrams for a viscous material. This stress-strain curve is uninformative and even somewhat misleading because of time dependence of strain in viscous materials. The stress-strain curve for a viscous material is meaningful only if the stress is applied instantaneously.

Plastic (Saint Venant) Behavior

Ideal ***plastic*** behavior involves permanent (nonrecoverable) deformation that affects the entire rock mass. It is rarely observed either in nature or in the laboratory. Ideal plastic behavior is also known as ***Saint Venant behavior*** in honor of Adhemar-Jean-Claude Barre de Saint-Venant (1797–1886), a French physicist who in the nineteenth century wrote mathematical expressions for plastic behavior in an attempt to relate various stress components in materials exhibiting plastic behavior.

In natural materials, plastic behavior is generally preceded by elastic behavior, plastic behavior

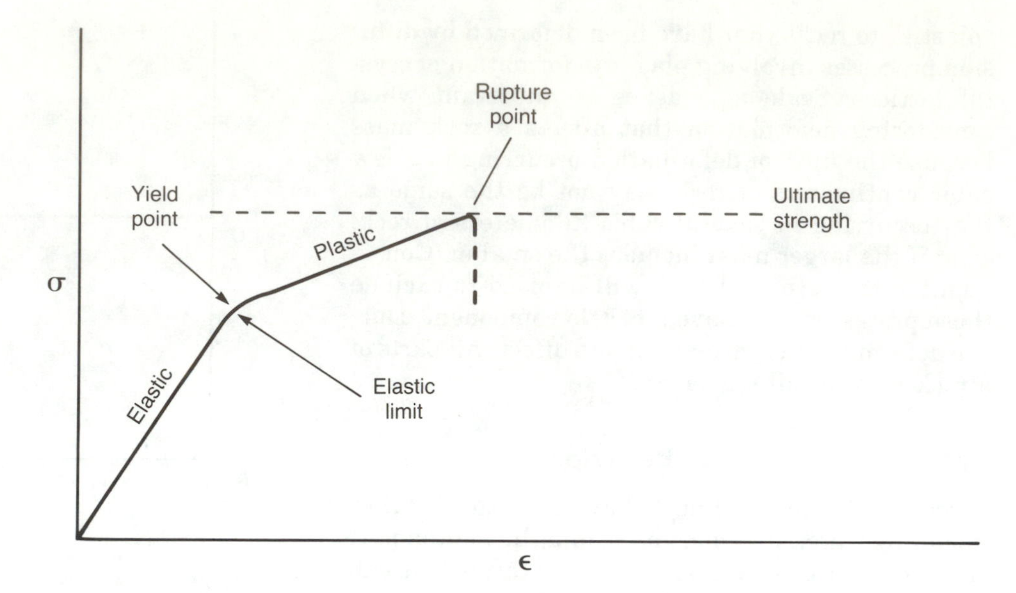

FIGURE 6–7
Stress-strain curve showing general properties of an elastic-plastic material (with strain hardening) and inflection points in the curve.

beginning at the elastic limit of the material being deformed (Figure 6–7). Total strain in a plastic material also depends on the *deformation path*. That is, it depends on the sequence of stresses applied to the body, rather than the final stress, as in an elastic material.

Once the yield point is passed, the material flows at a constant stress unless one of two things occurs. The first is ***strain hardening*** (increased resistance to strain as strain increases). The second, and opposite, is ***strain softening*** (Figure 6–8). Materials that either strain-harden or strain-soften should probably be described as ductile materials because they exhibit a more general property of nonideal behavior. Stress-strain curves for the limestone in Figure 6–8b indicate that, at relatively low confining pressure, the limestone deviates from ideal elastic behavior at moderate differential stress values before rupturing; at higher confining pressure, it still deviates but enters a realm of strain softening beyond the ultimate strength. At still higher confining pressure, the limestone exhibits almost ideal plastic behavior beyond the yield point. Plastic behavior is thought to dominate at depths in the crust and mantle where temperature reaches several hundred degrees Celsius and confining pressure reaches several kilobars. Presence of water or other fluid lowers threshold temperatures and pressures for ductile behavior.

Recrystallization processes, grain-boundary sliding, and some effects of pressure solution in metamorphic rocks (Chapter 7) are commonly associated with ductile deformation on the microscopic scale. The effects pervade the entire rock mass, or, in some instances, are confined to tabular zones of ductile deformation (ductile shear zone; Chapter 10).

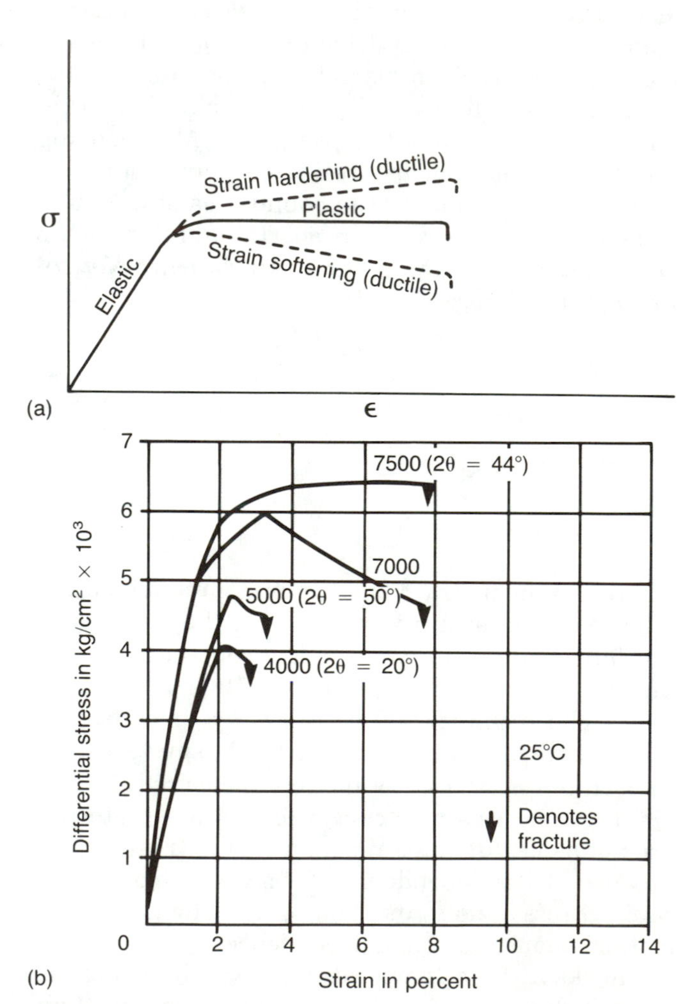

FIGURE 6–8
(a) Strain hardening and strain softening in an elastic-plastic material. (b) Experimental deformation of Solnhofen Limestone. Curves represent deformation at different confining pressures. (From H. C. Heard, 1960, Geological Society of America *Memoir* 79.)

FIGURE 6–9
Strongly foliated Late Proterozoic biotite gneiss containing feldspar layers on the Yadkin River near Siloam, North Carolina. The entire mass was ductilely deformed during the early Paleozoic. (RDH photo.)

Laboratory experiments designed to duplicate ductile deformation must be carried out at high temperature and pressure if ordinary rock materials are used. Structures in metamorphic and ductile fault rocks are thought to form under ductile conditions (Figure 6–9). In glacial ice and rock salt, ductile behavior may be initiated under surface conditions.

Several models have been proposed to explain plastic behavior, but none fully succeed in quantifying it because of the complexity inherent in the process. The mathematics describing plastic deformation is also complex, particularly that describing the plastic behavior of anisotropic crystalline solids like most rocks (Nicolas and Poirier, 1976). The two models outlined here were chosen because of their simplicity and historical significance.

Tresca's criterion states that plastic yield will begin under extensional conditions when the maximum shear stress (τ_{max}) reaches a critical value (C_y, a specific constant for the material) of half the initial stress difference of the material, or

$$\tau_{max} = C_y = (\sigma_1 - \sigma_3)/2 \quad \textbf{(6–8)}$$

where σ_1 and σ_3 are the principal stresses at yielding. This criterion was intended to account for some properties observed in metals where the constant maximum stress difference, $\Delta\sigma$ ($= \sigma_1 - \sigma_3$), indicates that yield stresses in both tension and compression have approximately the same magnitude. This relationship probably can be used to explain the behavior of metals in experiments involving initiation of plasticity in a metal rod by extension (Figure 6–10), but it does not adequately describe the plastic behavior of rocks.

Another model is based on the *Von Mises criterion,* which is obtained by taking the maximum deviatoric stress value as a constant. This constant is assumed to be related to the strain energy of distortion of the material and is commonly expressed as

$$(\sigma_1 - \sigma_2)^2 + (\sigma_2 - \sigma_3)^2 + (\sigma_3 - \sigma_1)^2 = 6C_y \quad \textbf{(6–9)}$$

where C_y is the yield constant. The Von Mises criterion was also intended to explain the initiation of plastic behavior in metals and does explain some plastic deformation in metals, but not in rocks.

CONTROLLING FACTORS

Many factors affect and control the behavior of rock materials, including composition, texture, temperature, total pressure, fluid pressure, rate of stress increase and strain rate, and character and spacing of anisotropies such as bedding or foliation. All affect the ways in which rocks deform, but particular variables—including temperature, composition, strain rate, pressure, and fluid content—have a greater effect than others. Given proper conditions, any of these variables may play a major role in the behavior of a rock mass being deformed. For example, most rocks deform elastically at low temperature, but will deform ductilely at high temperature and low strain rate. At high strain rate, the same

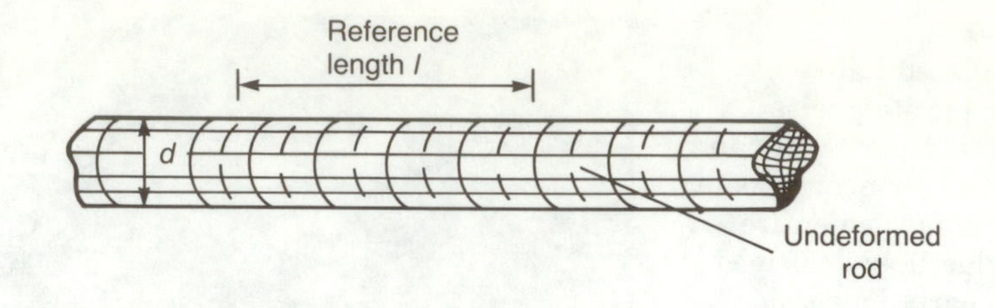

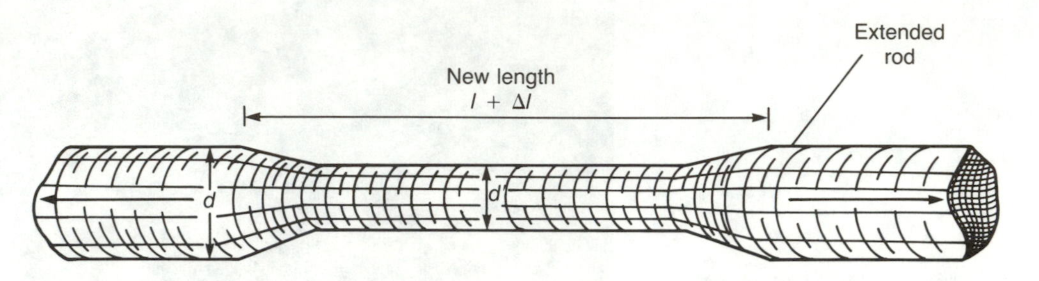

FIGURE 6–10
Extensional ductile necking of a metal rod.

rocks may deform brittlely. Increased fluid pressure may produce ductile behavior at lower temperatures than such behavior would be produced in a dry rock mass.

BEHAVIOR OF CRUSTAL ROCKS

The ideal-material behavior models discussed so far approximate the behavior of rock materials in the Earth. Some geologic materials exhibit almost ideal-model behavior (Figure 6–2). Thin sheets of muscovite and biotite exhibit elastic behavior as they are flexed and spring back to their original shapes. Muscovite and biotite sheets also have an elastic limit that can be exceeded. Some fluids upon and within the Earth approximate ideal viscous behavior, and we have abundant evidence for occurrence of nonideal plastic (or ductile) deformation. Transitional behavior—such as viscoelastic or elastic-plastic—is also prevalent at proper temperature and pressure.

Ideal-behavior models (Figure 6–1) are defined in large part by the shapes of stress-strain or strain-time curves (or both). Further discussion of deformation mechanisms in Chapter 7 will help in understanding these modes of behavior in rocks.

Ductile-Brittle Transition

Brittle behavior generally dominates in the upper crust, with elasticity providing rigidity and permitting faulting and jointing to occur at the elastic limit. A transition zone occurs in the lithosphere where brittle behavior is inhibited because softening occurs as a result of increases in temperature and pressure with depth. This is the ***ductile-brittle transition*** (DBT) or, properly, the plastic-brittle transition (Figure 6–11), according to Rutter (1986). Direct evidence of ductile behavior in crustal rocks is shown by particular types of folds that could form only in the ductile realm. Laboratory studies of minerals in these folded rocks indicate that temperatures of several hundred degrees Celsius and pressures of several kilobars were required to form them deep in the crust.

The actual depth to the DBT in the lithosphere is determined by the thermal gradient (rate of increase of temperature with depth determined by the amount of radioactive heat-producing elements U, Th, and K), intrusion of magma into the crust, and mantle heat flux. It is also controlled by the quantity of fluid present and other variables that determine pressure gradients. In tectonically inactive parts of the continents such as the eastern United States, where thermal gradients are very low (15° to 25°C/km), depth to the DBT may be as much as 15 km, but in areas of high heat flow (30° to 40°C/km) in the continents the depth to the DBT may range from 8 to 12 km. Depth to the DBT may be estimated from the maximum focal depth of most earthquakes in an area. In the eastern United States, most earthquakes occur at depths of less than 15 km; in the Great Basin in Utah and Nevada, where the thermal gradient is steeper, the foci of most earthquakes are within a few kilometers of the surface.

Large active faults that break the entire crust, like the San Andreas in California, unquestionably exhibit brittle behavior in the upper crust but probably involve movement in the fault zones by ductile flow below the DBT (Figure 6–11). The large faults also exhibit a kind of elastic-plastic behavior—the

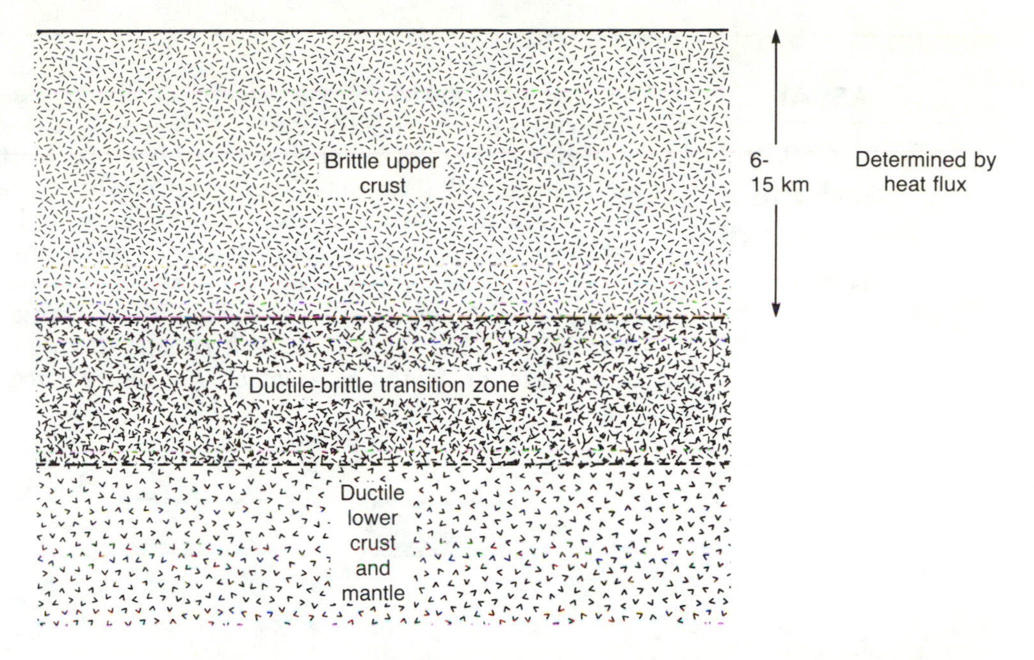

FIGURE 6–11
Ductile-brittle transition in the lithosphere. The depth to the transition is determined by the amount of heat produced in that part of the lithosphere, the nature and amount of fluid present, and the pressure variables.

upper crustal segments exhibit pure elastic behavior, and the lower crustal segments exhibit pure plastic behavior—but segments of the fault within the DBT (and sometimes below it) may exhibit elastic-plastic behavior (Figures 6–7 and 6–8). High strain rates may force the behavior back into the elastic realm, but rapid strain rates at high confining pressure may permit plastic behavior along the fault.

Natural models for the DBT occur in glaciers, salt, and unconsolidated sediment. A definable brittle zone occurs in the upper parts of a glacier where crevasses and other brittle fractures form. The fractures close at depths rarely exceeding 60 m, giving way to a zone of ductile flow (Figure 6–12). Direct observations in the ductile zone have been made by excavating a shaft through the upper brittle zone into the lower ductile deformation zone. Observers can enter the shaft and make observations because ductile flow closes the shaft at a very slow rate. Salt structures afford similar opportunities to observe ductile behavior occurring near the surface in natural materials (Chapter 2). The unconfined parts of the salt yield brittlely; the confined parts yield ductilely (Jackson and Talbot, 1986). Other analogies of ductile or viscous behavior exist where folds have formed in slumped water-saturated sediments (Stone, 1976) and in glacial silts deformed by slumping or ice movement (Figure 2–2a; Stone and Koteff, 1979).

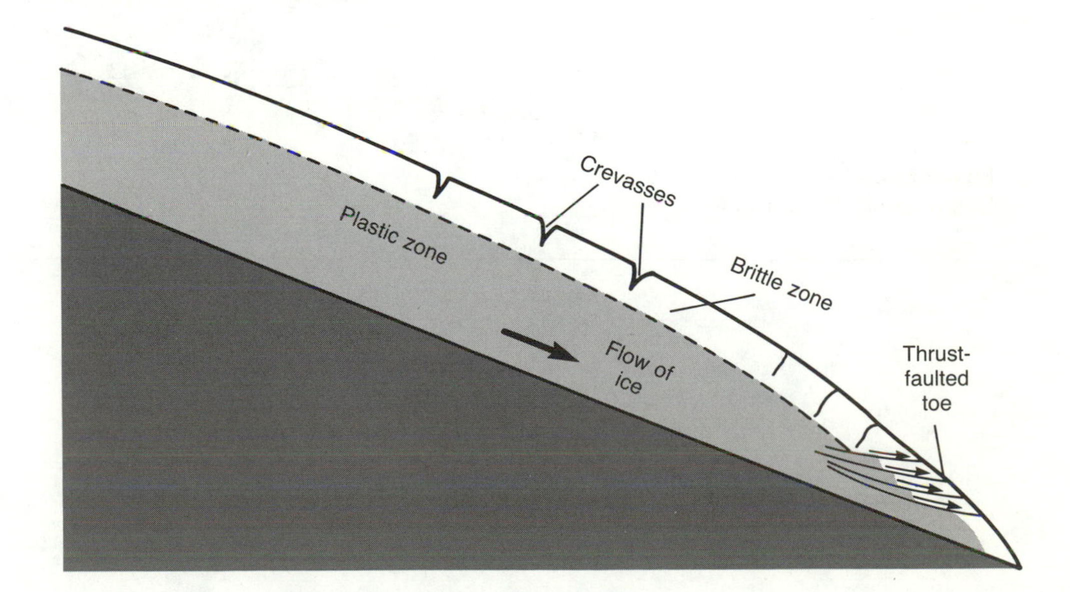

FIGURE 6–12
Cross section of a glacier showing deformation zones.

ESSAY

Silly Putty® and the Behavior of Mantle Rocks

An everyday model for viscoelastic or elastic-plastic behavior is a silicone material known as Silly Putty. It behaves as a ductile (probably viscous) material if it is deformed slowly and exhibits brittle elastic behavior if it is strained rapidly (Figure 6E–1). Pulled slowly, it stretches; hit with a hammer, it shatters. Crustal materials, except magma, probably do not exhibit viscous behavior, except perhaps over geologically long periods of time, but we can argue that viscous behavior can serve as a model for flow in the mantle (Chapter 1). A phenomenon in the mantle confounds any attempt to model the mantle as a homogeneous ideal viscous (or plastic) material: earthquakes occur here, at depths as great as 700 km (Figure 6E–2). Most geoscientists

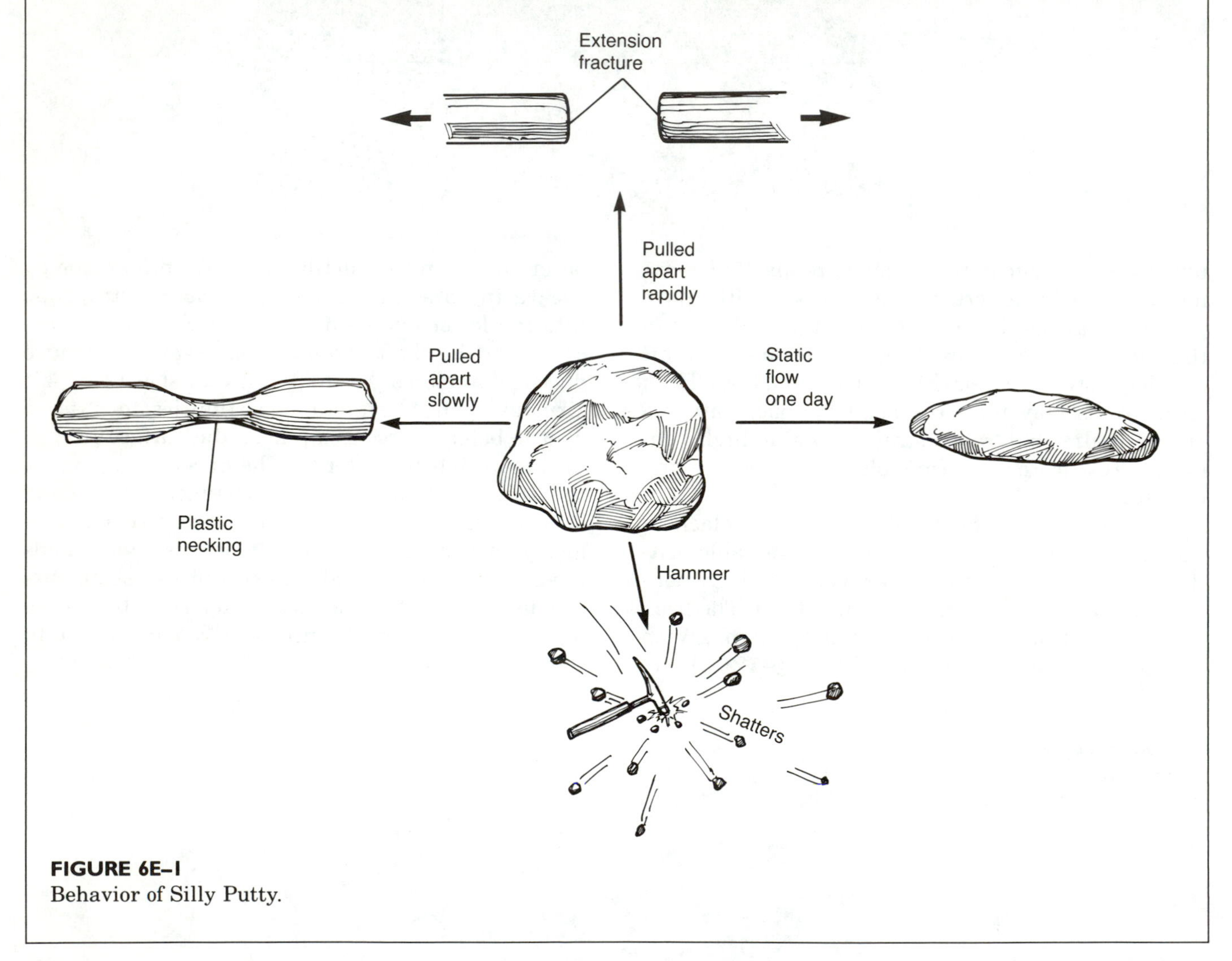

FIGURE 6E–1
Behavior of Silly Putty.

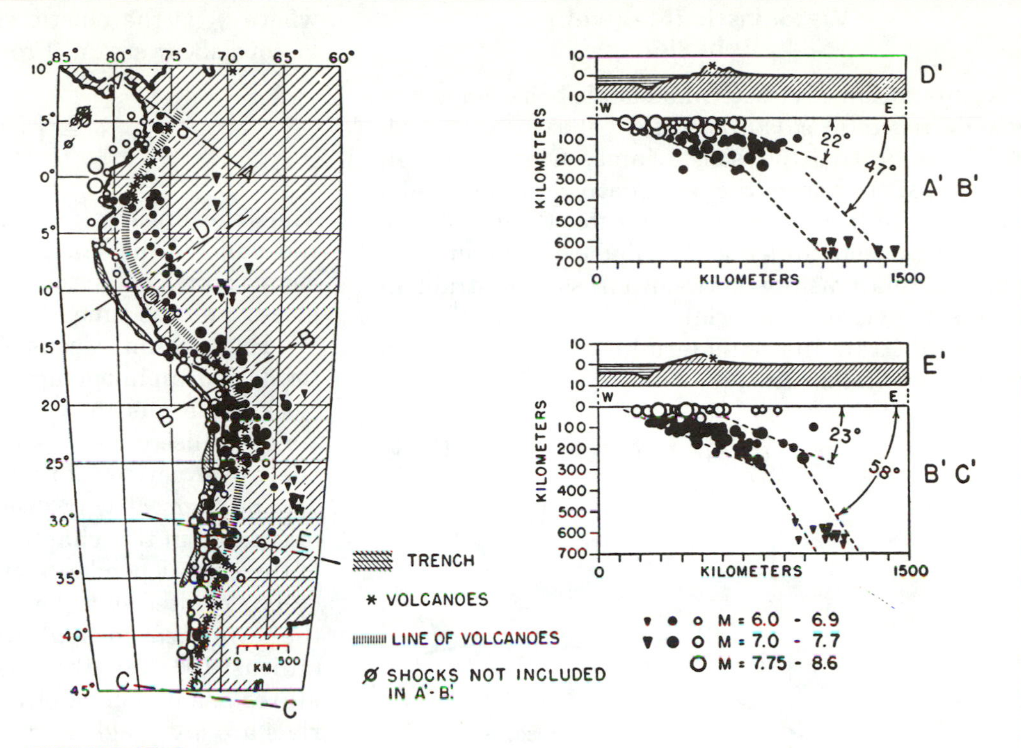

FIGURE 6E–2
Depth distribution of earthquakes beneath western South America. Note that few occur at depths greater than 250 km. (From Hugo Benioff, 1954, Geological Society of America *Bulletin.*)

agree that earthquakes involve an elastic-rebound mechanism that produces a suite of surface and body waves—elastic waves that travel on the surface of and inside the Earth. The largest deep-focus earthquakes in the mantle never produce as much energy as the largest shallow-focus earthquakes generated in the upper crust, but the same general mechanism—elastic rebound—must account for their origin. We also know that some of the brittle crust may be subducted into the mantle as cold upper lithosphere. Even if this alternative explanation is correct, generation of earthquakes requires sudden release of elastic strain energy—a rapid strain rate. Ductile flow may be occurring in the mantle much of the time, but rapid accumulation of strain energy may provoke transformation to elastic behavior that produces fractures and deep-focus earthquakes. Thus the behavior of Silly Putty subjected to markedly different strain rates may be a useful comparative model for understanding the behavior of mantle rocks and also help answer a question: Why do earthquakes occur there at all?

Viscoelastic (Maxwell) Behavior

A combination of viscous and elastic behavior called ***viscoelastic*** (or *elasticoviscous*) behavior was studied by a Scottish physicist, James Clerk Maxwell (1831–1879)—hence the alternative name; it depends on both stress and strain rates (Figure 6–13). Thus, equations 6–1 and 6–7 must be combined to express relationships between stress and strain in an elasticoviscous material.

Total strain in the material is given by the sum of incremental strains,

$$\gamma = \gamma_e + \gamma_v \tag{6–10}$$

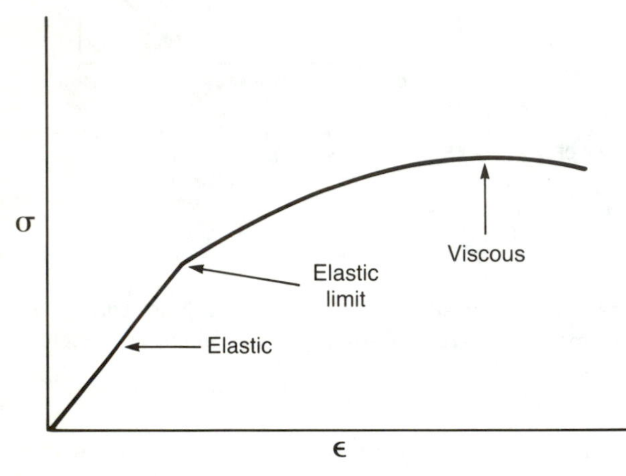

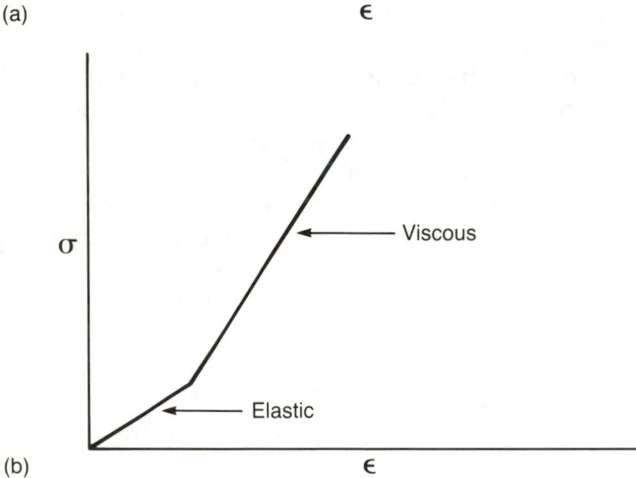

FIGURE 6–13
Stress-strain (a) and strain-time (b) diagrams for a viscoelastic material. The same problems exist here with stress-strain curves as for ideal viscous materials (Figure 6–6).

where γ_e is the elastic shear strain and γ_v is the viscous shear strain. From equations 6–1 and 6–6,

$$\dot{\gamma} = \left(\frac{\tau}{G}\right) + \left(\frac{\tau t}{\eta}\right) \tag{6–11}$$

Thus the linear viscoelastic strain rate ($\dot{\gamma}$) equals the sum of the elastic strain and viscous strain components. G is the shear modulus, τ is shear stress, and t is time.

Here we must consider time-dependence of deformation, an important factor in geologic processes. Rock materials that deform brittlely over a short time (milliseconds to a few years) will deform ductilely over periods of thousands to millions of years. The problem of why deep-focus earthquakes occur (discussed in this chapter Essay), is a good example of the time-dependence of geologic processes. Carey (1953) pointed out the relationship and suggested the term *rheid* for materials that deform like fluids (at least 10^3 times the elastic component) at temperatures below their melting points. Most rock materials are *not* rheids under strains of short duration, but more *are* rheids if deformation occurs over a geologically long time.

STRAIN PARTITIONING

Behavior of rocks may vary in space as well as in time; the separation into different mechanisms in the same rock mass is called ***strain partitioning.*** This phenomenon may be related to differences in flow rate in a ductilely deforming mass (Lister and Williams, 1983), the change in behavior type resulting from different physical properties or (occasionally) conditions (Figure 6–14). Plastic strain may be localized in narrow ductile deformation zones along deep crustal faults, leaving adjacent rocks mostly unaffected.

Different strains are affected by the bulk properties of the rocks being deformed. Relatively weak rocks, like shale, salt, and schist, commonly exhibit styles of deformation that contrast with those of stronger rocks, like sandstone, gneiss, and amphibolite, in the same deforming mass. Layers of different thickness in the same rock type may also cause partitioning of mechanical behavior. Shapes and wavelengths of folds are strongly influenced by layer thickness (Chapter 15). For example, folding of thinly layered sandstone/shale layers in a sequence of massively bedded sandstone will result in strong contrasts in mechanical behavior. In massive layers, folds of longer wavelength and less curvature will be produced; in the thinly bedded weaker layers, folds will have very short wavelengths (Figure 6–15). This may be explained by differences in the original anisotropic character of the sequence or by tectonically induced anisotropy (such as fractures and foliations) in the rock mass (Latham, 1985). Other factors, including local variations in fluid pressure, confining pressure, temperature, and strain rate, may create conditions under which strain will be partitioned in different parts of a rock mass.

Now the principles of mechanical behavior of rocks should be coming clear—not only in ideal materials but as strain occurs in the crustal rocks. We can turn to a discussion of the theory and observation of microstructures in rocks.

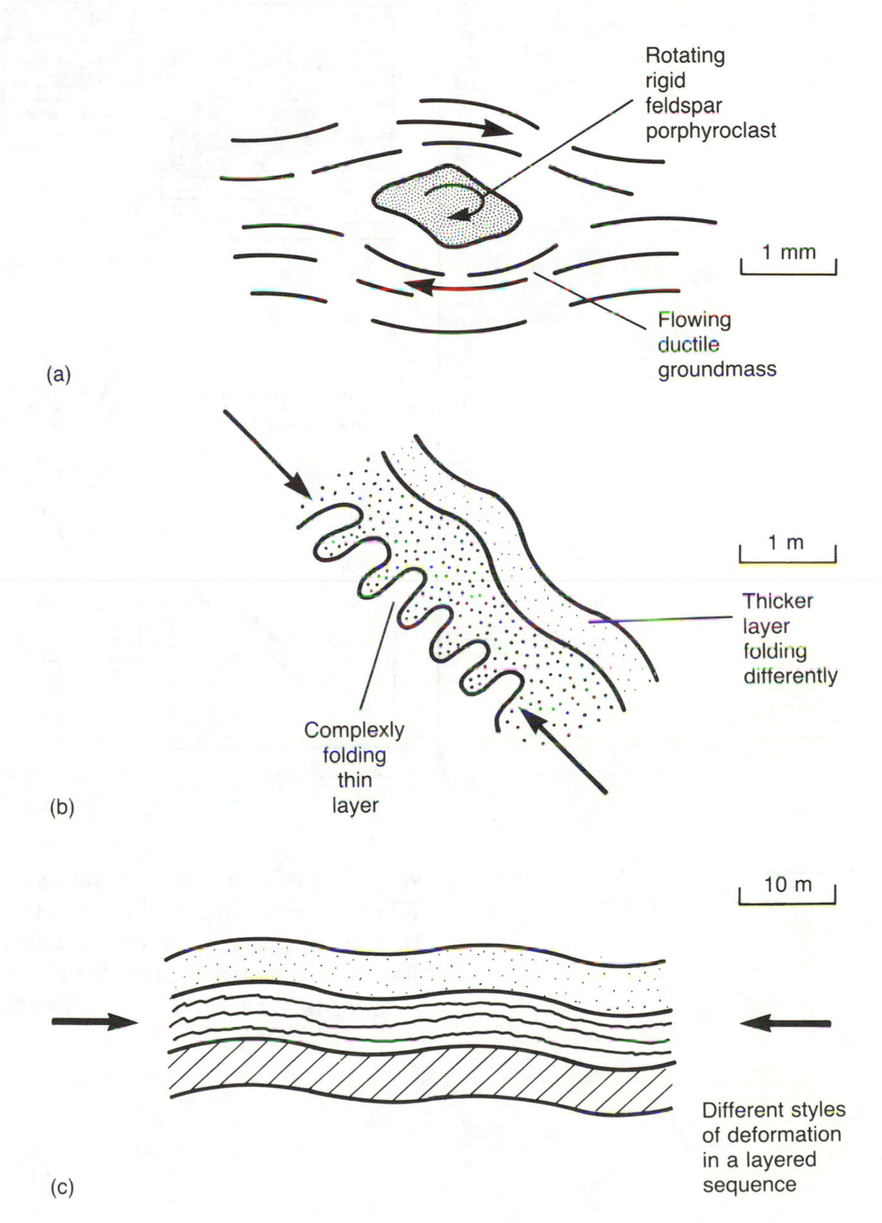

FIGURE 6–14
Strain partitioning at different scales and involving different variables of rigidity, layer thickness, and other properties. (a) Microscale deformation involving a strong object in a weak groundmass. (b) Outcrop-scale folding of layers with differing strength in a ductile groundmass. (c) Road-cut exposure of a sedimentary sequence of several rock types.

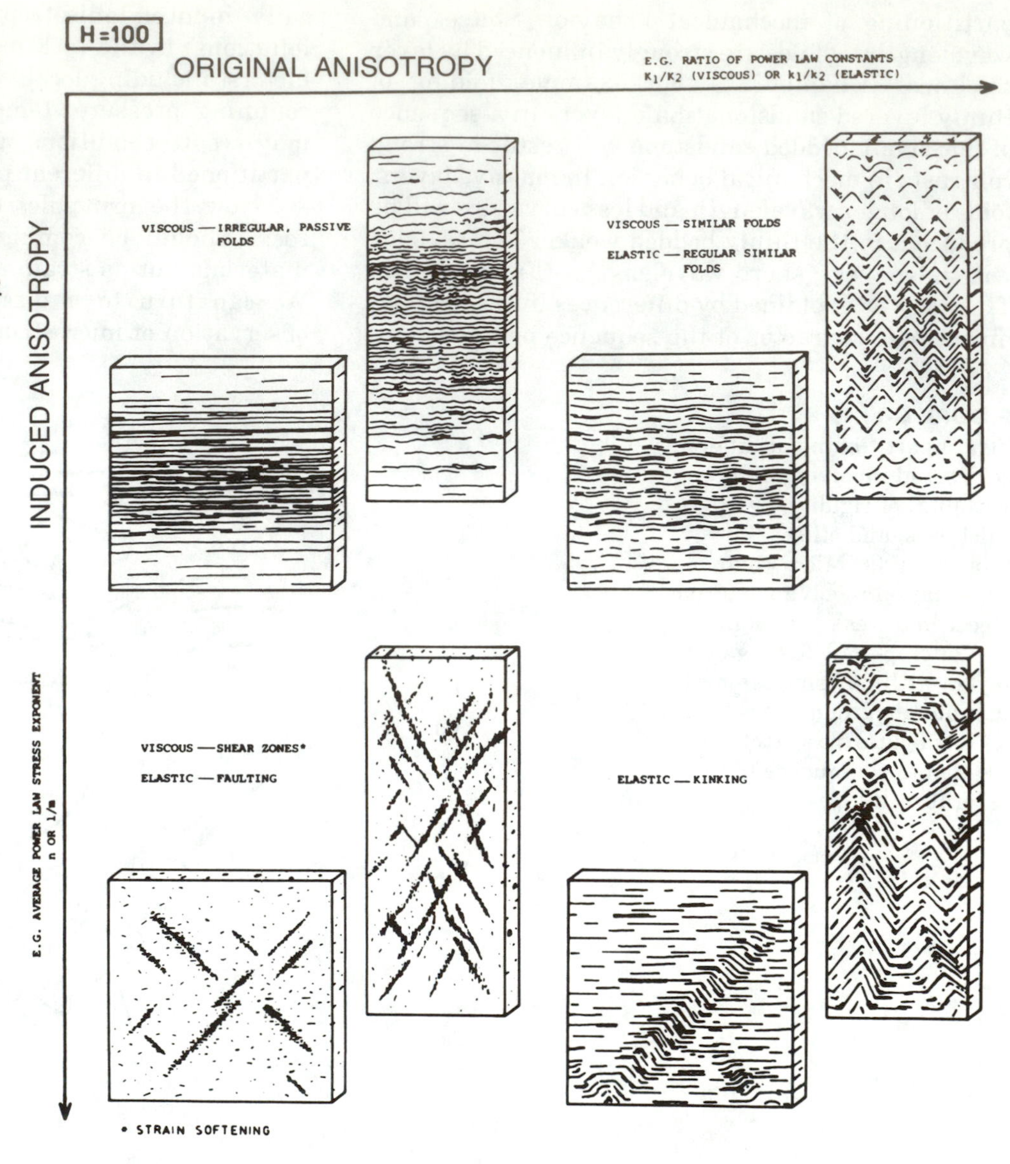

FIGURE 6–15
Relationships between original (intrinsic) anisotropy and tectonically induced anisotropy, producing different kinds of structures in rocks—and an opportunity for strain partitioning. (Reprinted with permission from *Journal of Structural Geology*, v. 7, J. P. Latham, A numerical investigation and geological discussion of the relationship between folding, kinking and faulting. © 1985, Pergamon Journals Ltd.)

Questions

1. Why do most rock materials exhibit nearly ideal behavior?
2. What factors control the depth to the ductile-brittle transition in the crust? How do these factors accomplish this?
3. The stress-strain data set tabulated here is from a loading experiment made by adding successive weights to the end of a 1-m length of steel wire 3 mm

σ (kg)	Δl	σ (kg)	Δl
0.00	0.00	3.84	4.63
0.63	0.15	4.11	5.30
1.23	0.24	4.37	6.23
1.88	0.33	4.40	7.34
2.45	0.60	4.36	8.08
2.79	0.47	4.17	8.70
2.39	2.45	3.88	9.17
3.17	3.33	3.51	9.49
3.55	4.00		

in diameter, producing an elongation strain (Δl). Using the data, calculate the stresses in MPa, plot the data on stress-strain axes, and interpret the kinds of behavior exhibited by the wire throughout the experiment. Note that the load needed to produce the next increment of strain may actually decrease at some stages. Calculate Young's modulus for the wire from the elastic region of the curve.

4. Poisson's ratios for most natural and synthetic materials are positive. Predict the properties a material must have to exhibit a negative Poisson's ratio. See Roderic Lakes, 1987, *Science,* v. 235, p. 1038–1040, for a discussion of synthetic materials with negative Poisson's ratios.
5. Why do so many rock cylinders break along planar shear planes in uniaxial experiments although the shear surfaces in a cylinder compressed from the ends are expected to be conical? (Think about the way the experiments are designed, with $\boldsymbol{\sigma}_1 \neq \boldsymbol{\sigma}_2 = \boldsymbol{\sigma}_3$. Are the principal stresses really so if planar shear surfaces form?)

Further Reading

Donath, F. A., 1970, Some information squeezed out of rock: American Scientist, v. 58, p. 54–72.

Relationships between experimental rock mechanics and structures we commonly observe in the field are presented, clearly, at the level of elementary structural geology. Laboratory techniques are explained in a way that clarifies the operation of rock-mechanics experiments, the construction of stress-strain curves from experimental data, and the formation of many common structures.

Jaeger, J. C., and Cook, N. G. W., 1976, Fundamentals of rock mechanics: London, Chapman and Hall, 585 p.

An excellent summary of the mechanical properties of rock materials presented somewhat as seen by an engineer. It contains numerous practical examples, as well as clear discussions of behavior models for rock materials.

Nicolas, A., and Poirier, J. P., 1976, Crystalline plasticity and solid state flow in metamorphic rocks: New York, John Wiley & Sons, 444 p.

This is a mathematically rigorous treatment of plastic behavior in rocks, but sections may yield some understanding of the plasticity of rocks without resort to the mathematical statements. A feeling will also be gained of the inherent complexity of plastic deformation processes.

Patterson, W. S. B., 1981, The physics of glaciers, 2d ed.: New York, Pergamon Press, 380 p.

Shumskii, P. A., 1964, Principles of structural glaciology, D. Kraus trans. of 1955 ed.: New York, Dover Publications, 497 p.

Both books survey glaciers and glacial ice, and each contains chapters on the ductile and brittle properties of ice. Shumskii includes photographs of glacial folds and other ice structures, the counterparts of which in rocks form only at great depths.

Turcotte, D. L., and Schubert, G., 1982, Geodynamics: Applications of continuum physics to geological problems: New York, John Wiley & Sons, 450 p.

Mathematical treatment of elasticity and viscous behavior, written at a suitable level and with many worked problems.

7

Microstructures and Deformation Mechanisms

Although the presence of dislocations in crystals was first proposed 30 years ago, it was only in the last 15 years that a general realisation of their importance was developed. Today, an understanding of dislocations is essential for all those concerned with the properties of crystalline materials.

DEREK HULL, *Introduction to Dislocations*

OUR DISCUSSION OF IDEAL MECHANICAL DEFORMATION MODELS in Chapter 6 provides a basis for exploring the microstructures and deformation mechanisms we actually observe in rocks (Figure 7–1), as well as some theory of how observed mechanisms operate on the microscopic scale. We now recognize the importance of microscopic-scale mechanisms to a complete understanding of deformation in the crust and mantle: rock deformation depends largely on scale; for example, brittle fracturing on the microscopic scale that produces small displacements of a layer may appear on the map scale as folds produced by ductile deformation.

Theory of dislocations and deformation mechanisms developed in metallurgy, materials science, and solid-state physics enhances our understanding of rock deformation. Several deformation mechanisms found in metals and ceramics have also been observed in rocks. Most metals, however, are isotropic and homogeneous, and many ceramic materials frequently include glass as a major constituent. Additionally, many deformation mechanisms observed in metals and ceramics may not occur in rocks because of anisotropy. As we learn more about deformation from studying simple monomineralic rocks (dunite, limestone, and quartzite) and turn to studying deformation of polymineralic rocks, we should not be much surprised if we discover new deformation mechanisms and document unanticipated behavior.

In this chapter we will see how rocks deform on a microscopic to submicroscopic scale and then learn how microscopic-scale deformation is related to larger structures. When we understand the conditions that produce various dislocations and deformation mechanisms, we will be able to better deduce the conditions that affected an ancient rock mass that was deformed at great depth before erosion exposed it for us to study at the present surface. For example, some fault zones contain crushed and ground-up wall rocks (called *cataclasite*), but another fault zone, quite different, may contain ductilely deformed rocks that were also derived from the walls: the latter contains evidence of flow and reduced grain size (*mylonite*) and may even be recrystallized. What different physical conditions produced fault zones that differ so markedly?

LATTICE DEFECTS AND DISLOCATIONS

We usually think of crystals in terms of groups of spheres in a convenient packing arrangement or as a series of lines whose intersections are occupied by atoms, ions, or groups of atoms such as CO_3 and SO_4. Regular geometric arrangement of atoms, ions, or groups in a compound is a ***crystalline solid,*** with a ***crystal lattice*** making up the internal array (Figure 7–2). We can isolate small parts of crystal

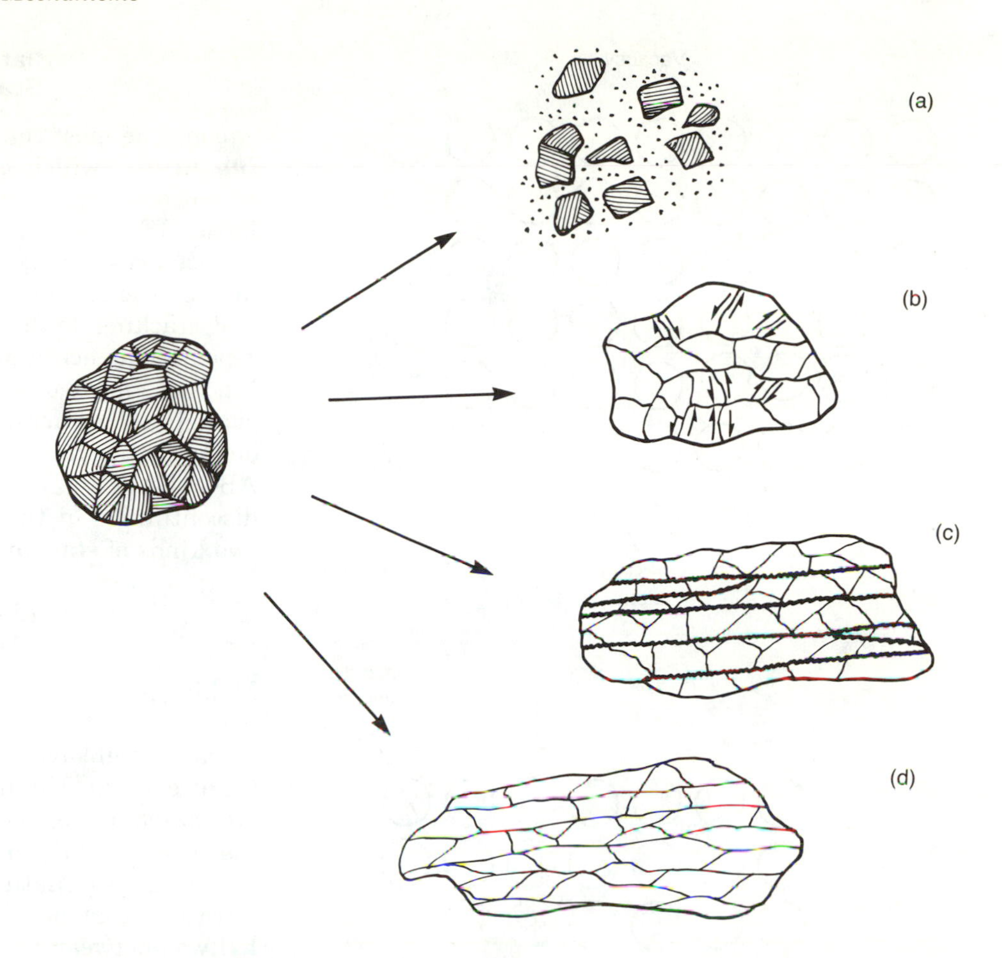

FIGURE 7–1
Common modes of deformation in rocks. (a) Cataclasis, involving granulation and fragmentation of the original rock. (b) Grain-boundary sliding, producing deformation by grains sliding past each other. (c) Pressure solution, forming zones of residues where the rock has been dissolved. (d) Plastic deformation, involving complete recrystallization of the rock, here producing preferred orientation of minerals.

lattices as building blocks called ***unit cells.*** Simple models like that relate some of the processes affecting crystal lattices on the atomic scale to larger structures we observe in minerals and rocks. A crystal lattice where all the sites are filled with the right atoms, ions, or groups, and which contains no *interstitial atoms* between lattice sites, is often called a ***perfect crystal*** and is a stack of identical atoms or ions. Ideally, perfect crystals exist only at 0°K (—273°C), where no thermal or other disturbances can affect the lattice. A halite lattice in which all the Na^+ sites are occupied by Na^+ ions and all the Cl^- sites are occupied by Cl^- ions is a perfect crystal. Such crystals are rare or nonexistent in minerals and rocks; most contain imperfections called ***defects.*** Introducing other ions into the structure (like K^+ into Na^+ sites or Br^- into Cl^- sites in halite), crystallizing lattices or finding natural lattices with ***vacancies*** (unoccupied sites), changing the stacking order of layers, or otherwise distorting the lattice—any of those produces defects. Several kinds of defects have been described and categorized as *point, line*—***dislocations***—*surface* (*planar*), and *volume* defects. All locally disturb the otherwise regular arrangement in a crystal lattice.

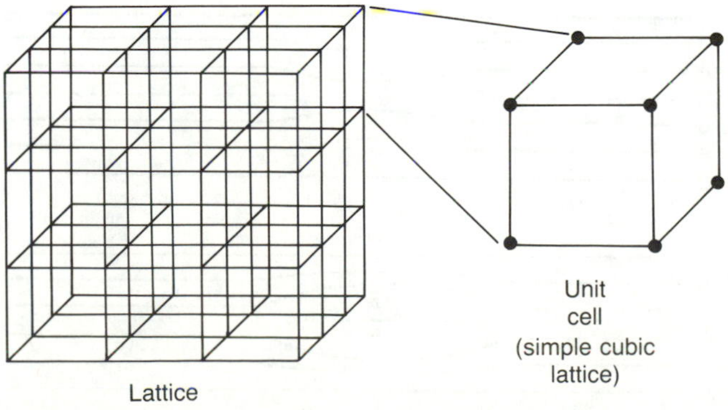

In the last few decades, defects in crystal lattices have profoundly influenced the way we live. Mass-produced semiconductors made of high-purity silicon crystals manufactured (or "doped") with carefully added compositional defects (like phosphorus or boron) have led to widespread use of transistors, microprocessor chips, and other components of computers, telephones, and other electronic systems. In the natural world, knowledge of lattice defects helps us understand the varied ways rocks deform on microscopic and larger scales.

Lattice

Unit cell (simple cubic lattice)

FIGURE 7–2
Crystal lattice (without atoms) and simple cubic unit cell.

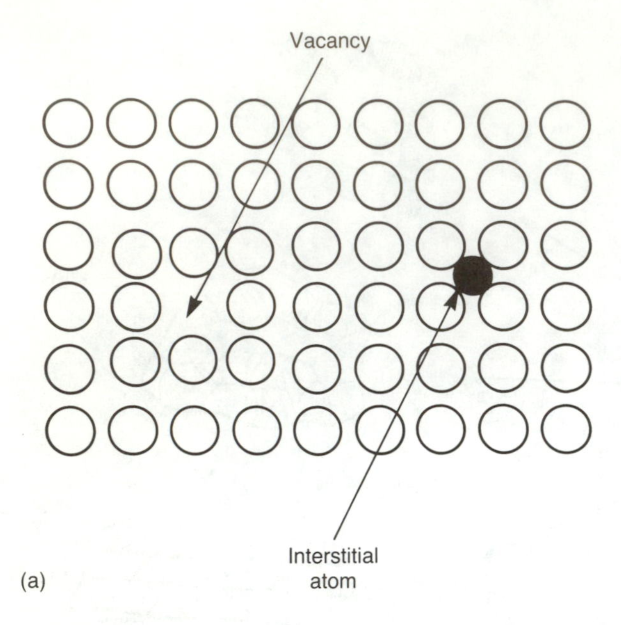

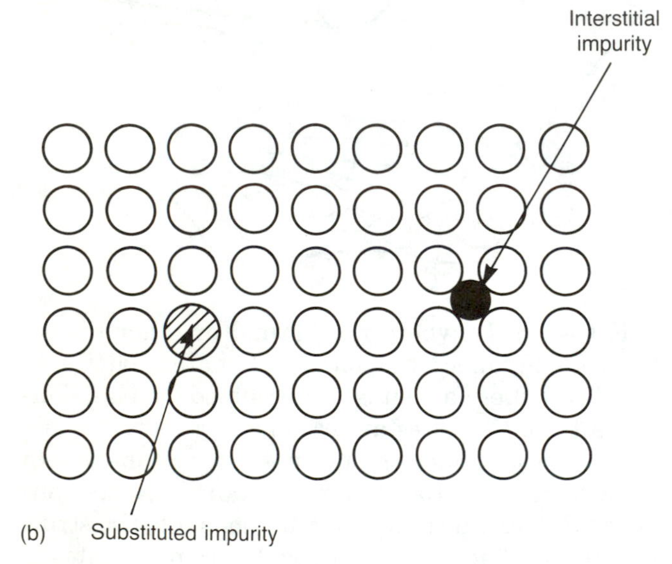

FIGURE 7–3
(a) Vacancies and interstitial atoms. (b) Impuritiesin crystal lattices.

Now we will examine lattice defects in more detail.

Point Defects

Substitution or *interstitial* defects involving a foreign atom or ion in a lattice site, or a normal atom out of its proper place, are introduced while a lattice is forming. Vacancies in a lattice may also be produced during ductile deformation, by irradiation by high-energy particles, and by sudden cooling of a crystal from high temperature (Figure 7–3). An equilibrium concentration of point defects will always occur at any temperature above absolute zero.

Planar Defects—Stacking Faults

Among the most common planar defects are ***stacking faults,*** which consist of irregularities in the repeat order in a series of layers in a close-packed lattice (Figure 7–4). If a lattice has a repeat sequence between layers of ABAB (only two kinds of layers), the sequence can alternate in only one way and stacking faults will not occur. If the normal repeat sequence among layers is ABCABC (three kinds of repeat units), possibilities exist for locally changing the stacking order. If part of a layer is left out so that part of the lattice has the stacking order ABCABABC . . . or ABCABACABC . . . , a local discontinuity in the lattice results. These are the two kinds of stacking faults (Figure 7–4).

Line Defects—Dislocations

Making an incision in a hypothetical crystal lattice, partly separating two layers, and then inserting an extra partial layer produces an ***edge dislocation*** (Figure 7–5a). The inserted layer is called an *extra half plane*. In real crystals it is attained by subjecting a lattice to simple shear (which is frequently encountered in nature), leading to distortion of the lattice so that part of a layer of atoms is isolated halfway between two normal layers. The edge dislocation begins as a small area and propagates outward as the lattice is progressively deformed by simple shear (Figure 7–5b).

The term line or edge dislocation is derived from the location of the edge of the extra layer separating the dislocation from the undisturbed lattice. Edge dislocations are designated $\perp$ if the dislocation is positive (the extra half plane lies above the reference line) and $\top$ if the dislocation is negative (the extra half plane lies below the reference line).

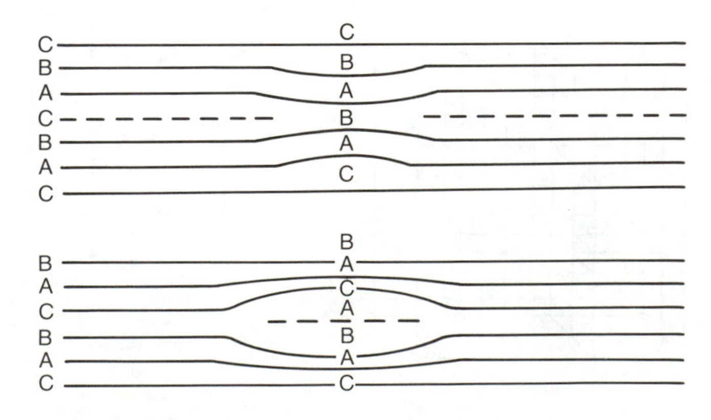

FIGURE 7–4
Two kinds of stacking faults.

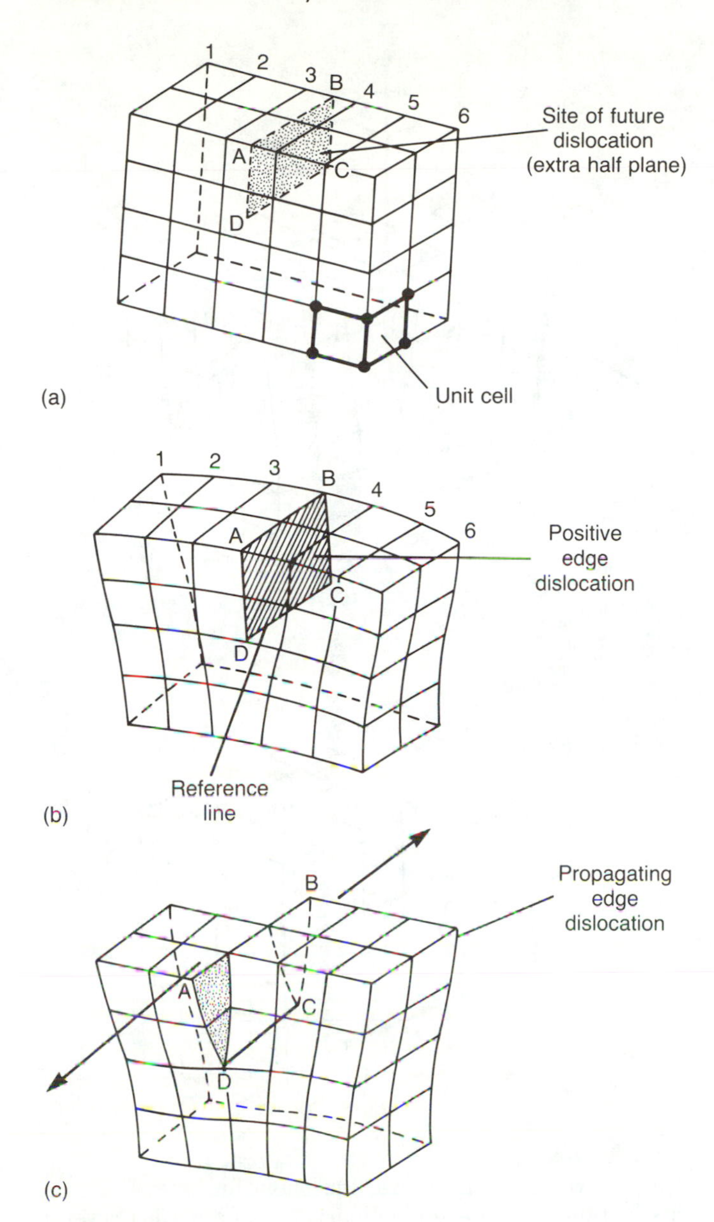

FIGURE 7–5
(a) Lattice before deformation showing the site of the future dislocation and a unit cell. (b) Positive edge dislocation. (c) Propagation of an edge dislocation by simple shear.

Another type, called a ***screw dislocation,*** involves helical rotation within the dislocation plane. The net result of a 360° rotation is that the dislocation moves down one lattice spacing to the next layer (Figure 7–6). Screw dislocations are *right-handed* if the sense of rotation is clockwise (as seen looking down the screw axis) and *left-handed* if the rotation is anticlockwise.

A convenient means of describing either edge or screw dislocations is by ***Burgers vector*** and ***Burgers circuit.*** Imagine a layer in a perfect crystal where a loop traverse is made from one site in the layer so that it returns to that site (Figure 7–7a). In a perfect crystal, the traverse should close precisely at the starting point. If the traverse crosses a dislocation, it will not close in the same distance as in the perfect crystal. The traverse is known as a Burgers circuit, and the magnitude of the failure of the traverse to close in the same loop that crosses a dislocation is a Burgers vector, **b** (Figure 7–7b). The Burgers vector of an edge dislocation is normal to the line of dislocation; that of a screw dislocation (Figure 7–8) is parallel to the line of the dislocation (Hull, 1975). Most dislocation lines are located at an angle to the Burgers vector, giving them a mixed character, with an edge component and also a screw component.

Dislocation lines can end at grain boundaries, but never within a crystal. They must either join (or branch into) other dislocations or form closed loops in the same crystal. Three or more dislocations meet at a point called a *node,* and the sum of the Burgers vectors for each dislocation is zero (Figure 7–9). Generalized,

$$\mathbf{b}_1 + \mathbf{b}_2 + \mathbf{b}_3 + \ldots = \sum_{1}^{n} \mathbf{b}_n = 0 \qquad (7\text{–}1)$$

Dislocations in minerals may be observed under the microscope by studying etched surfaces of crystals (Figure 7–10). They were actually observed—but not understood—during the nineteenth century

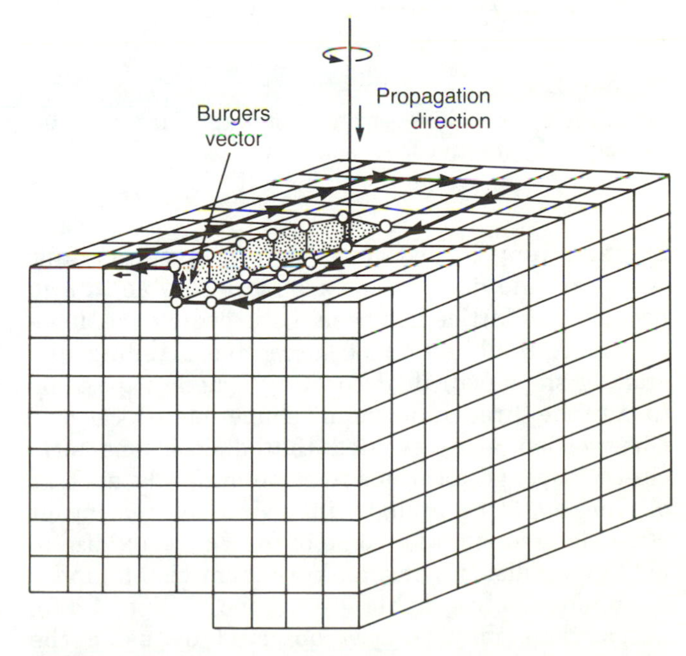

FIGURE 7–6
Left-handed screw dislocation, propagating downward.

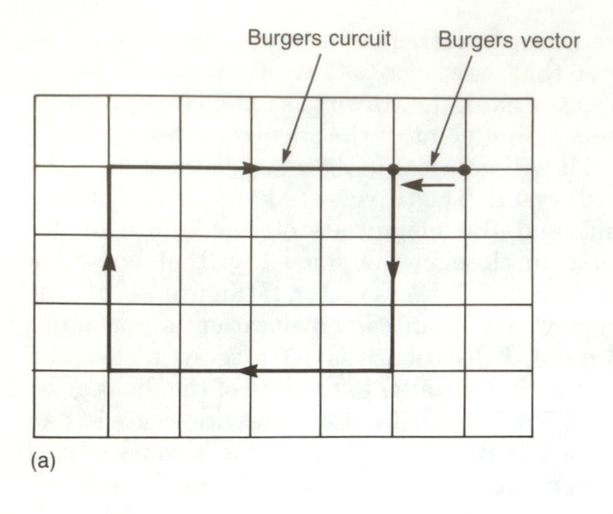

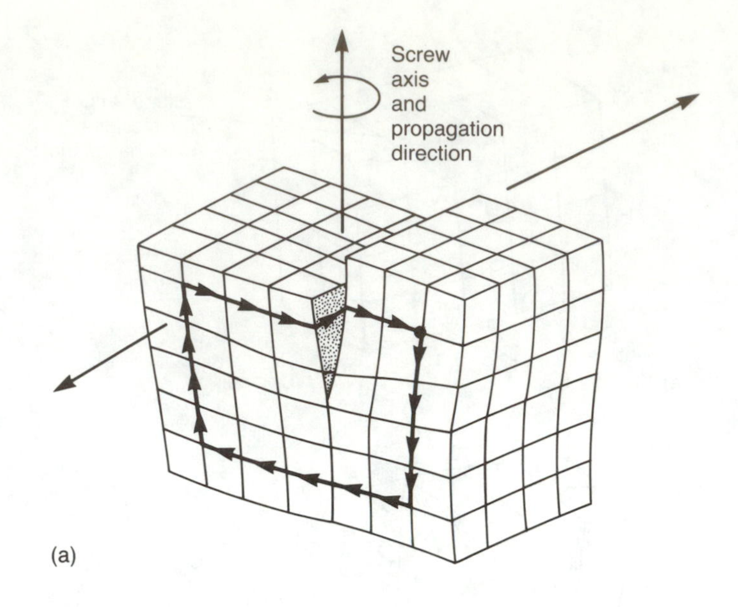

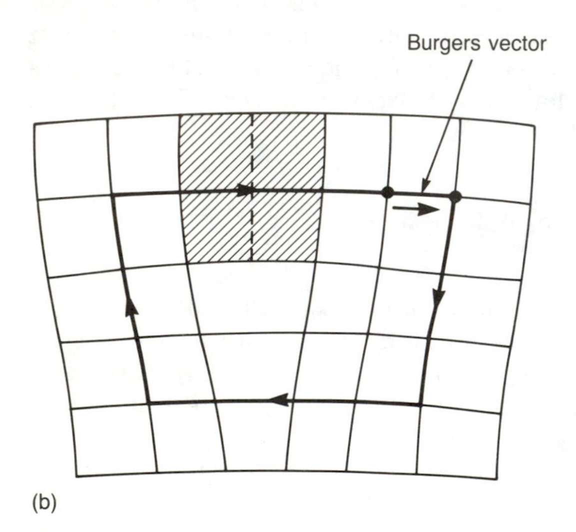

FIGURE 7–7
(a) Burgers vector in a perfect crystal. (b) Burgers vector in a crystal with an edge dislocation.

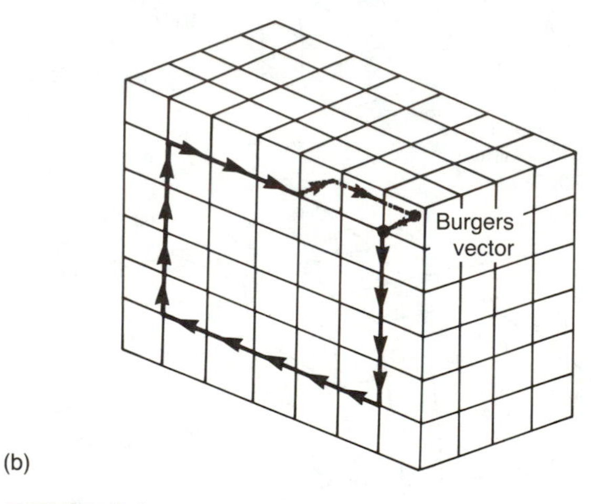

FIGURE 7–8
Burgers vector for an upward-propagating screw dislocation (a), and—for comparison—in a perfect crystal (b).

by examining etched surfaces of cleaved crystals. Areas around dislocations are more soluble in acid because the lattice has been disturbed by the dislocation. A combination of successive grinding and etching steps (called "decorating") made it possible to map the three-dimensional shapes of dislocations. Observation also revealed that dislocations form closed loops. Dislocations in some minerals, such as olivine, could be brought into view by heating at 900°C long enough for some of the Fe^{II} to oxidize to Fe^{III} along dislocations, making them visible under a standard petrographic microscope (Suppe, 1985). Dislocations may be best observed by using the transmitted electron microscope because of its greater magnification.

Measuring ***dislocation density*** in a series of grains indicates the degree of deformation of the crystal lattice. Sometimes it is useful to construct *dislocation maps* of distributions and kinds of dislocations in crystalline solids (Figure 7–11). Saturation limit can be reached in most lattices; then dislocations from several different slip systems or planes become entangled and that part of the lattice will accept no more dislocations, but dislocations may be thermally annealed out of a lattice by nucleation of a new and unstrained lattice. Annealing is accomplished by the ***dislocation glide and climb*** mechanism, in which material diffuses away from the deformed area and out of the extra half plane of the dislocation from one slip plane to the

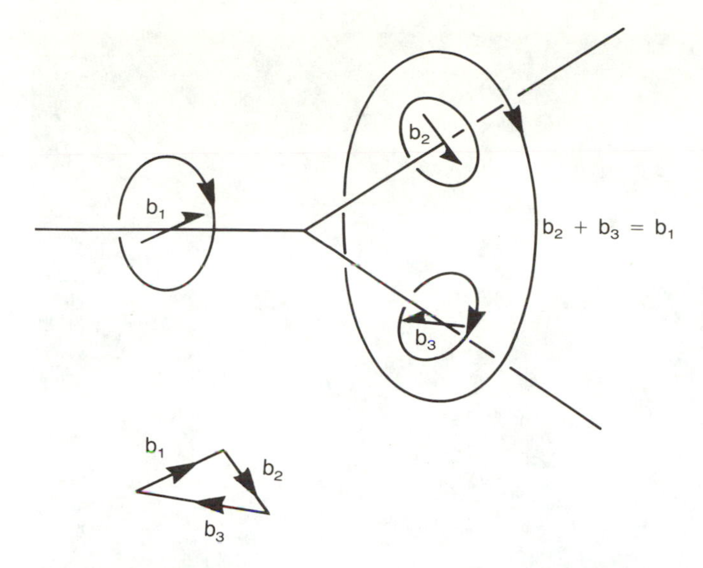

FIGURE 7–9
Several oppositely oriented Burgers vectors meet at a node and sum to zero. (Reprinted with permission from Derek Hull and D. J. Bacon, *Introduction to Dislocations,* 2d ed., © 1984, Pergamon Books Ltd.)

FIGURE 7–10
Dislocations in experimentally deformed synthetic quartz under the transmitted electron microscope. Width of field is approximately 2 μm. Note the concentration (pinning) of some dislocations around fluid inclusions. (John W. McCormick, SUNY College at Plattsburgh.)

next. (Such recrystallization mechanisms are discussed in the next section.) *Tangles* and *pileups* of dislocations at impurities in the lattice may prevent dislocation glide and climb from completely annealing the lattice.

DEFORMATION MECHANISMS

Factors that determine which deformation mechanism will dominate are temperature, total stress, fluid pressure, differential stress, composition, grain size, texture, and strain rate. Deformation mechanisms (Figure 7–1) will be discussed here in order of increasing temperature.

Cataclasis

Brittle deformation is concentrated in a rock mass along surfaces of movement and no cohesion—surfaces that separate undeformed regions—producing the rock *cataclasite*. This type of deformation, called ***cataclasis,*** is characterized by megascopic breccias, gouge, fractures and faults, microbreccias, and microfractures (Figure 7–12). Cataclasis consists of brittle granulation of rock at low temperature and low to moderate confining pressure, but may be accelerated by high fluid pressure. Cataclasis occurs mostly at low temperature within a few kilometers of the surface, but feldspars, garnets, and a few other minerals undergo cataclasis at high temperature. Where strain rate is high, failure occurs under brittle conditions by exceeding the elastic limit of the material. Under these conditions feldspars and garnets fracture, but quartz, calcite, and some other minerals deform ductilely. Total volume of rock may be reduced by closer packing of grains (that closes larger open spaces) and by fracturing and granulation. Fracturing has also been observed in rocks deformed at moderate to high temperature where the strain rate was very high.

Some map-scale fault zones contain rocks that resemble ductilely formed fault rocks (*mylonite*), but microscopic examination reveals that deformation was brittle and that the rock was granulated to a very fine powder of unaltered grains that flowed as deformation continued. Flow of fine-grained material produced by brittle deformation is called ***cataclastic flow.*** The process may also describe large-scale changes in shape due to small-scale brittle fracturing. Apparent megascopic ductile behavior shown to be brittle on the microscopic scale is called *triboplastic* behavior; it has been observed at shallow depth where rocks are readily ground to fine

FIGURE 7–11
Dislocations piled up at a barrier in experimentally deformed synthetic quartz under the transmitted electron microscope. Width of field is approximately 2 μm. (John W. McCormick, SUNY College at Plattsburgh.)

grain size during motion along a fault (Figure 7–13). Carbonate or other rocks of high mechanical strength on both sides of a fault may produce tribo-plastic cataclasite at low temperature.

Pressure Solution

The phenomenon of ***pressure solution*** was first recognized in the mid-nineteenth century by Henry C. Sorby (1826–1908), the English geologist who pioneered microscopic petrography and made several important contributions to structural geology. Pressure solution consists of dissolution, under stress, of soluble constituents such as calcite or quartz and is generally active at low to moderate temperature in the presence of water. The process is limited by the requirement of a water film between grains. Once the water is lost, pressure solution ceases to occur.

Parts of quartz, calcite, or other soluble mineral grains may be dissolved at points of greatest stress difference, frequently at points of high stress (or stored elastic strain energy) where one crystal touches another. These points of high stress are the most soluble and so dissolution starts there. The remaining surfaces are able to withstand the applied stress, and therefore the rate of pressure solution decreases with distance away from points or zones of high stress (Figure 7–14a). Minerals dissolved by pressure solution (like quartz and calcite) are sometimes precipitated in zones of lower pressure as overgrowths and fibers in pressure shadows, deposited in veins, or remain in solution and are completely removed from the rock mass. Parts of original grains, along with any reprecipitated material that occurs in optical discontinuity with original grains, may survive as evidence of pressure solution (Figure 7–14b). In addition, residues of insoluble materials (such as clays, micas, and opaque minerals) are commonly left behind on pressure-solution surfaces. Pressure-solution surfaces appear to develop best in slightly impure carbonate rocks rather than in pure limestone (Marshak and Engelder, 1985). Fine grain size and impurities nucleate pressure solution more readily than coarser and purer carbonate rocks.

FIGURE 7–12
Cataclastic materials and behavior.

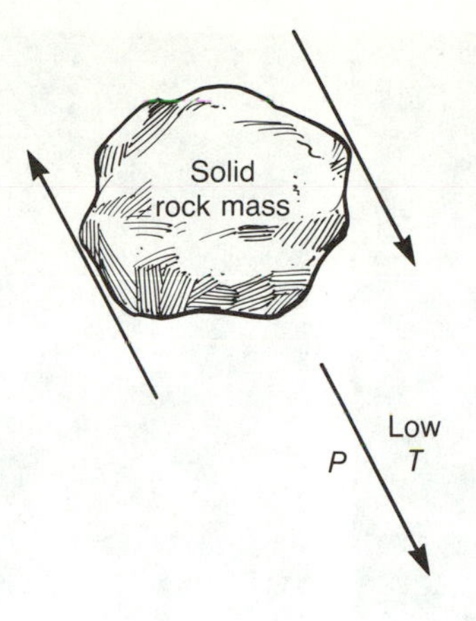

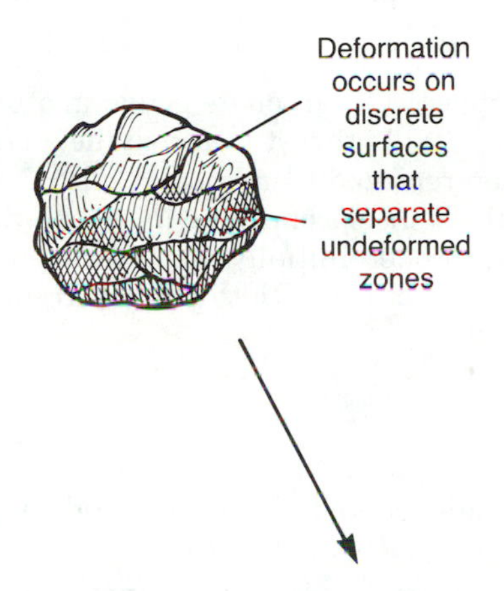

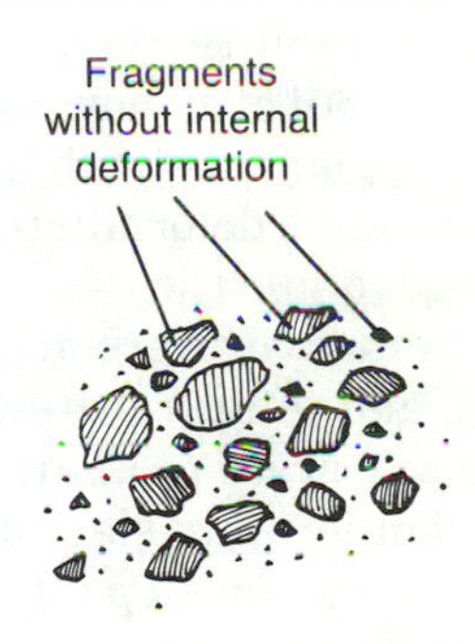

Pressure solution occurs during diagenesis, particularly during compaction and cementation of carbonate sediments where ***stylolites***—irregular surfaces coated with insoluble minerals—may form parallel to bedding (Figure 7–15), indicating that the maximum stress (*overburden pressure*) was vertical. Stylolites also form during deformation of carbonate rocks and sandstones normal to the direction of applied tectonic stress. The dark lines that mark stylolites are insoluble minerals that remain after soluble minerals have been dissolved by pressure solution.

Pressure solution may proceed grain by grain if the rock mass is water-saturated and is sufficiently porous. Most porosity in a rock mass tends to concentrate along bedding and fracture planes; pressure solution begins there and stylolites form along existing surfaces. Preferential dissolution will occur along favorably oriented surfaces, normal to the maximum principal stress.

Pressure solution may occur at high confining pressure and elevated temperature in both carbonate rocks and pelites where carbonate minerals and quartz are dissolved. It can be the dominant process up to 300°C (Rutter, 1976, 1983) and can reduce rock volume by as much as 30 percent if the system is open and dissolved constituents can escape (Kerrich, 1978). Even greater volume loss—on the order of 40

(a)

(b)

FIGURE 7–13
(a) Fine-grained triboplastic(?) carbonate gouge in a limestone breccia from cataclasite along the Saltville thrust in Knoxville, Tennessee. Small dark dolomite fragments were broken repeatedly and healed with white calcite. (b) Photomicrograph of the same specimen as in (a) showing flow banding and stylolites in cataclasite that was broken initially, then healed with veins of white calcite, then broken again by brittle fractures. Width of photograph is approximately 5 mm. Plane light. (RDH photos.)

to 60 percent—has been reported (Wright and Platt, 1982). In the past, it has been hard to understand the transport mechanism of dissolved constituents, given the solubilities of quartz and calcite at surface conditions. But it is now known that during formation of orogenic belts huge volumes of fluid are fluxed through deforming crustal rocks and transport large quantities of dissolved quartz, calcite, metals, and hydrocarbons to regions of decreased pressure and temperature (Oliver, 1986).

In many areas pressure solution is the dominant mechanism forming slaty cleavage. Irregular dissolution surfaces develop into discrete cleavage planes at increased pressure and temperature as deformation progresses. These surfaces form the well-developed slaty cleavage common in zones of low metamorphic grade. Direct evidence of pressure solution may be observed in thin sections as mica or quartz "beards" on larger grains (Figure 7–16), on partial grains (Figure 17–21b), and in zones of insoluble residues developed roughly parallel to each other (Figure 17–19). Pressure solution as a mechanism of cleavage formation will be discussed further in Chapter 17.

Material dissolved in zones of compression is frequently precipitated in veins that fill fractures in zones of extension. Open fractures are commonly the sites for deposition of dissolved constituents, forming veins—particularly where systems are open. Direct correlation generally exists between the composition of veins and the temperature—the metamorphic grade—of the enclosing rocks during deformation, except in strongly mineralized areas. Zeolite or calcite veins (frequently with prehnite) occur at the lowest temperature. Calcite-quartz, and then quartz-epidote and quartz-feldspar, veins occur at progressively higher temperatures. At the upper end of the temperature range, deformation by diffusion processes (to be discussed) dominates in the rock mass, but water still moves and transports materials in solution that may be precipitated in zones of low pressure (Figure 7–17). Composite veins containing fibrous crystals and multiple mineral assemblages may document a history of repeated extensional strain (Chapter 5) and a range of crystallization temperatures.

Translation and Twin Gliding

Deformation may occur in a crystal lattice by slip along an existing crystallographic plane (Figure 7–18). This kind of deformation is called ***translation gliding*** and the specific planes along which slip occurs are called *slip systems*. Slip may occur on the 0001, $10\bar{1}1$, and other planes in quartz (Christie and others, 1964; Carter, 1971); the $10\bar{1}1$, $02\bar{2}1$, and other planes in calcite (Turner and others, 1954); the

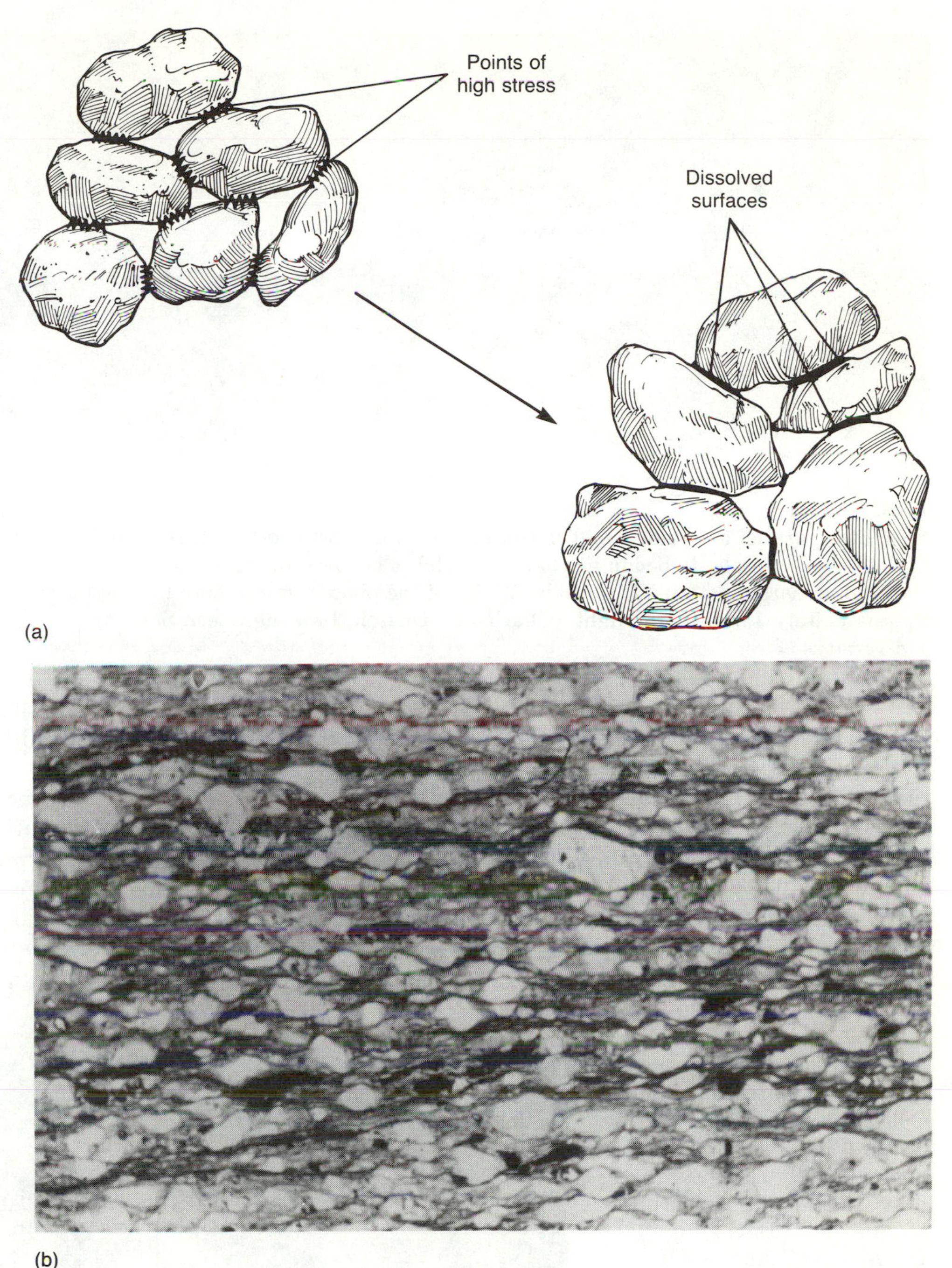

FIGURE 7–14
(a) Grains dissolve first at points of high stress. (b) Partly dissolved grains along slaty cleavage surfaces produced by pressure solution in Martinsburg Slate, Delaware Water Gap, New Jersey. Width of photograph is approximately 2 mm. Plane light. (Thin section courtesy of Timothy L. Davis, University of Tennessee.)

010 plane in plagioclase (Seifert, 1965); the 001 plane in the micas (Etheridge and others, 1973); and the 0kl, 110, and 010 planes in olivine (Raleigh, 1965; Carter and Avé Lallemant, 1970).

Deformation of crystal lattices may also occur by slip of a segment of the lattice along crystallographic planes, producing strain-induced twinning of the lattice—another form of translation, called ***twin gliding*** (Figure 7–19). Translation and twin-gliding mechanisms are each a type of planar dislocation and each is an important deformation mechanism in rocks at low pressure and temperature (subgreenschist facies conditions). Twin gliding may be produced artificially by applying pressure to an Iceland spar calcite crystal by inserting a knife blade parallel to the $01\bar{1}2$ plane of the negative rhombohedron (Figure 7–20). Mechanical twinning also occurs in quartz twinned on the Dauphiné and Brazil systems (McLaren and others, 1967) and in plagioclase twinned on the albite and pericline systems (Borg and Handin, 1966; Borg and Heard, 1970). For a much more complete discussion of mechanical twinning and slip systems, see Hobbs, Means, and Williams (1976).

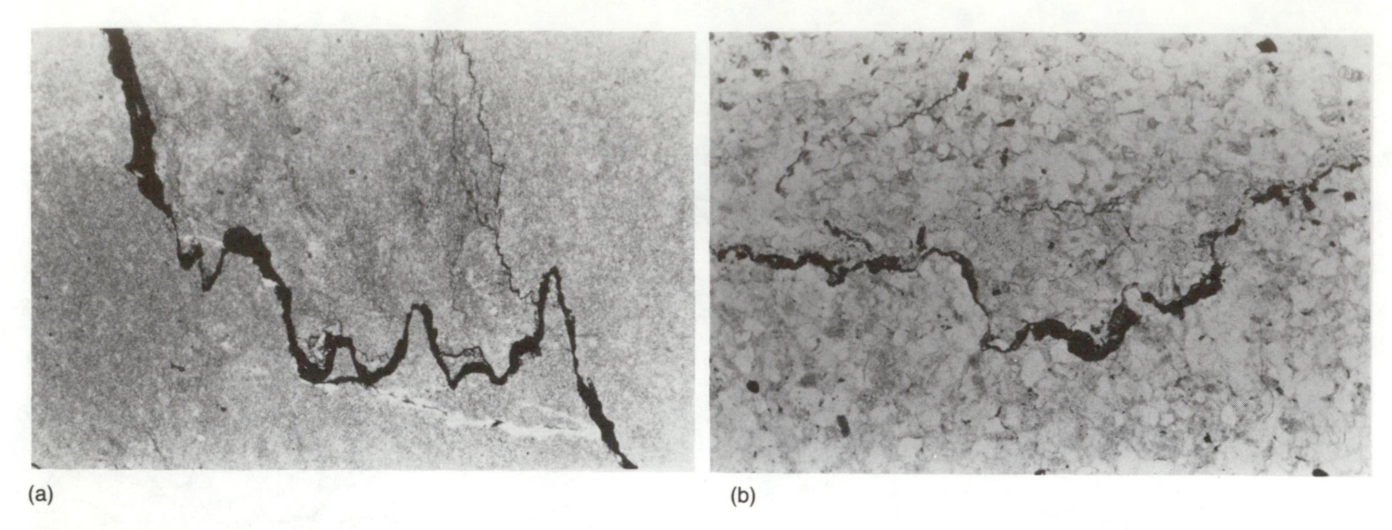

FIGURE 7–15
Stylolites parallel to bedding in Tymochtee Dolomite, northwestern Ohio. Smaller stylolites are oblique to bedding (a) and parallel to bedding in Massanutten Sandstone, northwestern Virginia (b). Width of the photograph in both (a) and (b) is approximately 7 mm. Plane light. (Charles M. Onasch, Bowling Green State University.)

Creep Processes

Creep processes are rate-dependent, but strain is not limited by deformation rate. Creep mechanisms may involve mass transport or diffusion of atoms or ions at grain boundaries, glide and climb of dislocations within a lattice, and diffusion of point defects through lattices. Each is a separate mechanism that is most efficient over a particular range of temperature and pressure. These processes also overlap and compete with each other and with pressure solution under the right conditions so that more than one mechanism may occur simultaneously, although one usually dominates. Because these mechanisms occur over a range of conditions, it is useful to plot their occurrence on *deformation maps* (Figure 7–21), which show experimentally determined range of occurrence of several mechanisms for particular rock types or minerals. These mechanisms, along with pressure solution, are sometimes referred to as

FIGURE 7–16
Mica beards parallel to slaty cleavage in Martinsburg Slate, Delaware Water Gap, New Jersey. Width of field is approximately 2 mm. Crossed polars. (Thin section courtesy of Timothy L. Davis, University of Tennessee.)

FIGURE 7–17
Veins in amphibolite in the Kapuskasing zone, south-central Ontario, filled with quartz and feldspar. Near-vertical vein in the center of the photograph probably formed later than the more gently inclined folded veins. (RDH photo.)

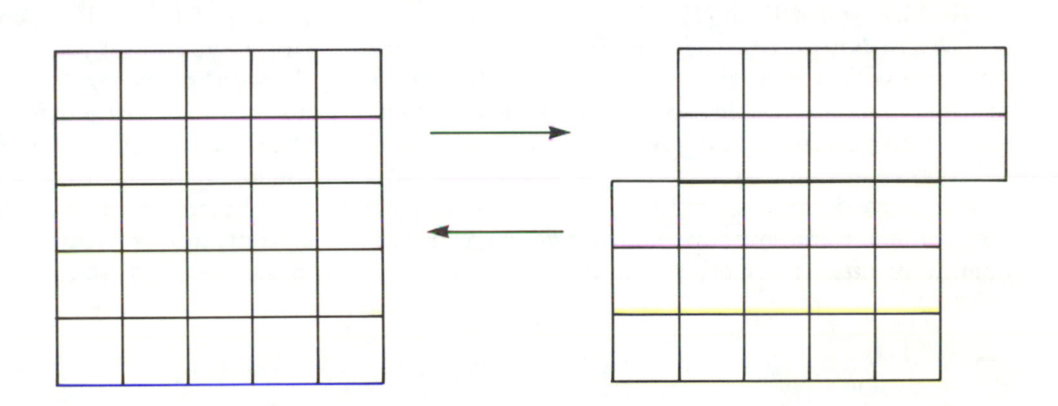

FIGURE 7–18
Translation gliding.

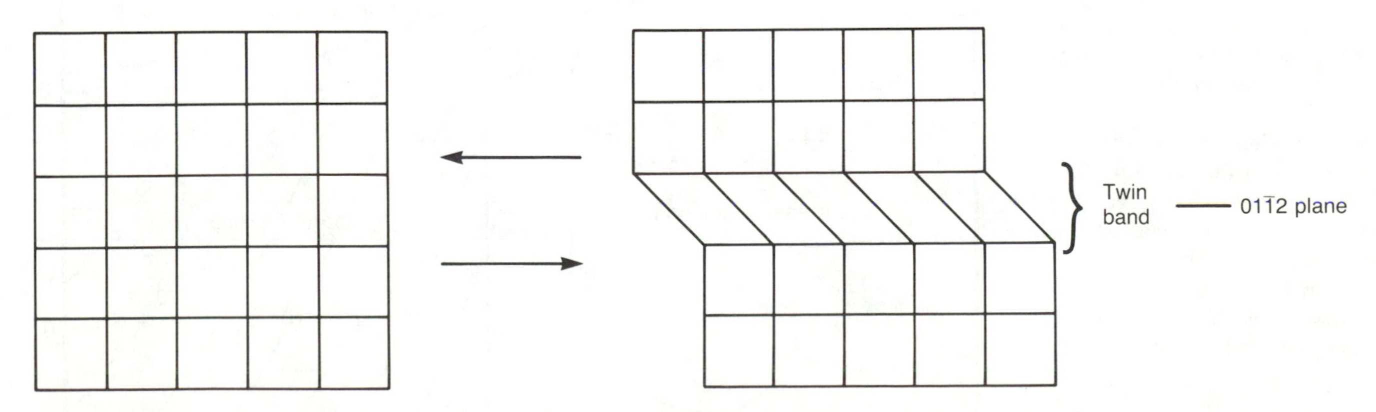

FIGURE 7–19
Twin gliding.

FIGURE 7–20
Mechanical twin of a cleavage rhomb of calcite produced by inserting a knife blade parallel to the $01\bar{1}2$ plane. The twin is the small upturned corner close to the pencil point. (Specimen courtesy of T. C. Labotka, University of Tennessee.)

steady-state deformation mechanisms. Much of the coming discussion is derived from a summary article by Kerrich and Allison (1978).

Grain-boundary diffusion creep (Coble creep) involves mass transport through diffusion at grain boundaries. Coble creep occurs at low to moderate temperature and overlaps the pressure-solution realm up to about 400°C (Figure 7–21). As temperature increases, diffusion processes become more efficient and pressure solution no longer is the dominant mass-transport mechanism.

Dislocations are produced by strain and are removed through a combination of glide and climb motion of the dislocations through the crystal lattice, occurring at moderate to high temperature as ***dislocation creep.*** Dislocations are constantly produced and migrate through the lattice as the rock is strained, causing each crystal to change shape. At the low-temperature end of the process, dislocations may become pinned at grain boundaries and other discontinuities within the rock. As a result, the rock mass becomes strain-hardened (Chapter 6) and

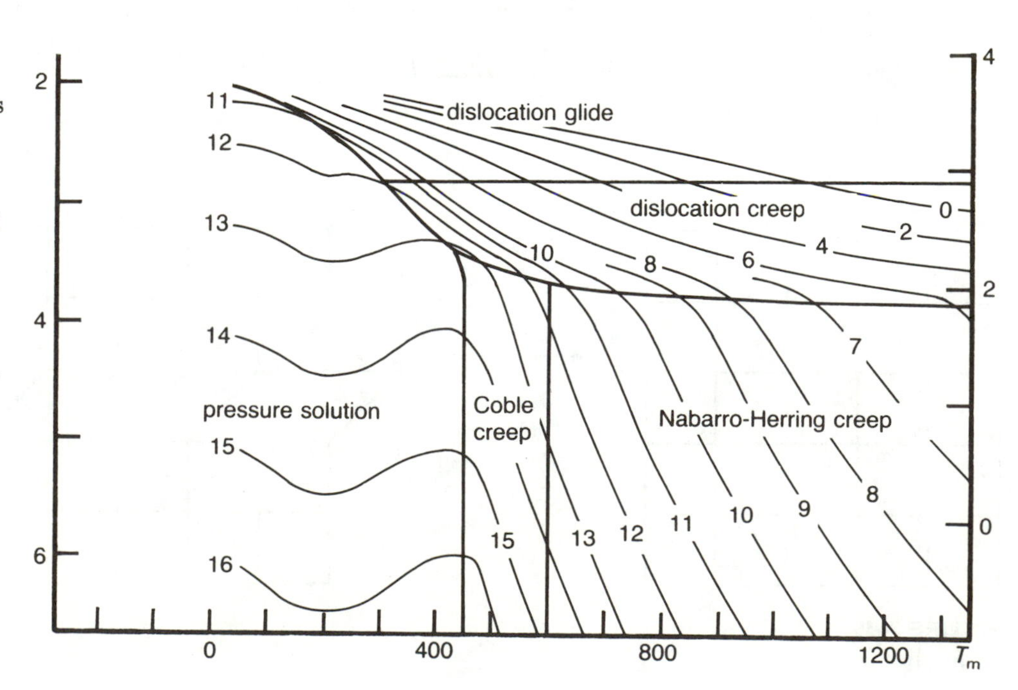

FIGURE 7–21
Deformation mechanism map for calcite, showing the various mechanisms that affect it at various temperatures and stresses. Contours are log strain rate per second. Vertical axis is in units of differential stress divided by the square root of 3 times the shear modulus (left axis) or divided by 10^5 Pa at 500°C (right axis). Both are logarithmic scales. Differential stress was defined in Chapter 3. (After E. H. Rutter, 1976, *Philosophical Transactions of the Royal Society of London*, v. 283.)

greater stress is required to increase the amount of strain further. At higher temperature (greater than half the melting temperature) the amount of energy present permits dislocations to bypass obstacles by climbing or slipping from one lattice plane to another. Dislocations may be annealed by the same process, thereby reducing the number of dislocations and strain in the lattice. Dislocation creep produces an unstrained lattice and is the most common plastic-flow process in metamorphic rocks at moderate to high temperature (Figure 7–22).

Another process involving creep of lattice defects is ***volume diffusion*** or ***Nabarro-Herring creep.*** It involves diffusion of point defects through crystals and occurs at high temperature and low stress. It would overlap and compete at lower temperature with grain-boundary diffusion creep, but it is a very slow process even at high temperature and cannot be considered an important deformation mechanism in crustal rocks. This process is separate and distinct from recrystallization.

Grain-Boundary Sliding

Creep may be compared to sliding. The slip of grains past one another, as sand grains or ball bearings move when a stick is pushed into them, is an important deformation mechanism in rocks. This mechanism, called ***grain-boundary sliding,*** occurs in association with Coble creep at moderate to high temperature. Cataclastic flow at low temperature is an analogous (but *not* identical) mechanism occurring in fault zones where grains move past one another under a strong component of simple shear. Cataclastic flow probably involves processes similar to grain-boundary sliding, but requires accompanying pressure solution to permit sliding along irregular grain boundaries at low temperature.

Superplastic Flow

Other mechanisms may be more complex. Superplasticity results from gliding along grain boundaries and probably involves both Coble creep and grain-boundary sliding (Nicolas, 1987). It was first observed as a phenomenon of extreme ductility in certain alloys undergoing deformation in high-temperature uniaxial tensile experiments. ***Superplastic flow*** in metals is thought to occur by a combination of grain-boundary sliding and flow (a kind of nonuniform flow). A single grain being deformed by a creep mechanism will assume the shape of the entire mass of grains; if the mechanism is grain-boundary sliding, a single grain will maintain its initial shape even though the shape of the aggregate changes significantly.

Superplastic flow is thought to occur at temperatures greater than half the melting point of coarse-grained equigranular rocks (Boullier and Gueguen, 1975). In contrast, the mechanism has also been reported in fine-grained rocks (average grain diameter 10 μm or less) at low temperature under a strong component of inhomogeneous simple shear, as occurs in fault zones (White and White, 1980).

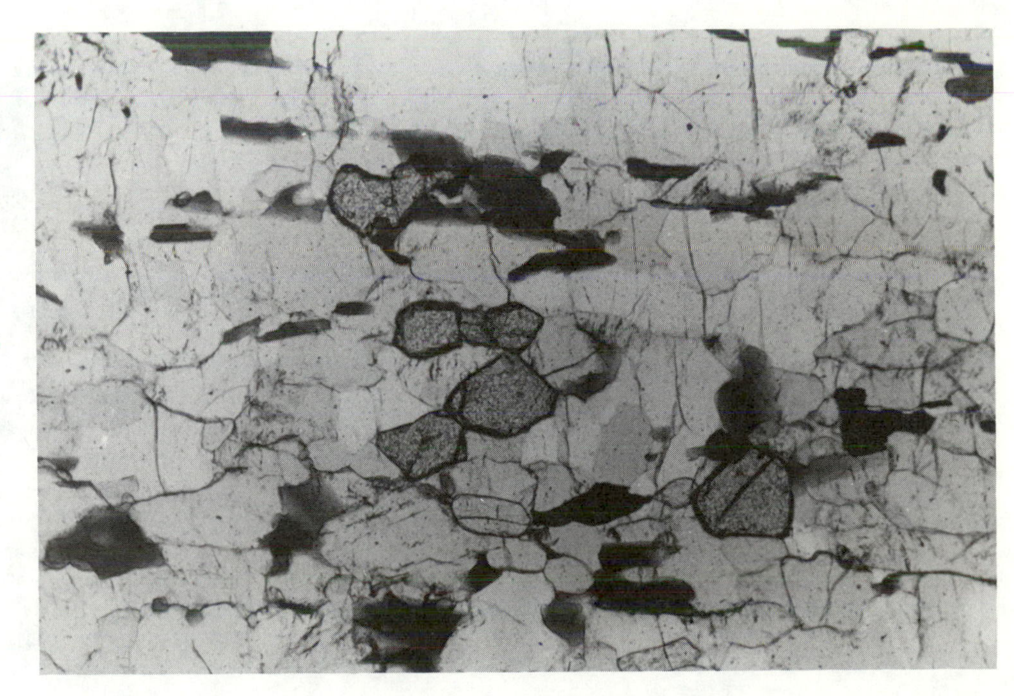

FIGURE 7–22
Annealed unstrained mineral assemblage in a thin section of Late Proterozoic quartzite from near Franklin, North Carolina. The grains with high relief are garnet. Dark grains are biotite and magnetite (note alignment). Clear grains are mostly quartz with a few grains of plagioclase. Note that many quartz grain boundaries meet at 120° angles. Width of photograph is approximately 3 mm. Plane light. (RDH photo.)

As indicated by the diversity of opinions just cited, superplastic flow of geologic materials may occur under a variety of conditions. A decrease in grain size at low temperature to the critical 10-μm size may enable strain rate to increase dramatically as grain-boundary sliding is initiated with no change in temperature. The obvious contrast in conditions under which the process is thought to occur suggests that the low-temperature process is more like cataclastic flow than superplasticity. Grain-size reduction may result from mylonitization or cataclasis (at lower temperature) or from syntectonic recrystallization.

UNRECOVERED STRAIN, RECOVERY, AND RECRYSTALLIZATION

Processes that produce microstructures as a mass of rocks is deformed and recrystallized involve formation of most structures to be discussed in later chapters. We have examined deformation mechanisms and how they affect a rock mass; the remaining task is to clearly distinguish between a strained lattice and the processes of recovery and recrystallization that restore the lattice to a condition less strained or even unstrained.

On the microscopic scale, deformation of a crystal lattice is indicated by unrecovered strain. *Undulatory extinction* in quartz grains (Figure 7–23) is caused by a difference in optical-extinction properties caused in turn by dislocations distorting the lattice. ***Subgrains*** are small parts of grains with lattice orientations differing in adjacent parts of the same grain. They begin to form as strain is relieved in the lattice through the mechanism of dislocation glide and climb and represent the first visible stage of reorganization of the lattice to an unstrained or lowest-energy state (Figure 7–24a). Boundaries of subgrains within old strained grains are ***low-angle boundaries,*** which indicate only a slight lattice misorientation across the boundary. At higher thermal or mechanical energy states, the crystal lattice will form sites for nucleation of entirely new unstrained crystals. In a crystal being deformed by dislocation glide and climb, steady-state flow may be accommodated by one of two mechanisms: ***dynamic recovery*** and ***dynamic recrystallization.*** The term ***recovery*** is used to describe processes that reduce dislocation density and dislocation interaction and increase the rate of glide and climb in a lattice being deformed by dislocation creep (Nicolas, 1987). The main difference between the two mechanisms is that dynamic recrystallization involves strain softening and dynamic recovery does not (Tullis and Yund, 1985). Recovery processes occur mostly at high strain rates (or lower temperatures); recrystallization, which produces high-angle grain boundaries, occurs at lower strain rates and higher temperatures (Figure 7–25). Recovery involves formation of subgrains (low-angle boundaries), but nucleation of new strain-free grains at high temperature is fundamentally a recrystallization process (Figure 7–24b). The newly formed grains contain

FIGURE 7–23
Undulatory extinction in quartz in cataclasite from the Homestake shear zone near Leadville, Colorado. Width of photograph is approximately 1 mm. Crossed polars. (RDH photo.)

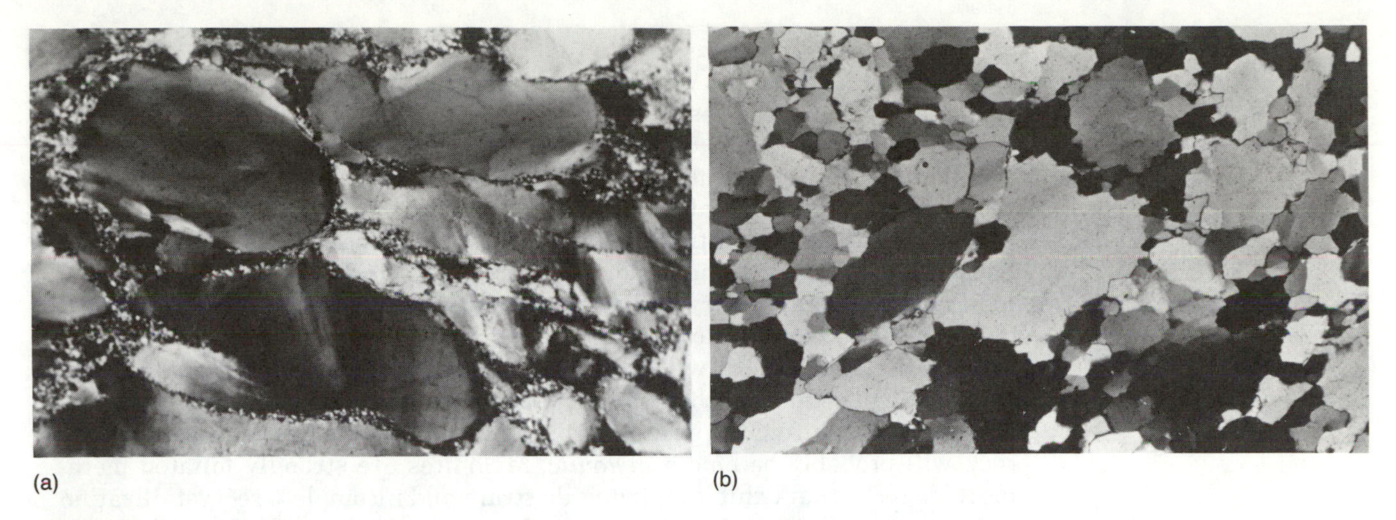

FIGURE 7–24
(a) Plastically deformed quartz grains in Cambrian Weaverton Quartzite from South Mountain near Harpers Ferry, West Virginia. Some grains have been strongly flattened into the foliation; a few maintain vestiges of originally round shape. All have new unstrained grains at the edges of original undulose sand grains. Field is approximately 2 mm wide. (b) Completely recrystallized fabric in Setters Quartzite, Baltimore, Maryland. The entire rock is a mosaic of unstrained polyhedra; note 120° grain-boundary intersections. Width of field is approximately 3 mm. Both (a) and (b) were photographed under crossed polars. (Both photos courtesy of Charles M. Onasch, Bowling Green State University.)

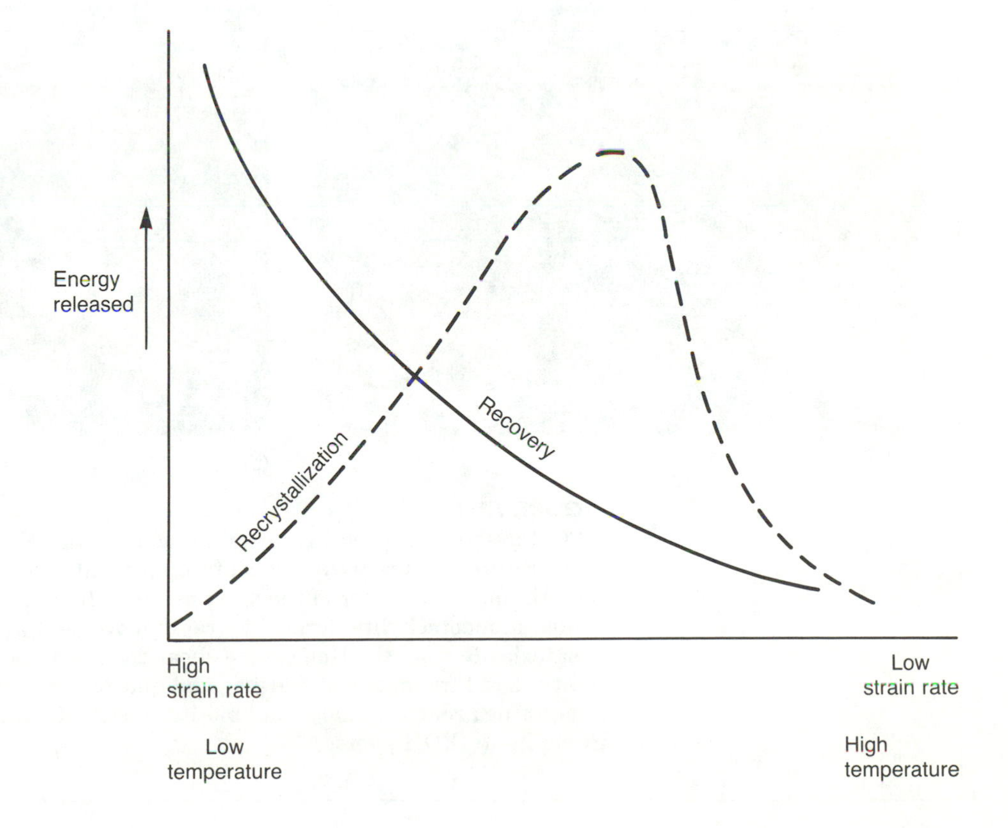

FIGURE 7–25
Relationships between strain rate, temperature, and energy released for recovery and recrystallization processes.

ESSAY

Cataclasites, Mylonites, and Metamorphic Rocks

If recrystallization occurs while a rock mass is being strained, the recovery and recrystallization processes in the lattices, which tend to remove strain, will compete with the tendency for brittle failure to occur as unrelieved strain accumulates in the grains. Competition also exists between strain rate and recovery or recrystallization rate within the rock mass. The competing processes and the resulting products differ in subtle ways (Hatcher and Hooper, 1981). On the one hand, rocks are being deformed very rapidly and may not recover and recrystallize very rapidly—or may not recover at all. If deformation occurs at low temperature and low confining pressure, the deformation mechanism will be brittle and the rock type will be cataclasite (Figure 7E–1). On the other hand, if the temperature is somewhat elevated, fluid is available, and the strain rate is high enough (as along a fault zone), the rock will probably become a *mylonite*. Mylonites are strongly foliated metamorphic rocks that exhibit high ductile strain and incomplete recrystallization or recovery. Larger grains are flattened and extended into the foliation, and ribbon quartz (which may be internally recrystallized) is common. Reduced grain size characterizes mylonitization (Bell and Etheridge, 1973; Hatcher, 1978).

Relationships between mylonites and metamorphic rocks depend on both strain rate and temperature. Sometimes dynamic recrystallization takes place in a rock mass, but the mass cannot recover rapidly enough to keep up with the strain rate—a strain-hardening process. Also, if the mylonitic texture is to be preserved on the microscopic scale, it must be frozen in by rapid cessation of strain and cooling; otherwise, recrystallization may coarsen the grains and

(a)

(b)

FIGURE 7E–1
(a) Cataclasite from the Brevard fault zone, South Carolina, formed at low temperature and pressure and containing highly strained constituents. The dark zone through the center of the specimen may have been a partially melted zone (called pseudotachylite) formed by rapid movement along the fault. (b) Thin section of cataclasite from the Homestake shear zone near Leadville, Colorado. Large, light-colored fragments of feldspar and quartz are fragmented and separated by very fine-grained zones of quartz and biotite. Width of field is approximately 16 mm. Plane light. (RDH photos.)

eliminate dislocation patterns characteristic of mylonite (Figure 7E–2a). Finally, at moderate to high temperature and low strain rate, intercrystalline elastic strain is recovered or recrystallized faster than the strain rate can increase it by multiplication of dislocations (Wise and others, 1984). Enough thermal energy remains after the stress is removed that static recrystallization eliminates strain, resulting in either a "normal" metamorphic rock

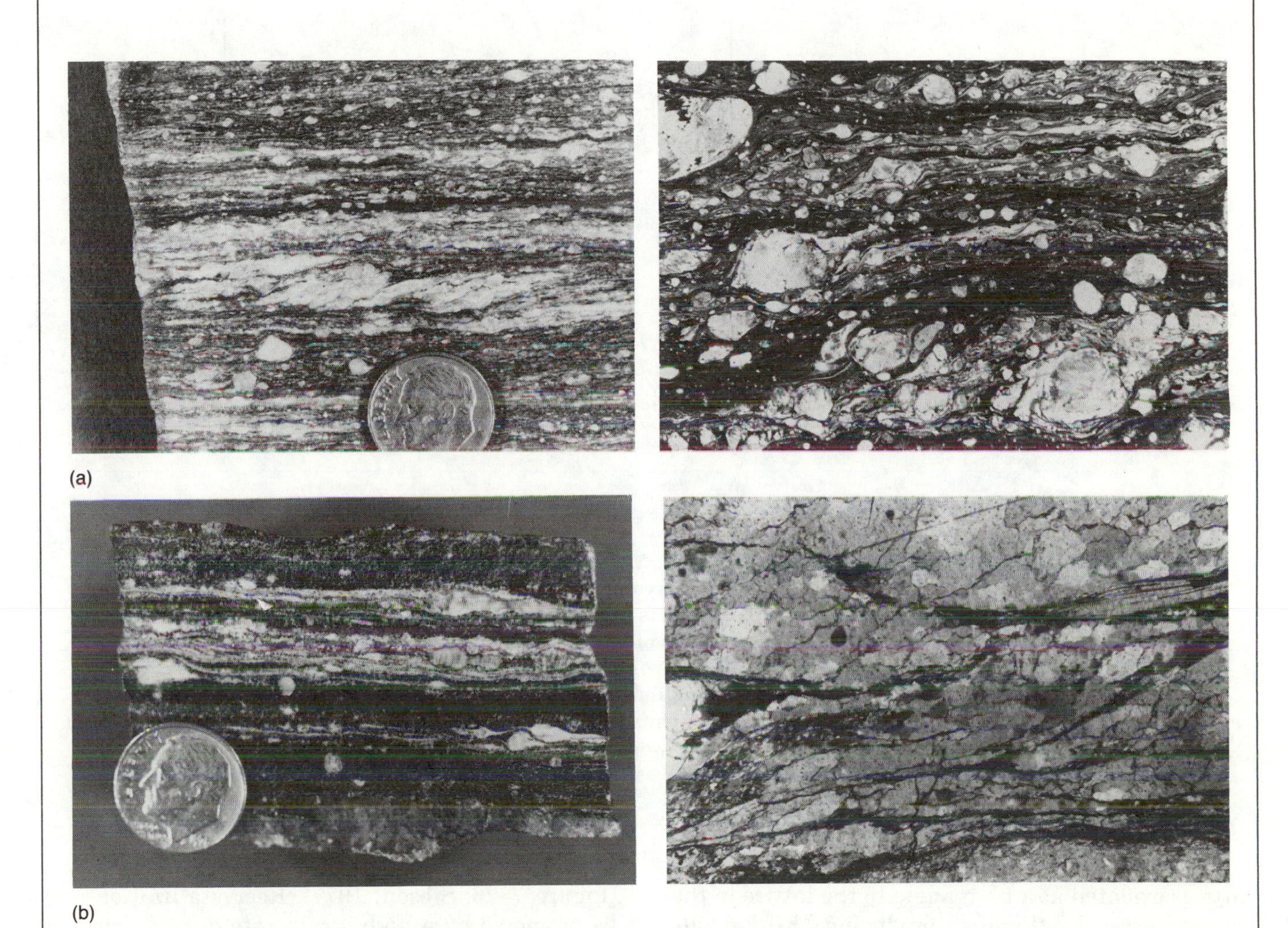

FIGURE 7E–2
(a) (*left*) Dynamically formed (nonannealed) mylonite from the Bartletts Ferry fault zone north of Auburn, Alabama. Note the S-C fabric that indicates sinistral shear sense and the filamentous quartz ribbons. (RDH photo.) (*right*) Thin section of nonannealed mylonite from the Carthage-Colton mylonite zone, Adirondack Lowlands, New York. Field is approximately 2 cm wide. Plane light. (Thin section courtesy of Teunis Heyn, Cornell University.) (b) (*left*) Mylonite from the Goat Rock fault zone near Forsyth, Georgia, formed dynamically but statically annealed. The salt-and-pepper texture of the biotite-rich portion of this rock is a metascopic clue that the microstructure is annealed. Feldspars with tails preserve the original mylonitic texture. (RDH photo.) (*right*) Thin section of an annealed mylonite from the Box Ankle fault in central Georgia. Quartz ribbons are totally internally recrystallized, but the original shape of the ribbons is preserved. Width of field is approximately 3 mm. Partially crossed polars. (RDH photo.)

(Figure 7E–3) or (possibly) a mylonite. In the latter case, the rock mass megascopically preserves the mylonitic texture but microscopically it is entirely recrystallized (Figure 7E–2b).

Mylonite and cataclasite commonly occur along faults. In nature, the competing processes (just described) overlap to a great degree and mylonites

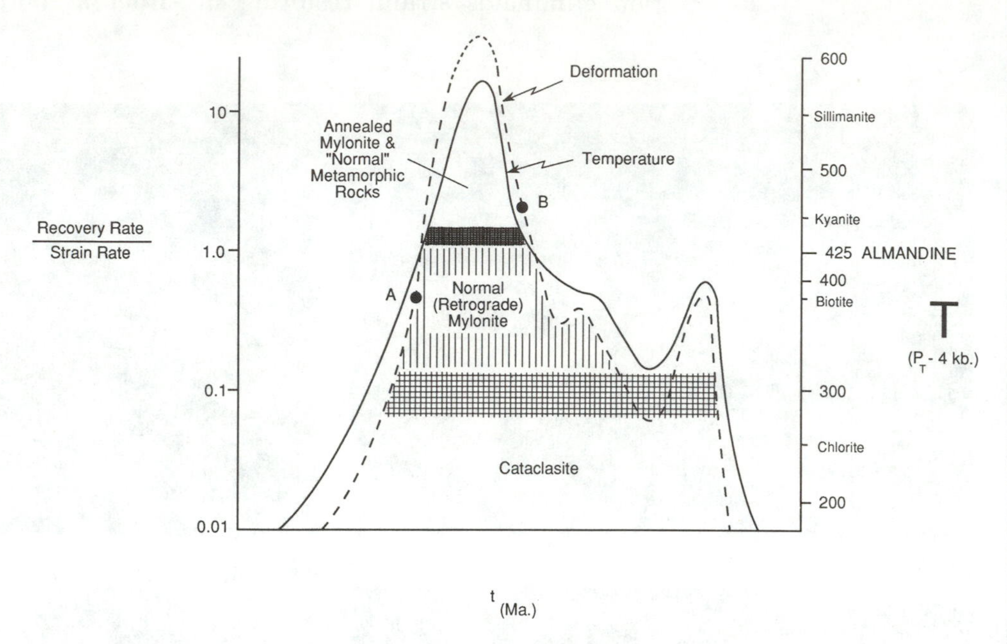

FIGURE 7E–3
Relationships between recrystallization-recovery rate, strain rate, temperature, and total strain and the products of deformation. Point A represents a rock that was deformed under conditions of increasing temperature. Mylonitic texture would probably be annealed following cessation of deformation, unless the rock mass was cooled without increasing temperature further so the existing textures would be preserved. Point B represents a rock deformed following the thermal peak. Mylonitic (or metamorphic) texture would be preserved, and much of the strain in the rock may not be annealed out.

lattices oriented at a high angle to the lattice in the parent crystal, thereby producing ***high-angle boundaries.*** As a result, very small and completely unstrained crystals form inside or at the boundary of an older large strained single grain. The original strained lattice may have formed subgrains to attain a lower energy state and then later recrystallized to form small new crystals with high-angle boundaries. The newly crystallized grains, polyhedra, or ***polygons,*** tend to meet at angles near 120°, an arrangement that has the lowest energy state and therefore the greatest stability for an unstrained lattice (Figure 7–24c).

Recrystallization may occur as ***dynamic*** or ***syntectonic recrystallization*** while stress is being applied to the rock mass, or it may occur as ***static recrystallization*** after stress has been removed (Figure 7–26; Sibson, 1977). Recrystallization may be promoted by a high strain rate during deformation, but it is less evident because new and unstrained grains are strained rapidly. Strain removed by recrystallization is the elastic strain of individual dislocations. *Strain energy plus grain-boundary energy drives recrystallization.*

The question of how preferred orientation is produced by deformation and recrystallization in rocks has evaded a precise answer for many years. Obviously, formation of preferred orientations, cleavage, and foliations in metamorphic rocks, and in many fault rocks as well, depends on deformation. Slip and rotation, in addition to recrystallization, are thought to be the common mechanisms leading to preferred orientation because preferred orientation has been produced experimentally without accompanying re-

should not result from every occurrence of high strain rate. Mylonitic texture may develop in ordinary metamorphic rocks in areas where there are no faults because of extremely localized development of high shear strain. Rocks in highly strained and thinned fold limbs undergoing ductile flow might contain mylonitic texture because a large amount of strain can be concentrated there. On the other hand, faults that form before the metamorphic-thermal peak in a rock mass will produce mylonite, but the mylonite will probably be overprinted by metamorphism. Mylonite in the fault zone will exhibit recrystallized microtextures common to metamorphic rocks, rather than a mylonitic texture, but will preserve some of the megascopic textures of mylonite (Figure 7E–2b). This relationship is important because few clues remain concerning the origin and nature of these early-formed fault rocks in the internal parts of mountain chains. As a consequence, other criteria, such as stratigraphic data (Chapter 9), must be used to distinguish between stratigraphic and fault contacts separating rock bodies.

To sum up, products of strain—including strain recovery and recrystallization—and the competing processes that form cataclasites, mylonites, and normal metamorphic rocks are all gradational. They involve a delicate balance between the thermal regime of the rock mass, pressures and fluids within the rock body, the strain rate, and the timing relative to the thermal peak.

References Cited

Bell, T. H., and Etheridge, M. A., 1973, Microstructure of mylonites and their descriptive terminology: Lithos, v. 6, p. 337–348.

Hatcher, R. D., Jr., 1978, Eastern Piedmont fault system: Reply: Geology, v. 6, p. 580–582.

Hatcher, R. D., Jr., and Hooper, R. J., 1981, Controls of mylonitization processes: Relationships to orogenic thermal/metamorphic peaks: Geological Society of America Abstracts with Programs, v. 13, p. 469.

Wise, D. U., Dunn, D. E., Engelder, J. T., Geiser, P. A., Hatcher, R. D., Jr., Kish, S. A., Odom, A. L., and Schamel, S., 1984, Fault-related rocks: Suggestions for terminology: Geology, v. 12, p. 391–394.

crystallization (Tullis, 1971). Hobbs, Means, and Williams (1976) discussed at length the mechanisms for crystallographic preferred orientations.

In Chapter 17, we will examine the mechanisms of cleavage formation and their relationship to recrystallization, but meanwhile we will turn to a consideration of laboratory studies of recrystallization processes.

LABORATORY MODELS OF DEFORMATION PROCESSES

Many simple geologic materials have been deformed in the laboratory. Mixtures of quartz and feldspar (the common constituents of granitic rocks) and pure mineral aggregates such as quartz sandstone, pure marble (calcite), and dunite (olivine) have been studied extensively. Mylonite can be produced experimentally in pure quartz sandstone only at temperatures in the range of 800° to 1,000°C and at high strain rates in the range of 10^{-5} to 10^{-7}/s (Figure 7–27; Jan Tullis, John Christie, and David Griggs, 1973). Design of experiments requires high strain rates, and as a result high temperatures are necessary to produce mylonites. Both temperature and strain rate greatly exceed the geologic conditions normally encountered by rocks being deformed in the core of a mountain chain or toward the flanks. But such laboratory conditions are necessary to create and study these materials during the investigator's lifetime.

Mixtures of quartz and feldspar in various proportions permit study of two materials of markedly

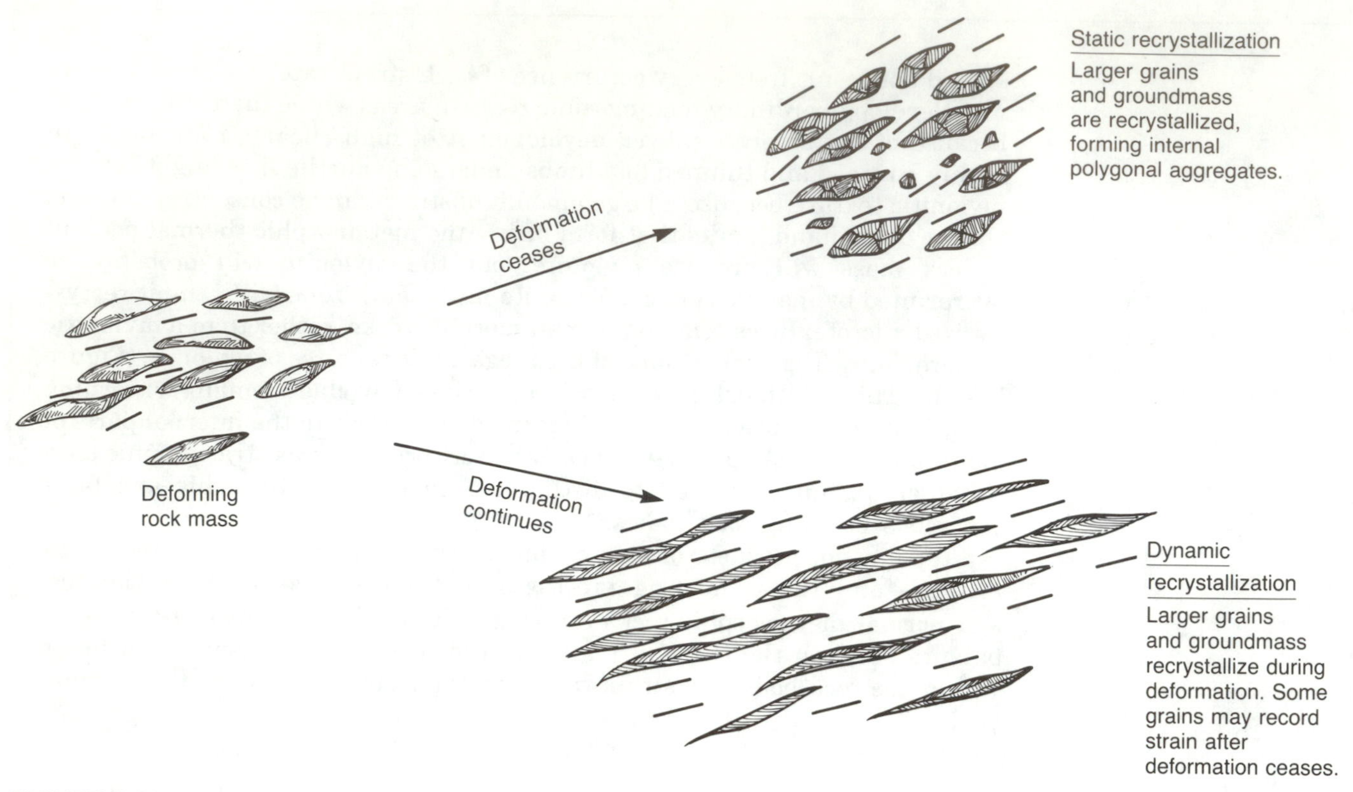

FIGURE 7–26
Relationships between static and dynamic recrystallization.

different mechanical properties in geologically realistic compositions. Quartz deforms ductilely at much lower temperature than feldspar, particularly in the presence of a small quantity of water. Feldspar generally remains brittle until temperature and pressure are very high, even with water present. Consequently, coexisting quartz and feldspars, as in granitic rocks, deform differently in the deforming rock mass. Experimental and field studies indicate that feldspar does recrystallize at higher temperature around the edges of large grains with the

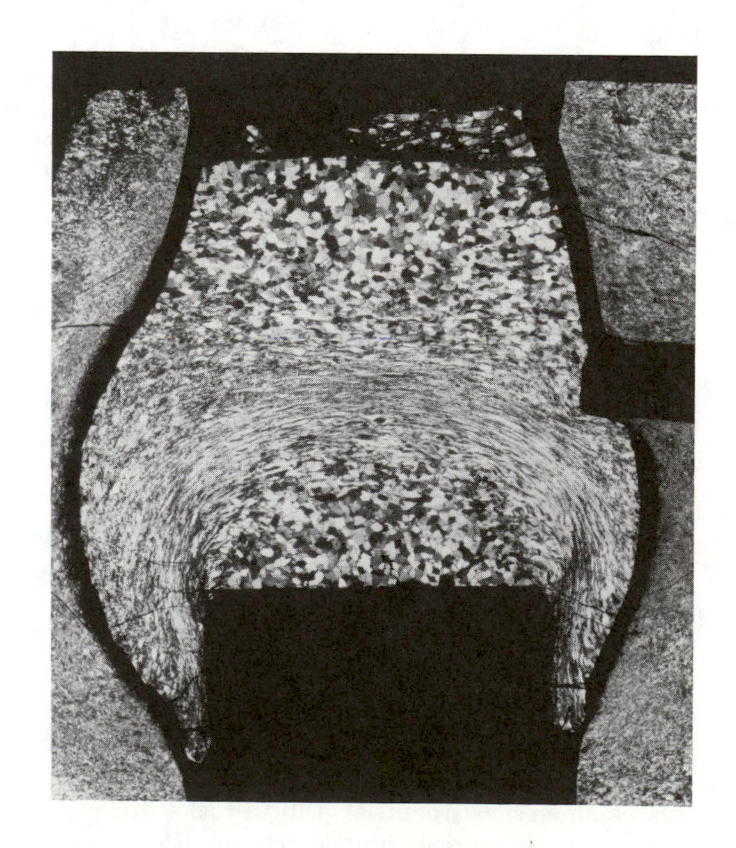

FIGURE 7–27
Thin section of an experimentally produced quartzite mylonite. Curved diffuse area of flattened quartz grains flowed around the sides of the dark obstacle, bulging the container as the ram compressed the sample. Annealed polygonal quartz is present on the ends of the sample. The important observation to be made here is that mylonite formed by both pure and simple shear. Crossed polars. (From J. Tullis, J. M. Christie, and D. T. Griggs, 1973, Geological Society of America *Bulletin*.)

(a)

(b)

(c)

FIGURE 7–28
Low-melting-point organic compounds that deform under controlled conditions on the stage of a petrographic microscope and exhibit various recrystallization and deformation processes.
(a) Deformation-induced twin lamellae (narrow lines) in crystals of para-dichlorobenzene. Width of field is 0.7 mm. Crossed polars. (b) Undeformed (or annealed) mass of interlocking octochloropropane crystals. Note that junctions of most crystals make ~120° angles. Width of field is 0.5 mm. Crossed nicols. (c) Abundant subgrains developed along grain boundaries following slight deformation of crystals in (b). Field width is 0.5 mm. Crossed polars. (W. D. Means, 1986, *Journal of Geological Education,* v. 34.)

appearance of small and unstrained polygons, leaving the old large grains largely unstrained (Tullis and Yund, 1985). As recrystallization progresses, the size of the old grains decreases and new and smaller grains appear. Presence of micas in a quartz-feldspar aggregate also increases the rate of deformation under the same temperature and pressure. Water, in even small quantities, also increases the deformation rate and lowers the temperature threshold for most deformation mechanisms.

A series of experiments designed by Winthrop D. Means (1986) involves three organic compounds with low melting points: monoclinic biphenyl, hexagonal octachloropropane, and triclinic para-dichlorobenzene. These compounds strain and recrystallize under minimum strain at room temperature so their behavior can be observed directly under a petrographic microscope (Figure 7–28)—something impossible with most rock materials, metals, and ceramics. The organic compounds exhibit a variety of deformation properties by kinking (Chapter 15), slip, recrystallization, and even phase change.

We now turn from rock mechanics to a study of fractures and faults.

Questions

1. Why are deformation processes in metals not good models for rock deformation?
2. Of all the factors that control the rates and thresholds of deformation processes, which single factor do you think is most important? Why?
3. How could you tell a cataclasite from a mylonite in the field? With the aid of a petrographic microscope in the laboratory?
4. Given a thin section of a metamorphic rock, how could you determine the extent to which strain in the rock has been relieved?

5. How would you determine whether a rock has been deformed by pressure solution or by volume-diffusion creep?
6. What is the difference between subgrain formation and formation of crystal polyhedra in a strained lattice?
7. Describe the sequence of deformation mechanisms that would occur with increasing temperature, in the presence of water, in an impure sandstone under stress starting at 100°C and ending at 450°C.

Further Reading

Groshong, R. H., Jr., 1988, Low-temperature deformation mechanisms and their interpretation: Geological Society of America Bulletin, v. 100, p. 1329–1360.

Reviews pressure solution and other low-temperature deformation mechanisms, with a discussion of how these are recognized in natural rocks.

Hull, Derek, 1975, Introduction to dislocations: New York, Pergamon, 271 p.

Defects and dislocations in crystals are explained clearly with good illustrations and without a great deal of mathematics.

Kerrich, R., and Allison, I., 1978, Flow mechanisms in rocks: Geoscience Canada, v. 5, p. 109–119.

A concise, readable summary of rock-deformation processes with good examples.

Kerrich, R., 1978, An historical review and synthesis of research on pressure solution: Zentralblatt für Geologie und Paleontologie, Part 1, p. 512–550.

A useful review of our knowledge of pressure solution, from its earliest recognition in the nineteenth century to modern concepts of how it works.

Means, W. D., 1986, Three microstructural exercises for students: Journal of Geological Education: v. 34, p. 224–230.

Presents a technique for observing a variety of microdeformational processes using a petrographic microscope, simple glass-slide apparatus, and organic compounds that deform at room temperature.

Means, W. D., 1989, in press, Kinematics, stress, deformation, and material behavior: A review of essentials for students: Journal of Structural Geology.

Reviews basic continuum mechanics and discusses applications to deformation processes and material behavior. Mathematics is used but at a level understandable to junior-senior geology students.

Nicolas, A., and Poirier, J. P., 1976, Crystalline plasticity and solid state flow in metamorphic rocks: New York, John Wiley & Sons, 444 p.

A rigorous book providing a comprehensive survey of plastic-deformation processes affecting rocks.

Schedl, Andrew, and van der Pluijm, B. A., 1988, A review of deformation microstructures: Journal of Geological Education, v. 36, p. 111–120.

Reviews the different lattice-scale deformation processes, subgrain formation, recovery and recrystallization processes, and some of the terminology of shear zones.

Tullis, Jan, and Yund, R. A., 1985, Dynamic recrystallization of feldspar: Geology, v. 13, p. 238–241.

Tullis, Jan, and Yund, R. A., 1987, Transition from cataclastic flow to dislocation creep of feldspar: Mechanisms and microstructures: Geology, v. 15, p. 606–609.

These papers clearly distinguish between dynamic recrystallization and recovery processes and the deformation mechanisms that occur in feldspar under various physical conditions.

Wise, D. U., Dunn, D. E., Engelder, J. T., Geiser, P. A., Hatcher, R. D., Jr., Kish, S. A., Odom, A. L., and Schamel, S., 1984, Fault-related rocks: Suggestions for terminology: Geology, v. 12, p. 391–394.
Undertakes to clarify the confused terminology related to fault rocks and the processes by which both ductile and brittle faults produce a variety of rock types. Discusses the differences and similarities of cataclasites and mylonites and their relationships to metamorphic rocks.

PART THREE

Fractures and Faults

OUTLINE

8 Joints and Shear Fractures
9 Fault Classification and Terminology
10 Fault Mechanics
11 Thrust Faults
12 Strike-Slip Faults
13 Normal Faults

WE NOW TURN TO A DISCUSSION OF ONE OF THE LARGE GROUPS of tectonic structures. The description and formation of joints and shear fractures, predominantly brittle structures, is discussed in Chapter 8. In the other chapters in this section, we will investigate the class of geologic structures—faults—that we once thought form dominantly by brittle deformation in the upper crust; but, as we learn more about them, we find that ductile faults exist in the lower crust. The geometry, terminology, and classification of faults is considered in Chapter 9, and the mechanics of faults and shear zones are discussed in Chapter 10. Here the relationships between brittle and ductile faults are outlined. The nature of the three major fault types and their interrelationships are discussed in Chapters 11, 12, and 13.

8

Joints and Shear Fractures

> Joints are the most ubiquitous structure in the Earth's crust, occurring in a wide variety of rock types and tectonic environments. . . . They control the physiography of many spectacular landforms and play an important role in the transport of fluids. . . . Today there is no doubt that joints reveal rock strain accommodated by brittle fracture. Establishment of reliable relationships between joints and their causes can provide the structural geologist with important tools for inferring the state of stress and the mechanical behavior of rock.
>
> D. D. POLLARD and ATILLA AYDIN, Geological Society of America *Bulletin*

FRACTURES ALONG WHICH THERE HAS BEEN NO APPRECIABLE movement parallel to the fracture and only slight movement normal to the fracture plane are ***joints*** (Figure 8–1). There is no offset parallel to the fracture plane. This kind of fracture is extensional, with the fracture plane oriented parallel to σ_1 and σ_2 and perpendicular to σ_3. As such, joints form in a plane not subject to shear. Joints are the most common of all structures, present in all settings in all kinds of rock, as well as in partly consolidated to unconsolidated sediment. We expect to see joints in rock, but they have also been found in unconsolidated glacial sediments in New England and in unconsolidated sediments of the Gulf and Atlantic coastal plains. Fractures also form in association with faults and other shear zones. Many of these are

FIGURE 8–1
Several orientations—sets— of near-vertical joints produced this irregular topography, in flat-lying sandstone, Canyonlands National Park, Utah. (L. P. Arnsbarger, National Park Service, and S. W. Lohman, U.S. Geological Survey.)

extensional, and are joints, but some may form parallel to shear planes in the strain ellipsoid.

Three types of fractures have been defined by Byron Kulander, Chris Barton, and Stuart Dean (1979), each type formed by a separate kind of motion (Figure 8–2). Mode I fractures are joints formed by opening the fracture; mode II fractures by sliding; and mode III fractures by a tearing motion. Mode II and III fractures are ***shear fractures.***

Study of joints and shear fractures is useful in both pure and applied science. Studying the sequence of fracturing and fracture filling in rocks to understand structures bracketed by fracturing events is applied directly in documenting the antiquity of deformation that produced fracturing (or lack of it) in the rocks beneath dams, bridges, and power plants (Figure 8–3). Fracture sequences help in understanding the nature of brittle deformation and the relationships of joints to faults and folds. That valuable minerals may be found in joints and shear fractures has been known for centuries.

Joints serve as the plumbing system for ground-water flow in most areas. They are the only routes by which ground water can move through igneous and metamorphic rocks at an appreciable rate, and they dominate as conduits in well-cemented sedimentary rocks. Consequently, fracture porosity (and permeability) produced by joints is important for the water supply in any area except those underlain by the classic regional aquifer systems, such as the unconsolidated sediments of the Gulf and Atlantic coastal plains and the Great Plains of the North American midcontinent.

Joint orientations in road cuts greatly affect both construction and maintenance. Fractures oriented parallel or dipping into a highway cut become hazardous during construction because they provide potential movement surfaces, particularly as the rock weathers. Rock-bolting is frequently employed to stabilize large blocks that cannot be removed; cement grout or rock-bolting may be employed if the rock is weak and the fracture density high. Undercutting and oversteepening of slopes during construction may lead to sliding and increase the cost of maintenance. In mines, subhorizontal joints in the roof must be rock-bolted to prevent falls.

Joints are generally planar structures and may be characterized as *systematic* or *nonsystematic*. ***Systematic joints*** have a nearly parallel orientation and a regular spacing. Joints that share similar orientations in the same area are referred to as a ***joint set.*** Two or more joint sets in the same area comprise a ***joint system.***

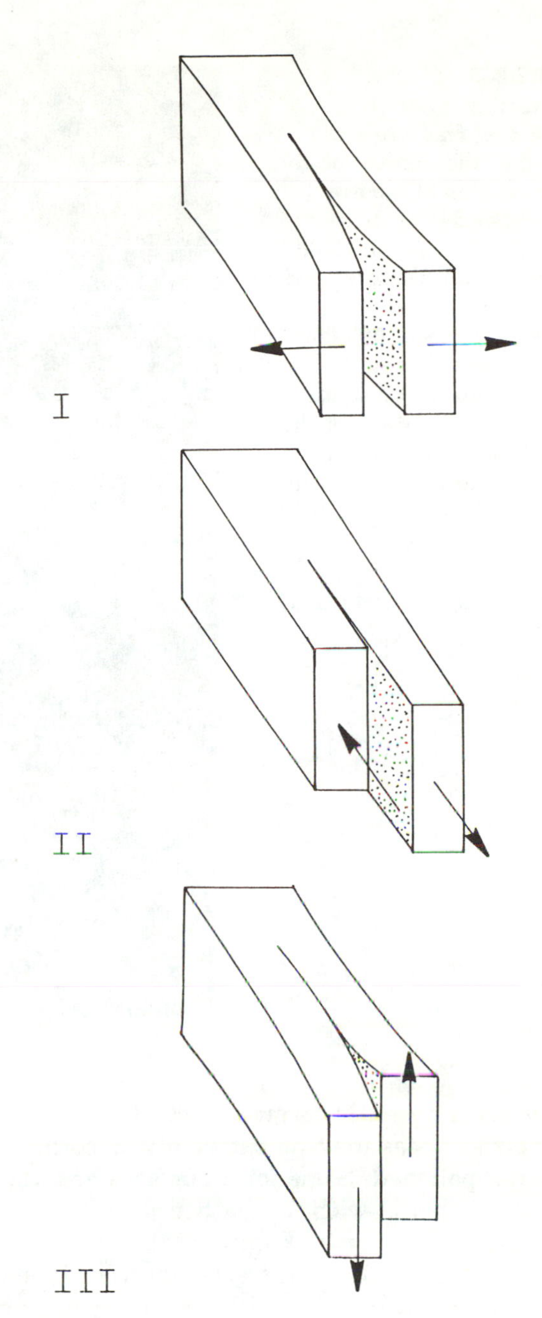

FIGURE 8–2
Three modes of fracture formation.

Joints that do not share a common orientation and in which fracture surfaces may be highly curved and irregular are called ***nonsystematic joints.*** They occur in most areas but do not lend themselves to standard means of fracture analysis. As a result, nonsystematic joints do not provide obvious evidence for a consistent relationship to a recognizable stress or strain field. Occasionally it can be shown that systematic and nonsystematic joints formed at the same time, but nonsystematic joints usually abut systematic joints—indicating the nonsystematic joints formed earlier (Figure 8–4).

Joints may be *unfilled;* that is, the fractures may be open and not filled with minerals. Generally they

FIGURE 8–3
Near-vertical view of intersecting fracture sets exposed in the excavation for the foundation of a power plant in South Carolina. Before the foundation could be constructed all fractures had to be demonstrated to be inactive for the most recent 500,000 years and not capable of movement during the future life of the plant. Many of the fractures are filled with minerals that have not been deformed. Scale is indicated by construction workers. (Courtesy of Malcolm F. Schaeffer, Duke Power Company.)

are the most recently formed fractures in an area and their surfaces may be extremely smooth, even appearing polished. Some joint surfaces are highly irregular; others, with a feathered texture, are termed ***plumose joints*** (Figure 8–5).

Filled joints (and shear fractures) occur in many areas (Figure 8–6) and the fillings range in composition from feldspar and quartz as high-temperature fillings (pegmatite and aplite veins) to lower-temperature quartz, calcite, adularia, chlorite, and epidote (and combinations) as well as ore minerals. Fractures may also be filled with combinations of zeolites, calcite, and other low-temperature minerals such as prehnite. Unraveling a sequence of cross-cutting relationships as well as the compositions of fillings, or lack thereof, and the orientations of fractures may lead to resolution of the chronology of brittle deformation.

Both filled and unfilled fractures may occur in ***conjugate*** (paired) systems. Conjugate systems may form parallel to the shear planes of the strain ellipsoid and intersect at an acute angle. They are shear fractures. For paired sets to be conjugates, they must form at the same time. It is difficult to prove the conjugate nature of shear fractures unless intersections at outcrop scale emphasize their relationships to each other and to associated structures such as folds or faults (Figure 8–7). Many intersecting fracture sets are first taken to be conjugates but detailed resolution of their movement histories shows they were formed at different times as joints later reactivated as shears.

Curved fractures occur frequently and may be related to "refraction" of joints from one rock type to another or by grain size or compositional changes within a thick bed or larger rock mass (Kulander and others, 1979). Many systematic joints may be curved on the mesoscopic scale, but they are planar regionally.

FRACTURE ANALYSIS

Study of joints in an area reveals the sequence and timing of formation—information with both pure and applied scientific value. Fracture studies in the

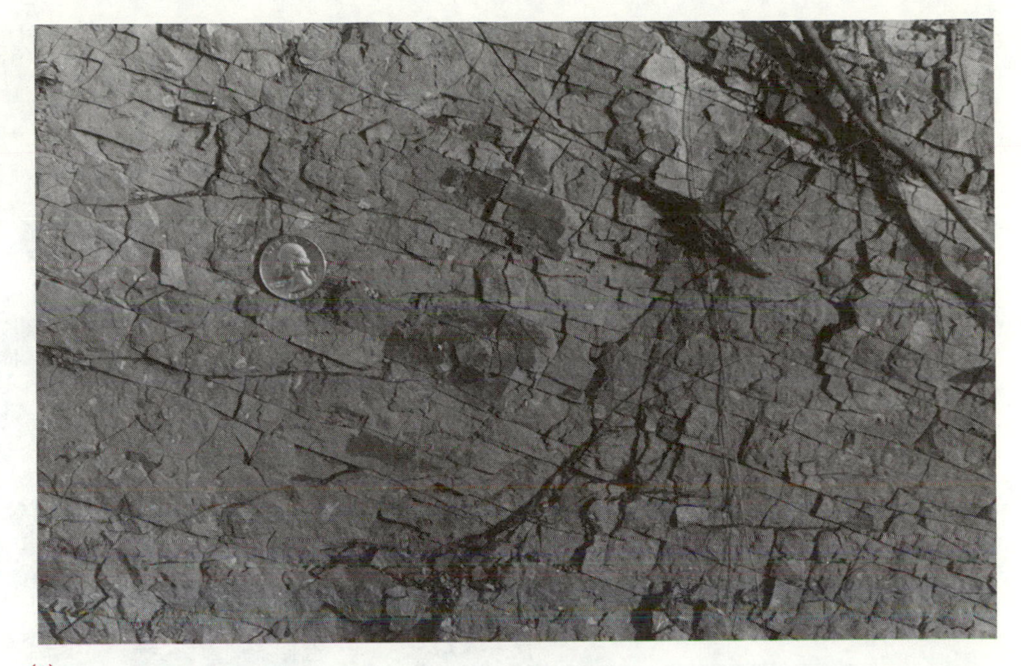

(a)

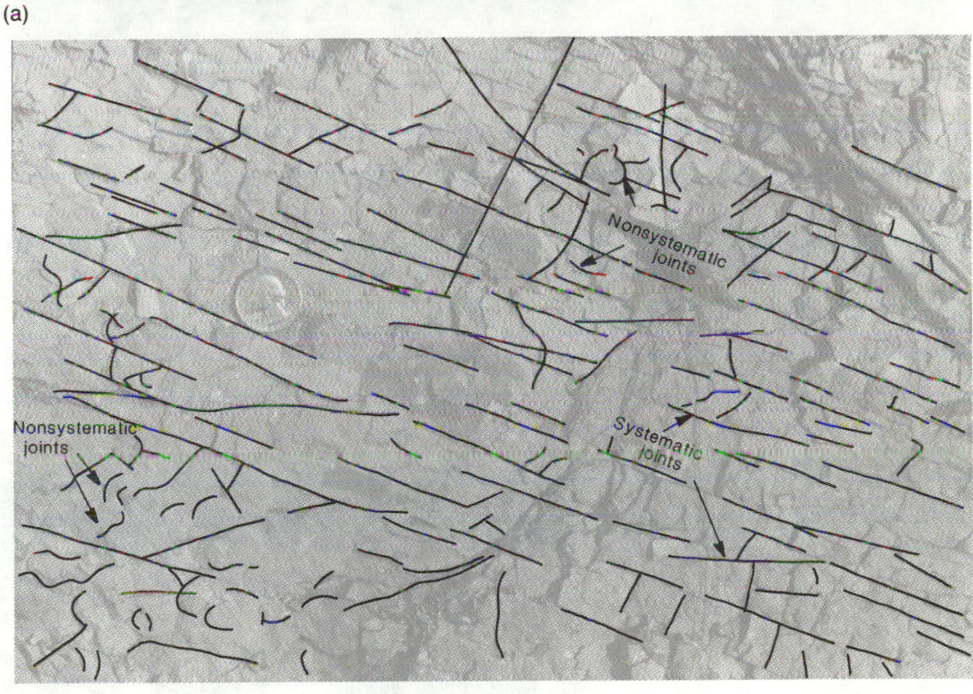

(b)

FIGURE 8–4
(a) Systematic and nonsystematic joints in Middle Cambrian Conasauga Shale, Oak Ridge, Tennessee. Note that systematic joints are linear and form parallel sets; nonsystematic joints are irregular. (RDH photo.) (b) Line drawing showing relationships in (a).

field provide information on the timing and geometry of brittle deformation of the crust and the way fractures propagate through rocks. Laboratory study of fractures in rocks, artificial materials, and glacial ice enables comparison with natural fractures and leads to understanding of the mechanism of brittle failure.

Significance of Orientation

Study of orientations of systematic fractures provides information about the orientation of one or more principal stress directions involved in brittle deformation.

Regional joint-orientation patterns may be determined by measuring strike and dip of mesoscopic-scale joints over a wide area. A good estimate of regional joint orientation is sometimes obtained by measuring the strike of linear stream segments on satellite imagery, topographic maps, or aerial photos (Figure 8–8). Data gathered either on the ground or from maps and photos related to orientation and spacing of lineaments may be analyzed to help understand relationships between joints and their

FIGURE 8–5
Plumose joints in Devonian siltstone near Glen Eldridge, New York. Note the plumose structure on both layers in the center of the photo indicates the fracture propagated left to right. (W. H. Bradley, U.S. Geological Survey.)

relationship to drainage and other topographic features. Data may be plotted using an equal-area net or rose diagrams, which show only the strike and frequency of orientations (Figure 8–9). Rose diagrams are as useful as equal-area plots for joints in most areas because most joint sets dip steeply. Shallow-dipping sets are better shown on equal-area plots. Joint sets may also be analyzed statistically.

Studies of joint and fracture orientations from LANDSAT and other satellite imagery and photography have a variety of structural, geomorphic, and engineering applications. As one example, Wise and others (1985) have used LANDSAT data to interpret relationships between fractures and the Alpine tectonics of Italy (Figure 8–10). They were able to improve delineation of the boundaries of widely

FIGURE 8–6
Several crossing sets of intersecting veins in limestone at Highgate Springs, Vermont, formed at different times and were filled with calcite. (C. D. Walcott, U.S. Geological Survey.)

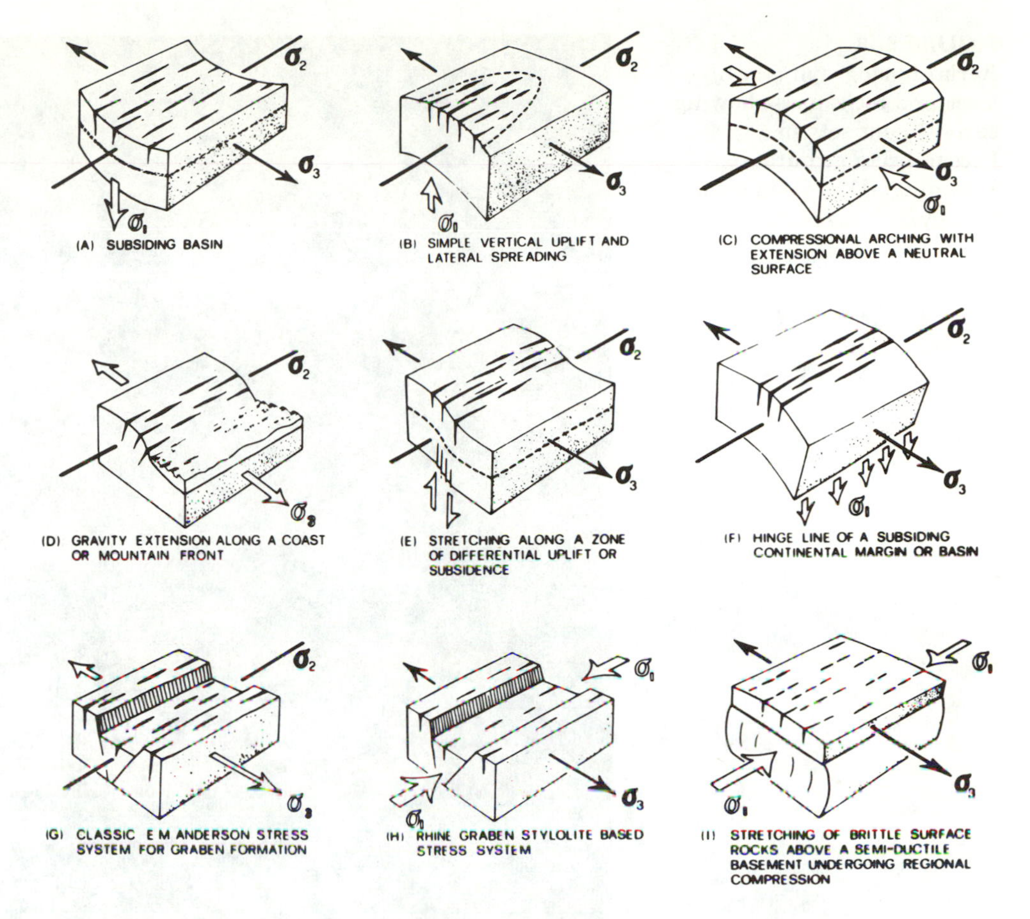

FIGURE 8–7
Possible joint orientations related to principal stress axes and larger structures. (From D. U. Wise, Renato Funiciello, Parotto Maurizio, and Francesco Salvini, 1985, Geological Society of America *Bulletin*, v. 96.)

known structural features and, because the region is tectonically active, drew conclusions about the orientation of the present-day stress field and the interrelationships of major structural features.

Strain-ellipsoid analysis of joints in an area may help determine dominant crustal extension directions. First, the fractures must be shown independently to be joints by finding evidence of extension along one or more sets. If it can be shown that two or more crossing sets formed at the same time, at least two sets may be related to one or more shear planes in the strain ellipsoid. If a third set is oriented properly, it may be related to the *YZ* extension plane (Figure 8–11).

Fracture Formation in the Present-Day Stress Field

It has been suggested that joint orientation in the sedimentary cover may be controlled by stress in the crystalline basement beneath. Joints may propagate from older crystalline basement rocks into overlying previously unfractured sedimentary rocks (Hodgson, 1961). The mechanism is poorly understood, but propagation into cover rocks may occur during reactivation of basement fractures—provided the basement fractures and stresses are suitably oriented.

Engelder (1982) has hypothesized that some regional joint sets in eastern North America are a product of the present-day stress field—*not* inherited patterns from older crust. The joint sets thought to have formed in the contemporary stress field are those *not* related to any older stress regime. After considering their orientations, surface markings (such as smooth or plumose), and information about stress orientation derived from hydraulic-fracturing measurements and focal-mechanism solutions, Engelder concluded that joints form normal to present-day σ_3.

Fold- and Fault-Related Joints

Joints may form during folding by a brittle-fold mechanism. They may form normal, parallel, and oblique to the fold axis and axial surface, depending on local stress conditions (Figure 8–12).

In one practical application, Richard P. Nickelsen (1979) made a detailed study of a well-exposed series of late Paleozoic deformed sandstone, coal, and shale at the Bear Valley strip mine in Pennsylvania. He was able to separate early jointing events from later faulting, cleavage formation, folding, later folding that produced new joint sets, and fi-

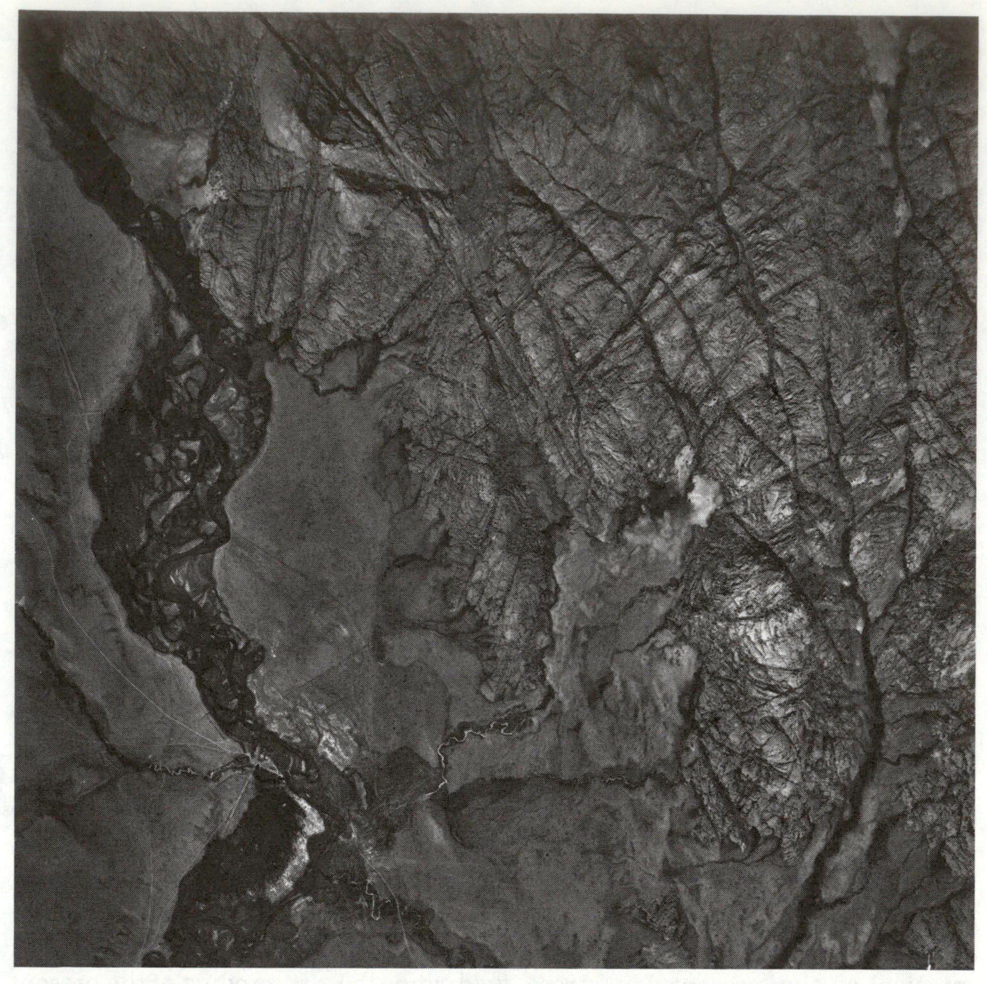

FIGURE 8–8
Aerial photograph of a highly fractured rock mass showing several joint sets in Precambrian granitic rocks, Wyoming. (U.S. Geological Survey).

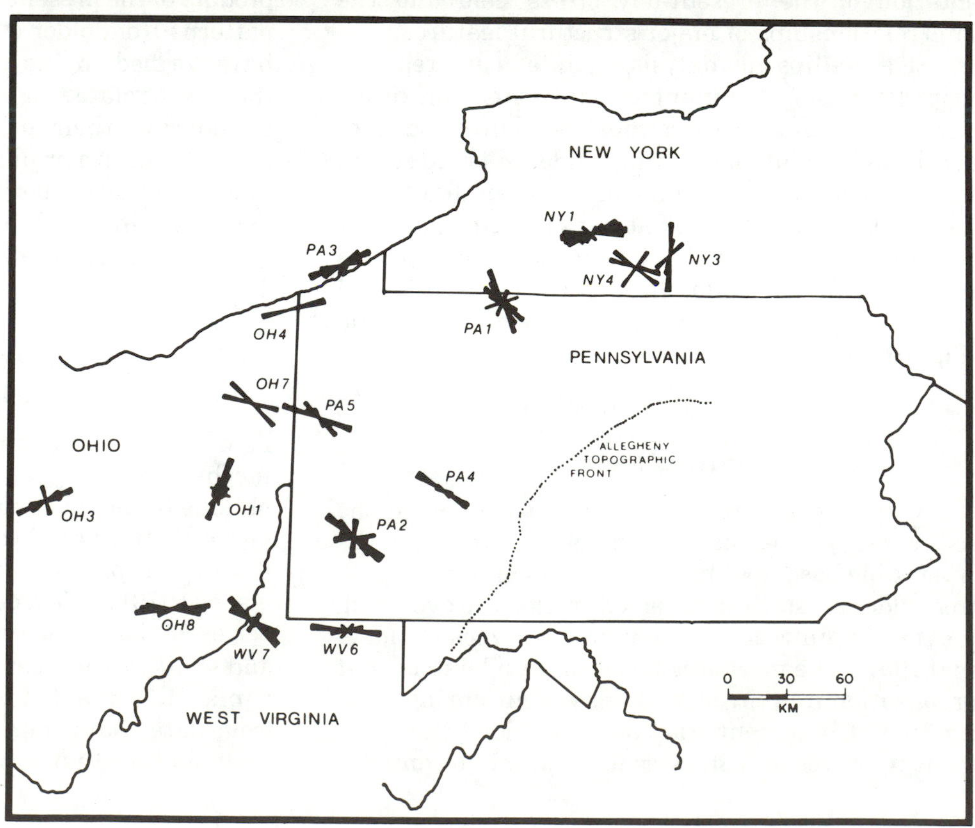

FIGURE 8–9
Rose diagrams of orientations of joints in drill cores from wells in Ohio, West Virginia, Pennsylvania, and New York. (From Terry Engelder, 1987, in *Fracture Mechanics of Rock,* B. Atkinson, ed., Academic Press.)

FIGURE 8–10
Fracture and lineament orientations in Italy. (From D. U. Wise, Renato Funiciello, Parotto Maurizio, and Francesco Salvini, 1985, Geological Society of America *Bulletin*, v. 96.)

nally, more faulting. The last faulting stage, he found, also produced new joint sets.

Joints frequently form adjacent to brittle faults. Movement along faults commonly produces a series of systematic fractures in which spacing increases closer to the fault zone, but the number of sets increases (Pohn, 1981). Most are probably joints, but some form as shear fractures.

JOINTS AND FRACTURE MECHANICS

For years, geologists have debated the origin of joints and other fractures in rocks. Much of the evidence favors origin of joints by extension, because many joints display no offset, except normal to the fracture surface, and are clearly related to an extensional strain field. Jointing may be geometrically related to resolution of shear fractures, particularly in some folds and near some fault surfaces, indicating that extensional strain may be present in rock masses undergoing simultaneous compression and shear.

Experiments can produce joints by forcing fluid into rock pores to lower the effective normal stress (normal stress minus pore pressure) until the tensile strength of the rock is exceeded. Theoretical and field studies indicate that fluid probably plays an important role in formation of many joints. Studies

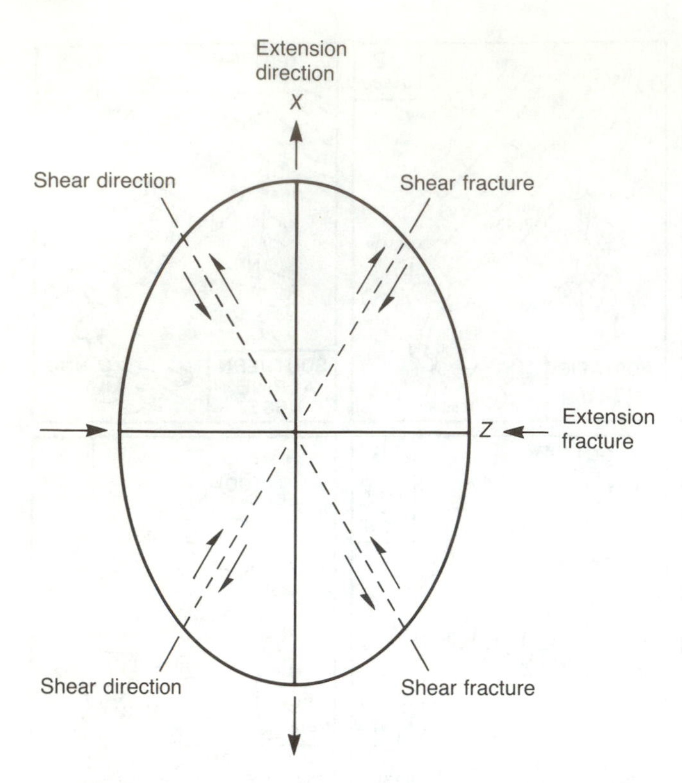

FIGURE 8–11
Jointing related to the strain ellipsoid.

by Secor (1965), Pollard and Holzhausen (1979), and Segall and Pollard (1983) indicate that joints form more readily in the presence of fluid, generating fractures by hydraulic fracturing. Secor (1965) considered effective stress ($\sigma - P$, where P is fluid pressure) and concluded that joints could form in a compressive environment if fluid pressure is high. Fluid under pressure (compression) is forced into the fracture and promotes continued fracture propagation. This process occurs in the upper crust where water is abundant and, interestingly, may also be the mechanism for forcible emplacement of dikes in the lower crust, with magma as the fluid (Odé, 1957; Pollard and Muller, 1976). Fractures in rock may begin as Griffith cracks (elliptical microcracks in grains), then coalesce into more continuous oriented fractures and ultimately form joints. One consequence of Griffith theory is the prediction that stress difference in extension can be no more than four times the tensile strength of the rock (Jaeger and Cook, 1976). If the stress difference exceeds the theoretical limit, the Mohr envelope is reached and fracturing occurs.

Terzaghi (1923) has suggested that shear resistance of soils (or rock) may be calculated by a modified version of the Coulomb criterion,

$$\tau = \tau_0 + \mu(\sigma - P) \qquad \textbf{(8–1)}$$

where τ is shear strength, τ_0 is inherent shear strength, μ is the coefficient of internal friction, σ is normal stress, and P is pore pressure, pressure exerted by a fluid against the walls of a microscopic opening in a rock. This idea has been applied extensively to shear fractures formed in rocks that contain a fluid (Figure 8–13).

Most joints form by extensional fracturing of rock in the upper few kilometers of the Earth's crust. The limiting depth for formation of extension fractures is a function of the tensile strength of the rocks present, which is influenced by rock type, fluid pressure, and the stress difference at a particular time.

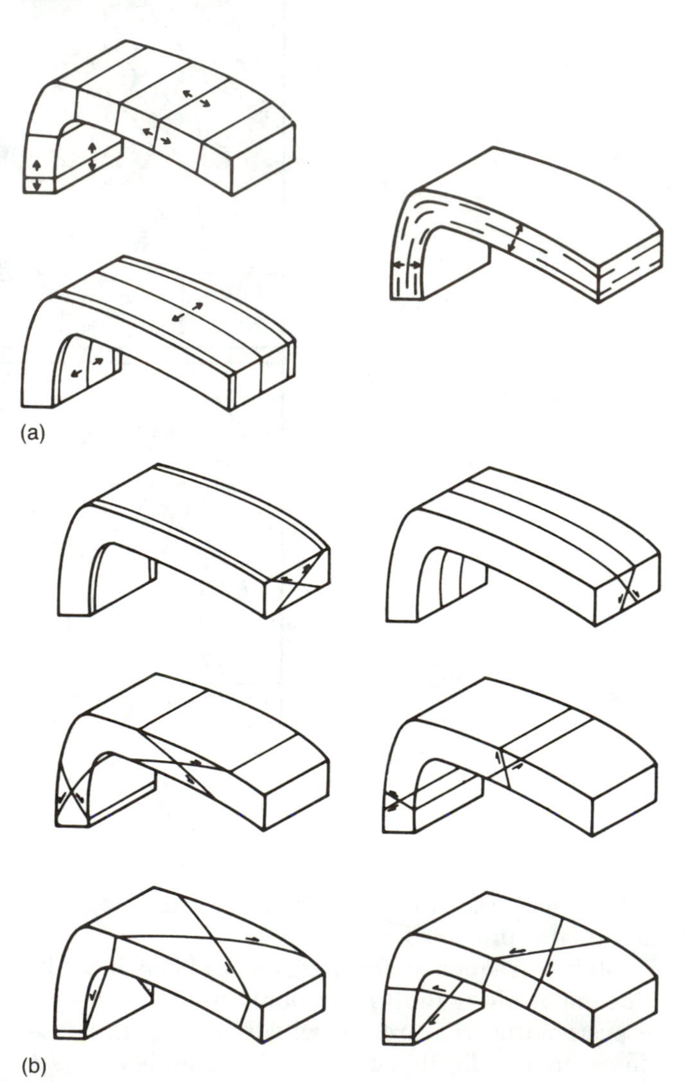

FIGURE 8–12
Fold-related joints and shear fractures. (a) Joints forming at different orientations in a fold. (b) Six possible shear-fracture orientations in a fold. (Reprinted with permission from P. L. Hancock, Brittle microtectonics: Principles and practice, © 1985, Pergamon Journals Ltd., *Journal of Structural Geology,* v. 7.)

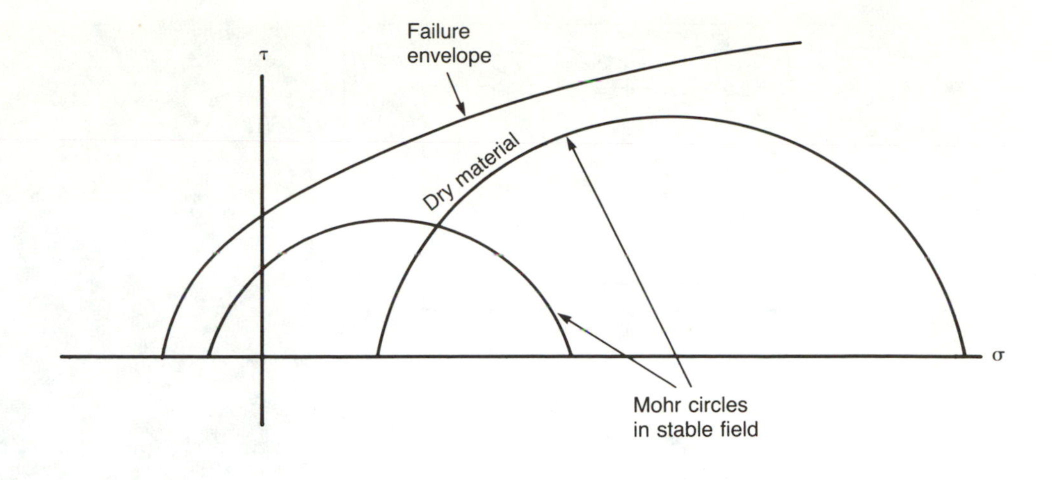

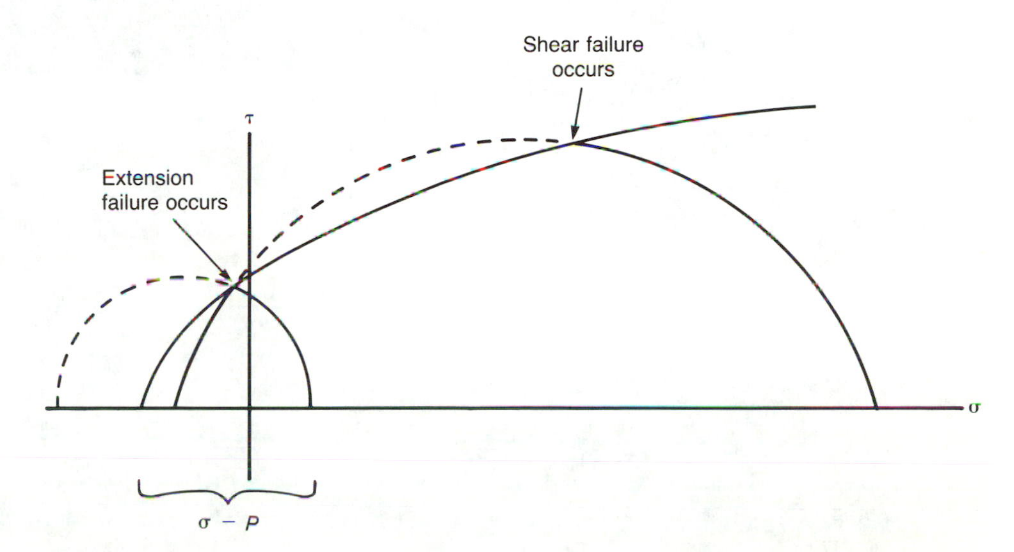

FIGURE 8–13
Mohr circles for stress, showing the effect of a fluid and pore pressure on effective normal stress.

Features on joint surfaces, such as plumose structure (Figure 8–5), provide clues to the direction of joint propagation, and other features provide details of the mechanics of formation of extension fractures and information on the rate and direction (Figure 8–14) of propagation of individual joint planes. Referring to Figure 8–14, *hackle marks* indicate zones where the joint propagated rapidly, and the *arrest line* indicates direction of propagation (Hodgson, 1961; Kulander and others, 1979). Kulander, Barton, and Dean (1979) have suggested that propagation of a joint always begins at a preexisting flaw in the rock mass, a grain of atypical size or hardness, a fossil, a concretion, a pore space, an irregularity in bedding, or other mechanical discontinuity. The joint may propagate from the flaw under ideal extensional conditions and form a symmetrical plumose surface with symmetrical twist hackle, Wallner lines, and arrest lines. Asymmetric plumose structure, twist hackle, and other structures require stress orientation to change as the joint propagates (Kulander and others, 1979). The explanation is reasonable if the asymmetric plume occurs in a single bed and if the beds above and below are symmetrical, but, if the same sense of asymmetry occurs on plumes in successive layers, the fracture is likely to be a hybrid shear (mixed-mode fracture) and not a joint (Hancock and Engelder, 1989). Such observations are important to understanding the mechanisms of fracture formation and determining whether a series of fractures consists of joints, shear fractures, or hybrid shears.

Bedding and foliation planes in coarser-grained rocks constitute barriers to joint propagation. Bedding in uniformly fine-grained rocks, such as shales and volcaniclastic rocks, appears to be less of a barrier or none at all. Joints frequently propagate through a sandstone bed and are slightly offset from those in the next layer above or below. Variation in bedding thickness also affects propagation direction. Differences in stress magnitude between layers produces a barrier to joint propagation through a section of alternating sedimentary rock types. For example, joints will not propagate from sandstone into

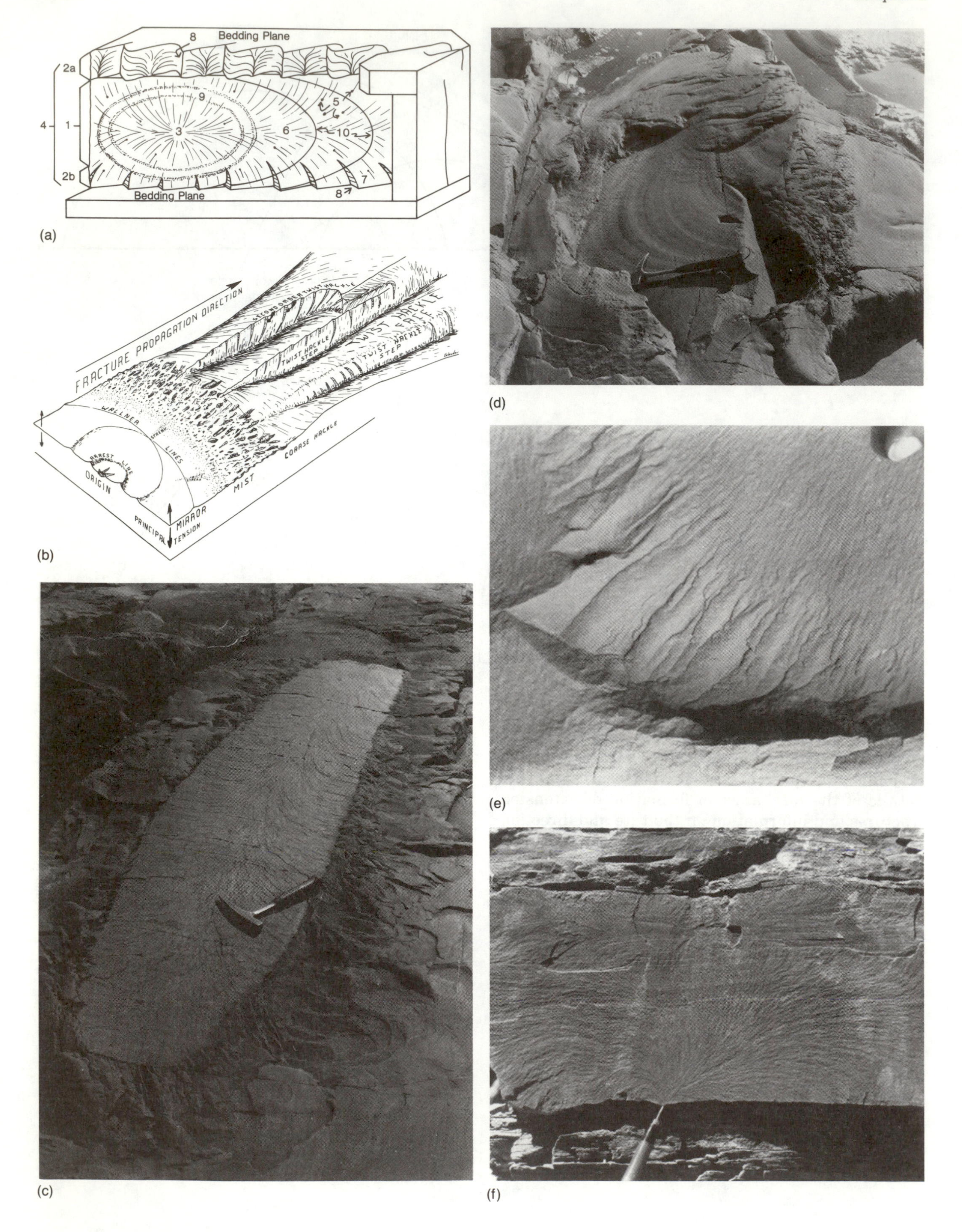

Bedding Plane
8
2a
4
1
2b
3
9
5
6
10
7
8
Bedding Plane
(a)
FRACTURE PROPAGATION DIRECTION
TWIST HACKLE STEP
TWIST HACKLE STEP
WALLNER LINES
ARREST LINE
ORIGIN
MIRROR
MIST
COARSE HACKLE
PRINCIPAL TENSION
(b)
(c)
(d)
(e)
(f)

shale if the least horizontal stress in the shale is greater than that in the sandstone. Fractures will terminate at the contact between the two rock types (Engelder, 1985).

Terry Engelder (1985) undertook to characterize both the environment and the mechanisms of joint formation and described four categories of joints as end-member paths for increase of stress (Figure 8–15). They are *tectonic, hydraulic, unloading,* and *release joints*. All categories involve the assumption (based on many observations) that the failure mechanism is tensile. ***Tectonic*** and ***hydraulic joints*** form at depth in response to abnormal fluid pressure and involve hydrofracturing. Hydraulic joints form during burial and vertical compaction of sediment at depths greater than 5 km, where escape of fluid is hindered by abnormal pore pressure. This process is static, whereas tectonic joints form essentially by the same mechanism but the stresses originate tectonically, and horizontal compaction occurs. Thus tectonic joints may form at depths of less than 3 km.

Unloading and ***release joints*** form near the surface as erosion removes overburden and thermal-elastic contraction occurs. Unloading joints begin to form when more than half the original overburden has been removed from a rock mass. Contemporary tectonic or remaining ancient stresses may serve to orient these joints. For vertical unloading joints to form, the effective stress in the horizontal plane necessary to exceed the strength of the rock mass must become tensile. This is thought to occur during cooling and elastic contraction of a rock mass as it is exhumed by erosion and may occur at depths of 200 to 500 m. Release of horizontal stress provides a similar opportunity for release joints to form.

Orientation of release joints is controlled by the existing rock fabric, in contrast to the three other types recognized by Engelder, which are stress-controlled. Release joints form late in the history of an area and are ultimately oriented perpendicular to the original tectonic compression that formed the dominant fabric in the rock. Tectonic stress may further increase the stress normal to future joint planes as burial depth and degree of lithification increase. Once erosion begins, and the rock mass begins to cool and contract, these joints begin to propagate parallel to an existing tectonic fabric, such as a prominent cleavage. In that way, release joints may also develop parallel to fold axes.

Beach (1980) has suggested that filled veins in low-grade metamorphic rocks are one of the best lines of evidence for hydraulic fracturing. Most of these fractures form by extension, but Beach showed that if the initial orientation of microcracks is parallel to shear directions, then cracks will continue to propagate parallel to shear directions or perhaps terminate as asymmetrically forked veins. Shear fractures may be straight or slightly curved but commonly cannot be traced over as great a distance as single extension fractures. Beach found that the length-to-width ratio of tensile veins is ordinarily greater than 500, whereas that of shear veins is less than 100. Hydraulic shear fractures are thought by Beach to form where differential stress is low, but hydraulic tensile fractures may form over a wide range of effective stress values. His ideas remain somewhat controversial.

Paul Hancock (1985) has defined a joint as *any* fracture discontinuity in rock, even a shear fracture. He concluded that joints and shear fractures may form by extension or shear, or by a hybrid mechanism, and suggested that the history of jointing may be pieced together by using surface markings (smooth, plumose) and parallelism to nearby kinematic indicators (shear zones, folds)—if all can be shown to have formed by the same deformational event. Symmetry with respect to other structures, joint refraction, curviplanar joints (sometimes indi-

FIGURE 8–14
(a) Detail of plumose joint surface showing primary surface structures: 1. Main joint face. 2. Twist hackle fringe. 3. Origin. 4. Hackle plume. 5. Inclusion hackle. 6. Plume axis. 7. Twist-hackle face. 8. Twist-hackle step. 9. Arrest lines. 10. Constructed fracture-front lines. (From B. R. Kulander and S. L. Dean, 1985, Proceedings of the International Symposium on Fundamentals of Rock Joints.) (b) Detail showing other features associated with propagation of the joint. (From Kulander, Barton, and Dean, 1979, *The application of fractography to core and outcrop fracture investigations,* U.S. Department of Energy METC/SP-79/3.) (c) Subhorizontal plumose joint in Middle Proterozoic Killarney Granite on Georgian Bay near Killarney, Ontario. The plumose pattern here indicates the fracture propagated from top to bottom of the photo. The irregular pattern at the edge of the smooth fracture is twist hackle. This joint was created as a Pleistocene glacier moved over the rock surface and plucked a mass of bedrock. (RDH photo.) (d) Glacially produced joint surface in Killarney Granite at the same locality as (b) with arrest lines that are concentric to the origin of the fracture. The joint therefore propagated right to left. Faint plumose structure is also visible on the joint surface. (RDH photo.) (e) Twist hackle at the edge of a joint in Jurassic Aztec Sandstone, southeastern Nevada. (f) Irregular plumose structure on a joint surface in Devonian Genesee siltstone near Watkins Glen, New York, indicating either a change of propagation direction during fracturing or a component of simple shear was present. (Parts e and f courtesy of Atilla Aydin, Purdue University.)

ESSAY

Mesozoic Fracturing of Eastern North American Crust—Product of Extension or Shear?

The present-day Atlantic Ocean probably opened along the coast of eastern North America during Early to Middle Jurassic time, after a period of rifting of the continental crust formed the Triassic-Jurassic basins along the axis of the Paleozoic Appalachian orogen. The basins and the adjacent Piedmont were intruded by diabase dikes (Figure 8E–1): from Virginia southward, they are oriented dominantly northwest; from Virginia northward, they trend more northerly to northeasterly (Ragland and others, 1983). A similar pattern occurs on both sides of the present Atlantic, interpreted by May (1971) as indicating extensional deformation along the line where spreading of the continents began. An alternative explanation, by de Boer and Snider (1979), is that Mesozoic dikes in eastern North America formed as a product of doming over a mantle hotspot in the Carolinas. The dikes examined by May and by de Boer and Snider have since been shown to be cross-cut by a younger set of north-south trending Mesozoic diabase dikes in the Carolinas and Virginia; this set converges toward a point near Charleston, South Carolina (Ragland and others, 1983; Figure 8E–1). Both sets of dikes appear to be extensional.

Fracture zones filled with siliceous cataclasite occur in the same region. In the central and southern Appalachians, they are oriented approximately north-south and east-west (Conley and Drummond, 1965; Birkhead, 1973); in some places they appear to cut the diabase dikes but in others the siliceous cataclasites are cut by the dikes. Thus the diabase dikes and siliceous cataclasite fracture zones may have formed about the same time in the Early Jurassic. If so, several extension directions may have existed at the same time, but that is mechanically impossible. Note that the siliceous cataclasite zones are filled with remobilized quartz (with or without feldspar and prehnite). Many contain open spaces, boxwork, and vuggy structure into which quartz crystals have grown, further indicating their extensional nature. They also contain abundant evidence of reactivation, with ground-up and fragmented early quartz recemented by later quartz veins. These zones rarely exhibit evidence of offset parallel to their lengths: rock-unit boundaries between bodies that trend into them are only minimally offset from one side to another.

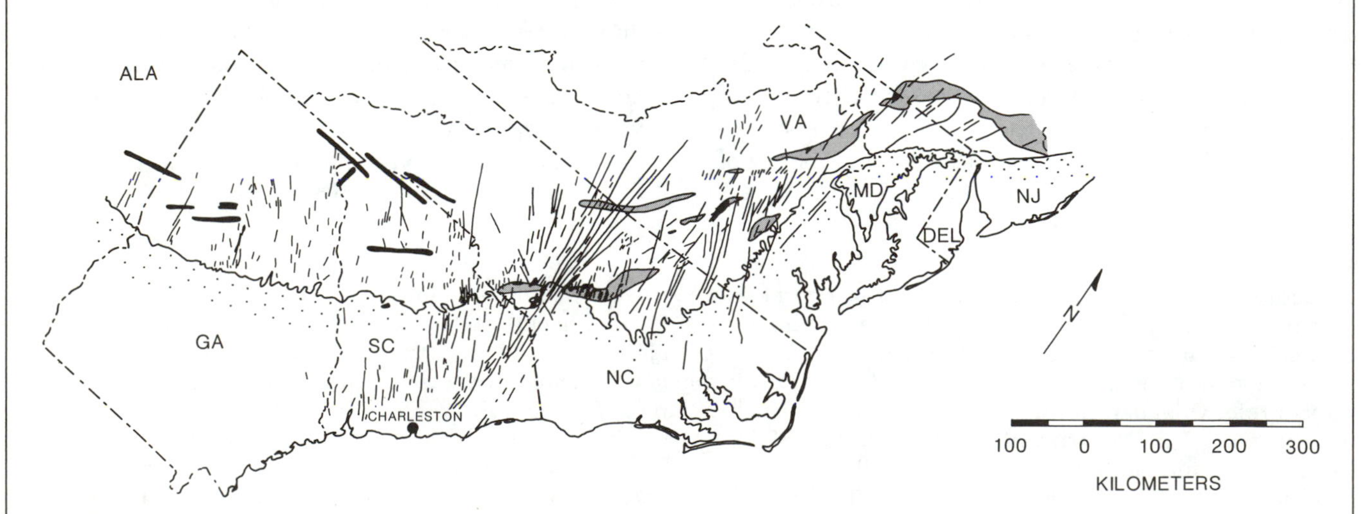

FIGURE 8E–1
Distribution and orientations of diabase and some siliceous cataclasite dikes in the eastern United States. (Modified from P. C. Ragland, R. D. Hatcher, Jr., and David Whittington, 1983, *Geology*, v. 11.)

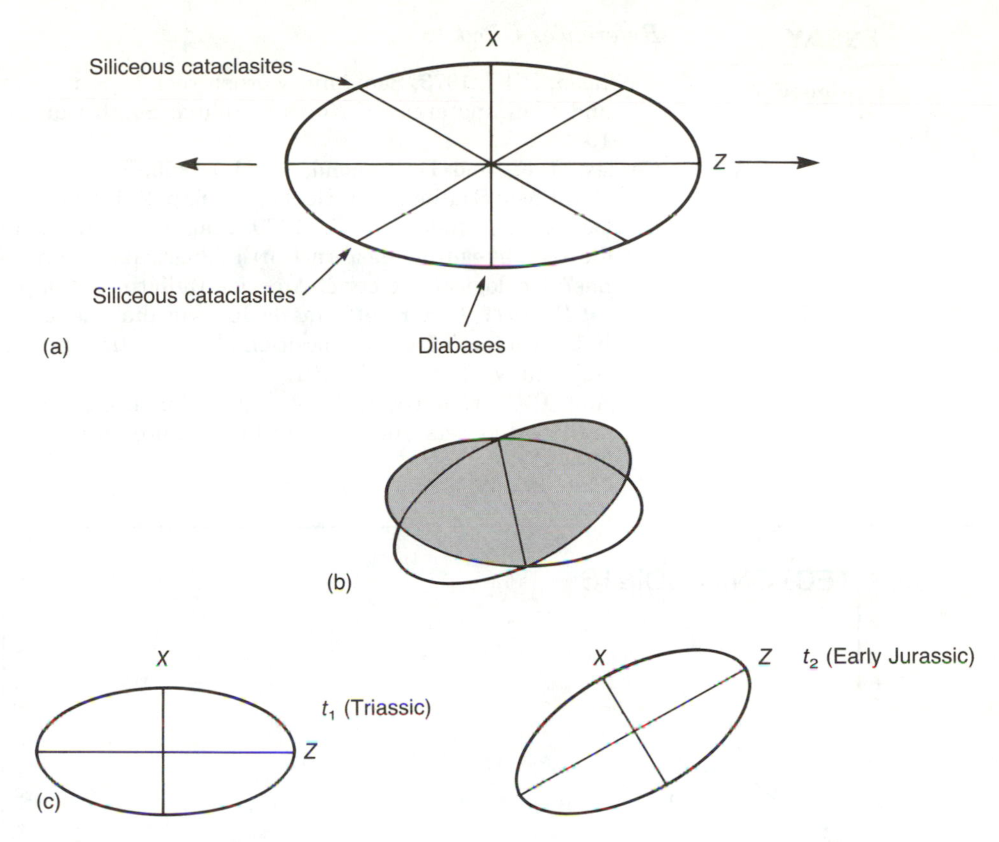

FIGURE 8E–2
Possible strain-ellipsoid orientations for siliceous cataclasite and diabase dikes in the eastern United States. (a) Model 1—Diabase dikes with a N 45° W orientation form by extension; siliceous cataclasite dikes form at the same time by shear. N–S diabases formed later by reorientation of the extensional stress field rotation of the strain ellipse. *Against the model:* Because cataclasites contain open-boxwork quartz, extension is required for at least part of their history. They also have little or no displacement. (b) Model 2—All diabases and siliceous cataclasites formed by extension, produced by multiple orientations of the strain ellipsoid at about the same time. The N–S diabases formed later. *Against the model:* It requires simultaneous and mechanically impossible extension directions and is also unlikely over short periods of geologic time. (c) Model 3—N 45° W diabases formed by extension; siliceous cataclasites were initiated by shear without much accompanying displacement. Siliceous cataclasites were reactivated by later extension and filled with hydrothermal quartz. N–S diabases formed by later extension involving a strain ellipsoid with a different orientation.

One area in the southern Piedmont of Georgia contains siliceous cataclasite dikes that have a left-lateral displacement of a few kilometers (R. J. Hooper, unpublished data).

Is it possible that the early diabase dikes are purely extensional features and that the siliceous cataclasite zones are conjugate shears (Figure 8E–2)? If so, the siliceous cataclasite bodies are shears with no displacement parallel to the shear zone and, once formed, were followed by repeated opening normal to the shears and filling with quartz. If not, simultaneous crustal extension in several directions is required.

ESSAY

continued

References Cited

Birkhead, P. K., 1973, Some flinty crush rock exposures in northwest South Carolina and adjoining areas of North Carolina: South Carolina Geologic Notes, v. 17, p. 19–25.

Conley, J. F., and Drummond, K. M., 1965, Ultramylonite zones in the western Carolinas: Southeastern Geology, v. 6, p. 201–211.

de Boer, J., and Snider, F. G., 1979, Magnetic and chemical variations of Mesozoic diabase dikes from eastern North America: Evidence for a hotspot in the Carolinas?: Geological Society of America Bulletin, v. 90, p. 185–198.

May, P. R., 1971, Pattern of Triassic-Jurassic diabase dikes around the North Atlantic in the context of predrift position of the continents: Geological Society of America Bulletin, v. 82, p. 1285–1291.

Ragland, P. C., Hatcher, R. D., Jr., and Whittington, D., 1983, Juxtaposed Mesozoic diabase dike sets from the Carolinas: A preliminary assessment: Geology, v. 11, p. 394–399.

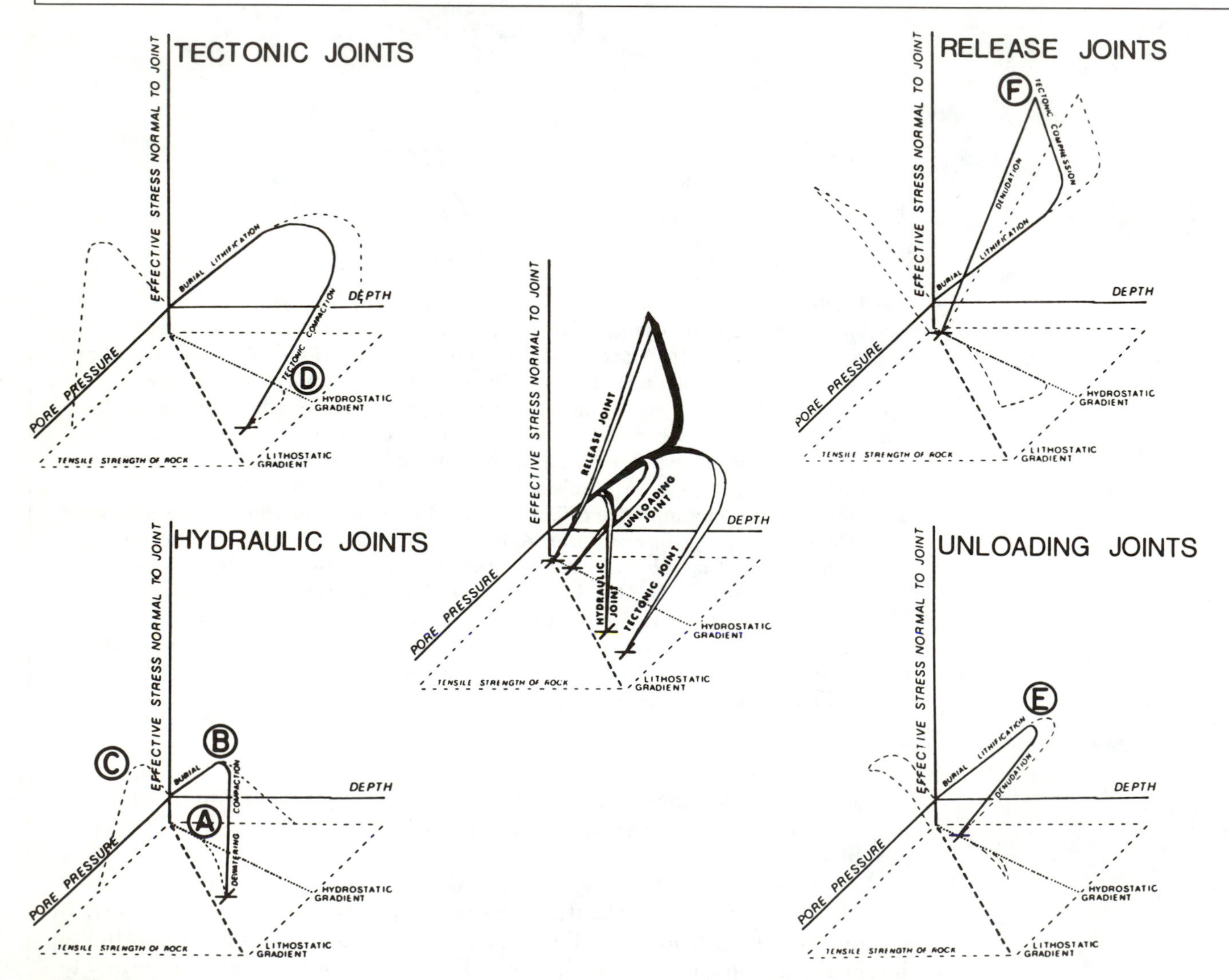

FIGURE 8–15
Loading paths for tectonic, hydraulic, unloading, and release joints. (Reprinted with permission from *Journal of Structural Geology,* v. 7, Terry Engelder, Loading paths to joint propagation during a tectonic cycle: An example from the Appalachian Plateau, © 1985, Pergamon Journals, Ltd.)

cating a transition from extension to shear), spatial relationships in the entire fracture system, and angular relationships among conjugate joints also may be used to decipher joint and shear fractures.

JOINTS IN PLUTONS

Fractures form in plutons in response to cooling and later tectonic stress (Figure 8–16). Orientations of joints that form in a cooling pluton may be influenced by the boundary of the pluton or by its general shape and internal structure. Some large stocks contain joints that are more or less concentric to the internal structure of the pluton and may also be related to the shape of the contact with the country rock. Fracture density and orientation may vary with changes in rock texture or fabric as well as proximity to the boundary of the pluton. Columnar joints (to be discussed shortly) form in flows and shallow plutons as a response to cooling and solidification of magma.

Robert Balk (1937) studied fractures in igneous bodies and recognized that fracturing occurs both as a result of flow related to emplacement and as a product of later regional stress fields. Early joints form in plutons during the final stages of crystallization of magma. Many joints are filled with hydrothermal minerals or crystallization products from late-stage fluid-rich magmas, crystallizing pegmatite, aplite, and quartz veins. Balk recognized that many early fractures form in relation to flow banding in the pluton, either across, parallel, or diagonal to banding. He concluded that the cross fractures are joints and that the diagonal fractures, which form as conjugate sets, are shear fractures. The parallel or longitudinal fractures are more difficult to relate to an extensional or shear-stress field.

Balk also found that joints may be more common and more closely spaced near the margins of a pluton. He suggested that the processes related to magma emplacement and cooling of the margins may fracture both the early-cooled parts of the magma and also the country rock (Figure 8–16), forming both joints and faults.

NONTECTONIC AND QUASITECTONIC FRACTURES

Sheeting

A kind of joint that forms subparallel to surface topography, generally in massive rocks, and corresponds to the unloading joints of Engelder (1985) is called ***sheeting.*** Most often, these fractures may be observed in igneous rocks but they also form in metamorphic rocks. The spacing between sheeting fractures decreases toward the surface and increases

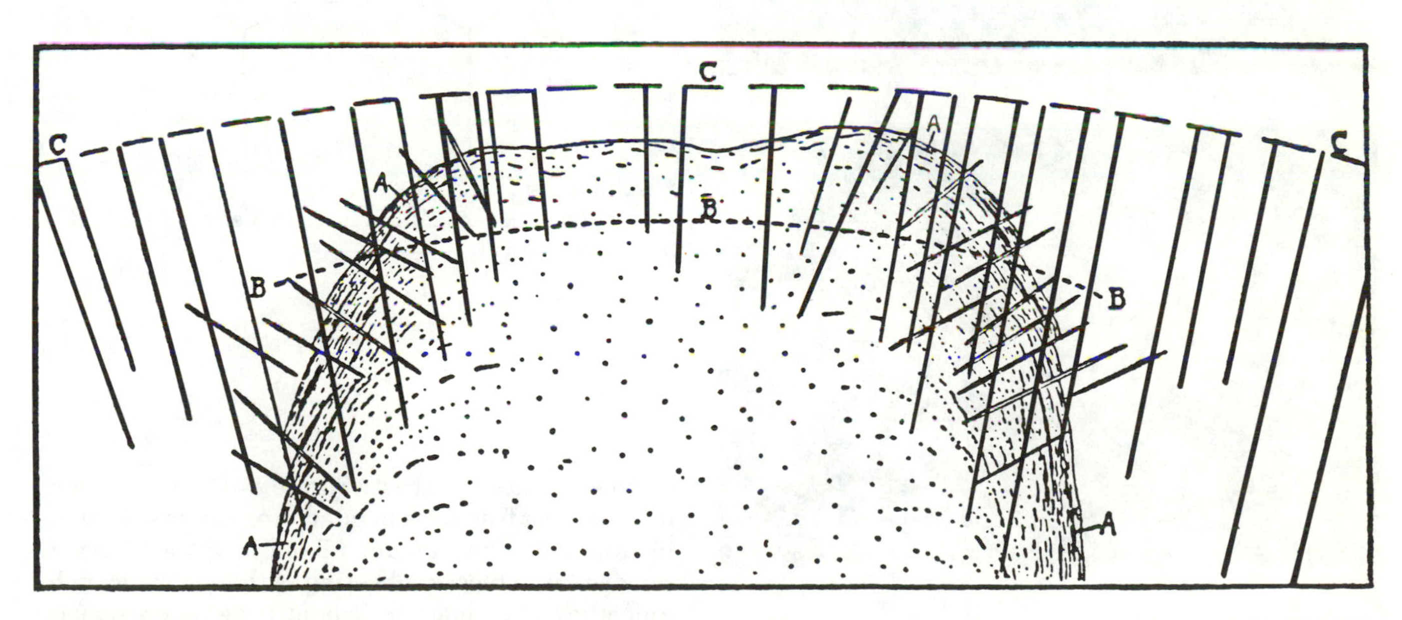

FIGURE 8–16
Joint patterns developed in a pluton. Note the greater joint density near the pluton margins, C–C is the present erosion surface. B–B represents sheeting joints developed parallel to C–C. Joints labeled A form perpendicular to flow directions. Near-vertical joints both outside and inside the pluton form by regional bending of the crust, possibly related to intrusion of the pluton. (From Robert Balk, 1937, *Structural behavior of igneous rocks:* Geological Society of America *Memoir* 5.)

FIGURE 8–17
Sheeting joints in a granite quarry in Massachusetts. Note that the joints are subparallel to the surface and that spacing increases with increased depth to a maximum of 6 m at the lower left. (L. C. Currier, U.S. Geological Survey.)

(a)

(b)

FIGURE 8–18
(a) Columnar joints in Quaternary basalt near Whistler, British Columbia. (b) Closeup of columnar basalt in (a). Columns are about 1 m across. Note fractures perpendicular to the long axes of the columns. (RDH photos.)

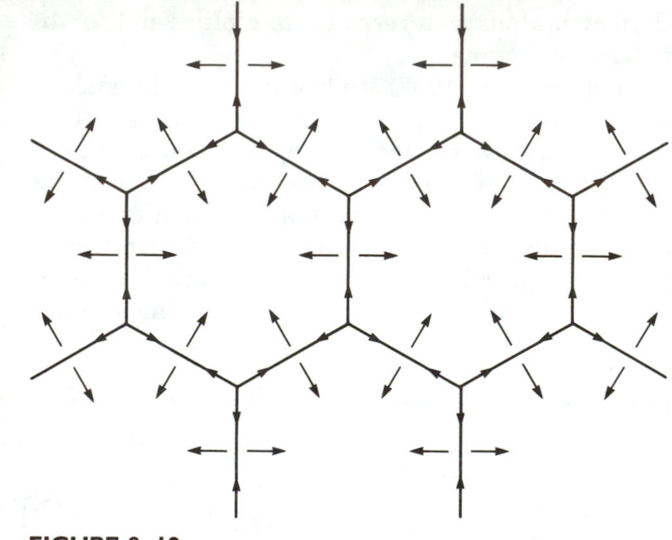

FIGURE 8–19
Formation of hexagonal fractures in a stress field that is nearly cylindrical.

downward into the crust (Figure 8–17). Quarrymen use sheeting fractures in quarrying dimension stone to minimize the amount of blasting necessary to remove large blocks. Sheeting is thought to form by unloading over long periods of time as erosion removes large quantities of overburden from a rock mass. The mass expands normal to the Earth's surface (σ_3 vertical) so that extension fractures form normal to the expansion direction and parallel to the surface (Johnson, 1970).

Columnar Joints and Mud Cracks

Columnar joints form in flows, dikes, sills, and (occasionally) in volcanic necks in larger plutons in response to cooling and shrinkage of congealing magma (Figure 8–18). Thermal gradients and contraction processes in the magma control the orientation of columnar joints. The orientation of columns is generally normal to the sides of a pluton. They may become curved if the thermal gradients and contraction processes are nonuniform or if the magma is still moving slightly as the joints form. Columnar joints commonly have five or six planar sides. Ideally, they would form hexagonal prisms if the stress fields, cooling rates, and thermal gradients were perfectly uniform throughout the cooling magma body (Figure 8–19). A highly symmetrical cylindrical stress field would form in which extension begins as the cylinders contract, forming hexagonal extension fracture patterns in the congealing magma. Mud cracks (Chapter 2) form on the Earth's surface by a similar mechanism—shrinkage and evaporation of water in unconsolidated sediment (Figure 2–7); they appear as four-, five-, six-, or seven-sided polygons, which frequently curl up at their edges.

Questions

1. How can we distinguish between systematic and nonsystematic joints?
2. What does asymmetrical twist hackle structure on a joint surface tell you about the way the joint formed?
3. How would systematic orientation of joints be established by a hydraulic-fracturing mechanism?
4. How could filled joints and unfilled joints form at the same time with differing orientations?
5. What is the best evidence that most joints form by extension?
6. How do columnar joints form?
7. What evidence tells us that some regional joints in eastern North America may have formed in the present-day stress field?
8. Suppose joints formed at a locality where the earliest set was filled with quartz-feldspar (pegmatite), then cut by a set containing quartz-epidote, which was in turn crossed by a quartz-calcite-prehnite set, then by a set of calcite- and laumontite-filled joints, and finally by an unfilled joint set. (Each set has a different orientation.) What do you conclude about the conditions producing joints there?
9. Are sheeting joints and Engelder's release joints related?
10. How can tectonic stresses tens or hundreds of millions of years ago influence orientations of joints that formed in the past 5 million years?

Further Reading

Engelder, Terry, 1985, Loading paths to joint propagation during a tectonic cycle: An example from the Appalachian Plateau, USA: Journal of Structural Geology, v. 7, p. 459–476.
Classifies joints and relates them to variables of effective stress normal to the joint, depth, pore pressure, tensile strength of the rock, and hydrostatic/lithostatic gradients through time.

Engelder, Terry, 1987, Joints and shear fractures in rock, *in* Atkinson, B., ed., Fracture mechanics of rock: London, Academic Press, p. 27–69.
Reviews concepts of joint and shear fracture formation, as well as features that occur on joint surfaces and information that can be derived from these surfaces.

Hancock, P. L., 1985, Brittle microtectonics: Principles and practice: Journal of Structural Geology, v. 7, p. 437–457.
Analyzes brittle structures as a way to solve tectonic problems. Joints are assumed to result either from extension or shear.

Nickelsen, R. P., 1979, Sequence of structural stages of the Alleghany orogeny, at the Bear Valley Strip Mine, Shamokin, Pennsylvania: American Journal of Science, v. 279, p. 225–271.

A classic study of a well-exposed series of rocks deformed during a single event recording a sequence of extension followed by compression and possibly later extension. Nickelsen related each stage of deformation to the strain ellipsoid and attempted to determine stress orientations for several stages.

Pollard, D. D., and Aydin, A., 1988, Progress in understanding joints over the past century: Geological Society of America Bulletin, v. 100, p. 1181–1204.

Reviews development of ideas on joint formation, including tectonic and nontectonic joints.

Secor, D. T., 1965, Role of fluid pressure in jointing: American Journal of Science, v. 263, p. 633–646.

Formation of joints under influence of pore fluids under pressure is a model that follows directly from Hubbert and Rubey's model for thrust faults (Chapters 10 and 11).

Wise, D. U., 1982, Linesmanship and the practice of linear geo-art: Geological Society of America Bulletin, v. 93, p. 886–888.

A tongue-in-cheek look at the ways lineaments are misinterpreted. Wise listed 32 rules derived from "logic" used in interpreting and misinterpreting lineaments.

Wise, D. U., Funiciello, Renato, Maurizio, Parotto, and Salvini, Francesco, 1985, Topographic lineament swarms: Clues to their origin from domain analysis of Italy: Geological Society of America Bulletin, v. 96, p. 952–967.

Lineaments in Italy from LANDSAT and other data were analyzed in an attempt to relate them to systematic joints in a tectonically active region.

9

Fault Classification and Terminology

The occurrence of faults and dykes must have been known from the earliest days of mining. It was only gradually, however, that they became an object of scientific study.

ERNEST M. ANDERSON, *The dynamics of faulting*

FAULTS ARE THE GENERATORS OF SEISMIC ACTIVITY TODAY; OUR interest in them is practical as well as scientific and aesthetic. Understanding faults is useful in the design for long-term stability of dams, buildings, bridges, and power plants, as well as for their effect on population distribution. Study of faults helps in understanding mountain-building and deformation processes—studies that many times turn out to have practical value. Faults have produced some of the Earth's most spectacular scenery (Figure 9–1), including the Grand Tetons in Wyoming and the Canadian and Montana Rockies. Faults are largely responsible for the great mountain chains on the Earth.

A ***fault*** is a fracture having appreciable movement parallel to the plane of the fracture. Sometimes it is difficult to specify what "appreciable movement" is and what we may or may not choose to call a fault—a somewhat scale-dependent judgment. We have no difficulty at all identifying a very large fault that records many kilometers of movement; on the other hand, a fault having a few centimeters offset in one study may be considered only a large fracture or a joint with slight offset in another because of the different scales of each study. Even small offsets become significant, however, when we weigh the impact of an active fault on nearby buildings or dams.

Faults occur in many forms and dimensions. They may be hundreds of kilometers long or only a few centimeters. Their outcrop traces may be straight or sinuous. They may occur as knife-sharp boundaries or as *fault* or *shear zones* millimeters to several kilometers thick (Figure 9–1). Fault or shear

FIGURE 9–1
Aerial view of the San Andreas fault, San Luis Obispo County, California. (R. E. Wallace, U.S. Geological Survey.)

zones may consist of a series of interleaving anastomosing brittle faults and crushed rock (cataclasite) formed near the surface or of mylonitic rocks produced by ductile faulting at great depth. Movement on ductile faults is distributed over a zone in contrast to brittle faults, where movement may be confined to a single plane. Most of our discussion in this chapter will deal with the descriptive properties of brittle faults; we will discuss shear zones in Chapter 10.

ANATOMY OF FAULTS

We need to understand the general descriptive anatomy of faults (Figure 9–2) before discussing the specific kinds of faults in detail. The most obvious feature related to faulting is the displacement of some marker, most commonly bedding. Displacement occurs along the ***fault plane,*** the actual movement surface. If the fault plane is not vertical, the rock mass resting on the fault plane is called the ***hanging wall,*** and the rock mass beneath the fault plane is called the ***footwall.*** We can measure the dip and strike of the fault plane and make other measurements of structures (such as slickensides, to be discussed shortly) that indicate relative movement along the fault plane.

Two end-member fault types exist: *dip-slip* and *strike-slip* faults (Figure 9–2). The ***slip*** along a fault describes the movement parallel to the fault plane. We speak of ***dip-slip*** where movement is down or up parallel to the dip direction of the fault. The term ***strike-slip*** applies where movement is parallel to

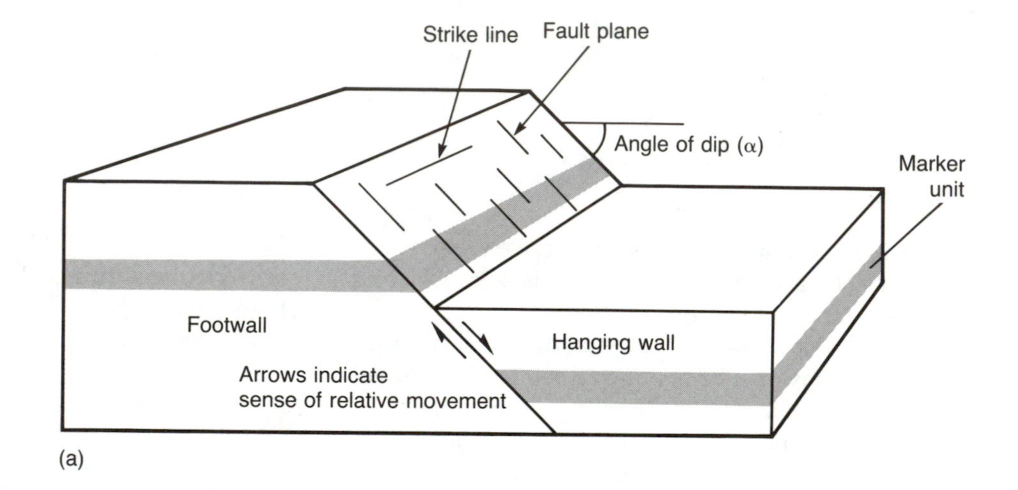

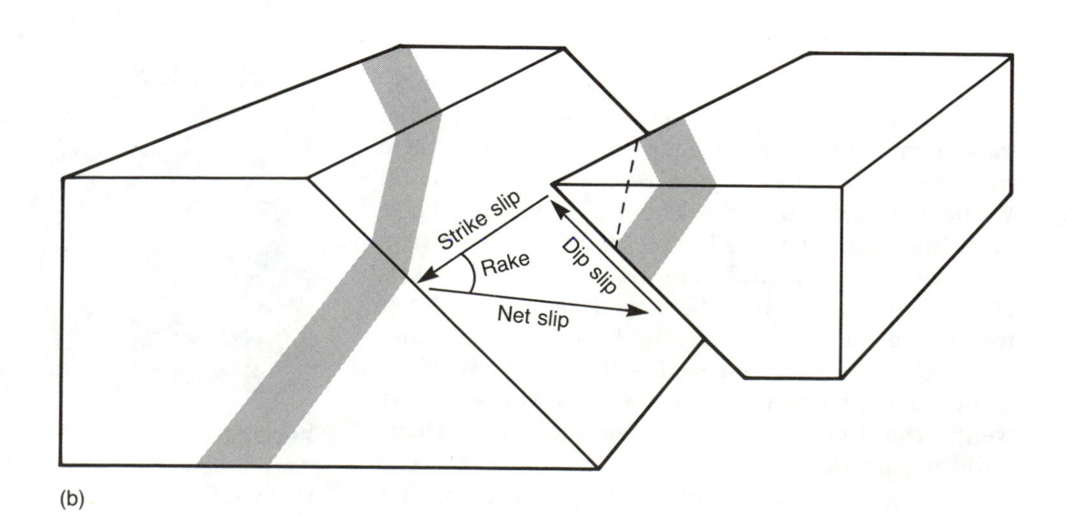

FIGURE 9–2
(a) Anatomy of faults.
(b) Oblique-slip fault showing the components of net slip and the rake of net slip.

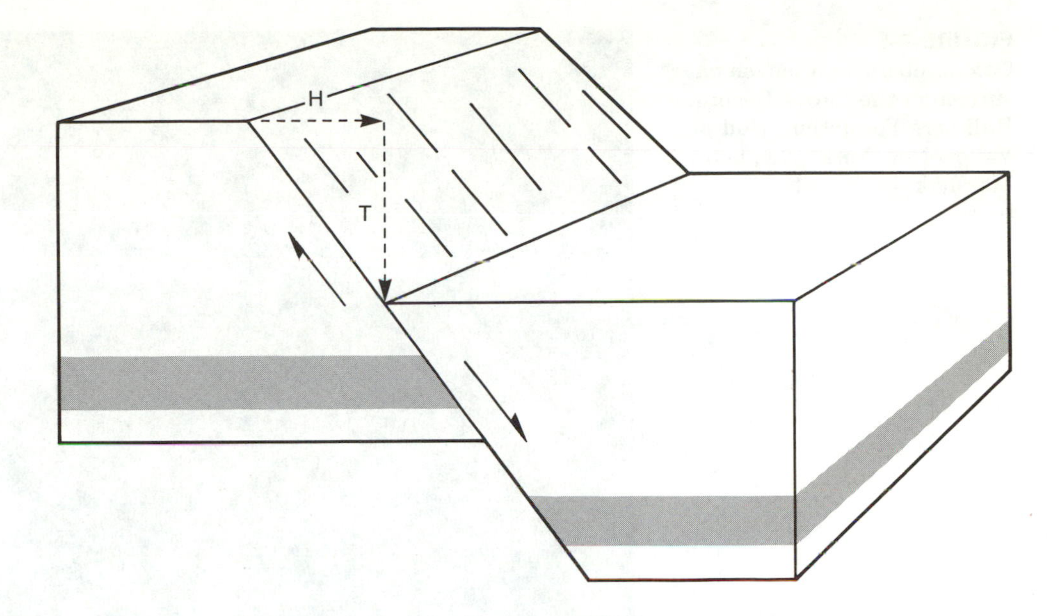

FIGURE 9–3
The components of dip-slip—heave (H) and throw (T).

the strike of the fault plane. Most faults exhibit dominantly dip-slip or dominantly strike-slip motion. A combination is referred to as ***oblique-slip.*** The ***net slip,*** or ***displacement,*** is the total amount of motion measured parallel to the direction of motion. We actually cannot determine the absolute motion sense on a fault without some primary criterion, as by establishing benchmarks before an earthquake and surveying them afterward, but the *rake of net slip* is often a useful measurement (Figure 9–2b). ***Heave*** describes the horizontal component of displacement and ***throw*** is the vertical component (Figure 9–3).

Another term is ***separation***—the amount of apparent offset of a faulted surface, such as a bed or a dike, measured in a specified direction (Figure 9–4). We can speak of ***strike separation, dip separation,*** and the total, or ***net, separation*** of a fault. The terms *heave* and *throw* are sometimes used to describe horizontal and vertical components of separation.

A common feature on fault surfaces is growth of fibrous minerals or grooves aligned parallel to the movement direction. They are frequently arranged in a series of steps, with the down side in the movement direction of the opposing fault block. Striated surfaces are called ***slickensides,*** and the striations on them are called *slickenlines* (Figure 9–5). Aligned fibrous minerals on a movement surface are *slickenfibers*. All indicate the trend of relative movement (such as north-south or northeast-southwest) along a surface, but the small steps may also be used to determine movement direction. The direction of down-stepping is commonly the movement direction. Slickensides most frequently record only the last movement event on the fault.

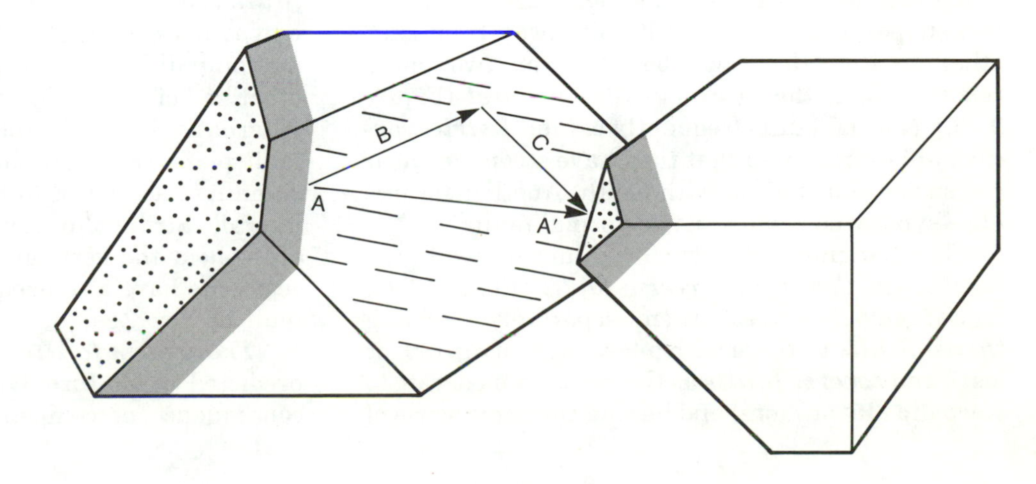

FIGURE 9–4
Oblique-slip fault showing the separation (A-A′) of a marker layer and its components, strike-separation (B) and dip-separation (C), measured in the fault plane.

FIGURE 9–5
Calcite fibers on a movement surface in the Lower Devonian Kalkberg Formation, Hudson Valley, New York. (Stephen Marshak, University of Illinois.)

ANDERSONIAN CLASSIFICATION

The fault classification devised by Ernest M. Anderson (1942), a British geologist, defines three fundamental categories: *normal faults, thrust faults,* and *wrench (strike-slip) faults* (Figure 9–6). (Elements of his classification may have been in use before Anderson's work, but the history is difficult to trace.) Thrust and normal faults are dip-slip faults on the basis of sense of motion of the hanging wall and footwall relative to each other.

Normal faults are dip-slip faults in which the hanging wall has moved down relative to the footwall. A ***graben*** consists of a block that has been dropped down between two subparallel normal faults that dip toward one another (Figure 9–7a). If two subparallel normal faults dip away from each other so that the block between the two faults remains high, the block is called a ***horst*** (Figure 9–7b). Normal faults frequently exhibit ***listric*** (concave up) geometry so that they have steep dips near the surface but flatten with depth. Another term is *lag*—synonymous with listric normal fault.

The hanging wall has moved up relative to the footwall in *thrust* and *reverse faults* (Figure 9–6). Some geologists treat them separately, defining ***thrust faults*** as those with a low angle of dip (30° or less), and ***reverse faults*** as those with a moderate to steep dip (45° or more) and having the same sense of motion as thrusts. These two fault classes are mechanically similar and we will assume them to be the same in our discussion here, although Anderson's theory of faulting (Chapter 10) does not predict the formation of reverse faults. Both high- and low-angle segments occur on many thrust faults. Low-angle segments may likewise be found along many reverse faults.

Anderson described two categories of ***strike-slip faults*** on the basis of direction of relative motion as seen by an observer looking along the fault plane (Figure 9–8). Strike-slip faults, where the right side moves toward the observer, are called ***dextral*** or ***right-lateral strike-slip faults.*** Those in which the left side moves relatively toward the observer are called ***sinistral*** or ***left-lateral strike-slip faults.*** Note that if the observer stands on one end of a hypothetical fault block, the same sense of relative motion will be observed if he (or she) moves to the other end of the block, turns around, and looks back. Alternatively, if the observer looking across the fault plane sees that the side opposite where he is standing has moved to the right, the fault is right-lateral. Strike-slip faults commonly have steep dips—near the vertical—but moderate and shallow segments have also been noted, as along the Alpine fault in New Zealand.

Transform faults are a type of strike-slip fault predicted by J. Tuzo Wilson (1965) as a necessary consequence of compensating for relative motion

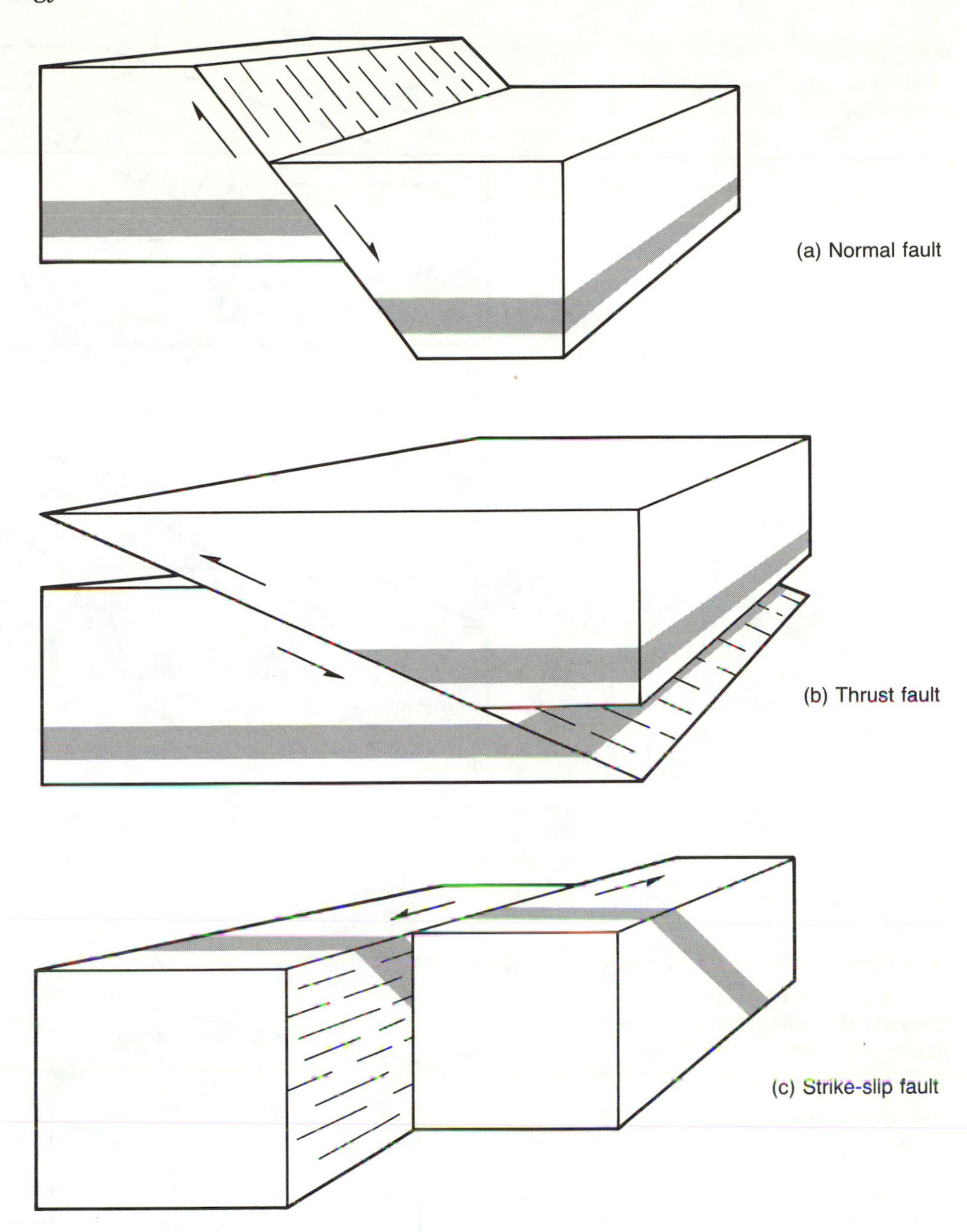

FIGURE 9–6
Anderson's fault classification.

between lithospheric plates (Figure 9–9a). His prediction was confirmed by seismologist Lynn Sykes (1967) from first-motion studies of earthquakes along the Mid-Atlantic Ridge. Because they connect parts of plates or are themselves plate boundaries, several kinds of transforms have been defined. They include ***ridge-ridge, ridge-arc,*** and ***arc-arc*** transform faults (Figure 9–9b).

Among terms that describe faults are *en echelon faults,* which roughly parallel one another but occur in short segments, sometimes overlapping; *radial* faults, which converge (or project) toward a single point; and *concentric faults,* which form concentric to a point (Figure 9–10; see also Figure 2–32). *Bedding faults* or *bedding-plane faults* follow bedding or occur parallel to the orientation of bedding planes (Figure 9–11). Many thrust and normal faults are bedding faults because the fault propagates along a weak zone parallel to bedding (Chapters 11 and 13).

CRITERIA FOR FAULTING

Existence of a fault at a particular locality is commonly demonstrated by the recognition of certain features and diagnostic characteristics. If all faults

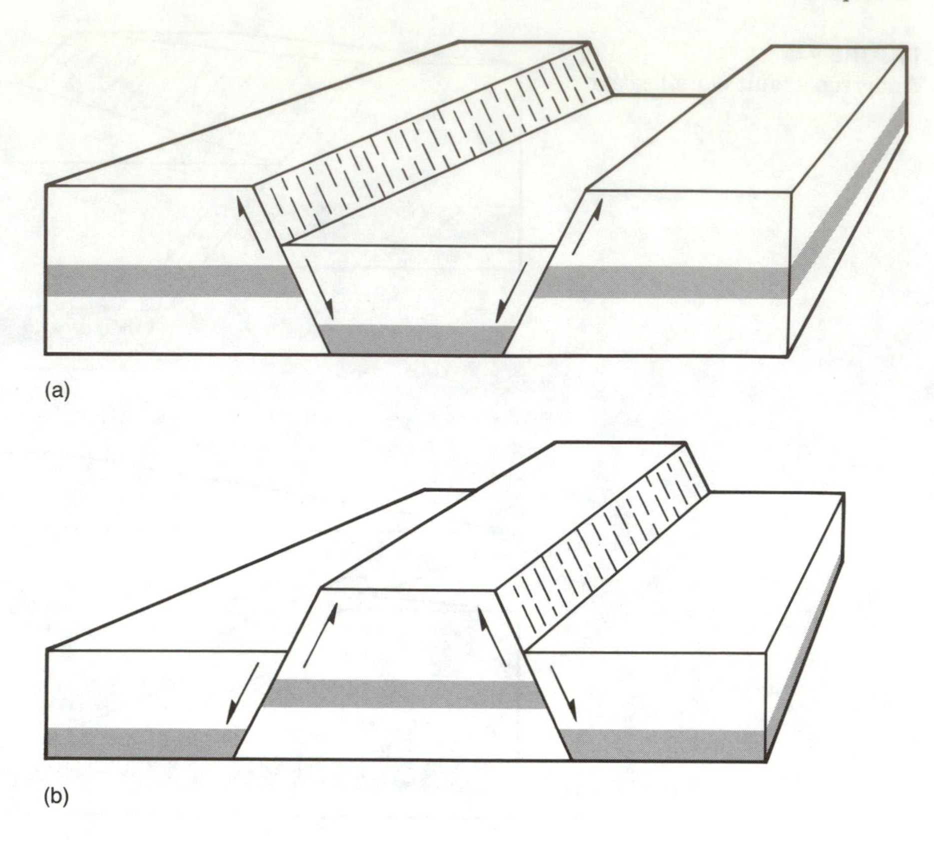

FIGURE 9–7
Graben (a) and horst (b) structures.

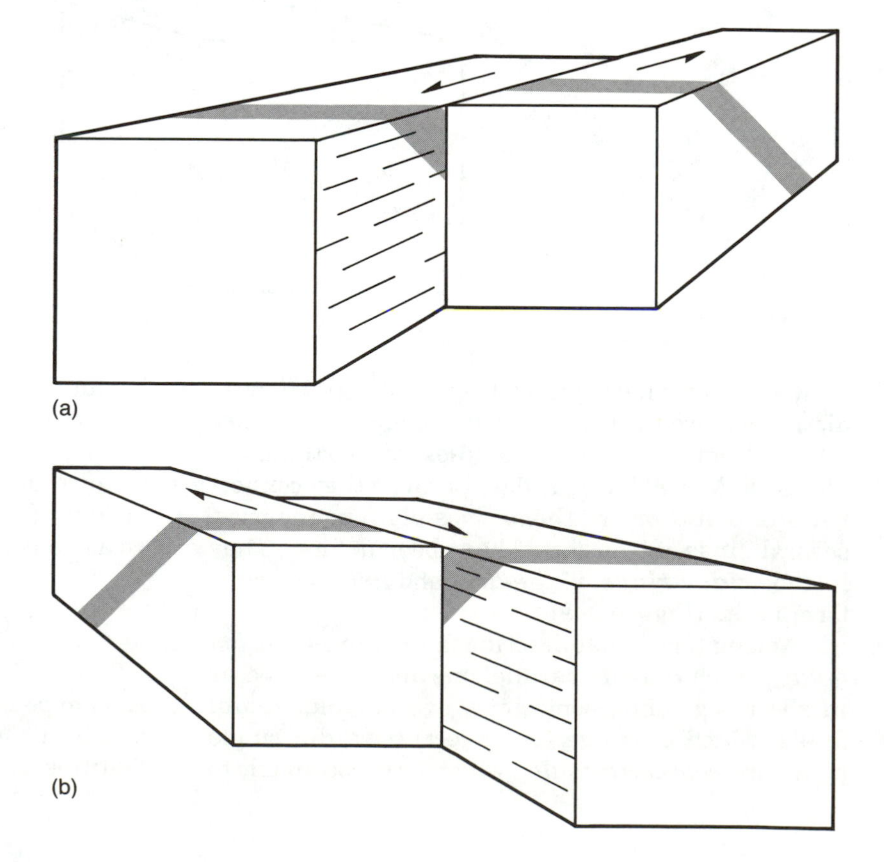

FIGURE 9–8
(a) Sinistral (left-lateral) and (b) dextral (right-lateral) strike-slip faults.

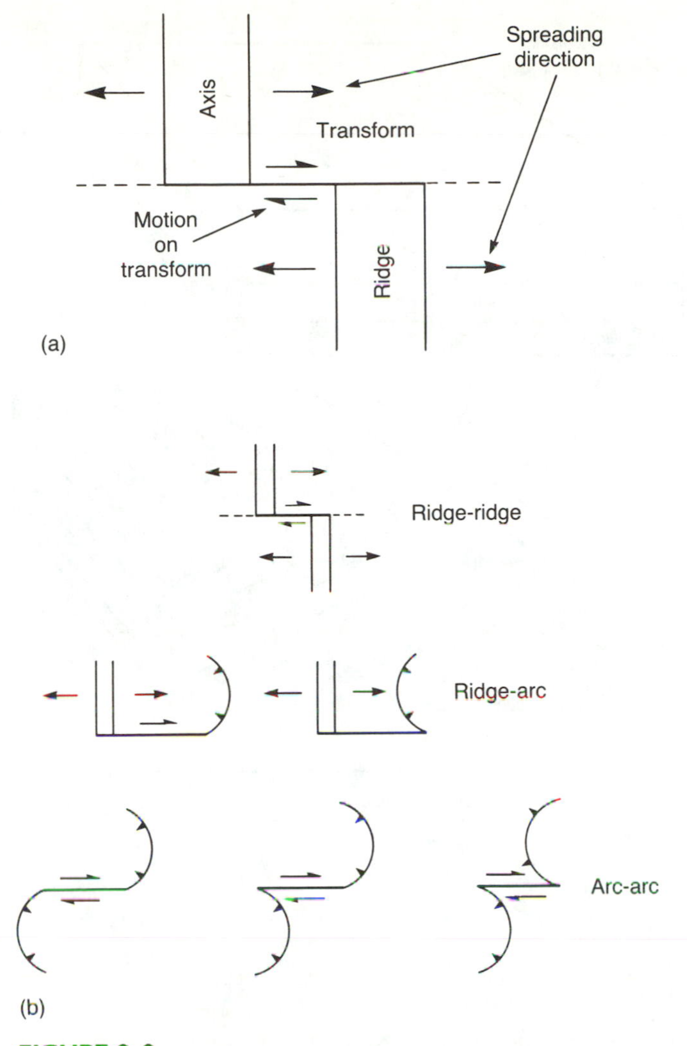

FIGURE 9–9
(a) Motion on transform faults. (b) Kinds of transform faults.

were active, we would need only one criterion: earthquakes and associated ground breakage. Most faults are inactive, however, and so we must use other means of recognizing them. Unfortunately, in many regions we cannot directly observe the fault surface because of younger deposits and soil or vegetation cover.

Probably the most fundamental way to demonstrate a fault is present is ***repetition*** or ***omission of stratigraphic units,*** or the ***displacement (offset) of a recognizable marker*** (Figure 9–12). Demonstrating that a rock unit or sequence has been repeated should prove the existence of a fault if it can be readily shown that repetition is not related to folding. Repetition by faulting commonly produces asymmetric repetition; repetition by folding commonly produces symmetric repetition. Likewise, units may be omitted from a sequence wherein continuity has been established elsewhere. Either repetition or omission can be demonstrated by recognition of marker units or by some other physical characteristic that allows us to make correlations and show that displacement or omission of rock units has actually occurred. Guide (index) fossils, color, texture, composition, or other distinctive features in a rock unit may set it apart so that it can serve as a marker unit.

The ***truncation of structures, beds,*** or ***rock units*** against some feature is another criterion for faulting (Figure 9–12), but truncation must be used with care because structures may be truncated by unconformities and intrusions (Chapter 2). Therefore a second criterion should be used to demonstrate that the boundary in question is a fault and not an unconformity. Truncation of structures on

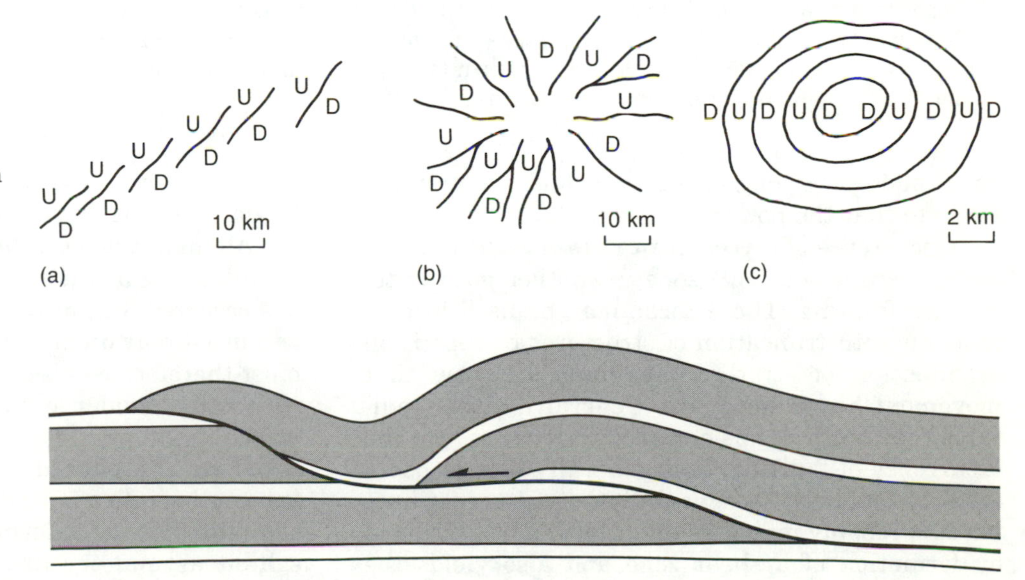

FIGURE 9–10
En echelon faults with sinistral steps (a), radial (b), and concentric faults (c) in map view. U indicates upthrown side, D downthrown side. (See Figure 2–32 for radial and concentric faults occurring in an impact structure.)

FIGURE 9–11
Bedding (plane) fault in cross-section view.

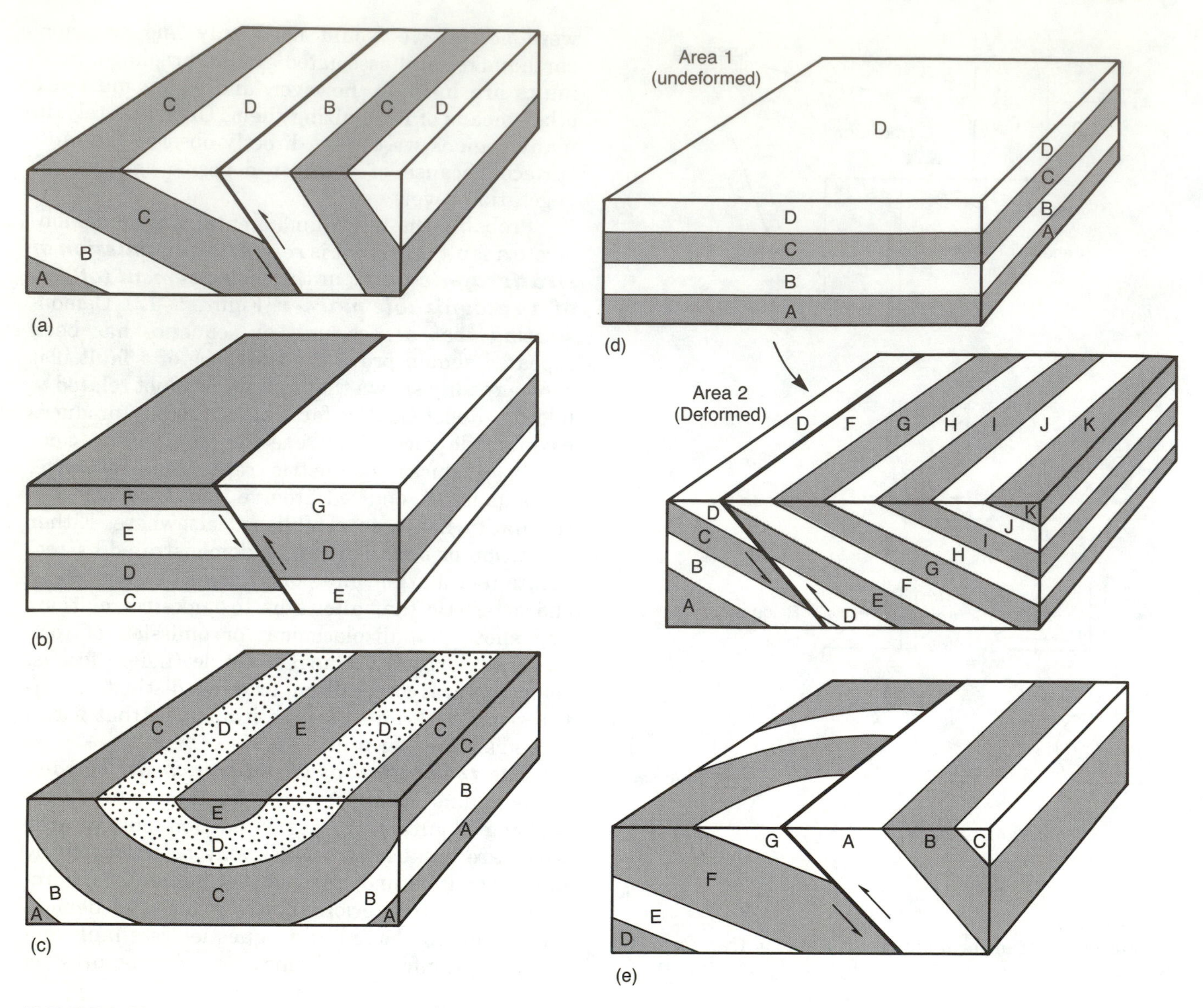

FIGURE 9–12
(a) Repetition or omission of stratigraphic units. (b) Offset of stratigraphic markers. (c) Repetition by folding. Note the symmetry in the outcrop pattern. (d) Omission of a unit by faulting in one area that can be shown to be present in a continuous sequence elsewhere. (e) Truncation of structures.

both sides of a discontinuity would most likely indicate that the boundary is a fault.

Occurrence of ***mylonite*** or ***cataclasite*** (or both) along a suspected fault zone is another good criterion for faulting. Those rocks may be used in conjunction with truncation of structures or repetition or omission of stratigraphic units to show that movement has taken place. Generally, either indicates that faulting has occurred. But to confirm the occurrence of faulting, it is generally best to use a displaced marker. Presence of S-C structures (Chapter 10) probably indicates that faulting has occurred by formation of a shear zone and also yields the sense of displacement. Other criteria are required for determining the amount of displacement.

Abundant *veins, silicification,* or other *mineralization* along a fracture zone may suggest faulting has occurred, but mineralization is a poor criterion because it may occur in fractures having no offset. It must therefore be used carefully and in conjunction with other evidence to prove whether a fault is present.

Drag is produced along a fault as units appear to be pulled into a fault during movement. The sense of motion may be determined because the drag folds exhibit asymmetry in the direction of movement.

Drag may occur on thrust faults so that the layers appear to have been pulled down-dip in the hanging wall or up-dip in the footwall (Figure 9–13a), opposite to the sense of relative motion. ***Reverse drag*** occurs along some listric normal faults (Figure 9–13b) where the layering appears to have been dragged down-dip parallel to movement along the fault. Reverse drag is produced by motion along the fault as it flattens, creating a rotational displacement up-dip on the more steeply dipping segment and apparent drag in the up-dip region.

Slickensides along a fault surface (Figure 9–5) often indicate movement but do not prove faulting because they may form along bedding surfaces during certain types of folding or may actually form on joints that have a small amount of movement. Slickensides help confirm only that movement has occurred.

Certain characteristics of topography may indicate faulting has occurred. Drainage may be controlled by faults (Figure 9–14). If active faulting is present, drainage may be successively offset over a few thousand—or a few hundred thousand—years, thereby providing a major criterion that can be observed on the surface as an indication of offset along a fault. ***Fault scarps*** form where a topographic surface is offset by dip-slip motion along a fault and directly indicate the movement sense of the fault (Figure 9–14a; Figure 9–15). A fault scarp may evolve through time into a ***fault-line scarp*** by differential erosion along the fault, lowering the level of the topographic surface and removing a

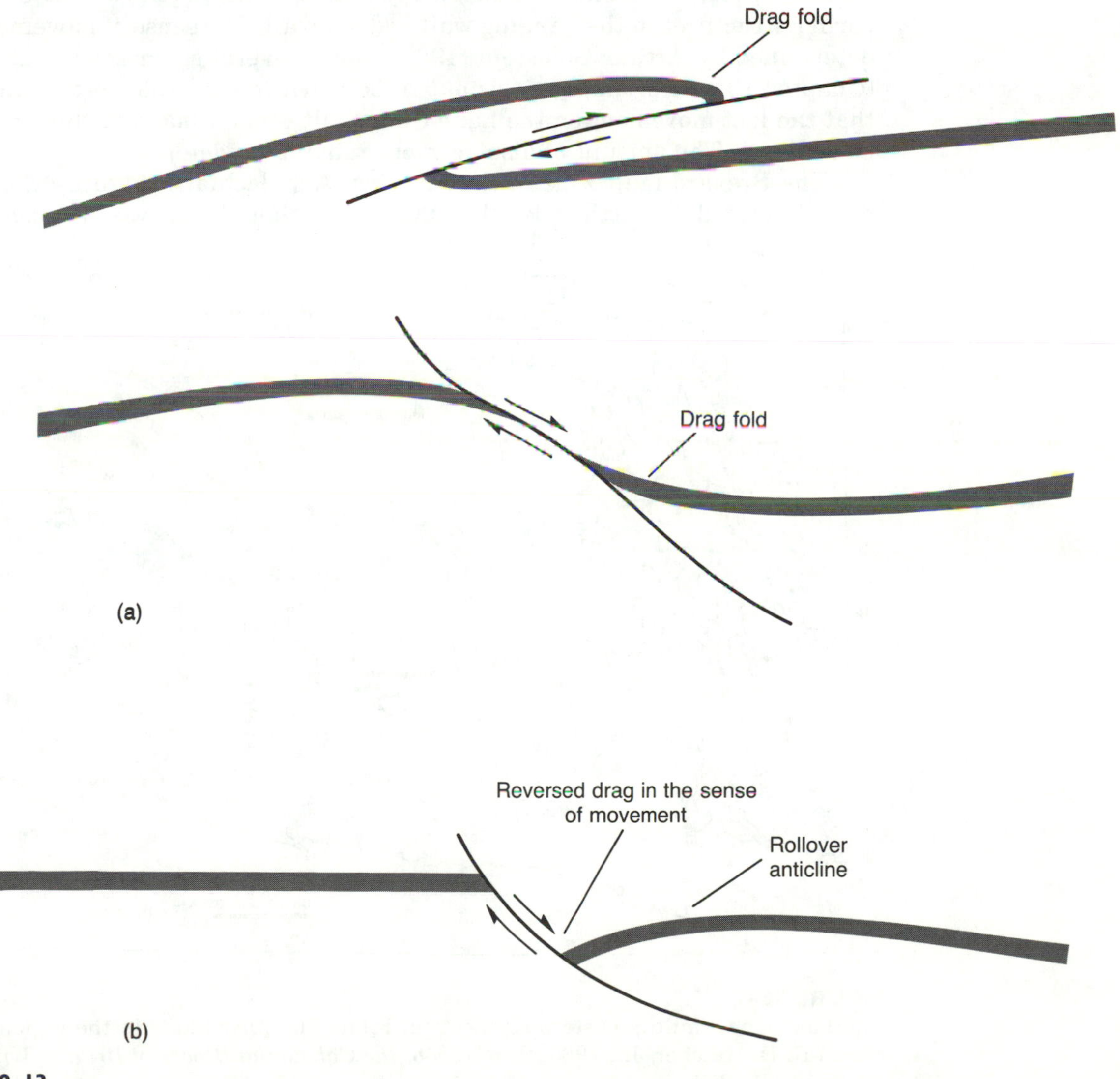

FIGURE 9–13
(a) Drag on thrust and normal faults. (b) Reverse drag on a normal fault. Cross-section view.

ESSAY

Existence and Displacement Sense of Large Faults

Some of the very large inactive faults in ancient mountain chains were discovered by recognition of one or more of the criteria for faulting discussed in this chapter. Yet the sense of motion and displacement on some of these faults has been debated for many decades because elements of the puzzle remain missing, or the history of movement is complex, or both.

The Lake Char fault (Char is short for Chargoggagoggmanchauggagogg-chaubunagungamaugg) in the southeastern New England Appalachians has been described by Roberta Dixon and Lawrence Lundgren (1968) as part of an extensive fault complex. It is a west-dipping low-angle fault in Connecticut that steepens northeastward and joins the more steeply dipping Clinton Newbury–Bloody Bluff fault system in Massachusetts (Figure 9E–1). The low-angle character of the Lake Char fault had been recognized earlier by the gentle dip along its outcrop trace, and then Robert Wintsch (1979) discovered that the fault reappeared in the core of the Willimantic dome farther west. The Lake Char fault has been assumed to be a thrust that moved rocks from the west in the hanging wall over the more easterly rocks now in the footwall. Unfortunately, there are no easily recognized marker units to help measure the displacement. Existence of the fault had been confirmed by Dixon and Lundgren primarily on the basis of mylonites along its trace that truncate earlier structures in the hanging wall and footwall. The sense of movement, as determined by Arthur Goldstein (1982), using several shear-sense indicators (Chapter 10), surprisingly turns out to be down toward the west, indicating that the last movement on the Lake Char fault was normal and thus may not be a thrust. The amount of displacement remains undetermined.

The Brevard fault zone in the southern Appalachians (Figure 9E–2) was first described by Arthur Keith (1907), who thought it was a syncline of

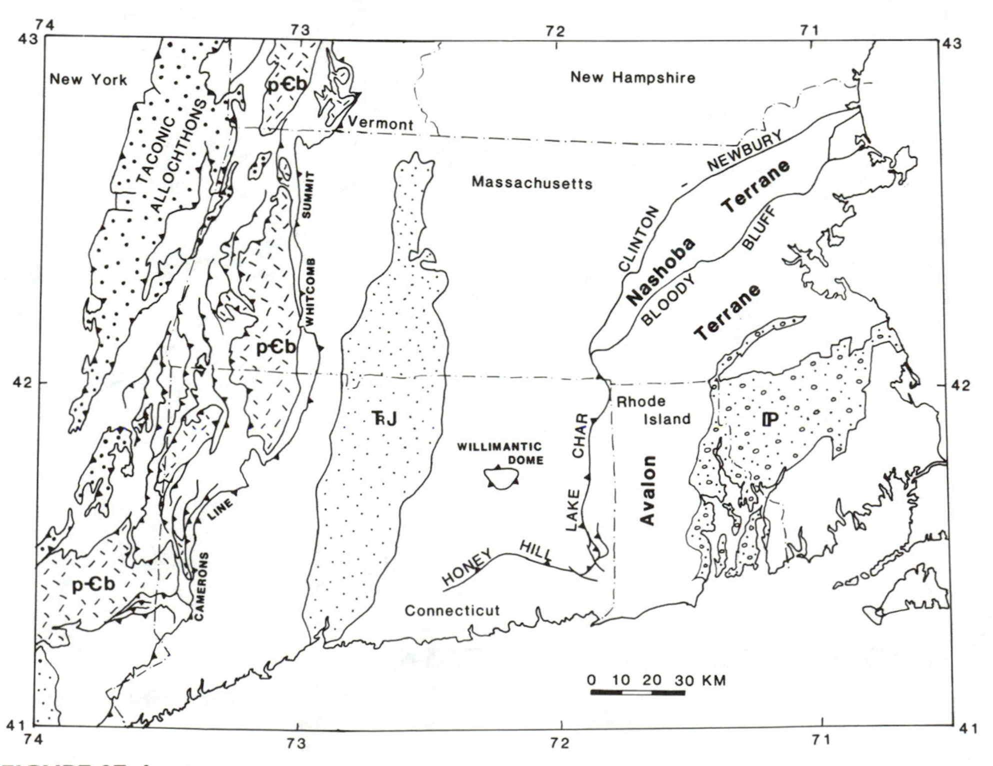

FIGURE 9E–1
The Lake Char fault in eastern Connecticut related to other faults in the region. (From R. D. Hatcher, Jr., 1986, *Synthesis of the Caledonian Rocks of Britain*, Kluwer Academic Publishers.)

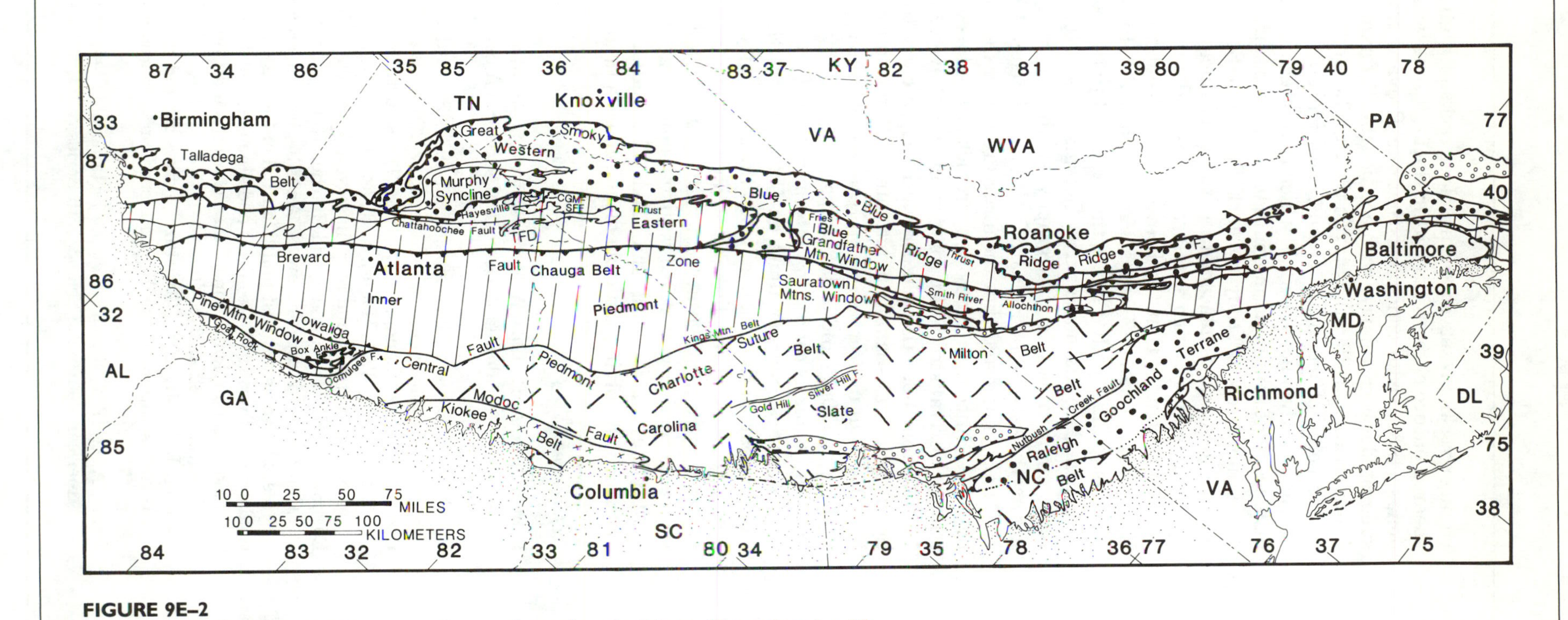

FIGURE 9E–2
Location of the Brevard fault zone in the southern Appalachians. (Reproduced, with permission, from the *Annual Review of Earth and Planetary Sciences,* Vol. 15. © 1987 by Annual Reviews Inc.)

ESSAY

continued

younger rocks preserved within older rocks of high metamorphic grade. The sheared nature of the rocks within the fault zone was first recognized by Anna Jonas (1932), who concluded that the Brevard fault zone is a thrust and suggested it is correlative with a major thrust farther north—in Maryland and Pennsylvania—called the Martic thrust. The Brevard fault, like many others in the cores of mountain chains, does not offset easily recognized markers throughout the known length, which is greater than 300 km. Consequently, its displacement sense and magnitude remain obscure.

A more recent study by Jack Reed and Bruce Bryant (1964) used the orientations and displacement patterns of linear structures along the North Carolina segment (such as quartz *c*-axes, and mineral lineations—see Chapter 18) and concluded that the Brevard is a left-lateral strike-slip fault. In 1970, Reed, Bryant, and Myers suggested that the Brevard was a right-lateral strike-slip fault with a major thrust (dip-slip) component. Then, after studies in South Carolina and nearby Georgia, I reported exotic blocks of footwall rocks and a traceable stratigraphy within and southeast—but unfortunately not northwest—of the Brevard fault zone (Hatcher, 1971). My studies (Hatcher, 1971) also suggested that the structure was a thrust, as originally concluded by Jonas. Since then, Andy Bobyarchick (1984) has found that the Brevard had a major dextral strike-slip component, and Steven Edelman, Angang Liu, and I (1987) concluded that the Brevard had an early thrust component, followed by later dextral strike-slip and later still moved by thrusting. This structure has also been interpreted as an Alpine root zone by Clark Burchfiel and John Livingston (1967) and as a transported suture by Douglas Rankin (1975).

Why so many conflicting interpretations for one fault? Part of the problem lies in critical marker units on both sides of the fault: they do not exist. Another part of the problem lies in the complexity of the structure. It probably has experienced both dip- and strike-slip movement. Yet another part of the problem is that different geologists have worked on different parts of the structure, which is exposed well in some areas but poorly in others. Large faults like the Brevard and Lake Char often undergo a long history of movement involving multiple displacements in different directions.

References Cited

Bobyarchick, A. R., 1984, A late Paleozoic component of strike-slip in the Brevard zone, southern Appalachians: Geological Society of America Abstracts with Programs, v. 16, p. 126.

Burchfiel, B. C., and Livingston, J. L., 1967, Brevard zone compared to Alpine root zones: American Journal of Science, v. 265, p. 241–256.

Dixon, H. R., and Lundgren, L., 1968, The structure of eastern Connecticut, *in* Zen, E-an, White, W. S., Hadley, J. B., and Thompson, J. B., Jr., Studies of Appalachian geology: Northern and Maritime: New York, Wiley-Interscience, p. 261–270.

Edelman, S. H., Liu, Angang, and Hatcher, R. D., Jr., 1987, The Brevard zone in South Carolina and adjacent areas: An Alleghanian orogen-scale dextral shear zone reactivated as a thrust fault: Journal of Geology, v. 95, p. 793–806.

Goldstein, Arthur, 1982, Geometry and kinematics of ductile faulting in a portion of the Lake Char mylonite zone, Massachusetts and Connecticut: American Journal of Science, v. 282, p. 378–405.

Hatcher, R. D., Jr., 1971, Structural, petrologic and stratigraphic evidence favoring a thrust solution to the Brevard problem: American Journal of Science, v. 270, p. 177–202.

Jonas, A. I., 1932, Structure of the metamorphic belt of the southern Appalachians: American Journal of Science, 5th Series, v. 24, p. 228–243.

Keith, Arthur, 1907, Description of the Pisgah quadrangle, North Carolina–South Carolina: U.S. Geological Survey Geologic Atlas Folio 147, 8 p.

Rankin, D. W., 1975, The continental margin of eastern North America in the southern Appalachians: The opening and closing of the Proto-Atlantic Ocean: American Journal of Science, v. 275-A, p. 298–336.

Reed, J. C., Jr., and Bryant, Bruce, 1964, Evidence for strike-slip faulting along the Brevard zone in North Carolina: Geological Society of America Bulletin, v. 75, p. 1177–1196.

Reed, J. C., Jr., Bryant, Bruce, and Myers, W. B., 1970, The Brevard zone: A reinterpretation, *in* Fisher, G. W., Pettijohn, F. J., Reed, J. C., Jr., and Weaver, K. N., eds., Studies of Appalachian geology: Central and southern: New York, Wiley-Interscience, p. 241–256.

Wintsch, R. P., 1979, The Willimantic fault: A ductile fault in eastern Connecticut: American Journal of Science, v. 279, p. 367–393.

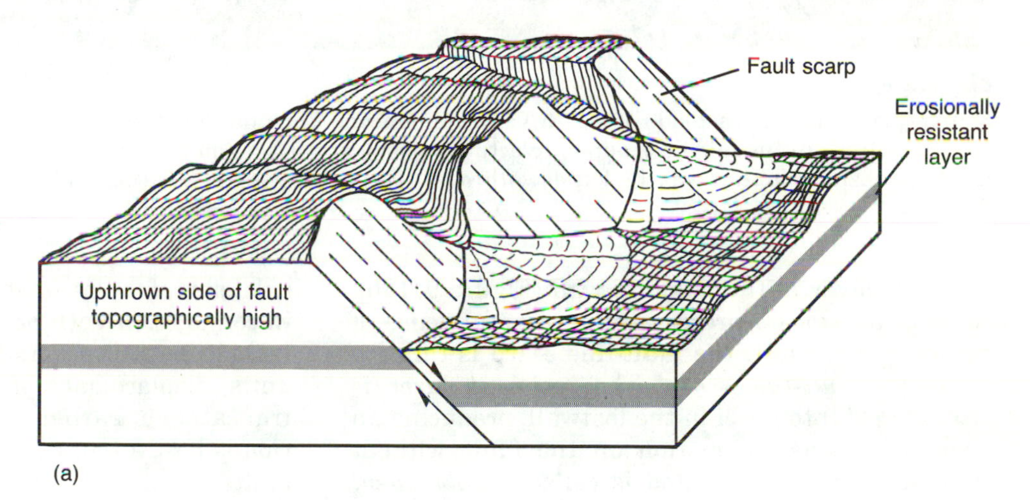

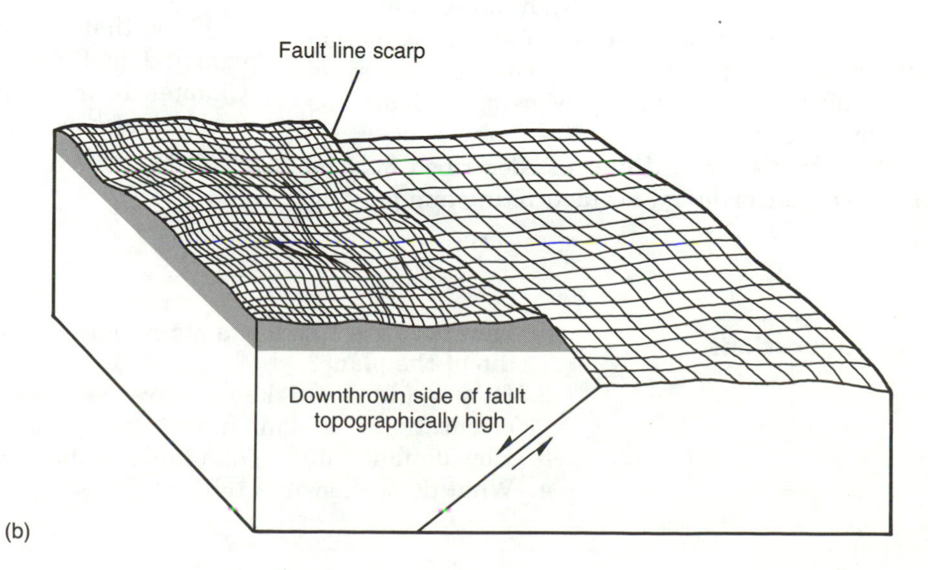

FIGURE 9–14
(a) Fault scarp. Note that the upthrown side of the fault is topographically high here. (b) Obsequent fault-line scarp with the downthrown side of the fault topographically high.

FIGURE 9–15
Fault scarp along the east side of Death Valley, California. The uniform slope on the mountain front in the middle ground are actual movement surfaces on the fault. The right (east) side is upthrown, the left downthrown. (Harold Drewes, U.S. Geological Survey.)

resistant layer in the hanging wall (Figure 9–14b). As long as erosion preserves the original sense of motion of the fault, the fault-line scarp is called a *resequent fault-line scarp.* The resistant layer is later etched into relief in the footwall, producing an anomalous sense of motion on the fault without subsequent movement, and is called an *obsequent fault-line scarp.* Consequently, a fault-line scarp may provide an opposite sense of apparent motion along a fault (Figure 9–14b). The real sense of motion must be determined by using at least one other major criterion.

The criteria for faulting just described may be listed as either primary, for identifying and proving existence of faults, or secondary, which must be used in conjunction with other criteria. Primary criteria include repetition and omission of stratigraphic units, displacement of a recognizable marker, and truncation of earlier structures—but the last criterion is best if truncation occurs on both sides of the fault.

Now that we have assembled the tools for recognizing and describing faults, we can go on to Chapter 10 on fault mechanics and deal with the problem of how brittle and ductile faults form and move.

Questions

1. How does the strike of a plane constrain the orientation (direction) of the dip of the plane?
2. How would you make measurements on the two limiting special cases of orientation of a plane in the Earth, horizontal and vertical surfaces?
3. How do faults differ from fault or shear zones?
4. What do slickensides tell us?

5. If you found a limestone-sandstone contact in the field in which both units dip 40° in the same direction, how would you determine whether it was a fault, a stratigraphic contact, or some other feature such as an unconformity?
6. What actually determines whether a fracture is a fault or a joint?
7. Most faults—regardless of whether they are thrust, normal, or strike-slip faults—when studied carefully turn out to be oblique-slip faults with a dominant thrust, normal, or strike-slip component of motion. Why?
8. Why does drag occur on some faults? Reverse drag?
9. Which criteria are most reliable for recognizing faults? Why?
10. Would fault and fault line scarps form where rocks on either side of the fault have equal resistance to erosion? Explain.
11. How are net slip and separation alike? How are they different?

Further Reading

Several other elementary textbooks on structural geology discuss fault classification and terminology from a different point of view. Some, such as Davis, include fault mechanics in their treatment of faulting.

Billings, M. P., 1973, Structural geology, 3d ed.: New York, Prentice-Hall, 606 p.

Davis, G. H., 1984, Structural geology of rocks and regions: New York, John Wiley & Sons, 492 p.

Hills, E. S., 1963, Elements of structural geology: New York, John Wiley & Sons, 483 p.

Park, R. G., 1983, Foundations of structural geology: London, Chapman & Hall, 135 p.

10

Fault Mechanics

If these principles hold in the case of rocks under pressure, rupture by shearing would be expected to occur along planes oblique to the axis of greatest and least intensity of compressive stress but (if the material is isotropic) inclined at angles of less than 45° to the axis of greatest stress. In such a case, two intersecting sets of planes of rupture may develop, cutting each other at an oblique angle, the greatest pressure bisecting the acute angle.

L. M. HOSKINS, U.S. Geological Survey Annual Report

WE DO NOT KNOW WHO FIRST RECOGNIZED A FAULT, OR WHEN OR where. Probably it was many centuries ago, when someone noted ground breakage after an earthquake. The word "fault" may have been coined by eighteenth-century English coal miners when they recognized offset coal seams and considered this a defect or "fault" in the perfect deposition they expected in nature. In the eighteenth century, Coulomb formulated the failure criterion that bears his name (Chapter 3), recognizing that shearing and normal stresses are interdependent with respect to the cohesive shear strength and the coefficient of internal friction of a material. In the 1800s, Playfair, Lyell, and other geologists wrote about faults much in the modern sense. In 1882, Otto Mohr related the Coulomb failure criterion to both shearing stress and normal stress, so the criterion is frequently referred to as the Coulomb-Mohr criterion.

During the late 1800s, Bailey Willis and H. M. Cadell experimented with some of the first geologic models (Figure 10–1), exploring the nature of faults and other structures, but it was not until E. M. Anderson (1942, 1951) systematically examined the three fundamental types of faults that modern study of *fault mechanics* began. Anderson's clear and simple—and rigorous—assumptions form the basis for most later study of the mechanics of faulting. In this chapter we will outline the modern concepts of fault mechanics from the standpoint of both laboratory experiments and field observations and will also look briefly at shear-zone mechanics and the criteria for determining shear sense in deformed rocks.

FORMATION OF FRACTURES: GRIFFITH THEORY

We can easily observe ruptures in the Earth's crust in the form of faults and joints. Many faults are traceable for hundreds of kilometers; joint sets (Chapter 8) are easily traced for tens of kilometers using the techniques of lineament analysis, along with measurement of joint orientations on the ground. What is the ultimate source of these fractures? How do they begin? In 1921, Alan Arnold Griffith, an English engineer, reported that elliptical microscopic cracks exist in glass and suggested that stress is concentrated and magnified by several orders of magnitude at crack tips. *Griffith cracks* in glass have since been confirmed, their dimensions being on the order of 100 μm wide by 1000 μm long. Their existence in other brittle materials, and the applicability of Griffith theory to propagation of fractures through rocks, have been debated for many years. Griffith cracks are generally accepted as a feature of rock materials despite the anisotropic nature of rock. The size of Griffith cracks in rocks must be on the order of the maximum grain diameter.

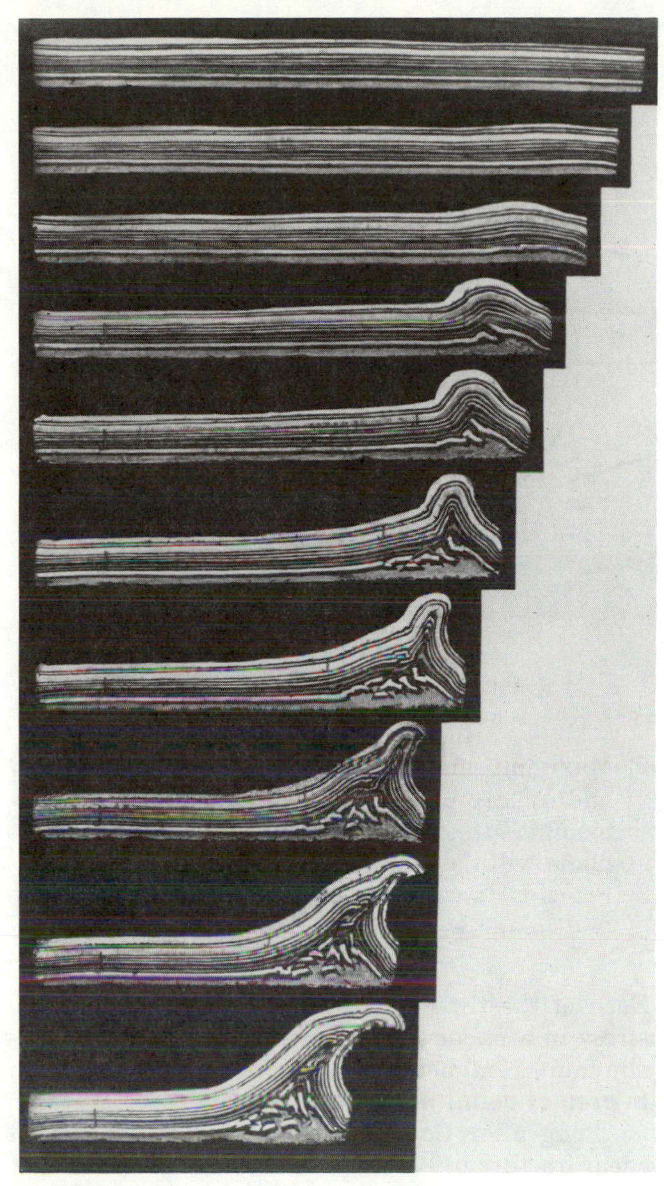

FIGURE 10–1
Pressure-box model of Bailey Willis consisting of layers of differing strength that produced a faulted fold as they were compressed from the ends. (U.S. Geological Survey Annual Report, 1893.)

Just how far Griffith theory is applicable to rocks is uncertain, but the theory helps us understand the effect of the cracks on elastic properties of rock materials, particularly if the theory is applied to large rock bodies where the entire mass can be assumed to behave homogeneously (Jaeger and Cook, 1976).

A homogeneous stress state exists in a rock mass if no Griffith cracks or discontinuities are present. Griffith cracks magnify stress concentration at crack tips and edges, raising it by several orders of magnitude. The amount of increase depends on the orientation and dimensions of each crack. Larger and suitably oriented cracks would undergo greater stress magnification than smaller and less suitably oriented cracks. The cracks with greater stress magnification would propagate into larger fractures; those with less would remain as microcracks. The tensile stress, σ_T, at the tip of an elliptical crack in a brittle material under tension, σ_0, perpendicular to the crack, would be

$$\sigma_T \sim \frac{2}{3}\sigma_0\frac{(2l)^2}{d} \qquad \textbf{(10–1)}$$

where l is the length of the crack and d is the width (Suppe, 1985). Solving this equation predicts that crack propagation may occur at low stress because amplification results from the relationship between length and width. For example, if the ratio of length to width of the crack is 10 to 1, stress amplification is about 270. If the ratio is 40 to 1, stress amplification is about 4300.

Griffith theory predicts that, for principal stresses σ_1 and σ_3, failure will occur under uniaxial compression if

$$(\sigma_1 - \sigma_3)^2 = 8T_0(\sigma_1 + \sigma_3) \qquad \textbf{(10–2)}$$

and under triaxial compression ($\sigma_2 = \sigma_3$) if

$$(\sigma_1 - \sigma_3)^2 = 12T_0(\sigma_1 + 2\sigma_3) \qquad \textbf{(10–3)}$$

where T_0 is the tensile strength of the material. The importance of these last two equations is that *Griffith theory predicts that the compressive strength of most materials will be 8 or 12 times the tensile strength.* Measured ratios of 7 samples range from 9 to 17, with an average near 14 (Jaeger and Cook, 1976). Another important conclusion to be drawn from these equations is that the stress difference ($\sigma_1 - \sigma_3$) must be 8—or 12—times the compressive strength before the Coulomb-Mohr failure envelope is reached in compression and 4 times the tensile strength before it is reached in tension.

ANDERSON'S FUNDAMENTAL ASSUMPTIONS

E. M. Anderson in 1942 rigorously defined three groups of faults that may be related to principal and shear stresses in the Earth. In doing this, he first defined a "standard state" as

a condition of pressure which is the same in all directions at any point, and equal to that which would be caused by the weight of the superincumbent material, across a

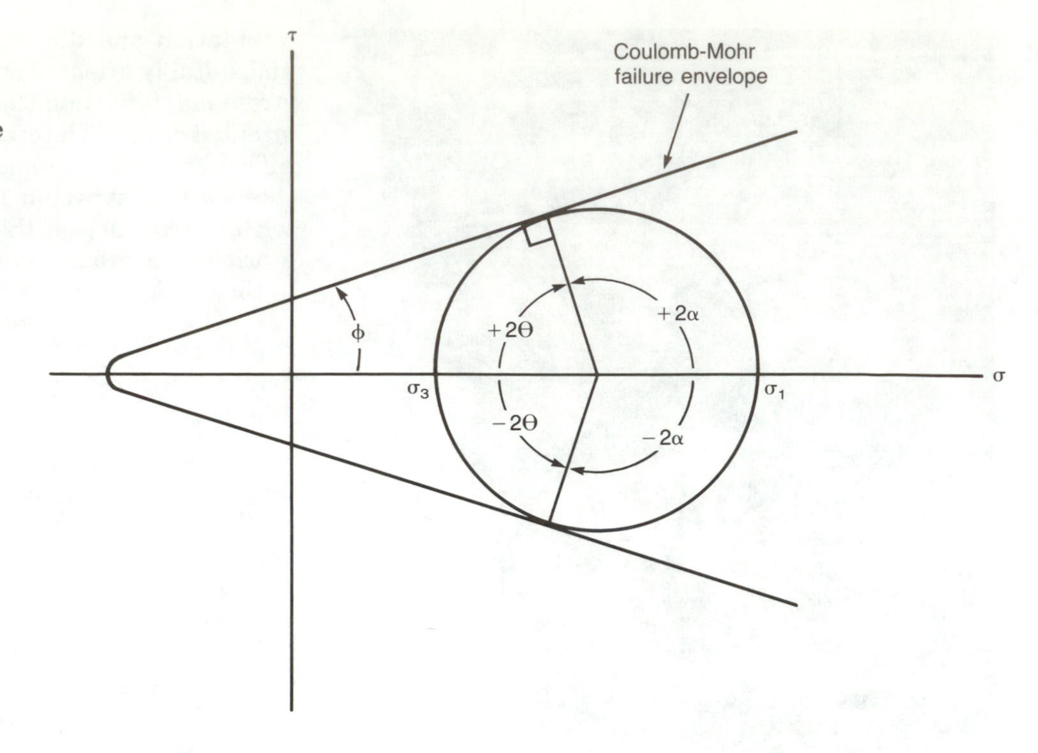

FIGURE 10–2
Mohr diagram showing relationships among ϕ, θ, α, and the Coulomb-Mohr failure envelope.

horizontal plane, at the particular level in the rock. It is assumed in this definition that the surface is flat and that the strata are of uniform specific gravity.

Accordingly, his standard state represents a stress system made up of two parts: (1) the force of gravity and (2) a superposed horizontal stress that remains constant in any horizontal plane but increases with depth (Hafner, 1951).

Anderson (1942) made several assumptions regarding stress in the Earth, assumptions still used in fault mechanics: (1) the crust was initially "intact" and unfractured; (2) one principal stress direction is vertical; (3) the two other principal stresses, which are normal to the first and to each other, must be horizontal and probably maintain a constant orientation over millions of years; and (4) forces must balance each other in a body in equilibrium. Assumption 2 actually recognizes that the Earth's surface is a surface of no shear; so, at least one principal stress must be vertical. Anderson also recognized that stress orientation in the upper crust can differ from one province to another. His assumptions make possible rigorous statements about fault mechanics, thus clarifying relationships between faulting and stress in the Earth.

Following up on his assumptions, Anderson suggested three possibilities for deviation from hydrostatic stress conditions:

1. Principal stresses $\boldsymbol{\sigma}_1$ and $\boldsymbol{\sigma}_2$ are horizontal. The least principal stress, $\boldsymbol{\sigma}_2$, is vertical, and the dip on the fault plane will be 45° + ϕ/2—about 30°.
2. Maximum and minimum principal stresses, $\boldsymbol{\sigma}_1$ and $\boldsymbol{\sigma}_3$, are horizontal, implying that the intermediate stress, $\boldsymbol{\sigma}_2$, is vertical and that the fault plane will have a near-vertical dip.
3. Principal stress, $\boldsymbol{\sigma}_1$, is vertical. Therefore the fault plane will dip 45° + ϕ/2—about 60°.

Each of the three sets of conditions describes the stress in a major fault group: thrust faults, strike-slip faults, and normal faults. We will discuss them in greater detail in the next section.

Long after Coulomb formulated the theory of shear fracture in 1776, Navier observed in 1833 that rupture does not occur on the shear planes that bisect the angles between the greatest and least principal stresses. He suggested that "internal friction" was responsible for the difference and proposed using a constant he called the *coefficient of internal friction* to account for it. The observation that $\boldsymbol{\sigma}_1$ (maximum principal stress) bisects the acute angle between the conjugate shear planes of the stress ellipsoid is called ***Hartman's rule.*** This angle (2α) commonly is closer to 60° than to 90° because the shear angle depends on the angle of friction, ϕ, in the Mohr diagram (Figure 10–2) so that

$$\pm\boldsymbol{\alpha} = 45° - \phi/2 \qquad \textbf{(10–4)}$$

From the Mohr diagram in Figure 10–2, it is also evident that when 2θ is 90°, $\boldsymbol{\tau}_{max}$, the maximum shear angle, is 45° to $\boldsymbol{\sigma}_1$. Alternatively, the angle 2θ (the supplement to α) may be used to obtain similar results. The significance of these relationships lies in the ability to predict the angles produced by faults and other shear fractures.

ANDERSON'S FAULT TYPES

Separation of stress regimes into the three fundamental possibilities just outlined gives rise to Anderson's three basic fault groups. The mechanics is easy to understand if his scheme is followed.

Type 1. Thrust Faults

Ideal thrust faults may be defined mechanically as faults produced by maximum and intermediate principal stresses oriented horizontally, with the minimum principal stress oriented vertically (Figure 10–3). A state of *horizontal compression* is thus defined for thrust faulting. The strain ellipsoid corresponding to this stress configuration has shear planes oriented at 45° or less to the horizontal and striking parallel to the intermediate stress and strain axes. The fault planes commonly form at smaller angles because of inhomogeneities in the rocks. Anderson's theory does not allow for the steep dips (>45°) of reverse and high-angle thrust faults.

The two shear planes are potential fault planes; ideally, a fault should form parallel to each shear plane, but commonly only one dominant shear plane becomes a fault. The alternate shear plane may be used to form a fracture or series of fractures with minor displacement. The result is a thrust fault with the upper block (the hanging wall) moved up relative to the lower block (the footwall).

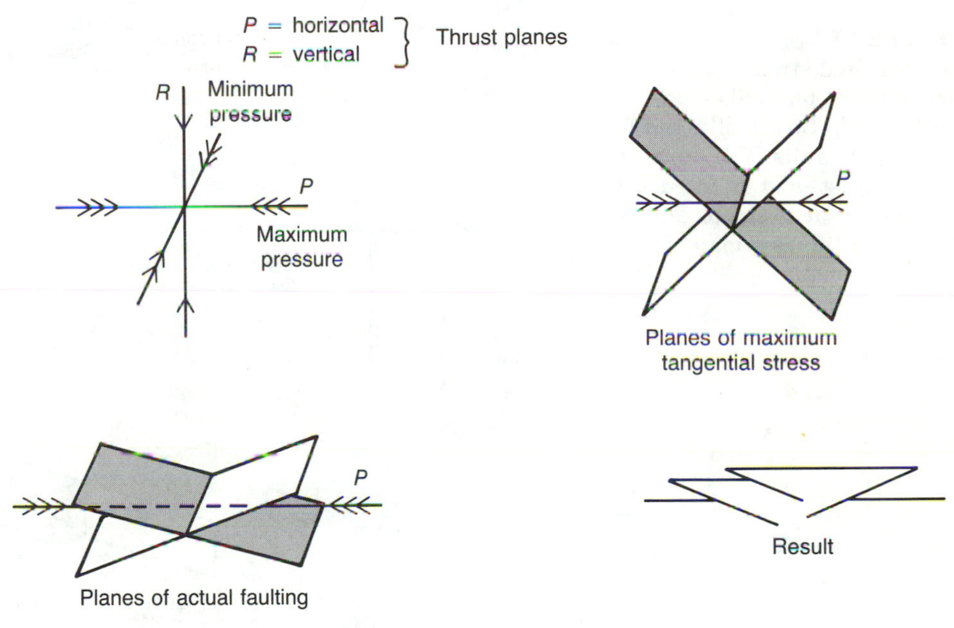

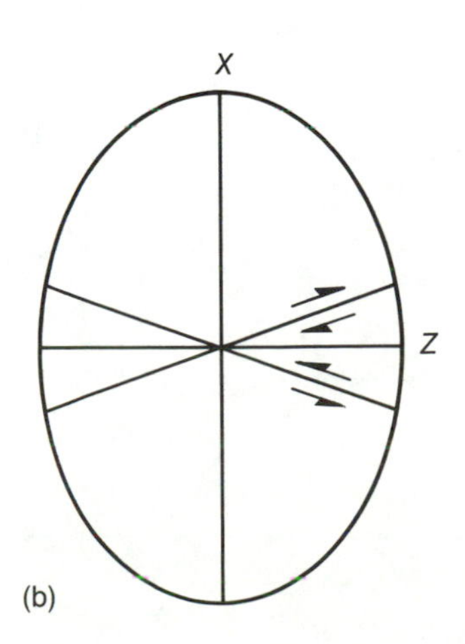

FIGURE 10–3
(a) Orientation of principal stresses for thrust faulting. (b) Strain-ellipsoid configuration for thrusting. Note that Y is horizontal and perpendicular to the plane of the page. View is parallel to strike of faults.

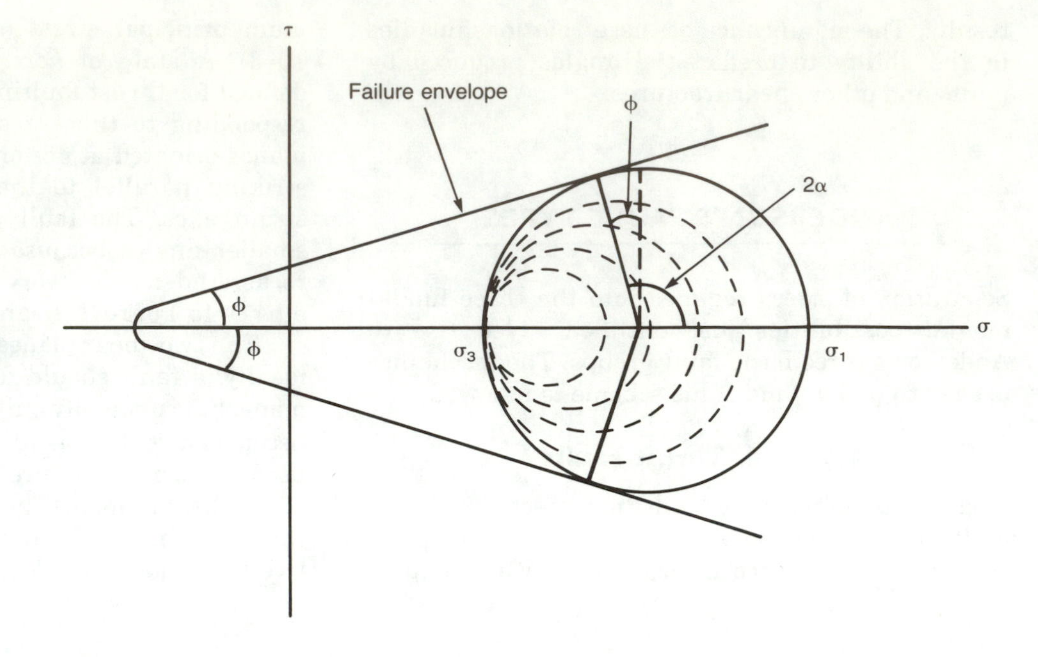

FIGURE 10–4
Mohr-circle analysis of thrust faulting. As σ_3 remains constant, σ_1 increases to failure (dashed circles).

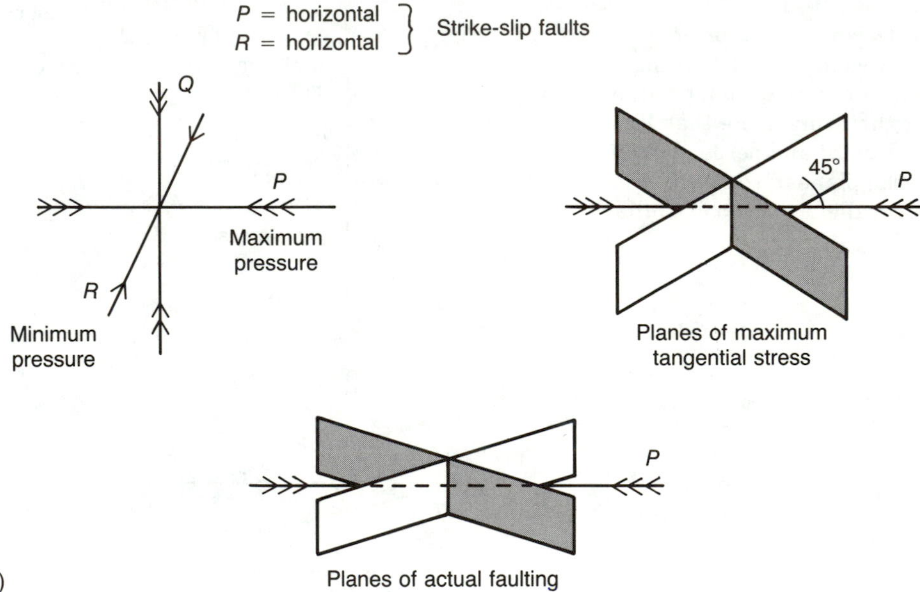

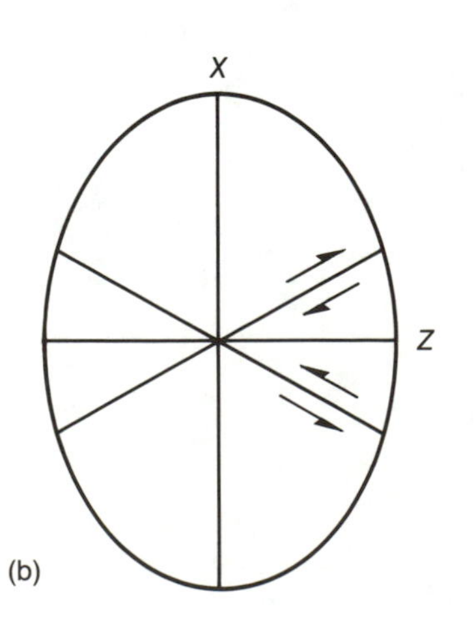

FIGURE 10–5
(a) Principal stress orientations for strike-slip faulting. (b) Strain ellipsoid for strike-slip faults. *Y* is vertical (in the crust) and oriented perpendicular to the page here. Note that this is a map view.

Mohr-circle analysis of thrust faulting (Figure 10–4) might involve successively increasing values of σ_1, producing circles of increasing radii until the failure envelope is reached. At that point rupture occurs.

Type 2. Strike-Slip Faults

If the maximum and minimum principal compressive stress is horizontal and the intermediate principal stress axis vertical, conditions are right for strike-slip faulting (Figure 10–5). As with all faults, any shear plane that forms will include the principal intermediate stress axis and, as a result, must be vertical. The corresponding strain ellipsoid for strike-slip faulting also contains vertical shear planes, which in theory should lie at 45° or less to the axes of maximum and minimum strain; they commonly form at angles more acute and symmetrically about the axes of minimum strain (and maximum stress).

Mohr-circle analysis for strike-slip faults can be the same as for thrusts increasing σ_1 to failure because both are compressional.

Type 3. Normal Faults

Normal faulting involves extension in one horizontal direction, with the maximum principal stress vertical (Figure 10–6). Shear planes may be expected at 45° or less to the axes of maximum and minimum principal stress and strain. In reality,

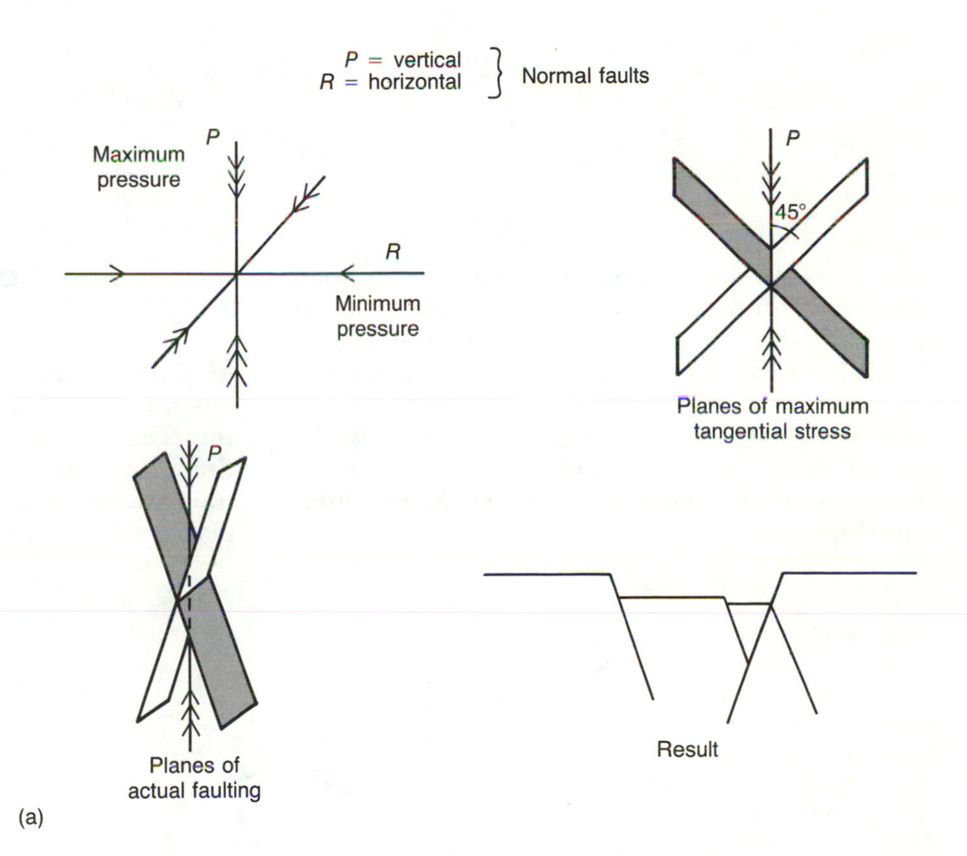

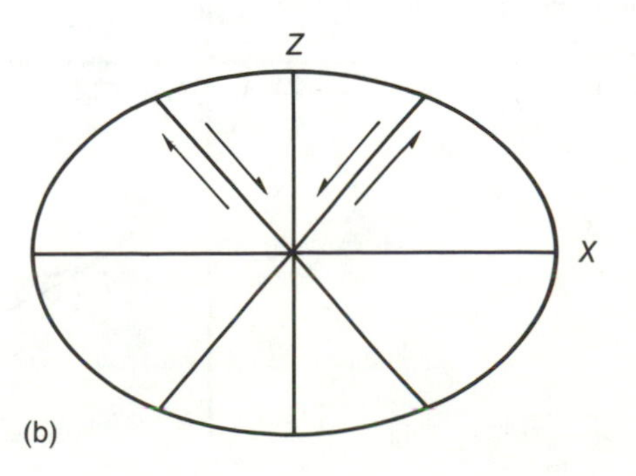

FIGURE 10–6
(a) Orientation of principal stress axes for normal faulting. (b) Strain ellipsoid for normal faults. *Y* is horizontal and perpendicular to the page. View is parallel to strike of faults.

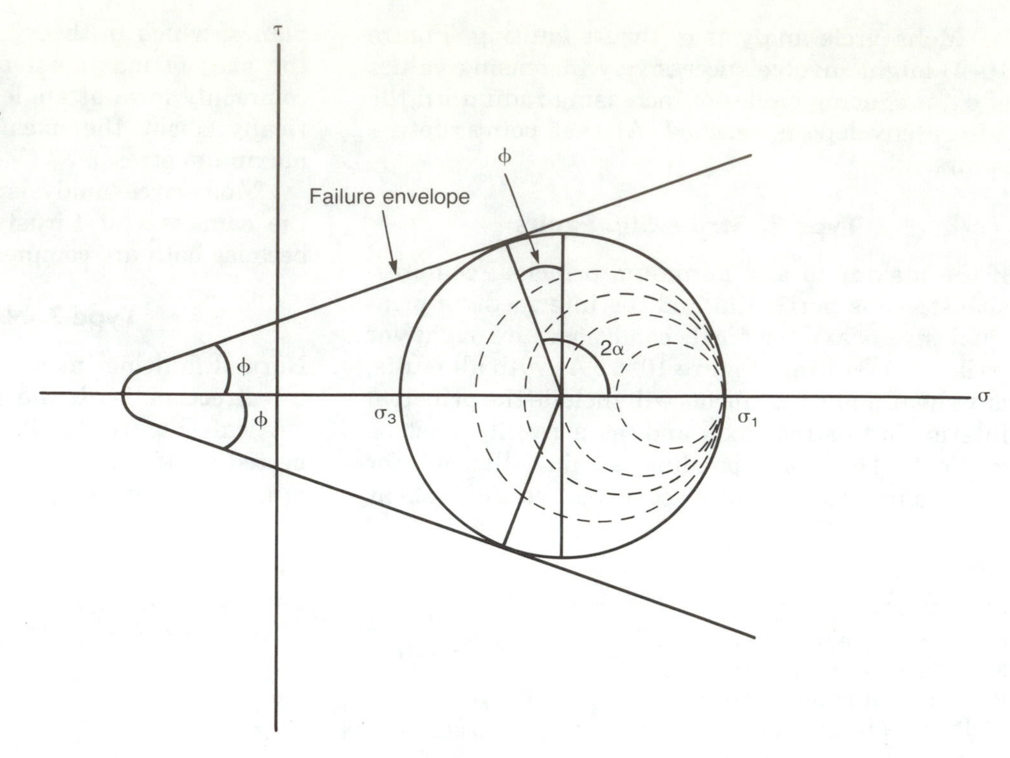

FIGURE 10–7
Mohr-circle analysis of normal faulting: σ_3 decreases with σ_1 constant (increasing the size of the Mohr circles—dashed circles) until the Mohr circle intersects the failure envelope. Note: the linear failure envelopes in Figures 10–4 and 10–7 indicate that the experimental rock material is brittle.

they form steeply dipping fault planes symmetrically about the vertical axis (σ_1). Both shear planes frequently develop in complex normal fault zones.

Mohr-circle analysis of normal faulting may be thought of as starting at a large confining pressure, with σ_3 decreasing (producing extension) until the size of the circles produces a stress difference ($\sigma_1 - \sigma_3$) large enough to reach the failure envelope (Figure 10–7).

ROLE OF FLUIDS

M. King Hubbert and William W. Rubey (1959) demonstrated that fluid plays an important role in faulting. They recognized that the lubricating effect of fluid in a fault zone is really due to buoyancy as the fluid pressure reduces the normal stress on the fault plane. Frictional resistance is thereby reduced and

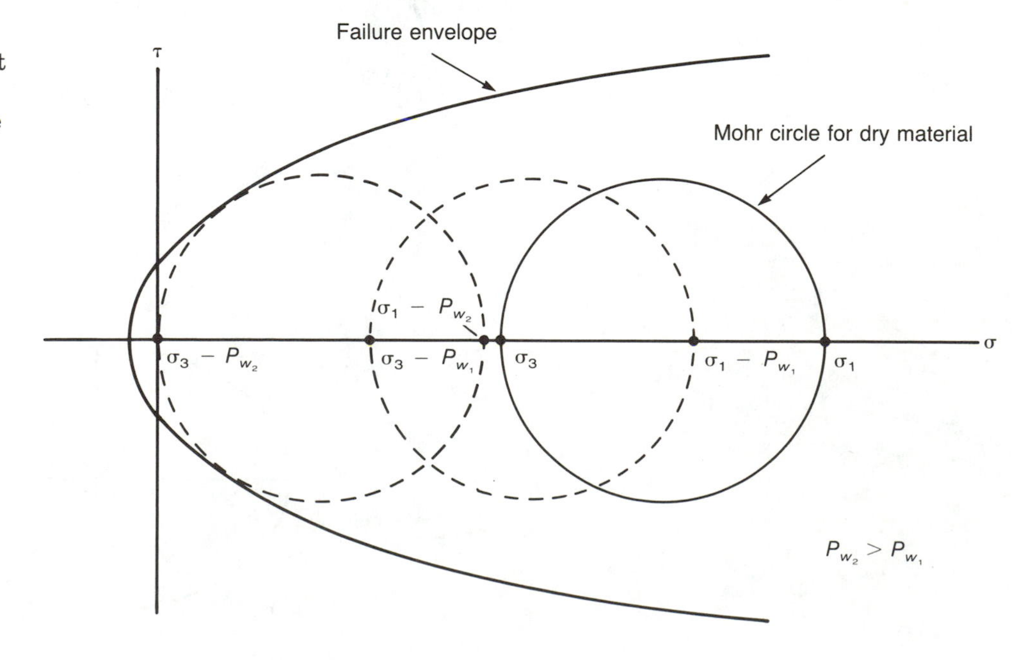

FIGURE 10–8
Mohr circles showing the effect of fluid pressure (P_w) on effective normal stress and the failure envelope. Increasing fluid pressure reduces the strength of the material and forces it into the unstable region above the failure envelope. Note the curved envelope, indicating that the rock behaves ductilely.

movement facilitated. The resulting stress is *effective normal stress,* **S,** and we can determine it by subtracting the fluid pressure, **P,** from the normal stress, σ. Mohr circles are shifted to the left toward the failure envelope and then across it, leading to rupture and movement along the fault (Figure 10–8).

The effect of fluid on movement is illustrated vividly by landslides and snow avalanches. Conditions favoring landslides may exist for centuries in an area without initiating slides (Chapter 2); then influx of a large amount of water into the susceptible materials may buoy up the mass and landsliding results. Snow avalanches travel at high velocity on a cushion of air, another fluid that lowers effective normal stress. In unconsolidated sediment in accretionary wedges and deltas, movement along faults is made easier by overpressured sections containing fluid that cannot escape. Stress is borne by the fluid rather than by the sediment, facilitating movement.

FRICTIONAL SLIDING MECHANISMS

Some movement will occur on any fracture produced by shear. If appreciable movement is to occur on an existing fault surface, the coefficient of friction along the fault must be overcome. Two laws concerning friction were probably first discovered by Leonardo da Vinci, then rediscovered by Guillaume Amontons (1663–1705), a French physicist who first presented them to a scientific gathering in 1699 (Suppe, 1985). ***Amontons' first law*** states that the tangential force parallel to the fault surface necessary to initiate slip is directly proportional to the force normal to the fault surface. The proportionality constant, μ, is the coefficent of friction. Dividing force by area, Amontons' first law may be stated as

$$\tau = \mu\sigma \qquad \textbf{(10–5)}$$

where τ is tangential (shearing) stress and σ is normal stress. *Amontons' second law* states that the frictional resistance to motion is independent of the contact area. This law has an important bearing on the nature of fault surfaces and resistance to movement.

If water occurs in the fault zone, normal stress will be decreased by the quantity P_w, which is the water pressure. Equation 10–5 then becomes

$$\tau = \mu\,(\sigma - P_w) = \mu \mathbf{S} \qquad \textbf{(10–6)}$$

where $(\sigma - P_w)$ is effective normal stress, **S.** Therefore water lowers the shearing stress required for failure and fault motion and allows movement to begin at lower values of τ.

In 1959, Hubbert and Rubey illustrated the effective stress relationship, using only an empty beer can on a glass plate (Figure 10–9). First, they placed the can (open end down) on the dry plate and lifted one end until the can moved—an angle greater than 15°. Then they lowered the plate and wetted it with water, chilled the can, and again placed it on the wet surface. This time, when they raised one end of the plate, only a slight incline—about 2°—was needed for the can to begin sliding down the slope. A similar model used a concrete block weighing several hundred kilograms. The block was smooth and rested on a smooth surface. As long as the surfaces remained dry, the block could be moved only with great difficulty, but when a film of water separated the two smooth surfaces, the block moved so easily that Hubbert and Rubey said it was dangerous to stand near the heavy and unstable block.

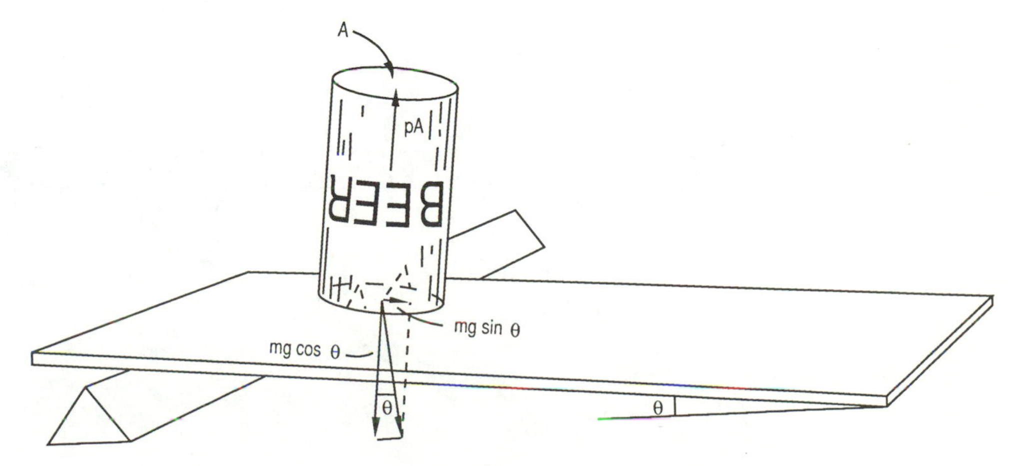

FIGURE 10–9
The Hubbert and Rubey beer-can experiment. If the glass plate is dry, $\theta > 15°$; if wet, $\theta < 2°$; *m*—mass of can, *g*—acceleration due to gravity, *A*—area of base of can, *p*—excess pressure of air inside can over that outside. (From M. K. Hubbert and W. W. Rubey, 1959, Geological Society of America *Bulletin,* v. 70.)

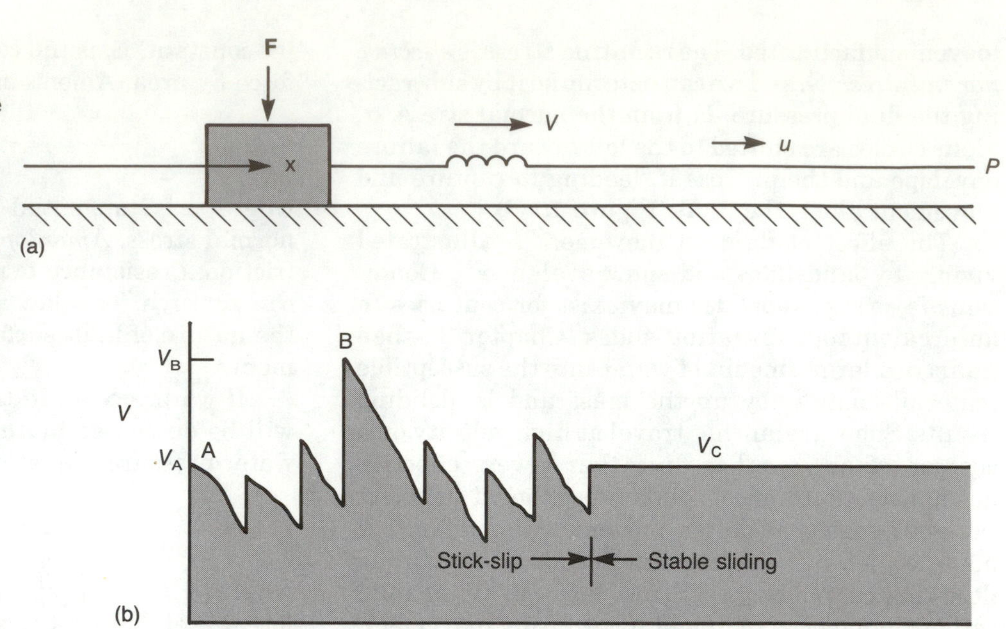

FIGURE 10–10
(a) To move an object connected by a spring by a line load (P) (e.g., a strong cable), friction (F) must be overcome before the object will move. The screen will stretch as the line is tightened at velocity (u) until friction with the surface is overcome and the object will move at the velocity as the spring relaxes. (b) Diagram of time (x) versus displacement (V) illustrating stick-slip and continuous-creep mechanisms. (From James Byerlee, 1977, *Friction of rocks,* U.S. Geological Survey, Office of Earthquake Studies.)

MOVEMENT MECHANISMS

Movement on faults occurs in at least two different ways: by ***stick-slip*** (unstable frictional sliding) and by ***continuous creep*** (stable sliding). The stick-slip mechanism (Figure 10–10) involves sudden movement on the fault after long-term accumulation of stress (Turcotte and Schubert, 1982). This mechanism and the accompanying *elastic rebound* (Chapter 3) is what we think causes earthquakes. The continuous-creep mechanism involves uninterrupted motion along a fault so that strain is relieved continuously and does not accumulate. The differ-

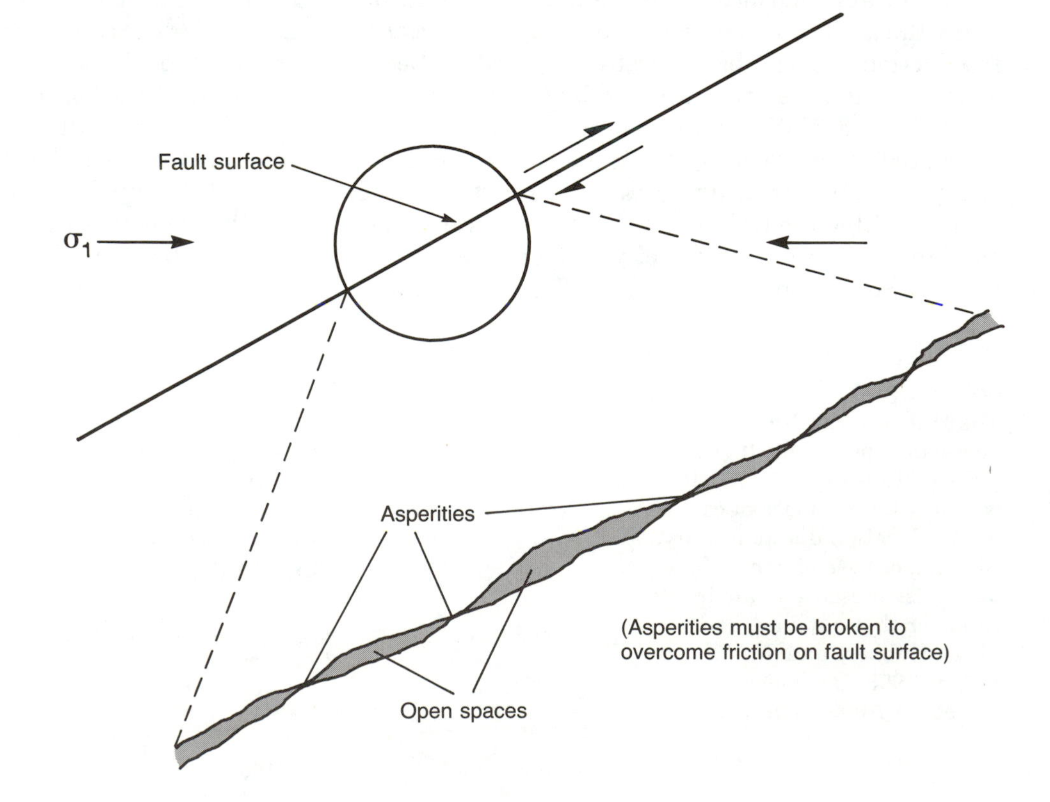

FIGURE 10–11
Asperities and contact-surface area along a fault.

ence in behavior may be produced along segments of the same active fault undergoing continuous creep where ground water is abundant, but other segments, with less ground water, may move by stick-slip. Some segments of the San Andreas fault in California exhibit stick-slip behavior; others, continuous creep. Other and more complex factors may determine the mechanism involved.

It has been suggested that withdrawal of ground water may cause active faults to switch mechanisms, from continuous creep to stick-slip, thereby increasing the earthquake hazard. The converse may also be true, but there are difficulties. Pumping fluid into a fault zone has been proposed as a way to relieve accumulated elastic strain energy and reduce the likelihood of a large earthquake, but the rate at which fluid should be pumped into a fault remains unknown. Pumping fluid into a locked fault zone does raise the fluid pressure and lowers the effective normal stress, but if pumping is too rapid, it could trigger earthquakes artificially by allowing the stick-slip mechanism to go to completion rather than slowly bleeding off accumulated stress by creep.

Fault Surfaces and Frictional Sliding

Fault surfaces between two large blocks are never perfectly smooth and planar, especially on the microscopic scale. Microscopic irregularities and imperfections, called *asperities* (Figure 10–11), increase the resistance to frictional sliding because of the overall irregular-shaped fault surface; they also reduce the percentage of surface area actually in contact. F. Bowden, a metallurgist, first suggested that because of asperities even the best-prepared surfaces will not fit together and that breaking asperities increases the contact area. The initial contact area may be as little as 10 percent but, once motion begins, irregularities on the surface are removed and as the displacement increases the contact area decreases (Teufel and Logan, 1978). Because asperities provide the surface with its frictional strength (normal strength) and resistance to movement, they must be broken through in order for movement to occur. Rock-mechanics experiments on fault surfaces demonstrate that resistance to frictional sliding increases after movement begins, suggesting an increase in strength of asperities. The increase may result from the greater contact area, requiring that more asperities be broken to initiate movement, but the strength of the fault zone decreases again with further movement. Thus, despite the increased contact area, continued movement reduces the asperity population and hence also the strength of the fault surface.

James Byerlee (1977, 1978) has shown experimentally (Figure 10–12) that for low effective normal stress, **S** (<0.2 GPa)—the usual condition in the upper crust—

$$\frac{\tau_{max}}{\mathbf{S}} = 0.85 \qquad \textbf{(10–7)}$$

where τ_{max} is maximum frictional shearing stress, and 1 Pa, or pascal, is 1 kg $m^{-1}s^{-2}$ (so an MPa, megapascal, is 1 million pascals, and a GPa, gigapascal, is a billion pascals). The relationship has been called *Byerlee's law of rock friction* and is an important statement regarding faulting and the brittle behavior of most rock materials under stress in the upper crust. That Byerlee's ratio for most rock types is 0.85 (Figure 10–12) indicates that the coefficient of friction in Amontons' laws is independent of rock type and depends solely on values of shearing and normal stress. Byerlee also showed that for values of **S** between 0.2 GPa and 2 GPa,

$$\tau_{max} = 50 \text{ MPa} + 0.6\ \mathbf{S} \qquad \textbf{(10–8)}$$

These conditions probably occur at greater depth in the crust. There is some doubt that the 0.85 value for the coefficient of friction is valid, particularly because earthquakes occur in the shallow crust at ambient stress much below that suggested by Byerlee's experiments (Alsop and Talwani, 1984). Fluid, clay minerals, or anything else that weakens rocks could cause failure at values lower than predicted.

BRITTLE VERSUS DUCTILE FAULTS

So far, most of our discussion of fault mechanics has been directed toward brittle faults in the upper 5 to 10 km of the Earth's crust. Faults in the upper crust consist of either a single movement surface or an anastomosing complex of fracture surfaces (Figure 10–13). An individual fault may have knife-sharp contacts (Figure 10–14) or it may consist of a zone of cataclasite (Chapter 7).

The ductile-brittle transition (Chapters 4 and 6) is thought to occur 10 to 15 km deep in continental crust. Faults that penetrate the entire crust, or most of it, must involve both ductile and brittle behavior. Large active strike-slip faults, like the San Andreas fault in California, the Anatolian fault in Turkey, and the Alpine fault in New Zealand, probably penetrate most of the lithosphere, for they are plate

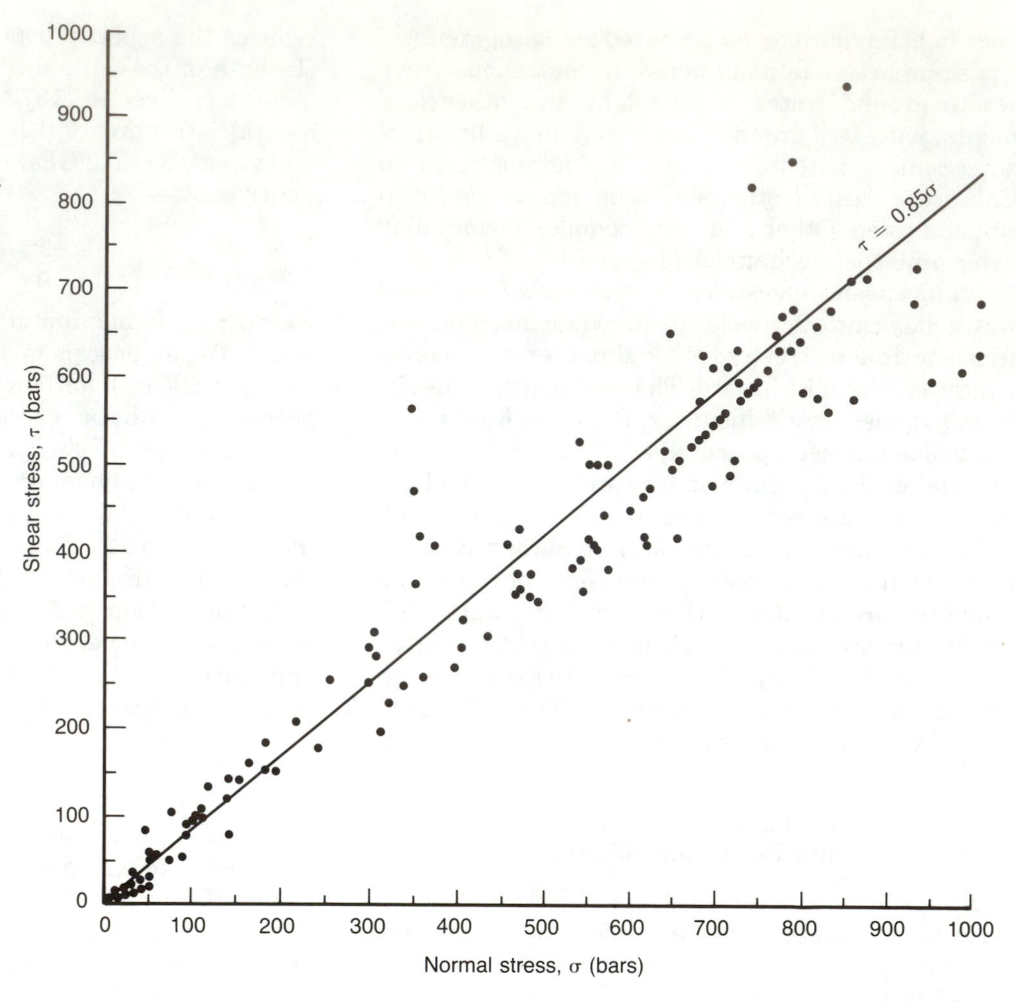

FIGURE 10–12
Least-squares fit of points derived from experimental measurement of maximum shear stress (τ) and effective normal stress (σ), producing a curve with the equation $\tau = 0.85\sigma$. (From James Byerlee, 1977, *Friction of rocks,* U.S. Geological Survey Office of Earthquake Studies.)

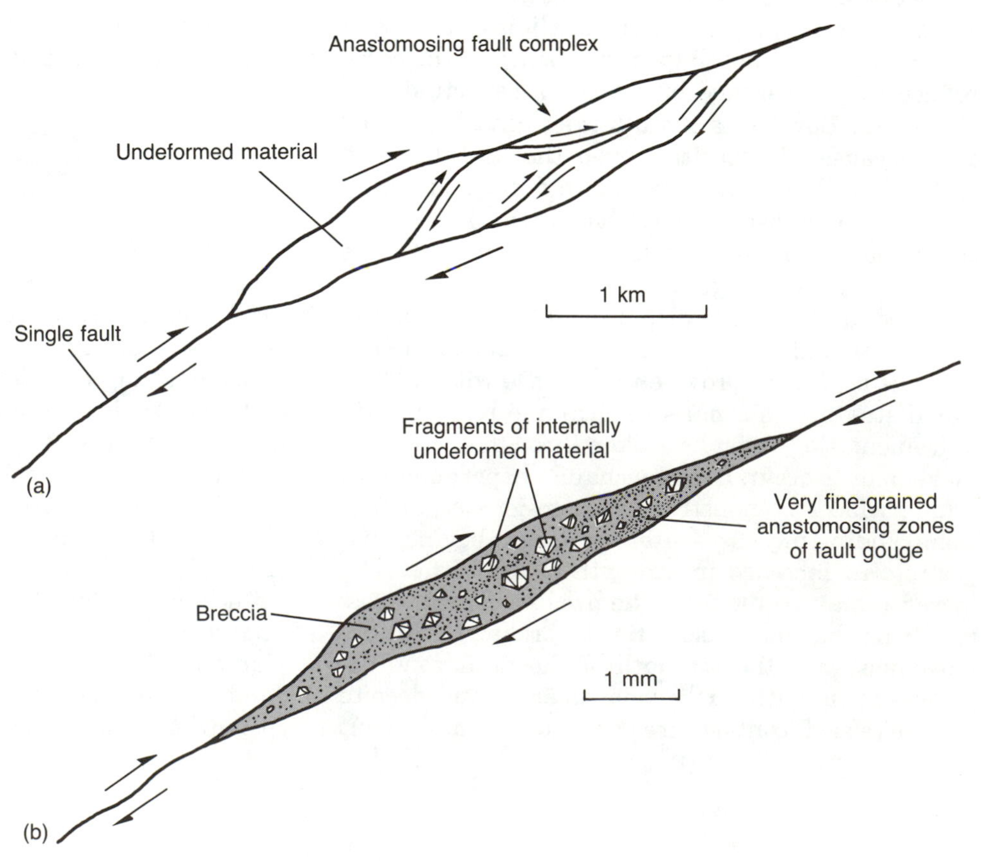

FIGURE 10–13
Characteristics of brittle faults in the upper crust. (a) Map view of a fault zone showing distribution of movement surfaces and the way a fault zone may change from a single fault to an anastomosing fault zone. (b) Close-up view of one of the faults in (a). Note that fragments of internally undeformed material survive to the microscopic scale.

(a) (b)

(c)

FIGURE 10–14

(a) Knife-sharp fault contact along the Highgate Springs thrust (part of the Champlain thrust system) at Line Rock Point near St. Albans Bay, Vermont. Lower Cambrian Dunham Dolomite in the hanging wall is thrust over Middle Ordovician Iberville Shale. (RDH photo.) (b) Zone of cataclasite along a brittle fault, southern England. Note the dark rocks to the left of the lighter cataclasite in the fault zone (center of photo). (William E. Dunne, University of Tennessee.) (c) Cataclasite from the Homestake shear zone near Leadville, Colorado. An older ductile fabric of quartz and feldspar was fractured under brittle conditions. Width of field is approximately 7 mm. Plane light. (RDH photo.)

ESSAY

Artificial Earthquakes

Should we try to artificially relieve accumulated elastic strain energy on large faults otherwise certain to cause sizable earthquakes? The San Andreas fault is likely to produce an earthquake of magnitude 7 or 8 in southern California early in the next century or before. One proposal is that we pump fluid into a locked segment of the fault zone so the strain energy will be relieved by very small earthquakes and thus minimize damage and loss of life. But no one has ever successfully carried out such an experiment. So, should southern California cities be evacuated before pumping? Who will pay for damages produced by the small earthquakes? What if pumping triggers a large earthquake and devastates the region anyway? Who bears legal responsibility? Such questions discourage attempts to try the experiment.

Artificial earthquakes were produced unintentionally in the early 1960s as a byproduct of disposal of toxic chemicals in a 12,045-foot well at Rocky Mountain Arsenal near Denver, Colorado. In 1965, geologist David M. Evans suggested that occurrence of earthquakes after disposal of fluid there from 1962 until 1965 showed that it was possible to influence the mechanism and timing of release of accumulated elastic strain (Evans, 1966). He plotted the occurrence of earthquakes from March of 1962 to November of 1965 and correlated them with the pumping of fluid into the well at the arsenal. His data showed that earthquakes began within a month after pumping started, decreased markedly from September of 1963 through September of 1964 when no fluid was injected—and resumed with greater frequency when pumping was restarted in September of 1965 (Figure 10E–1).

The waste fluid was pumped into Precambrian crystalline basement rock beneath the Paleozoic to Cenozoic sedimentary cover, with disposal in fractures that did not communicate with the present-day surficial ground-water system. The fluid was initially injected under pressure, the pressure being relieved by flow along existing fractures enlarged in the crystalline rocks. New fractures were expected to form by hydrofracturing during injection.

Although seismic activity was not unknown, no earthquakes had been felt or recorded in the Denver area between 1882 and the beginning of pumping in 1962. From April of 1962 until September of 1965, local seismic stations recorded 710 earthquakes with epicenters in the vicinity of the Rocky Mountain Arsenal. Magnitude ranged from 0.7 to 4.3 on the Richter scale, and about 75 of the earthquakes were felt. Most foci were determined to be from 4.5 to 5.5 km deep (Evans, 1966)—about the same depth as the bottom of the well. Low-level elastic strain energy accumulated and was being relieved by fluid pressure, reducing the effective normal stress so that elastic strain could be relieved by shear on appropriately oriented fracture surfaces.

No one doubts the Denver earthquakes were produced by deep-well injection. Public opinion brought an end to the pumping, and contractors at Rocky Mountain Arsenal devised a chemical means of breaking down the toxic waste so it could be disposed of conventionally.

The events in Denver proved that we can induce earthquakes, but to enter a major active fault zone and attempt to relieve stored elastic strain energy

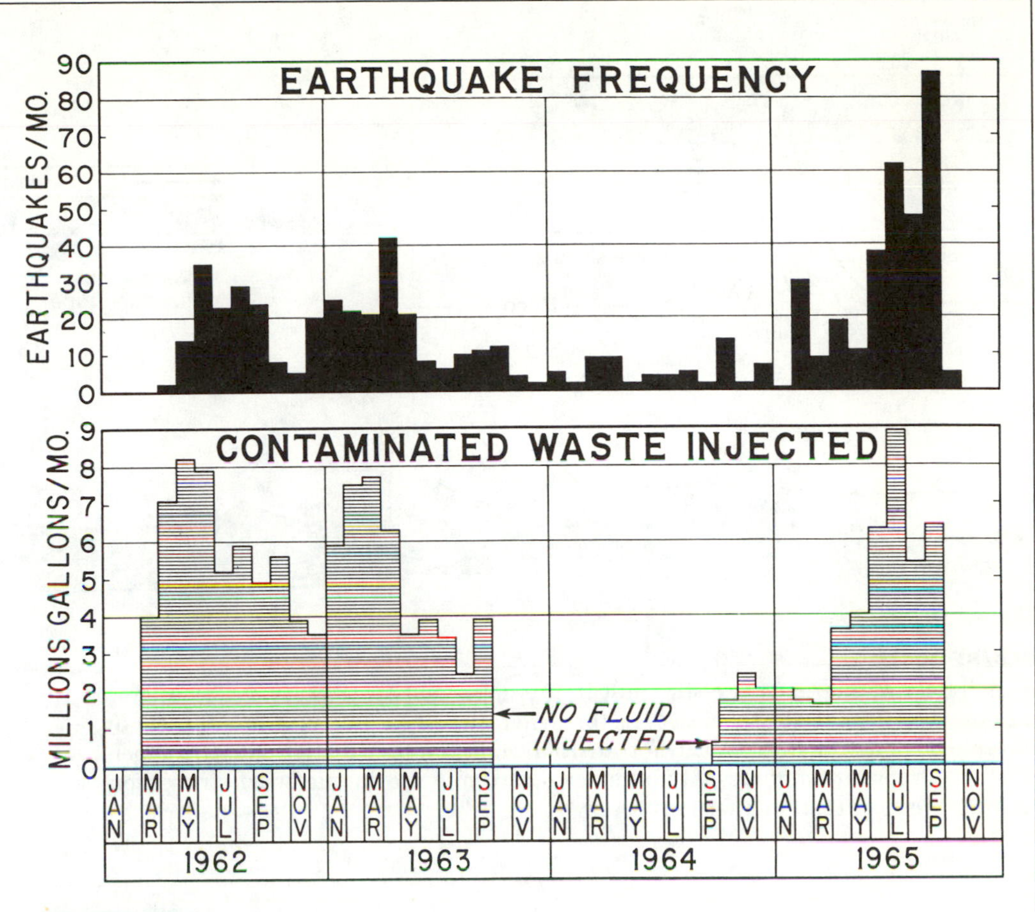

FIGURE 10E–1
Correlation of earthquake frequency with the injection of liquid waste at Rocky Mountain Arsenal from 1962 to 1965. (From D. M. Evans, 1966, *Geotimes*.)

artificially—before nature releases it suddenly and without warning—requires detailed knowledge of the fracture system associated with the fault at great depths. We do not have the data necessary to predict how much fluid should be pumped, where it should be pumped, and how rapidly it should be pumped into the system to lower the effective normal stress and bleed off the excess energy. Apparently the crust in the Denver area did not contain a great excess of accumulating and stored elastic strain energy, as does the crust along the San Andreas fault in southern California. Before we can attempt to drain off energy from a large and active fault zone, we must answer many questions—scientific, sociological, political, and legal.

Reference Cited

Evans, D. M., 1966, Man-made earthquakes in Denver: Geotimes, v. 10, no. 9, p. 11–18.

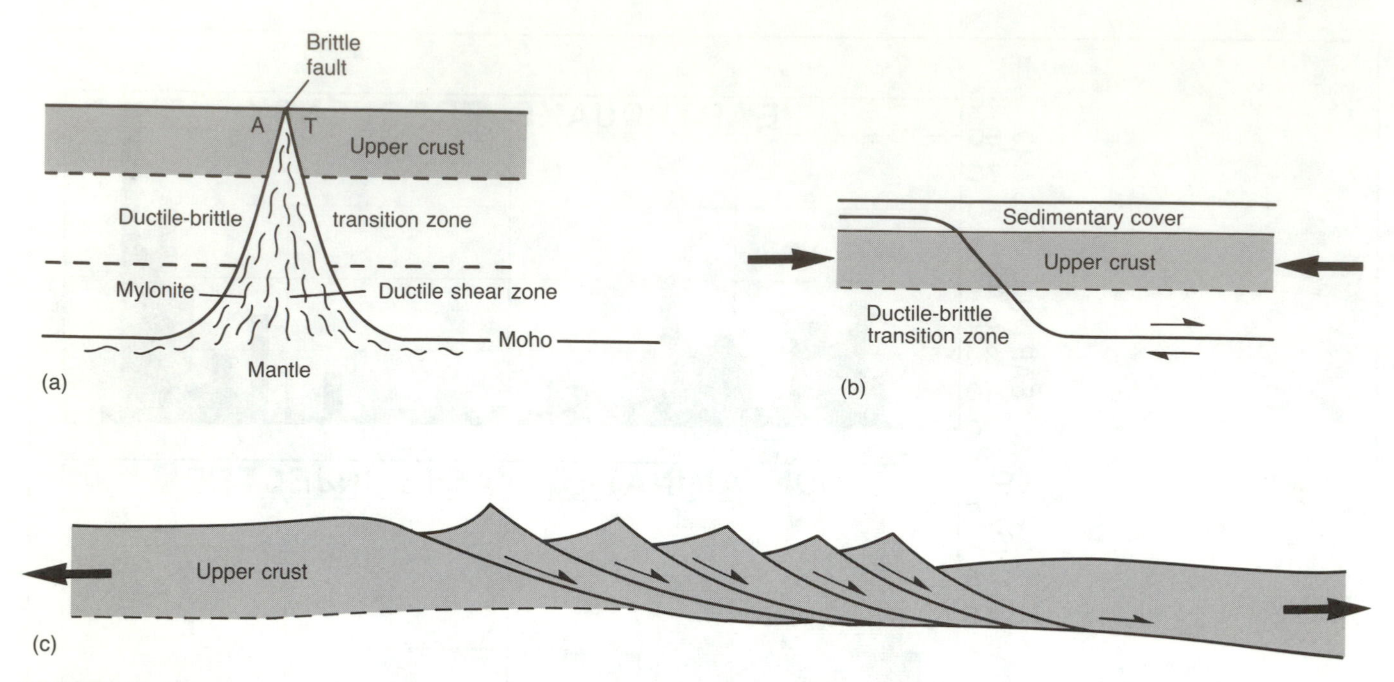

FIGURE 10–15
(a) Behavior of a large strike-slip fault in the upper and lower crust. A—away from the observer; T—toward the observer. (b) Generation of large crystalline thrusts by movement along the ductile-brittle transition zone and propagation into the upper crust. (c) Propagation of normal faults downward into the ductile-brittle transition zone in zones of crustal extension.

boundaries (Figure 10–15a). They consist of brittle fault zones at the surface, but at great depth probably are generating mylonite. Direct evidence for the transition may be found at the surface along some faults like the Alpine in New Zealand (Reed, 1964). Thrust faults that displace large blocks of crystalline rock may be generated within or below the ductile-brittle transition and propagate into the brittle upper crust (Figure 10–15b). Ultimately, the entire thrust sheet may be displaced onto upper crustal rocks. Crustal extension, as in the Basin and Range province, may create normal faults that on regional scale propagate downward into the ductile-brittle transition (Eaton, 1979).

SHEAR ZONES

Our previous discussions of ductile and brittle faults lead directly to consideration of the characteristics of shear zones. They are produced by inhomogeneous simple shear and are thought of as zones of ductile shear (Figure 10–16), although the term has been used to describe fault zones where brittle behavior dominates. John Ramsay (1980) has classified them as brittle, brittle-ductile, and ductile shear zones.

Attributes of ductile shear zones have been listed by J. Carreras and Stanley H. White (1980) and are slightly modified here:

1. *Shear zones on all scales are zones of weakness and represent localized strain softening.*
2. *Strain softening in ductile shear zones is associated with formation of mylonite and may result from several processes.*
3. *Transient brittle behavior (rate-dependent?) can occur in ductile shear zones.*
4. *Sheath folds [Chapter 15] are common in mylonite zones and apparently can form in both steady-state and nonsteady-state flow regimes.*
5. *There is little evidence to indicate that shear heating is common in ductile shear zones.*
6. *Shear zones may act both as closed and open geochemical systems with respect to movement of fluids and elements, irrespective of size of the zone.*

Ramsay (1980) added these boundary conditions as geometric features of ideal shear zones: (1) shear zones generally have parallel sides, and (2) displacement profiles along any cross section through a

ESSAY

Differences in Behavior of Crustal Faults

We can directly observe large faults that produce earthquakes and break the present-day surface. All such faults are brittle. Speculation about the behavior of large faults at depth in the crust and mantle is based on geophysical models and on direct observation of fault zones interpreted as having been formed at great depths but after millions of years of erosion are now exposed. We believe that faulting deeper in the crust (more than 10 to 15 km) occurs by ductile flow, which produces mylonite zones that range in width from less than a meter to several kilometers.

Studies of inactive broad mylonitic fault zones in present-day exposures in western Greenland led John Waterson (1975) and J. Grocott and Waterson (1980) to suggest that these broad zones may represent the lower-crustal equivalents of stick-slip brittle faults in the upper crust. The mylonite zones contain mineral assemblages indicating that they formed at upper amphibolite to granulite-facies metamorphic conditions. Mineral assemblages in these fault zones, along with the textures of fault rocks, provide information about the conditions and depths where the zones formed. It would be useful if we could directly trace a fault from the lower crust, where it is now moving, to the surface, where it is undergoing brittle deformation. Seismic-reflection profiles across active island arcs, like the Barbados arc in the Caribbean (Westbrook and Smith, 1983), image faults that pass from the surface, where evidence for brittle deformation may be observed, through the crust and into the mantle, where earthquake activity diminishes but the faults appear to continue. Ductile deformation would be expected at these greater depths, but direct observation is impossible. Thus, we must rely on indirect information from geophysical images and direct observations of ancient fault zones thought to have formed in the deeper crust.

References Cited

Grocott, J., and Waterson, J., 1980, Strain profile of a boundary within a large shear zone: Journal of Structural Geology, v. 2, p. 111–117.

Waterson, J., 1975, Mechanism for the persistence of tectonic lineaments: Nature, v. 253, p. 520–522.

Westbrook, G. K., and Smith, M. J., 1983, Long décollements and mud volcanoes: Evidence from the Barbados Ridge complex for the role of high pore-fluid pressure in the development of an accretionary complex: Geology, v. 11, p. 279–283.

shear zone should be identical. His second condition implies that small-scale structural features and strain profiles across a shear zone in any cross section should be identical, except near its ends. This should permit study of finite strain and the geometry of any shear zone along several representative cross sections. Most real shear zones approximate these features. By considering the various types of shear zones in light of these boundary conditions, Ramsay concluded that three kinds of displacement can occur in shear zones: (1) heterogeneous simple shear, (2) heterogeneous volume change largely by pressure solution, and (3) combinations of the first two.

SHEAR-SENSE INDICATORS

Determination of shear sense or direction of movement along shear zones is an important factor that later may help estimate displacement. Sense of shear may not be immediately obvious in many larger shear zones, and so indicators of shear sense on the mesoscopic and microscopic scales may be used ultimately to determine the sense of motion of shear zones that are kilometers wide.

Various criteria may help determine shear sense, as summarized by Carol Simpson and Stefan Schmid (1983), Gordon Lister and Arthur Snoke

FIGURE 10–16
Near-horizontal ductile shear zones in augen gneiss near Vålån in the Swedish Caledonides. Two well-defined ductile shears are visible just below the coin. Note how porphyroblasts are rotated dextrally (clockwise) by the shear zones. (RDH photo.)

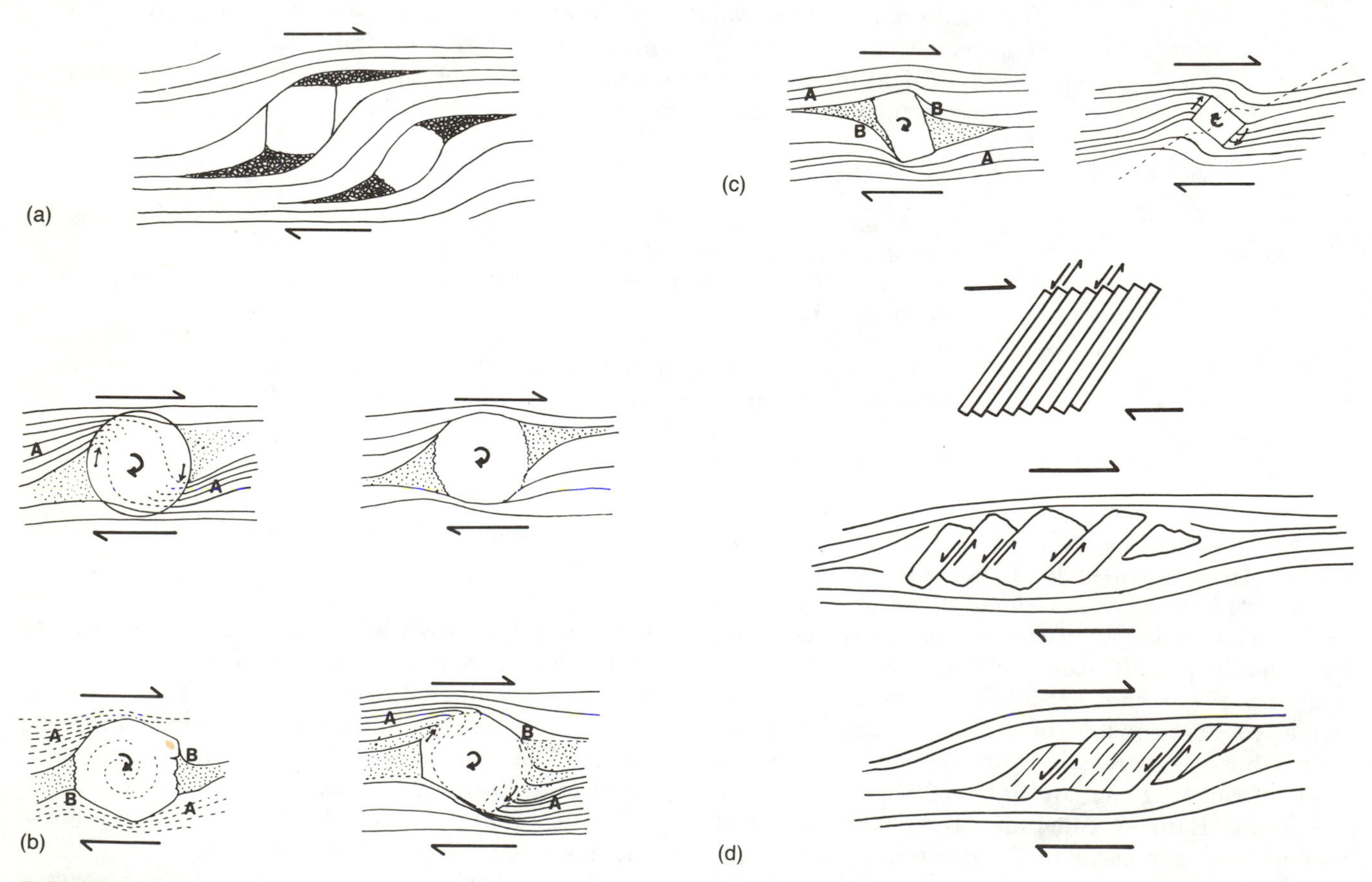

FIGURE 10–17
Rotated porphyroclasts (a), crystals (b), pressure shadows (c), and fractured grains (d). Arrows indicate movement sense. Note that fractured porphyroclasts frequently yield an opposite movement sense. (From Carol Simpson and Stefan Schmid, 1983, Geological Society of America *Bulletin,* v. 94.)

(1984), C. W. Passchier and Simpson (1986), and Simpson (1986). The discussion here is based largely on their compilations.

Asymmetric Porphyroblasts, Porphyroclasts, and Pressure Shadows

Porphyroblasts are large grains that have grown in the rock mass during deformation and metamorphism, and *porphyroclasts* are relict earlier large grains. If they are rotated in a more ductile matrix—commonly finer grained—the rotation produces structures associated with these more brittle objects that provide clear indicators of shear sense (Figure 10–17). The associated structures are examined in the following sections.

Asymmetric Augen and Boudins. Augen gneisses and mylonitic gneisses deformed ductilely by a strongly asymmetric simple-shear component commonly develop "tails" on porphyroclasts; the tails are asymmetric in the direction of ductile flow (Figures 10–17, 10–18a). Smaller grains making up

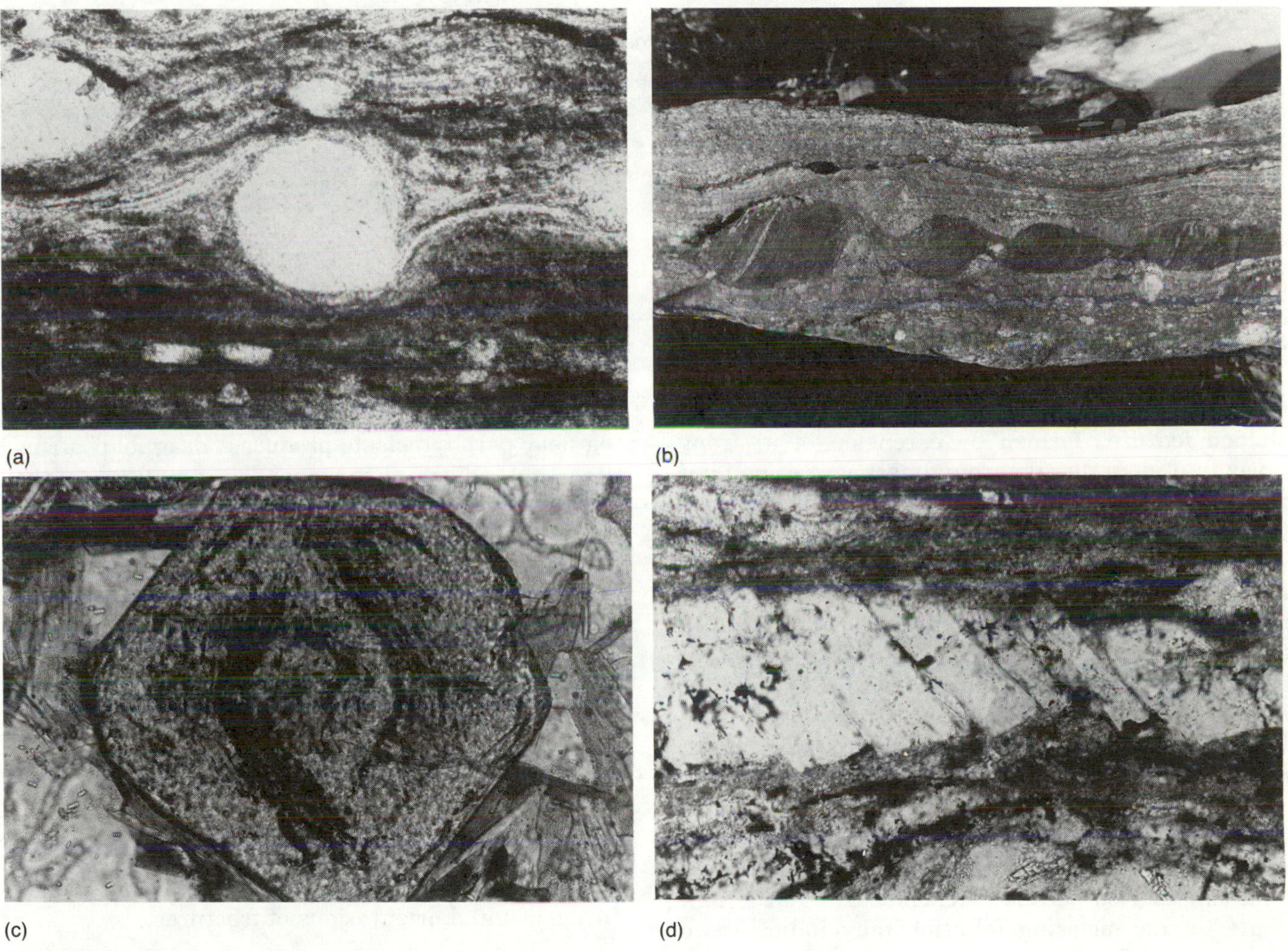

FIGURE 10–18
(a) Feldspar porphyroclast with asymmetric tails indicating (dextral) movement. Towaliga fault zone near Forsyth, Georgia. Width of photo is approximately 8.5 mm. Plane light. (From R. J. Hooper and R. D. Hatcher, Jr., 1988, *Tectonophysics,* v. 152, Elsevier Science Publishers.) (b) Amphibolite boudins in biotite gneiss showing dextral rotation of original layering at Woodall Shoals, South Carolina. (RDH photo.) (c) Rotated inclusion trails of graphite in garnet in Poor Mountain metasiltstone near Walhalla, South Carolina, indicating sinistral shear. This garnet has a spiral "snowball" structure. The garnet is approximately 0.7 mm wide. Plane light. (RDH photo.) (d) Fractured and rotated quartz porphyroclast from the Homestake shear zone near Leadville, Colorado, indicating apparent dextral but actual sinistral shear. The quartz grain is approximately 1 mm long. Plane light. (RDH photo.)

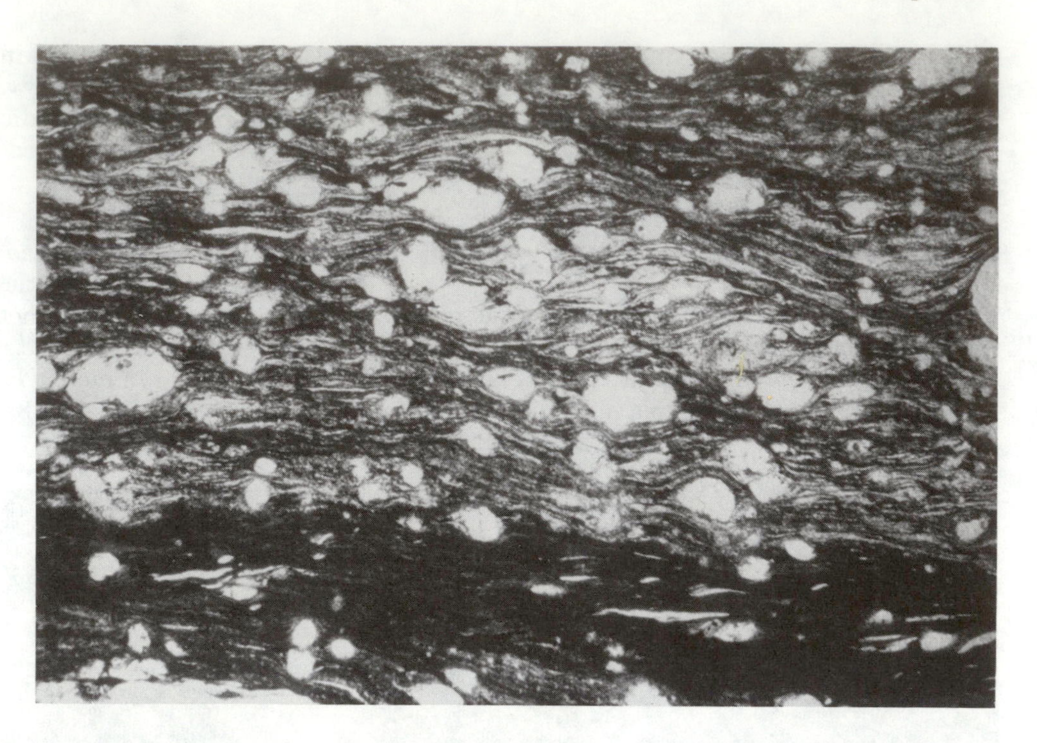

FIGURE 10–19
The acute angle between the older subhorizontal S-surface and new inclined C-surfaces indicates the shear direction. Mylonite from the Towaliga fault zone near Forsyth, Georgia. The C-surfaces parallel the sides of the shear zone. Width of field is about 4 mm. Plane light. (From R. J. Hooper and R. D. Hatcher, Jr., 1988, *Tectonophysics,* v. 152, Elsevier Science Publishers.)

the tails may be derived both from the porphyroclasts and from the recrystallized groundmass. Compositional similarities to the porphyroclasts suggest that most of the material in the tails is derived from the porphyroclasts. Boudins, which are sausage-shaped features formed by extension of a strong layer in a more ductile groundmass, are sometimes rotated, producing pressure shadows or wrapped foliations that may indicate shear sense.

Inclusion Trails. Inclusion trails in garnet, staurolite, cordierite, feldspars, and several other minerals may have a "snowball," an S, or a spiral asymmetric pattern (Figures 10–17, 10–18) that indicates shear sense. These crystals were rotated during crystallization in a more ductile matrix, recording the sense of rotation. Most inclusion trails record prograde deformation rather than retrograde. Care should be exercised in using inclusion trails in minerals, like garnet, that record growth during a long time period because inclusion trails at a high angle to the enclosing foliation may indicate two directions of flattening and not simple shear (Bell, 1985; Vernon, 1988). According to T. H. Bell (1985), even asymmetric inclusion trails can be produced by different orientations of pure shear strains at different times. Spiral and snowball inclusion trails probably can be used to indicate shear sense.

Asymmetric Pressure Shadows. Pressure shadows formed adjacent to grains or crystals in a more ductile matrix may be used for determination of shear sense if the pressure shadows are asymmetric (Figure 10–17). They function in the same way as asymmetric tails on porphyroblasts, but display an opposite sense of asymmetry.

Distorted Layering. Geometry of distorted layering near porphyroclasts produces "drag fold" structures that indicate shear sense. The amplitude of the foldlike structures may also provide a measure of shear strain (Bjornerud, 1989).

Fractured Grains

Large grains of some minerals like feldspars, amphiboles, pyroxenes, and micas continue to deform brittlely, even in a ductilely deforming matrix. During progressive deformation in a zone of heterogeneous simple shear, they may fracture and become offset (Figure 10–18d). Simpson and Schmid (1983) have pointed out that these fractures indicate a sense of motion opposite that of actual movement (Figure 10–17). In fact, the sense of motion depends on the original orientations of fractures.

Composite Foliations

S- and C-Surfaces. Foliations (parallel alignment of platy minerals or bands in a metamorphic rock) of several types develop in shear zones related to progressive simple shear. A foliation that D. Berthé, P. Choukroune, and P. Jegouzo (1979) first called a ***C-surface*** (the C is from the French *cisaillement,* meaning shear) is related to shearing forms in the shear zone during progressive simple shear (Figure 10–19). Depending on the initial orientation, slip

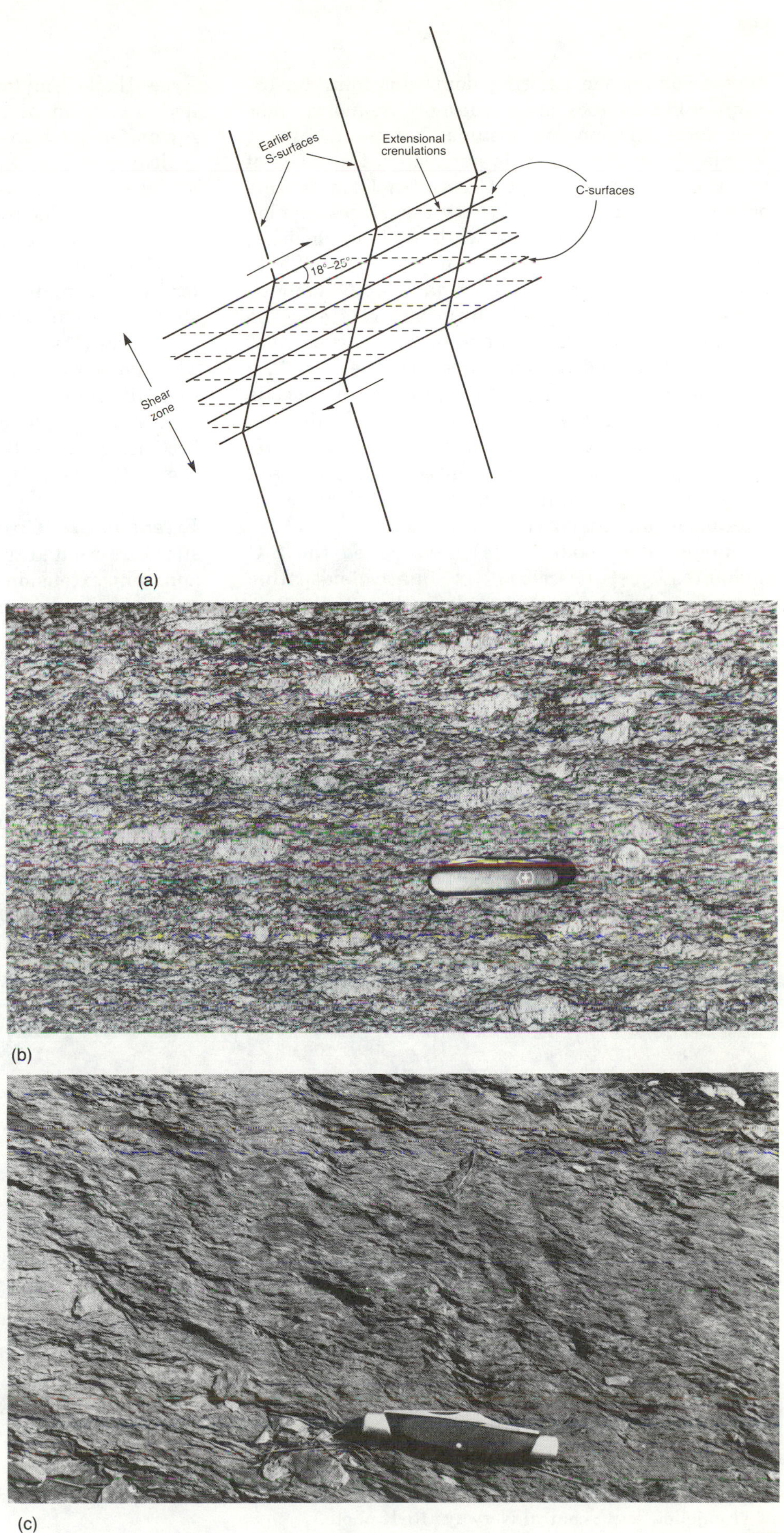

FIGURE 10–20
(a) Generalized shear-zone geometry showing development of foliations relative to the shear components and the walls of the shear zone. (b) Ductilely deformed late Mississippian Clouds Creek Granite in the Pennsylvanian Modoc fault zone near Batesburg, South Carolina. The large fractured microcline phenocrysts are aligned parallel to the dominant foliation (S), and the dextral C-surfaces are inclined about 20° to the S-surfaces. This is a Type I S-C mylonite. (Arthur W. Snoke, University of Wyoming.) (c) Phyllonite in the Brevard fault zone, Mountain Rest, South Carolina, a Type II S-C mylonite. (RDH photo.)

may occur on the existing dominant foliation (***S-surface***) in the rock mass, but more frequently new C-surfaces develop. We assume that the dominant simple shear component is parallel to the walls of the shear zone and that C-surfaces form in this orientation (Figure 10–19). New C-surfaces develop parallel to the shear-zone walls and as shearing continues may be rotated to an angle of 18° to 25° to the shear-zone walls (Weijermars and Rondeel, 1984). S-surfaces may also form and rotate in the shear zone; sometimes they exhibit drag structure at the walls that indicate motion sense. The C-surface foliation in quartzofeldspathic or quartzitic rocks is the same as ***shear band foliation*** (S. H. White and others, 1980) that forms in more micaceous rocks. Microscopically, both are recognized as thin layers of fine-grained recrystallized aggregates of minerals like micas and quartz (Figure 10–20a).

Lister and Snoke (1984) have called the S-C mylonites (just described) in quartzofeldspathic rocks Type I mylonites. They defined another class, Type II S-C mylonites, as those resulting from mylonitization of quartz-mica rocks. Type II S-C mylonites contain flame-shaped buttons (or fish scales) of micas, which they called "mica fish." C-surfaces are defined by trails of mica, whereas S-surfaces are defined by an oblique foliation in the adjacent quartz aggregate.

Subgrains in quartz may form at angles of 10° or less to the main lattice orientation and indicate shear sense in the same relationship as C- and S-surfaces (Lister and Snoke, 1984; Simpson, 1986). They appear as a part of a crystal having an extinction different from that of the larger host grain. Where the angle between the subgrain and host becomes greater than about 15°, recrystallization begins (Chapter 7).

Extensional Crenulation Cleavage. Another structure related to S- and C-surfaces originates by apparent extension and rotation as the shear zone moves; it is called ***extensional crenulation cleav-***

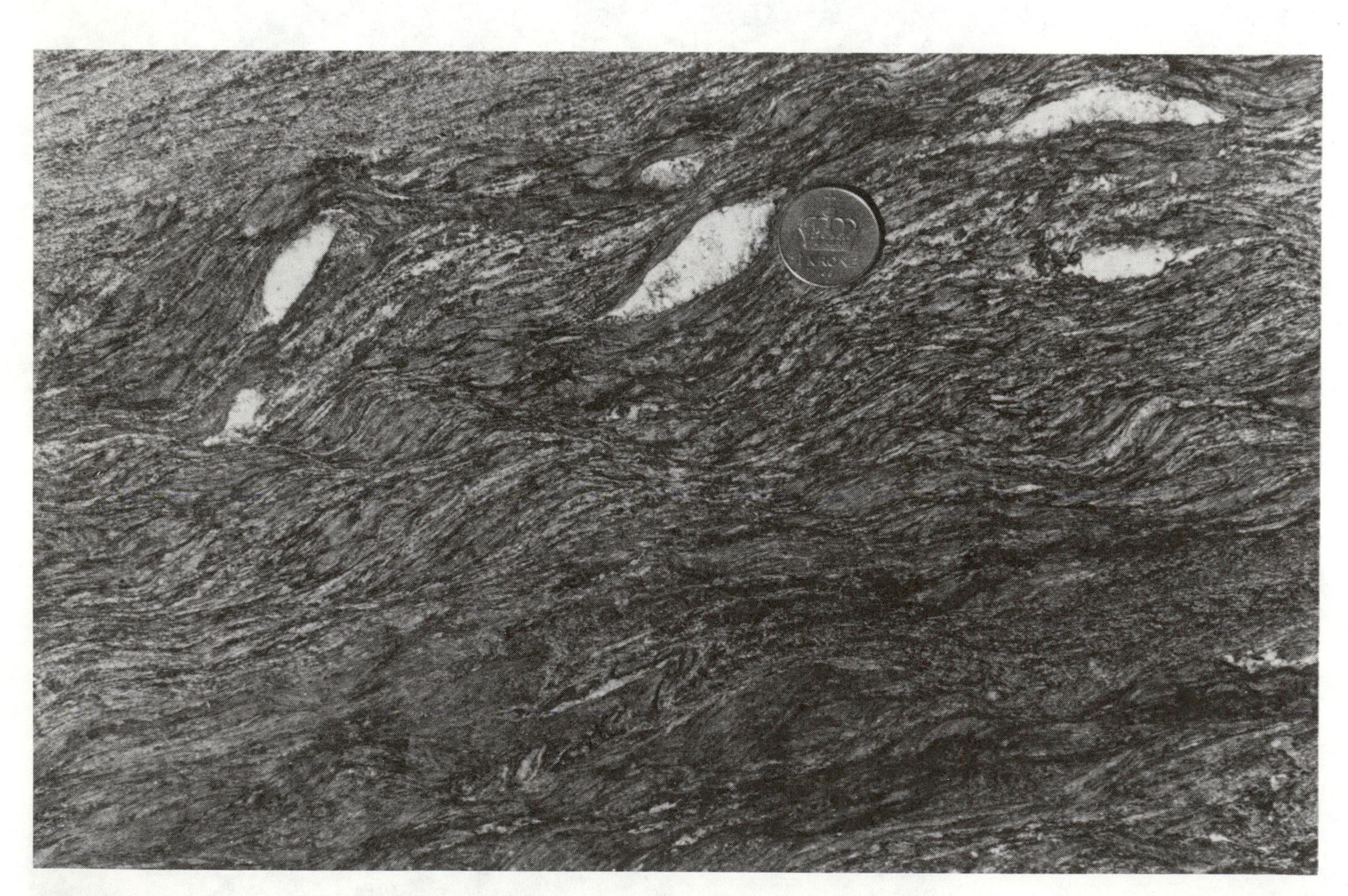

FIGURE 10–21
Shear-band C-surfaces deforming earlier S-surfaces in a fine-grained Cambrian phyllonite from the fault zone at the base of the Seve thrust sheet on the west flank of Tronfjellet, south-central Norway. (RDH photo.)

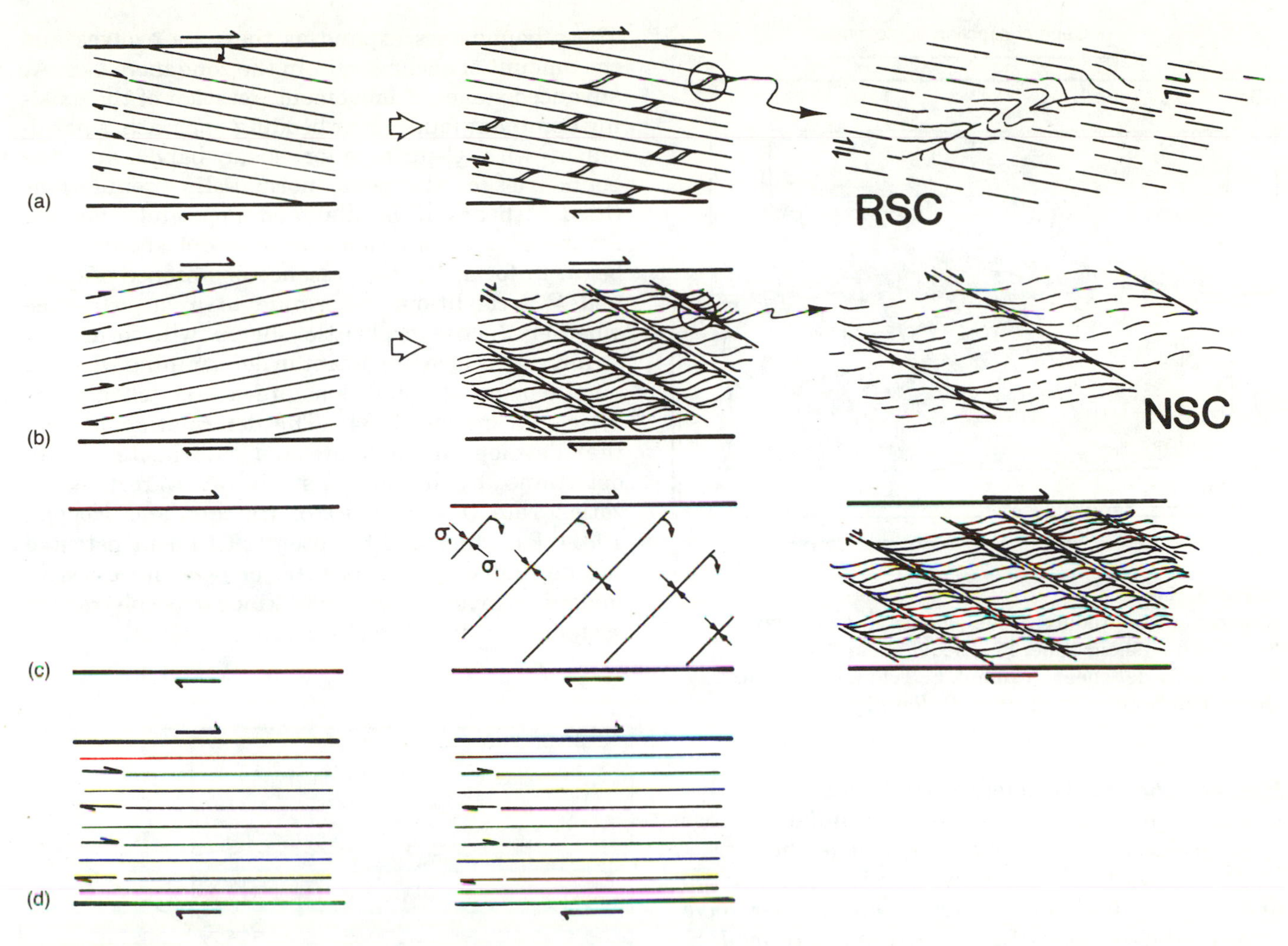

FIGURE 10–22
Development of C-surfaces and oppositely vergent structures related to conveniently oriented earlier S-surfaces that became reactivated. (a) Development of reverse-slip crenulations (C) compensates for movement normal to shear-zone walls (S) where the original foliation is gently inclined clockwise to shear-zone movement direction. (b) Normal slip crenulations compensate for normal movement where the original foliation (S) is inclined gently counterclockwise to the movement direction. (c) If shearing of the foliation perpendicular to the cumulative flattening direction occurs so that the foliation is rotated into an easy slip orientation and slip is an important mode of strain in the shear zone, normal-slip crenulations will develop. (d) If the original foliation is parallel to the shear-zone walls, no compensating mechanism is required. NSC—normal-slip crenulations; RSC—reverse-slip crenulations. (Reprinted with permission from A. J. Dennis and D. T. Secor, *Journal of Structural Geology,* v. 9, A model for the development of crenulations in shear zones with applications from the southern Appalachian Piedmont, © 1987, Pergamon Journals Ltd.)

age (Platt and Vissers, 1979; Figure 10–21). Structures resembling intrafolial folds (folds lying within a foliation) appear between conveniently oriented active S-surfaces inclined opposite to the shear direction and to the C-surface orientation (Figure 10–21). They may result from shearing motion on an existing S-surface becoming locked, then forced to "ramp" to the next higher detachment S-surface on the hand-specimen to microscopic scale—a sequence similar to the detachment-ramp behavior of thrust faults at the road-cut to crustal scale (Figure 10–22). Formation of extensional crenulation cleavage requires a preexisting layering or layering produced during motion of the shear zone. An earlier foliation (S-surface), or new C-surfaces, provides the layering for formation of extensional crenulations (Dennis and Secor, 1987). Allen Dennis and Donald T. Secor (1987) noted that extensional crenulations may form as either *normal-* or *reverse-slip* crenulations (Figure 10–22).

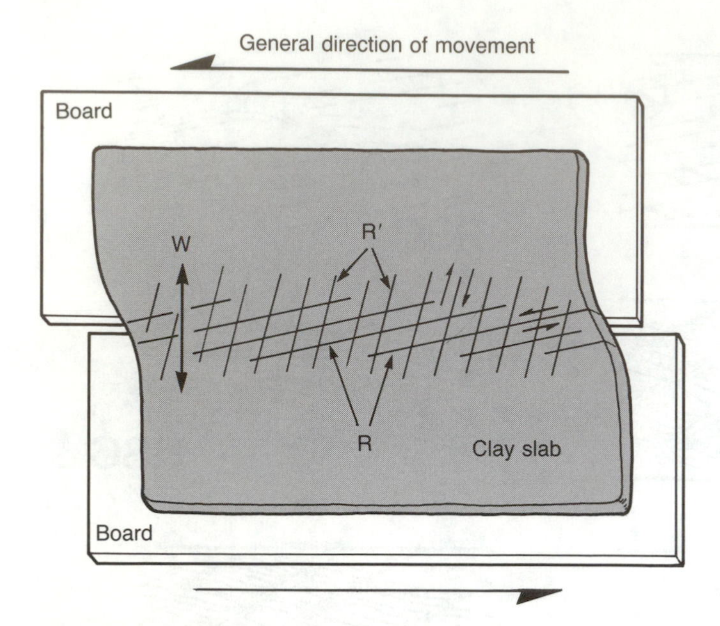

FIGURE 10–23
Experiment to produce Riedel shears in a clay layer. W—width of shear zone; R—Riedel shear; R′—anti-Riedel shear. (From J. S. Tchalenko, 1970, Geological Society of America *Bulletin*, v. 81.)

Riedel Shears. Geometric similarity exists between normal- and reverse-sense crenulations that form in ductile shear zones and brittle shear fractures that were first recognized by W. Riedel (1929) and form in near-surface fault zones. They have been produced experimentally in dry clay by J. S. Tchalenko (1970; Figure 10–23). ***Riedel shears*** and the oppositely moving structures called ***anti-Riedel shears*** form initially at very low strain, but movement is immediately transferred to low-angle shears that form subparallel to the shear-zone boundaries. Interesting comparisons can be made between the through-going low-angle shears that bisect the Riedel and anti-Riedel shears and the C-surfaces that form in ductile shear zones. In contrast to normal and reverse-slip crenulations that form in ductile shear zones and function to thicken, thin, or maintain its thickness (Dennis and Secor, 1987), Riedel and anti-Riedel shears appear to form early in the movement history of a brittle shear zone, then are either destroyed or cease moving.

SHEAR-ZONE KINEMATICS

Regardless of whether a shear zone forms under brittle or ductile conditions, the kinematics appears similar and is also independent of scale. Once simple shear is initiated and the rock mass begins to deform, the boundaries of the zone are defined. These boundaries expand as the zone evolves and the amount of shear strain in the zone increases. At advanced stages of movement, rotation of the existing dominant foliation or bedding approaches parallelism with C-surfaces or shear bands and the boundaries of the zone; normal-slip crenulations (Riedel shears if brittle) and oppositely moving reverse-slip crenulations (anti-Riedel shears) also begin to form. Depth, presence or absence of fluid, and P-T conditions determine to a large degree whether ductile or brittle shears will form, but strain rate is also important in determining whether ductile or brittle conditions prevail (Hatcher, 1978; Wise and others, 1984). The deformation process that develops in mylonite and cataclasite is also determined by the ratio of strain rate to recrystallization rate (Hatcher, 1978; Hatcher and Hooper, 1981). Passchier and Simpson (1986) demonstrated the dependence on this ratio of flow processes in mylonites involving different kinds of porphyroclast systems that indicate shear sense.

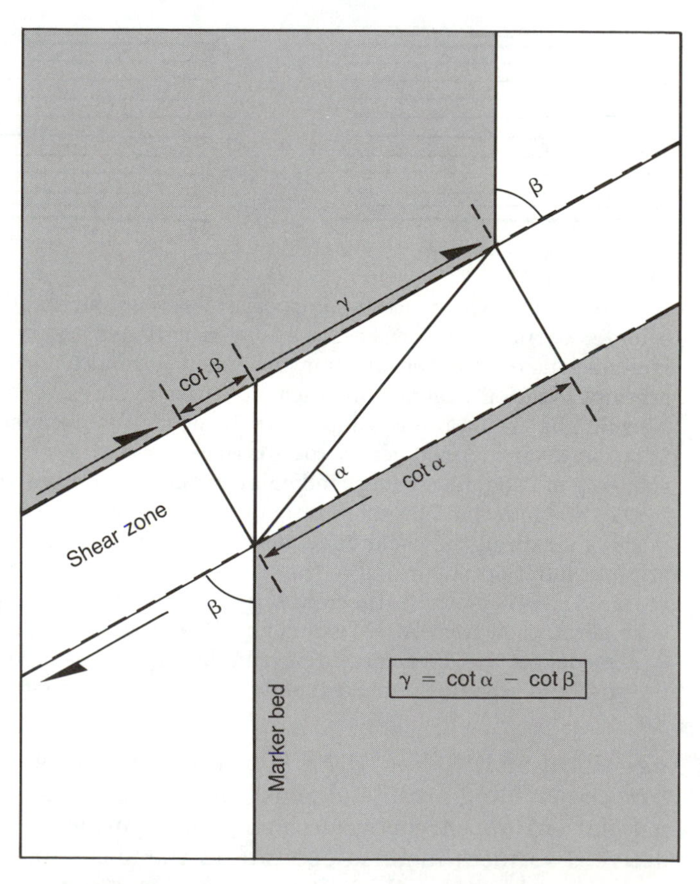

FIGURE 10–24
Displacement of a marker in a shear zone. (After J. G. Ramsay, 1967, *Folding and fracturing of rocks:* McGraw-Hill Book Company, and Ruud Weijermars and H. E. Rondeel, 1984, *Geology,* v. 12.)

Strain-ellipsoid analysis of shear zones is initially unclear, because in a zone of heterogeneous simple shear, the ellipsoid has no constant orientation. Net flattening does occur roughly perpendicular—and with elongation parallel—to the walls of the shear zone (Figure 10–20); but the C-surfaces must parallel the shear planes.

Displacement of some marker by a shear zone may be estimated by a technique derived by Ramsay (1967). Lines making an original angle α with the shear-zone boundary (Figure 10–24) are displaced to a new angle α'. The shear displacement, γ, may be obtained from

$$\cot \alpha' = \cot \alpha + \gamma \qquad (10\text{–}9)$$

or

$$\gamma = \cot \alpha' - \cot \alpha \qquad (10\text{–}10)$$

The displacement, D, on a shear zone may also be estimated from the simple integral

$$D = \int_0^x \gamma dx \qquad (10\text{–}11)$$

derived by John Ramsay and Rod Graham (1970), where x is the width of the shear zone. So, if we can estimate the magnitude of shear strain on a shear zone from either equation 10–11 or equation 4–5, we can obtain an estimate for displacement by measuring the width of the zone in the field. For example, the Brevard fault zone in the southern Appalachians is about 5 km wide. Dextral shear strains, determined using equation 4–5, are near 1. Substituting and solving the integral produces $D = 1$ (5 km) = 5 km as the displacement estimated for the fault zone.

With that ends our discussion of fault mechanics. Now we turn to more detailed consideration of each of the three kinds of faults, beginning with thrust faults.

Questions

1. Why are Mohr diagrams for normal and thrust faults constructed of nested circles of oppositely increasing radii?
2. How is Amontons' first law related to Byerlee's law?
3. How do Anderson's fundamental assumptions relate to real faults? Can you suggest an alternative scheme of orientation of principal stress (strain) axes that will work?
4. How does water lower the coefficient of sliding friction on a fault?
5. Describe the transformation of part of a fault from a stick-slip mechanism to a continuous-creep mechanism. What are the reasons for the transformation?
6. If a fault in a gneiss has a maximum shear stress of 68 MPa and an effective normal stress of 77 MPa, what is the coefficient of sliding friction for the fault?
7. A dextral fault zone, 2 km wide, exposed in the central part of the West African craton, consists entirely of mylonite. You can show (using zircon Pb-U geochronology) that it was formed 2.3 b.y. ago in rocks that were in the upper amphibolite facies (20 km deep, temperature above 500°C). What can you deduce about the behavior of the fault in the upper crust before erosion removed the upper crustal rocks?
8. Why do fractured grains commonly yield an opposite shear sense?
9. How can extensional crenulation cleavage form in a ductile shear zone that formed in a homogeneous granite?
10. How could Riedel shears developed on a crustal scale be analogous to microscopically developed ductile shear bands? (For example, see S. Mosher, 1983, Kinematic history of the Narragansett Basin, Massachusetts and Rhode Island: Constraints on late Paleozoic plate reconstructions: Tectonics, v. 2, p. 327–344.)
11. Explain in terms of fault mechanics the mechanism by which earthquakes were produced in the Denver area soon after deep injection of fluid.

Further Reading

Anderson, E. M., 1951, Dynamics of faulting and dyke formation: Edinburgh, Oliver and Boyd, 191 p.
A classic and readable work discussing the characteristics and mechanics of the three major types of faults.

Simpson, Carol, 1986, Determination of movement sense in mylonites: Journal of Geological Education, v. 34, p. 246–260.
Discusses both shear-sense indicators and properties of fault rocks.

11

Thrust Faults

Eight great faults [occur in East Tennessee as] ribbon-like masses or blocks . . . crowded one upon another, like thick slates or tiles on a roof, the edge of one overlapping the opposing edge of the other.

JAMES M. SAFFORD, *A Geological Reconnaissance of the State of Tennessee*

THRUST FAULTS ARE FASCINATING STRUCTURES, CONSISTING OF thin sheets of rock that have huge areal extents. Many sheets have moved horizontally tens to hundreds of kilometers. Chief Mountain, Montana, is a remnant of a formerly more extensive thrust sheet (Figure 11–1). Thrusts form some of the largest structures in mountain chains, comparable in size to large accreted terranes or microcontinents—hundreds of kilometers long. Thrusts affect migration of ground water, oil and gas, and ore-bearing fluids and also form traps for hydrocarbons and metallic minerals. Overthrusting may force fluid migration in sedimentary rocks beneath thrust sheets, forming traps for accumulation of hydrocarbons and metallic minerals (Figure 11–2).

Thrust faults are found in mountain chains—orogenic belts—and, because of the stratigraphy present, are most easily recognized on the continentward side. A belt of thrust faults occurs between the undeformed craton and the metamorphic core of nearly every mountain chain (Figure 11–3). This zone is called a ***foreland fold-and-thrust belt,*** and most examples contain stratigraphic sequences from former stable trailing margins like the present-day East Coast of the United States. Thrusts also occur in the metamorphic cores of mountain chains, where some thrusts form under ductile conditions where temperatures are moderate to high. Other thrust faults in orogenic interiors form later, at lower temperatures and under more brittle conditions.

The idea of large-scale horizontal transport of great slabs of rock was conceived in the 1840s by Arnold Escher von der Linth, a Swiss geologist. Escher was the first to correctly interpret the Glarus overthrust in Switzerland and called these structures *Uberschiebung,* which in German means "overpushing." Shortly afterward, James M. Safford (1856) recognized the Appalachian thrusts in Tennessee and visualized the fault geometry. In Switzerland—as related by Edgar Bailey (1935)—Roderick I. Murchison reexamined the Glarus thrust and confirmed Escher's interpretation, but refused to agree with J. Nicol (1861)—a Scot known for the optical prism of Iceland spar that bears his name—that thrusts exist in the Scottish Highlands. In fact, some British geologists sought to disprove the existence of Scottish thrusts by testing a supposed example in the Northwest Highlands near Loch Eribol and Assynt. First, Charles Lapworth, representing Her Majesty's Geological Survey, went to the Assynt District, where several months of study convinced him that the Moine thrust did exist. (He also recognized the importance of mylonite and coined the name.) Unfortunately, he fell mentally ill, believing the Moine thrust was still moving and threatening his cabin near the base of Knockan Crag, and so he never finished his work there. Later in the nineteenth century, several members of Her Majesty's Geological Survey, including B. N. Peach, J. Horne, C. T. Clough, and H. M. Cadell, mapped

FIGURE 11–1
Chief Mountain, Montana—a klippe of the Lewis thrust that transported Middle Proterozoic Belt Series rocks over Upper Cretaceous sedimentary rocks. The Lewis thrust is the prominent inclined boundary traceable from the lower left of the mountainside and is truncated by the nearly horizontal dark boundary—another thrust carrying Precambrian rocks—near the top of the mountain. Boyer and Elliott (1982) concluded the upper—nearly horizontal—boundary forms the roof of a duplex structure in the Belt Series rocks. (Bailey Willis, U.S. Geological Survey.)

the same region in an attempt to dispose of any notion of large-scale horizontal transport of a thin sheet of rock in the British Isles. Instead, they published in 1884 compelling evidence that at least 16 km of transport had occurred along a low-angle movement surface. The usefulness of their geologic maps of the Assynt District (Peach and others, 1888), including the Moine thrust (Figure 11–4), is likely to outlast the stone monument to their work that now stands weathering on a low hill near Inchnadamph.

As shown by studies of the Glarus and Moine thrusts, thrust faults have long provoked controversy. First, their very existence was disputed; later, involvement of *basement* in foreland thrusts was debated; more recently, geologists have focused on whether the motive force for thrusting is gravity or compression. (Basement generally consists of crystalline rock that is the product of an earlier orogenic cycle and underlies a less deformed and less metamorphosed cover.) Great strides were made during the 1970s and 1980s toward understanding the geometry and mechanics of thrust faults, but even recent mechanical models are not accepted by all. Thus controversy continues.

This chapter is more detailed than the chapters on strike-slip and normal faults partly because so much knowledge has accumulated on the geometry of thrusts. The petroleum industry has long explored thrust-faulted terranes, gathering extensive drilling and seismic-reflection data to help understand the geometry and behavior of thrusts. Regions of normal and strike-slip faults are also much explored, but information—and controversy—about thrust faults remain more abundant.

NATURE OF THRUST FAULTS

Thrust faults are gently dipping faults in which the hanging wall has moved up relative to the footwall. Several kinds of thrusts exist, and their behavior varies with the types and strengths of rocks involved, the ambient temperature at the time of formation, and the degree of involvement by water during movement.

Thrust faults are usually found in mountain chains or their eroded remnants. They occur most commonly in a continentward-thinning wedge of sedimentary rocks above an undeformed basement in a foreland thrust belt (Figure 11–5). The undeformed mass of continental-margin sediments has a characteristic wedge shape—a shape maintained throughout the deformational history of the belt—and is also a feature inherited from the shape of the original continental margin on which the thrust belt is developed because trailing margin sediments thicken oceanward. Wedge geometry in thrust belts was first recognized by William M.Chapple (1978) as a fundamental property of deforming thrust belts. After studying structures in the still-active Taiwan thrust belt and formulating dynamic geometric models of them, John Suppe (1981) concluded that the shape is preserved during deformation through a combination of surface erosion and slumping. These processes tend to conserve the original angles—equilibrium angles—and thus the shape of the wedge. During thrusting, the principal change is that the wedge is shortened and thickened until it reaches a steady-state condition (Chapple, 1978; Suppe, 1981). Nicholas Woodward (1987) has con-

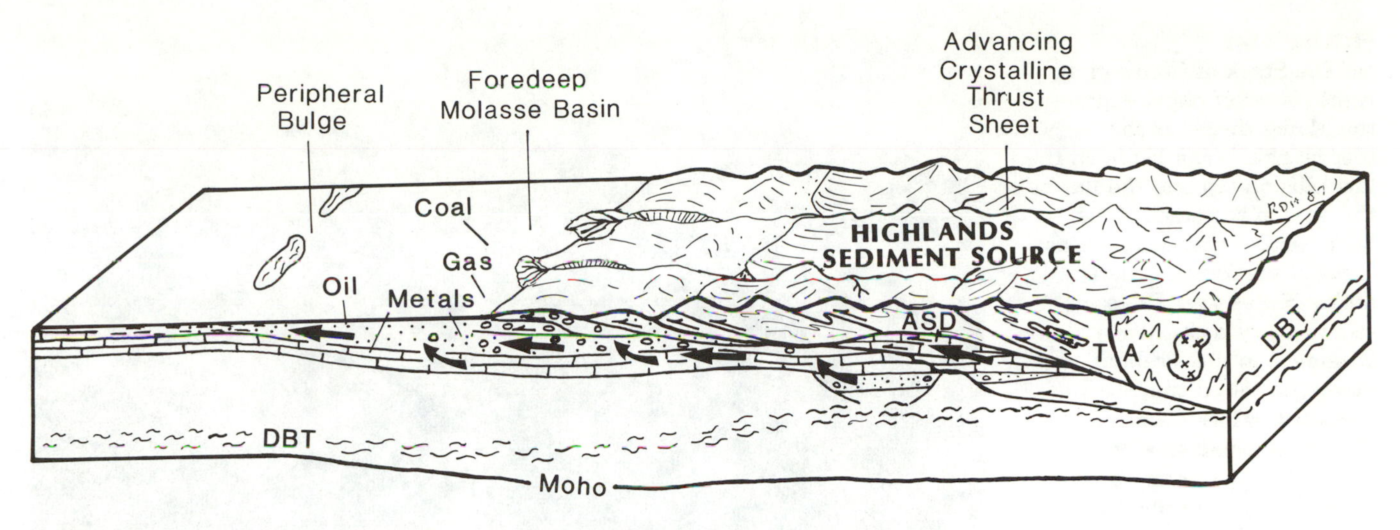

FIGURE 11–2
Effects of the arrival of a large thrust sheet on a continental margin. Fluids carrying hydrocarbons and metals are expelled. Loading of the crust by the sheet produces a depression beneath and in front of it (foredeep) and a crustal arch (peripheral bulge) farther inland. DBT—ductile-brittle transition zone; ASD—antiformal stack duplex; T—movement toward observer; A—movement away from observer. (Modified from J. E. Oliver, 1986, *Geology*, v. 14.)

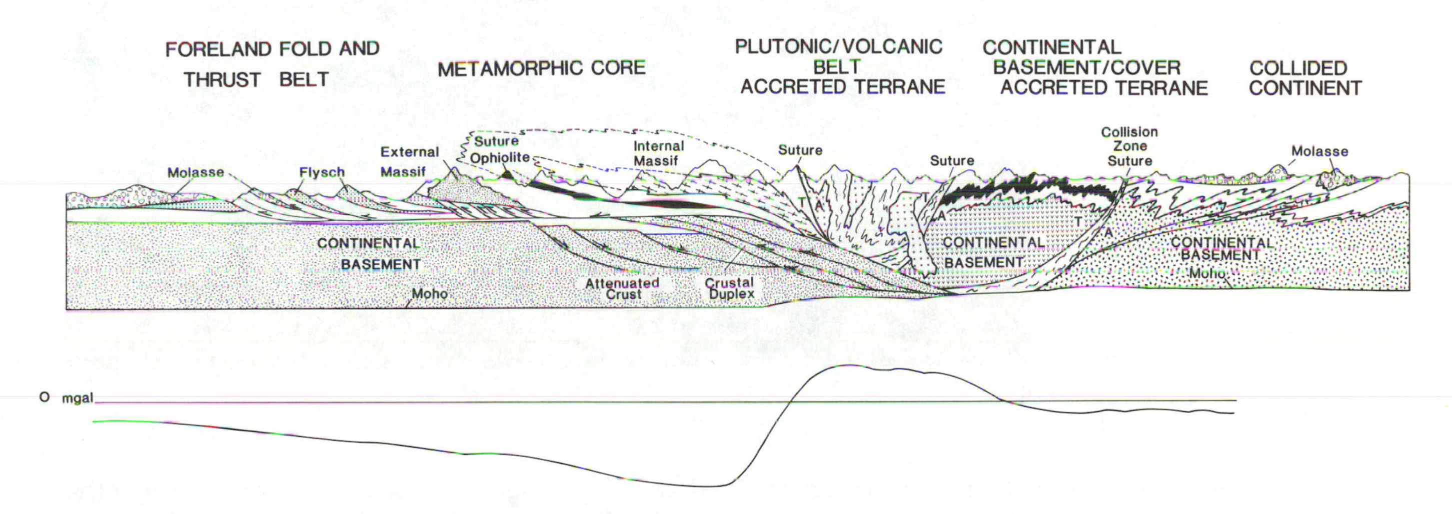

FIGURE 11–3
Ideal mountain chain. Masses of continental or oceanic crust or island arcs may be accreted by either head-on or oblique collision, producing a variety of deformational styles, metamorphism, and plutonism. T—movement toward observer; A—movement away from observer. Curve is a hypothetical gravity profile of the kind observable in most mountain chains. (From R. D. Hatcher, Jr., and R. T. Williams, 1986, Geological Society of America *Bulletin*, v. 97.)

cluded that the wedge must be deformed internally as new material is added to the front if the shape of the wedge is to be preserved. Evidence is lacking for internal deformation in foreland-type wedges; the wedges probably become stronger overall during deformation, though fault zones probably weaken them. Woodward suggested that the critical wedge models developed primarily from study of modern accretionary wedges may not survive rigorous application to foreland thrust belts because of differences in mechanical properties of the rocks and in the amount of fluid present.

Thrusts have historically been debated as either ***thin-skinned,*** with no basement involved, or ***thick-skinned,*** with basement involved (Figure 11–6). Because basement is not commonly observed at the

(a)

(b)

FIGURE 11–4
(a) The Stack of Glencoul at right center of photo exposes the Moine thrust at the topographic break between the low hills on top and the flat at the base of the stack. This fault carried metasedimentary rocks of the Precambrian Moine Series over Paleozoic sedimentary rocks and basement in the foreland. (b) Moine thrust on the Stack of Glencoul. White Cambrian Pipe Rock Sandstone was overthrust by darker-colored metasandstone of the Late Proterozoic Moine Series. (RDH photos.)

surface in foreland thrust belts, many geologists came to believe that basement was also not involved in the subsurface. The thin-skin concept originated with A. Buxtorf (1916), who worked in the Jura Mountains. Later, John Rich (1934) discovered most of the principles of thin-skinned deformation during his classic study of the Pine Mountain block in the southern Appalachians (Figure 11–7a). Rich based his detailed analysis on surface geologic mapping and a few shallow drill holes; data gathered since have modified his conclusions very little. Long after Rich, Shankar Mitra (1988) confirmed his conclusions, but used seismic-reflection and existing drill data to show that the subsurface structure was more complex than Rich had thought (Figure 11–7b).

Rich proved that thin-skinned deformation worked for the Pine Mountain overthrust sheet, and that set the stage for similar discoveries around the world. Even so, John Rodgers (1949, 1964) and Byron Cooper (1964) took opposing sides in the thin- and thick-skin debate in the Appalachians, and A. W. Bally, P. L. Gordy, and G. A. Stewart (1966) used seismic-reflection and surface geologic data to prove that the deformation mechanism in the Canadian Rockies fold-and-thrust belt was thin-skinned. Later studies, using seismic-reflection profiling in other foreland thrust belts, confirmed Rich's conclusion that thin-skinned deformation is the dominant style, but thick-skinned thrusts have also been shown to exist.

Metamorphic rock—either as basement slices or as progressively metamorphosed sedimentary rocks—appear on the thickened side of the wedge in the transition to the metamorphic core of a moun-

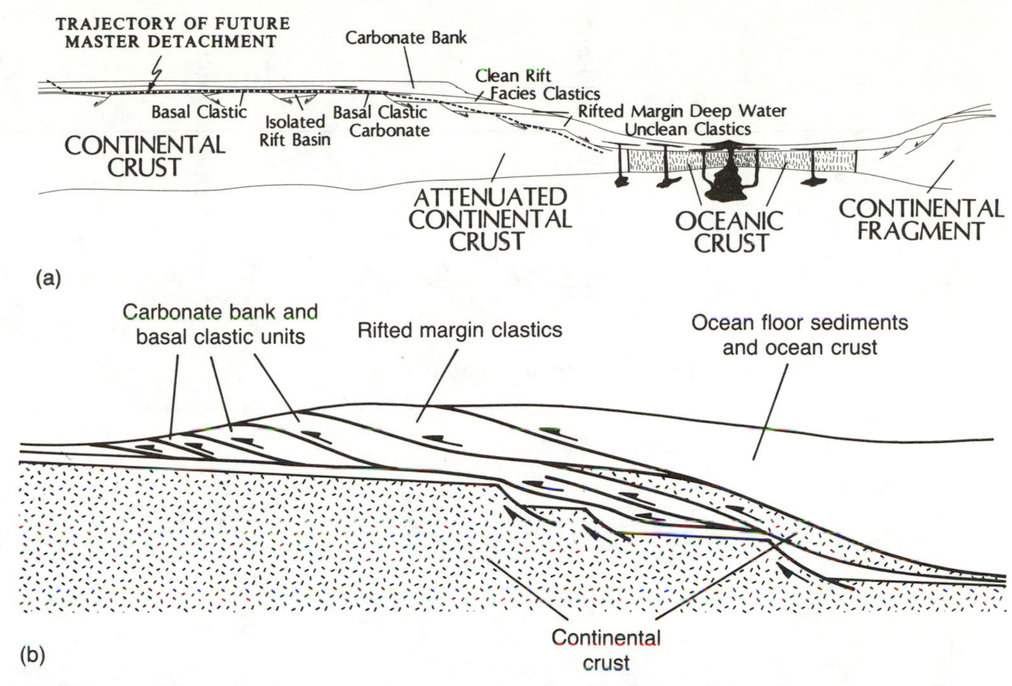

FIGURE 11–5
Attributes of the wedge before and after deformation. (a) Continental platform margin showing many commonly associated features and a slight thickening toward the ocean. Dashed line is the trajectory of a future major thrust. (b) Deformed continental platform assemblage. Note that deformation has greatly thickened and shortened the wedge.

tain chain (Figure 11–8). Thrusts involving crystalline (metamorphic and igneous) rocks are called ***crystalline thrusts.*** Those closer to the foreland form under brittle conditions; those farther into the metamorphic core form under conditions ranging from ductile to brittle, depending on temperature and strain rate during movement (Hatcher and Williams, 1986). Thermal or strain softening of the crystalline mass may be required before initiation of faulting if low-angle thrust faults are to propagate through crystalline rocks. The thrust plane may be initiated by local ductile behavior, but the mass may be transported by brittle translation. The temperature during movement depends on when it begins relative to one or more of the thermal/metamorphic peaks affecting the orogen. A source far inside an orogen may produce warmer thrust sheets than one on the outer flanks. Hatcher and Odom (1980) observed a definite correlation between movement of large faults (mostly thrusts) and thermal/metamorphic peaks in the southern Appalachians, noting that older thrusts formed in the internal parts early in the history of the chain, but the youngest faults formed on the flanks, suggesting the orogen had an inside-out deformation plan. A similar relationship exists in the Alps, the Canadian Cordillera, British and Scandanavian Caledonides, and—probably—other mountain chains. In contrast, we also noted the Appalachians of New England and the Canadian Maritime provinces contain older thrusts along the western flank and younger faults toward the east, indicating an outside-in deformation plan. This also reflects a contrasting late Paleozoic plate tectonic history for these regions—head-on collision in the southern Appalachians, oblique collision in New England.

Earlier concepts of thrust faults implied significant differences in the modes of origin, propagation, and motion of crystalline and foreland thrusts. For many decades since, we have been able to study their geometry in eroded surface exposures. The phenomenon of low-angle thrusting of crystalline rocks was first noted by Törnebohm (1872) in Norway and Sweden and by Peach and others (1888) when they first mapped the Moine and Arnabol thrusts in Scotland. More recently, it has been shown that crystalline thrusts share many properties with foreland thrusts (Elliott and Johnson, 1980; Hatcher and Williams, 1986). The powerful combination of detailed surface geologic studies and geophysical data gathered in metamorphic cores of orogens, particularly in the Appalachians (Bryant and Reed, 1970; Hatcher, 1971, 1972; Cook and others, 1979; Hatcher and Zietz, 1980), the British Caledonides (Butler and Coward, 1984), and the Alps (Panza and Müller, 1979; Laubscher, 1983), has confirmed many of the similarities. Still, as the ensuing discussion will make clear, important differences remain.

Many thrust faults tend to form in a concave-up spoon-shaped geometry. Thrusts with this concave-up geometry are called ***listric*** thrusts. Normal faults frequently exhibit the same geometry, indicating mechanical similarities to thrusts.

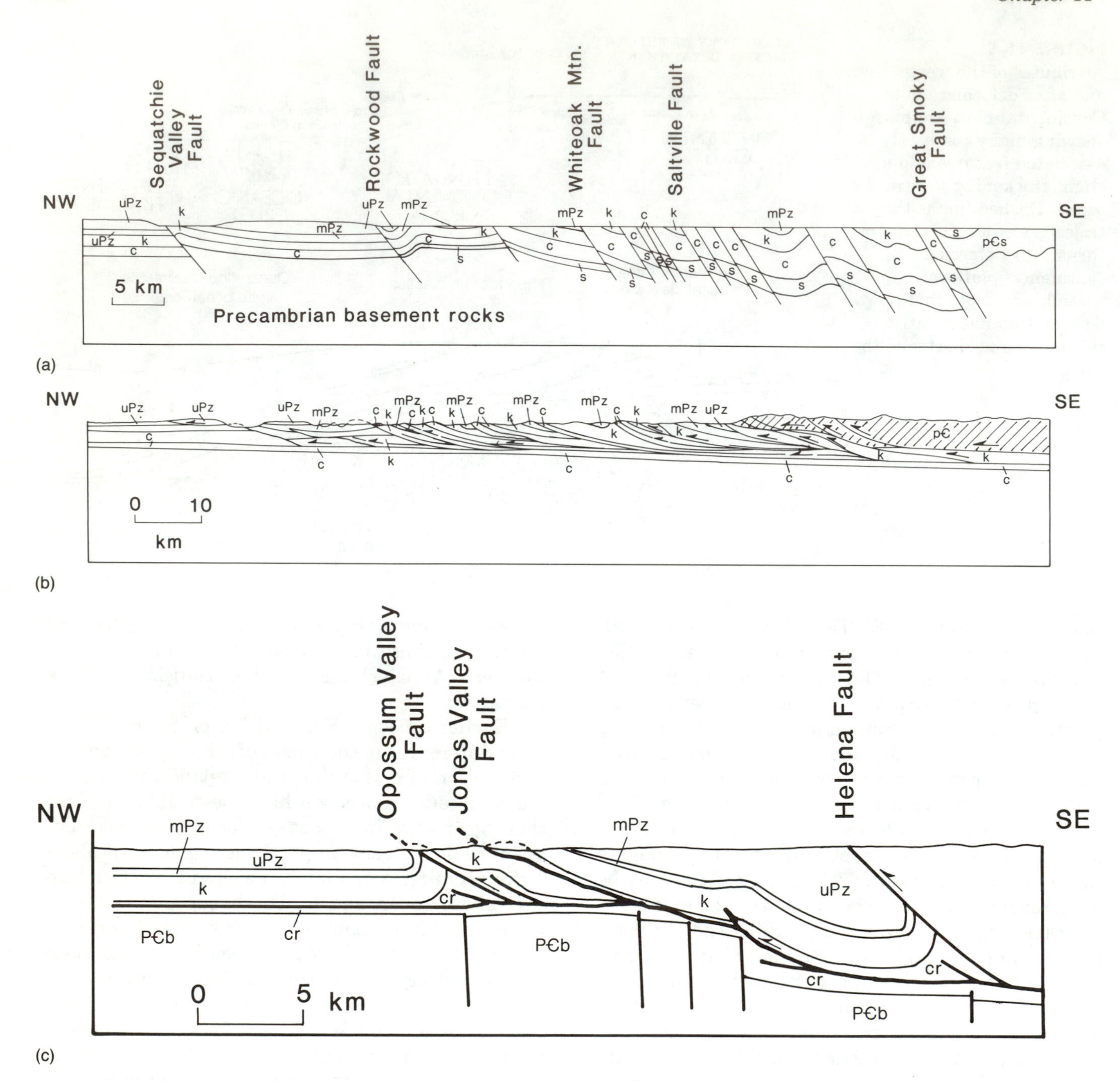

FIGURE 11–6
(a) Thin-skinned (no basement) thrusting versus (b) thick-skinned basement-involved thrusting along a cross section through the Appalachians in Tennessee and western North Carolina. (Part a from John Rodgers, 1953, Kentucky Geological Survey Series 9, Special Publication 1; part b from a section by RDH.) (c) Cross section from the Appalachian Valley and Ridge in Alabama showing control of thin-skinned structures by older faults in the basement. uPz-upper Paleozoic rocks; mPz—middle Paleozoic rocks; k—Knox Group carbonate rocks (Cambrian-Ordovician); c—Cambrian shale and carbonate; cr—Conasauga Group and Rome Formation (Cambrian); p€s—Precambrian sediments; p€ and P€b—Precambrian basement rocks. (From W. A. Thomas, 1986, Virginia Tech Geological Sciences Memoir 3.)

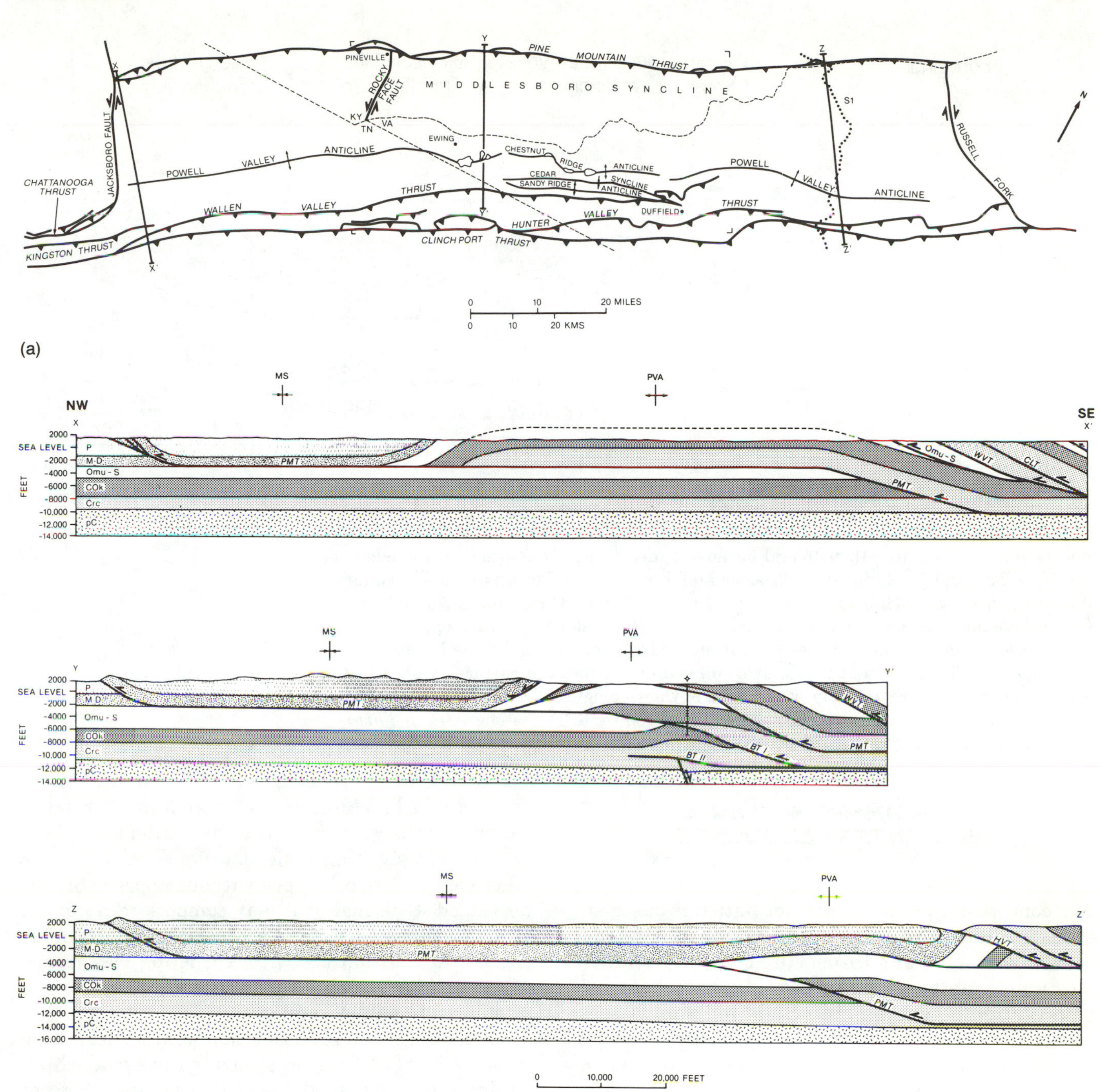

FIGURE 11–7
(a) Attributes of the Pine Mountain thrust sheet, Tennessee, Virginia, and Kentucky. The three lines X-X′, Y-Y′, and Z-Z′ locate the sections in (b). (b) Sections through the Pine Mountain block showing the fault-bend fold character of the Pine Mountain thrust (PMT). MS—Middlesboro syncline; PVA—Powell Valley anticline; WVT—Wallen Valley thrust; CLT—Clinchport thrust; P—Pennsylvanian rocks. M-D—Mississippian and Devonian rocks; Omu-S—Middle to Upper Ordovician and Silurian rocks; COk—Cambro-Ordovician carbonate rocks; Crc—Cambrian clastic rocks; pC—Precambrian basement rocks. The COk, Omu-S, and P are strong rock units; Crc and M-D are weak units. (From Shankar Mitra, 1988, Geological Society of America *Bulletin,* v. 100.)

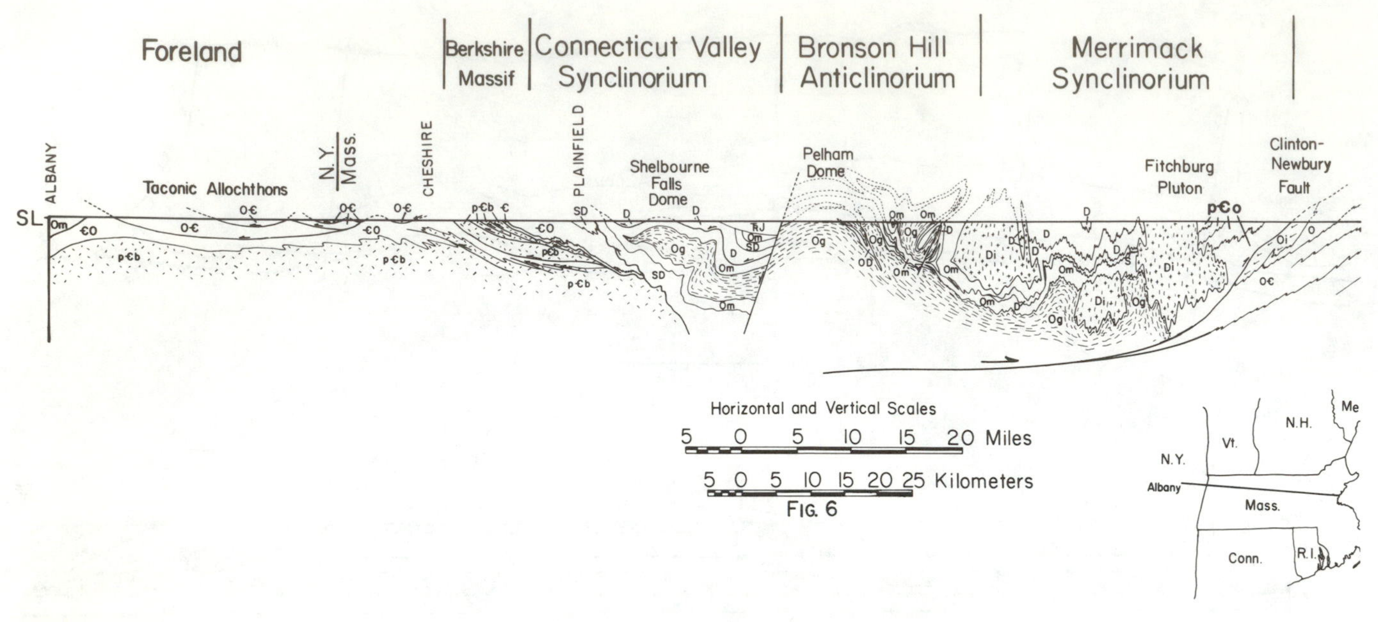

FIGURE 11–8
Formation of crystalline thrusts and basement involvement in thrusts at the edge of the New England foreland. ₱J—Triassic and Jurassic; D—Devonian; Di—Devonian intrusive rocks; SD—Silurian and Devonian; S—Silurian; Om—Middle Ordovician; O—Ordovician; Oi—Ordovician intrusive rocks; Og—Ordovician granite; O€—Ordovician and some Upper Cambrian; €O—Cambrian and some Lower Ordovician; p€b—basement rocks; p€o—Precambrian rocks in eastern Massachusetts. (Reproduced by permission of the Geological Society from Thrust and nappes in the North American Appalachian Orogen by R. D. Hatcher, Jr., in *Thrust and Nappe Tectonics,* Geological Society of London Special Publication No. 9, 1981.)

DETACHMENT WITHIN A SEDIMENTARY SEQUENCE

The concept of thin-skinned deformation is based on certain common properties of thrusts in foreland thrust belts. Foreland thrusts are characteristically low-angle faults that thrust older rocks over younger rocks. They may include high-angle segments and (very rarely) may bring basement rocks to the surface; they commonly repeat the same oldest rock unit across the entire belt. The oldest unit most commonly is a shale, evaporite, coal, or other weak (*incompetent*) rock type, or it may be a strong (*competent*) unit, one susceptible to strain softening. Thrusts generally propagate along these layers and are called *detachments* or *décollements*. Because they represent translation along a bedding plane, they are also called *bedding thrusts*. Décollements (French for "ungluing") may occur as the result of either folding or faulting (Figure 11–9), but the net result is relative displacement of the upper layers.

High-angle segments exist where a thrust cuts across a strong ***(competent)*** unit from one detachment to another (Figure 11–10). A high-angle segment is called a ***ramp*** and may occur on both small (outcrop) or larger scales as units a kilometer thick or more. Massive limestone, dolostone, sandstone, or other competent rock types without appreciable interbedded weak material may compose ramp units. Their thickness also limits the amplitude of folds formed in association with thrusting, as ramps make up the strong unit, or ***strut,*** in a thrust-faulted sedimentary sequence.

Folds may form in association with thrust faults. Bailey Willis (1893), after studying the Appalachian foreland, distinguished *break thrusts*—which form by thrusting the connecting limb of an anticline-syncline, overthrust the *hanging-wall anticline,* and preserve a *footwall syncline*—from *shear thrusts,* formed independently of folding. Shear thrusts do not have a footwall syncline. John Suppe (1983) has defined *fault-bend folds* (Figure 11–11) as those formed as the thrust surface changes from steeper dip to shallower in an up-dip direction. A thrust sheet wherein dip flattens as it passes over a ramp would produce a fault-bend fold. Suppe (1985) also recognized *fault-propagation folds,* which form as layers fold during propagation of a thrust through a sedimentary sequence; they are similar to Willis's break thrusts.

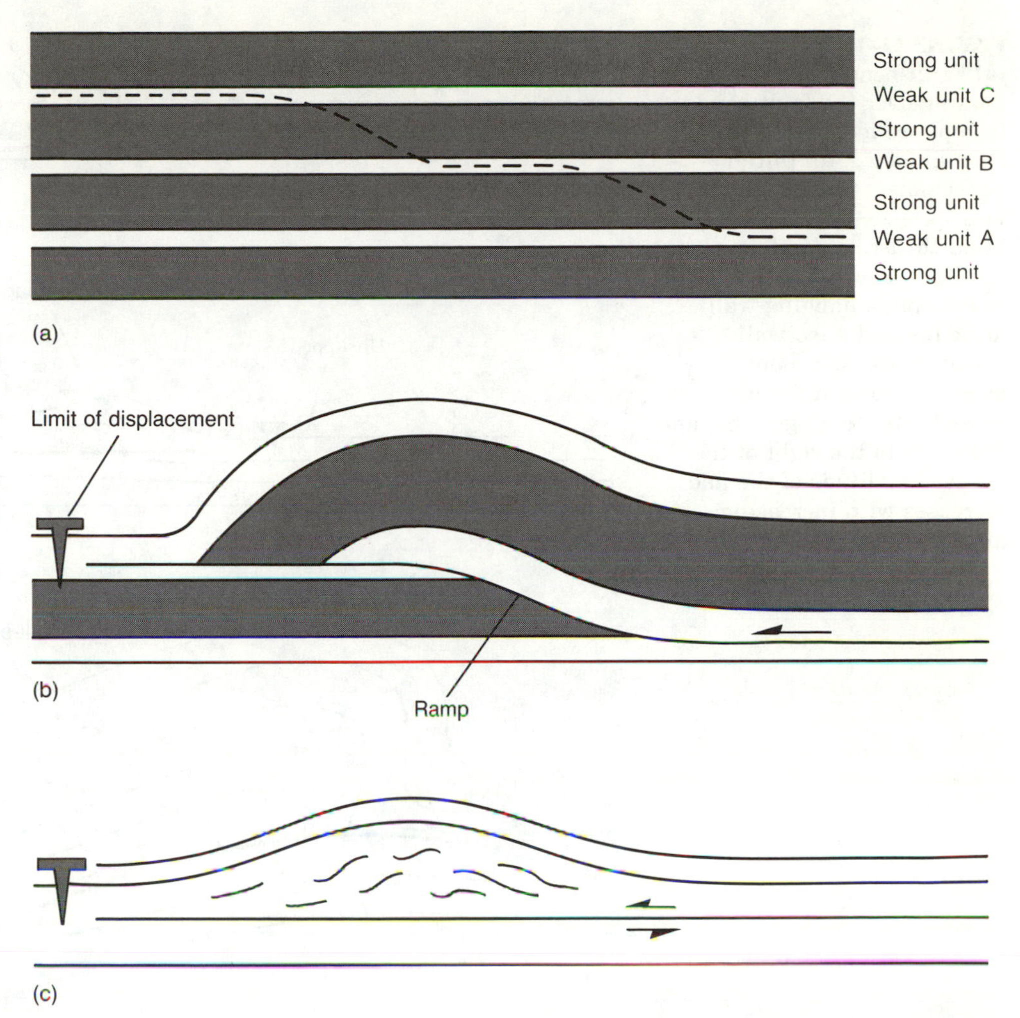

FIGURE 11–9
Formation of a detachment in a sequence of alternating weak and strong rock units. The dashed line in (a) is the path of a future detachment. The detachment follows weak units and ramps (or refracts) across strong units. Ramping occurs because of a change in the physical properties of a weak unit, as in weak unit in (a), or thinning of a weak unit, as in weak unit in (b). The entire sheet moves over the undeformed footwall, forming a ramp anticline (or fault-bend fold; see Figure 11–11).
(c) Folds may also form by this mechanism above a detachment.

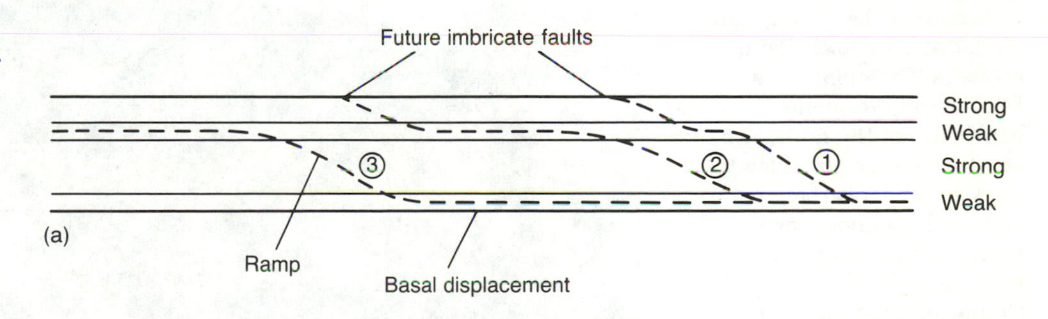

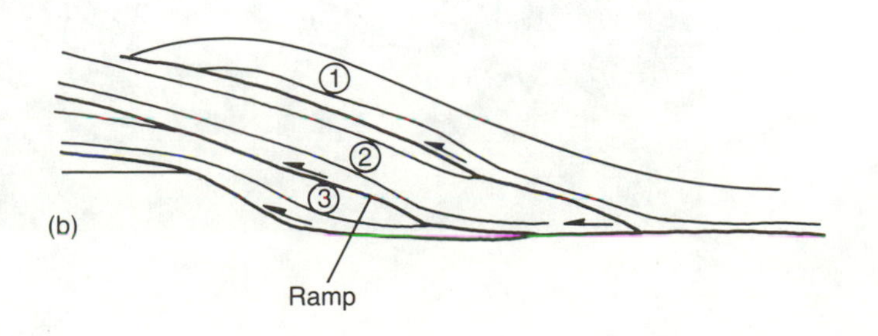

FIGURE 11–10
Properties and propagation of thrusts in cross section. (a) Undeformed. (b) Deformed.

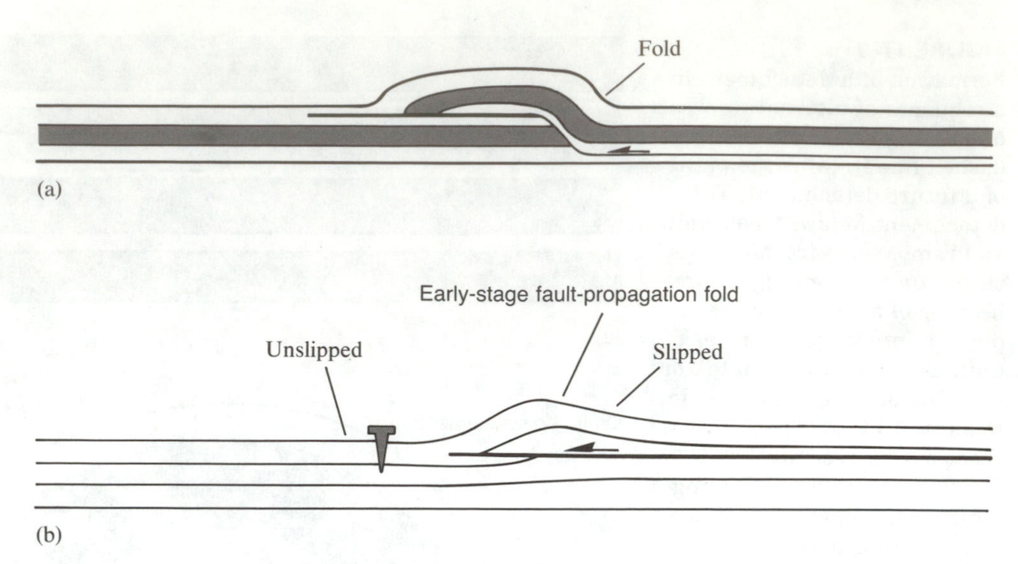

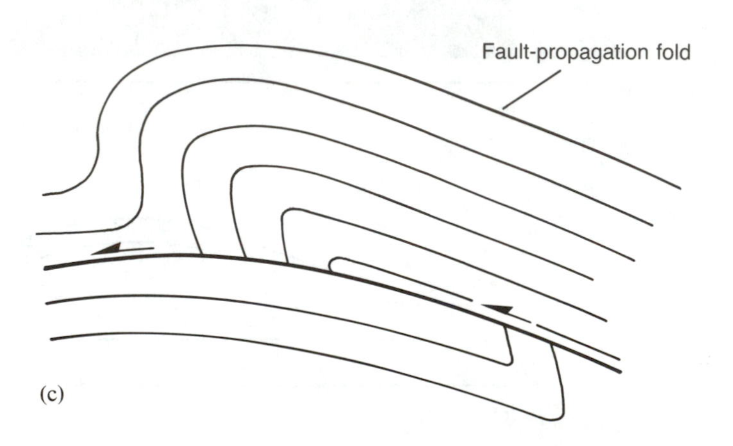

FIGURE 11–11
(a) Fault-bend folds. Rocks passing over the fault-bend (ramp) are ideally folded, then unfolded. (b) Initial development of a fault-propagation fold as a blind thrust. (c) Well-developed fault-propagation fold containing a hanging-wall anticline and a footwall syncline. Displacement decreases toward the left, away from the hinge zone, and increases to the right of the hinge. Amplitude of the fold increases with increasing displacement.

FIGURE 11–12
Exposure of the Champlain thrust at Low Rock Point on Lake Champlain near Burlington, Vermont. Note the gentle dip of the fault. A horse of white Lower Ordovician Beekmantown carbonate on right side of photo rests in the fault zone between the hanging wall of gray Cambrian Dunham Dolomite and footwall of black Middle Ordovician Iberville Shale. (RDH photo.)

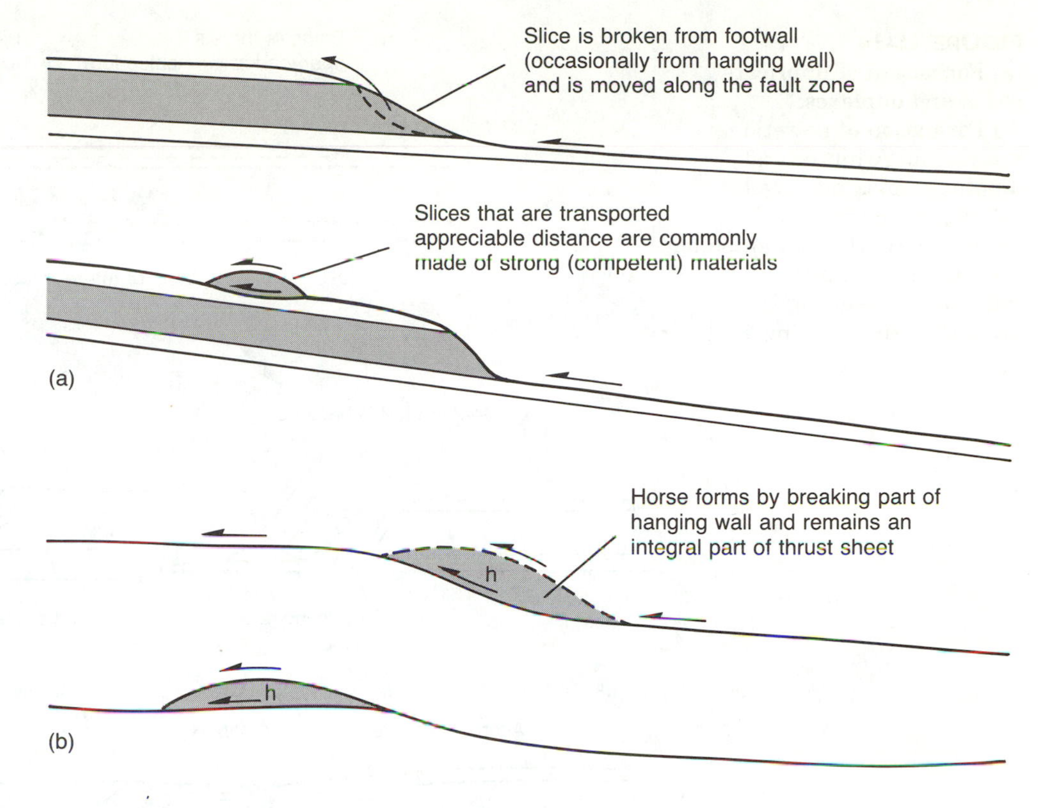

FIGURE 11–13
Formation of horses (slices, h) in cross section. Slices are commonly derived from the footwall of the fault; horses may be derived from the hanging wall.

Folds that form near the thrust surface during movement are called ***drag folds*** (Figure 9–13a). Drag folds are a direct product of friction along the thrust surface and may range from open to tight folds, depending on the competence of the rocks being deformed. The greater the amount of incompetent material in a section, the tighter the drag folds.

Thrust faults commonly break off and grind material along the fault surface. If the deformation mechanism is dominantly brittle, fracturing is more likely. Pieces of material called *horses,* or *slices,* ranging up to several kilometers in maximum dimension, may be transported within the fault zone beneath a thrust sheet. Here a general rule: *the age of the material composing slices is usually intermediate between the ages of the hanging-wall and footwall rocks* (Figure 11–12). An exception occurs where the slice consists of crystalline basement. Slices may roll or slide along and, in the process, be folded and fractured internally. As a result of the mode of formation and transport, these blocks are commonly made of competent materials and are most frequently derived from the footwall, but slices derived from the hanging wall do exist. Obviously, fault surfaces completely surround a slice (Figure 11–13). Less commonly, they are composite competent/incompetent materials, with the competent dominating. A *horse* may also be thought of as a slice that remains an integral part of the thrust sheet and is not transported far from its point of origin, but many geologists use the terms slice and horse synonymously.

Thrust faults may occur as isolated faults, but most occur in groups consisting of a master fault with several associated smaller faults; the smaller ones are called ***imbricate thrusts*** (Figure 11–14). The term *imbricate zone* may also be used to describe the entire series of thrusts making up a foreland thrust belt. Imbricates probably form by branching from a master décollement as the thrust sheet moves forward. As one segment "sticks," another propagates beneath it to form an imbricate. In most cases propagation is toward the foreland. A few imbricates may propagate in the opposite direction, if the overburden is thin. Repetition of the sequence results in an ***imbricate stack.*** An ***accretionary wedge,*** which forms as the sediments in a subduction zone along an active margin are scraped off the descending slab, consists of an imbricate stack of thrusts (Figure 11–14b).

Thrusts are known to occur where two parallel to subparallel subhorizontal thrusts of roughly equal displacement are separated by a deformed

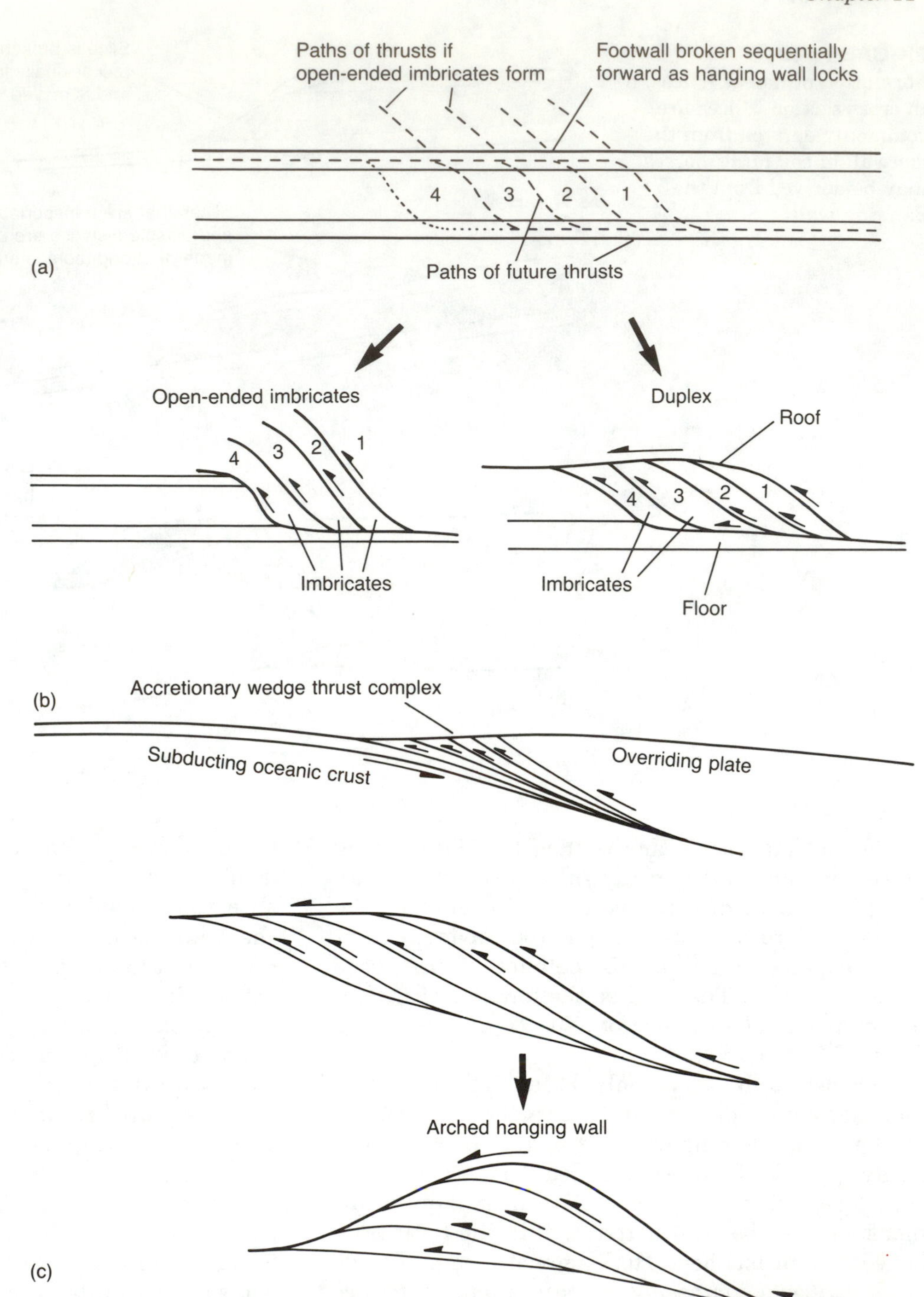

FIGURE 11–14
(a) Formation of imbricate zones and duplexes.
(b) Formation of accretionary wedges. (c) A hinterland-dipping duplex may evolve into an antiformal stack duplex by continued motion on the thrust sheet and imbricates, resulting in arching of the hanging wall.

interval that is thin relative to their total areal extent. Structures of this kind are called ***duplexes*** (Figure 11–14). The upper of the two master faults is the ***roof thrust,*** and the lower is the ***floor thrust.*** Smaller imbricates connect the two master faults. The imbricates dip more steeply than the roof or floor thrusts and gradually merge with the master faults. Duplexes occur in most thrust-faulted terranes and are probably more common than once thought, even though they were recognized more than a century ago by Peach and others (1888) along the Glencoul and Moine thrusts in the Assynt District of Scotland. They form as the roof thrust ramps and then locks, and successive imbricate faults form, ramp, and lock, allowing formation of the floor thrust and continued movement of the entire thrust complex.

Antiformal stack duplexes may be produced as the imbricate stack forms and rotates toward the horizontal (Figure 11–14c). Examples include the

Mount Crandell structure in the Canadian Rockies (Dahlstrom, 1969; Price and Mountjoy, 1970) and the Limestone Cove window in the Appalachians of Tennessee (Diegel, 1986; Figure 11–15). Duplex development is a fundamental property of the interiors of foreland thrust belts in mountain chains. It has recently been discovered, using a combination of surface geologic and seismic-reflection data that a duplex, although formed in response to movement of a thrust sheet, frequently arches the thrust sheet as the duplex is built by duplication of rocks beneath it. Duplexes generally form beneath an appreciable overburden and with a ratio of strong rocks to weak between 1:1 and 4:1 (Costello and Hatcher, 1985). In some imbricate thrust structures, a roof thrust exists with only minor displacement, but in others the roof thrust is easily identified and has considerable displacement. Banks and Warburton (1986) and Morley (1986) have called them *passive* and *active roof duplexes*.

Synthetic and *antithetic normal faults* are normal faults that dip either subparallel or oppositely to join a larger fault. They commonly form within a moving thrust sheet as the sheet passes over a ramp, and extensional stresses result as layers are extended and produce normal faults (Figure 11–16). They may also form as the entire sheet is extended as a single layer or by relaxation of the sheet. The strike of the faults generally parallels the strike of the more gently dipping thrusts. Dip of the normal faults is typically greater than 45°, but the dip may also become shallow so that the faults are listric.

Using fold geometry and deformation mechanisms, Peter Geiser has classified thrusts in three major categories (Figure 11–17): *layer shortening, fracture,* and *fold thrusts.* Geiser's fracture thrusts are the same as Willis's shear thrusts and Suppe's fault-bend folds, and his fold thrusts are the same as Willis's break thrusts and Suppe's fault-propagation folds. Combinations are also possible. The layer-parallel shortening mechanism would account for only 15 to 25 percent strain if it depended solely on pressure solution as the dominant strain mechanism in the thrust sheet. But over a thrust belt 100 km wide, this mechanism could produce 15 to 25 km of displacement (decreasing with distance from the direction of thrusting). Other layer-parallel shortening mechanisms, such as contraction faulting (see next section), may result locally in 50 to 60 percent shortening. The mechanism related to pressure solution has been well documented by Geiser and Terry Engelder (1983) in the Appalachian Plateau of New York and by Walter Alvarez, Engelder, and William Lowrie (1976) in the northern Apennines of Italy. Very large thrusts in the internal parts of foreland thrust belts may be initiated as a low-strain variety.

SMALL-SCALE FEATURES OF THRUST SHEETS

Some predominantly small-scale structures form in association with thrusts. Because they provide important clues to thrust mechanics, we will discuss them here.

D. K. Norris (1958) described *extensional faults,* in which bedding undergoes layer-parallel extension, and *contractional faults,* in which bedding is shortened, in coal mines of the foreland thrust belt of the Canadian Cordillera (Figure 11–18). Later, Ernst Cloos (1964) described contractional faults in the central Appalachians but called them *wedges* or wedge faults. They have since been reported in other thrust-faulted terranes. Frequently associated with extensional and contractional faults is a *broken formation* zone. A broken formation generally develops beneath a detachment, perhaps by hydrofracturing under high pore pressure. It may also form where the ratio of competent material to incompetent is low (less than 1) and the rocks are too weak to support formation of a duplex. Steven Wojtal (1986) has interpreted broken formations as brittle shear zones.

PROPAGATION AND TERMINATION OF THRUSTS

A thrust moves until the energy responsible for its motion is spent or until something stops it from moving and its energy is transferred to another existing fault or to a new fault. A thrust belt rarely consists of a single thrust, so there must be some reason for motion to be transferred from one thrust to another or for the unbroken sequence to form a new thrust.

Movement along a thrust may become difficult or diminish where the thrust passes over a ramp. Therefore the angle between the shearing stress and the fault plane differs from that along the main detachment or higher detachments. It can be shown mathematically that frictional resistance to movement is greater in the high-angle segment of the

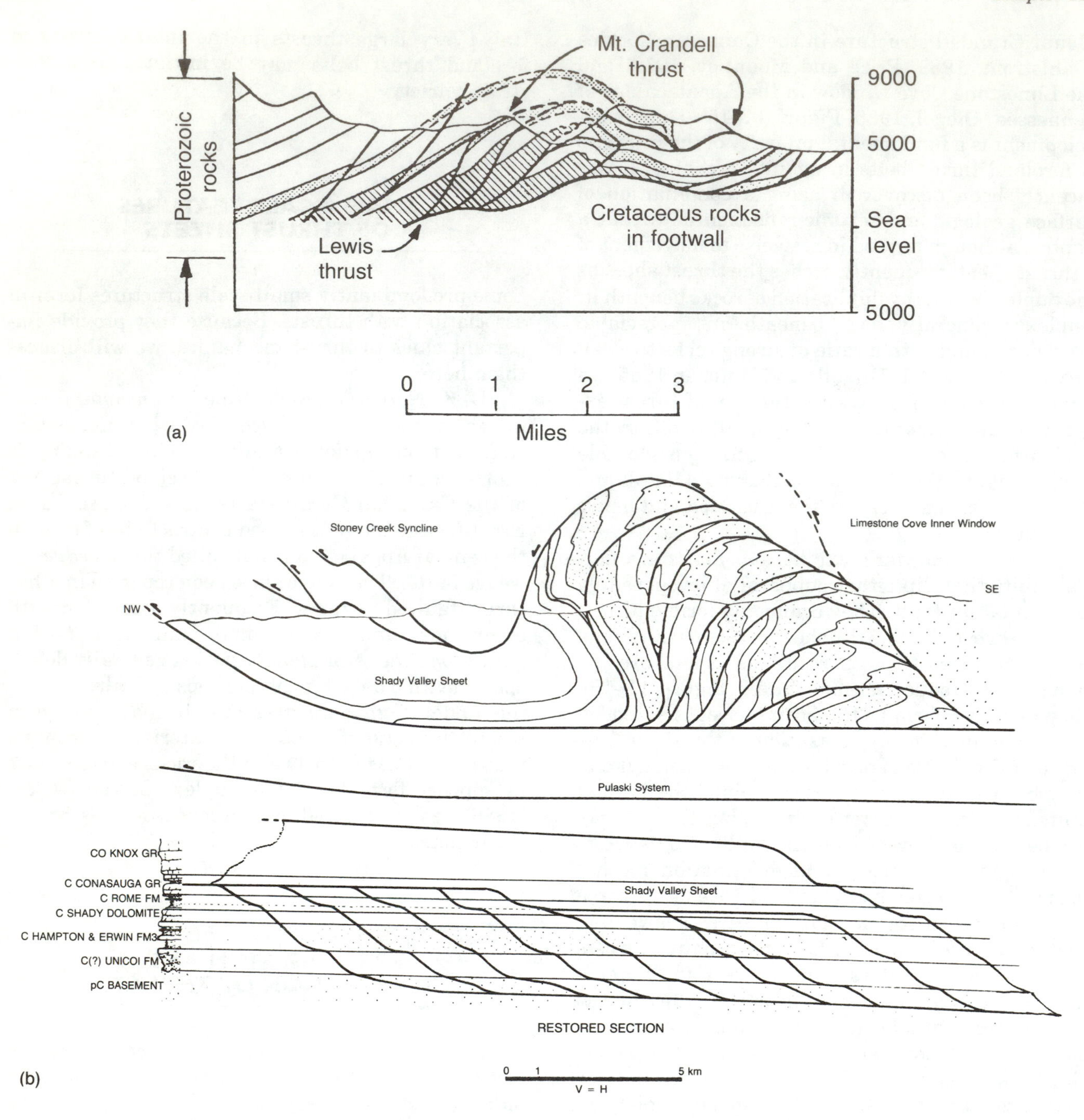

FIGURE 11–15
(a) Cross section of the Mt. Crandell structure in southern Alberta, a hinterland-dipping duplex. The Lewis thrust forms the floor, and the Mt. Crandell thrust forms the roof. (After R. J. W. Douglas, 1952, Geological Survey of Canada, *in* C. D. A. Dahlstrom, 1969, Canadian Society of Petroleum Geologists *Bulletin,* v. 18.) (b) Cross section of the Limestone Cove duplex, Tennessee, a foreland-dipping duplex. (From F. A. Diegel, 1985, Geological Society of America Southeastern Section Guidebook.) (c1) Explosure of a small duplex in Lower Devonian limestones in a quarry at Fuera Bush, south of Albany, New York. Dm—Manlius Formation; Dc—Coeymans Formation; Dk—Kalkberg Formation. (c2) Relationships between different elements in the duplex. (c1 and c2 courtesy of Stephen Marshak, University of Illinois-Urbana.)

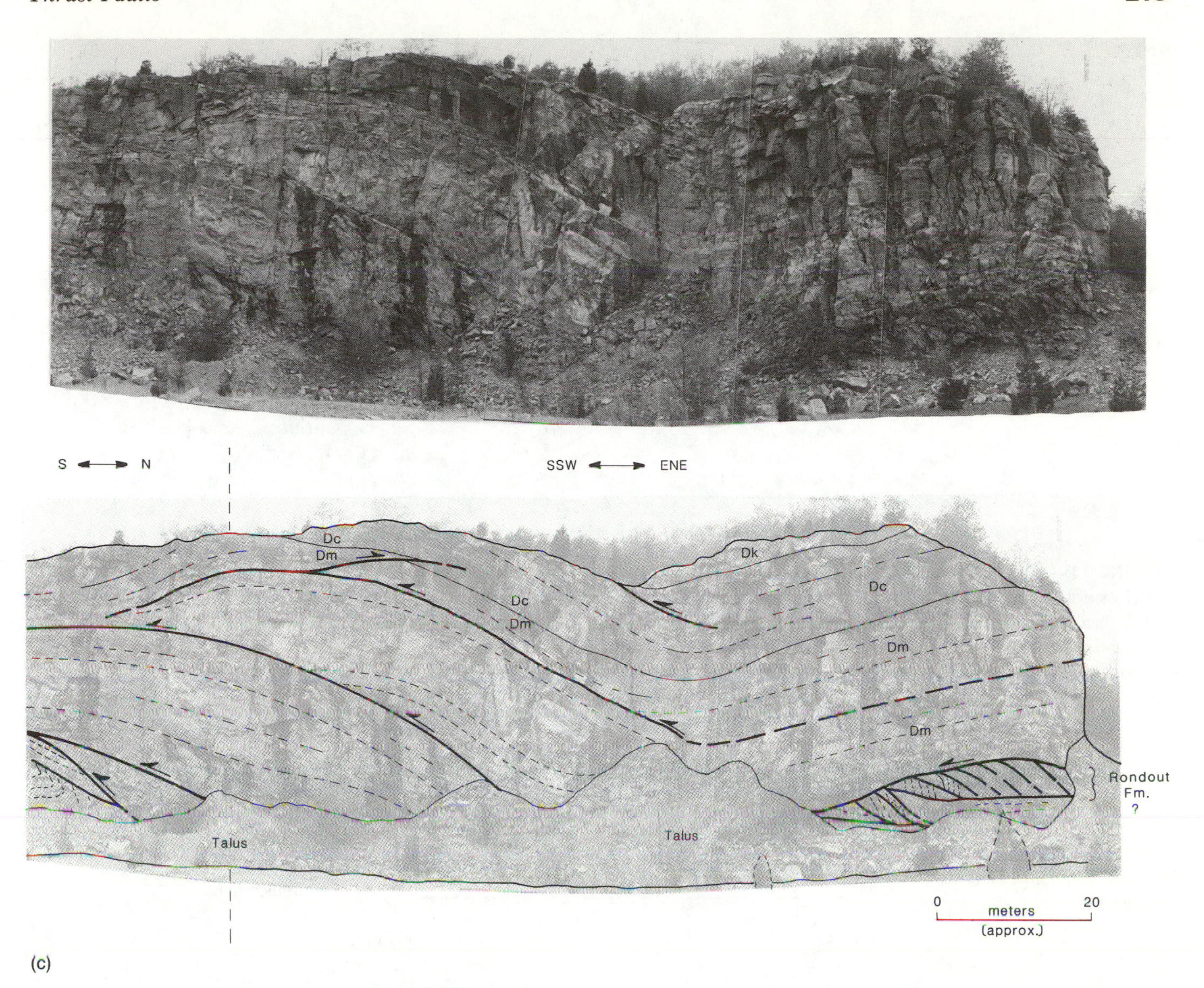

(c)

fault plane than along low-angle segments (Figure 11–19). If the geometry of the fault changes enough that the force necessary to overcome the frictional resistance exceeds the strength of the footwall rocks, a new thrust may form and propagate through the footwall from an existing detachment. New thrusts may form at any level, depending on where the older thrust surface becomes locked and a suitable detachment is present for continued propagation of the thrust. This is also part of the duplex-forming mechanism described earlier.

A thrust may be forced to ramp to a higher detachment by stratigraphic pinchout of the detachment unit, or by facies changes within the detachment (reducing the amount of weak material), or by basement irregularities. Basement irregularities may change the angular relationships with the dominant shear strain, causing the fault to ramp to a higher detachment rather than continue to propagate along the lower detachment. Facies changes may also weaken the ramp unit.

Thrust faults project down-dip until they join a larger fault. The join of two surfaces is called a ***branch line*** (Figure 11–20). Imbricates have only one branch line; a slice or horse is terminated at each end by leading and trailing branch lines.

Thrusts may terminate up-dip in several ways. A thrust may terminate at the surface as an ***erosion*** (or ***emergent***) ***thrust*** (Figure 11–21). There, surficial materials may be caught up and overridden by the advancing thrust sheet. The amount of movement that occurs once the surface is reached depends on the available energy. Smaller thrusts may terminate in the sedimentary section, either within a

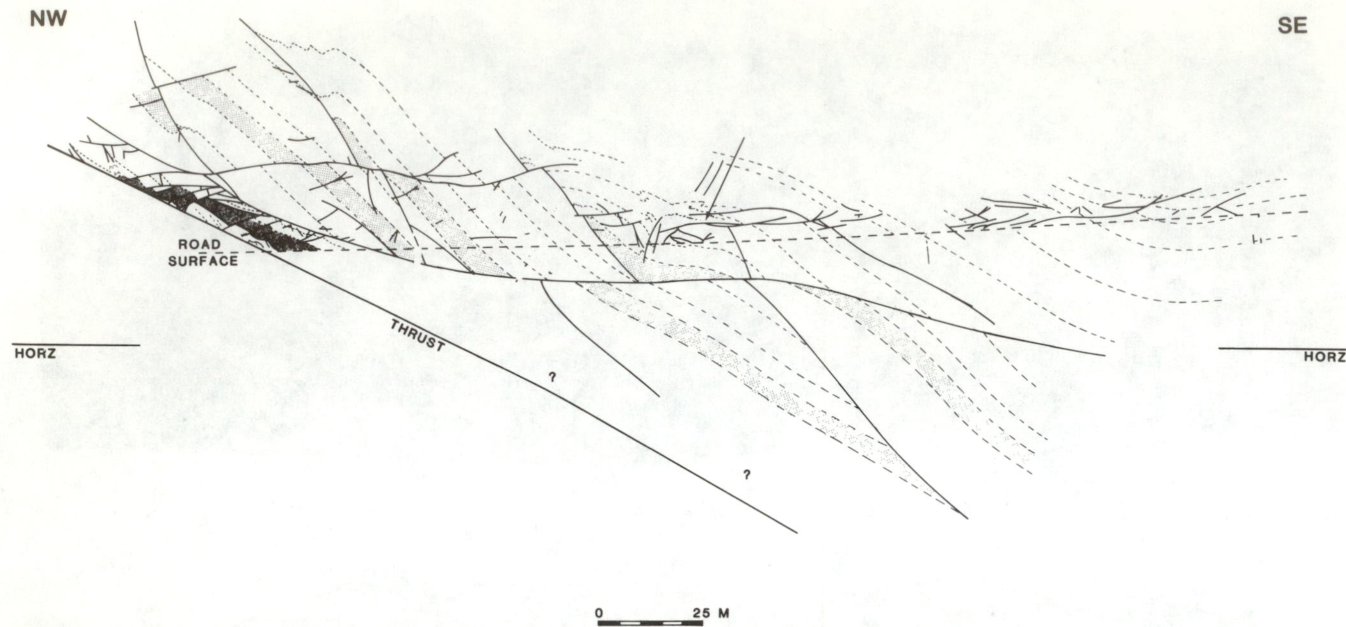

FIGURE 11–16
Synthetic and antithetic normal faults in sandstone and shale of the Lower Cambrian Rome Formation in the hanging wall of the Copper Creek thrust at Diggs Gap on Interstate 75 north of Knoxville, Tennessee. (Reprinted with permission from *Journal of Structural Geology*, v. 8, Steven Wojtal, Deformation within foreland thrust sheets by populations of minor faults, © 1986, Pergamon Press PLC.)

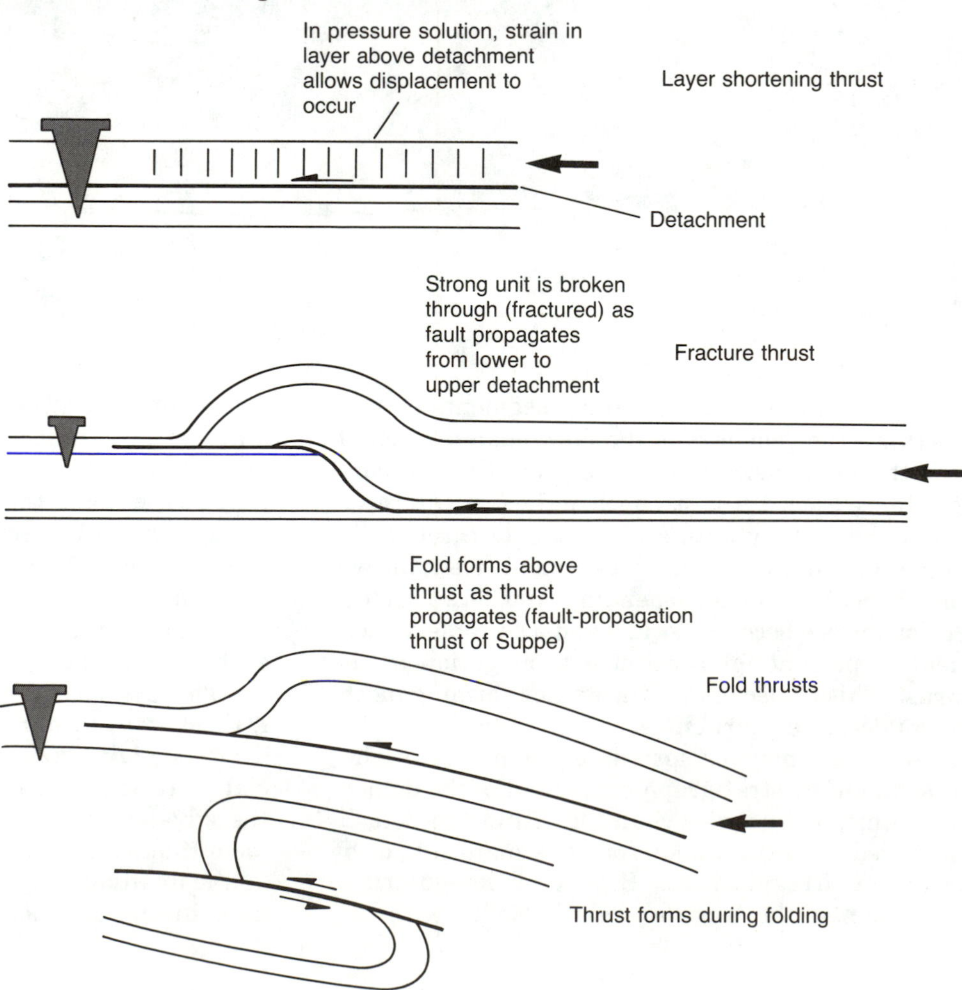

FIGURE 11–17
Geiser's thrust categories. Note the similarities in the second and third categories to Suppe's fault-bend and fault-propagation folds in Figure 11–11.

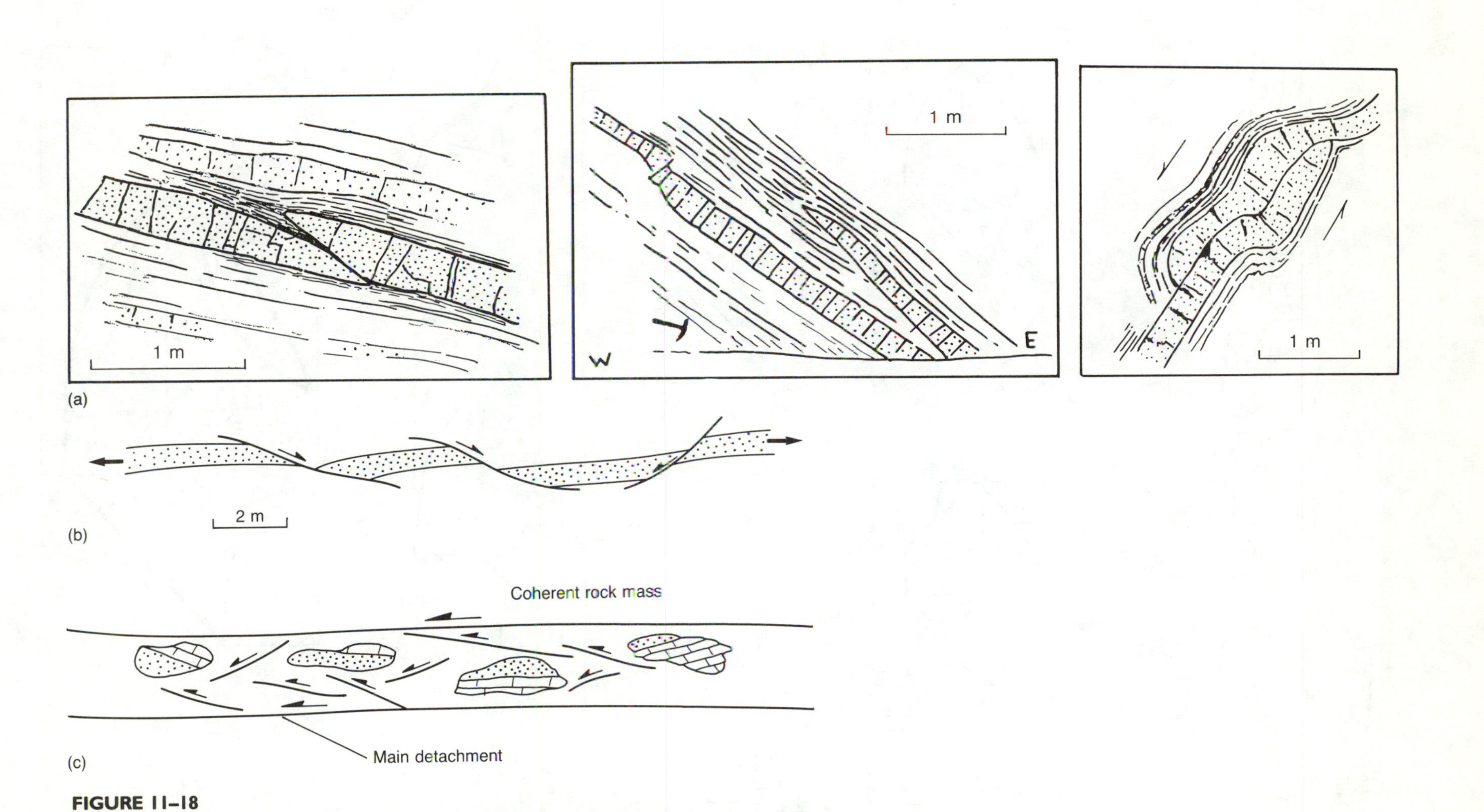

FIGURE 11–18
(a) Contractional faults shown in initial, intermediate, and advanced stages of development. Drawn from photographs by Ernst Cloos of structures in Maryland. (From Ernst Cloos, 1964, Virginia Tech Department of Geological Sciences Memoir 1.) (b) Extensional faults. (c) Broken formation zone. All are shown in cross-section view.

FIGURE 11–19
Propagation sequence in the development of a thrust belt (1, oldest; 7, youngest).

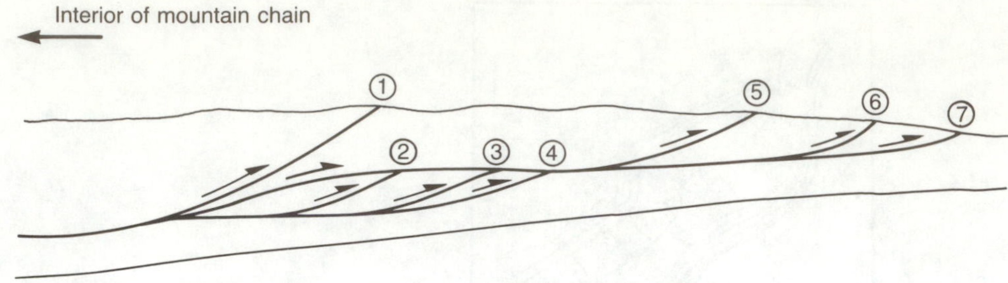

FIGURE 11–20
Branch lines. (a) Diverging splay (S), which has a single tip line (T) and a map termination T′, and one branch line (B) that intersects the erosion surface at B′. (b) Rejoining splay with a single branch line B that intersects the map at two branch points B'_1 and B'_2. (c) Two major faults Q and R with connecting splay S. Two branch lines have surface terminations B and B′. (d) Horse surrounded by fault surfaces (left). Two fault surfaces meet at a single closed branch line B with two cusps U. Diagram at right illustrates half of horse cut along section X-X′. (e) Two fault surfaces meet along two branch lines, B_1 and B_2; the map pattern resembles a diverging splay and the cross section resembles a horse. (From S. E. Boyer and David Elliott, AAPG *Bulletin,* v. 66. © 1982. Reprinted by permission of American Association of Petroleum Geologists.)

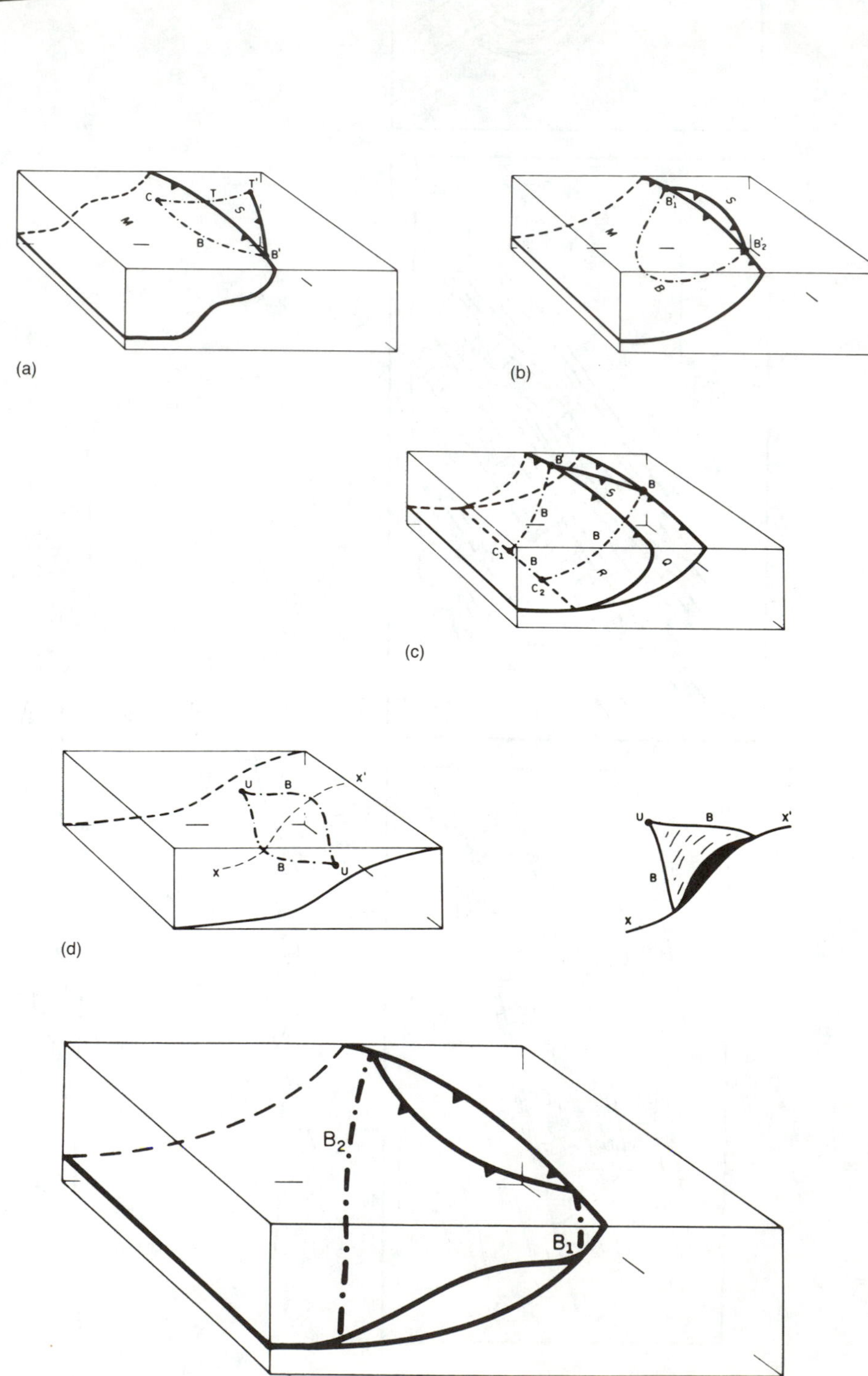

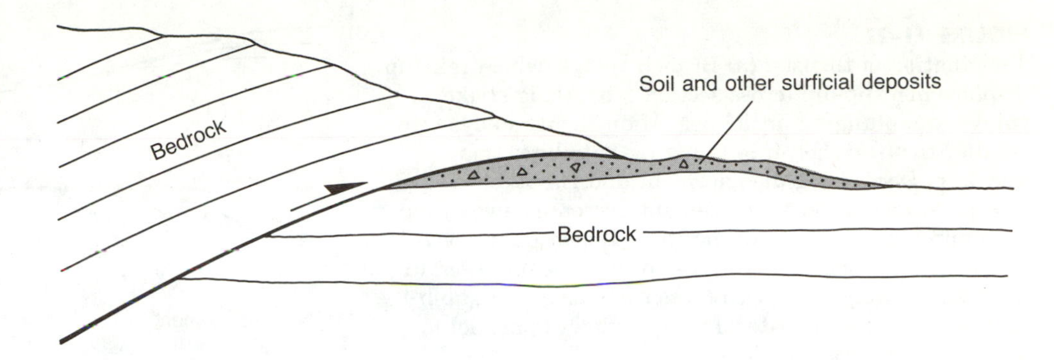

FIGURE 11–21
Emergent thrust breaks the surface and involves surficial materials.

detachment or as a series of branching splays that distribute motion on the smaller faults (Figure 11–22). Erosion thrusts have been documented by Declan DePaor and David Anastasio (1987) in the Pyrenees in Spain, where unconformities—formed beneath debris eroded from the advancing thrust sheet—had been deformed by the sheet. Several thrusts that branch and transfer displacement as they terminate in the sedimentary section or overlap one another have been described by B. C. Burchfiel, R. J. Fleck, D. T. Secor, R. R. Vincelette, and G. A. Davis (1974) in the Spring Mountains of southern Nevada. The opposite of an emergent thrust is a ***blind thrust,*** in which displacement decreases upward within the sedimentary section; it never reaches the surface.

Terminations of thrusts along strike sometimes yield information about subsurface up-dip behavior. A thrust may splay along strike and distribute movement among several smaller thrusts before terminating (Figure 11–22). A single thrust or a set of splays may strike into the axial zones of anticlinal folds that plunge and die along strike but farther from the main thrust from which they were derived (Figure 11–22b). Folds and splays frequently strike at angles of 15° to 30° into the hanging wall of the thrust sheet from the strike of the main fault, a result that can be predicted by orientation of the principal and shear planes in the strain ellipsoid.

Some thrusts terminate along strike and up-dip by decreasing displacement until the fault terminates in bedding. The *bow-and-arrow rule,* formulated by David Elliott (1976a), states that the length of a normal line at the bisectrix of the chord connecting the terminal ends of a thrust sheet is a measure of the maximum displacement of the thrust sheet (Figure 11–22c). Elliott's rule works only if displacement is roughly symmetrical to the ends of the sheet.

Movement of a thrust sheet may result in the ends terminating in strike-slip faults ***(tear faults),*** which at depth become part of the thrust sheet (Figure 11–22). Along-strike changes in detachment level of a thrust may also occur where the thrust is forced to ramp to a higher level—a ***lateral ramp.***

Fault length and displacement are directly related, as noted by Elliott (1976b). He concluded that maximum displacement, u, may be approximated by

$$l \cong 14u \qquad \textbf{(11–1)}$$

where l is fault length as obtained by his bow-and-arrow rule. As Elliott noted further, longer thrusts must be part of stronger thrust sheets. If so, we can predict that the largest of all thrust sheets—composite crystalline thrust sheets, to be discussed soon—should be the strongest and have the greatest displacements. If we plot length versus known displacement for several thrust sheets (Figure 11–23), we find his conclusion right. Elliott's equation may underestimate displacement by as much as 50 percent for very large thrust sheets unless independent criteria, such as seismic-reflection data, are available to help constrain cross sections.

Many geologists believe thrust faults are produced by outward propagation from the internal parts of a mountain chain. These are *normal-sequence thrusts*. Sometimes a moving thrust sheet will lock or achieve a geometry that causes the sheet to break back of the leading edge, producing thrusts that are *out of sequence* as compared with normal inside-out geometry (Figure 11–24).

FEATURES PRODUCED BY EROSION

Thrust sheets characteristically have a large areal extent, and their fault surfaces commonly dip at low angles; also, many thrust sheets are folded after emplacement. This produces several important erosion-related features.

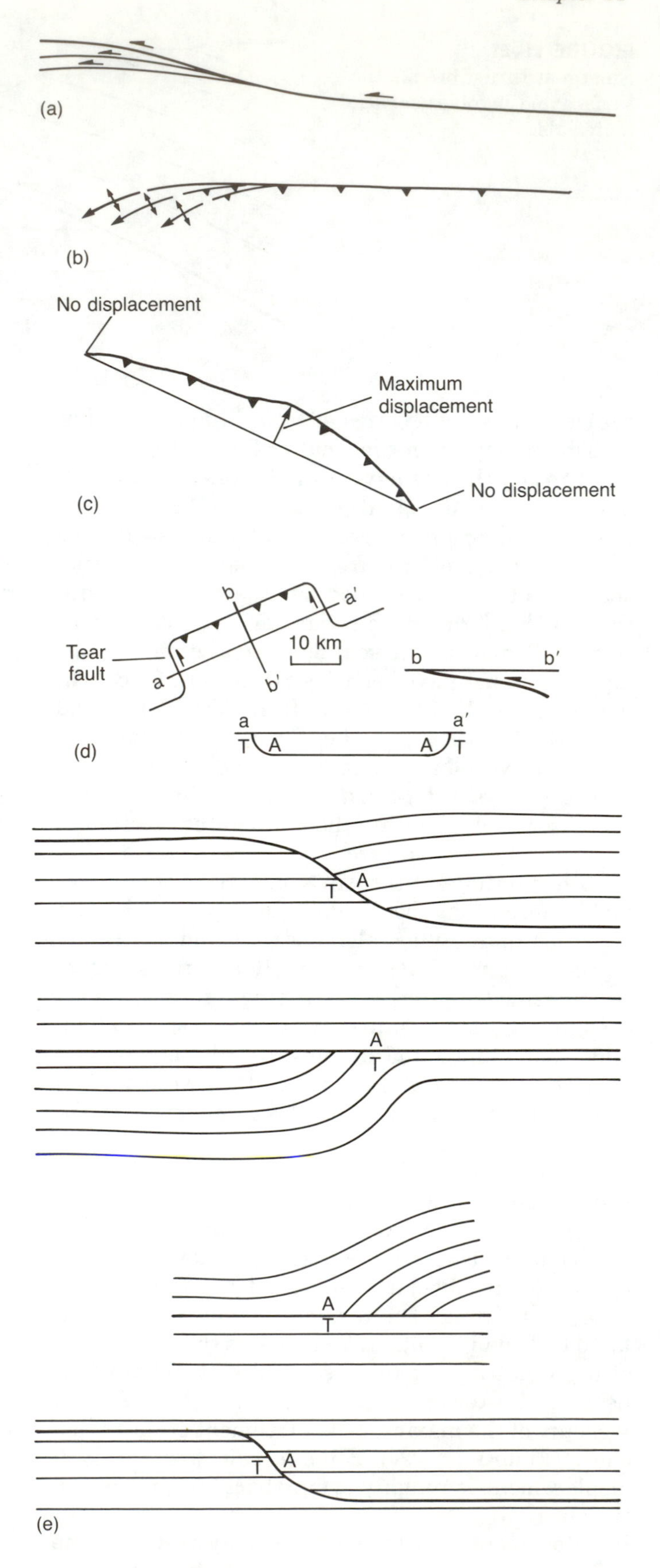

FIGURE 11–22
Termination of thrusts. (a) Branch splays, which take up displacement up-dip (cross section). (b) Along-strike splays into plunging anticlines. Main thrust splays into smaller thrusts, then into plunging anticlines (map view). (c) Decreasing displacement and Elliott's bow-and-arrow rule. Displacement decreases away from the approximate center of the thrust sheet (map view). A—away from the observer; T—toward the observer. (d) Tear faults (map view with cross sections a-a′ and b-b′). (e) Configurations of lateral ramps, viewed parallel to strike; all are cross sections.

Erosion may make holes in a thrust sheet, exposing the footwall rocks. Footwall rocks are thus completely surrounded by the hanging wall, producing a feature called a ***simple window,*** or ***fenster.*** A window usually develops where a thrust sheet has been folded, resulting in part of the sheet having a higher elevation than adjacent parts (Figure 11–25). The higher part of the sheet may also have a greater fracture density.

A thrust sheet may undergo renewed movement during or after folding. An antiformal fold thus produced may prevent the entire thrust sheet from moving and lead to imbrication of the sheet along the crest of the antiform. Then erosion may expose a more complex window into the footwall rocks, but the window is opened along the trace of a fault and will appear to be partly closed because the elongate outcrop pattern parallels the fault trace (Figure 11–25). Steven S. Oriel (1950) first used the term ***eyelid window*** to describe a complex window geometry in the North Carolina Blue Ridge.

Many windows in overthrust terranes expose footwall duplexes (Figure 11–15b), suggesting an interactive relationship between the growth of duplexes and the arching of a thrust sheet during thrust emplacement (Figure 11–26). Many examples of isolated domes occur in large thrust sheets; breached windows that expose duplexes in the footwall include the Grandfather Mountain window in the southern Appalachians (Boyer and Elliott, 1982) and the Assynt window in the northwestern Scottish Highlands (Elliott and Johnson, 1980).

Some geologists use the term window to describe any erosionally exposed rock unit that is surrounded laterally by the unit above it. If the contact between the two is not a fault, but a normal contact or unconformity, the proper term is *inlier* because the rocks are still in stratigraphic order, with the oldest on the bottom. True windows involve younger rocks exposed by erosion through a sheet of older rocks.

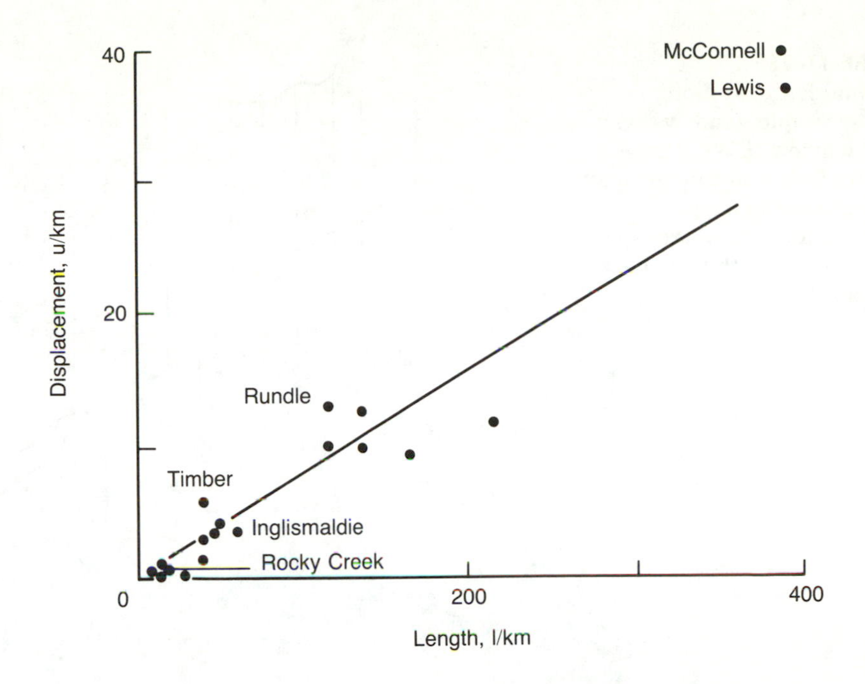

FIGURE 11–23
Plot of along-strike length versus displacement of thrust sheets. (Modified from D. Elliott, 1976, *Journal of Geophysical Research*, v. 81, p. 949–963. Copyright American Geophysical Union.)

Erosion also frequently isolates parts of a thrust sheet. Dismemberment of a thrust sheet by erosion may leave a large remnant called a ***klippe*** (Figure 11–25). A klippe sometimes results from folding of the thrust sheet, where the preserved part remains in a syncline, but erosional dissection of a nearly flat thrust sheet may also preserve parts of the sheet as klippes. Klippes are *erosional outliers* of thrust sheets, but not all outliers are klippes.

Windows and klippes are useful in measuring minimum distance of transport of a thrust sheet (Figure 11–25). The minimum horizontal transport is the distance measured across strike from the outcrop trace of the thrust on the down-dip side of a window to the outcrop trace of the thrust or to the outcrop trace of the thrust on the up-dip side of a klippe. Displacement is better measured on a map than on a cross section because a cross section contains more interpretation.

CRYSTALLINE THRUSTS

Crystalline thrusts, as remarked earlier in this chapter, involve transport of metamorphic or igneous rocks, or both, as part or all of a thrust sheet (Figure 11–27). They have been known for many decades in the Alps, the Appalachians, the Scandinavian and British Caledonides, and other orogens. The large thrusts first identified in Sweden and Scotland are crystalline thrusts. Geologists once thought their propagation and motion involved a process totally different from that in thin-skinned foreland thrusts in a sedimentary sequence, but many attributes of thin-skinned thrusts may also be identified in crystalline thrusts, leading to the conclusion that the processes may be nearly the same. That they are low-angle thrusts indicates an important similarity to thin-skinned thrusts. But how

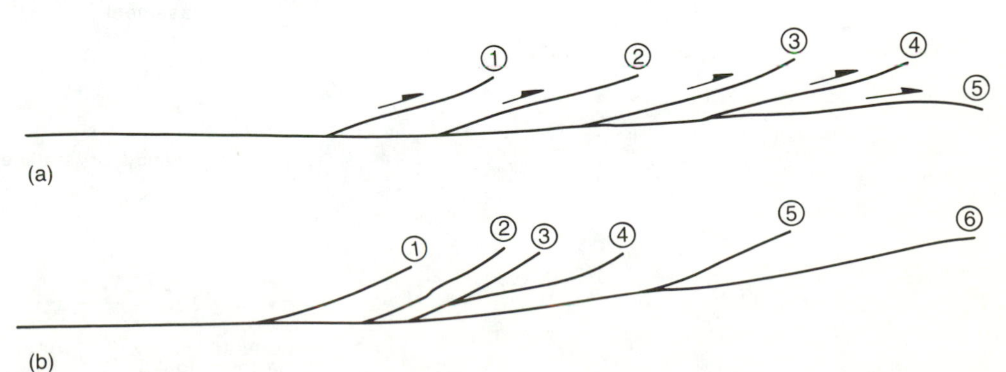

FIGURE 11–24
In-sequence (a) and out-of-sequence (b) thrusts. A normal sequence involves progression of thrusts from one side of a deformed zone to another. Out-of-sequence thrusting involves thrusts that form within a pile of already-formed thrusts.

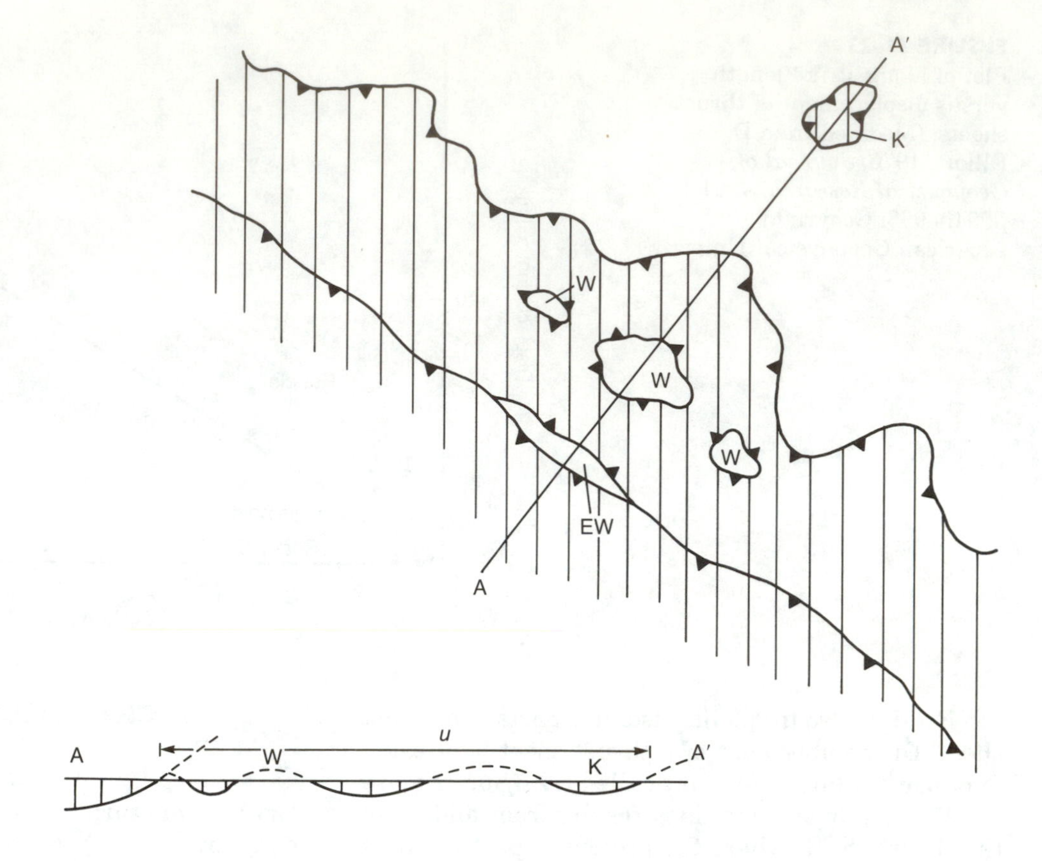

FIGURE 11–25
Erosional features along thrusts: Simple windows (W); eyelid window (EW); klippe (K). Windows and klippes may be used in estimating minimum amount of transport (u). Rocks of the thrust sheet are striped.

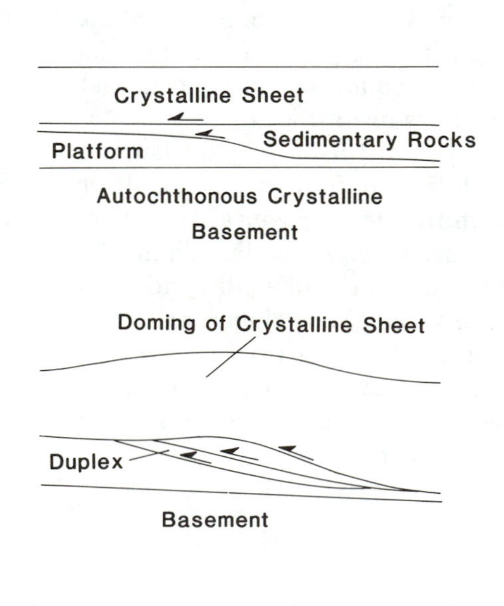

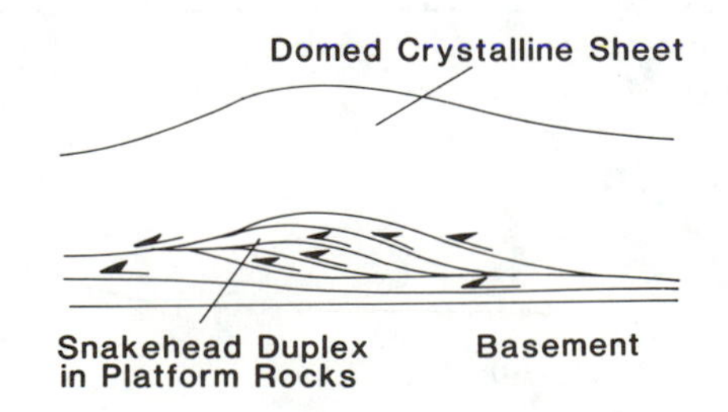

FIGURE 11–26
Possible interactive relationships between formation of a duplex during emplacement of a thrust sheet, arching of the thrust sheet by the growing duplex, and subsequent erosional unroofing of the domed area. The result is a window that exposes a duplex.

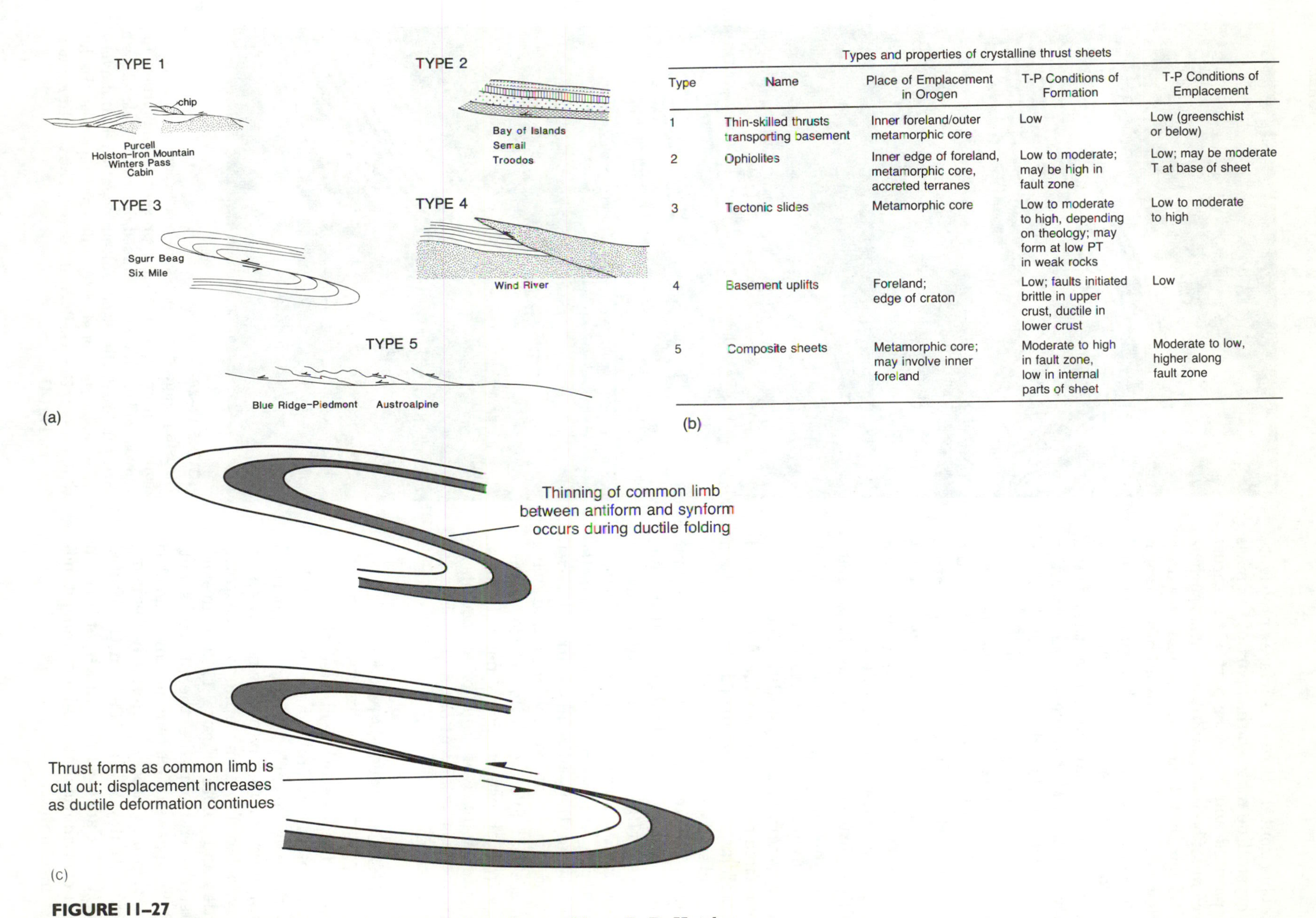

Types and properties of crystalline thrust sheets

Type	Name	Place of Emplacement in Orogen	T-P Conditions of Formation	T-P Conditions of Emplacement
1	Thin-skilled thrusts transporting basement	Inner foreland/outer metamorphic core	Low	Low (greenschist or below)
2	Ophiolites	Inner edge of foreland, metamorphic core, accreted terranes	Low to moderate; may be high in fault zone	Low; may be moderate T at base of sheet
3	Tectonic slides	Metamorphic core	Low to moderate to high, depending on theology; may form at low PT in weak rocks	Low to moderate to high
4	Basement uplifts	Foreland; edge of craton	Low; faults initiated brittle in upper crust, ductile in lower crust	Low
5	Composite sheets	Metamorphic core; may involve inner foreland	Moderate to high in fault zone, low in internal parts of sheet	Moderate to low, higher along fault zone

(b)

FIGURE 11–27
Types (a) and generalized properties (b) of crystalline thrusts. (From R. D. Hatcher, Jr., and R. T. Williams, 1986, Geological Society of America *Bulletin*, v. 97.) (c) Formation of a tectonic slide.

does a subhorizontal thrust surface propagate through a crystalline mass so that part of the mass becomes separated and moves as part of the thrust sheet? There must be important differences, such as the initially horizontal layering in sedimentary rocks, in view of the different material properties of sedimentary and crystalline rocks. Hatcher and Williams (1986) have shown that thrusts follow zones of weakness in crystalline rock, as they do in a sedimentary section, and that the shared properties with thin-skinned thrusts led to a general rule of crystalline thrusts: *The largest crystalline thrust sheet in an orogen is always larger than the largest foreland thrust.* The rule no doubt partly reflects the greater inherent strength of crystalline thrust sheets. After studying the Moine thrust zone, Elliott and Johnson (1980) came to a similar conclusion about propagation of thrusts in crystalline rocks.

Crystalline thrusts may form as large slabs wherein materials appear to exhibit brittle to semibrittle behavior. They may also form during ductile folding where large crystalline isoclinal recumbent folds (fold nappes) are produced. Continued transport attenuates the overturned limb between an antiform and synform, finally producing a ductile thrust (Figure 11–27c). Such thrusts were first called *slides* by Edgar B. Bailey (1910), but Michael J. Fleuty (1964) has urged that they be called *tectonic slides* to prevent ambiguity. Tectonic slides are common on all scales in high-grade metamorphic rocks, but they may also form at lower grades in less-competent assemblages that have undergone isoclinal folding. Bailey (1934) distinguished *lags* from thrusts as listric normal faults, which cut out the *upright limb* of a recumbent anticline, from thrusts that cut out and replace the inverted limb.

THRUST MECHANICS

Ever since their discovery in the nineteenth century, the mechanics of thrust faults have been studied intensively. Some of the early insights were derived from work with pressure boxes in which a piston deformed layers of different materials with contrasting properties and colors (Figure 11–28). In such experiments, the end opposite the piston is buttressed and the upper surface remains unconfined so that deformation may proceed internally. By using pressure boxes, Bailey Willis (1893) in the United States and H. M. Cadell (1890) in Great Britain made several contributions toward our understanding of thrust mechanics.

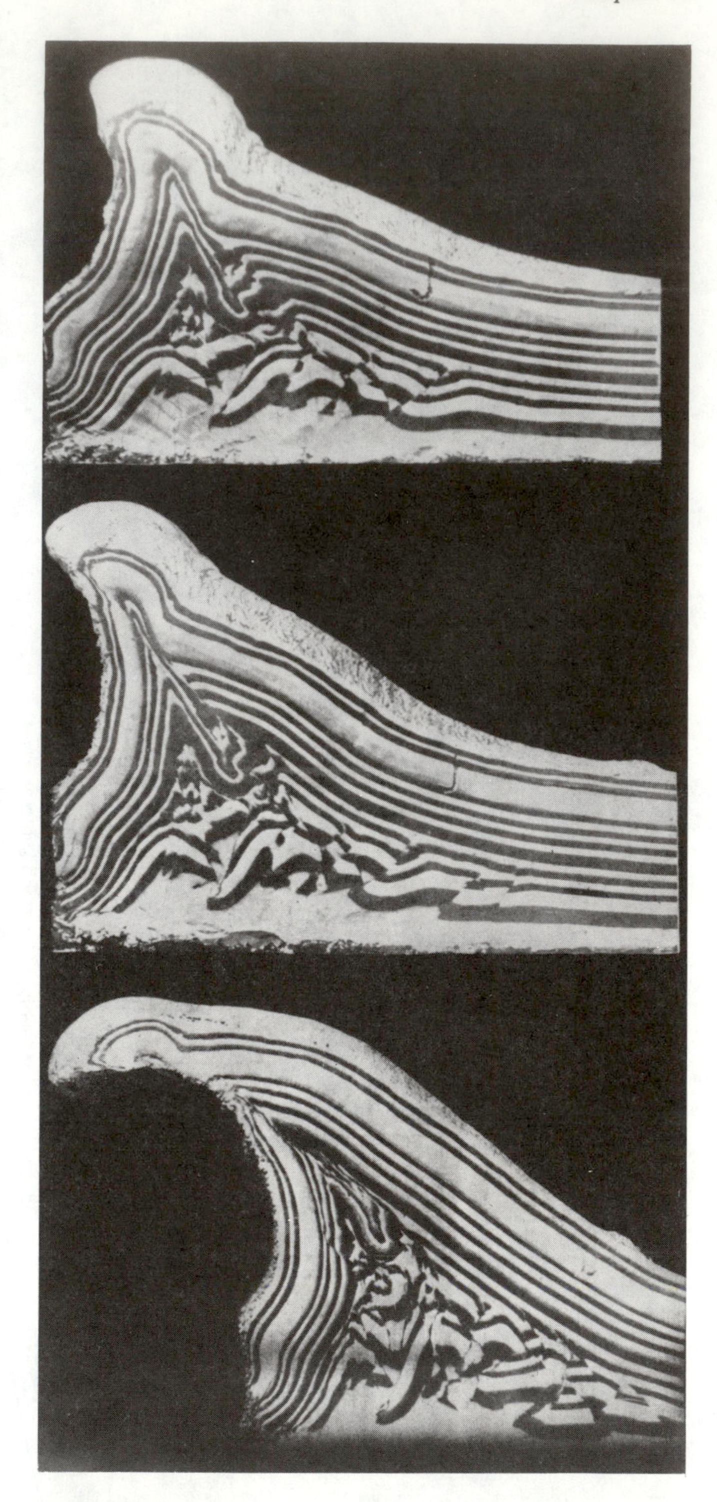

FIGURE 11–28
Thrusts produced in a pressure-box experiment. The sequence from top to bottom represents a fold formed in a pressure box that on continued deformation develops a thrust (a break thrust or fault-propagation fold). Layers are made of clays of slightly different properties. (Bailey Willis, 1893, U.S. Geological Survey, Thirteenth Annual Report, Part II.)

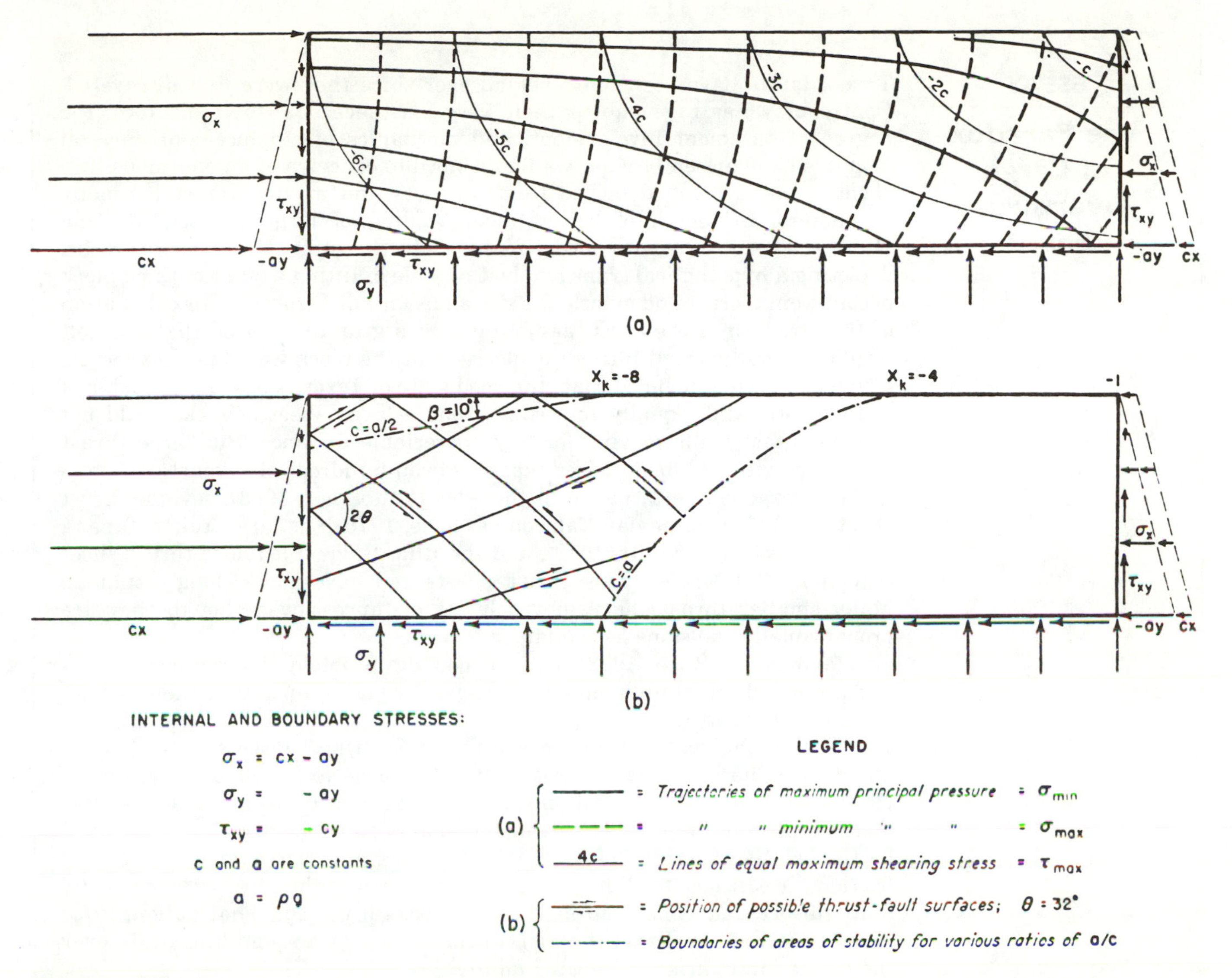

FIGURE 11–29
Hafner model of stress distribution in a buttressed block. Stress is applied horizontally and is assumed to be constant with depth and with a constant lateral gradient. (From W. Hafner, 1951, Geological Society of America *Bulletin,* v. 62.)

Attempts to apply the principles of continuum mechanics to thrusting began in the 1940s and 1950s with the work of Jean Goguel (1965), W. Hafner (1951), and M. K. Hubbert (1951). Models of stress distribution in a thrusted block by Hafner are still used (Figure 11–29). Probably the most significant paper on thrust faulting published in midcentury was one by Hubbert and William W. Rubey (1959), who showed that fluid pressure along a zone of weakness facilitates motion with expenditure of much less energy than if the zone is dry (Chapter 10). Their calculations of the force required to move a thrust sheet (such as the Keystone thrust sheet in Nevada) always exceeded the inherent strength of the sheet if they assumed it dry—indicating that it would fragment before motion could occur and therefore could not exist in nature. But they concluded that the buoyant effect of fluid in a mass of rocks would lower the frictional resistance to fracturing and subsequent motion—lower it enough to show how a very large thrust sheet can exist in nature. Once fracture is initiated, they suggested, the Mohr-Coulomb criterion may be modified to

$$|\tau_c| = \tau_0 + (S - P_w) \tan \phi \qquad \textbf{(11–2)}$$

where τ_c is the critical shear stress at failure, τ_0 is

ESSAY

The Paradox of Large Overthrust Faults

Thrust faults have been controversial ever since they were first discovered. Controversy has raged about their very existence, transport distance, the degree of basement involvement, and mechanics of emplacement. Several disputes have arisen because we find it hard to conceive of an enormous thin sheet of rocks being detached from its roots and moving intact for many kilometers. In search of better understanding of large thrusts, M. King Hubbert (1945) devised a model to simulate the strength of crustal rocks, depicting a hypothetical crane and bolt capable of lifting a 600-km-thick block of continental crust and mantle the size and shape of Texas. But his calculation of the strength of the crust, assuming it was granite, showed that the bolt would pull out without lifting the block—that the block would not support its own weight. He concluded that "the good state of Texas is utterly incapable of self-support" and, equally important, that a large sheet of rock would not remain intact while moving for any appreciable distance. But large thrust sheets do exist and have moved tens and even hundreds of kilometers.

The large Silvretta nappe in the Alps (Laubscher, 1983), several thrust sheets in the Scandinavian Caledonides (Gee, 1978), the large Yukon-Tanana sheet (Templeman-Kluitt, 1979), and the Blue Ridge–Piedmont thrust sheet (Hatcher, 1981) are all huge thrust sheets and have moved long distances. Many smaller thrusts have moved lesser distances, even though they are proportionally the same as the larger thrusts.

Raymond A. Price (1973) has outlined the problem of large overthrusts and discussed constraints placed by mechanics on the initiation and propagation of large overthrusts.

"The mechanical paradox in overthrust faulting," Price says in his abstract, "originates in the conceptual models chosen for mechanical analysis." He then summarizes "the notion that the maximum dimensions of a tabular mass of rock that may undergo translation along an overthrust fault are fixed by the relative magnitudes of: (1) the stress required to overcome the *total* frictional resistance to sliding (and the cohesion or adhesion?) over the *entire* fault surface and, (2) the strength of the rocks involved." That notion, Price says, "tacitly assumes that sliding is initiated, and occurs simultaneously, over the entire fault surface." He went on to say:

> A realistic model for overthrust faulting must be compatible with the nature of actual overthrust faults, the way in which displacements occur on active faults, and the focal mechanisms of earthquakes. Most large overthrusts are discrete slip surfaces within a

the inherent cohesive shear strength of the material, **S** is normal stress, P_w is the pore pressure [$(\mathbf{S} - P_w) = \boldsymbol{\sigma}$, effective normal stress], and ϕ, the internal friction angle, may be modified. The equation then becomes

$$|\tau_c| = (\mathbf{S} - P_w) \tan \phi \qquad \textbf{(11–3)}$$

Effective normal stress is an important factor in movement along thrusts and other faults (Chapter 10), as well as in the general process of rock deformation. Pore pressure reduces the effective normal stress and in effect reduces the overall strength of the rock mass. Fluid also enables renewed movement by lessening frictional resistance along existing faults. Hubbert and Rubey assumed that once movement begins, cohesive shear strength along the fracture becomes negligible and concentrates stress along the fracture; stress is then transferred to the fluid, producing a buoyant effect and greatly enhancing motion. Earlier, we saw a useful analog to the buoyant-fluid concept—Hubbert and Rubey's beer-can experiment (Figure 10–9). This experiment demonstrates the importance of buoyant force in

coherent mass of rock that is physically continuous around the ends. The net slip, which changes along the fault surface, can be viewed as the cumulative effect of innumerable incremental displacements, each of which is initiated as a local shear failure that propagates as a dislocation on a scale and at a velocity that is small in comparison with the total area of the fault surface. The Coulomb-Navier failure criterion and the law of sliding friction, as modified for the effects of pore pressure, are empirical relationships derived from experiments on small rock specimens and may be applied over the area of a dislocation, but should not be extrapolated, in the same way, over a linear scale of up to six orders of magnitude to encompass an entire fault surface. The shape and dimensions of overthrust faults are controlled by the stress field, mechanical properties, and anisotropy of the rock mass, but not by the total frictional resistance to sliding over the entire fault surface.

Here Price advocates application of physical laws to a complex structure by considering the real properties of the structure. For several decades now, we have been attempting to do so. One barrier to understanding why overthrust faults exist is our inability to make direct measurements on active thrusts buried several kilometers beneath the surface. The barrier is likely to remain, but we can obtain a good idea of the shapes and properties of thrusts by using geophysical techniques (Chapter 20) to image the crust and by studying samples we can obtain from fault zones.

References Cited

Gee, D. G., 1978, Nappe displacement in the Scandinavian Caledonides: Tectonophysics, v. 47, p. 393–419.

Hatcher, R. D., Jr., 1981, Thrusts and nappes in the North American Appalachian orogen, *in* McClay, K. R., and Price, N. J., Thrust and nappe tectonics: Geological Society of London Special Publication 9, p. 491–499.

Hubbert, M. K., 1945, Strength of the Earth: American Association of Petroleum Geologists Bulletin, v. 29, p. 1630–1653.

Laubscher, H. P., 1983, Detachment, shear, and compression in the central Alps, *in* Hatcher, R. D., Jr., Williams, Harold, and Zietz, Isidore, Contributions to the tectonics and geophysics of mountain chains: Geological Society of America Memoir 158, p. 191–212.

Price, R. A., 1973, The mechanical paradox of large overthrusts: Geological Society of America Abstracts with Programs, v. 5, p. 772.

Templeman-Kluitt, D. J., 1979, Transported cataclasite, ophiolite and granodiorite in Yukon: Evidence of arc-continent collision: Geological Survey of Canada Paper 79-14, 27 p.

decreasing effective normal stress.

As pointed out by Kenneth J. Hsü (1969), Hubbert and Rubey's simplification applies *only* if renewed movement occurs on an existing fracture. In all other situations, we must consider the cohesive shear-strength term (τ_0).

Study of thrust sheets in the Muddy Mountains of Nevada led William G. Brock and Terry Engelder (1977) to conclude that thrust sheets (or other fault blocks) may be too permeable to maintain fluid pressure of the kind Hubbert and Rubey postulated, except in a few places such as Taiwan and possibly the Gulf Coast. Fibers of calcite and other minerals on many fault surfaces also suggest that deformation occurs on thrusts by a creep mechanism and may further indicate that the buoyancy mechanism is not viable for all thrusts. Fibers and mineral growth along faults indicate that fluid was abundant and not trapped.

Mechanical analysis by David Elliott (1976a, 1976b) suggested that a thrust propagates as fractures, with displacement related to the length of the outcrop trace. Elliott also suggested that much of the energy expended in emplacement is dissipated

FIGURE 11–30
Ideal Coulomb wedge: α is the dip of the basal thrust; β is the topographic slope angle; α + β is the angle at the apex of the wedge.

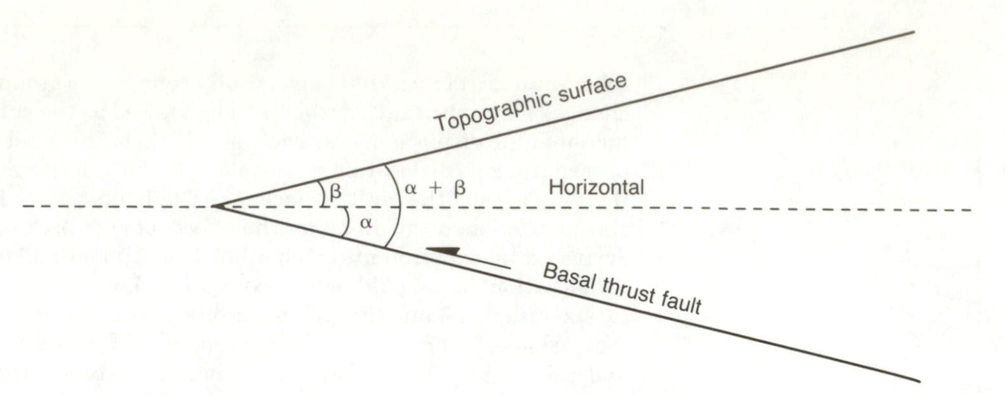

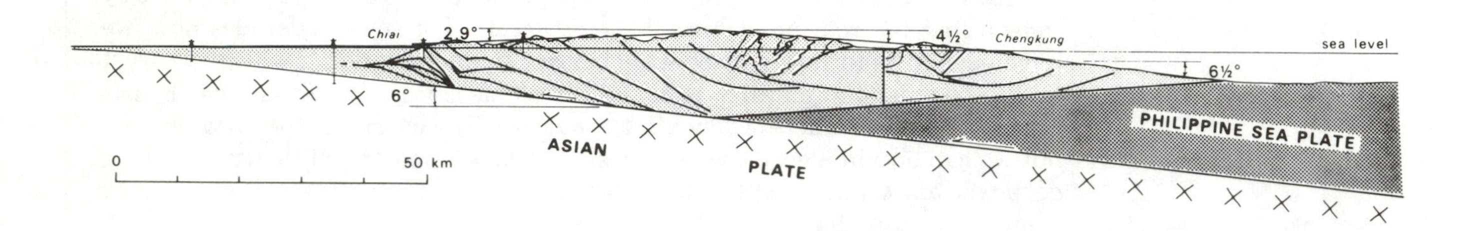

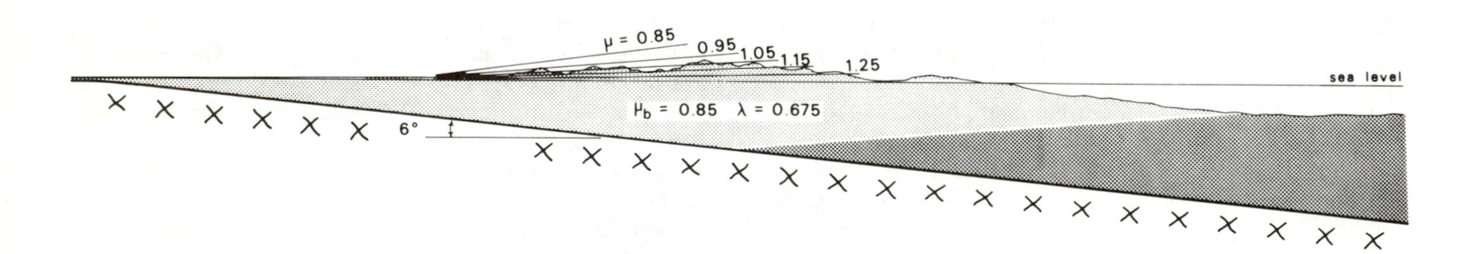

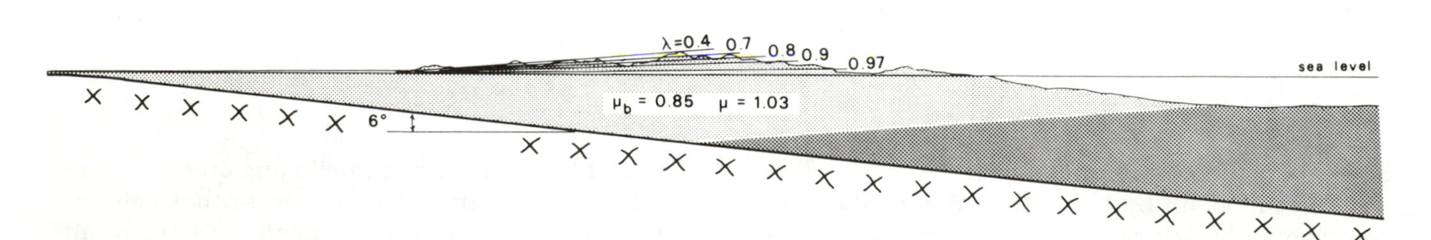

FIGURE 11–31
Propagation of thrusts in Taiwan maintaining the wedge shape and equilibrium surface angle of 2.9°. μ—coefficient of friction, μ_b—coefficient of friction along base of wedge; λ—pore fluid pressure, obtained from a ratio of pore fluid pressure minus the product of water density, acceleration of gravity, used directly in the form of gravity divided by vertical normal traction σ_z, minus the product of the same quantities. (From Dan Davis, John Suppe, and F. A. Dahlen, *Journal of Geophysical Research,* v. 88, p. 1153–1172, 1983, copyright by the American Geophysical Union.)

within the thrust sheet. In the McConnell thrust sheet in the Canadian Rockies, energy is dissipated by frictional sliding only in the uppermost five kilometers. Pressure-solution slip along individual surfaces and penetrative pressure solution with cleavage development dominate at greater depths, but have also been found at shallow depths—within hundreds of meters of the surface. Where pressure-solution slip has been observed, slip velocity appeared to vary with magnitude of shear stress.

Mechanical analysis of foreland thrusting by William M. Chapple (1978) may prove as significant as the earlier study by Hubbert and Rubey. Chapple recognized five characteristics now considered fundamental: (1) Foreland thrust belts are thin-skinned, and all deformation occurs above a particular layer without basement involvement. (2) The basal layer in the thrust sheet consists of a weak material such as shale, coal, or evaporite. The weak material occurs in a stratigraphic section composed mostly of materials of higher strength, so movement is restricted to the weak layer. (3) Before deformation, the entire stratigraphic section of a foreland belt is wedge-shaped. The wedge thins toward the interior of a continent, with depth to basement increasing toward the sea (or an ancient sea); the surface of the wedge, too, slopes toward the continent. (4) As deformation proceeds, the wedge is continuously shortened and thickened, but it maintains its shape throughout deformation (Figure 11–30). (5) On a regional scale, the entire deforming foreland thrust belt behaves plastically.

Dynamic thrust models developed by John Suppe (1981) from study of the active thrust belt in Taiwan suggest that an equilibrium angle is maintained throughout deformation of the wedge and that materials composing thrust sheets are Coulomb (brittle) materials (Figure 11–31). Thus Dan Davis, Suppe, and Tony Dahlen (1983) have called the deforming wedge a ***Coulomb wedge.*** The basal angle is maintained by the dip of the basement surface, and the surface angle is maintained by erosion, which removes the steepened part of the shortened end of the wedge. These angles are also related to the coefficient of internal friction, μ, which equals tan ϕ for the rock mass. In addition, the surface-slope angle is partly a function of climate; for example, in an arid climate the surface equilibrium angle would be maintained by increased landsliding. Even so, Davis and his colleagues concluded that the rock mass does not begin to move forward until the *critical taper* of the wedge is achieved.

To formulate an ideal mechanical model, we must see how materials involved in thrusting will behave under stress (that is, their rheology), as well as knowing their strengths. It is also essential that we understand the degree to which body (penetrative) deformation (such as cleavage formation) affects the thrust sheet during initiation and emplacement. Different ideal-behavior modes—such as elastic, viscous, and plastic—may be used in model experiments to duplicate natural behavior, but it may not be possible to describe material behavior exactly in terms of an ideal rheology. Consequently, we must make certain assumptions about rheology in order to devise a mechanical model for foreland thrusting.

The mechanical model formulated by Chapple (1978) embodies both elastic and plastic behavior. Chapple assumed that the basal weak layer and the total wedge behave as ideal plastic materials. As a result of his analysis, thrust faults are considered a manifestation and consequence of shortening of the wedge. A second conclusion is that plastic flow under compression is an attribute of foreland fold-and-thrust belts and that it provides most of the shear stress along the basal weak layer to drive deformation of the wedge.

Elliott (1976a) concluded from his mechanical analysis that gravity was primarily responsible for deformation of foreland thrust belts. He did not assume a specific rheology but did make preliminary assumptions about stress distribution within thrust sheets. He assumed that σ_1 is oriented at 45° to the basal layer in thrust sheets and so did not incorporate a weak basal layer and its attendant properties in his analysis. Thus the fundamental strength of the entire thrust sheet must be the same as that of the basal layer. In contrast, Chapple assumed that the strength of the remainder of the thrust sheet was much greater than that of the weak basal layer. A consequence of Elliott's approach is that horizontal compression is not necessary to form a foreland thrust belt, leading to the additional consequence that high shearing stress is not produced in the basal layer—contrary to Chapple's belief.

Several major thrusts, such as the Keystone in Nevada, appear to have involved only strong rocks, such as massive carbonates, so the basal detachment may not have been involved or was not available when the thrusts formed. A surface slope, though present, may not be necessary in Chapple's model, but it is essential in Davis, Suppe, and Dahlen's model, for critical taper must be achieved in the wedge and maintained.

GRAVITY VERSUS COMPRESSION

Now let us compare gravity with compression as two possible primary motive forces producing foreland thrust belts. Gravity is an appealing mechanism. Hubbert and Rubey (1959) argued that it would be very difficult to push a thin sheet of rock very far horizontally without breaking it up because its internal strength is too low; they reasoned that the magnitude of shearing stress necessary to move a mass of rock would be greatly reduced if the base of the mass were under high pore-fluid pressure. But they maintained that body forces—gravity—were necessary to move a thrust sheet.

We can argue geometrically that gravity is the primary force responsible for moving thrust sheets. Robert C. Milici (1975) has pointed out that the pattern of overlap suggested by outcrop traces of thrust faults in part of the Appalachian foreland thrust belt yields a deformation plan in which the oldest thrusts form on the outer fringe of the belt and successively younger thrusts form deeper in the core of the orogen (Figure 11–32). The logical mechanism for producing this geometry is gravity, but the geometry may be a local aberration, produced during the latter stages of emplacement of out-of-sequence thrust sheets where the footwall sequence locks and the hanging wall breaks and moves over the deformed footwall (Morley, 1986). Experiments with models by Hans Ramberg (1967) and field work, including studies of accretionary wedges in subduction zones (Stockmal, 1983; Dahlen, Suppe, and Davis, 1984), have shown that most thrust belts result from propagation of thrusts outward from the core of the orogen toward the continent. It has also been noted that similar map patterns exist in foreland thrust belts where it can be shown independently that the deformation plan is inside-out. Independent evidence has also shown that the oldest thrusts—not the youngest—are in the interior of the orogen and that deformation there generally occurred long before deformation in the foreland. Such evidence led Hans Stille to suggest in the 1930s that an orogen, including the foreland thrust belt, is deformed from the inside out.

Another objection to the gravity mechanism concerns the direction of slope of basement beneath the foreland. Nowhere in the world is there a foreland thrust belt with the basement surface sloping *away* from the interior of an orogen: all are inclined *toward* the interior. The original shape of the foreland wedge also argues against a reverse slope. It is hard to envisage gravity propagating thrusts *up*-slope. Raymond A. Price (1974) attempted to resolve the problem with his alternative mechanism of *gravitational spreading,* based partly on earlier observations and studies of models by Walter Bucher and Ramberg. Price suggested that as the core of an orogen is shortened and buoyantly uplifted during compression, metamorphism, and plutonism, the orogen compensates for the uplift by flowing laterally under the influence of gravity, as a large ice sheet may push lobes uphill (Price, 1974; Elliott, 1976a). Lateral spreading provides stress that deforms the foreland and pushes a thrust up the otherwise inward-dipping basement surface, but the amount of uplift actually needed to

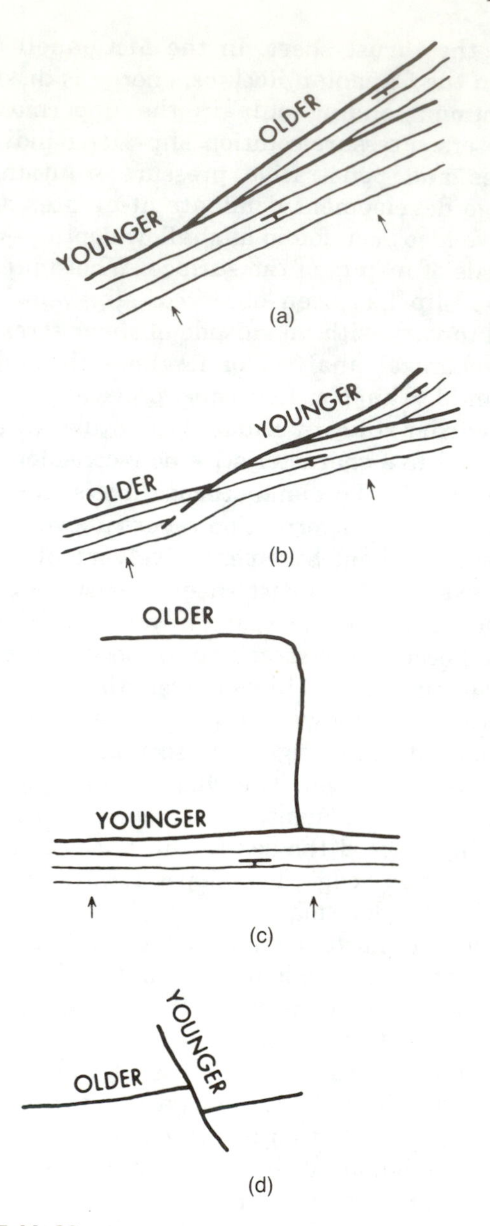

FIGURE 11–32
Milici's geometric relationship suggesting development of thrusts by a break-back mechanism. (From R. C. Milici, 1975, Geological Society of America *Bulletin,* v. 86.)

produce the observed deformation greatly exceeds most estimates. For this reason alone, Price's mechanism seems unworkable. Add the requirement of an internal zone of stretching and attenuation of structures (none has been discovered) and gravitational spreading becomes difficult to accept as a primary mechanism of foreland deformation. Price (1981), recognizing the problem, has recommended that the concept of gravitational spreading be abandoned.

The deformed wedge, the weak basal layer, and the overall geometry of the foreland thrust belt all suggest that Chapple and Davis, Suppe, and Dahlen are right. Their mechanical models—despite initial objections to compression as the primary mechanism of foreland deformation—best explain how foreland thrusts form.

MECHANICS OF CRYSTALLINE THRUSTS

The mechanics of crystalline thrusts remain a partly unsolved mystery. Crystalline thrusts share many properties with thin-skinned thrusts. They are transported along low-angle faults, occur in sheets up to a few kilometers thick, and may follow zones of weakness in the footwall sequence, but there are important differences: (1) They are among the largest structures in orogenic belts. (2) There is no obvious lithologic plane of weakness along which detachment may begin. (3) Crystalline thrust sheets may form as part of the thick foreshortened wedge, but can also form as plastic structures, either as detached thrust sheets or as products of folding (Hatcher and Williams, 1986).

Understanding the formation of detachments along which crystalline thrusts propagate is a difficult problem. E. Ronald Oxburgh (1972) has suggested that crystalline thrusts form as "tectonic flakes" when two pieces of continental crust collide. Richard L. Armstrong and Henry J. B. Dick (1974) have suggested that propagation and detachment of crystalline thrusts occur at the ductile-brittle interface in zones of high heat flow, as in back-arc basins. Oxburgh's mechanism may account for single large crystalline-thrust complexes, such as the Austroalpine sheets of the Alps, which can be directly related to a collision zone. Moreover, Armstrong and Dick's mechanism may in part apply directly to generation and emplacement of ophiolite thrust sheets, but many crystalline thrusts do not occur in the situations described by the two models. They explain some crystalline thrusts, but they do not explain the majority, which occur in the metamorphic cores of most mountain chains. Hatcher and Williams (1986) have described five basic types of crystalline thrust sheets—thin-skinned basement-cover sheets, ophiolites, tectonic slides, basement uplifts, and composite sheets—and postulated a mechanical model to explain formation of the composite, ophiolite, and thin-skinned types (Figure 11–27). Their model embodies Armstrong and Dick's idea of detachment within the ductile-brittle transition zone and also relates across-strike width to compressive stress, friction and dip on the basal thrust, and thickness of the sheet.

Most geologists nowadays agree that faults beneath crystalline thrust sheets (and deep-seated listric normal faults) propagate within the ductile-brittle transition, which occurs at a depth controlled by the thermal properties of the crust—thereby also controlling the thickness of these sheets.

THE ROOM PROBLEM

Unfilled voids may seem possible in the construction of cross sections in foreland fold-and-thrust belts, suggesting too much room (or volume)—a statement of the ***room problem*** in structure and tectonics. Usually the problem really lies in the skills of the person constructing the section because nature does not allow large voids in the crust at depths of several kilometers, where most structures form.

Many thrust faults form in regions where the local deformational style occurs by slip of layers past each other (flexural-slip) and where ideally there is no ductile flow of material into voids, either potential or real. In a deformational realm dominated by horizontal compression, there should be no void space larger than pores or small fractures (openings of millimeters or less), even in rocks being brittlely deformed. Yet, given the geometric possibilities that exist in foreland thrust belts, voids of tens to hundreds of cubic meters might be expected; deep cuts, seismic-reflection profiles, and drill data tell us that is *not* so.

Thrusts may also form in the cores of brittle (flexural-slip buckle) folds—generally, anticlines—as the folds tighten and a room problem arises in the more tightly folded inner layers. A single thrust fault may take shape, or tighter folds may form, or two thrusts may dip in the same directions as the limbs of the fold and opposite to each other (Figure 11–33). The last is called a ***delta,*** or ***triangle, structure.*** Delta structures commonly form in the outer or upper parts of the foreland fold and thrust belt where the overburden is relatively thin. They also form where the ratio of competent to incompetent rocks is high, on the order of 1:1 to 4:1. Duplexes may form at greater depths and serve the

ESSAY

Natural Model Gravity Foldbelt

An ideal fold-and-thrust system can form on a small scale under the influence of gravity, if all the mechanical elements are present. In December 1974, I observed a fold-and-thrust belt of this kind in a frozen, moss-covered road cut in the Blue Ridge of North Carolina (Figure 11E–1). The soil beneath the thin moss was water-saturated and had frozen overnight. As it froze, it expanded, forming needle ice that lifted the moss layer, increasing the slope angle and forming a detachment between the more rigid moss layer above and the freezing soil below. The surface layer moved down the slope above the master detachment, perhaps lubricated by a film of water that had not yet frozen.

The sliding mass developed thrusts, drag structures, fault propagation folds, and tear faults as it moved downslope above the detachment (Figure E–1c). Both folds and thrusts may be traced to termination points, with many terminations overlapping the terminal zones of other folds and thrusts. As the sheet moved, it began to fragment, but mostly remained intact.

The only motive force that created the mossy fold-thrust system was gravity. The phenomenon is easily and frequently observed, and until recently downslope movement of a gravity-driven mass off a topographic high was the mechanism generally invoked to explain most foreland thrust belts in mountain belts. Edward Hansen (1971) has described a similar gravity-driven small-scale multiple-fold system in partly frozen surficial materials in south-central Norway.

The Helvetic Alps and the Prealps in Switzerland and France were long thought to have slid northward off the Aar and Gotthard basement massifs and

(a)

FIGURE 11E–1
Small-scale fold-thrust system developed on a slope covered by a thin layer of moss that froze; the moss detached, moved downslope under the influence of gravity, and produced thrusts, fault-propagation folds, tear faults, and other features characteristic of a fold-thrust belt. (a) View of needle ice–covered upper slopes from which a layer of moss has detached and slid downslope. (b) Closeup of the detached moss layers showing the variety of structures developed. (RDH photos.) (c) Sketch of structures in (b).

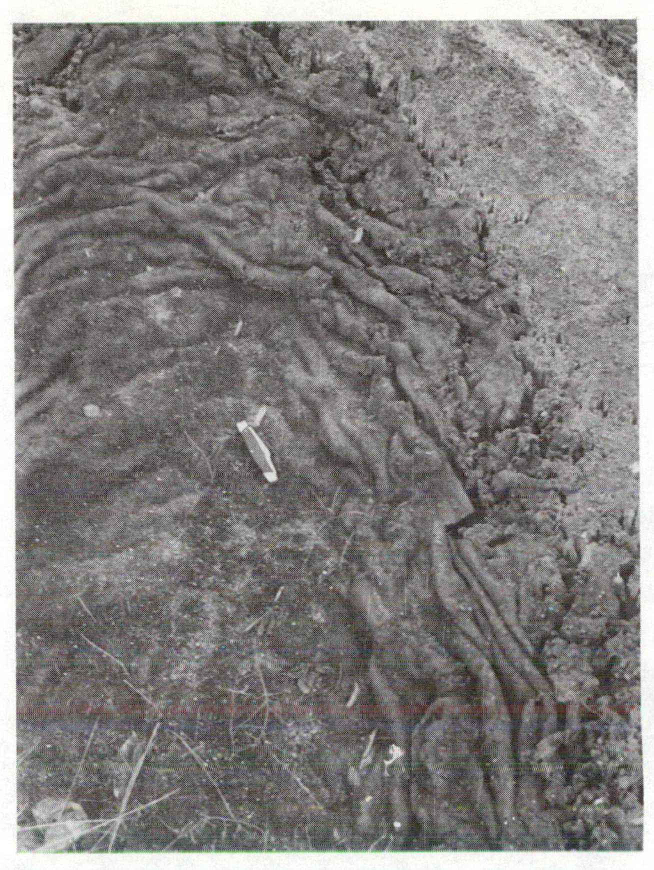

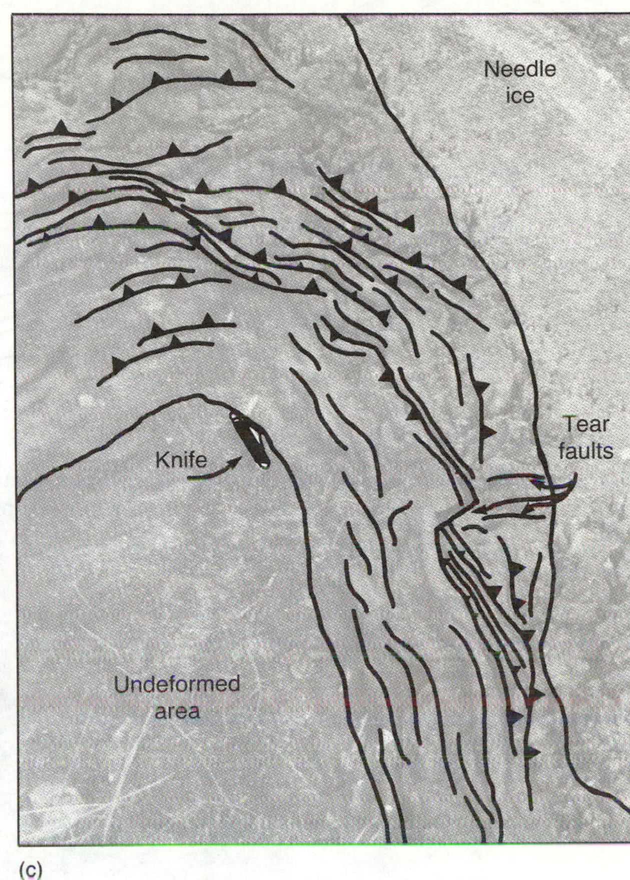

(b) (c)

FIGURE 11E–1 *(continued)*

the more internal Pennine Alps (Trümpy, 1960). At one time, many geologists believed the Appalachian foreland from Alabama to Pennsylvania had slid northwestward off the Blue Ridge (Gwinn, 1970) and that the Wyoming-Montana thrust belt had slid eastward from the internal higher parts of the Cordillera (Rubey and Hubbert, 1959). George H. Davis (1975) has documented gravity-driven folding and décollement from the gneissic basement of the Rincon Mountains near Tucson, Arizona.

References Cited

Davis, G. H., 1975, Gravity-induced folding off a gneiss dome complex, Rincon Mountains, Arizona: Geological Society of America Bulletin, v. 86, p. 979–990.

Gwinn, V. E., 1970, Kinematic patterns and estimates of lateral shortening, Valley and Ridge and Great Valley provinces, central Appalachians, south-central Pennsylvania, *in* Fisher, G. W., Pettijohn, F. J., Reed, J. C., Jr., and Weaver, K. N., eds., Studies of Appalachian geology: Central and southern: New York, Wiley-Interscience, p. 127–146.

Hansen, Edward, 1971, Strain facies: New York, Springer-Verlag, 207 p.

Rubey, W. W., and Hubbert, M. K., 1959, Role of fluid pressure in mechanics of overthrust faulting, II. Overthrust belt in geosynclinal area of western Wyoming in light of fluid pressure hypothesis: Geological Society of America Bulletin, v. 70, p. 167–206.

Trümpy, Rudolf, 1960, Paleotectonic evolution of the central and western Alps: Geological Society of America Bulletin, v. 71, p. 843–908.

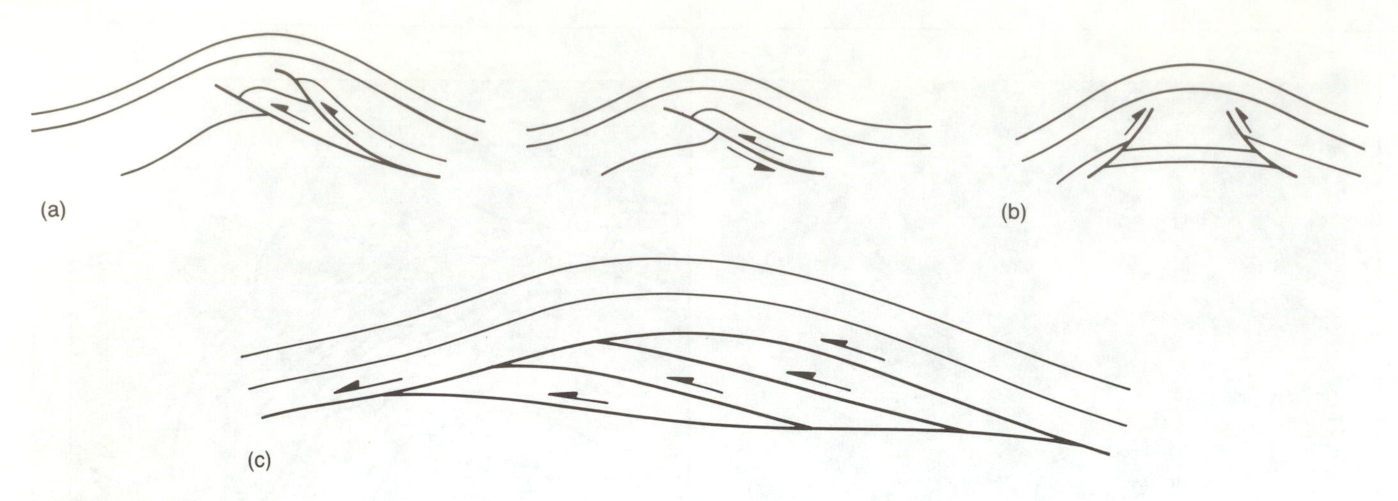

FIGURE 11–33
Possible solutions to the room problem in thrust-fold systems. (a) Single thrusts and splaying imbricates. (b) Delta or triangle structures. (c) Duplexes.

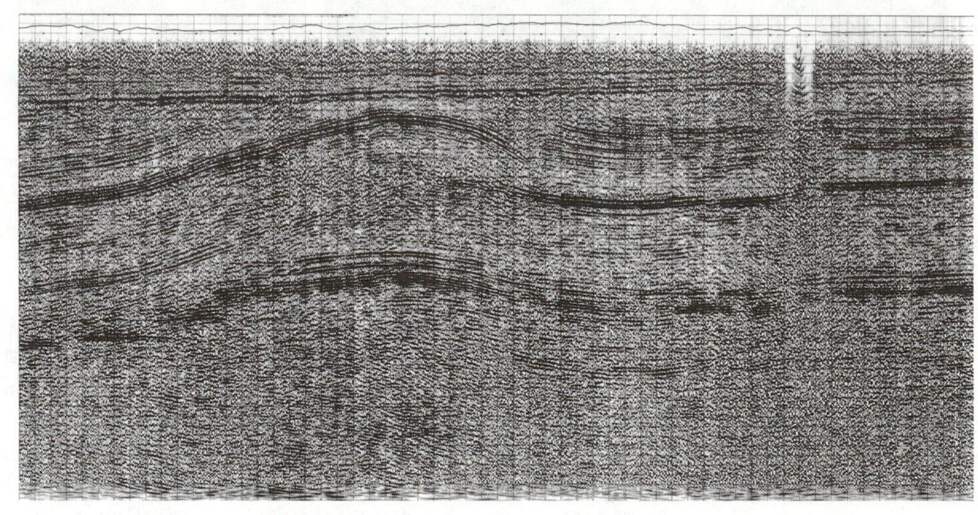

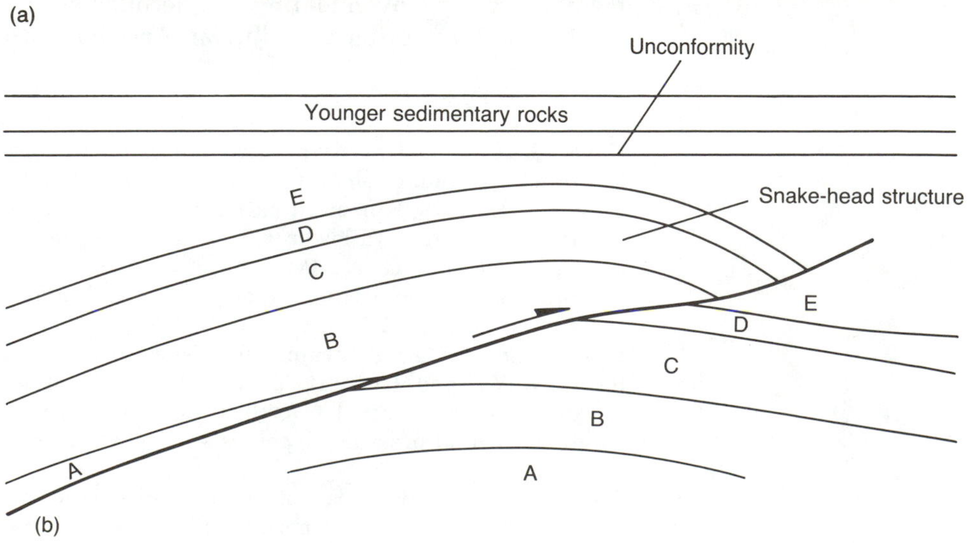

FIGURE 11–34
(a) Seismic-reflection profile of snake-head structure from the Appalachians buried beneath the Coastal Plain in Alabama. (b) Cross-section interpretation of the profile.

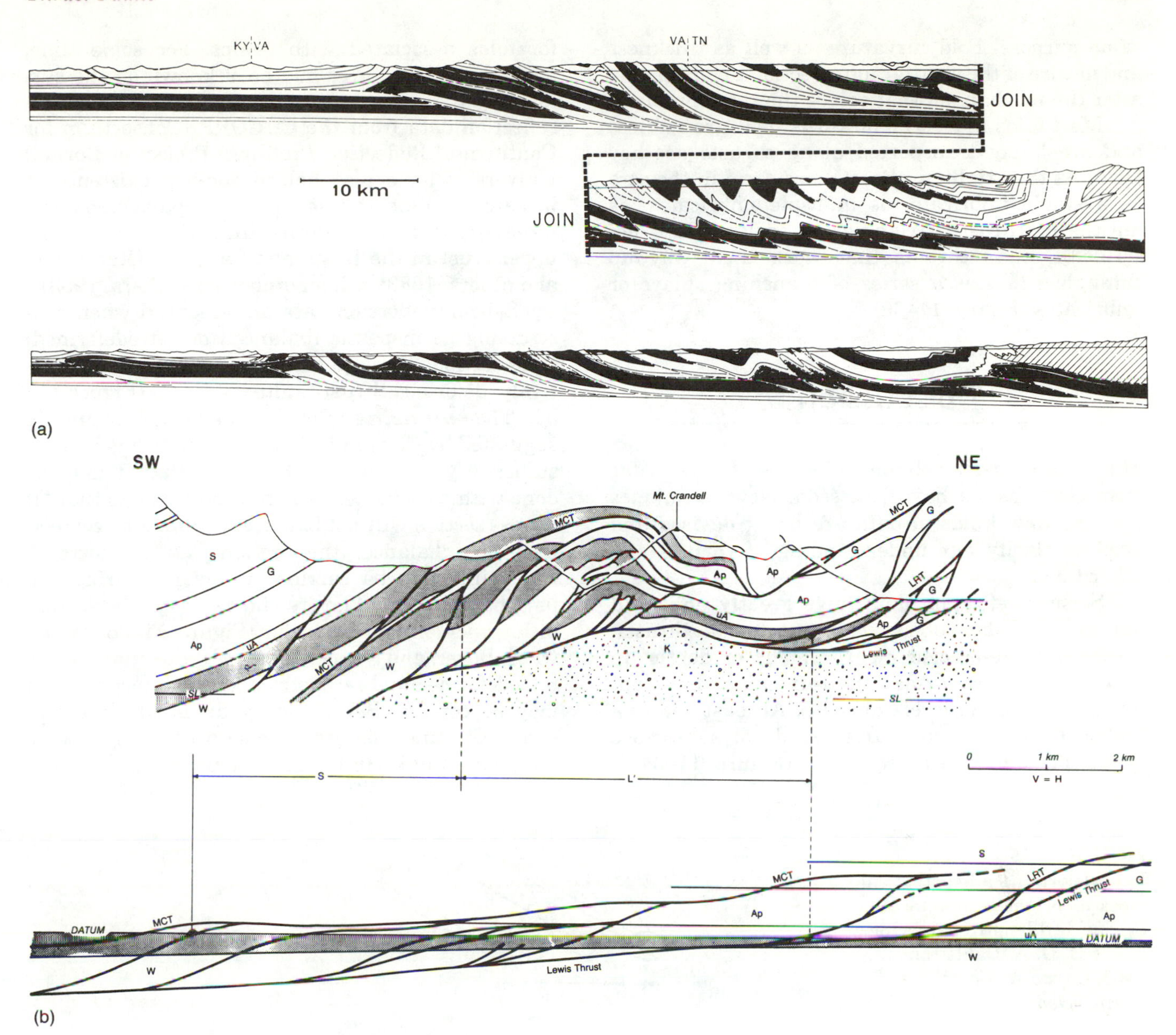

FIGURE 11–35

Cross sections through parts of a foreland fold-and-thrust belt. (a) Sections through the Appalachian Valley and Ridge. Black is the strong rock unit in the sequences. (From D. H. Roeder and W. D. Witherspoon, 1978, Palinspactic map of East Tennessee, *American Journal of Science,* v. 278, p. 543–55. Reprinted by permission of American Journal of Science.) (b) Section through part of the Canadian Rockies (above) and a restored (retrodeformed) version (below). MCT—Mt. Crandell thrust; LRT—Livingston Range thrust; S—Siych; G—Grinnell; Ap—Appekunny; fine stipple—lower Altyn; uA—upper Altyn; W—Waterton; K—Cretaceous. (From S. E. Boyer and David Elliott, AAPG *Bulletin,* v. 66. © 1982. Reprinted by permission of American Association of Petroleum Geologists.)

same purpose. Fold curvature as well as thickness and nature of the stratigraphic section involved may alter the mechanism that solves the room problem.

Most folds in a foreland thrust belt are rootless and are being transported along thrusts as they form. As they tighten, thrusts may form in the core as a single imbricate, as multiple imbricates, as antithetic thrusts (triangles or deltas, just described), or as thrusts that work their way up through a fold as a series of branching splays or imbricates (Figure 11–33).

DISCUSSION

The nature and mechanics of thrust faults is far from clear, as we have just seen. New techniques yielding new kinds of data are being devised and used to clarify our understanding of thrusts and other faults.

Seismic-reflection data have greatly enhanced our view of the subsurface geometry of thrusts, provided more-precise measurement of depths to basement, and confirmed the thin-skinned concept for deformation of foreland areas. At least one geometric term has arisen from study of subsurface geometry—*snake-head structure* (Figure 11–34)—for folds associated with ramps. For some time, snake-head structures had been known on the surface, but they were not called that. Seismic-reflection data from the COCORP (Consortium for Continental Reflection Profiling) Project at Cornell University have also helped confirm existence of large crystalline thrusts in the Appalachians and large listric normal faults through much of the upper crust in the Basin and Range in Utah (Cook and others, 1983; Allmendinger and others, 1983).

Seismic-reflection data are essential when constructing palinspastic (balanced or retrodeformed) cross sections through brittlely (mostly nonpenetratively) deformed thrust-faulted terranes (Figure 11–35). The ***balancing of cross sections,*** as originally suggested by Clinton Dahlstrom (1969), has become sufficiently iterative mechanically that it may be done with a computer. An oft-quoted rule is that "if a cross section will not balance, it cannot be correct; if it does balance, the section *may* be correct." Balancing may be carried out by measuring the lengths of deformed layers and restoring the section to an undeformed condition (Figure 11–36; Woodward, Boyer, and Suppe, 1985). If the section can be retrodeformed and reconstructed, the interpretation may be correct. Balancing by areas (thickness of rock units times length) is also possible by taking the areas of units in the section and restoring them

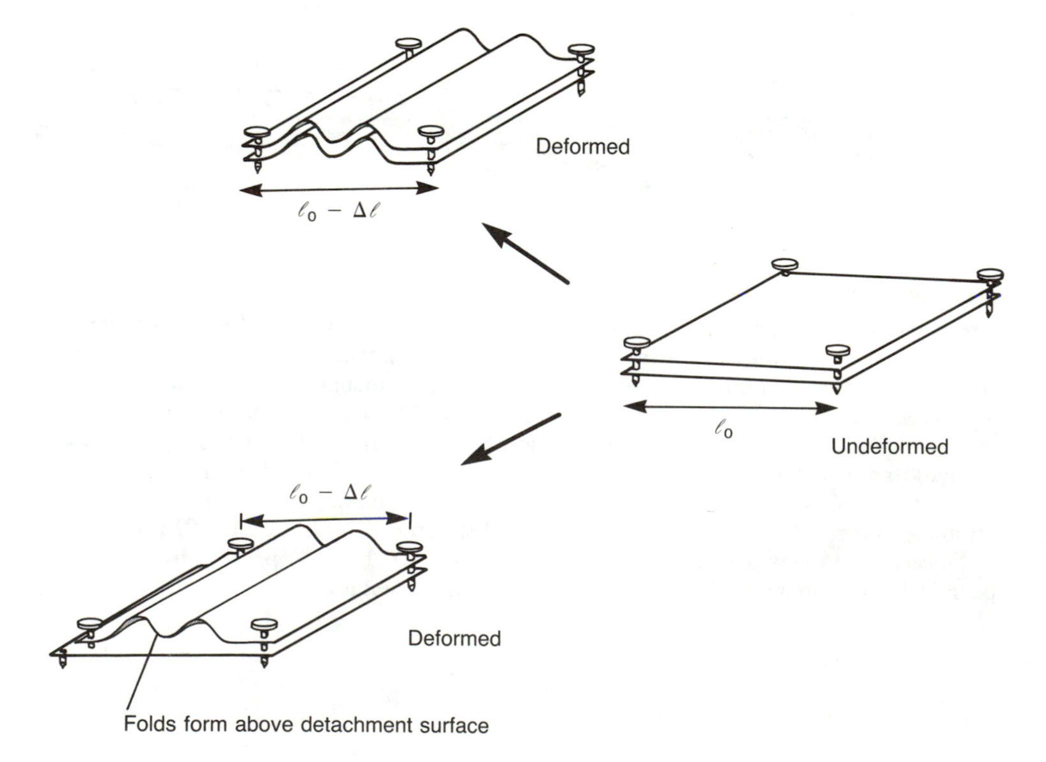

FIGURE 11–36
Retrodeforming a hypothetical block using the technique of line (or bed-length) balancing. (After C. D. A. Dahlstrom, 1969, *Canadian Journal of Earth Science,* v. 6.)

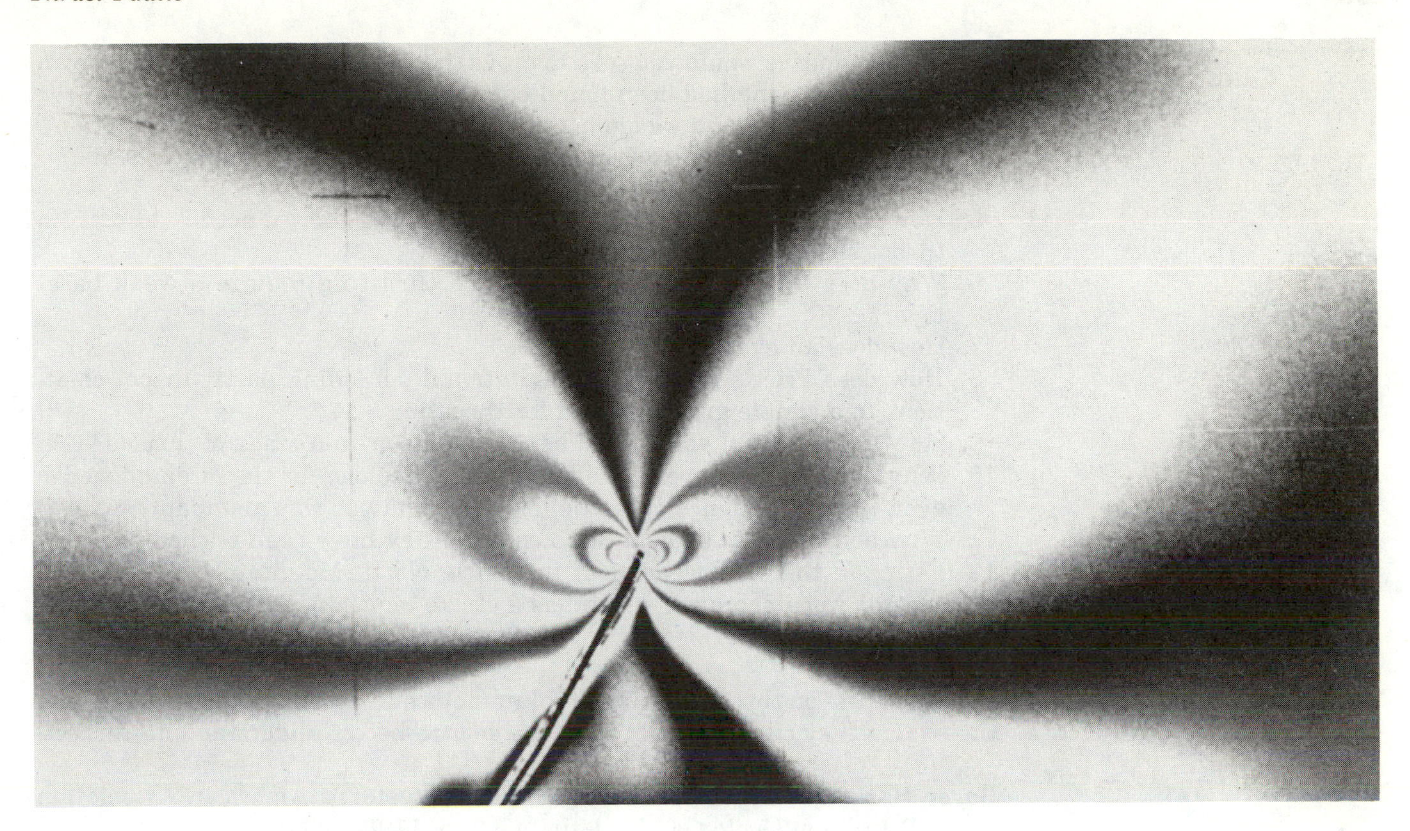

FIGURE 11–37
Photoelastic strain pattern developed around a slot in a stressed sheet of plexiglass. Note pattern at crack and at crack tip. (Courtesy of John J. Gallaher, Jr., Arco Oil and Gas Company.)

to an undeformed condition. Major problems arise with construction and balancing of any section if internal deformation within rock units occurs and if there is deformation out of the plane of the section. The problem of internal deformation may be partly solved by area balancing. Deformation out of the plane of the section requires construction of sections in several directions—a form of three-dimensional analysis. Accurate coupling of surface and subsurface data in cross-section analysis aids reconstructions that enhance our ability to construct accurate mechanical models. Note that in this kind of analysis the component of movement parallel to strike ultimately prevents balancing sections if the sections cross into the metamorphic core. Another problem with balancing sections in complexly deformed rocks lies in estimating the amount of penetrative strain (Woodward, Gray, and Spears, 1986). A poor estimate may prevent balancing a section otherwise correctly drawn.

A technique borrowed from engineering has found widespread use in mechanical modeling of thrust faults—*finite-element studies*. It is a numerical method for dividing a continuum into triangles or parallelograms with nodes on the boundaries where displacements are calculated by a computer program and a structure is defined. Boundary conditions of the finite-element method are either nodal displacements or forces at nodes. The computer program simulates the deformation by applying stress of various magnitudes at various orientations to thrust or other fault-fold geometries, but it has some difficulty in handling discontinuities such as faults.

Photoelastic studies of pressure-sensitive plastics in polarized light (Figure 11–37), another technique from engineering, allow model studies of various stress distributions to be made with different fault, ramp, and fold configurations. Coupled with finite-element analysis, such studies have helped elucidate simple foreland thrusts.

Now we can turn from thrust faults to the other class of compressional faults—strike-slip faults.

Questions

1. What evidence would you seek to prove that a large overthrust exists in an area where none had been found before?
2. Why does a foreland wedge maintain its shape throughout deformation?
3. Why do thrusts have steeper dips on ramps than within detachments?
4. How and why do slices (horses) form?
5. How does fluid pressure reduce the amount of energy needed to move a thrust sheet?
6. Why does Chapple's model for foreland thrusting require a weak basal layer?
7. How does an eyelid window form?
8. How does Price's concept of gravitational spreading partly dispel objections to a simple gravity model for thrusting?
9. How are potential voids filled beneath folds or in a stack of thrusts?
10. Why did Chapple choose an ideal plastic rheology for the foreland wedge even though evidence of brittle (elastic) behavior was abundant?
11. Why don't thrusts slide back down once they have been pushed up?
12. Estimate the displacement of the Little North Mountain thrust (in the central Appalachians), which has a strike length of 220 km.

Further Reading

The literature on thrust faulting is voluminous and growing. These works, and the papers they cite, will help open the door to better understanding thrusts.

Boyer, S. E., and Elliott, David, 1982, Thrust systems: American Association of Petroleum Geologists Bulletin, v. 66, p. 1196–1230.
Describes the terminology and geometry of thrusts, with examples from several mountain chains, in both maps and cross sections.

Chapple, W. M., 1978, Mechanics of thin-skinned fold and thrust belts: Geological Society of American Bulletin, v. 89, p. 1189–1198.
Lists the properties of a foreland fold and thrust belt, then derives a mechanical model. The first part of the paper and the conclusions will be useful to undergraduate students.

Dahlstrom, C. D. A., 1969, Balanced cross sections: Canadian Journal of Earth Science, v. 6, p. 743–758.
Principles of balancing cross sections (mainly line balancing) are explained clearly and concisely, with examples mostly from the Canadian Rockies.

Hatcher, R. D., Jr., and Williams, R. T., 1986, Mechanical model for single thrust sheets, Part I: Taxonomy of crystalline thrust sheets and their relationships to the mechanical behavior of orogenic belts: Geological Society of America Bulletin, v. 97, p. 975–985.
The first part of this paper outlines the properties and modes of occurrences of crystalline thrust sheets and shows how their behavior compares and contrasts with that of foreland thrusts.

Hossack, J. R., and Hancock, P. L., eds., 1983, Balanced cross sections and their geological significance: Journal of Structural Geology, v. 5, p. 98–223.
Special issue dedicated to David Elliott, containing 10 papers on balanced cross sections.

McClay, K. R., and Price, N. J., 1981, Thrust and nappe tectonics: Geological Society of London Special Publication 9, 539 p.
This compendium ranges from thrust mechanics through rock products of faulting to studies of thrusts in different parts of the world. Papers by Bally, Ramberg, Ramsay, Price, and Coward and Kim, in particular, will interest undergraduates, but others also contain excellent examples and problems related to thrust faulting.

Mitra, Shankar, 1986, Duplex structures and imbricate thrust systems: Geometry, structural position, and hydrocarbon potential: American Association of Petroleum Geologists Bulletin, v. 70, p. 1087–1112.
This study of duplex and imbricate thrusts provides more details than do Boyer and Elliott (1982). Cites good examples from several fold and thrust belts.

Moores, E. M., 1982, Origin and significance of ophiolites: Reviews of Geophysics and Space Physics, v. 20, p. 735–760.
Summarizes the structure and processes that form and emplace ophiolite sheets, with examples from various settings.

Perry, W. J., Roeder, D. H., and Lageson, D. R., 1984, North American thrust-faulted terranes: American Association of Petroleum Geologists Reprints Series 27, 466 p.
A compendium of papers, including several classic works, such as the 1934 AAPG Bulletin paper by J. L. Rich on the concept of thin-skinned thrusting, V. E. Gwinn's 1964 GSA Bulletin paper on thin-skinned structure of the Allegheny Plateau, the 1981 paper by R. A. Price on the Canadian Rocky Mountains foreland published in Geological Society of London Special Publication 9 (see McClay and Price, earlier in this list), and the 1982 AAPG Bulletin paper by Boyer and Elliott.

12

Strike-Slip Faults

The powerful dislocation which intersects Scotland along the line of the Great Glen has, in the past, been regarded by most geologists as a normal or dip-slip fault with a predominant vertical downthrow to the south-east. A reconsideration of the entire problem now suggests that this view is no longer tenable and that the dislocation is, in reality, a lateral-slip or wrench fault with a horizontal displacement of approximately 65 miles. . . .

WILLIAM Q. KENNEDY, Quarterly Journal of the Geological Society of London

MANY STRIKE-SLIP FAULTS ARE KNOWN WORLDWIDE AS SPECTACULARLY active earthquake generators. Some have repeatedly caused heavy damage and killed or injured millions of people. The earthquakes near Kansu (1920) and Tangshan (1976), China, that killed 100,000 and 500,000 people were produced by strike-slip faults. Such disasters justify the intensive study and attempts to predict major earthquakes. Routine prediction of earthquakes is not yet possible, but we have gathered an enormous amount of data and greatly increased our knowledge of strike-slip faults. Some of these large active strike-slip faults, like the San Andreas in California and the Alpine fault in New Zealand, are located on the Pacific margin as a result of plate motion. Others, like the Great Glen fault in Scotland, studied by Kennedy (1946), have been known for centuries because of their linear traces.

PROPERTIES AND GEOMETRY

It is relatively easy, using several different lines of evidence, to determine the sense of motion along a fault after an earthquake. One is accompanying ground breakage. Another is that topographic features, such as streams and ridges, may be offset by motion and so indicate movement (Figure 12–1). Yet another is that strike-slip faults are characterized by their steep dips. Most dip more than 60°, and many approach the vertical. Shallow-dipping segments do exist, particularly where a fault changes orientation.

Strike-slip faults may branch and splay into smaller segments and into other types of faults. They may contain *slices,* or *horses,* of material plucked from either side of the fault and carried along in the same way that blocks are carried along thrust faults. Slices may be derived from either block, so the age-relationship rule related to thrusts (Chapter 11) does not apply here.

Strike-slip faults are also called ***wrench, tear,*** and ***transcurrent*** faults. (Arthur Sylvester, 1988, has recommended the term wrench fault be abandoned because of the confusion generated by the name.) The term *accident* is used by French geologists to describe large, steeply dipping faults with uncertain motion sense. No similar term appears in English terminology for faults. Many faults called accidents by French geologists, like the South Atlas fault in Morocco, are probably strike-slip faults. The most recent class of faults, called *transforms* (J. T. Wilson, 1966), was added by consideration of plate motion. *Sinistral,* or *left-lateral,* and *dextral,* or *right-lateral,* strike-slip faults were defined in Chapter 9 (Figure 12–2).

Where a strike-slip fault abruptly changes strike, various things may happen to its geometry. The dip may change along with the sense of motion, and that segment may become a thrust or normal

FIGURE 12–1
Dextral offset of several streams along the San Andreas fault, San Luis Obispo County, California. View is toward the south. (Robert E. Wallace, U.S. Geological Survey.)

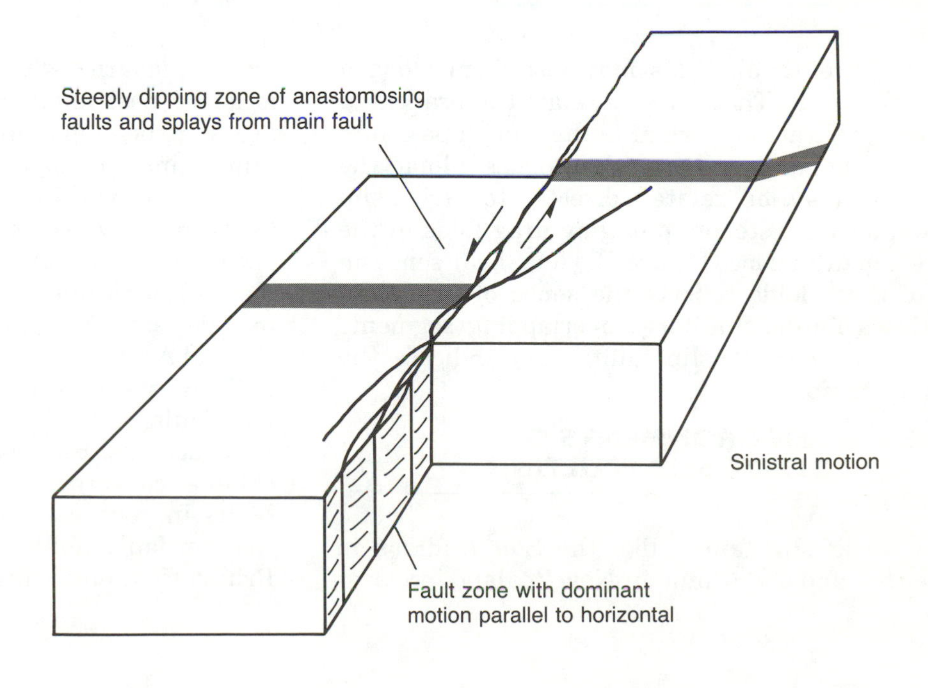

FIGURE 12–2
Geometry and anatomy of strike-slip faults.

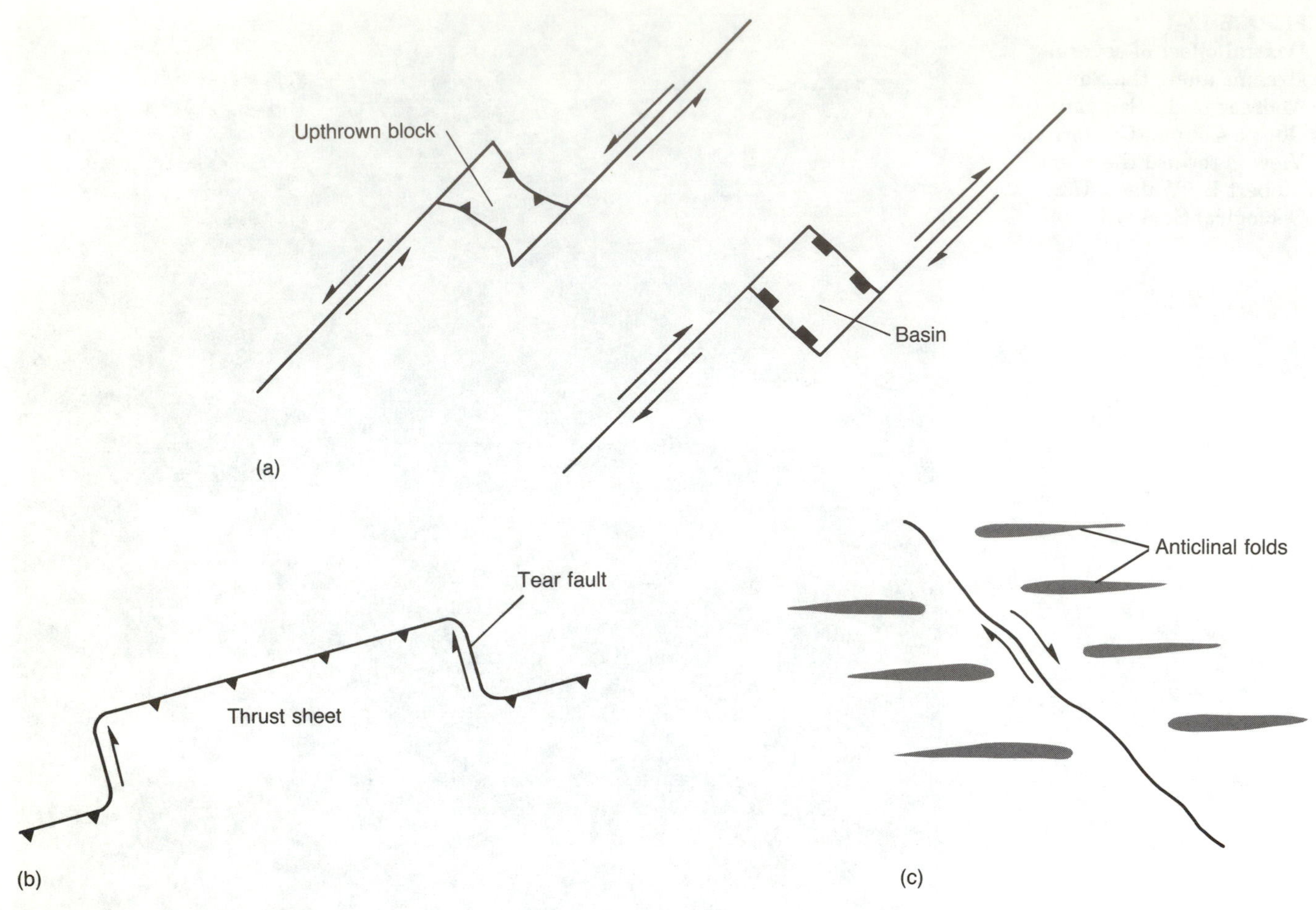

FIGURE 12–3
(a) Transfer of motion from strike-slip into thrust or normal slip motion. (b) Tear faults at the edges of a thrust sheet. (c) Folds related to strike-slip motion. (d) Map of the Pine Mountain block in Tennessee showing a thrust sheet bounded by tear faults. (After J. L. Rich, 1934, AAPG *Bulletin*, v. 18. Reprinted by permission of American Association of Petroleum Geologists.)

fault (Figure 12–3). Folds may also form along a strike-slip fault. They may be related to branching into either thrust or normal faults, which pass into folds as the displacement diminishes along the faults—or may be related directly to strike-slip motion, forming steeply plunging drag folds in the strike-slip fault zone (Figure 12–3c). Shear sense on any of these folds reflects the sense of motion of associated faults. Similarly, overlapping segments of *en echelon* strike-slip faults may indicate the sense of motion.

ENVIRONMENTS OF STRIKE-SLIP FAULTING

Many strike-slip faults, like the San Andreas in California and the Alpine in New Zealand, occur at plate boundaries where one large block of material is forced past another; they are considered *transforms*. Strike-slip faults may also be found in mountain chains, moving large blocks of material along or across strike within them. They form the boundaries between many accreted terranes and the continent (or other terranes) to which accretion is taking place. Other kinds of strike-slip faults (Figure 12–3) include large faults, such as the Motagua fault in Central America, and *tear faults* bounding the edges of thrust sheets, such as the Jacksboro and Russell Fork faults on the Pine Mountain thrust sheet in the Appalachians. Strike-slip faults also occur behind oblique convergent margins, like several large faults in southeastern Alaska (such as the Fairweather fault) and adjacent parts of the Yukon and British Columbia (the Tintina fault).

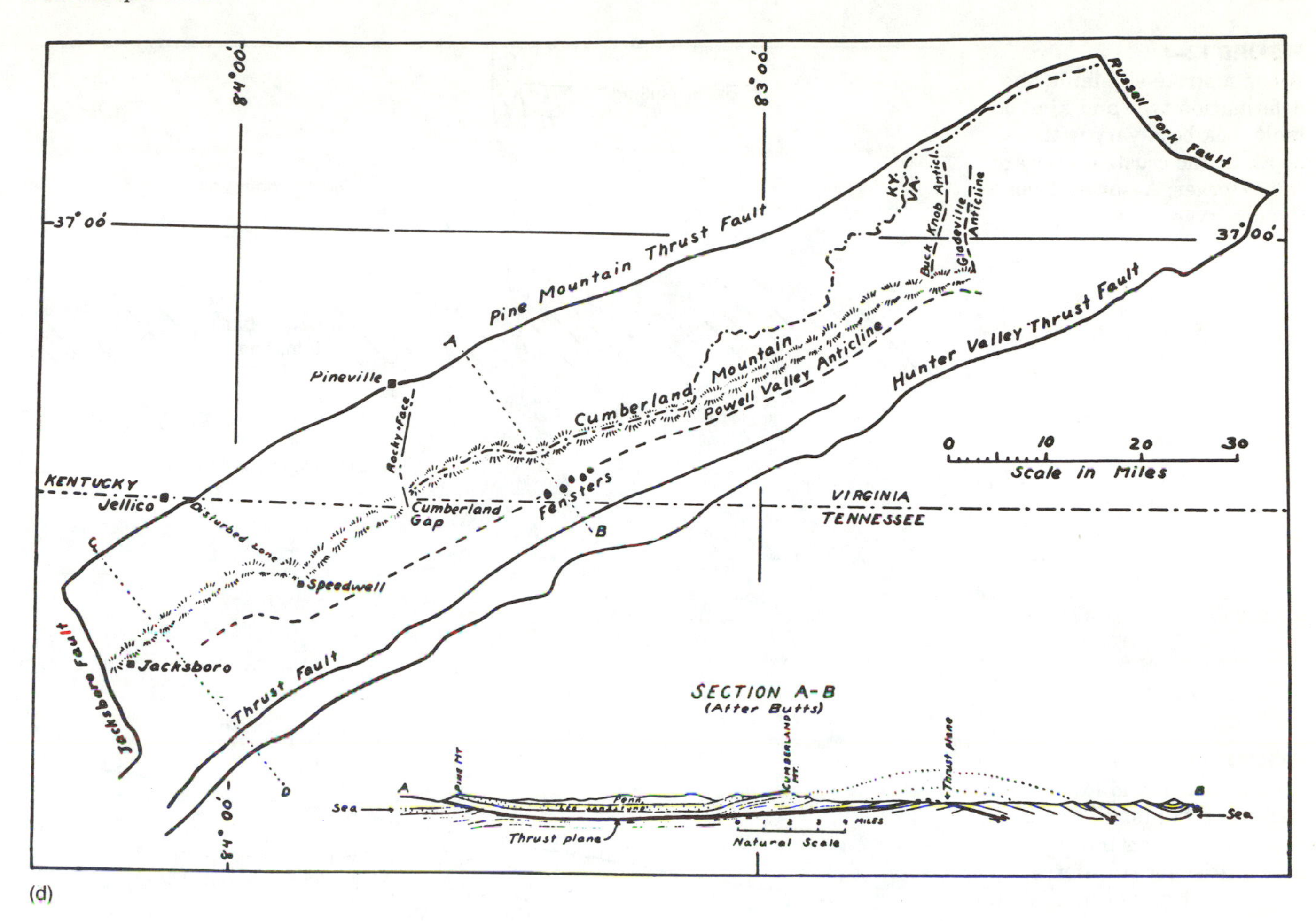

(d)

NATURE OF FAULT ZONES

A strike-slip fault is commonly thought of as a single brittle fault surface separating two blocks, and some do occur as simple planar fault surfaces. Generally, these simple planar faults form in upper crustal rocks, where multiple fractures do not tend to form, or in zones that contain abundant fluid. *Tear faults* bounding thrust sheets (Chapter 11) may have sharp planar boundaries, depending on the nature of the rocks and availability of fluid. Many strike-slip faults, like other faults in mountain chains, and at terrane and plate boundaries, are commonly *not* single clean brittle surfaces. They most frequently occur as cataclastic zones consisting of anastomosing faults containing gouge and breccia, or—if erosion has exposed the deeper parts of the zone—a considerable thickness of mylonite. These are the classic *shear zones* discussed in Chapter 10.

Evidence from inactive strike-slip faults and others eroded to expose lower crustal parts suggests a continuous transition from brittle deformation near the surface to ductile deformation at middle to lower crustal depths (Figure 12–4). At the surface, elastic strain energy would be relieved by instantaneous strike-slip rebound (Chapter 10), as indicated by cataclastic rocks (breccia and gouge), but at depth this motion may occur as ductile creep, continuously producing mylonite and other ductilely deformed rocks (Sibson, 1983). Also, comparison of brittle strike-slip faults and those in deeply eroded crust indicate that strike-slip fault zones near the surface are relatively narrow, but some ductile strike-slip fault zones in deeply eroded parts of high-grade Precambrian shields, such as western Greenland (Bak and others, 1975), may be several kilometers wide (Chapter 10). The mineralogy and texture in these zones exhibit abundant evidence of ductile deformation at high temperature and pressure. Those are conditions found in the lower crust.

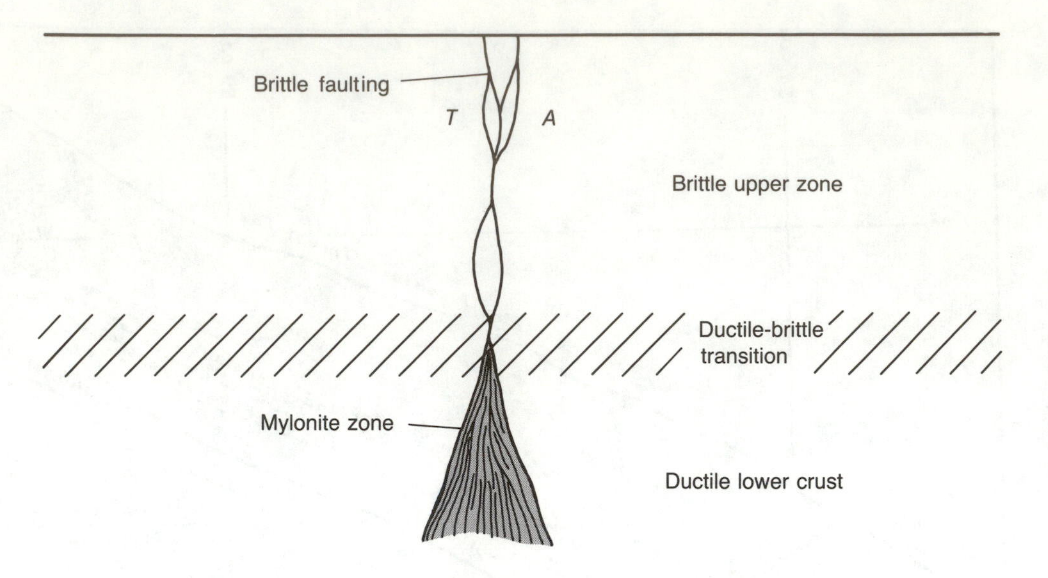

FIGURE 12–4
Along a strike-slip fault, the deformation type and kind of fault rock both vary with depth in the crust. T—toward the observer; A—away from the observer.

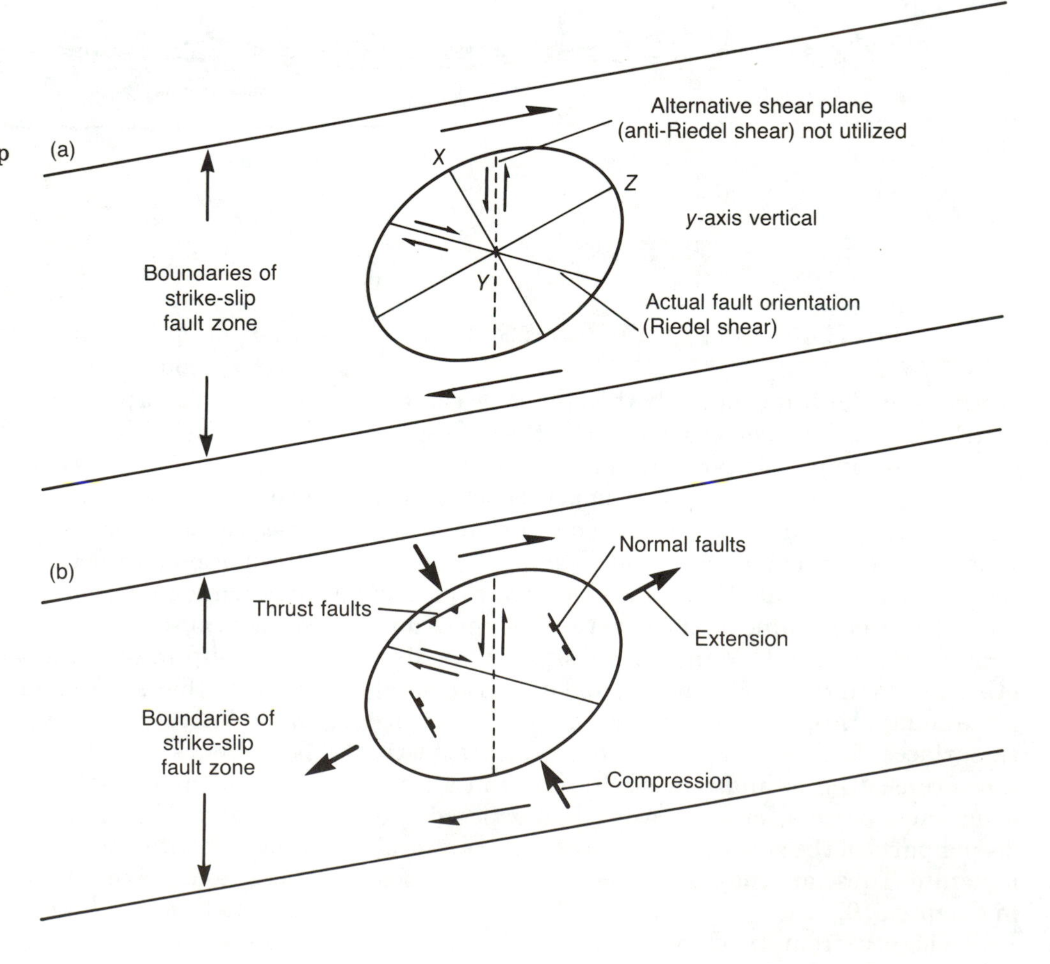

FIGURE 12–5
(a) Strain ellipsoid for strike-slip faults. (b) Strain ellipsoid showing the possibilities for transferred motion to thrust or normal-slip domains.

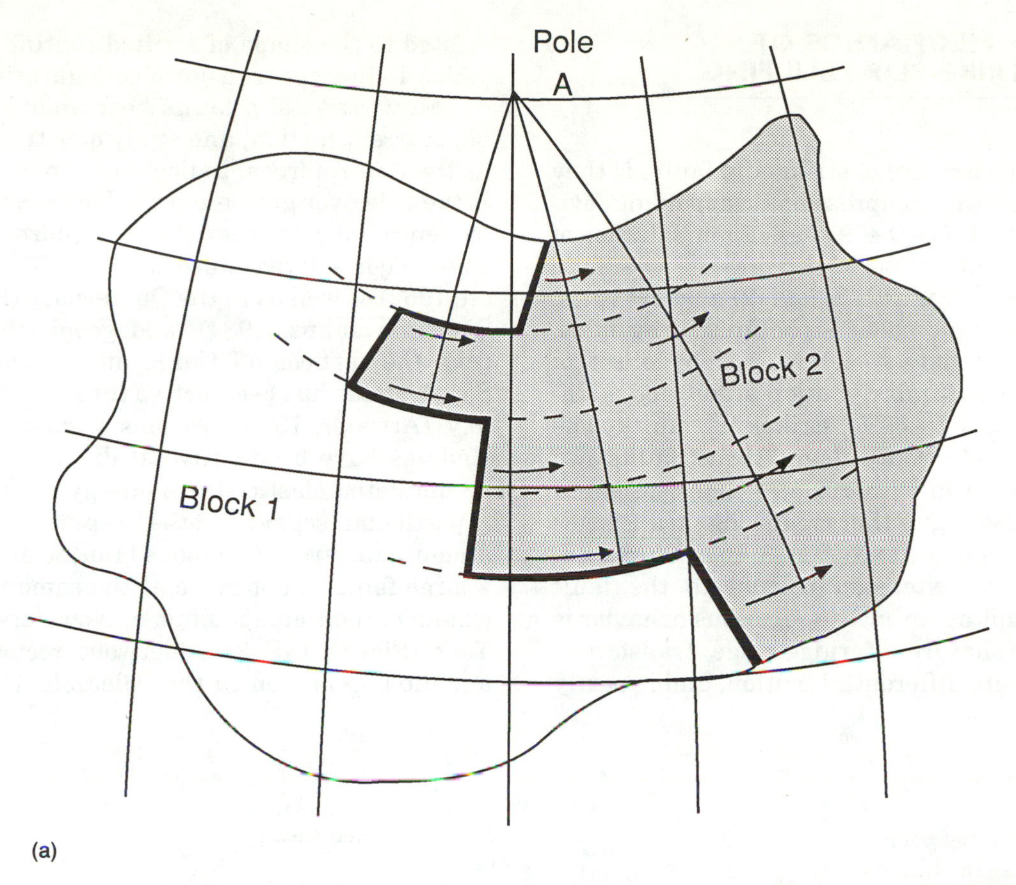

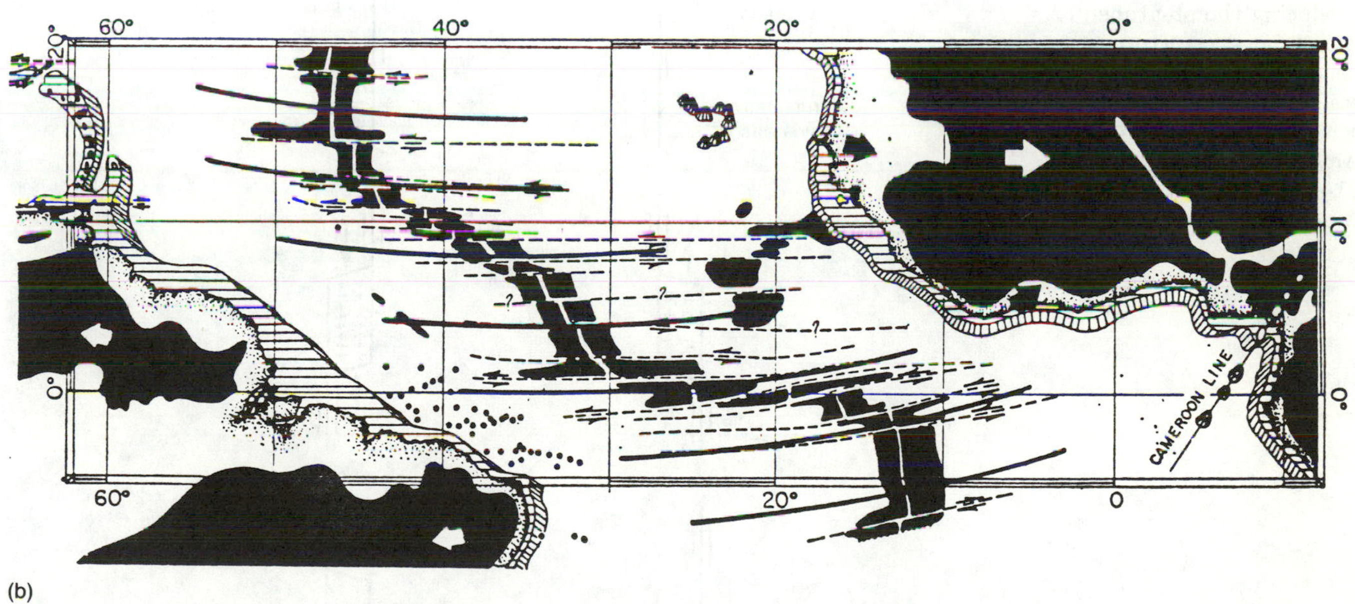

FIGURE 12–6
Small-circle geometry relationship of strike-slip and transform faults. (a) Generalized diagram showing the motion of two blocks on a sphere that must follow small circles on the sphere concentric to pole A. (b) Transforms (dashed faults) in the present-day Atlantic Ocean relative to small circles (heavy solid lines) concentric to a pole at 62° N 36° W. (From W. J. Morgan, *Journal of Geophysical Research,* v. 73, p. 1959–1982, 1968, copyright by the American Geophysical Union.)

MECHANICS OF STRIKE-SLIP FAULTING

A fundamental attribute of strike-slip faults is they form by horizontal compression. Supporting evidence is threefold: (1) the Andersonian solution of the strain ellipsoid for strike-slip faults—possibly a circular argument (Figure 12–5a); (2) field evidence, including mesoscopic shear structures, drag folds, and the sense of motion determined by offset of marker units; and (3) first-motion studies of earthquakes produced by active strike-slip faults. The close association of strike-slip and thrust faults also shows that they form by horizontal compression.

We could also argue that ridge-ridge transforms are not compressional, except that they do not exhibit evidence of extension normal to the fault plane. Part of this apparently anomalous behavior is related to the nature of ridge-ridge transforms, which compensate differential motion, and is partly related to the shape of a rifted continental margin, which influences transform configuration.

Most strike-slip faults bear abundant evidence of recurrent motion, and study of active faults, such as the San Andreas, indicate that recurrent motion is the rule over geologic time. For example, geologic evidence indicates that the San Andreas has undergone major activity since the early Tertiary (Crowell, 1962) as well as in the Quaternary (Figure 1E–1; Sieh and Jahns, 1984), and geophysical evidence from the seafloor off California indicates that the San Andreas has been active throughout the last 30 m.y. (Atwater, 1970). Various segments of the San Andreas have been active at different times. Once accumulated elastic-strain energy is relieved along a particular segment, other segments continue to accumulate strain and move later, so all segments of a large fault do not move at the same time. As one example, the Ramapo fault in New Jersey and New York (Figure 13E–2) underwent recurrent strike- and dip-slip motion in the Paleozoic, Mesozoic, and

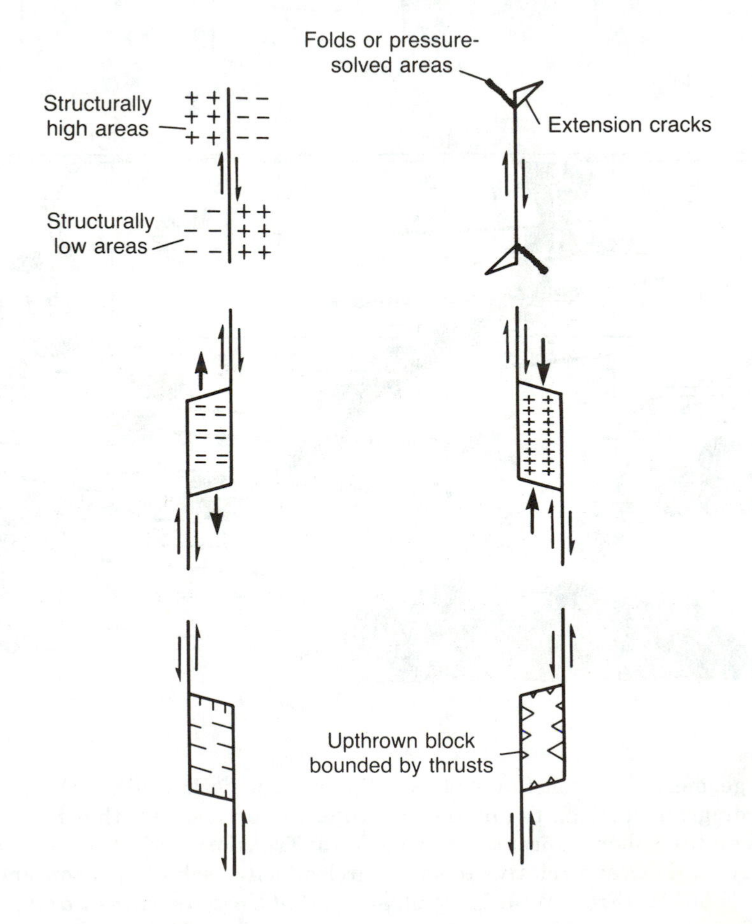

FIGURE 12–7
Transfer of motion between strike-slip fault segments producing rhomb-graben or rhomb-horst structures, depending on whether the step-over is in a right or left sense. (After Atilla Aydin and Amos Nur, *Tectonics,* v. 1, p. 91–105, 1982, copyright by the American Geophysical Union.)

Tertiary throughout 400 m.y. (Ratcliffe, 1971). The Ramapo may even be active today, for many small earthquakes have been located along it (Aggarwal and Sykes, 1978).

FAULT GEOMETRY AND OTHER FAULT TYPES

J. Tuzo Wilson (1965), D. P. McKenzie and R. L. Parker (1967), and Jason Morgan (1968) used the fact that Earth is nearly spherical and recognized that major strike-slip and transform faults must geometrically follow small circles on the Earth rather than great circles. We can predict, using spherical geometry, that with any deviation of a fault plane from a small circle the sense of motion on the fault will include components of either normal or thrust motion (Figure 12–6). Once orientation of the fault plane deviates significantly, it passes from the realm of strike-slip (shear) into the realm of normal or thrust motion (Figure 12–5b).

Relationships between *en echelon* or parallel segments of strike-slip fault systems that have no strike-slip connections have been studied by Atilla Aydin and Amos Nur (1982); they saw that motion may be transferred from one segment to another—by a *step-over*—through zones of extension or of compression, depending on the relative motion of the fault segments and the sense of each step-over (Figure 12–7). As a result, an *en echelon* dextral strike-slip fault with right step-overs produces *pull-apart,* or *rhomb-graben* (or *rhombochasm*), *basins.* The same result is produced by left-lateral strike-slip faults with left step-overs. Two sides of the basin will be bounded by segments of the strike-slip fault with a significant component of normal slip. The other sides will be bounded by normal faults oriented obliquely to the primary direction of horizontal strike-slip motion (Figure 12–7). J. C. Crowell (1974) and T. W. Dibblee (1977) had recognized the same relationships along segments of the San Andreas fault.

Dextral strike-slip faults with left step-overs or sinistral faults with right step-overs produce *rhomb horsts,* or *push-up ranges.* Strike-slip fault segments with significant thrust motion will bound two sides of a horst, and the other two sides will be bounded by obliquely oriented thrusts.

Small rhomb grabens and horsts have been identified at many places on the Earth. The Dead Sea and the San Francisco Bay area have been interpreted by Aydin and Nur (1982) and Aydin and Page (1984; Figure 12–8) as rhomb-graben structures. Similar conclusions were reached earlier for the Dead Sea structure by Freund (1965) and Freund, Zak, and Garfunkel (1968). The Ocotillo Badlands and Borrega Mountain along the Coyote Creek fault in California have been interpreted by Aydin and Nur (1982) as rhomb horsts. Each of these structures formed as part of a complex of interacting plate boundaries where relative motion has changed through time.

Complex fault geometry frequently produces large rhomb grabens and horsts (Figure 12–9). Then, as noted by Aydin and Nur (1982), more faults form, producing smaller segmented grabens and horsts, which together form larger composite basins or uplifts. Aydin and Nur also noted a consistent ratio of length to width of pull-apart basins, ranging from 2.4 to 4.3, with a mean of 3.2. The ratio appears to be independent of scale, because length/width ratios of basins and uplifts of all sizes fall within this range.

Transfer of motion on a strike-slip fault into thrust or normal slip or into smaller strike-slip faults may be considered formation of second- or third-order features. The master strike-slip fault does not terminate, but changes geometry so as to form the second- and third-order structures.

TERMINATIONS OF STRIKE-SLIP FAULTS

Strike-slip faults may terminate by splaying along strike into smaller faults, by changing strike and becoming thrust or normal faults (which may further splay into smaller faults or terminate in folds), or by diminishing displacement along the main fault until none is detectable (Figure 12–10a). Splays may develop as Riedel and anti-Riedel shears (Figure 12–6a) off a main fault, reducing the displacement.

In cross section, strike-slip faults frequently splay upward into outwardly branching segments that exhibit *flower, palm,* or *horse-tail* configurations (Figure 12–10b). These are more difficult to observe in the field unless a great deal of vertical relief exposes a cross section of the fault zone. Existence of flower structures has been confirmed and observed as offset rock units in seismic-reflection profiles. Some strike-slip motion involved

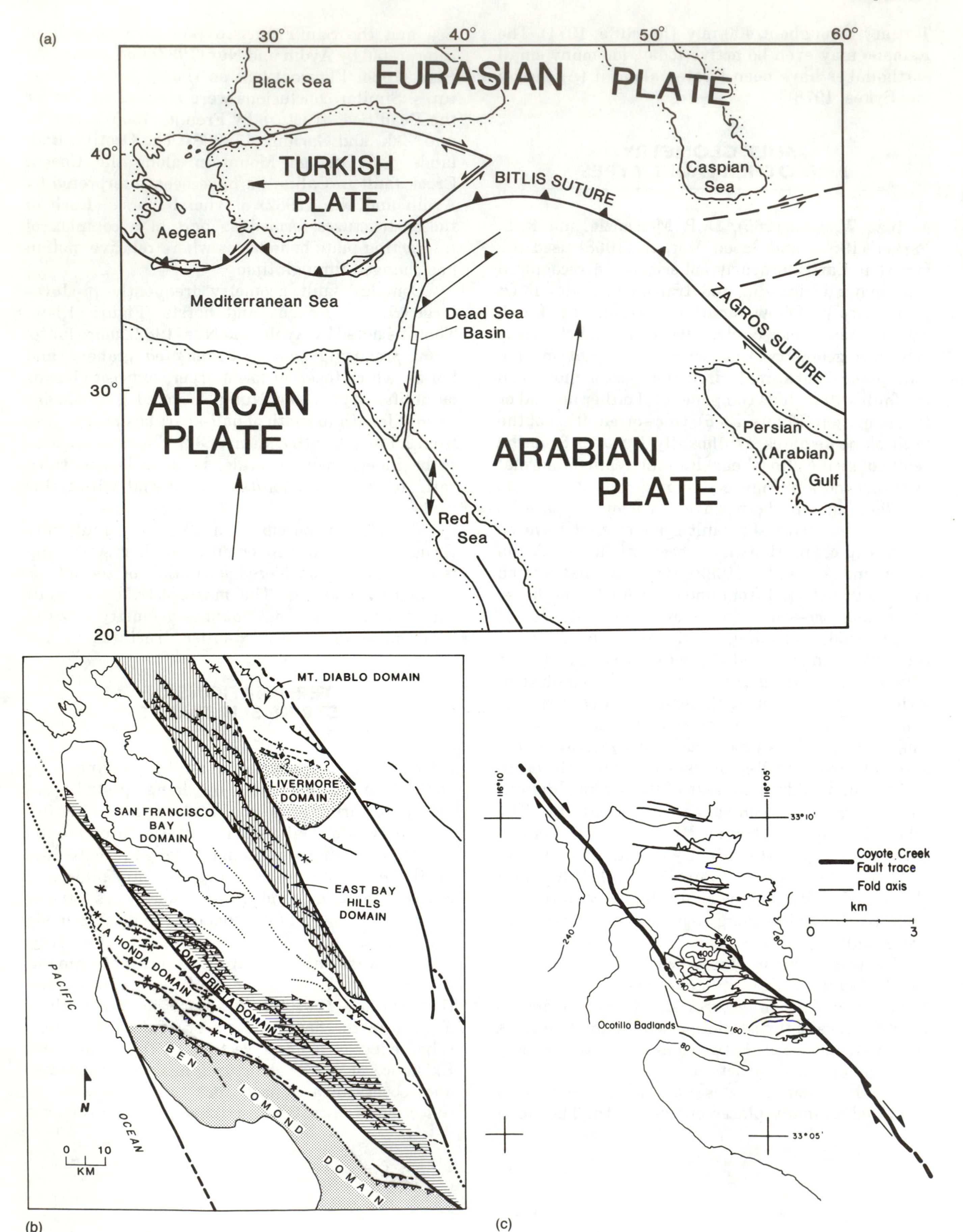
(a)
30°
40°
50°
60°
EURASIAN PLATE
Black Sea
TURKISH PLATE
BITLIS SUTURE
Caspian Sea
Aegean
Mediterranean Sea
Dead Sea Basin
ZAGROS SUTURE
AFRICAN PLATE
ARABIAN PLATE
Persian (Arabian) Gulf
Red Sea
20°
MT. DIABLO DOMAIN
LIVERMORE DOMAIN
SAN FRANCISCO BAY DOMAIN
EAST BAY HILLS DOMAIN
LA HONDA DOMAIN
LOMA PRIETA DOMAIN
BEN LOMOND DOMAIN
PACIFIC
OCEAN
N
0 10
KM
(b)
116°10'
116°05'
33°10'
33°05'
Coyote Creek Fault trace
Fold axis
km
0 3
Ocotillo Badlands
240
160
80
400
(c)

◊ **FIGURE 12–8**
(a) Arrangement of plate boundaries in part of the Middle East showing the Dead Sea rhomb-graben and transform fault system. (After M. R. Hempton, *Tectonics,* v. 6, 1987, copyright by the American Geophysical Union.) (b) Rhomb-graben fault system in the San Francisco Bay area. (After A. Aydin and B. M. Page, 1984, Geological Society of America *Bulletin,* v. 85.) (c) Rhomb horst along the Coyote Creek fault in California. (From R. V. Sharp and M. M. Clark, 1972, U.S. Geological Survey Professional Paper 787.)

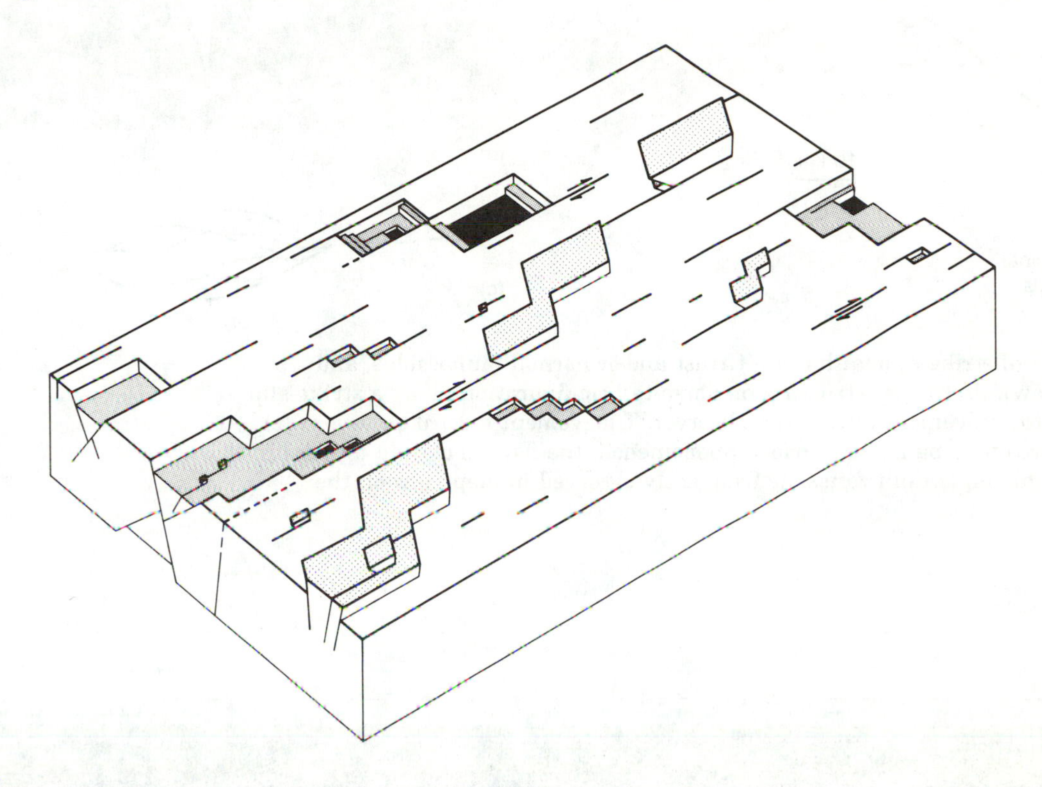

FIGURE 12–9
Complex rhomb-graben and -horst structure along a large fault system. (After Atilla Aydin and Amos Nur, *Tectonics,* v. 1, p. 91–105, 1982, copyright by the American Geophysical Union.)

in forming a flower structure must be resolved into dip-slip. Faults associated with the Paleozoic aulocogen of southern Oklahoma exhibit flower and *inverted rift structure* (Figure 12–11; Harding, 1974). It may be that flower structures open upward and then close again, just as anastomosing networks of fault splays sometimes develop along the surface trace of any fault type.

Resolution of strike-slip into significant compressional dip-slip motion is called *transpression,* which often results from oblique plate convergence where the dominant motion is strike-slip. Brian Harland (1971) was perhaps the first to describe transpression, in the course of his study of Caledonian structures in Spitzbergen. It has since been reported in many places, including the Mecca Hills near the Salton Sea in California (Sylvester and Smith, 1976). Transpression and the opposite, *transtension,* involve nothing more than complex oblique-slip motion resolved from dominant strike-slip. It is surprisingly common, for most faults exhibit one form or another of oblique-slip.

TRANSFORMS

We have discussed transform faults several times earlier (see Figures 1–11 and 9–9). Wilson (1966) first postulated their existence to compensate for differential motion between adjacent plates, whether at a spreading center (such as an oceanic ridge), or between segments of a plate moving past one another, or connecting segments in a subduction zone where some segments are consumed faster than others. Shortly after Wilson's prediction, studies by

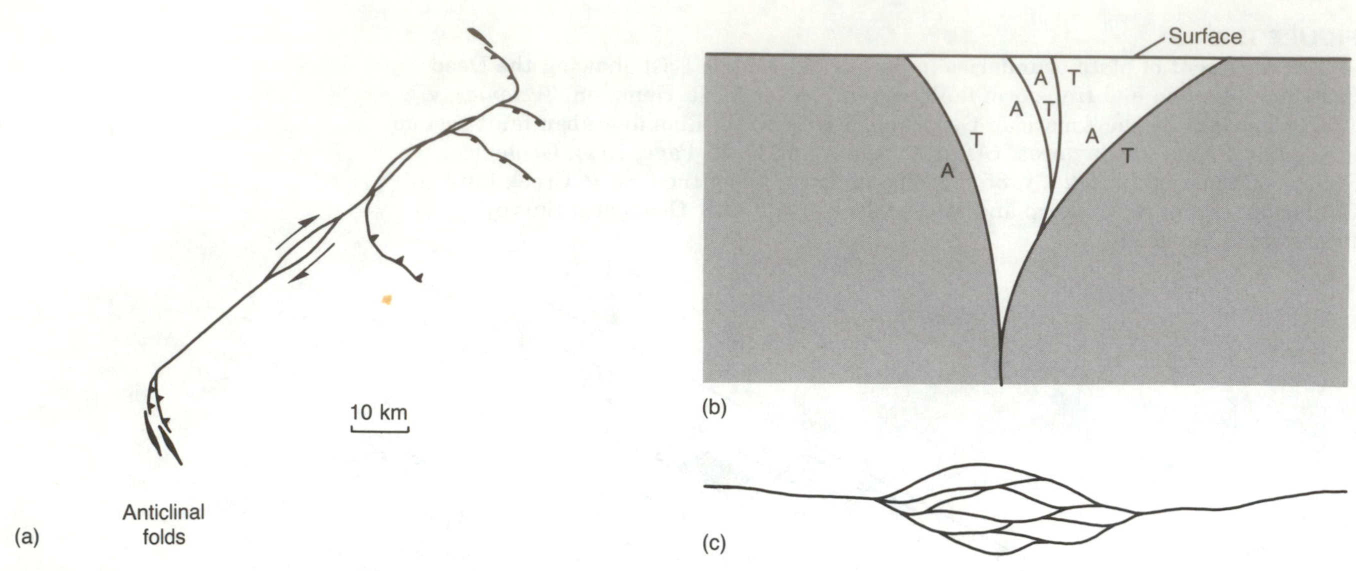

FIGURE 12–10
(a) Termination of strike-slip faults into thrust and/or normal faults, folds, and splays (map view). (b) Flower structure or horse-tail configuration along a strike-slip fault (A indicates movement away from observer, T movement toward observer). (c) Flower structures may be a near-surface phenomenon, the flowers closing upward in cross section into single fault zones, as frequently observed in map view on the surface.

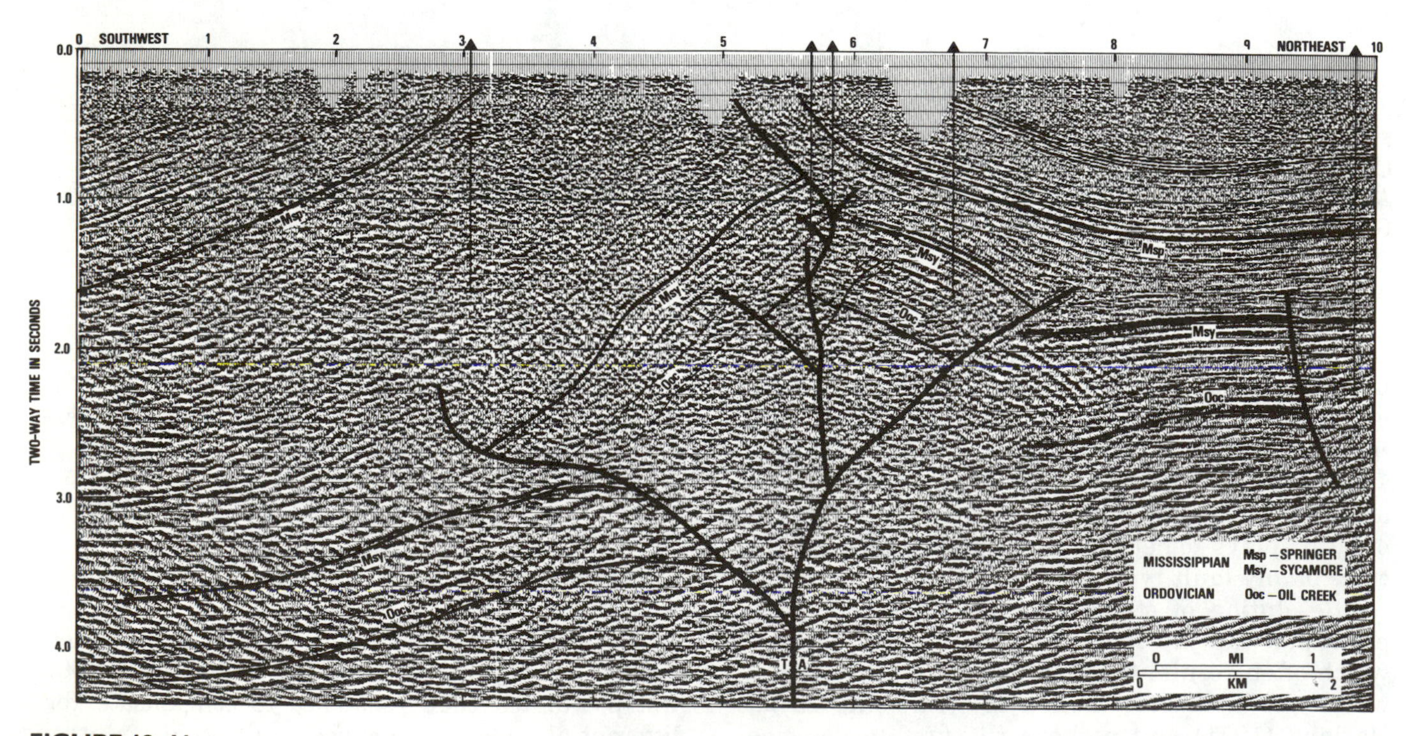

FIGURE 12–11
Structure section constructed from a seismic-reflection profile and drill data through the Ardmore basin in the Oklahoma aulocogen, illustrating flower and inverted- rift structures. Msp, Msy, and Ooc are Paleozoic rock units. (After T. P. Harding and J. D. Lowell, 1974, AAPG *Bulletin,* v. 58. Reprinted by permission of American Association of Petroleum Geologists.)

Lynn Sykes (1967) of earthquake focal mechanisms along the Mid-Atlantic Ridge verified the hypothesis. Several kinds of transforms have been described, among them ridge-ridge transforms, ridge-arc transforms, and arc-arc transforms (see Figure 9–9b). The classic ridge-ridge transform has an apparently simple strike-slip offset, but the real motion on the fault is opposite to the apparent offset sense.

One widely known transform is the San Andreas fault. It is a major *intracontinental transform,* connecting the ends of the East Pacific Ridge where it disappears beneath the North American continent in the Gulf of California and where it reappears off the coast of Oregon (Figure 12–12). As another example, Gregory A. Davis and B. Clark Burchfiel (1973) have suggested that the Garlock fault east of the San Andreas in southern California and westernmost Arizona is an intracontinental transform: they pointed out that two different tectonic styles are present on each side of its trace. Consider too, the Dead Sea fault in the Middle East, the Alpine fault in New Zealand, and the Great Glen fault in Scotland. All may be intracontinental transforms.

As we end our discussion of strike-slip faults, note that even the largest and most widely known examples still deserve intensive study, particularly because of the gradational nature of fault types.

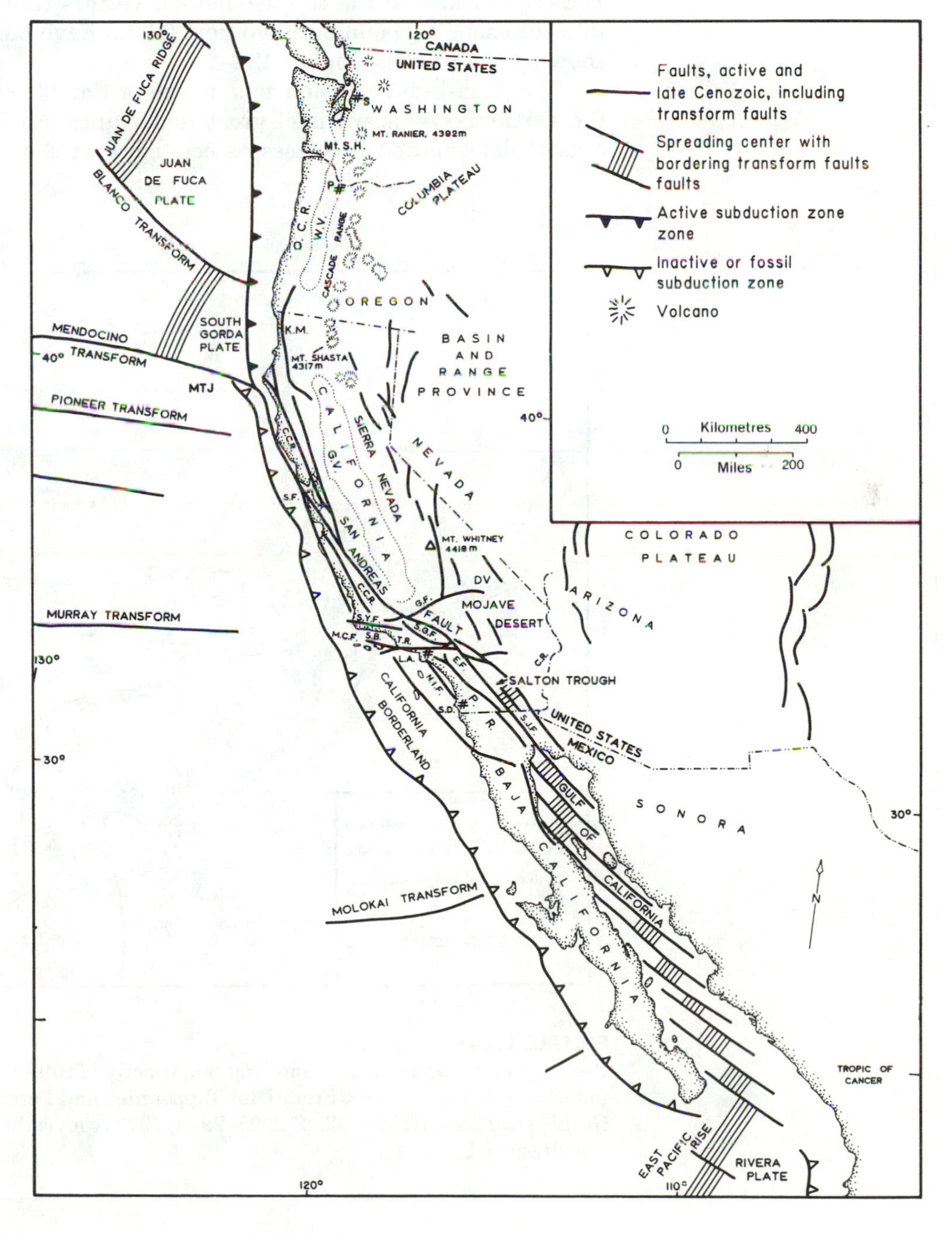

FIGURE 12–12
San Andreas and related fault systems in California and northern Mexico. CCR – California Coast Ranges; CR – Colorado River; DV – Death Valley; EF – Elsinore Fault; GF – Garlock Fault; GV – Great Valley; KM – Klamath Mountain; LA – Los Angeles; MCF – Malibu Coastal Fault System; MTJ – Mendocino Triple Junction; Mt. SH – Mount St. Helens; NIF – Newport-Inglewood Fault; OCR – Oregon Coast Ranges; P – Portland; PR – Peninsular Ranges; S – Seattle; SB – Santa Barbara; SD – San Diego; SF – San Francisco; SGF – San Gabriel Fault; SJF – San Jacinto Fault; SYF – Santa Ynez Fault; TR – Transverse Ranges; WV – Willamette Valley – Seattle Lowland. (After J. C. Crowell, 1987, *Episodes,* v. 110.)

ESSAY

Rigid Indenters and Escape Tectonics

A large system of intraplate continental strike-slip faults was first described by Peter Molnar and Paul Tapponnier (1975) in central Asia, where they found that known strike-slip faults interconnect (Figure 12E–1). As a result, Tapponnier suggested that India acted as a *rigid indenter* (Figure 12E–2) when it collided with Asia and that strike-slip deformation resulted when large parts of Asian crust moved laterally in response to the collision. The crust responded to the collision with the indenter by forming large strike-slip faults bounding blocks that moved laterally out of the collision zone. In Asia north of the indenter, dextral strike-slip faults ideally would have moved blocks westward from the west wide of the collision zone, and sinistral faults (again ideally) would have moved blocks eastward and out of the eastern part of the collision zone. Deviation of actual motion from ideality along the large faults was related to both the original shape of India as it collided with Asia (Figure 12E–1) and also to the relative-motion vectors that brought Asia and India into collision. Tapponnier and others (1982) have built accurate scale models showing the process (Figure 12E–2).

The Asia-India collision zone is one of Earth's most populous regions, so the earthquake hazard is of great importance. So is the incentive to study crustal deformation processes associated with the indenter mechanism and

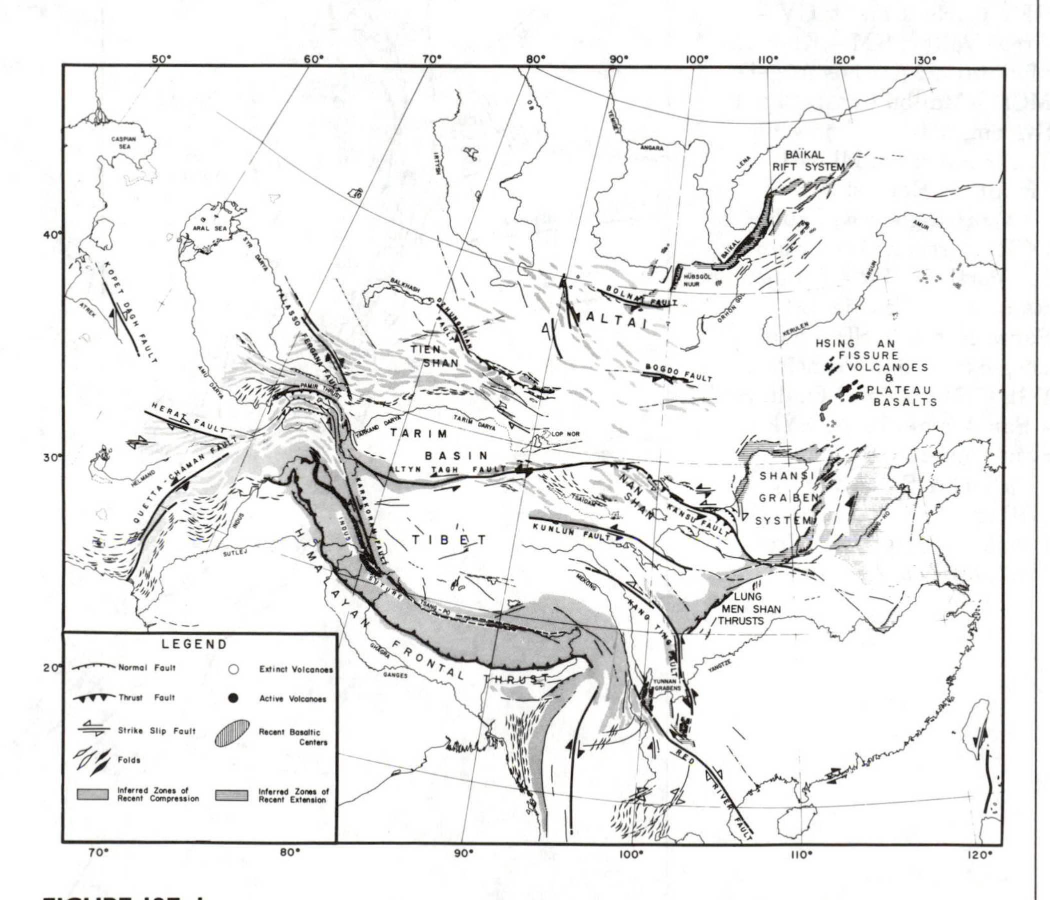

FIGURE 12E–1
Recent tectonic map of Asia showing major active faults (heavy lines). Arrows indicate sense of motion. (From Paul Tapponnier and Peter Molnar, *Journal of Geophysical Research,* v. 82, p. 2905–2930, 1977, copyright by the American Geophysical Union.)

collision-related faults (both large and small) and to apply the results to earthquake prediction.

Since the indenter mechanism was originally proposed, Brian Davies (1984) has applied it to explain the origin of the late Precambrian deformed belt in the Arabian shield, and Jean-Pierre LeFort (1984) has used it to explain the structure of large faults and the curvature of the central Appalachians.

(a)

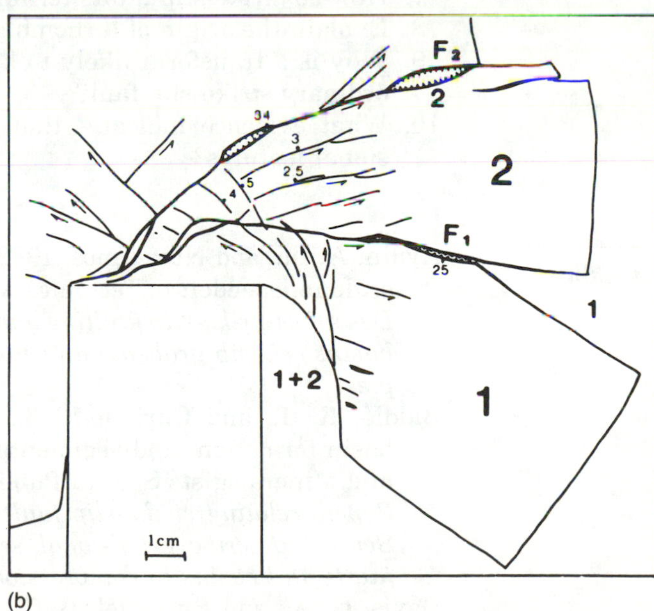

(b)

FIGURE 12E–2
Plasticine model of escape tectonics involving a rigid indenter and several laterally moving blocks. (a) Photo of one model. (b) Sketch of the deformation in the model. (From P. Tapponnier, G. Peltzer, A. Y. Le Dain, R. Armijo, and P. Cobbold, 1982, *Geology,* v. 10, p. 611–616.)

References Cited

Davies, Brian, 1984, Strain analysis of wrench faults and collision tectonics of the Arabian shield: Journal of Geology, v. 82, p. 37–53.

LeFort, J.-P., 1984, Mise en évidence d'une virgation carbonifère induite par la dorsale Reguibat (Mauritanie) dans les Appalaches du Sud (U.S.A.), Arguments géophysiques: Bulletin Societé Géologique de France, v. 26, p. 1293–1303.

Molnar, Peter, and Tapponnier, Paul, 1975, Cenozoic tectonics of Asia: Effects of a continental collision: Science, v. 189, p. 419–426.

Tapponnier, P., Peltzer, G., Le Dain, A. Y., Armijo, R., and Cobbold, P., 1982, Propagating extrusion tectonics in Asia: New insights from simple experiments with Plasticine: Geology, v. 10, p. 611–616.

Questions

1. Why do strike-slip faults characteristically dip steeply?
2. How can strike-slip, normal, and thrust faults—as well as folds—all form in the same stress system?
3. Why do some strike-slip faults contain cataclastic material along their extent (or are single planar structures) but others contain mylonite and consist of zones several kilometers wide?
4. Why do some strike-slip fault zones contain mylonite that is cut through by later cataclasite?
5. What evidence supports the statement that strike-slip faults are compressional structures?
6. Why must major strike-slip faults become some other type if they deviate from a small circle of the Earth?
7. How do strike-slip faults terminate?
8. Explain the origin of a rhomb graben.
9. Why is a transform likely to have a sense of motion opposite that of an ordinary strike-slip fault?
10. What evidence indicates that ridge-ridge transforms are not compressional features?

Further Reading

Aydin, Atilla, and Nur, Amos, 1982, Evolution of pull-apart basins and their scale independence: Tectonics, v. 1, p. 91–105.
Describes strike-slip faulting and its relationships to the origin of pull-apart basins (rhomb grabens) and uplift (rhomb horsts), with numerous examples.

Biddle, K. T., and Christie-Blick, Nicholas, 1985, Strike-slip deformation, basin formation, and sedimentation: Society of Economic Paleontologists and Mineralogists Special Publication 37, 386 p.
Papers relating strike-slip faulting and formation of sedimentary basins. Several describe extensional settings; others, development of strike-slip faults and basins in compressional settings.

Davis, G. A., and Burchfiel, B. C., 1973, Garlock fault: An intracontinental transform structure: Geological Society of America Bulletin: v. 84, p. 1407–1422.
The Garlock fault—part of the San Andreas system—is interpreted as an intracontinental transform, in contrast with the San Andreas, which has been interpreted as a ridge-ridge transform.

Sylvester, A. G., ed., 1984, Wrench fault tectonics: American Association of Petroleum Geologists Reprints Series 28, 374 p.

Contains several excellent papers informative for undergraduates. Several papers by J. C. Crowell, J. Tuzo Wilson's paper defining transforms, J. S. Tchalenko's paper on shear zones, and the comparative paper by R. E. Wilcox and others on wrench-fault tectonics all make good reading.

Sylvester, A. G., 1988, Strike-slip faults: Geological Society of America Bulletin, v. 100, p. 1666–1703.

An up-to-date review of the nature of strike-slip faults written for students, structural geologists, and geologists not conducting research on strike-slip faults.

Tapponnier, P., Peltzer, G., Le Dain, A. Y., Armijo, R., and Cobbold, P., 1982, Propagating extrusion tectonics in Asia: New insight from simple experiments with Plasticine: Geology, v. 10, p. 611–616.

Models the strike-slip tectonics of Asia, based on experimental work and the concept of a rigid indenter. The results are striking.

13

Normal Faults

The labors of Pennsylvania geologists have rendered so familiar the structure of the Appalachians, that it has been accepted as typical of all mountains, and a comparison will facilitate an understanding of the basin ranges. Indeed, I entered the field with the expectation of finding in the ridges of Nevada a like structure, and it was only with the accumulation of difficulties that I reluctantly abandoned the idea.

GROVE KARL GILBERT, *Explorations and Surveys West of the One Hundredth Meridian*

G. K. GILBERT'S RECOGNITION OF THE BLOCK-FAULTED STYLE OF the Basin and Range Province was one of the great discoveries in nineteenth-century geology. Though chief geologist of the Wheeler reconnaissance, Gilbert was under the command of an Army officer and had only three field seasons to explore the geology of large parts of Utah and Nevada. He was also conditioned to expect Appalachian geology in the Far West. Only his hard work, open mind, and careful observation led to the discoveries that provide some of the striking examples we will discuss in this chapter.

Normal faults are dip-slip faults in which the hanging wall has moved down relative to the footwall (Figure 13–1). In many parts of the world they are important as structural traps for hydrocarbon accumulations, and in the Houston area in the Texas Gulf Coast they cause subsidence problems, affecting roads and buildings. (Excessive withdrawal of ground water from regional aquifers contributes to the subsidence.) Normal faults are also the primary brittle mechanism for extending and thinning the crust before the opening of a new ocean basin. Only now are we beginning to understand the similarities and differences of extensional processes that result in formation of symmetrical rifts (like the Red Sea and the Gulf of California) and the ones that produce asymmetrical extension (like the Basin and Range province). These, too, involve normal faulting.

Normal faults have been called *gravity faults,* implying that the primary motive force is gravity. They have also been called *extensional faults* because they extend layering, or the crust. ***Listric faults*** (Figure 13–2) have concave-up surfaces, flatten with depth, and steepen toward the surface. They may either be normal or thrust faults, as "listric" describes the geometry and not the sense of motion. ***Growth faults*** involve simultaneous deposition and fault motion. Many growth faults are listric normal faults.

PROPERTIES AND GEOMETRY

Normal faults have moderate to relatively steep angles of dip and have long been thought of as propagating directly into crystalline basement beneath the sedimentary cover and terminating deep in the crust without change in geometry. With the accumulation of better field and geophysical data, it has been demonstrated that many normal faults flatten listrically into incompetent units at depth and display many of the detachment properties of thrust faults. Others pass from the sedimentary cover into the crystalline basement. Modern seismic-reflection profiles (Figure 13–3), coupled with surface studies, suggest that many flatten at

FIGURE 13–1
Normal faults in Pleistocene clay in northeastern Washington, rotated by landsliding and later overlapped by terrace deposits. (F. D. Jones, U.S. Geological Survey.)

great depth into the ductile-brittle transition (Figure 13–4). The geometry and evolution of normal-fault systems has been described in the Gulf Coast region by Ernst Cloos (1968), who modeled them in clay-block experiments. These experiments with models demonstrated the concave-up listric shape in the vertical dimension. Cloos also successfully modeled the map pattern of normal faults in Texas and Louisiana, where they are concave toward the Gulf of Mexico.

Normal faults also show many of the branching characteristics of thrust and strike-slip faults (Figure 13–5). ***Splays*** occur along normal faults. ***Synthetic*** faults dip in the same direction as the master fault and join the master fault at depth (Figure 13–5b). ***Antithetic*** faults (Figure 13–5b) likewise join the master fault at depth, but dip in the opposite direction.

Paired normal faults exist where related blocks flanked by parallel faults dip either away from the blocks or toward them (Figure 13–6). A ***graben*** involves two normal faults that dip toward each other, with a down-dropped block in between. A ***half-graben*** is a block bounded by a normal fault on one side; on the other side it passes into a gentle fold. A ***horst*** is a relatively upraised block flanked by parallel normal faults that dip away from the horst. The mountains and alluvium-filled valleys of the Basin and Range Province, which extends from southern Idaho into Mexico, may seem a classic graben-and-horst region, with the ranges as horsts and valleys as grabens, but crustal extension over geologic time has created a geometry much more complex. The Rhine graben region of Germany and Switzerland is dominated by a single large down-dropped crustal block, but within it are many smaller horsts and grabens.

Normal faults may also be related to folding. Some normal faults at depth pass into monoclinal

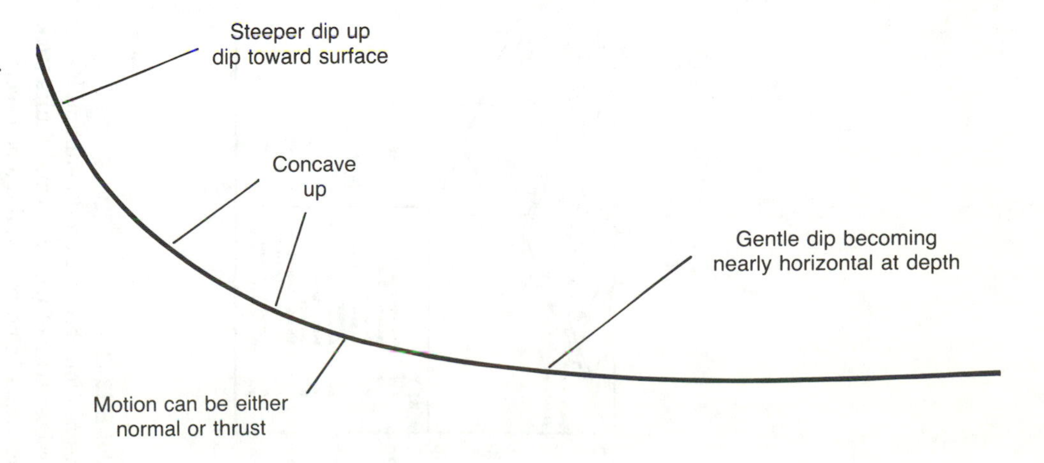

FIGURE 13–2
Characteristics of listric faults. Note that they apply to either normal or thrust faults.

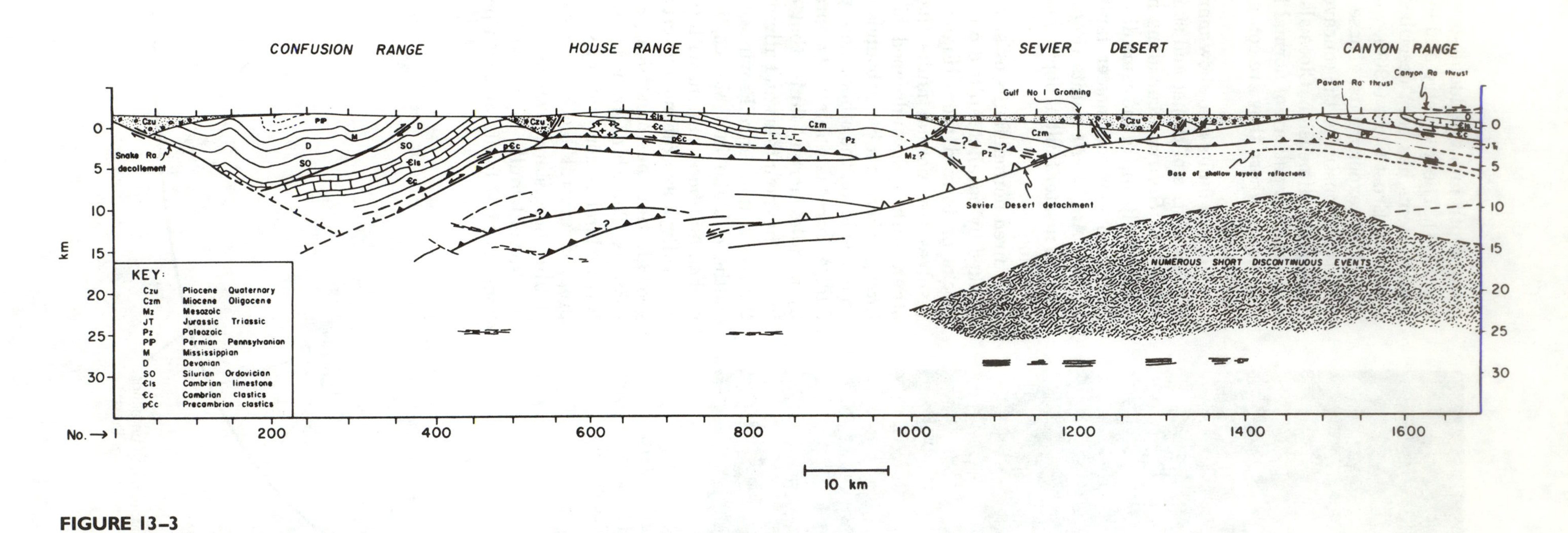

FIGURE 13–3
Cross section through the Basin and Range province in Utah showing flattening characteristics of listric normal faults (ticks), thrusts (solid teeth), and older thrusts reactivated as normal faults (open teeth with ticks). Section is based on the COCORP Utah seismic-reflection line. (From R. W. Allmendinger and others, 1983, *Geology.*) This section is derived from the seismic-reflection profile shown in Figure 20–15.

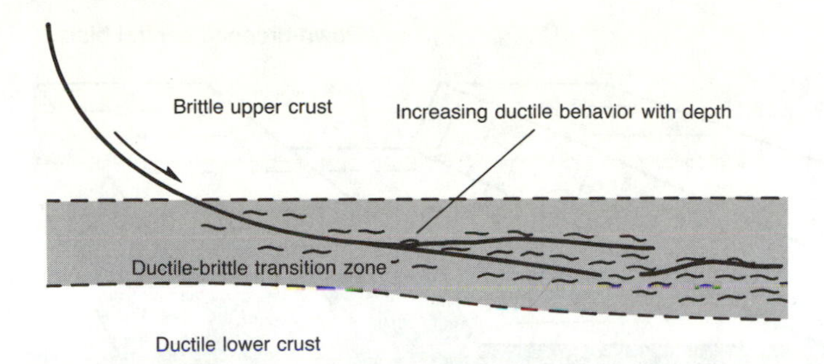

FIGURE 13–4
Listric normal fault flattens into the ductile-brittle transition.

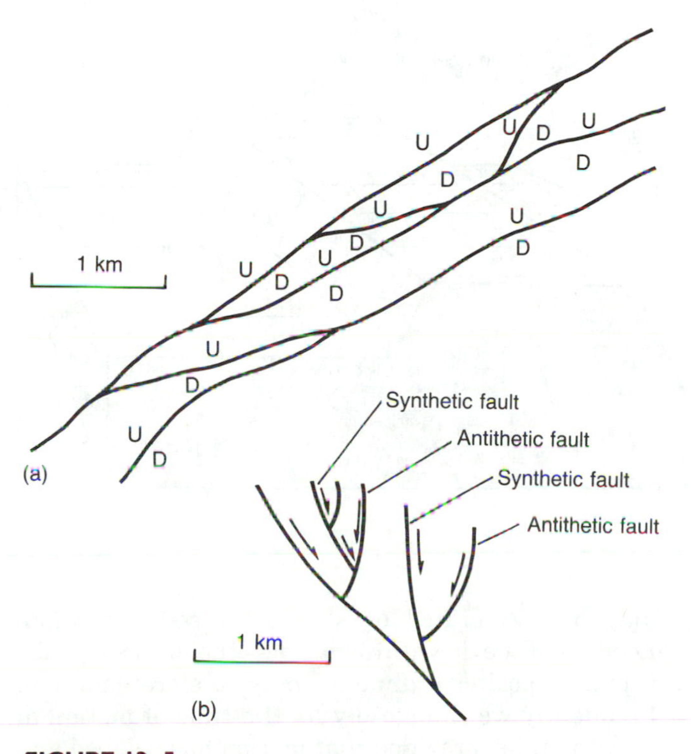

FIGURE 13–5
Branching characteristics of normal faults. (a) Anastomosing splays: U—upthrown side; D—downthrown side (map view). (b) Synthetic and antithetic faults (cross section).

folds near the surface or along strike (Figure 13–7). The shape of the fold is frequently related to the shape of the fault surface. Good examples occur in the Grand Canyon region, where normal faults pass into monoclines along strike.

A mass of basement rocks broken by normal faulting may only partly break the sedimentary cover, producing a ***drape fold*** in the sedimentary rocks. Drape folds also form in association with steeply dipping basement thrusts, as in the Rocky Mountain Front Ranges of Colorado and Wyoming (see Figure 11–27).

Drag folds (Chapter 9) form because of friction along the fault surface and occur along normal faults; they provide evidence of motion sense on a fault through observation of the sense of shear of the folds along the fault. ***Reverse drag folds*** and *roll-over anticlines* form along growth faults where the part of the downthrown block close to the fault is displaced downward more than the parts farther away (see Figure 9–13). These structures are related to the listric shape of the faults on which they commonly form. Rollover anticlines are targets for petroleum exploration and in the Gulf Coast have proved to be highly productive.

ENVIRONMENTS AND MECHANICS

Normal-fault zones are commonly described as brittle structures that develop only in the upper crust. Those formed near the surface commonly have sharp contacts, cataclastic zones, breccias, and rotated blocks (or horses) characteristic of brittle fault zones; if formed deep in the crust, they exhibit ductile features, notably mylonites. The large-detachment normal faults produced by crustal extension, as in the Basin and Range Province, probably formed in the ductile-brittle transition—in the middle to lower crust. They are characterized by the same features we see along faults formed by compression, including mylonite and related ductile structures.

The extensional character of normal faulting may be examined with aid of the strain ellipsoid (Figure 13–8). Ideally, normal faults dip more than 50° and form on shear planes that are bisected by a vertical (Z) axis, with the maximum principal strain (X) horizontal. The Y axis lies in the fault plane. The stress ellipsoid involves a vertical maximum principal stress (σ_1) and a horizontal σ_3, with σ_2 lying within the fault plane. In addition, the rules for fluid pressure, buoyancy, and reduction of frictional resistance (Chapter 10) apply to normal faulting, just as to other fault and fracture types.

Upward propagation of a normal fault—as argued by Neville Price (1977)—may result from hydrofracturing by overpressured fluid, generally water. Once pore-water pressure exceeds the tensile strength of a potential glide zone within a gently dipping part of a listric fault zone, hydraulic fracturing occurs. Frictional resistance to movement falls to zero, and the fault block tends to move, causing the fracture to propagate upward—still under high water pressure—until the entire block moves along the fault. The fault will steepen as the

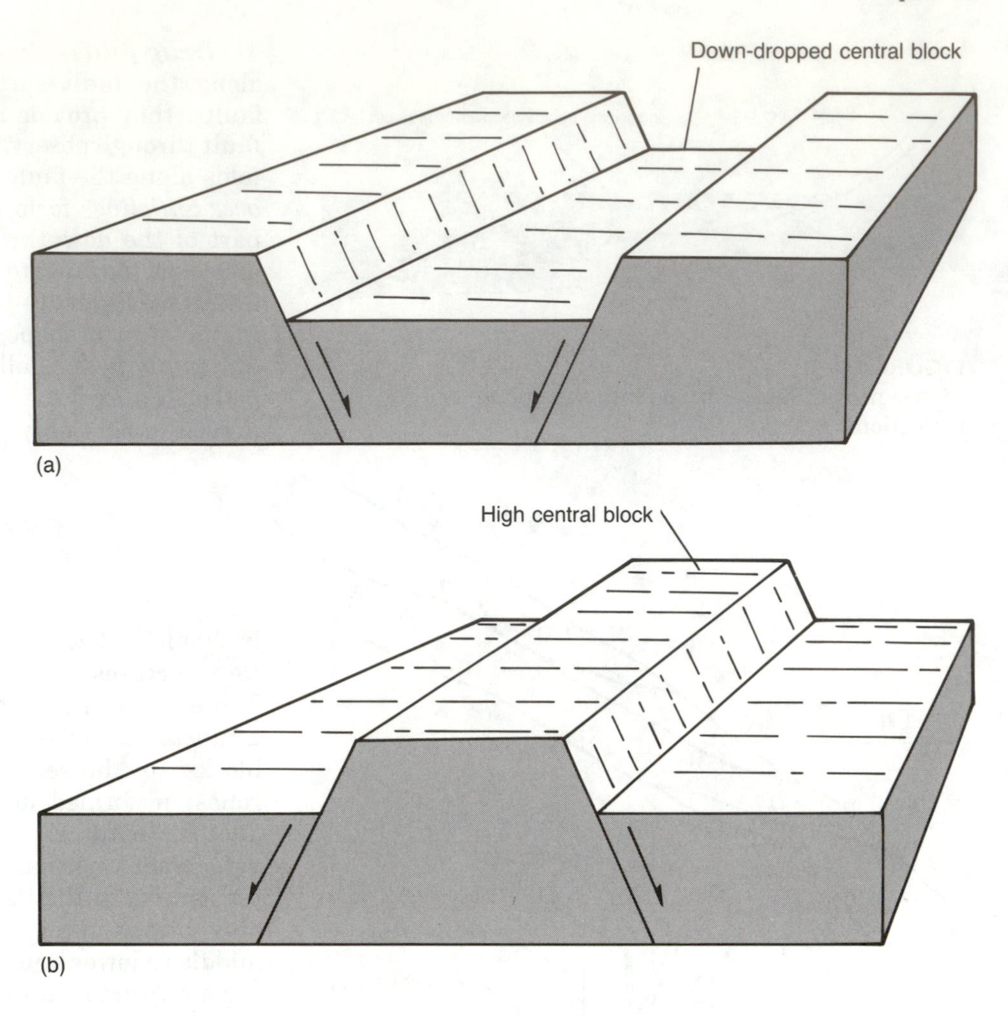

FIGURE 13–6
Graben (a) and horst (b) structures.

block tends to move downslope, forming a steeply dipping segment that propagates through the upper part of the block and permits it to move as a coherent unit. The block thus formed is bound by an ideal listric surface or is a rotational slide block.

Growth Faults

Growth faults commonly form in relatively unconsolidated sediments during deposition and thicken stratigraphic units in the downthrown block (Figure 13–9). They may be either normal or thrust faults, but the former are probably more common. The best-documented growth faults in North America (possibly the world) occur in the Gulf Coast region. There drilling has revealed many large listric normal growth faults, some with vertical displacements of a kilometer or more. They have also been observed in the Niger delta, the Bay of Biscay, and in other basin margins. High-quality seismic-reflection data have provided additional documentation of fault geometry and properties (Figure 13–10).

Listric normal growth faults are among the best examples of gravity faults because they can move only under the influence of gravity. Their movement may be accelerated or slowed by sedimentation. Loading of the downthrown side—the depocenter for rapidly deposited sediment—may accelerate motion. Frequently we can closely fix the time of motion of growth faults, provided that motion has stopped and that sediment has been deposited across the fault. Changes in relative thickness across the fault are therefore keys to timing of motion. The situation is unique because material is transported from one fault block to the other both during and after movement on the fault. Hong-Bin Xiao and John Suppe (1986) have presented evidence suggesting that the curvature of listric normal growth faults is related to compaction of unconsolidated sediments after deposition on the downthrown blocks of growth faults.

In the Gulf Coast, overall movement along growth faults is toward the Gulf of Mexico basin. As a result, the faults tend to form arcuate, steplike patterns roughly concentric to the Mississippi delta and the Gulf Basin (Figure 13–11; Cloos, 1968). The term ***down-to-basin faults*** is frequently used because the downthrown side is always toward the Gulf Basin. These faults are also classic listric faults, flattening downward into a detachment in

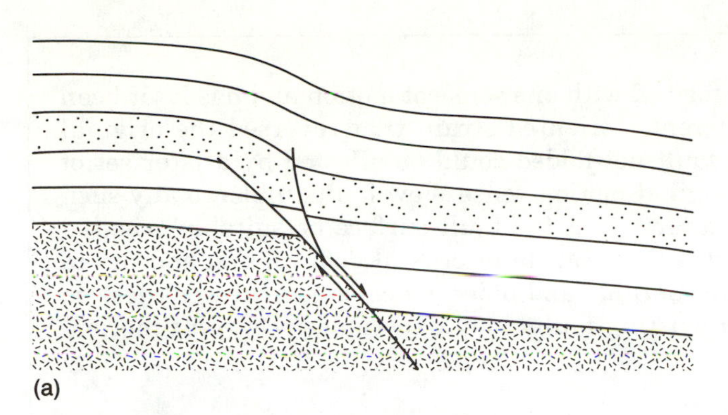

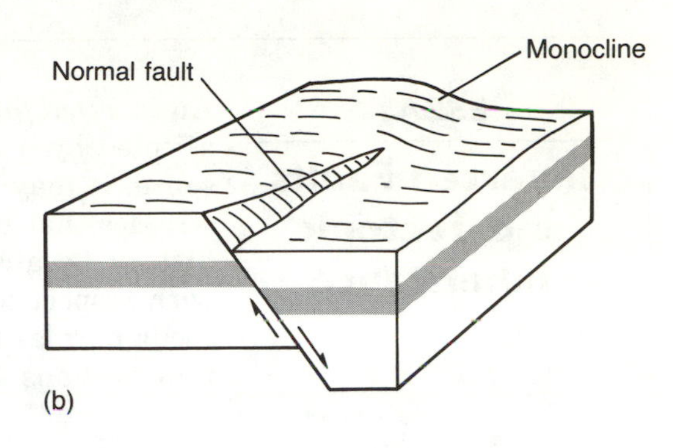

FIGURE 13–7
(a) Up-dip termination of a normal fault into a monoclinal drape fold. (b) Monocline grading along strike into a normal fault.

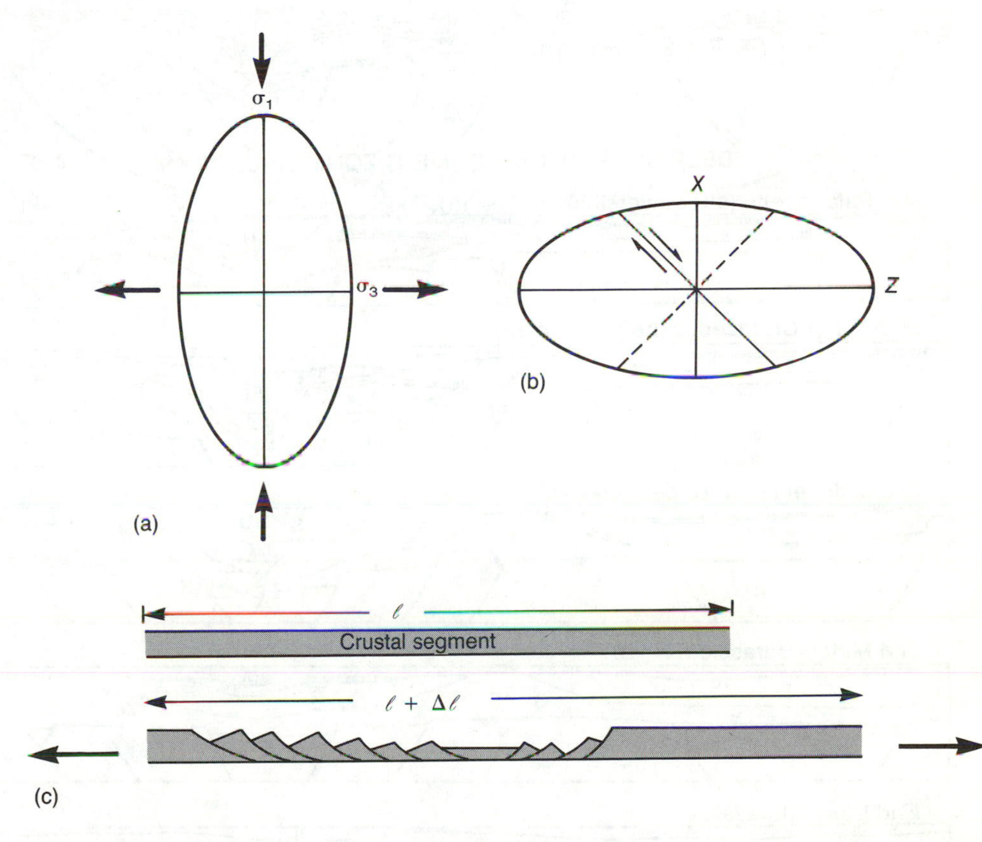

FIGURE 13–8
Stress (a) and strain (b) ellipsoids for normal faulting. (c) Sequential sections of the crust before and after normal faulting and extension, showing displacement taking place primarily as horizontal extension. A crustal segment of original length ℓ is extended to a new length $\ell + \Delta\ell$; the extended part is also thinned.

the sedimentary section. In the Gulf Coast, the detachments may be either weak "geopressured" shales (actually unconsolidated shales—muds—containing excess trapped pore water) or they may be in the extensive Jurassic salt unit. They may be analogous to rotational landslides and slump structures. Compression and thrusting occurs at the toes of these structures (Figure 13–12) because of the change during movement from extension at the rear to compression at the toe. (You may recall from Chapter 11 the arguments for and against gravity as the primary motive force for large-scale thrusting.) The characteristics of slump structures and down-to-basin faults must be found in large thrusts before we can argue convincingly that gravity was responsible for forming them.

Growth faults are aseismic. Areas where active growth faults are known to occur are commonly areas where no seismicity has been associated with faulting. Continuous and very recent movement on growth faults has been demonstrated in the Texas Gulf Coast, seismically one of the least active regions in North America. High fluid pressure in the fault blocks and the relatively unconsolidated nature of the sediments induces a state of continuous creep, and the blocks move aseismically toward the Gulf of Mexico.

Rift Zones

Rifts are narrow linear zones where the crust has been pulled apart, producing grabens, half-grabens,

ESSAY

Inverted Faults and Tectonic Inheritance

An *inverted fault structure* formed with one sense of motion and has later been affected by a new and differently oriented strain that reversed the original sense. It may be that any fault not folded could be affected by a later set of stresses that reverse the original motion sense. Few faults undergo any such history, because reversal requires that the fault surface be suitably oriented with respect to the new stress field. But later crustal deformation appears to occur normal to continental margins and older zones of crustal weakness so that perhaps 40 percent of existing faults may experience recurrent motion.

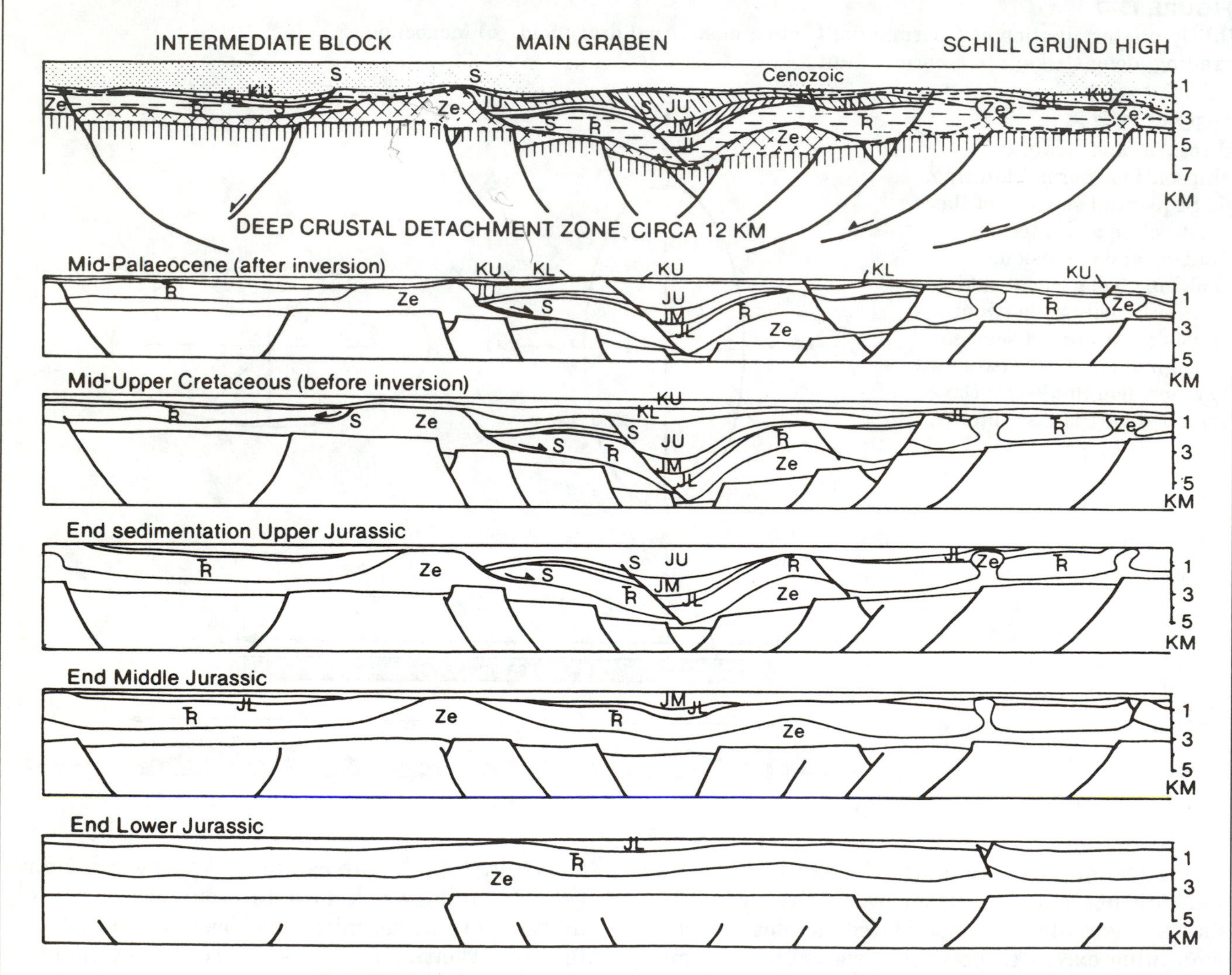

FIGURE 13E–1
Balanced regional cross section through part of the North Sea with successive back-stripping of sediments to the end of Early Jurassic (JL) time showing older faults that remain beneath the Zechstein (Ze) evaporite sequence. These faults were reactivated later in the Mesozoic and the Tertiary as the Viking graben formed. Ŧ —Triassic; K—Cretaceous; L—lower; M—middle; U—upper; S—fault action during sedimentation. (Reprinted with permission from *Journal of Structural Geology*, v. 5, A. D. Gibbs, Balanced cross-sections from seismic sections in areas of extensional tectonics, copyright 1983, Pergamon Journals Ltd.)

Consequently, the tendency for reactivation of old faults must be related to the alignment of newly formed plate boundaries in a later tectonic cycle.

Some geologists have speculated that the crust may undergo later extension and normal faulting in a direction exactly opposite to the direction of compression that earlier produced thrusting. One example of such a reversal is in the Viking graben, in the North Sea (Figure 13E–1). The Viking graben is a complex of extensional structures formed during the late Mesozoic and early

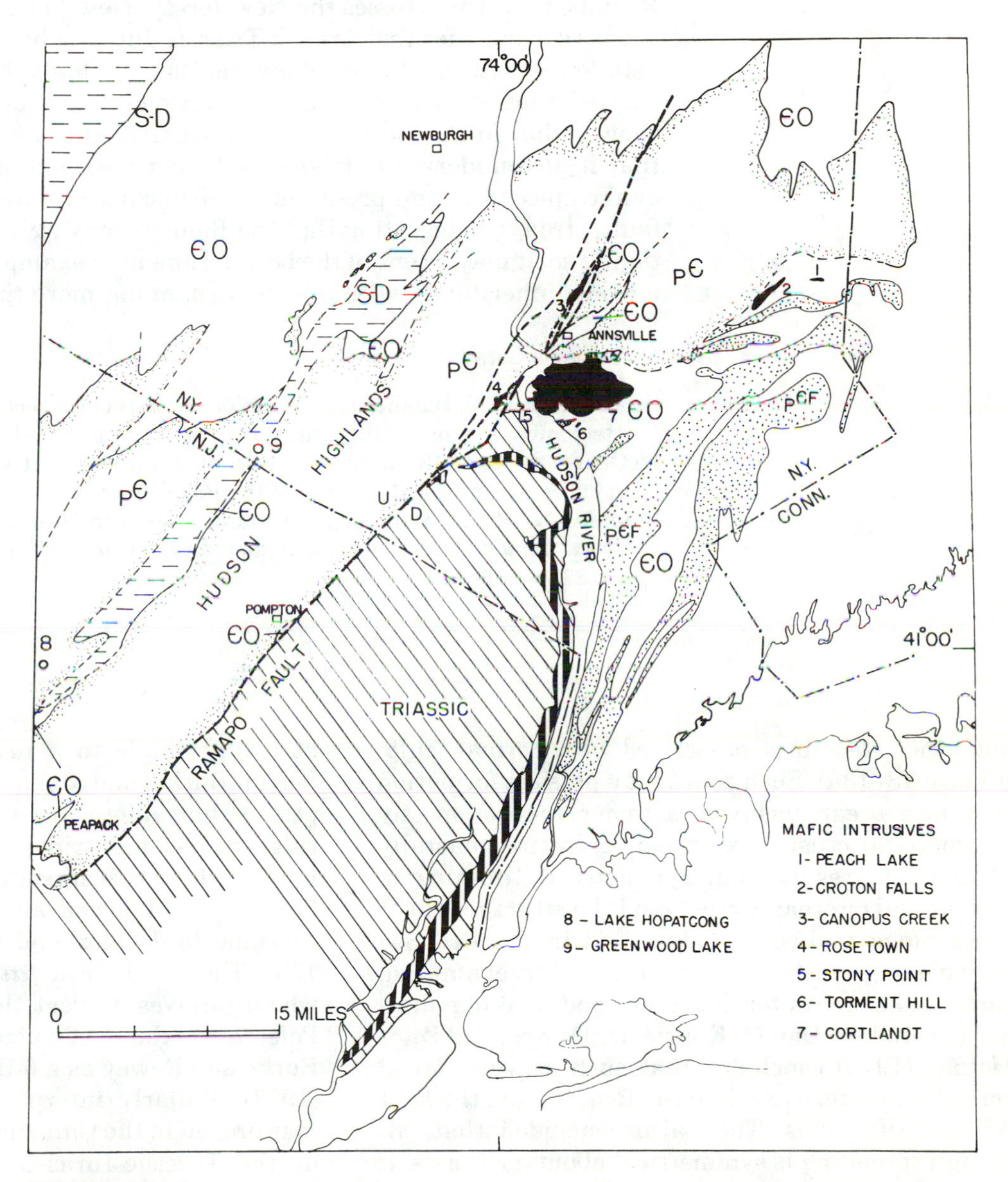

FIGURE 13E–2

Geologic map of part of northern New Jersey and southern New York showing the northern end of the Newark Triassic-Jurassic basin (striped) and the Ramapo fault and its extension northeast into the Hudson Highlands. p€ and p€f—Precambrian rocks; €O—Cambrian and Ordovician rocks; S-D—Silurian-Devonian. (From N. M. Ratcliffe, 1971, Geological Society of America *Bulletin,* v. 82.)

Cenozoic (Gibbs, 1983). Where high-quality crustal seismic-reflection data have been acquired, the later normal faults appear to have reactivated older Paleozoic thrusts originally formed during the Caledonian orogeny. Some thrusts had been reactivated earlier, during the Devonian, and formed the extensional Old Red Sandstone basins (McGeary, 1987).

Most of the exposed Triassic-Jurassic basins of the East Coast of the United States trend parallel to the strikes of major Appalachian faults, and most of the faults are compressional. Some basins are localized along the outcrop traces of older faults, indicating a further genetic connection. The Ramapo fault that crosses the New Jersey–New York border is the northwestern border fault for the Newark Triassic-Jurassic basin (Figure 13E–2). After detailed structural studies along the Ramapo fault, Nicholas Ratcliffe (1971) showed that it formed as a compressional fault during the Paleozoic. He showed that in the Mesozoic it was reactivated as an extensional feature and that it also underwent strike-slip during the same period. Historical earthquake epicenters and present-day seismic activity also are located along the fault, strongly suggesting that the Ramapo may still be undergoing reactivation. If so, it may be one of the best-documented examples of both inversion and tectonic inheritance, with a history spanning more than 350 m.y.

References Cited

Gibbs, A. D., 1983, Balanced cross-sections from seismic sections in areas of extensional tectonics: Journal of Structural Geology, v. 5, p. 153–160.

McGeary, S., 1987, Nontypical BIRPS on the margin of the North Sea: The SHET survey: Geophysical Journal of the Royal Astronomical Society, v. 89, p. 231–238.

Ratcliffe, N. M., 1971, The Ramapo fault system in New York and adjacent northern New Jersey: A case of tectonic heredity: Geological Society of America Bulletin, v. 82, p. 125–142.

and other structures associated with normal faults (Figure 13–13a). Such zones may presage formation of a new ocean basin or a major zone of mostly symmetrical crustal extension. They are generally assumed to result from symmetrical thinning of continental or oceanic crust, and the structure of rift zones along mid-ocean ridges and in the Red Sea region supports the assumption. After examining earthquake epicenter locations and making first-motion studies, Dan McKenzie, D. Davies, and Peter Molnar (1970) concluded that such a symmetrical spreading pattern exists in the Red Sea and the East African rift zones. They also concluded that, although spreading is symmetrical about a rift axis, it may decrease to zero along the axis at a pole of rotation. These ideas have since become known as the "McKenzie model" of rift formation.

Aulocogens are tectonic troughs, bounded by normal faults and formed within continental crust at a high angle to a nearby continental margin. Kevin Burke and John Dewey (1973) have suggested that aulocogens result from formation of a ridge-ridge-ridge triple junction above a mantle plume, where two spreading ridges have developed, but the third forms an initial rift and does not continue to develop and spread apart (Figure 13–13b). The result is a ***failed rift,*** or ***failed arm,*** which receives a great thickness of sediment. The Paleozoic Oklahoma aulocogen was interpreted by Burke and Dewey as a failed rift, and Paul Hoffman (1973) similarly interpreted the Proterozoic Wopmay orogen in the Canadian shield. Rift basins, such as the Triassic-Jurassic basins of eastern North America, may form as marginal rifts before the opening of an ocean. Formation of the basins just mentioned also involved a major strike-slip component (Swanson, 1982). In contrast, the Gregory Rift of the East African system has been shown to be an asymmetric rift structure (Bosworth and others, 1986); there, the master fault occupies the east side of the rift and is down toward the west for more than 100 km (Figure 13–13c), then changes to the west side and becomes down to the east for more than 100 km north and south.

Regional Crustal Extension

Large-scale extension of continental crust occurred in the Basin and Range province during middle to late Tertiary time (Eaton, 1979; Coney, 1980; Wernicke and Burchfiel, 1982; Wernicke, 1985). The crust there has been thinned by as much as 25 percent vertically during lateral extension of more than 100 percent. Gently dipping faults now exposed at the surface were at first interpreted as thrusts. Thrusts do exist here, but many of these faults have been reinterpreted as erosionally exhumed flat segments of very large listric normal faults on a crustal scale. They seem to flatten from steeply dipping upper crustal segments into the ductile-brittle transition during crustal spreading and thinning. Uplift and erosion expose old detachments at the surface. Because of a constant rate of heat flow, new levels of subhorizontal detachment form deep in the crust as the level of crustal extension and detachment within the ductile-brittle transition remains constant. Uplift and erosion cause existing detachments to become inactive, and, as extension continues, new detachments form beneath them (Eaton, 1979). In the Basin and Range, many older detachment zones are now exposed at the surface, cut by younger active normal faults—faults that presumably flatten at depth into the present ductile-brittle transition.

Several models were proposed during the 1970s and 1980s to explain regional crustal extension (Figure 13–14). All are based on the fact that many low-angle fault surfaces in the Basin and Range are erosionally exhumed normal faults. Models have been developed involving one of these: (1) Symmetrical extension and pure shear, suggested by John Proffett (1977), Gordon Eaton (1979), and Warren Hamilton (1982). (2) Asymmetric extension involving simple shear and a master detachment cutting through the entire lithosphere (Wernicke, 1985). (3) Asymmetric extension and simple shear involving delamination of the lithosphere (Lister, Etheridge, and Symonds, 1986). The symmetrical-extension model explains thinning of the crust but not the asymmetric structure of several extensional regions, like the Basin and Range; this model was undoubtedly derived after consideration of symmetrical rifts such as the Red Sea and the mid-ocean ridges.

Extensional detachments are intriguing structures. Any that can be closely examined in the Basin and Range (Figure 13–15) consist of a lower mylonitic zone overlain by a series of higher-level brittle faults that grade downward into the mylonitic zone (Coney, 1980; G. H. Davis, 1980). The Whipple Mountains detachment in southeastern California (Figure 13–16) is an excellent example: in many places a moderately to steeply dipping structure clearly has been truncated by the underlying detachment (G. A. Davis and others, 1980; G. A. Davis and Lister, 1988).

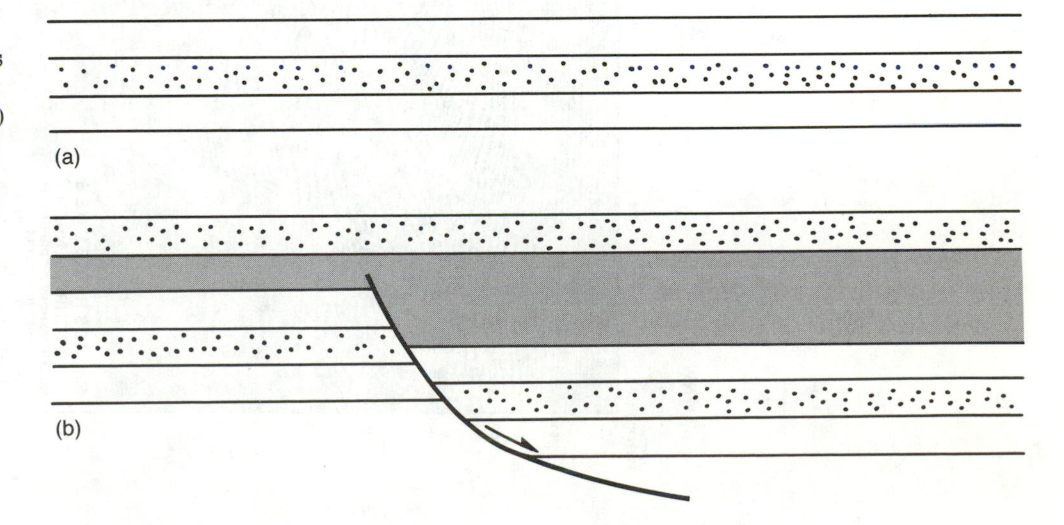

FIGURE 13–9
Geometry and characteristics of growth faults. (a) Uniform deposition before faulting. (b) Deposition continues during faulting, with excessive accumulation of sediment on the downthrown block of the fault. Movement ceases, but deposition continues on both sides of the fault.

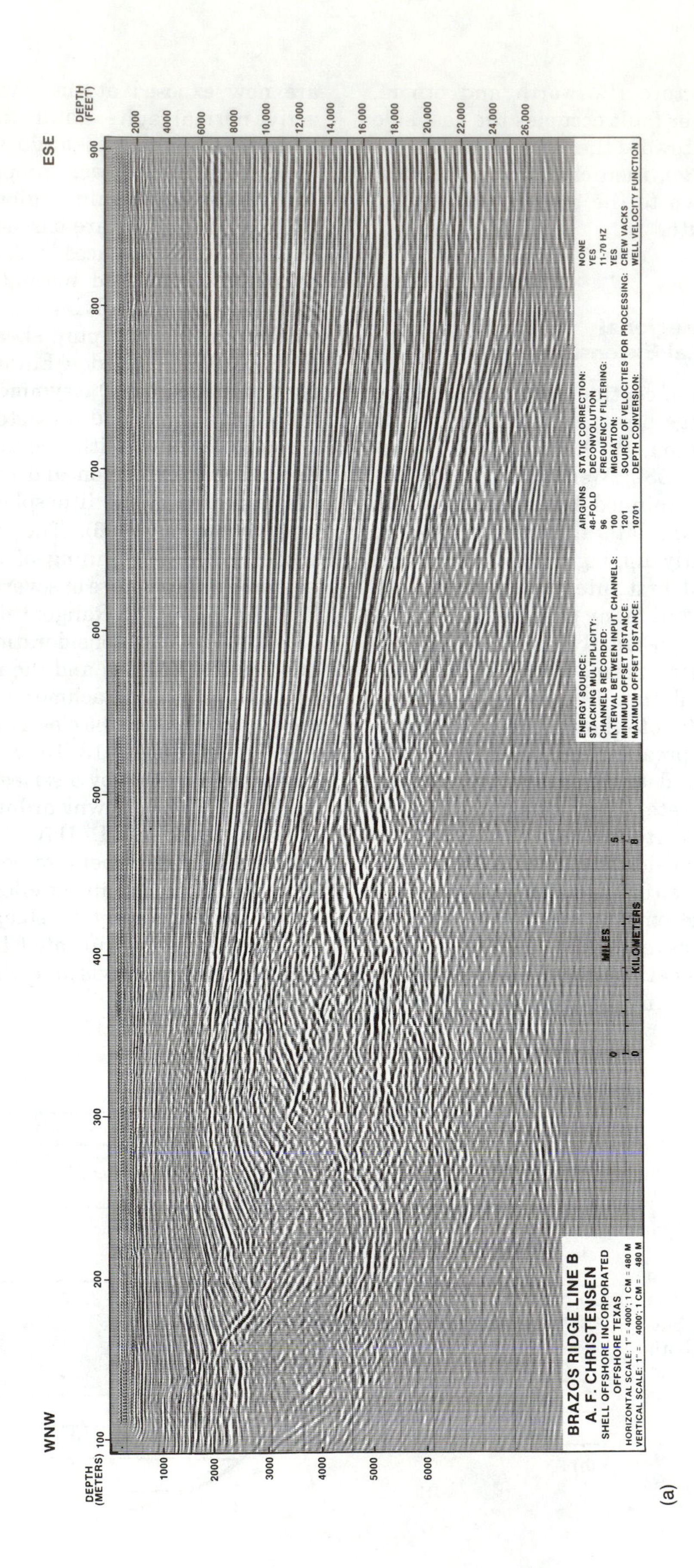
WNW
ESE
DEPTH (METERS)
1000
2000
3000
4000
5000
6000
DEPTH (FEET)
2000
4000
6000
8000
10,000
12,000
14,000
16,000
18,000
20,000
22,000
24,000
26,000
100
200
300
400
500
600
700
800
900
BRAZOS RIDGE LINE B
A. F. CHRISTENSEN
SHELL OFFSHORE INCORPORATED
OFFSHORE TEXAS
HORIZONTAL SCALE: 1" = 4000'; 1 CM = 480 M
VERTICAL SCALE: 1" = 4000'; 1 CM = 480 M
MILES
0
5
KILOMETERS
0
8
ENERGY SOURCE: AIRGUNS
STACKING MULTIPLICITY: 48-FOLD
CHANNELS RECORDED: 96
INTERVAL BETWEEN INPUT CHANNELS: 100
MINIMUM OFFSET DISTANCE: 1201
MAXIMUM OFFSET DISTANCE: 10701
STATIC CORRECTION: NONE
DECONVOLUTION YES
FREQUENCY FILTERING: 11-70 HZ
MIGRATION: YES
SOURCE OF VELOCITIES FOR PROCESSING: CREW VACKS
DEPTH CONVERSION: WELL VELOCITY FUNCTION

(a)

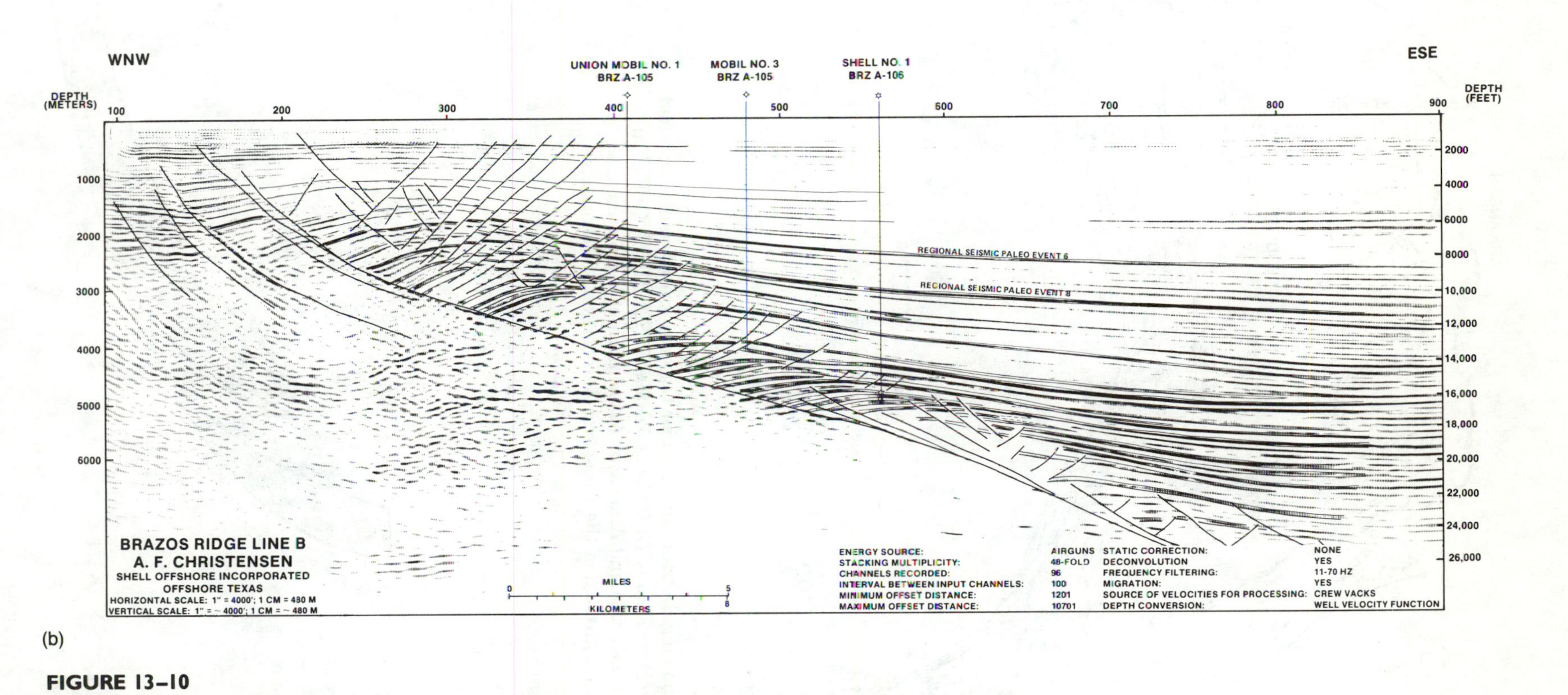

(b)

FIGURE 13–10
Seismic-reflection profile (a) and interpretation (b) showing a major growth fault and smaller synthetic and antithetic normal faults along the Corsair trend in the U.S. Gulf Coast. (Courtesy of A. F. Christensen and Shell Offshore, Inc.)

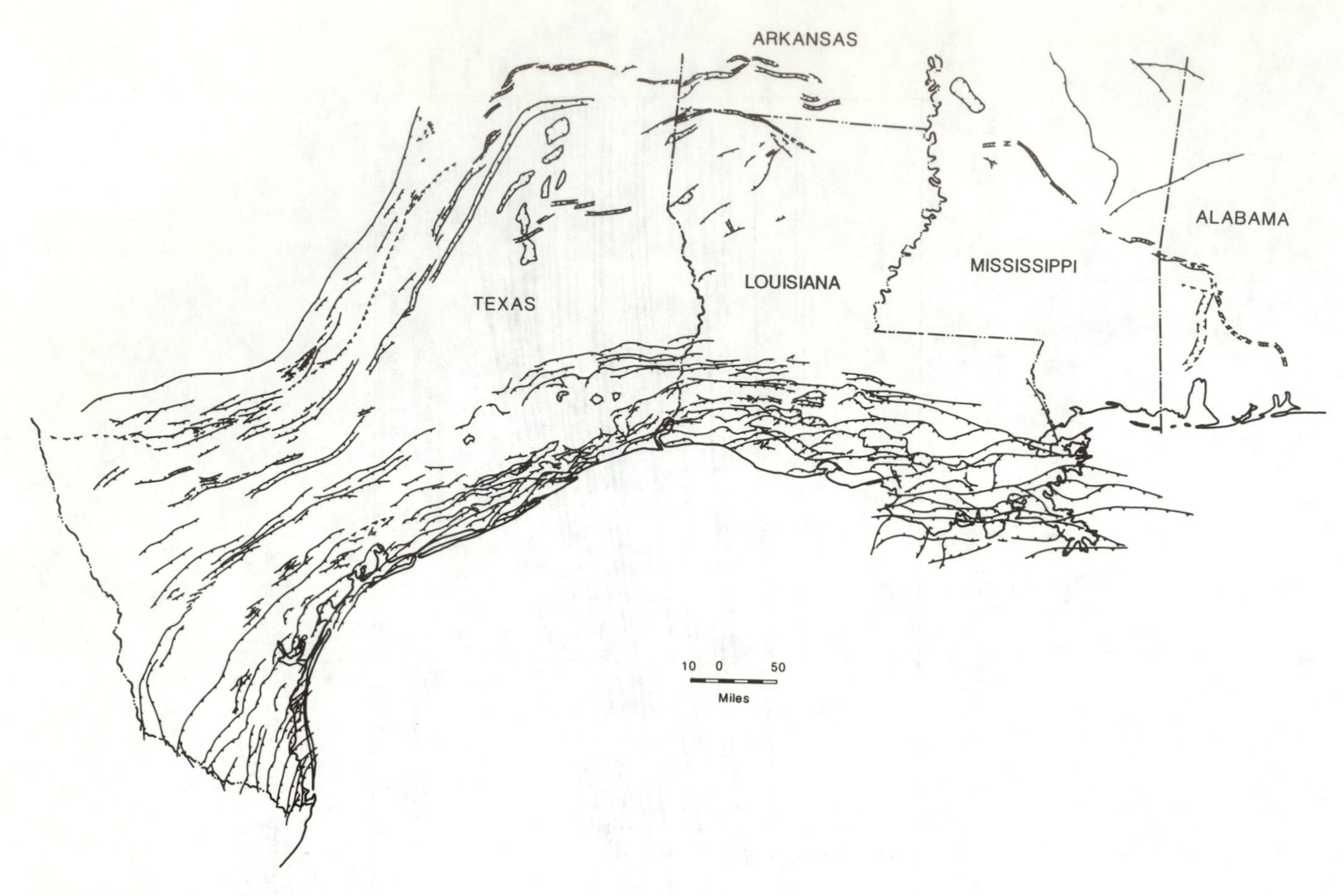

FIGURE 13–11
Map of the Gulf Coast showing distribution of major regional growth faults (toothed lines). (Modified from *Tectonic Map of Gulf Coast Region, U.S.A.*, 1972, Gulf Coast Association of Geological Societies and American Association of Petroleum Geologists. Reprinted by permission of the American Association of Petroleum Geologists.)

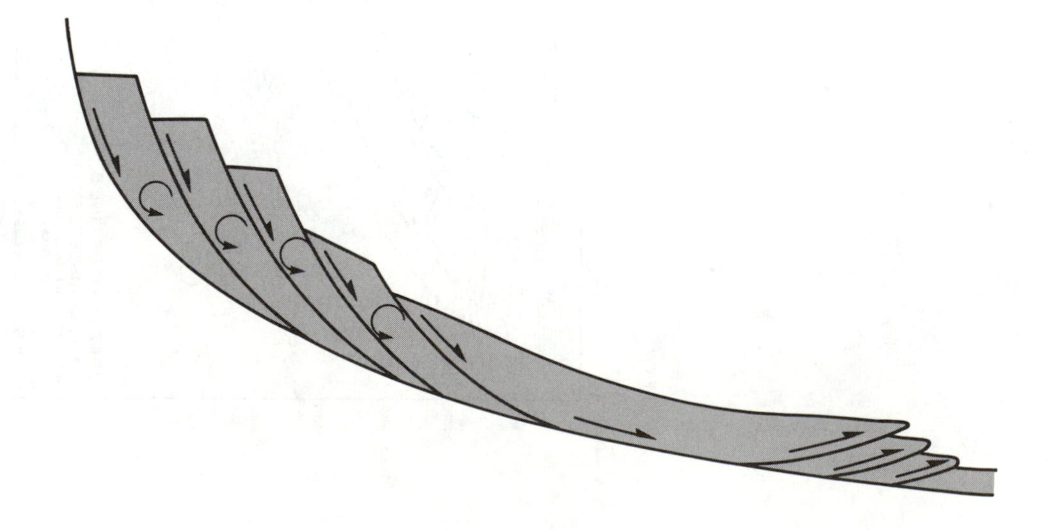

FIGURE 13–12
Cross section showing rotational character and thrusting of the toes of listric faults in sediments. Curved arrows indicate counterclockwise rotation of blocks.

Formation of crustal listric normal faults and resultant large-scale extension has been suggested to explain the opening of the Bay of Biscay, between France and the Iberian peninsula (de Charpal and others, 1978). Seismic-reflection profiling in the Bay of Biscay reveals large normal faults that dip steeply in the upper crust and appear to flatten into a major detachment zone in the lower crust. The detachment horizon is interpreted as the ductile-brittle transition. Such faults are thought to be extensional normal faults like those in the Basin and Range province.

J.-P. Brun and Pierre Choukroune (1983) have devised several models to explain the mechanics of extended crust. Their models (Figure 13–17) take into account several boundary conditions and behavior modes, including (1) vertical zoning of the crust into ductile (continuous) or brittle (discontinuous) domains of deformation; (2) the nature of discontinuities in the brittle realm; (3) presence—or absence—of preexisting gently dipping mechanical discontinuities; and (4) existence of an extensional (stretching) gradient across the zone. To sum up Brun and Choukroune's models:

Model I—Crustal thinning occurs only by ductile processes, like the extensional necking of a steel rod.
Model II—A vertical downward transition from brittle to ductile deformation develops whereby near-surface normal-fault displacement is related directly to thinning of the crust by ductile deformation at depth.
Model III—A pre-existing mechanical discontinuity is assumed within the crust as a zone of movement. This discontinuity may be the ductile-brittle transition.
Model IV—All deformation is brittle.
Model V—Crustal extension may be accomplished by intrusion of magma, separating and isolating large blocks of crust.

Of the five models, II and III seem to account for the most properties and known boundary conditions of zones of crustal extension. Geologic and geophysical evidence also favor II and III.

Collapse Structures and Related Features

Some normal faults occur as collapse structures related to the subsidence of magma chambers, and they may deform volcanoes and the tops of plutons (Figure 13–18). They form calderas and other structures and function in the overall process of magma emplacement; they also provide conduits for magma.

Salt domes also produce normal faults above and on the flanks of salt structures as the salt arches and intrudes overlying sediments (see Figure 2–30b). Some of these faults have large displacements (1 to 3 km) and in the Gulf Coast interact locally with regional down-to-basin faults.

The impact on the Earth of large bodies (*bolides*) from outer space produces rebound structures in which normal faults may develop concentrically or radially (see Figure 2–32). As a result, a series of concentrically arranged horsts and grabens may develop in a well-exposed and deeply eroded impact structure. Such structures, called *astroblemes,* or *impact structures,* are known in many parts of the world and are attributed to large meteors or even asteroids colliding with the Earth. Some, like Meteor Crater in Arizona, were formed as recently as the Quaternary. Others, like the Wells Creek and Flynn Creek structures in Tennessee, are Paleozoic or Mesozoic structures, and the Vredefort structure in South Africa was probably formed in the Precambrian.

Relationship to Strike-Slip Faults

The geometry and sense of motion of normal faults was at one time thought to be unique and invariably related to gravity or to crustal extension, but detailed study of fault systems has revealed that normal faults are related both to folds and to the compressional faults. Strike-slip faults (Chapter 12) have been shown to be intimately related to formation of isolated normal faults and of horsts and grabens (rhomb grabens and rhombochasms). Resolution of strike-slip faulting into an extensional sense results from the stepping over of strike-slip motion into that of extension, where displacement is transferred from one fault to another.

We have seen that the formation of normal faults may be linked with formation of other fault types at all scales and that their geometry is similar to that of thrusts, but with an opposite movement sense. With that we close our discussion of fractures and faults and turn to a discussion of folds and folding.

FIGURE 13–13
(a) Cross section through the Red Sea Rift. (From J. D. Lowell and G. J. Genik, 1972, AAPG *Bulletin*, v. 56. Reprinted by permission of American Association of Petroleum Geologists.)
(b) Failed rift and aulocogen.
(c) Interpretive map of faults in the Gregory Rift of the East African rift system; made from LANDSAT images. Note asymmetry along the length of the rift, indicating that even supposedly classic symmetrical rifts may prove to have been formed by asymmetrical extension. (From W. Bosworth, J. Lambiase, and R. Keisler, *EOS*, v. 67, July 1986, copyright by the American Geophysical Union.)

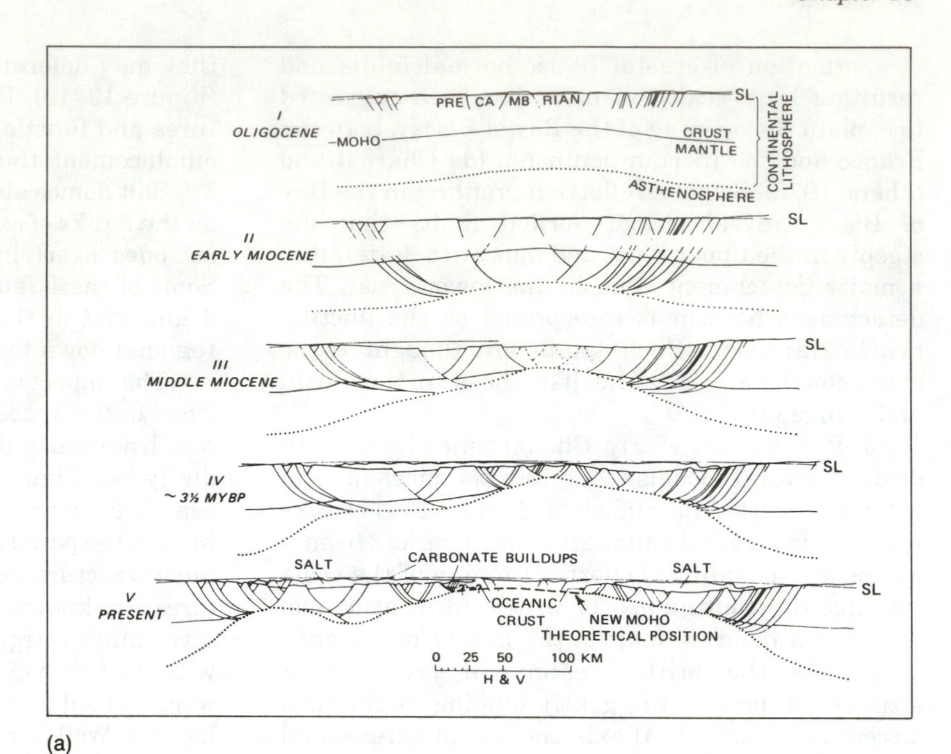

(a)

(b)

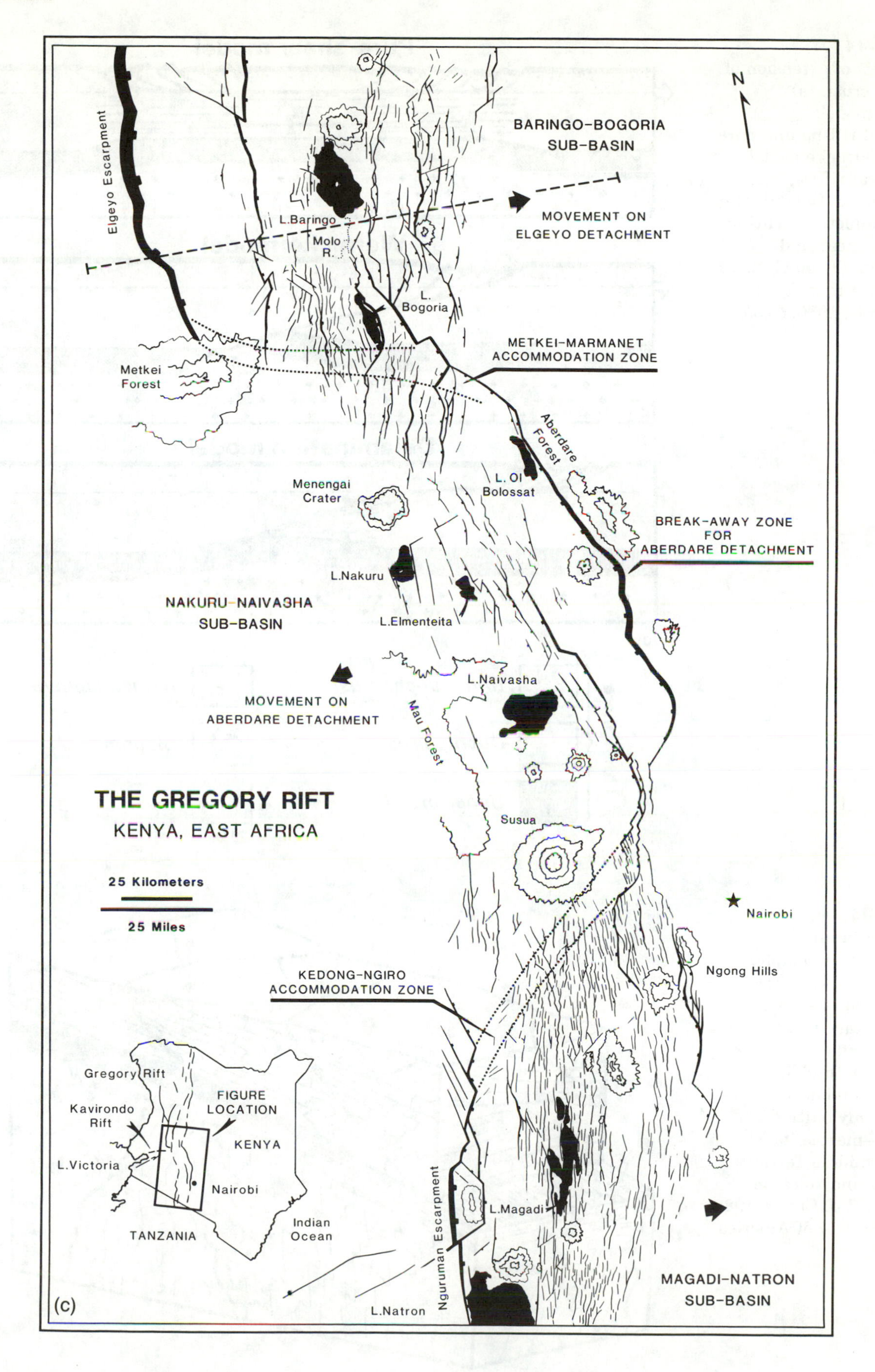
BARINGO-BOGORIA
SUB-BASIN
N
Elgeyo Escarpment
L.Baringo
Molo R.
MOVEMENT ON
ELGEYO DETACHMENT
L. Bogoria
METKEI-MARMANET
ACCOMMODATION ZONE
Metkei Forest
Aberdare Forest
Menengai Crater
L. Ol Bolossat
BREAK-AWAY ZONE
FOR
ABERDARE DETACHMENT
L.Nakuru
NAKURU-NAIVASHA
SUB-BASIN
L.Elmenteita
L.Naivasha
MOVEMENT ON
ABERDARE DETACHMENT
Mau Forest
THE GREGORY RIFT
KENYA, EAST AFRICA
Susua
25 Kilometers
25 Miles
Nairobi
Ngong Hills
KEDONG-NGIRO
ACCOMMODATION ZONE
Gregory Rift
Kavirondo Rift
FIGURE
LOCATION
KENYA
L.Victoria
Nairobi
TANZANIA
Indian Ocean
Nguruman Escarpment
L.Magadi
MAGADI-NATRON
SUB-BASIN
L.Natron
(c)

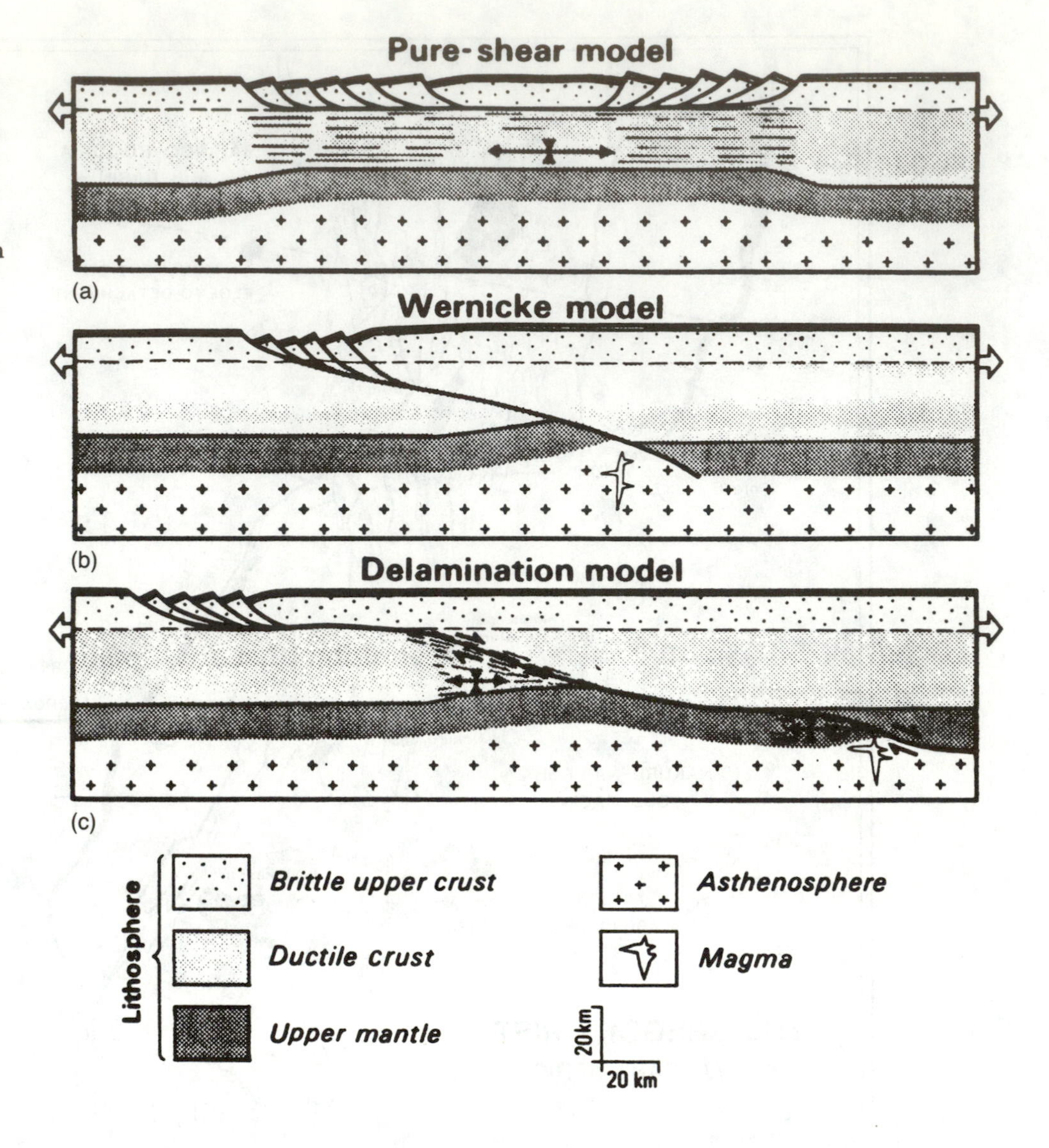

FIGURE 13–14
Three models of extension of continental crust. (a) Pure-shear model: Symmetrical rifting and pure shear. (b) Wernicke model: Asymmetrical rifting with simple shear. (c) Delamination model: Asymmetrical rifting with simple shear and delamination. (From G. S. Lister, M. A. Etheridge, and P. A. Symonds, 1986, *Geology,* v. 14.)

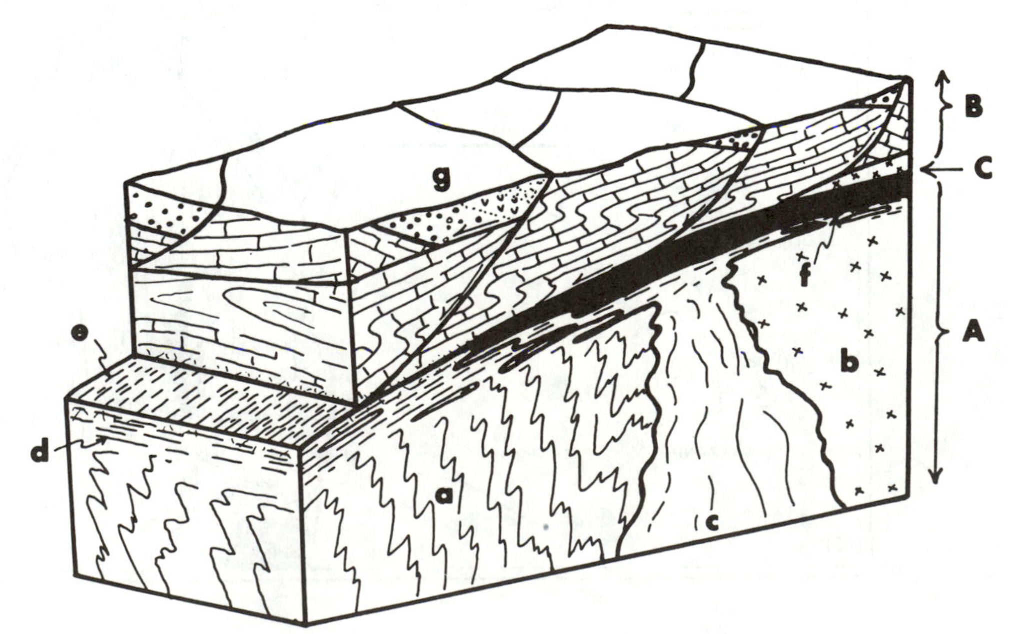

FIGURE 13–15
Typical Cordilleran metamorphic core complex. A—basement; B—cover; C—detachment; a—older metasedimentary rocks; b—older pluton; c—younger pluton (early to middle Tertiary); d—mylonitic foliation; e—mylonitic lineation; f—marble (black); g—lower to middle Tertiary sedimentary and volcanic rocks. (From P. J. Coney, 1980, Geological Society of America *Memoir* 153.)

(a)

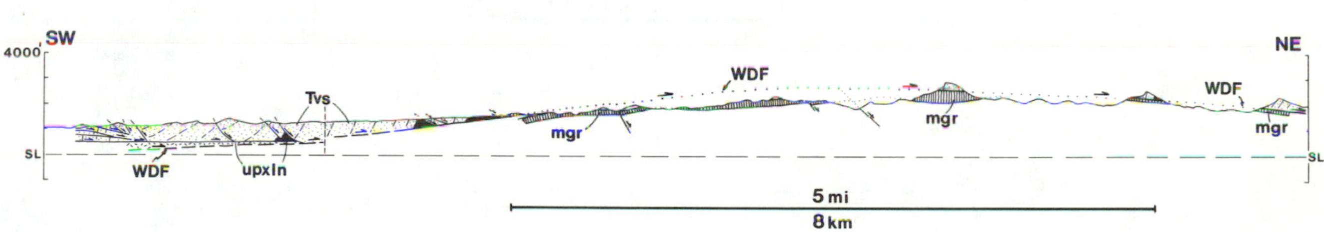

(b)

FIGURE 13–16
(a) View of the Whipple Mountains, California, looking southwest along the axis of the Whipple Peak antiform. The detachment fault is the light-dark contact that wraps around most of the range. Dark rocks in the upper plate consist of Miocene volcanic and sedimentary rocks, which were deposited on upper-plate crystalline rocks that form the light-colored area in the lowest portion of the photo. Light-colored rocks forming most of the range are lower-plate mylonitic gneisses that sit structurally beneath the nonmylonitic gneisses that compose the low-relief lower plate in the western part of the range (upper right part of photo). (E. J. Frost, San Diego State University.) (b) Cross section across the Whipple Mountains, illustrating middle Miocene geologic relations before domal uplift and warping of the Whipple detachment fault (WDF). The cross section illustrates evidence for two phases of rotational normal fault displacement along the detachment surface. Tvs—Tertiary sedimentary and volcanic rocks; mgr—mylonitic granitic rocks; upxln—upper plate crystalline rocks. (Unpublished section courtesy of G. A. Davis.)

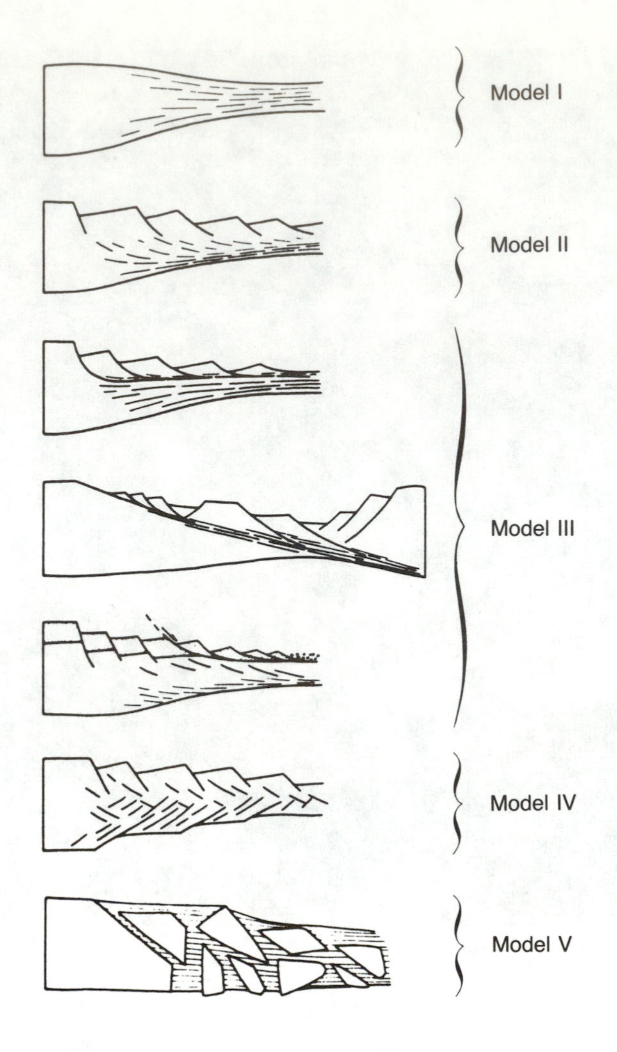

FIGURE 13–17
Models for stretched crust. (After J.-P. Brun and Pierre Choukroune, *Tectonics*, v. 2, p. 345–356, 1983, copyright by the American Geophysical Union.)

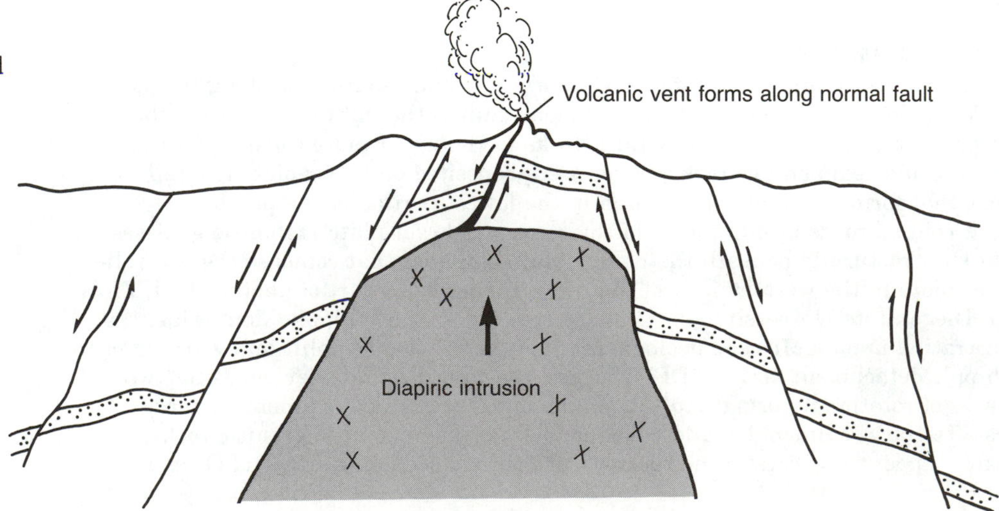

FIGURE 13–18
Normal faults developed in association with volcanoes and plutons.

Questions

1. Why are normal faults also called gravity faults?
2. Why do normal faults splay and form antithetic and synthetic faults?
3. How do reverse drags folds form? Why are they commonly associated with growth faults?
4. How do normal faults terminate?
5. If a listric normal fault zone flattens downward into a detachment in shale, and a listric normal fault forms in the upper crust and flattens into a detachment in the ductile-brittle transition zone, how would you expect them to differ in appearance and in fault rocks (if any)?
6. How would you document the timing of motion on a growth fault?
7. What kinds of evidence would help you decide whether a particular fault has been reactivated with a sense of motion differing from its original motion sense?
8. What kinds of data would you need to determine if rifting occurred by a symmetrical or asymmetrical mechanism in an area like the East African rift or the continental margins around the present Atlantic?

Further Reading

Brun, J.-P., and Choukroune, P., 1983, Normal faulting, block tilting and décollement in a stretched crust: Tectonics, v. 2, p. 345–356.
Outlines a series of models for crustal thinning—ductile, ductile-brittle transition, and mechanical boundaries, such as shear zones, brittle faulting, and magma intrusion. All relate to the normal faulting process.

Cloos, Ernst, 1968, Experimental analysis of Gulf Coast fracture patterns: American Association of Petroleum Geologists Bulletin, v. 52, p. 420–444.
Models the normal-fault system of the Gulf Coast by subjecting clay blocks to extensional deformation. The listric geometry was successfully modeled, as was the major regional pattern of down-to-basin faults.

Keen, C. E., Stockmal, G. S., Welsink, H., Quinlan, G., and Mudford, B., 1987, Deep crustal structure and evolution of the rifted margin northeast of Newfoundland: Results from LITHOPROBE East: Canadian Journal of Earth Sciences, v. 24, p. 1537–1549.
Discusses the crustal structure of part of the present-day Atlantic extensional margin and attempts to evaluate the applicability of both symmetric and asymmetric spreading models.

Wernicke, Brian, 1985, Uniform-sense normal simple shear of the continental lithosphere: Canadian Journal of Earth Science, v. 22, p. 108–125.
Examines models of pure and simple shear for crustal extension in the Basin and Range and elsewhere. Wernicke concluded that simple shear oriented in a single direction is the most common mechanism for crustal extension and likely involves much of the lithosphere.

PART FOUR

Folds and Folding

OUTLINE

14 Fold Geometry and Classifications
15 Mechanics of Folding
16 Complex Folds

WE NOW TURN TO A DISCUSSION OF A SECOND MAJOR GROUP OF structures—folds. They may be observed on all scales in crustal rocks and occur in a variety of shapes. Folds are described and classified in Chapter 14. Chapter 15 deals with the mechanics of how folds form, and Chapter 16 investigates more complex multiple-phase fold systems and sheath folds.

14

Fold Geometry and Classifications

Folds in layered rock commonly are nearly perfect geometric figures, approaching segments of circles in parallel or *concentric* folds, sinusoidal waves in some *similar* folds and sharp angles in some *kink* folds. Partly because of their geometric beauty, folds have attracted the attention of geologists almost from the time of the birth of geological science.

ARVID M. JOHNSON and STEPHENSON D. ELLEN, *Tectonophysics*

WAVELIKE STRUCTURES—FOLDS—IN ROCKS RESULT FROM DEformation of bedding, or foliation, or other originally planar surface. These structures occur in a variety of scales, ranging from those visible only under a microscope to those extending for kilometers (Figure 14–1). Folds form in all environments in the Earth's crust, from near-surface brittle to lower-crust ductile, and under conditions of simple shear to pure shear; they range from very broad and gentle folds to tightly compressed and attenuated structures. They occur singly as isolated folds and in extensive fold trains of different sizes. Rocks may be deformed by single folding events and by serially overprinted generations. Folds were the first structures recog-

FIGURE 14–1
Disharmonic folds in shale and thin limestone beds near Whitehall, New York. (RDH photo.)

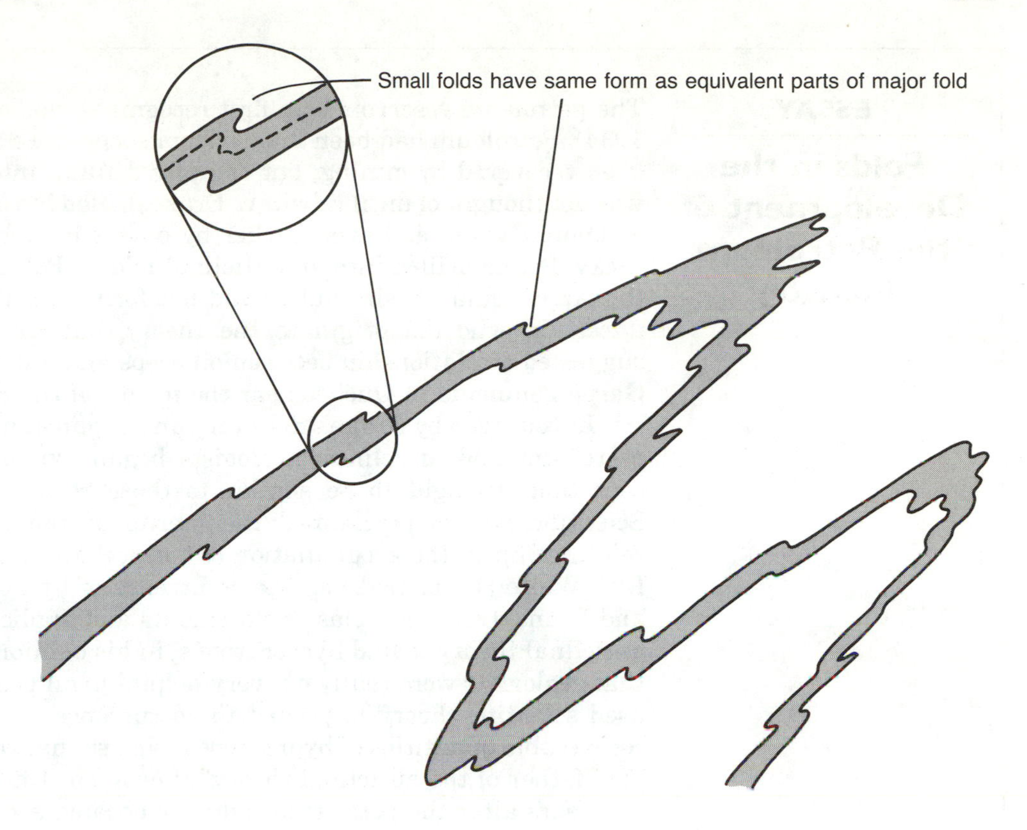

FIGURE 14–2
Pumpelly's rule relating small- and large-scale folds.

nized as hydrocarbon traps, leading to the *anticlinal theory* of petroleum accumulation and development of a major new industry about the end of the nineteenth century. (The petroleum industry is still the largest single employer of geologists.) Folds also play an important role in concentrating other valuable minerals. For example, the widely known ***saddle-reef deposits,*** first identified in Australia, contain minable concentrations of sulfide minerals localized in the hinges of folds.

In this chapter we will introduce folds, describe their properties, and discuss some of the classifications devised to help us communicate about folds and explore their nature. Chapter 15 will present current views on the mechanics of folding, and Chapter 16 will discuss development of multiple fold systems and other complex folds.

It is important at this point to reintroduce the concept of scale: structures small enough to require magnification to be seen are ***microscopic.*** Those whose sizes range from hand specimen to outcrop scale are ***mesoscopic,*** and those of map scale or larger are ***macroscopic.*** Most observations and measurements that provide information about orientations of the various fold elements (limbs, axes, and axial surfaces), and allow inferences about the folding process, are made on the mesoscopic scale; yet, microscopic- and map-scale studies also yield useful results. Conclusions regarding the orientation, shape, and size of folds apply on all scales.

Recall ***Pumpelly's rule*** (Chapter 1), which states that *small-scale structures generally mimic larger-scale structures formed at the same time.* If Pumpelly's rule holds true, mesoscopic folds should have the same style and orientation as macroscopic folds (Figure 14–2), enabling study of mesoscopic and microscopic folds to enhance our understanding of macroscopic folds and the structural history of a major tectonic zone.

DESCRIPTIVE ANATOMY OF SIMPLE FOLDS

Anatomy of Folds

Consider the simple upright fold pair in Figure 14–3. Any fold of this type has a ***crest,*** at its highest point (elevation), and a ***trough,*** at its lowest point. The straighter or least-curved segments are called ***limbs*** and connect the parts of the fold exhibiting the greatest curvature—the ***hinges.*** The line where layering changes orientation from dip in one direction to dip in the opposite direction across a crest or trough is called the ***hinge line.*** For ideal (cylindrical) reversal of dip, the hinge line is parallel to the

ESSAY

Folds in the Development of the Petroleum Industry

The petroleum reservoir trap first recognized was an anticlinal fold (Howell, 1934). Petroleum had been known from seeps since ancient times and tar had been recovered by mining, but actually drilling into the Earth to recover oil was not thought of until Edwin L. Drake drilled his first well in August of 1859 in Pennsylvania and even earlier by others in Germany, Canada, and Kentucky. Drake drilled into an anticlinal fold in Paleozoic sedimentary rocks of the Appalachian basin, but he did not formulate the anticlinal theory. It is uncertain who did originate the theory, but in 1846 Sir William Logan suggested a relationship between oil seeps and anticlinal folds in rocks of the Gaspé Peninsula in Québec near the mouth of the St. Lawrence River.

Encouraged by Drake's discovery and Logan's suggestion of a link between petroleum and anticlines, geologists began exploring for petroleum under conditions thought to be similar to those responsible for Drake's success. Scientific papers proclaimed the merits of the anticlinal theory and its relationship to the accumulation of both oil and natural gas. Two papers by I. C. White (1885, 1892) on the occurrence of hydrocarbons in West Virginia and Pennsylvania are classic statements and applications. White's use of the anticlinal theory to find hydrocarbons, in his opinion, helped dispel the notion that geologists were really not very helpful in oil prospecting. He had actually used scientific theory to predict the occurrence and location of economically recoverable quantities of hydrocarbons and so inadvertently became known as the "father of the anticlinal theory" (Levorsen, 1954).

Years after the petroleum industry became a major factor in the United States and world economy, geologists realized that many structures trapping petroleum were not anticlines and that an anticline cannot trap petroleum except under special stratigraphic conditions. A. W. McCoy and W. R. Keyte (1934) recommended that anticlines be considered one of a group of structural traps. Others proposed that a parallel group of stratigraphic traps also be recognized. By the 1930s, geologists applying scientific concepts had helped establish the petroleum industry, many structural and stratigraphic traps had been identified, and geologists had become an essential part of petroleum exploration.

References Cited

Howell, J. V., 1934, Historical development of the structural theory of accumulation of oil and gas, *in* Problems in petroleum geology: Tulsa, Oklahoma, American Association of Petroleum Geologists, p. 1–23.

Levorsen, A. I., 1954, Geology of petroleum: San Francisco, W. H. Freeman, 703 p.

Logan, Sir William, 1846, Report of progress, 1844: Ottawa, Geological Survey of Canada.

McCoy, A. W., and Keyte, W. R., 1934, Present interpretations of the structural theory for oil and gas migration and accumulation, *in* Problems of petroleum geology: Tulsa, Oklahoma, American Association of Petroleum Geologists, p. 253–307.

White, I. C., 1885, The geology of natural gas: Science, v. 5, p. 521–522.

______ 1892, The Mannington oil field and the history of its development: Geological Society of America Bulletin, v. 3, p. 187–216. (An appendix describes the anticlinal theory for natural gas.)

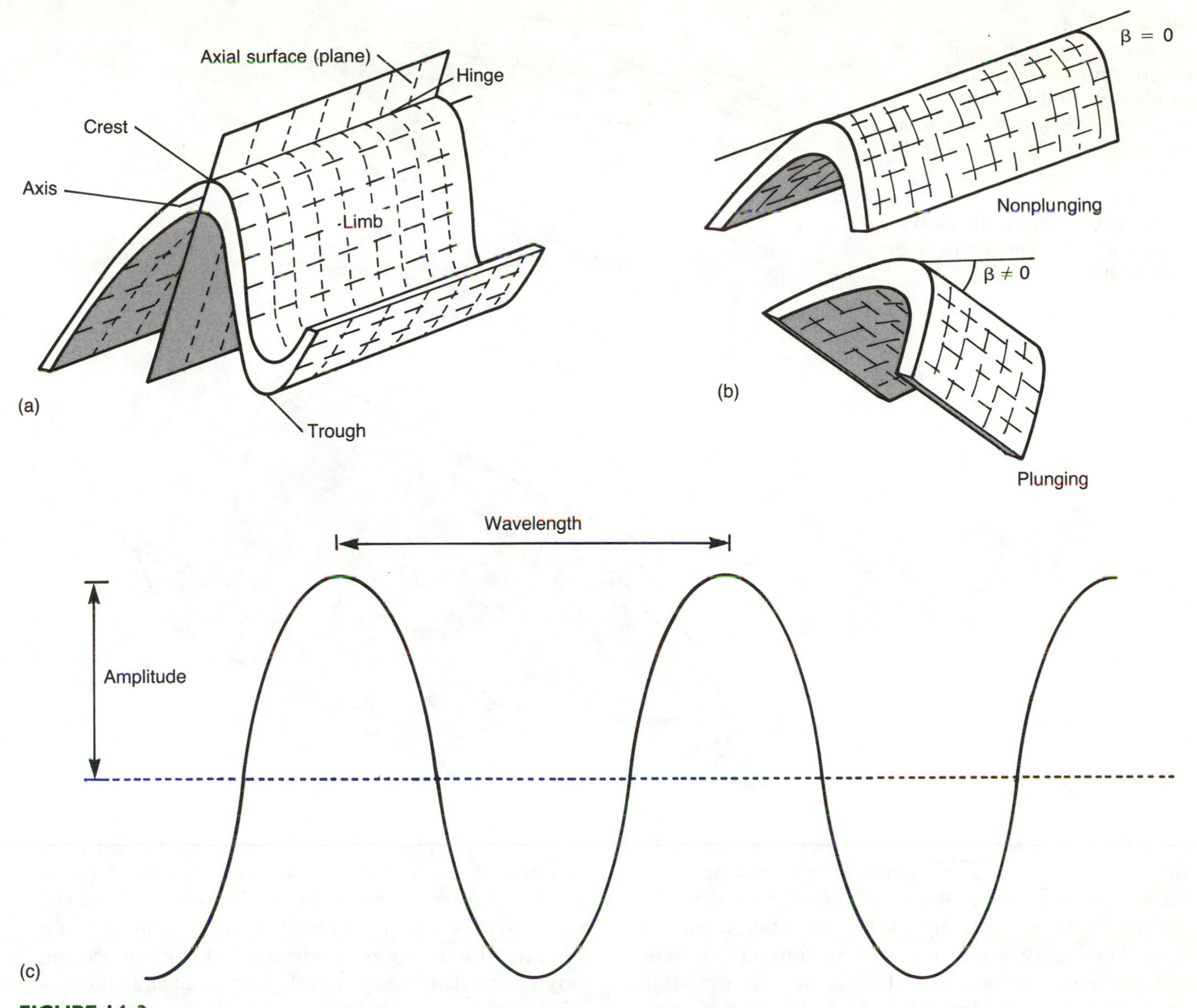

FIGURE 14–3
(a) Anatomy of a fold. (b) Plunging and nonplunging folds. (c) Relationships between wavelength and amplitude of folds. β is the angle of plunge.

fold axis, or ***axial line.*** The hinge and crest lines coincide only in symmetrical folds having vertical axial surfaces. If several fold axes are connected on successive folded surfaces of the same fold, they form the ***axial surface*** (***axial plane,*** if the surface is planar). Fold axes may be horizontal, but more often they are inclined to the horizontal and are said to ***plunge;*** folds with nonhorizontal axes are called ***plunging folds.***

The distance from crest to crest of adjacent anticlines—or trough to trough of adjacent synclines—is the *wavelength.* Half the distance from the crest of an anticline to the trough of an adjacent syncline, measured parallel to the axial plane, is the *amplitude* of the fold.

The direction of leaning (the opposite of dip direction) of the axial surface, sense of shear, or the direction of overturning (producing an inverted limb) is called the ***vergence*** of a fold (Figure 14–4). This property will work only on folds having one limb that dips more steeply than the other—an *asymmetric fold.* Vergence is not a property of *symmetrical folds,* but small folds may be asymmetrical on the limbs of a larger symmetrical fold. A related

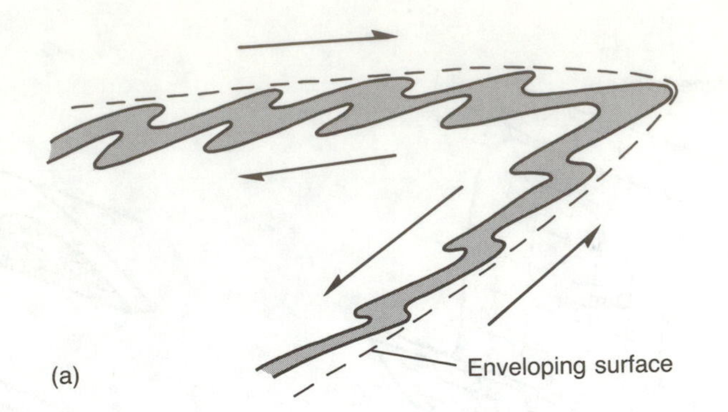

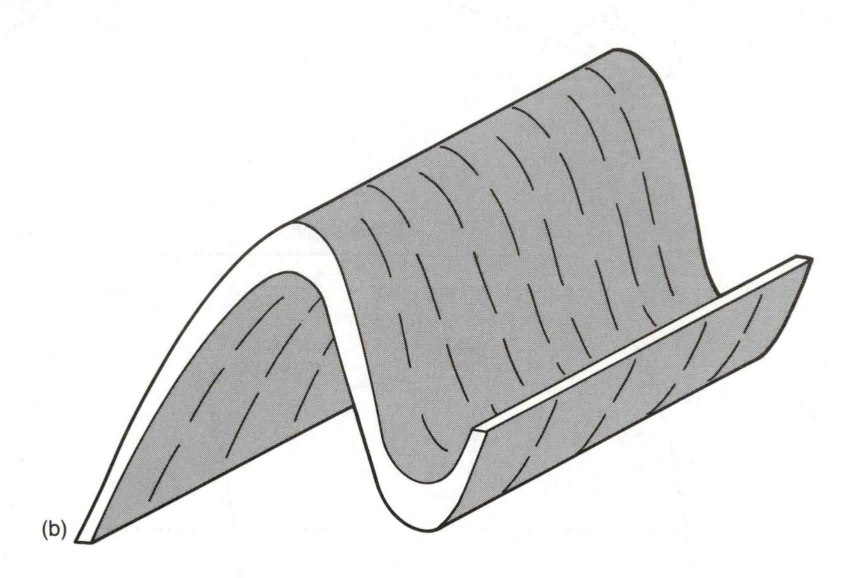

FIGURE 14–4
(a) Vergence (direction of overturning) of large and small folds (indicated by arrows) and relationships to an enveloping surface. (b) Slip lines illustrated by lines—e.g., fibers or slickensides—on a layer surface that indicate the direction of motion of one layer past another.

property of folds is *slip lines,* which are generally another indication of vergence. Slip lines are produced by relative motion of reference points in successive layers during folding or other deformation. They may occur on real-movement surfaces and produce slickensides, fibers, or other visible lines indicating motion (Figure 14–4b), or they may be imaginary lines deduced from vergence. Determination of vergence is useful in working out the overall direction of tectonic transport of all structures in an area, in addition to helping fix an observer's location on a large fold.

The largest folds in a given area are often called *first-order folds; second-order folds* are smaller folds on the flanks (Figure 14–4a). Folds of successively higher order are also possible. It may be feasible to relate the geometry of small- to large-scale folds by using an ***enveloping surface,*** which is constructed by connecting the crests or troughs of small folds by a line or surface tangent to one of the folded surfaces. Enveloping surfaces are useful for studying folds at outcrop scale or in cross section where many small folds occur on limbs of larger folds and the geometry of the larger folds is not clear.

While deciphering a fold, a geologist must keep in mind the relationship between the axial plane and the plunge of the fold axis (Figure 14–5) as the axial surface dips more or less steeply. The axis of a fold must lie in the axial surface, but the orientation may vary within the plane. The trend of the fold axis may vary considerably from the strike of the axial surface.

Kinds of Folds

In this section we will consider simple folds, and you will see that many of the terms are closely interrelated as opposites or as synonyms.

The term ***anticline*** applies to folds that are concave toward older rocks in the structure; an anticline contains older rocks in the center (Figure 14–6a). A fold that is concave downward and has the form of an anticline is called an ***antiform;*** the rocks may not be older in the middle, or the age of the rocks may be unknown. In the opposite sense, a ***syncline*** is a structure wherein layering is concave toward the younger rocks and, as a result, younger rocks are found in the central part of the structure

(Figure 14–6b). Similarly, a structure where layering is concave up and dips toward the center is a ***synform.*** A ***dome*** is a special kind of antiform wherein layering dips in all directions away from a central point, and a ***basin*** is a unique synform in which layering dips inward toward a central point (Figure 14–6c).

Where the sequence can be worked out such that the ages of the rocks are determinable, it may be possible to categorize antiforms and synforms as antiformal synclines or synformal anticlines, depending on the relative ages of rocks found in the center of the structures (Figure 14–6d). An ***antiformal syncline*** (*downward-facing syncline*) is a structure in which layering dips away from the axis, but the rocks in the center are younger. In contrast, a ***synformal anticline*** (*upward-facing anticline*) is a structure wherein layering dips inward as in a syncline, but the rocks in the center of the structure are older rather than younger. These structures are produced during multiple episodes of folding (Figure 14–7).

Rocks that dip uniformly in one direction may be described as ***homoclinal.*** A slight complication of this structure, involving a local steepening of an otherwise uniform regional dip, is called a ***monocline.*** A ***structural terrace*** is a local flattening of a uniform regional dip (Figure 14–8).

A fold is ***cylindrical*** (properly *cylindroidal,* or cylinder-like) if the fold can be generated by moving the fold axis parallel to itself. Folds may occasionally be sinusoidal. Generally, cylindrical folds are those in which the axes are everywhere parallel on successive folds of the same generation (Figure 14–9). *Aberrant folds* deviate slightly from ideal cylindricity (Ramsay and Sturt, 1973a). Most folds

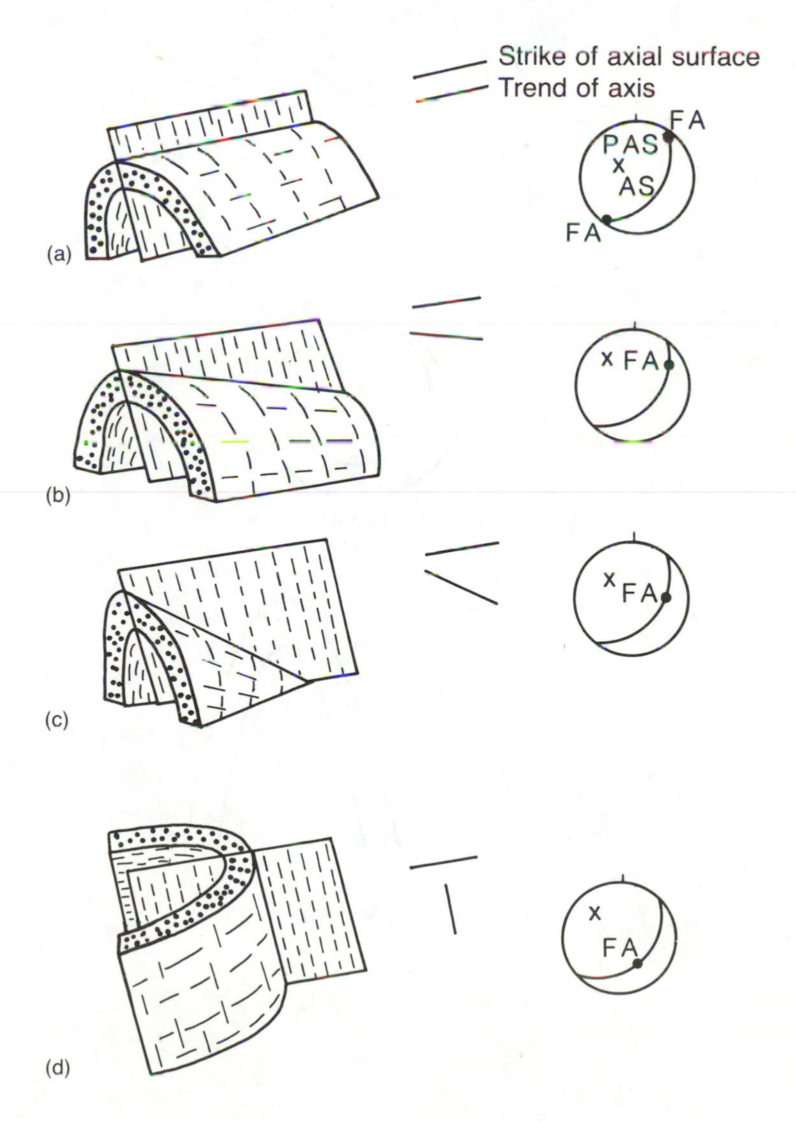

FIGURE 14–5
Axial-surface orientation and constraints on the orientation of the fold axis at zero (a), intermediate (b and c), and steep plunges (d). PAS—pole to axial surface; AS—axial surface; FA—fold axis.

FIGURE 14–6
(a) Anticline and syncline (cross section). (b) Antiform and synform (cross section). (c) Doubly plunging anticline and syncline (map view). (d) Dome and basin (map view). (e) Antiformal syncline and synformal anticline in cross section. Є, O, and S indicate Cambrian, Ordovician, and Silurian rocks.

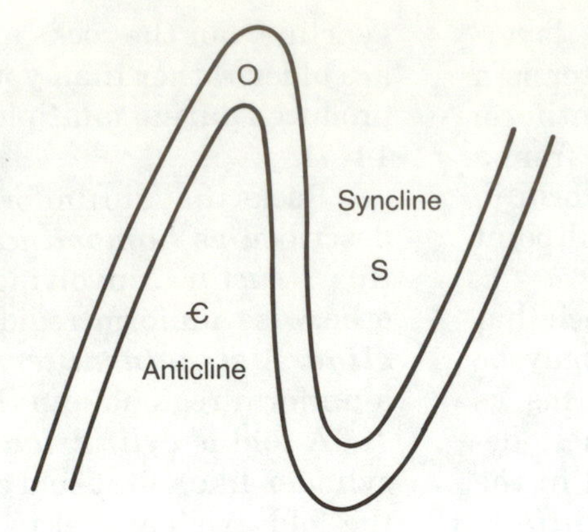

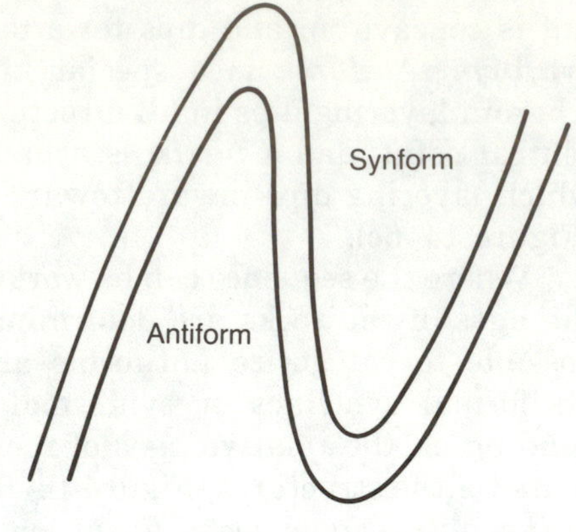

(a) Ages of rocks known (b) Ages of rocks unknown

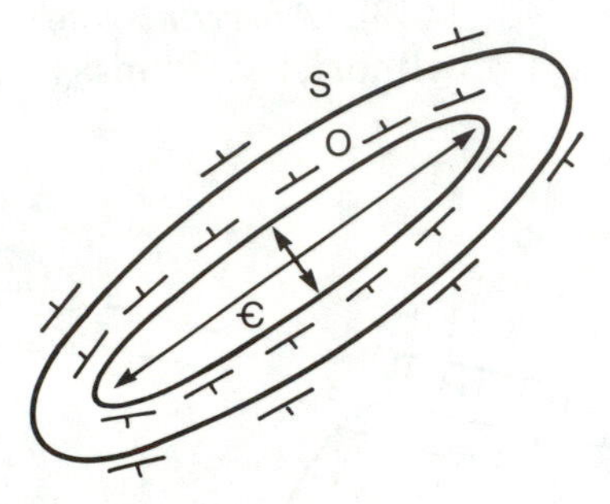

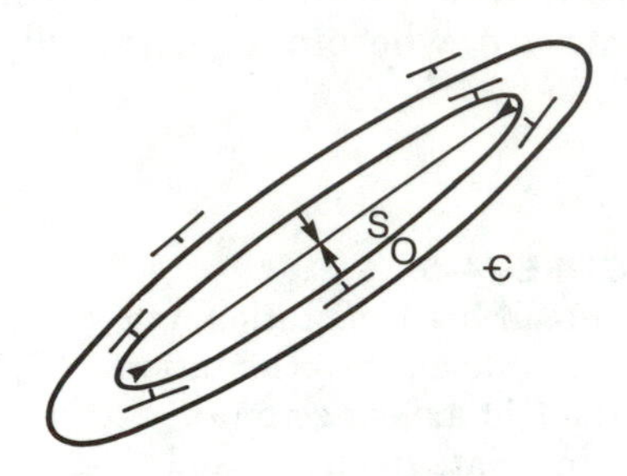

(c) Anticline (doubly plunging) Syncline (doubly plunging)

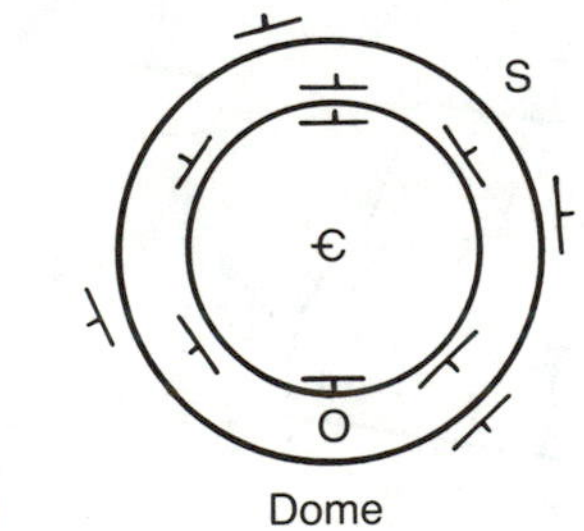

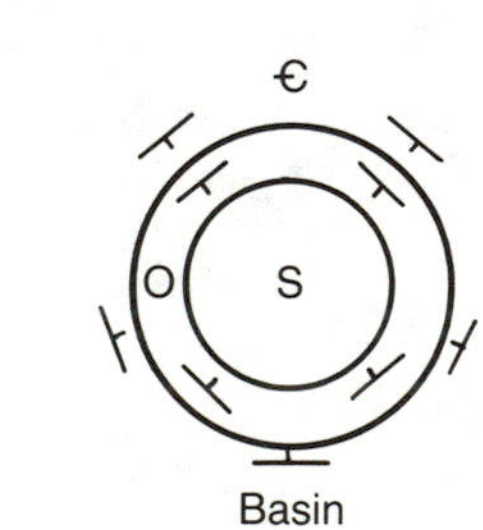

(d) Dome Basin

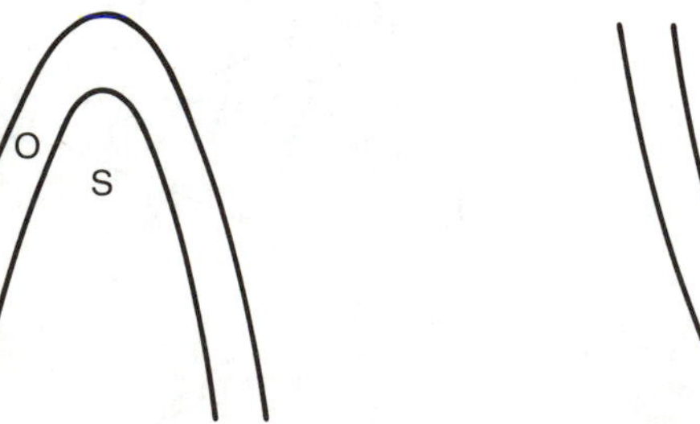

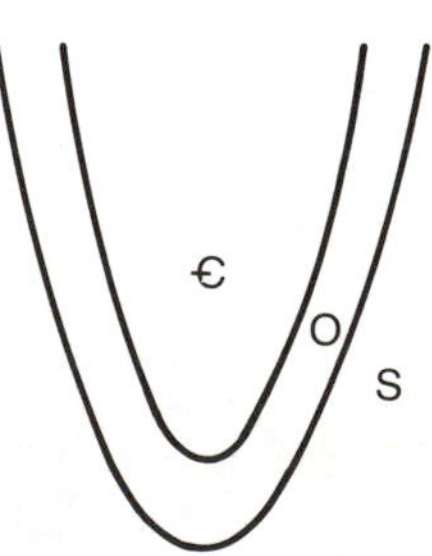

(e) Antiformal syncline Synformal anticline

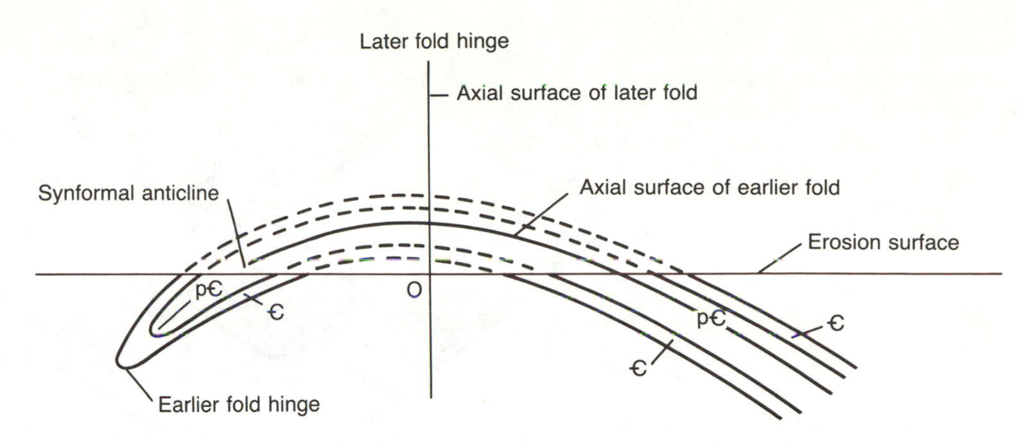

FIGURE 14–7
Synformal anticline as a product of multiple folding. Note axial surfaces of both earlier and later folds. pЄ, Є, and O indicate Precambrian, Cambrian, and Ordovician rocks.

probably fall into this class because the deviation from ideal cylindricity is only slight or may be detected only by tracing the fold a great distance along the trend. Aberrant folds are commonly not considered separately but grouped with cylindrical folds. On the other hand, ***noncylindrical folds*** contain axes that are not parallel on successive folds or in which axes in successive folds of the same group may converge toward a point rather than being parallel and are considered a separate group. Noncylindrical folds with convergent axes are called ***conical folds*** (Figure 14–9). ***Sheath folds*** are noncylindrical but tubular—closed at one end—and the fold axes curve within the axial surfaces (Figure 14–10). They commonly occur in zones of ductile deformation where rocks have been deformed by a strong component of inhomogeneous simple shear.

Upright folds have vertical axial surfaces (Figure 14–11). ***Overturned folds*** have one inverted limb. ***Reclined folds*** have plunging axes and axial surfaces that dip at low angles; in either case, generally the axis plunges normal—or at a high angle—to the strike of the axial surface. ***Recumbent folds*** have horizontal axes and axial surfaces. Folds may be classed as ***open folds*** if the limbs dip very gently away from or toward one another, or ***tight,*** if the fold limbs dip steeply toward or away from one another. ***Isoclinal folds*** are tight folds wherein axial surfaces and limbs are parallel; they are common in rocks that have been ductilely deformed (Figure 14–12).

To distinguish among many possible combinations of plunging, reclined, inclined, and recumbent folds, Michael J. Fleuty (1964) has constructed a diagram plotting the plunge of a fold hinge versus the dip of the axial surface (Figure 14–13). M. J. Rickard (1971) has constructed a similar plot with end members—vertical, horizontal, upright, and recumbent folds—that vary by pitch (or rake) of the axis, plunge of the axis, and dip of the axial surface.

Folds that maintain constant layer thickness are called ***parallel folds*** (Figure 14–14a). ***Concentric folds*** are parallel folds in which folded surfaces define circular arcs and maintain the same center of curvature. These characteristics restrict fold shape and require that both fold types die out upward or downward (sometimes both) from the zone of greatest deformation (Figure 14–15a). *Ptygmatic folds* are near-parallel folds formed by the folding of stronger

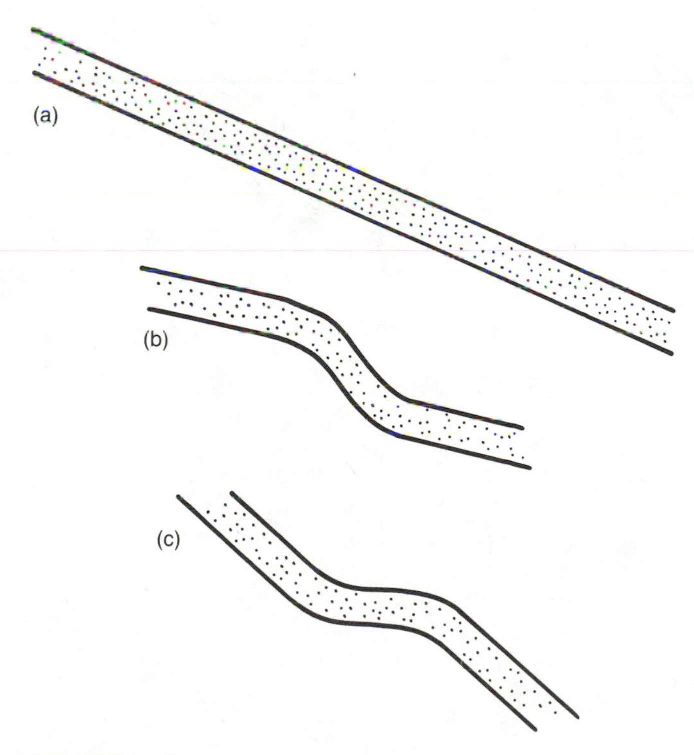

FIGURE 14–8
Homocline (a), monocline (b), terrace (c).

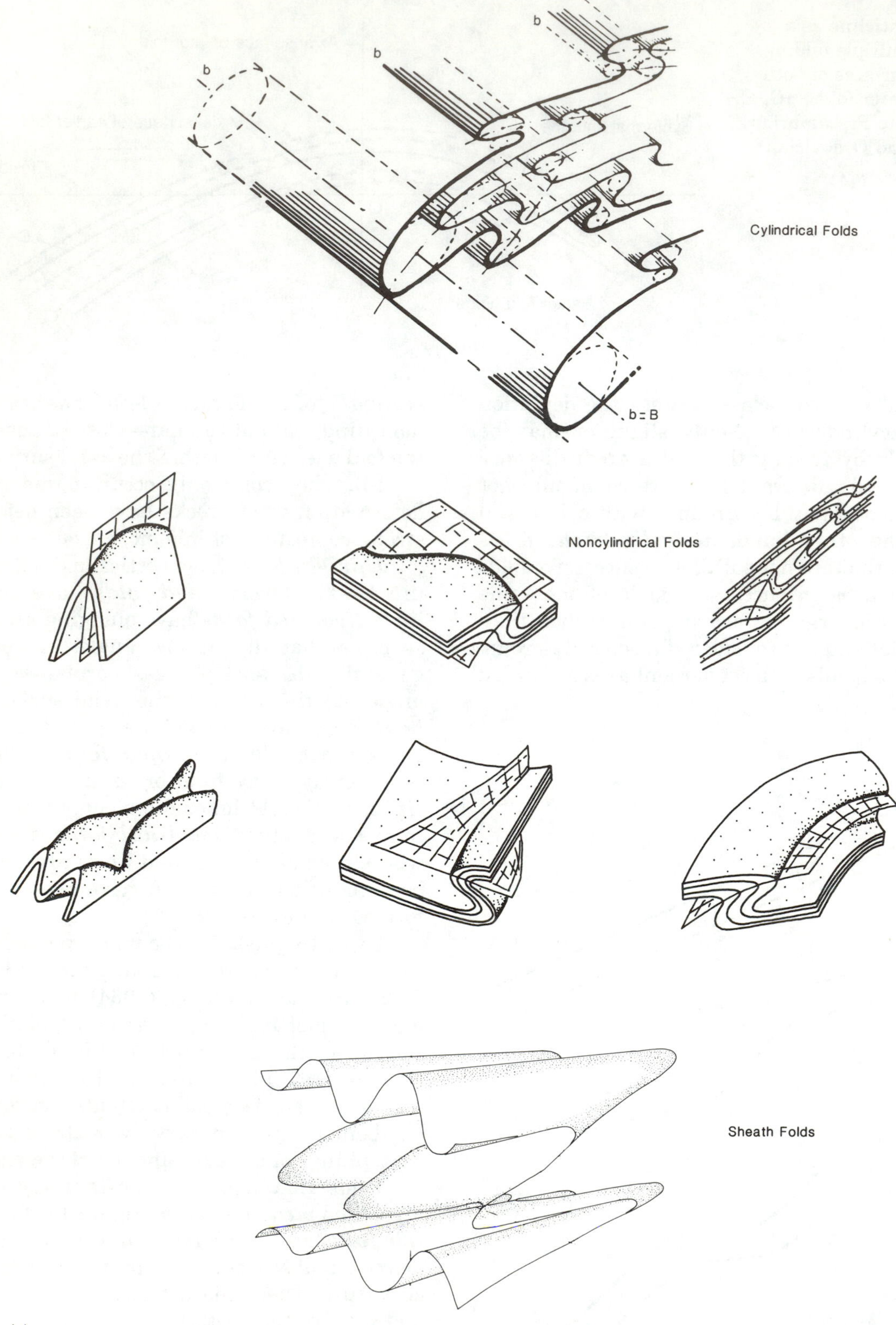
b
b
b
Cylindrical Folds
b = B
Noncylindrical Folds
Sheath Folds
(a)

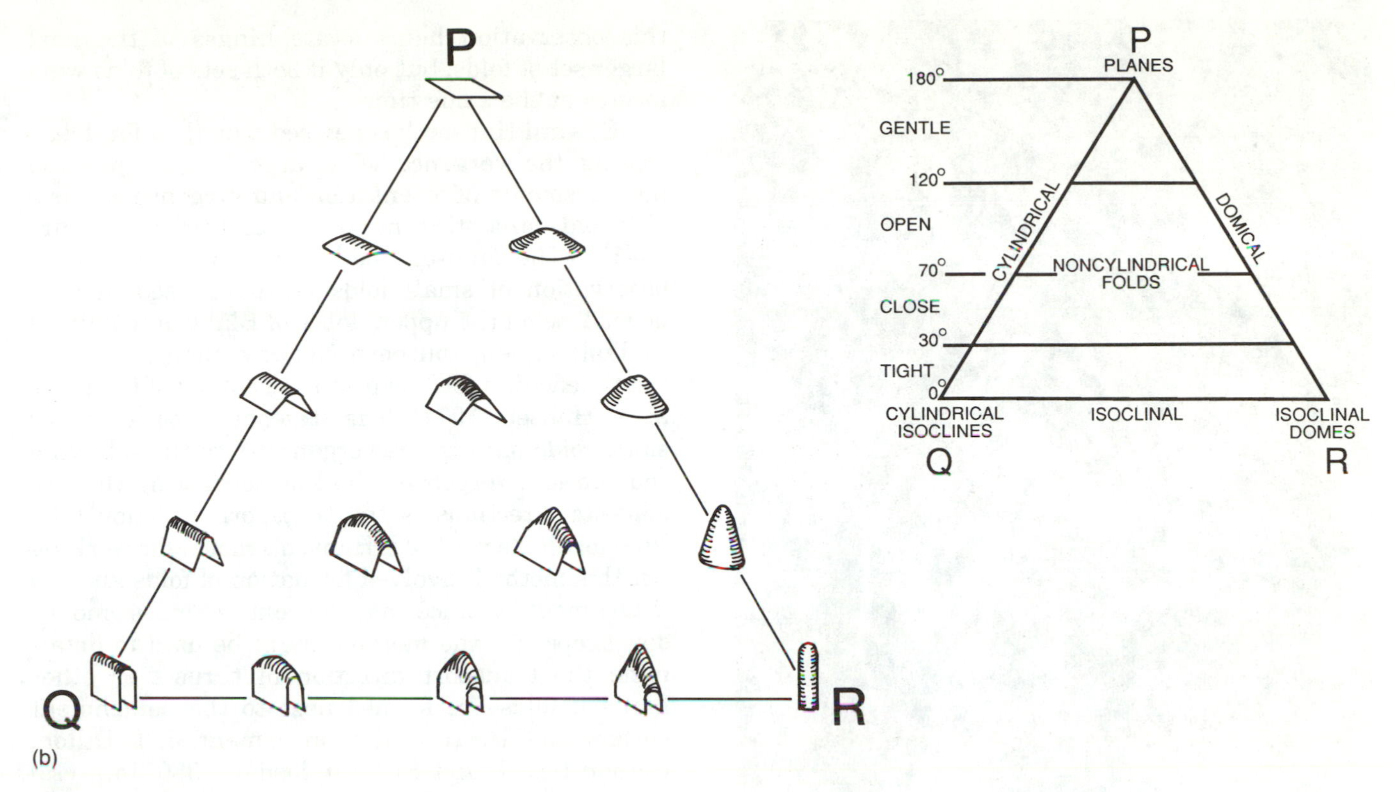

FIGURE 14–9
(a) (opposite page) Cylindrical, noncylindrical, and sheath folds. Compare the shapes of the hinges from top to bottom: the ideal cylindrical folds have very linear hinges, and the sheath folds have tightly curved hinges. (After Gilbert Wilson, 1982, *Introduction to Small Scale Geological Structures,* Unwin Hyman Ltd; D. M. Ramsay and B. A. Sturt, 1973, *Tectonophysics,* v. 18, Elsevier Science Publishers; and *Journal of Structural Geology,* v. 3, J. R. Henderson, Structural analyses of sheath folds with horizontal X-axes, Northeast Canada, copyright 1981, Pergamon Press PLC.) (b) End-member fold classification (PQR diagram) showing fold shapes and terminology. This classification scheme illustrates the gradual changes in shape from unfolded surfaces (planes, P) through ideal cylindrical isoclinal folds (Q) to ideal isoclinal domes (sheath folds, R). (Reprinted with permission from *Journal of Structural Geology,* v. 1, G. D. Williams and T. J. Chapman, The geometrical classification of noncylindrical folds, copyright 1979, Pergamon Press PLC.)

layers in a much more ductile matrix. Quartz or quartz-feldspar veins in weak schist frequently produce ptygmatic folds (Figure 14–15b). ***Similar folds*** maintain their shape throughout a section: they do not die out upward or downward, but maintain the same curvature in the hinges (Figures 14–14b and 14–15c). Layer thickness changes (measured perpendicular to layering) in similar folds, from thicker in the axial zones to thinner on the limbs. Folds with straight limbs and sharp angular hinges are ***chevron*** and ***kink folds.***

Folds in which shapes change from one layer to another are ***disharmonic*** (Figures 14–1 and 14–14c). Disharmonic folds can form during parallel folding by overtightening and crumpling of thinner layers between thicker layers (Figure 14–15d).

Folds wherein the synclines are thickened and the anticlines thinned are called ***supratenuous folds*** (Figure 14–14d). Many supratenuous folds form in unconsolidated sediments and so are nontectonic; others form where uplift occurs during deposition so that more sediment accumulates in structural lows and less accumulates over highs.

Fault-bend and ***fault-propagation folds*** that form in association with thrust faults were described in Chapter 11. Recall that fault-bend folds are generally open parallel folds, but fault-propagation folds may be tighter parallel folds.

FIGURE 14–10
Sheath fold of calcsilicate layers in Archean gneiss, Melville Peninsula, District of Franklin, Canada. (J. R. Henderson, Geological Survey of Canada.)

Use of Parasitic Folds in Determining Position in a Fold

Pumpelly's rule notes similarities between the shapes of folds of different sizes on the same major structure (Figures 14–2, 14–4a, and 14–16a), but does not take into account a useful difference. The sense of asymmetry (or vergence) of higher-order folds depends on which limb of the next lower fold they occur. Shear couples commonly form as a result of flow of material out of hinges or relative motion of oppositely moving layers. Figure 14–16a shows that on one limb small or parasitic folds have a clockwise (Z) sense of rotation, but on the opposite limb a counterclockwise (S) sense. Small folds in the hinge are symmetrical and have an M shape. In the field, this observation helps locate hinges of the next larger set of folds, but only if both sets of folds were formed at the same time.

Edward Hansen has devised a method for determining the vergence of a large fold by plotting measurements of orientation and vergence of parasitic folds on a stereonet (or equal area net; Figure 14–17). The Hansen method was worked out by observation of small folds in tundra sod moving downslope on the upper slopes of Blåhø, a mountain in Trollheimen, southern Norway, then applied to the Caledonian folds exposed in bedrock of the same area (Hansen, 1971). The stereonet plot separates small folds having one vergence from those having the opposite vergence. The line separating the two opposite directions is the transport direction (slip line) for the larger fold. Hansen's model for working out this method involved formation of folds above a detachment surface as the entire mass moved downslope. So, the method might be used to determine the transport direction of thrusts or other faults if folds are formed next to the detachment surface and are related to movement. J. T. Dillon, Gordon Haxel, and Richard Tosdal (1989, in press) have used this method to determine the transport direction on the Chocolate Mountains thrust in southeastern California.

Observations of overturning and vergence should always be made while viewing fold shape in the direction of plunge (Figure 14–4a). A simple experiment with a piece of folded paper will show that the sense of vergence changes when the same fold is viewed from different directions.

This technique is useful but risky. In areas with only one generation of folds, it works very well. Where more than one fold generation exists, the geologist must make sure observations and conclusions concern folds of the same generation. Vergence may be used to determine whether the geologist is on an upright or an overturned limb of a fold—if the vergence is related to the fold generation being considered. In many areas the stratigraphic section may be overturned (Figure 14–16b), but a set of later folds may indicate the sequence is upright. Close study of both stratigraphy and structure will usually reveal the correct relationships.

MAP-SCALE PARALLEL FOLDS AND SIMILAR FOLDS

Parallel folds and similar folds are the most common groupings of folds; they form two classes of contrast-

FIGURE 14–11
Upright, overturned, open, tight, reclined, recumbent, isoclinal, and isoclinal-recumbent folds. AS—axial surface.

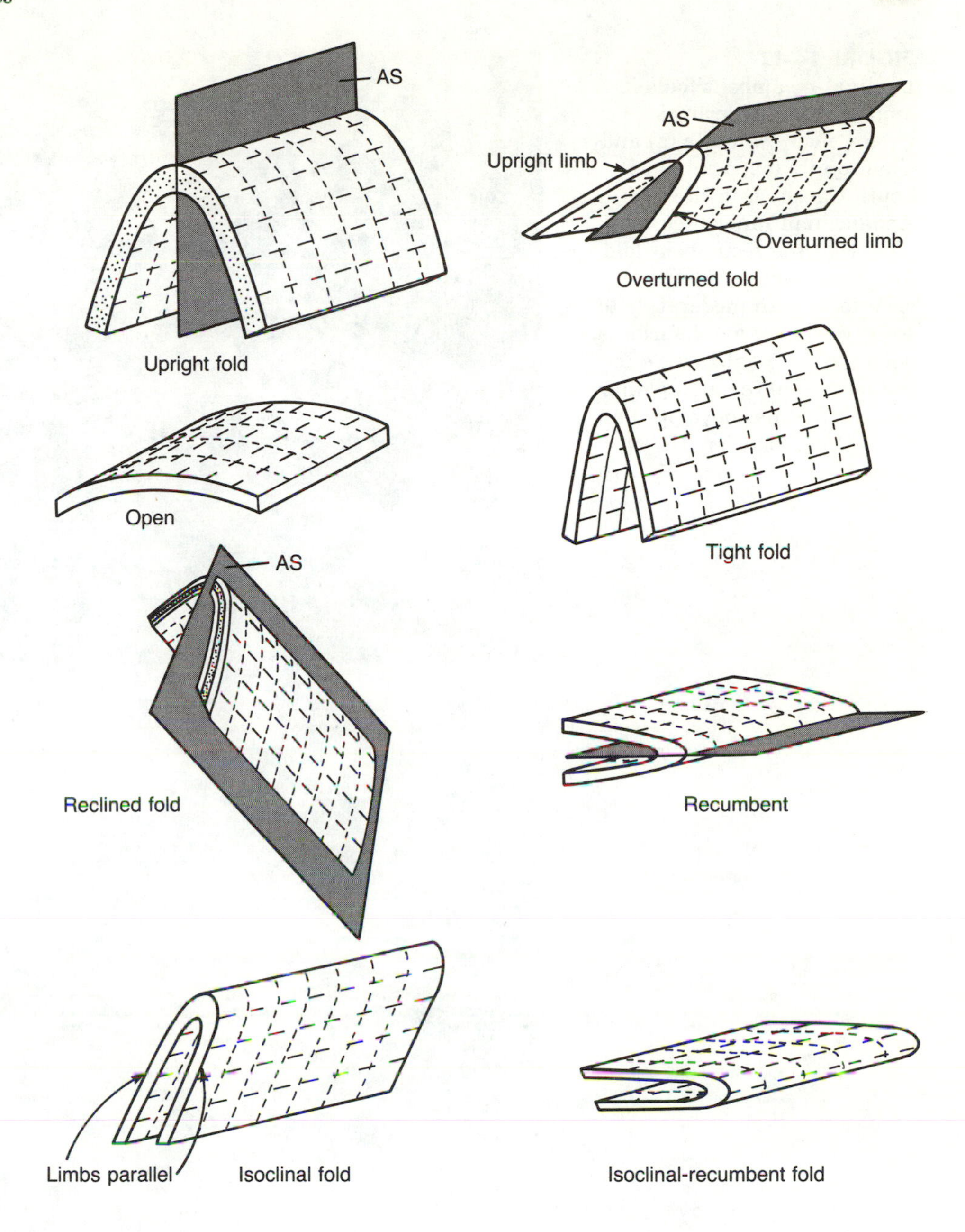

ing fold shape, but may not be immediately evident in map view (Figures 14–18a and 14–19a). They are clearly shown in cross sections constructed through regions folded by these different styles (Figures 14–18b and 14–19b). Determination of fold style is made by combining field observation of mesoscopic folds (and related structures) and careful plotting of dip-strike data on sections as they are constructed.

Several techniques have been devised for constructing geologic sections through regions deformed by parallel folds. H. G. Busk (1929) has invented a method for section construction, assuming parallel folds are also concentric. He realized that, in parallel-concentric folds, dip of bedding is parallel to circular arcs of folds. He extrapolated folded layers to depth using dip measurements made on the surface and, after identifying the center of curvature of each fold, constructed lines perpendicular to bedding to locate the centers of the folds and then connected them with circular arcs (Figure 14–20a). Then he eliminated overlapping segments of the arcs, producing smooth curves connecting anticlines and synclines. Busk's method works well if the rocks are deformed by parallel-concentric

FIGURE 14–12
Isoclinal-recumbent folds in amphibolite and granitic gneiss, near Walhalla (a) and Clemson (b) (opposite page), South Carolina, in the Appalachian Inner Piedmont. Note that the recumbent fold in (a) has been refolded by open folds with moderately to steeply dipping axial surfaces, whereas the axial surface in the fold in (b) remains planar and the fold has developed an axial-plane foliation. (RDH photos.)

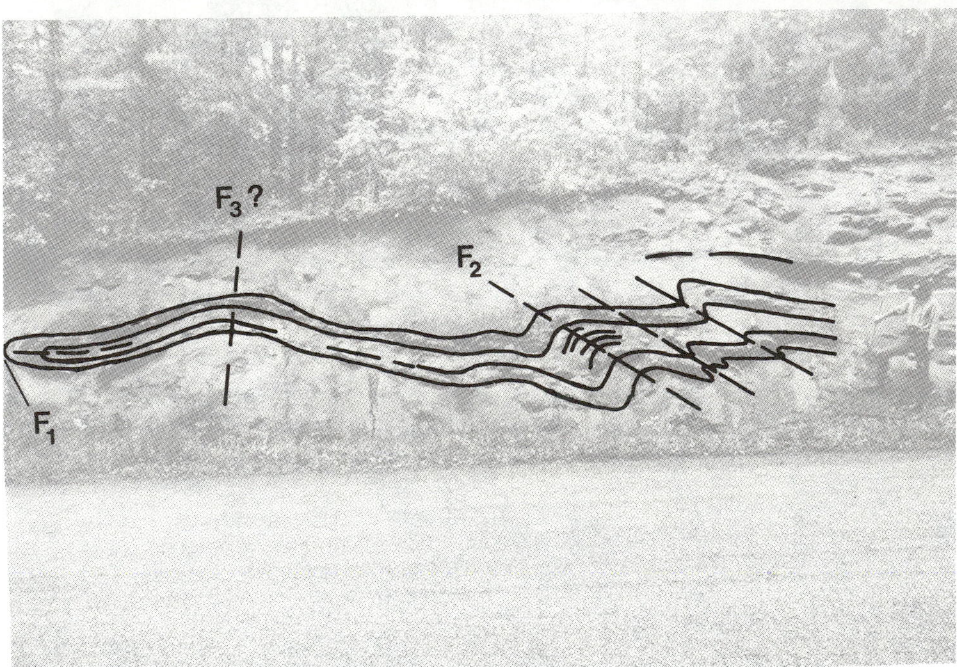

(a)

(b)

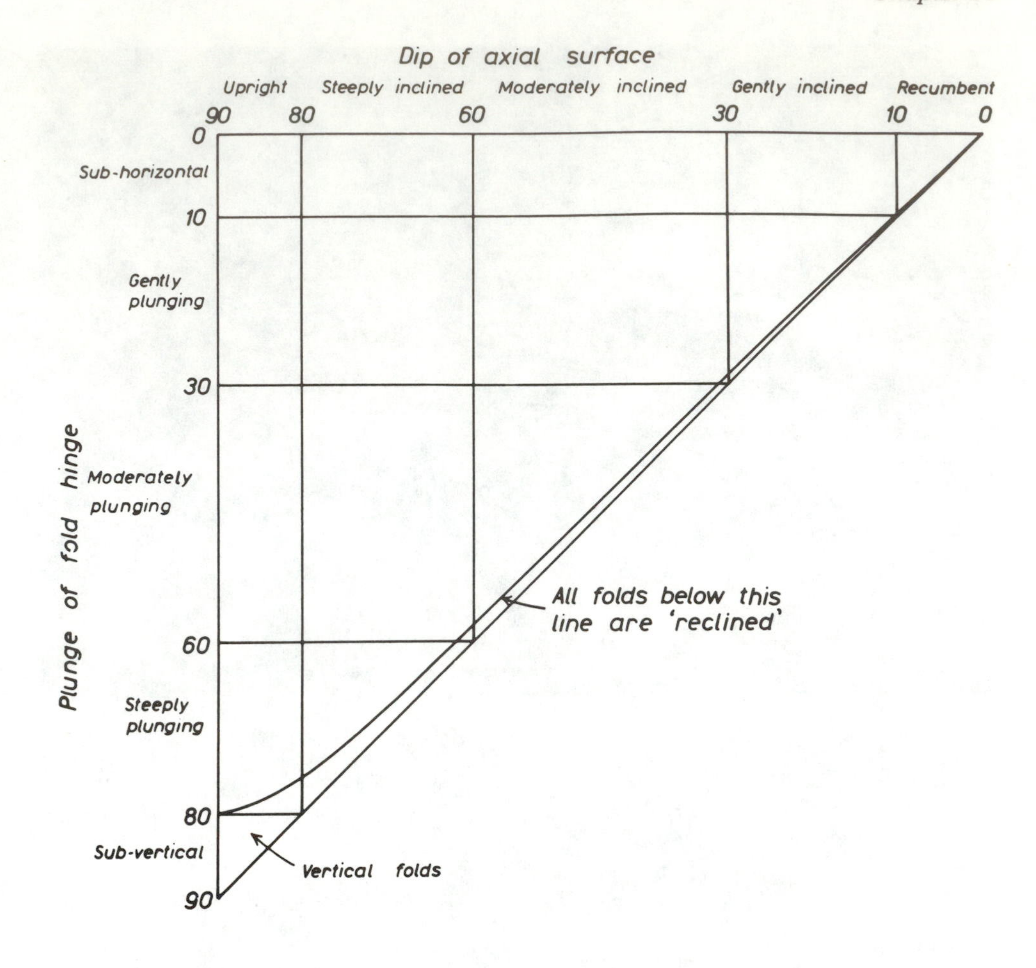

FIGURE 14–13
Plot for distinguishing between several fold types on the basis of relative plunge of the fold hinge and dip of the axial surface. (From M. J. Fleuty, 1964, *Proceedings of the Geologists' Association,* v. 75, p. 461–492.)

folds. Unfortunately, few regions are deformed in this manner—usually areas with thick sections of very strong, thickly bedded rocks (like massive sandstone, dolostone, or limestone) with few weak interlayers.

An alternative way to construct sections through parallel folded regions is based on the assumption that the rocks were deformed by parallel kinking rather than by concentric folding (Figure 14–20b). Kink folds have been observed for many years in layered rocks, and Roger Faill (1973), working in the Appalachians of Pennsylvania, was the first to suggest that small-scale kinks may be extrapolated to map scale. John Suppe (1983, 1985) has used this assumption and the kink style extensively because, once a section is constructed, this method, more than others, makes the section easy to balance. Kink folds do exist in foreland fold and thrust belts (Chapter 11), but are no more common than parallel-concentric folds and probably less common than parallel folds with curved (rather than angular) hinge zones. Therefore, anyone constructing sections through regions of parallel folding should try to combine the two techniques so as to interpret the structure more realistically.

Areas deformed by similar folding involve thinning of limbs and thickening of hinge zones of folds, and layers no longer are parallel; these are fundamental differences between parallel folds and similar folds (Figure 14–19). Cross sections are still a useful way to depict the structure of an area, but are harder to construct because similar folds also have a greater tendency to be noncylindrical, and extrapolation of layering to depth is less certain.

FOLD CLASSIFICATIONS

Folds have been classified in many ways, each intended to pigeonhole folds using one or more of the properties, but no classification takes into account all aspects of folds and folding. Every classification suffers from inherent defects, depending on what the classification was devised to accomplish. Some classifications attempt to categorize folds from a purely descriptive point of view; others determine and classify the fold-forming mechanism. Genetic classifications have been severely criticized because of their subjectivity and the requirement of interpretation by the user (Hudleston, 1973a). Some geologists argue that the best classification requires the least interpretation and is therefore the most objective, but drawbacks remain. Every classification, regardless of its basis, has some merit. In the next few sections, we will consider several fold classifications and discuss their advantages and disadvantages. Emphasis will be placed on learning John Ramsay's geometric and descriptive classification, along with the rationale for the others.

Classifications Based on Interlimb Angle

A fold classification devised by Michael J. Fleuty (1964) is based on the angle between the limbs of a fold, termed the *interlimb angle* (Figure 14–21). His classification is purely descriptive: the interlimb angle is measured and the tightness of the fold identified as *gentle,* having an interlimb angle of 180° to 120°; *open,* 120° to 70°; *closed,* 70° to 30°; *tight,* 30° to 0°; *isoclinal,* 0°; and *elasticas,* in which the angle is negative. Elasticas are rare in natural rocks, but do occur in ductilely folded stronger veins in a weaker matrix (ptygmatic folds; Figure 14–15b). Interlimb angle as a measure of fold tightness, as Fleuty defined it, depends not

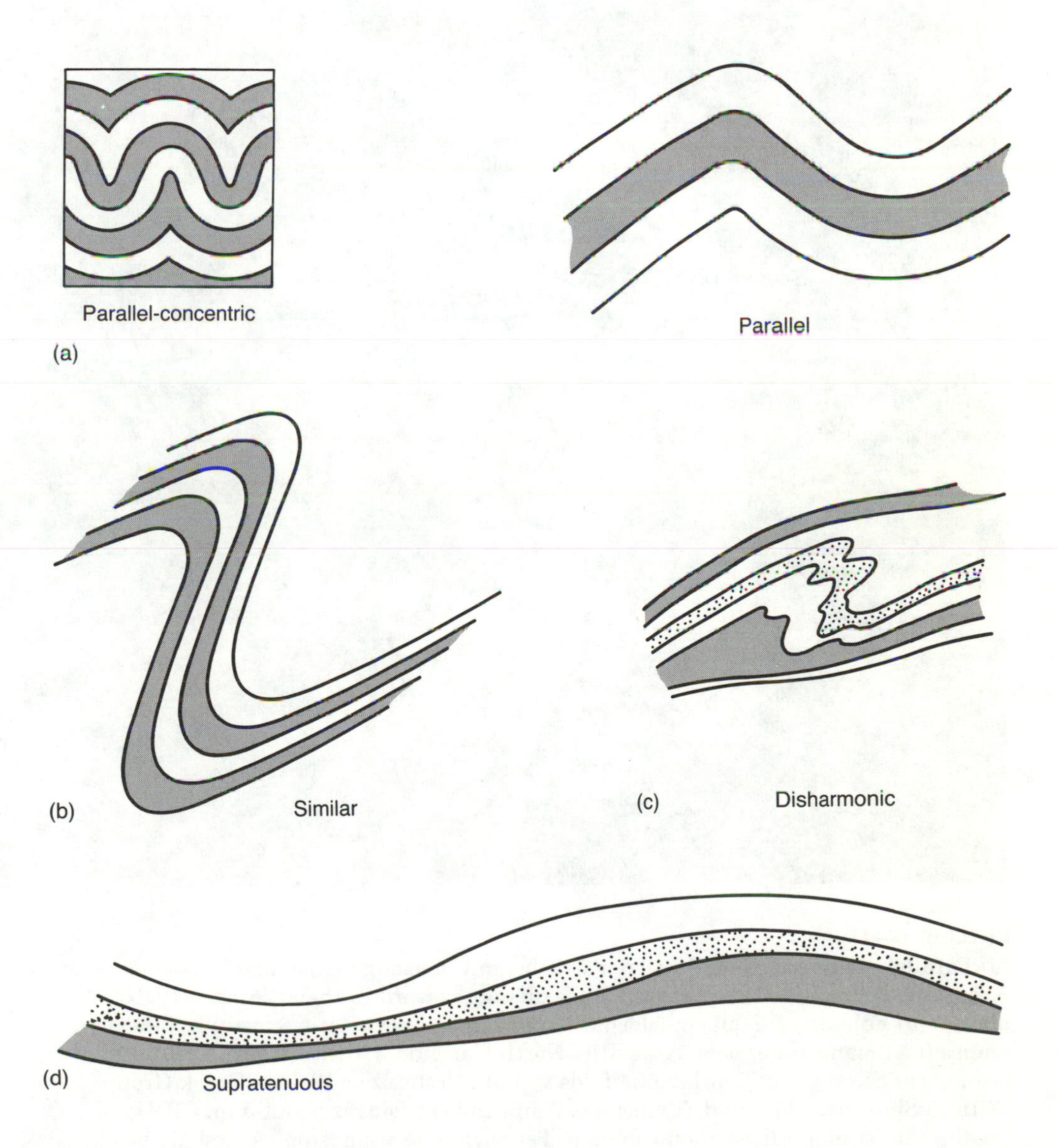

FIGURE 14–14
(a) Ideal parallel-concentric folds and parallel folds. (b) Ideal similar folds. (c) Disharmonic folds. (d) Supratenuous folds. All are shown in cross-section view.

(a)

(b)

FIGURE 14–15
(a) Parallel-concentric folds in Upper Cambrian Conasauga shale and limestone near Kingston, Tennessee. Note that the folds die out upward in the exposure. (RDH photo.) (b) Folding of a quartz-feldspar vein produced ptygmatic folds in biotite gneiss (metasandstone) near Asheville, North Carolina. (Arthur Keith, U.S. Geological Survey.) (c) Similar-like folds in Late Proterozoic Walden Creek Group Wilhite Slate near Walland, Tennessee. Amplitude of folds is about 5 m. (RDH photo.) (d) Strongly disharmonic folds in Pennsylvanian sandstone and shale near Rockwood, Tennessee. Note fault in left center. (RDH photo.)

(c)

(d)

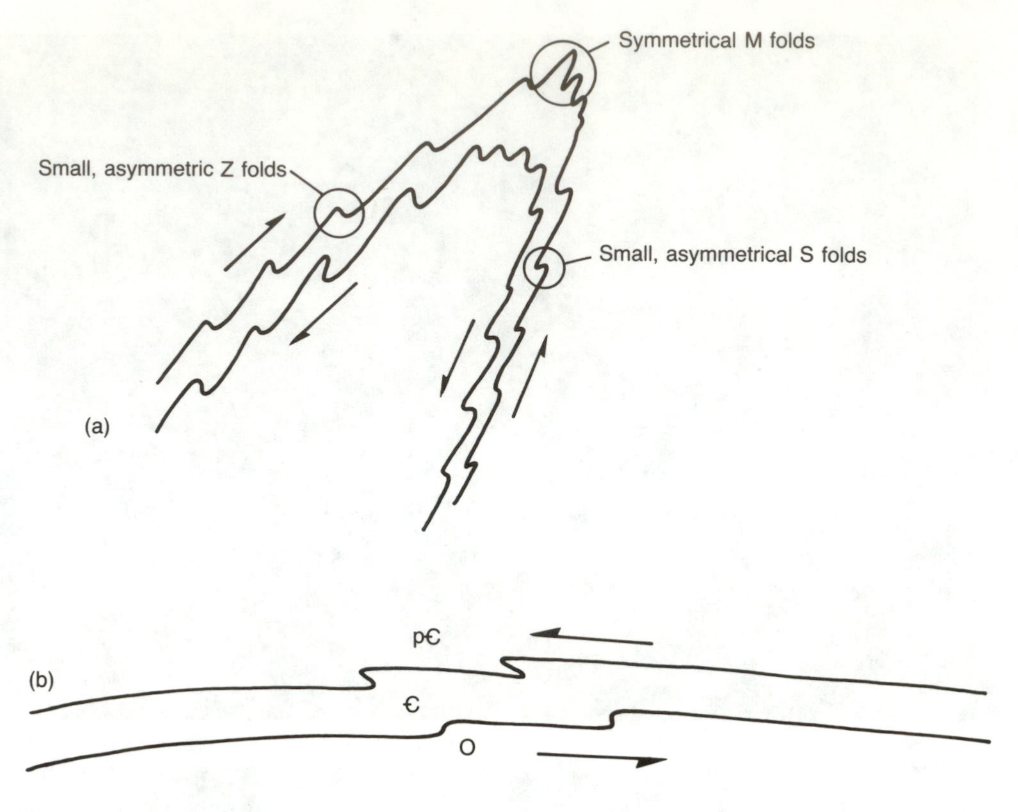

FIGURE 14–16
(a) Relationships of small S, Z, and M folds to the location of a large fold hinge. (b) Multiple deformation and Pumpelly's rule. Small folds are interpreted as an upright limb of an anticline, but stratigraphy (p€ oldest, O youngest) indicates the sequence is overturned and small folds were superimposed after the sequence was overturned.

only on this angle, but also on the rate of curvature of the fold, as John Ramsay (1967) pointed out. Therefore a working descriptive fold classification suitable for use with most folds must take into account the amount of curvature, or the curved area, of the fold.

One such classification (by Ramsay, 1967) relates variables of hinge area, or the length of the hinge, and the interlimb angle. An equation describes fold shape involving a parameter P_1 (Figure 14–21b):

$$P_1 = \frac{\text{Length of projection of limbs of join on } i_1 i_2}{\text{Length of projection of the hinge zone on the join of } i_1 i_1} \qquad \textbf{(14–1)}$$

i_1 and i_2 are inflection points on adjacent limbs of the same fold, and the hinge zone is where maximum curvature occurs. Ramsay's P_1 ratio may be useful in describing folds where a quantitative representation of the hinge and interlimb angle is important. Ramsay's modified interlimb-angle classification is purely descriptive; an observer can make measurements on a fold, calculate the parameter P_1, and categorize it without making any interpretation. Therefore, a geologist knowing very little about structural geology or the nature of folds can use this classification if he or she can make the required measurements and divide the one by the other.

G. D. Williams and T. J. Chapman (1979) have devised a classification of noncylindrical folds using a diagram with three end-members: planes, cylindrical isoclines, and isoclinal domes (Figure 14–9b). The classification is based on measurements of both interlimb angle and hinge angle. It permits classification of the entire spectrum of variations between the three end numbers.

Peter J. Hudleston (1973a) has devised a simple classification based on the assumption that an observer can recognize basic fold shapes and estimate amplitude (Figure 14–22). His scheme, called *visual harmonic analysis,* involves assigning to a given fold one basic shape (A to F) and one amplitude number (1 to 5). He recommended that diagrams of the fold shape and amplitude be taken to the field for direct comparison with natural folds in rocks.

Ramsay's Standard Classification

A geometric and descriptive classification used by many structural geologists is another devised by John Ramsay (1962, 1967). It is based on the profile

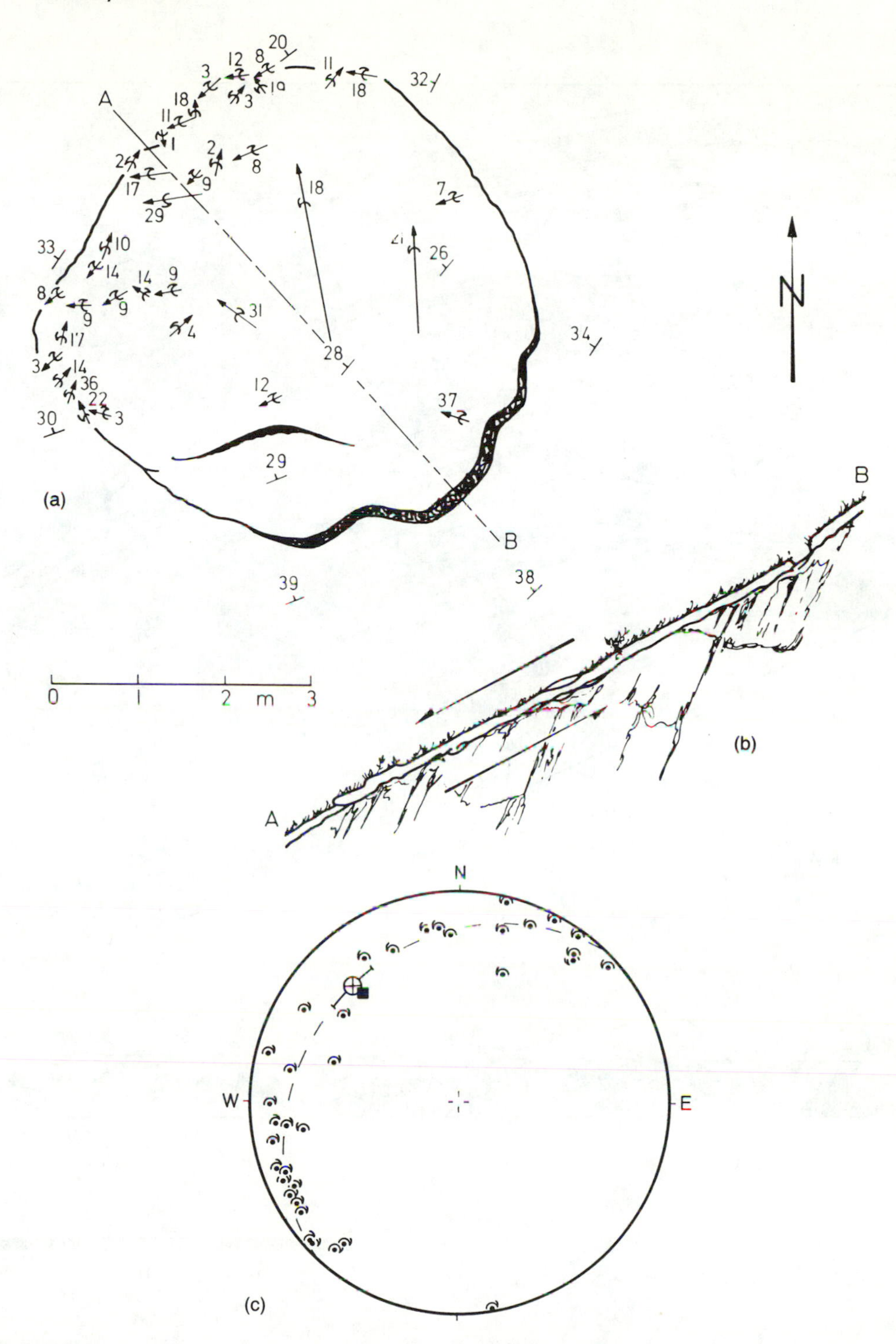

FIGURE 14–17
(a) Map of fold-axis orientations on a lobe of tundra sod on Blåhø, Trollheimen, southern Norway, moving downslope, that served as part of the model for development of Hansen's method. Each arrow represents a fold axis; the length of each arrow is proportional to the length of the hinge line measured. Semicircular arrows on each fold axis indicate the sense or asymmetry of each fold in a downplunge direction. Dip-strike symbols indicate the dip of the tundra surface where unaffected by the slide. (b) Cross section A-B; location indicated in (a). (c) Stereoplot of 36 fold axes from map in (a), each point showing the sense of overturning. Counterclockwise arrows are S-folds; clockwise arrows are Z-folds; dark square indicates transport direction. (From Edward Hansen, 1971, *Strain Facies,* Springer-Verlag.)

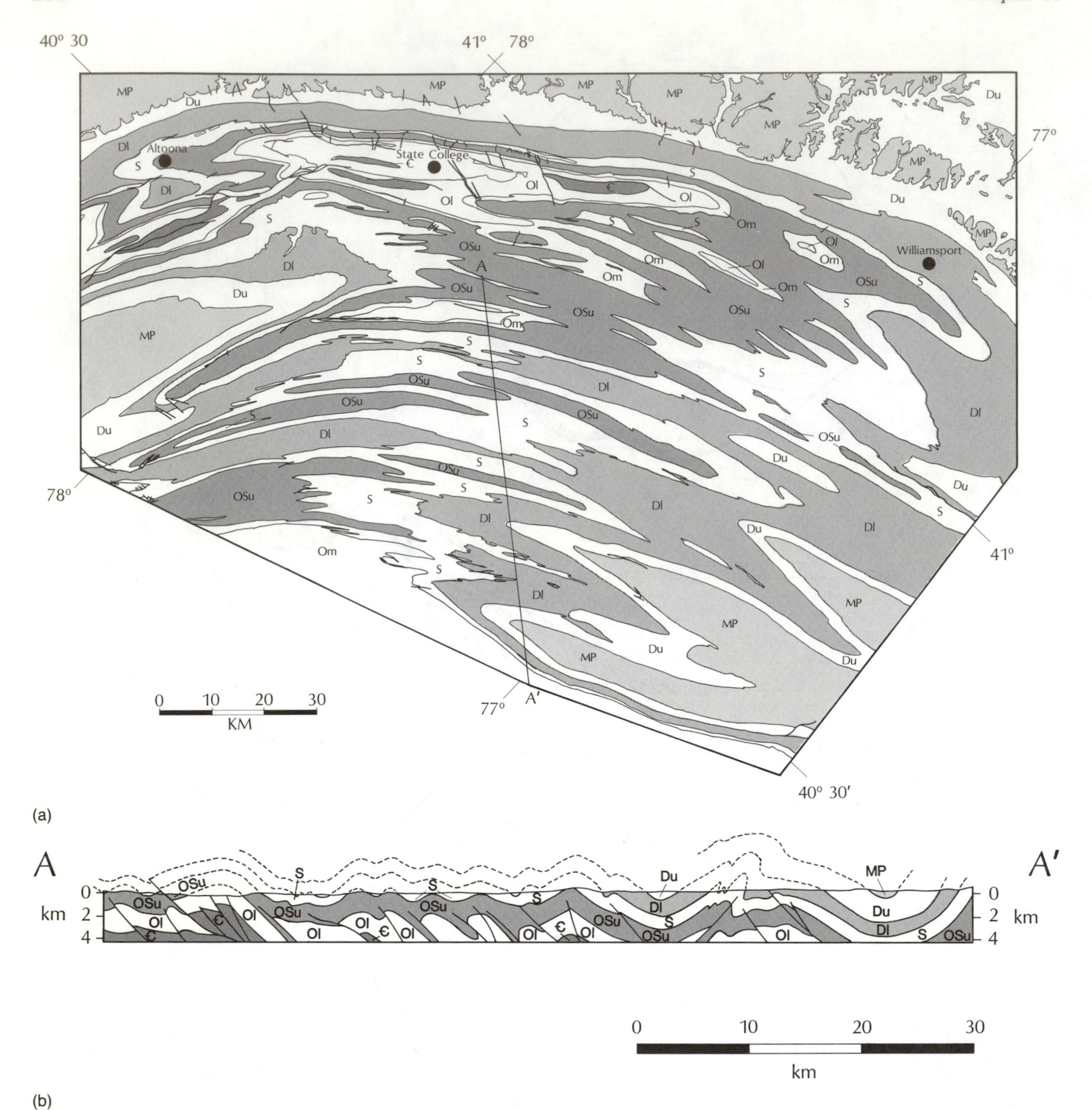

FIGURE 14–18
(a) Map-scale parallel folds in part of the Appalachian Valley and Ridge of Pennsylvania. (b) Cross section along line A-A′ showing the parallel style of folding. Note that the layers remain parallel and maintain constant thickness and that folds are faulted at depth. Є—Cambrian; Ol—Lower Ordovician; Om—Middle Ordovician; Osu—Upper Ordovician and Lower Silurian; S—Middle and Upper Silurian; Dl—Lower Devonian; Du—Upper Devonian; MP—Mississippian and Pennsylvanian. (From Pennsylvania Geologic and Topographic Survey, 1980, *Geologic Map of Pennsylvania*.)

perpendicular to the hinge of a fold. Several measurements enable the fold to be categorized without bias by any observer, even one uninitiated in fold mechanics. Thus it is one of the most useful of all fold classifications.

Ramsay's standard classification involves an indirect relationship between layer thickness both normal to layering (t_α) and parallel to the fold axial surface (T_α), and angle of dip (α) at different points on successive folded surfaces (Figure 14–23). The parameters are related by

$$T_\alpha \cos \alpha = t_\alpha \qquad (14\text{–}2)$$

The thickness at the fold hinge $t_0 = T_0$. The ratio

$$t'_\alpha = t_\alpha/t_0 \text{ or } T'_\alpha = T_\alpha/T_0 \qquad (14\text{–}3)$$

may then be calculated and plotted versus the angle of dip, α (Ramsay, 1967). Equation 14–3 expresses change in orthogonal thickness (t_α or T_α) with change in dip, α.

Lines connecting points of equal slope, or dip, are ***dip isogons.*** They may be connected between successive layers and in Ramsay's technique are constructed at regular intervals (10° is convenient) on successive layers within a fold (Figure 14–23a). The relative convergence, divergence, or parallelism of the dip isogons is a key to Ramsay's classification. Folds where the isogons converge toward the concave part of the fold are Class 1 folds. Folds with parallel isogons belong to Class 2, and folds with isogons that diverge from the concave part of the fold are in Class 3 (Figure 14–24). Class 1 is subdivided into three groups: Class 1A folds have strongly convergent isogons; Class 1B corresponds to parallel/concentric folds with convergent isogons; Class 1C folds are modified similar folds having weakly convergent isogons. Class 1C folds may correspond roughly to the flexural flow folds of Fred Donath and Ronald Parker (see the next section). Class 2 folds are ideal similar folds with parallel isogons, and Class 3 folds are similar-like folds having divergent isogons.

Ramsay's classification has a distinct advantage, being both descriptive and objective. But it classifies many folds—perhaps the majority—as 1C folds, and, as Ramsay (1962) pointed out, most similar folds as Class 2 and most parallel folds as Class 1B. Ramsay concluded that this is an attribute of the folding process rather than a defect in the scheme. His classification cannot be used readily in the field; either a photograph must be made facing parallel to the hinge of the fold, or the fold must be collected, sawed perpendicular to the hinge, and traced or photocopied so the isogons can be constructed.

Peter Hudleston (1973a) modified Ramsay's scheme by defining another measurable quantity, ϕ_α, to be used in conjunction with dip isogons. He defined ϕ_α for a folded layer as the angle between the normal to the tangents drawn to either fold surface at an angle of apparent dip, α, and the isogon (Figure 14–25). The quantity ϕ_α is either positive or negative, depending on whether the isogon is deflected clockwise (+) or counterclockwise (−) relative to the normal to the folded surfaces as the observer traces the isogon from the inner to the outer arc. The dip, α, of the right limbs of antiforms and left limbs of synforms is assumed positive and all other limbs negative. The fold in Figure 14–23a is plotted in Figure 14–26b as a plot of ϕ_α versus α. Figure 14–26a is a plot of t'_α versus α for the same fold to illustrate the standard Ramsay classification. Hudleston's modification of Ramsay's scheme permits limiting each class to a distinct field on the plot of ϕ_α versus α (Figure 14–26b). Table 14–1 summarizes relationships between Ramsay's fold classes and these parameters.

TABLE 14–1
Relationships between Hudleston's parameters and Ramsay's fold classes

Class	t'_α	ϕ_α
1A	> 1.0	< 0
1B (parallel)	1.0	0
1C	$\cos \alpha < t'_\alpha < 1.0$	$\alpha > \phi_\alpha > 0$
2 (similar)	$\cos \alpha$	α
3	$< \cos \alpha$	$> \alpha$

Values of ϕ_α are plotted for a positive α. (From P. J. Hudleston, 1972, *Tectonophysics,* v. 16, p. 1–46.)

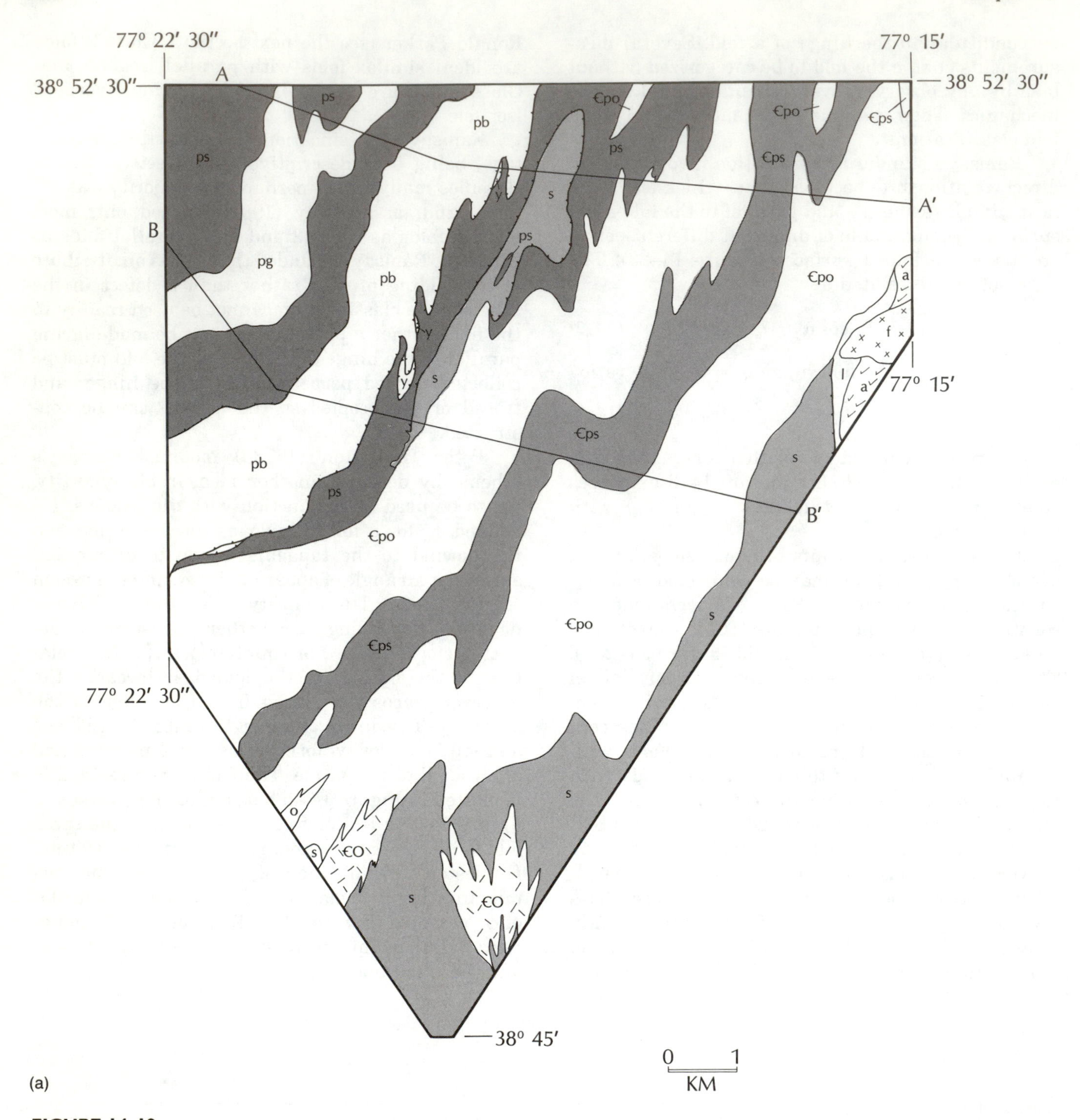

FIGURE 14–19
(a) Map of similar folds in northern Virginia. (b) (opposite page) Cross sections through the same area showing the similar folding style. Note that a thrust sheet was emplaced and later affected by similar folding. The rocks here are metamorphosed and folded sedimentary, volcanic, and plutonic rocks. Cambrian (€po, €ps, o) rocks are preserved in a syncline flanked by Precambrian rocks (ps, pg, pb, s, and y). Patterned units (a, €O, f, l) are igneous bodies. (From A. A. Drake, Jr., 1986, Geology of the Fairfax Quadrangle, Virginia: U.S. Geological Survey.)

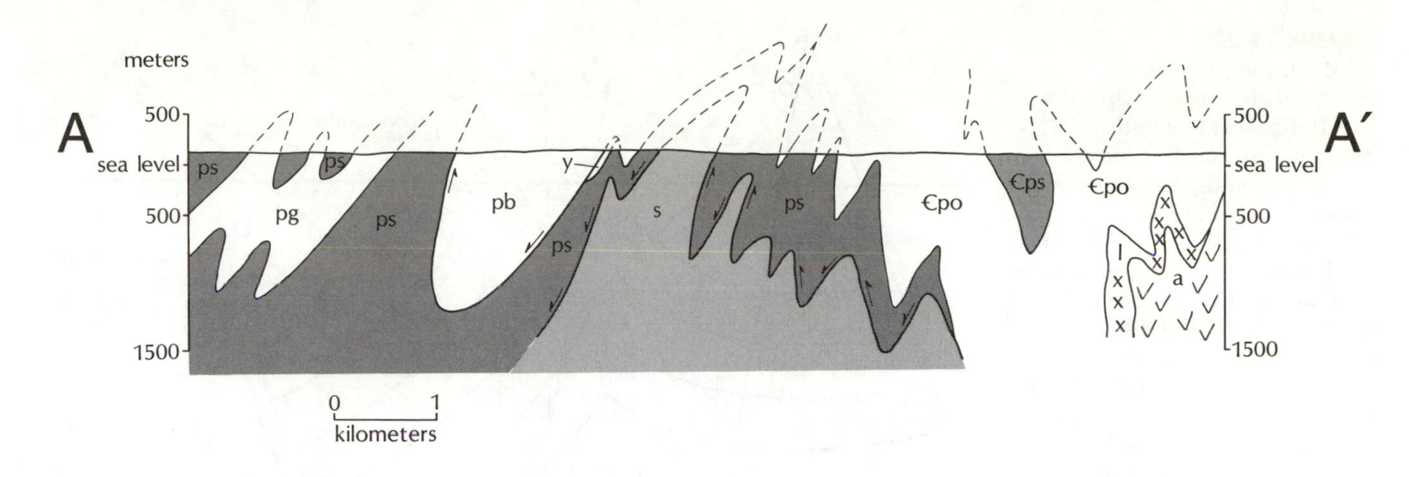

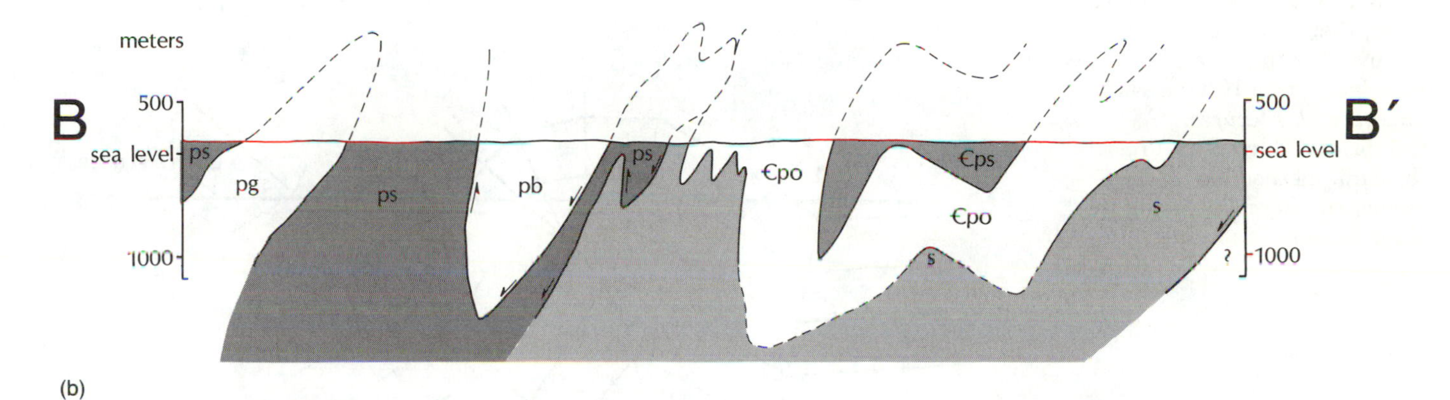

(b)

Donath and Parker Classification

A refinement of several older classifications (Knopf and Ingerson, 1938; Turner and Weiss, 1963) is one devised by Donath and Parker (1964). It is a genetic-mechanical scheme based on *mean ductility* and *ductility contrast* within the folded sequence (Figure 14–27), and thus some geologists argue that it does not belong in a discussion of fold properties and classifications. We will consider it here because it contrasts with the descriptive classifications and because it provides another option that can be used in the field.

According to the Donath and Parker scheme, two broad groups of folds exist. One, where shape is controlled by the layering in the rocks, is called ***flexural folds.*** The other, where layering serves only as a strain marker during folding, is called ***passive folds.*** A second broad, twofold subdivision within the Donath and Parker scheme is fundamentally a separation of brittle and ductile behavior. ***Slip*** along bedding, cleavage, or foliation planes is important in forming brittle folds. The process of *ductile flow* dominates in passive folds. Another category of folds that fit into neither twofold subdivision is called ***quasi-flexural*** and corresponds to disharmonic folds (Figure 14–1).

Flexural-slip folds correspond in shape and mode of formation to parallel or parallel-concentric folds (Figure 14–28). Flexural-slip folds form by buckling (being pushed from the ends of layers), bending (flexing across layering), and slip parallel to layering. Generally, these folds are not very tight except in less-competent rocks or in rocks with a strong contrast in competence (ductility). Chevron folds (Figure 14–29) form by flexural slip and commonly have tightened hinges and straight limbs. The temperature and pressure at which these folds form are low. Such folds may overprint earlier folds in rocks of high metamorphic grade, forming after the high-grade rocks have cooled and pressure has dropped considerably. Flexural-slip folds are easily recognized by slickensides, fibers, or other movement indicators (such as slip lines) on layer surfaces and by constant layer thickness.

Passive-slip folds (Figure 14–30) are a type of similar folds thought to form by shearing along planes inclined to the layering (Figure 14–27b),

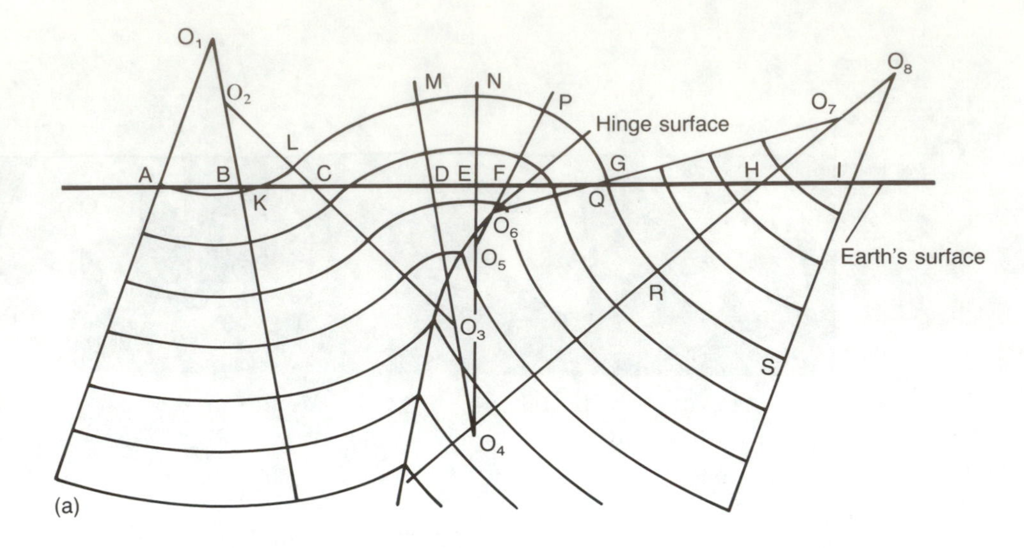

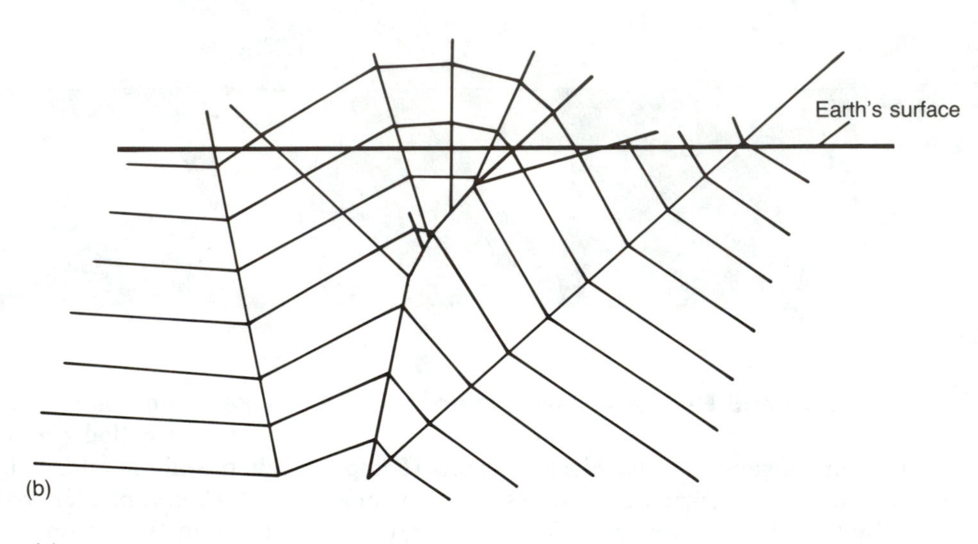

FIGURE 14–20
(a) Busk method for constructing sections through parallel-concentric folds. Normals are constructed through each dip measurement (labeled A through I). Normals intersect at points O_1 through O_8. Using O_1 as a center and O_1A as a radius, an arc is drawn connecting O_1A and O_1B. The process is repeated using O_2 as center and O_2K as radius, creating point L, and the process is continued until the fold system is constructed. Deeper layers are constructed using their thicknesses, and the overlapping arcs are rounded. (From H. G. Busk, 1929, *Earth Flexures,* Cambridge University Press.) (b) Kink method for reconstructing folds, using the same example as in (a).

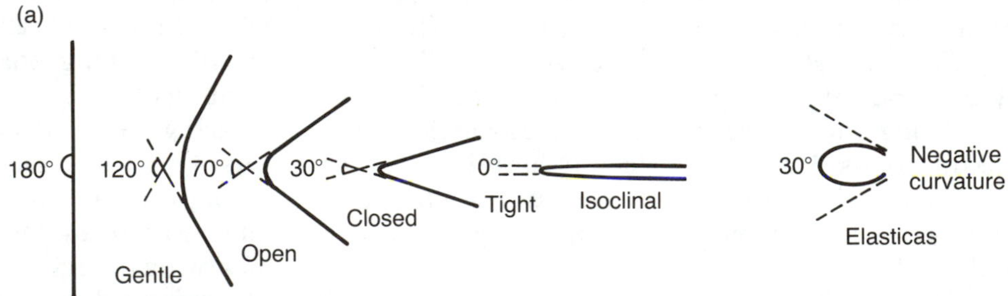

FIGURE 14–21
(a) Interlimb angle classification and categories. (After M. J. Fleuty, 1964, *Proceedings of the Geologists' Association,* v. 75, p. 461–492.) (b) Derivation of the parameters measured for calculation of Ramsay's P_1 (equation 14–1).

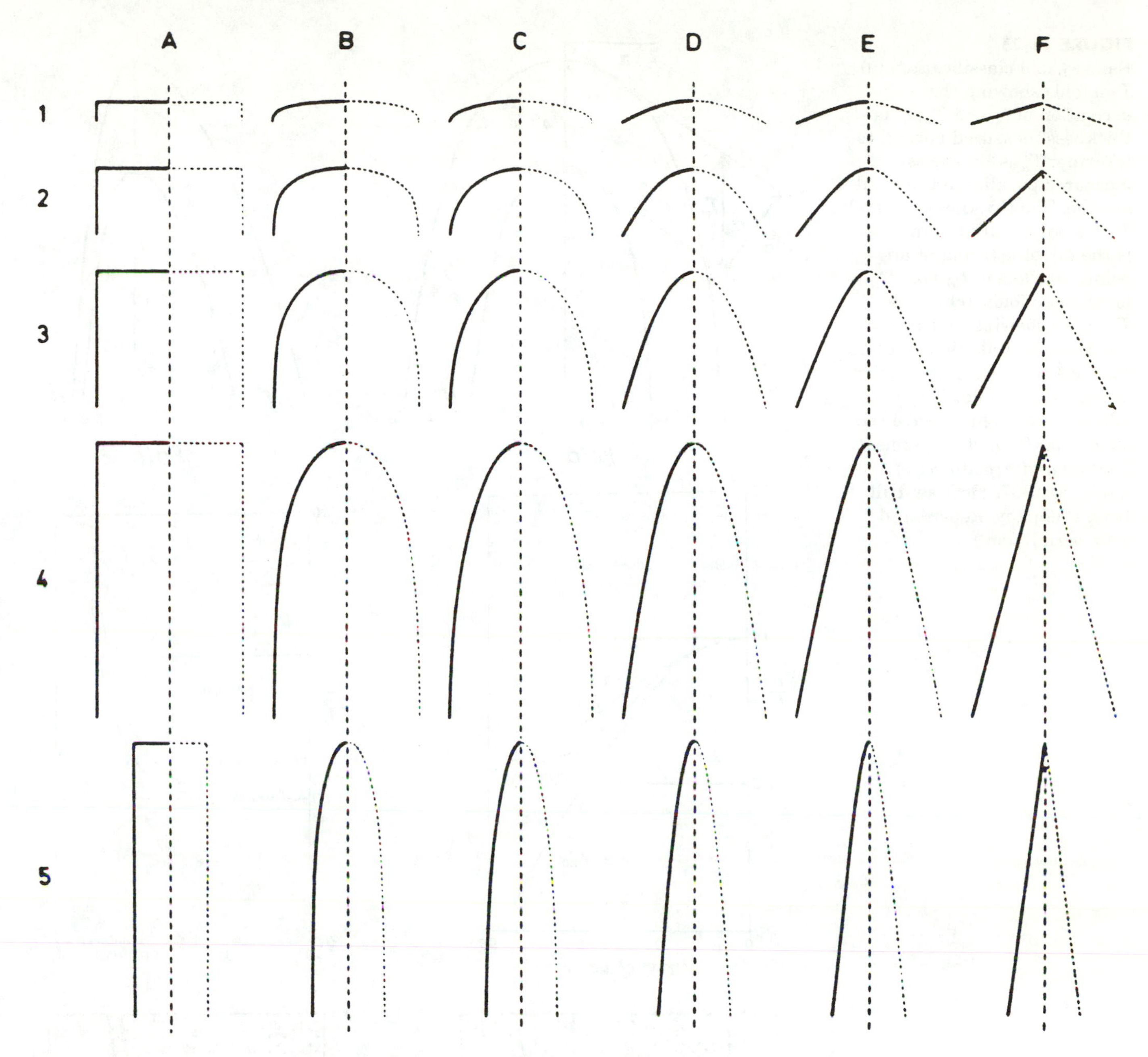

FIGURE 14–22
Hudleston's thirty folds having different combinations of shapes and amplitudes for visual harmonic analysis. (From P. J. Hudleston, 1973, *Tectonophysics,* v. 16, Elsevier Science Publishers.)

FIGURE 14–23
Ramsay fold classification. (a) Two folds showing the derivation of t_α and T_α; t_α is thickness measured normal to layering; T_α is thickness measured parallel to the axial surface. The thicknesses t_0 and T_0 are equal only in hinge; α is the dip of layering at any point. (b) Plots of t'_α and T'_α for the two folds. (c) t'_α and T'_α plots showing domains of the fundamental fold classes in Figure 14–17. (d) Construction of isogons in a fold. Points labeled 1 through 12 have the same dip. (From J. G. Ramsay, *Folding and Fracturing of Rocks*, © 1987, McGraw-Hill Book Company. Reproduced with permission.)

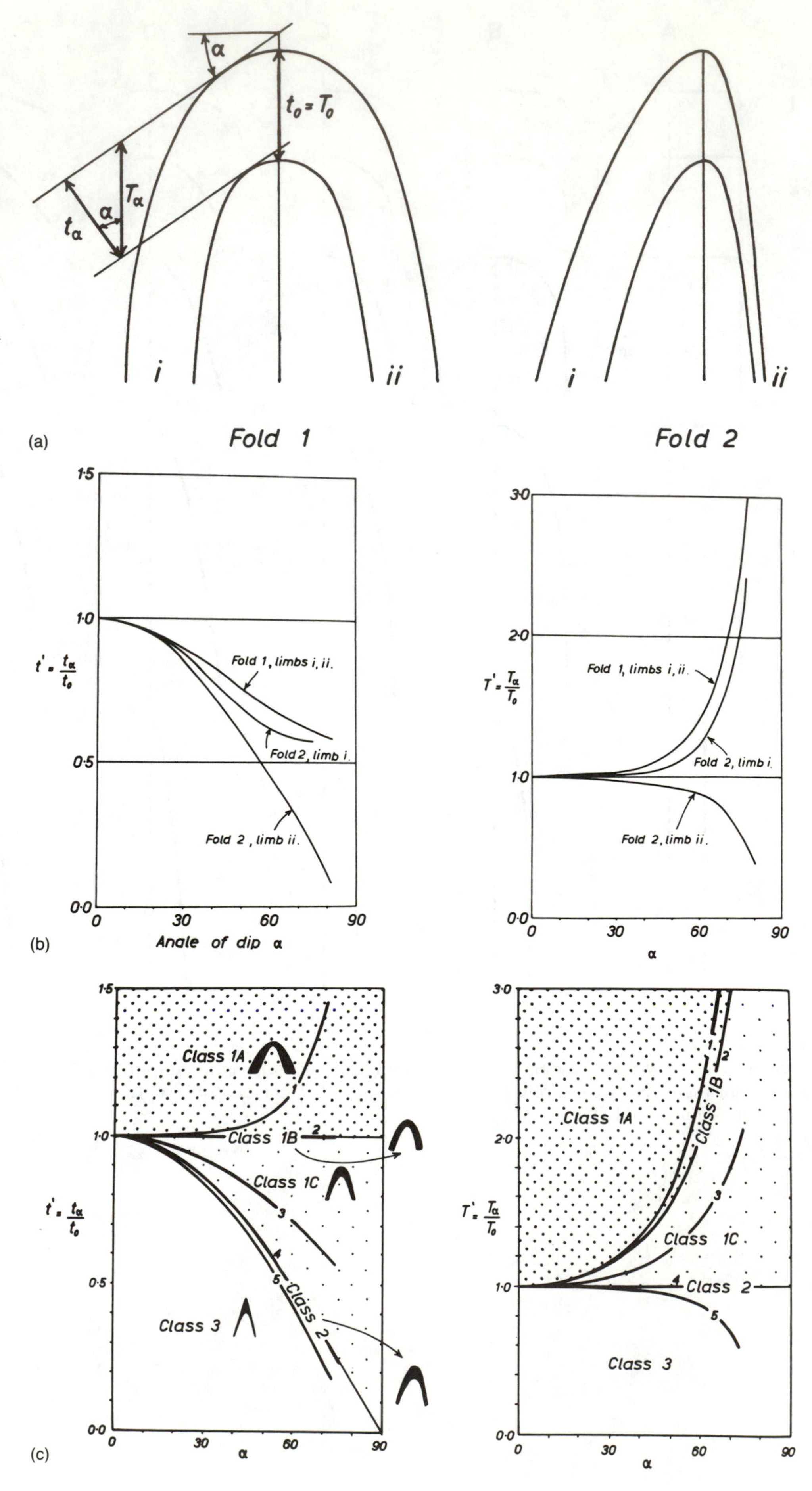

FIGURE 14–23 *(continued)*

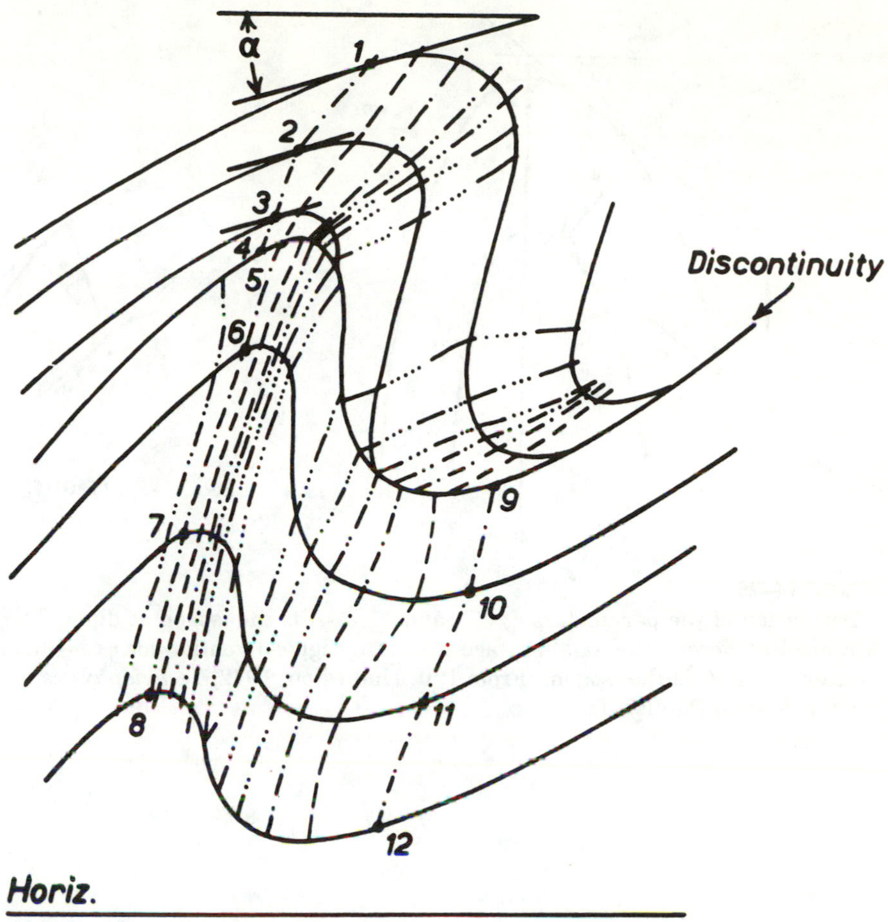

(d)

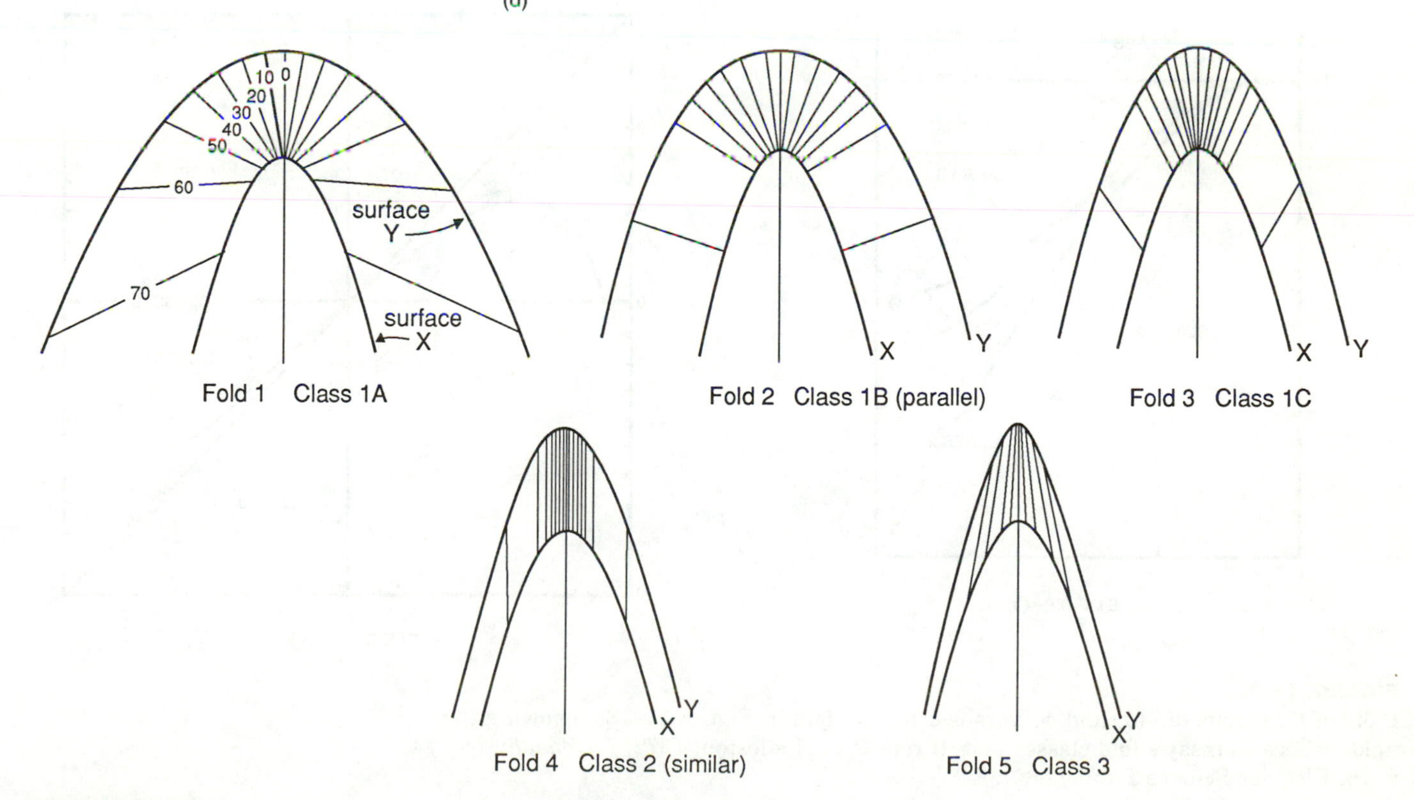

FIGURE 14–24
Ramsay fold classes. (From J. G. Ramsay, *Folding and Fracturing of Rocks,* © 1967, McGraw-Hill Book Company. Reproduced with permission.)

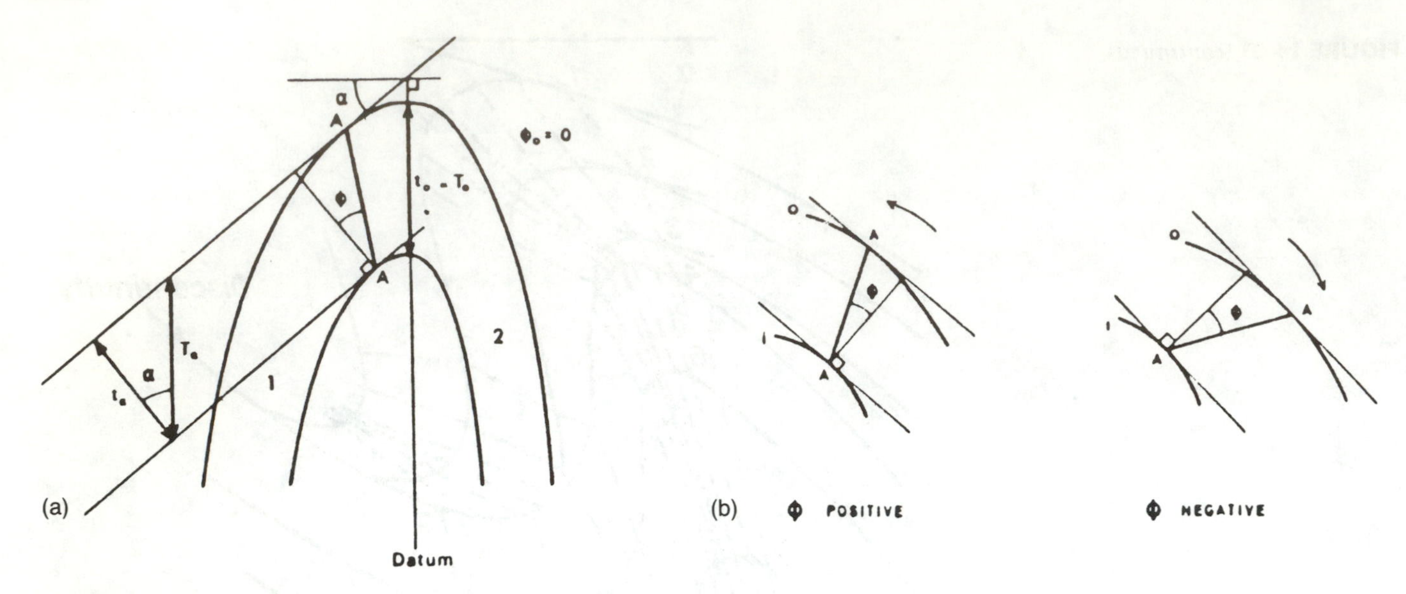

FIGURE 14–25
(a) Derivation of the parameters t_α, T_α, and ϕ_α. A-A is the isogon at dip α. The reference line here is the axial-surface trace. (b) Sign convention for ϕ: i—inner arc; o—outer arc. A-A is the isogon. (From P. J. Hudleston, 1972, *Tectonophysics,* v. 16, Elsevier Science Publishers.)

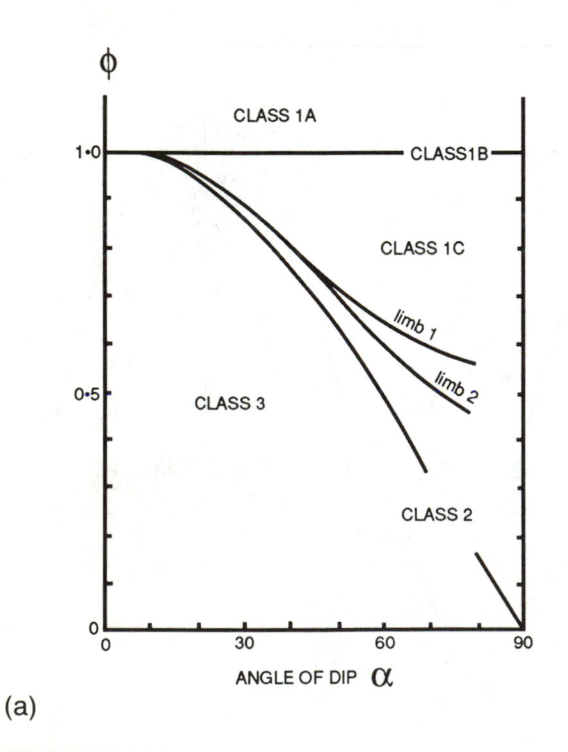

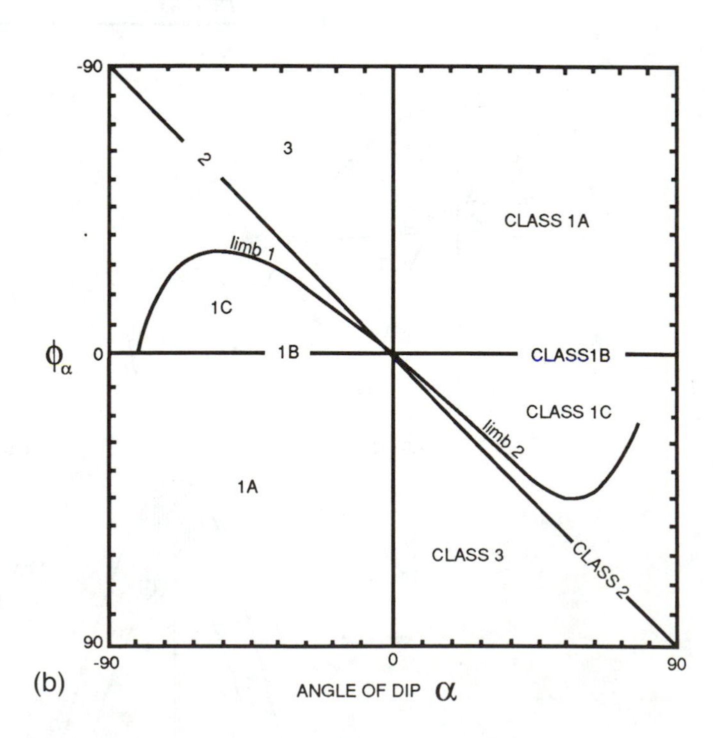

FIGURE 14–26
Plots of t'_α versus dip (α) and ϕ_α versus α for the fold in Figure 14–18a, showing the fields where Ramsay's fold classes plot. (From P. J. Hudleston, 1972, *Tectonophysics,* v. 16, Elsevier Science Publishers.)

FIGURE 14–27
(a) Basis of Donath and Parker classification. (From F. A. Donath and R. B. Parker, 1964, Geological Society of America *Bulletin*.) (b) Types and mechanisms of Donath and Parker fold types. In flexural-slip folds, layer thicknesses remain constant and folding is accomplished by slip along layers. In flexural flow, strong layers change thickness little or not at all, weak layers undergo appreciable thickness changes, and cleavage is strong in weak layers but poorly developed in strong layers. Passive slip folds are ideally developed by movement parallel to a strong cleavage, a mechanism that may not exist in nature. Passive-flow folds develop by ductile flow with limbs thinned (or relatively thickened) equally in all rock types.

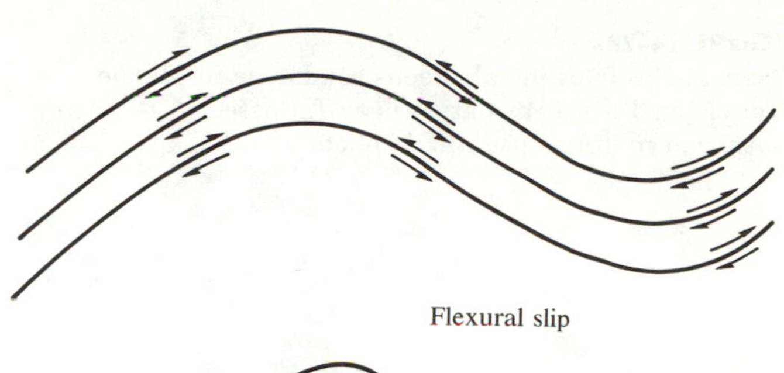

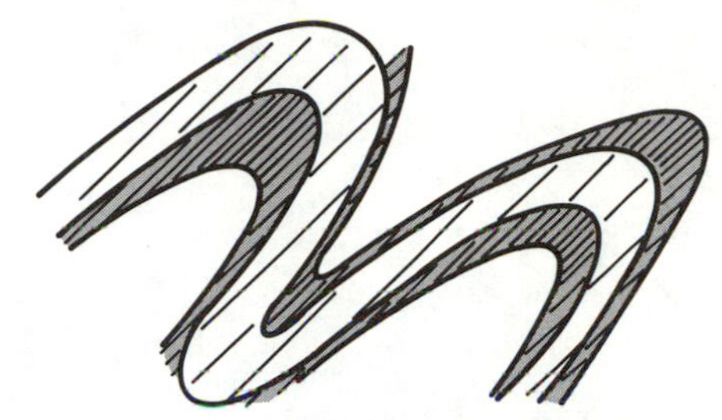

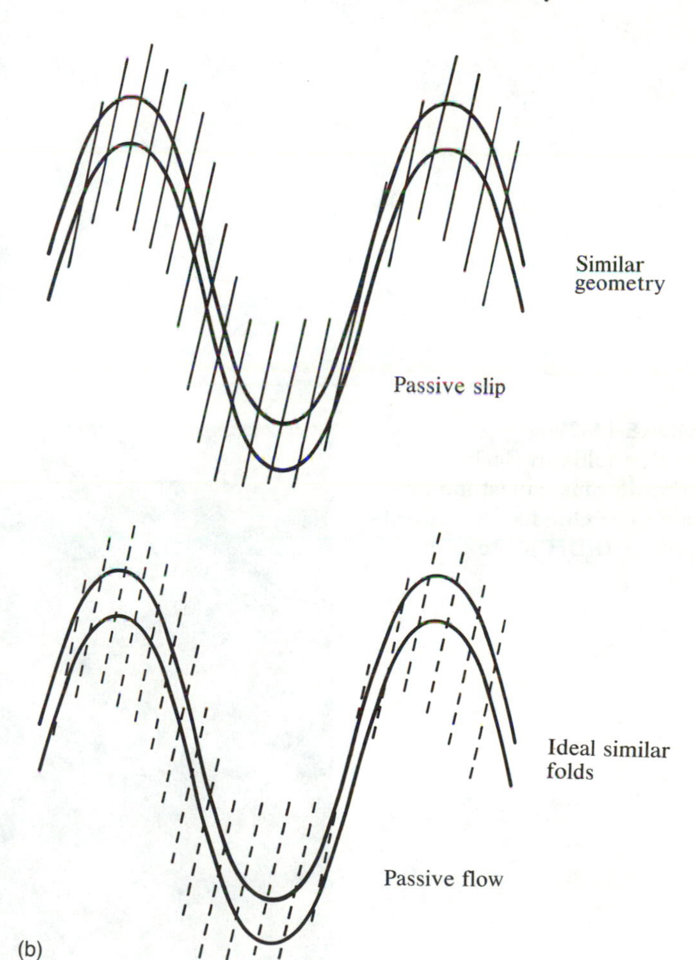

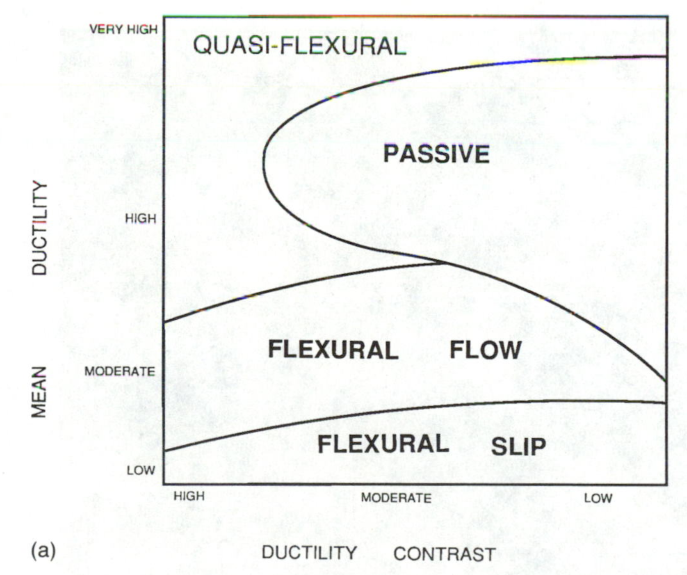

FIGURE 14–28
Flexural-slip folds in calcareous sandstone in Middle Ordovician Tellico Formation near Tallassee, southeastern Tennessee. (RDH photo.)

FIGURE 14–29
Chevron folds in Early Carboniferous sandstone and shale near Oued Zem, central Morocco. (RDH photo.)

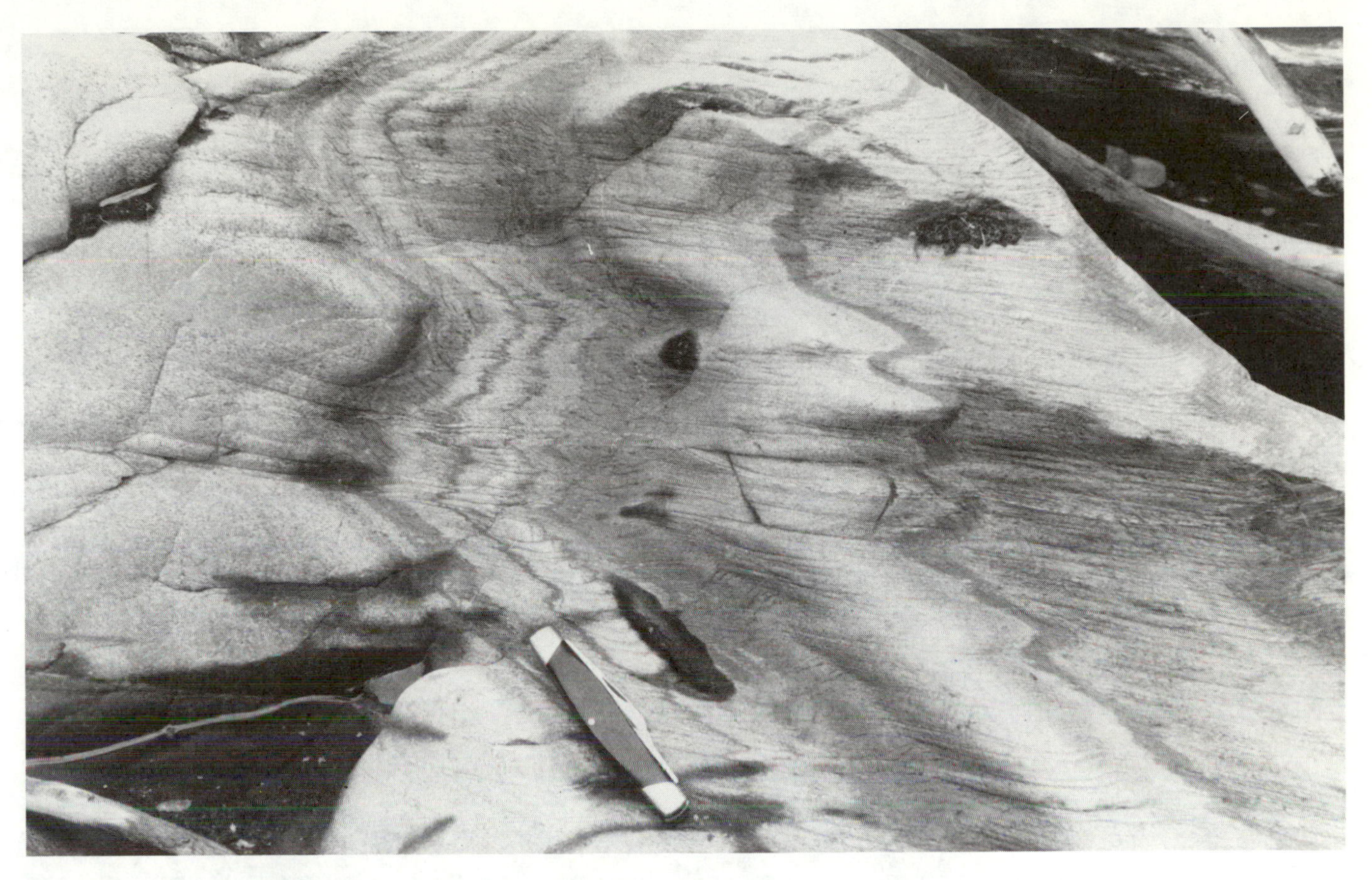

FIGURE 14–30
Folds in cleaved Lardeau Group phyllitic metasiltstone along the Illecillewaet River in southern British Columbia. Displacement of layers parallel to cleavage could result from passive slip or from pressure solution and buckling. (RDH photo.)

much as playing cards move past each other when pushed at one end of the deck. Folds of this kind would not form by pure shear.

Flexural-flow folds form in rocks from low to moderate metamorphic grade, depending on the nature of the rocks (Figure 14–31). Flexural-flow folds are mostly similar-like folds, but may also include some parallel folds. Some layers in single flexural-flow folds maintain constant thickness; others are thickened into axial zones and thinned into limbs as folding proceeds, indicating a higher contrast in internal ductility. As a result, flexural flow may be observed in shales in an interbedded limestone and shale sequence deformed at very low temperature. The shale layers change thickness appreciably as they undergo strongly plastic or semiplastic flow, and the limestone layers undergo brittle deformation and maintain constant thickness (Figure 14–31a). In much the same way, flexural-flow folds may occur in moderate to high-grade metamorphic rocks in which a layer, such as amphibolite with a relatively low mean ductility (high competence), is interlayered with a weak material like schist (Figure 14–31b). The schist will flow ductilely but the amphibolite will deform brittlely or have a slower rate of ductile deformation. As a result, the amphibolite may not change thickness much during folding. In flexural-flow folding, the layers that do not undergo appreciable thickness changes control the overall shape of the folds.

Passive flow folds are ideal similar folds (Figure 14–32) that involve plastic deformation. The layering acts only as a strain marker to record the effects of deformation. All layers are thinned or thickened equally. Passive-flow folds form in metamorphic rocks with low mean ductility and ductility contrast regardless of grade, and in salt, glacial ice, and water-saturated unconsolidated sediments wherein properties are uniformly ductile (Figures 2–2a, 2–30a, 2–31, and 15–2).

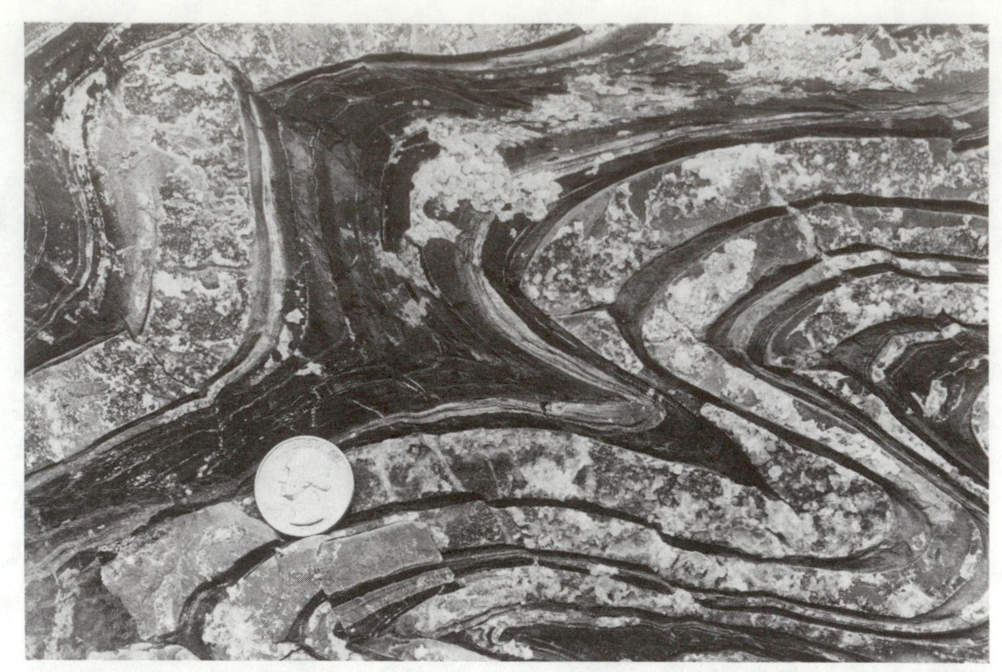

(a)

(b)

FIGURE 14–31
(a) Flexural-flow folds in shale and limestone near Whitehall, New York. Weak shale layers (dark) thicken and thin by flow; stronger limestone layers change thickness less in the short limbs and hinges. They tend to fracture in the long limbs of folds and where limestone beds are thick. (b) Flexural-flow folds in Lower Cambrian Hamill Group quartzite and schist, northern Selkirk Mountains, British Columbia. Darker schist layers undergo greater thickness changes than the lighter quartzite layers. (RDH photos.)

The Donath and Parker classification has one important shortcoming: it is genetic and is meant to explain fold mechanics, requiring the user to make a judgment each time a fold is categorized. The user must have more than a cursory understanding of fold mechanics and must be able to apply impartially the criteria for mechanical processes. The principal advantage of the classification lies in usefulness in the field. For example, to conclude that a fold is a flexural-slip fold, a geologist must observe the fold shape and bed thickness (whether thickness changes or remains constant), make a judgment about ductility contrast and mean ductility, and show that slip occurred parallel to bedding. That the

FIGURE 14–32
Passive-flow folds in Middle Proterozoic migmatitic biotite gneiss, Thor-Odin dome, Shuswap complex, British Columbia. These folds have a similar or near-similar geometry. Quartz-feldspar layers (light-colored) provide minimal control of the shape of folds.

observations can be made in the field is an advantage, but interpretations required by the observer may prove a disadvantage.

Which Classification to Use?

The brief comparison in this chapter points out advantages and disadvantages of descriptive and genetic classifications. No classification is ideal. One reason is that few folds are ideal and so few can be pigeonholed readily. The descriptive classifications may be too rigid. Genetic classifications are too subjective and may not describe real folds. Perhaps the ideal classification would be usable in the field and would yield information about the nature of the folding process without demanding much judgment by the observer. Folding is a complex process, and it is often hard to determine the boundary between brittle and ductile behavior at the time of deformation; some rocks may exhibit brittle behavior, and others in the same fold may exhibit ductile behavior. (The interbedded shale and limestone producing flexural flow is an example.) Rock composition, texture, and overall physical properties involve many more variables than can be accounted for in any simple classification.

Which fold classification should be used? It depends partly on what the observer needs to determine about the folds in an area. Probably the best scheme is Ramsay's standard classification—given an ability to distinguish between parallel folds and similar folds in the field.

In the next chapter, we will examine the mechanics of fold formation, further expanding our ability to understand the differences between folds and the mechanisms that form them.

Questions

1. What is the main pitfall in relating small and large scale folds using Pumpelly's rule?
2. If a fold axial surface has a strike of N 35° E (035) and dips 49° SE, what is the range of possible orientations for the trend and maximum magnitude of plunge of the fold axis?
3. What is meant by the statement, "These folds have a northeast vergence"?
4. Determine the positions and nature of major fold axial zones by reconstruction from the small structures represented in the outcrop sketches shown here. The layers in the second and fourth exposures are not the same; those in the first, third, and fifth exposures are the same layer.

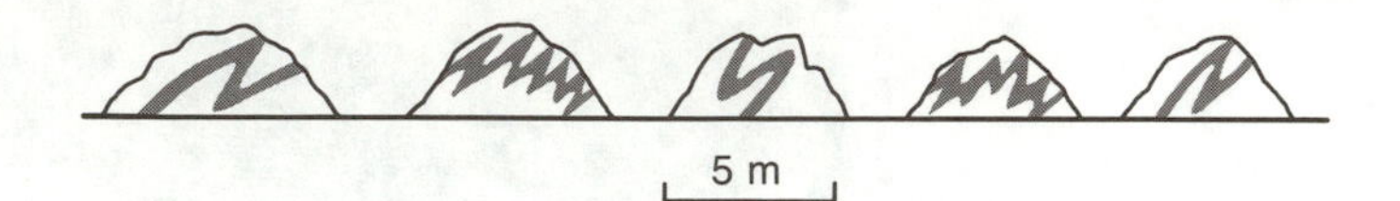

5. What kind of axis does an ideal dome or basin have?
6. How can a large fold structure be defined using the enveloping surface?
7. What are the relative merits of genetic and descriptive classifications? Shortcomings?
8. What is the basis of the Donath and Parker fold classification? Ramsay's classification?
9. What information would you need to determine whether a fold is a passive-flow fold? A class 1C fold? How and where would you obtain this information?
10. The following orientations (trend, plunge, sense of rotation—S or Z) of mesoscopic folds above the Chocolate Mountains thrust were collected in southeastern California by Gordon Haxel of the U.S. Geological Survey. Plot these data on a stereonet, indicating sense of rotation on each point, and determine the transport direction of the thrust.

N 8 E	19 NE	(Z)	N 32 E	4 NE	(S)	N 54 W	27 NW	(Z)
N 22 E	7 NE	(Z)	N 33 E	7 NE	(Z)	N 39 W	14 NW	(Z)
N 17 W	8 NW	(Z)	N 28 E	22 NE	(Z)	N 79 E	3 NE	(S)
N 1 W	12 NW	(Z)	N 68 E	4 NE	(S)	N 82 E	7 NE	(S)
N 28 E	31 NE	(Z)	N 48 E	34 NE	(S)	N 86 E	1 NE	(S)
N 52 W	4 NW	(Z)	N 86 W	4 NW	(Z)	N 89 W	27 NW	(Z)
N 19 E	22 NE	(Z)	N 31 W	17 NW	(Z)	N 88 E	4 NE	(S)
N 36 E	19 NE	(S)	N 72 W	19 NW	(Z)	N 89 E	12 NE	(S)
N 37 E	53 NE	(S)	N 69 E	12 NE	(S)	N 61 W	27 NW	(Z)
N 23 W	11 NW	(Z)	N 34 E	7 NE	(Z)	N 63 W	31 NW	(Z)
N 22 E	24 NE	(Z)	N 32 E	34 NE	(Z)	N 66 W	3 NW	(Z)
N 32 E	17 NE	(S)	N 24 E	33 NE	(Z)	N 70 W	3 NW	(Z)
N 53 W	32 NW	(Z)	N 24 E	3 NE	(Z)	N 73 W	5 NW	(Z)
N 87 E	28 NE	(S)	N 73 E	8 NE	(S)	N 66 W	8 NW	(Z)
N 63 E	12 NE	(S)	N 76 E	6 NE	(S)	N 59 W	6 NW	(Z)
N 82 E	33 NE	(S)	N 75 E	9 NE	(S)	N 56 W	7 NW	(Z)
N 79 W	18 SE	(S)	N 78 E	22 SW	(Z)	N 58 E	33 SW	(Z)
N 68 W	8 SE	(S)	N 58 W	27 SE	(S)	N 57 W	23 SE	(S)
N 74 W	23 SE	(S)	N 82 E	24 SW	(Z)	N 87 W	4 SE	(Z)
N 81 E	4 SW	(Z)	N 58 W	9 SE	(S)	N 47 W	28 SE	(S)
N 49 W	17 SE	(S)	N 43 W	8 SE	(S)	N 28 W	16 SE	(S)
N 32 W	11 SE	(S)	N 62 E	7 NE	(S)	N 69 W	7 NW	(Z)

Further Reading

Donath, F. A., and Parker, R. B., 1964, Folds and folding: Geological Society of America Bulletin, v. 75, p. 45–62.
A useful genetic synthesis and classification of folds based on mean ductility and ductility contrast. The classification is easy to follow and use in the field.

Fleuty, M. J., 1964, The description of folds: Proceedings of the Geological Association: v. 75, p. 461–492.
Critically reviews the descriptive terminology of folds and provides worthwhile comments about the terms Fleuty considers most useful.

Hansen, Edward, 1971, Strain facies: New York, Springer-Verlag, 207 p.
Describes a useful method for determining transport direction of a large fold from sense of overturning and orientation of parasitic folds.

Hobbs, B. E., Means, W. D., and Williams, P. F., 1976, An outline of structural geology: New York, John Wiley & Sons, 571 p.
The chapter on folds in this text is especially good and discusses fold mechanics.

Hudleston, P. J., 1973, Fold morphology and some geometrical implications of theories of fold development: Tectonophysics, v. 16, p. 1–46.
Takes the geometric classifications to a higher level with the modification of Ramsay's classification. It also presents a useful discussion of fold mechanisms.

Ramsay, J. G., 1967, Folding and fracturing of rocks: New York, McGraw-Hill, 568 p.
Discusses several of the fold classifications outlined in this chapter. It is an advanced text, but elementary structure students will find the mathematics simple.

Williams, G. D., and Chapman, T. J., 1979, The geometrical classification of noncylindrical folds: Journal of Structural Geology, v. 1, p. 181–185.
An easy-to-follow classification that should prove useful in the field, as it requires only observation of relationships between axes, hinges, and limbs and measurement of interlimb and hinge angles.

15

Mechanics of Folding

One of the intriguing features of layered rocks deformed during natural orogenic processes is that their surfaces are often curved to form folds. The methods of classification of these folds and the determination of their mechanisms of formation have long been controversial subjects in the study of rock structure.

JOHN G. RAMSAY, *Folding and Fracturing of Rocks*

FOLDED LAYERS IN ROCKS HAVE BEEN OBSERVED FOR MANY centuries. The ancient Greeks reported folded rocks in their exploration of the Mediterranean region. Leonardo da Vinci (1452–1519) was impressed by contorted strata preserving marine fossils high in the Alps, as indicated in Vallisnieri's early eighteenth-century sketch (Figure 15–1). Sir James Hall (1761–1832) in 1788 described folds in the sea cliffs on the coast of Scotland (A. M. Johnson and Ellen, 1974).

The geologists who first described folds doubtlessly speculated about how they formed. The phenomenon is readily observed in glacial ice (Figure 15–2) and in soft sediment deformed by gravity (Figure 2–2). The close association of folds and cleavage was first noted by several geologists before 1850, directly linking internal forces and folding. Da Vinci's uplifted and contorted limestones containing marine fossils, high above sea level, required tectonic forces to elevate them—forces unexplained for centuries.

The wide distribution of folds, their occurrence on all scales, and their great variety all provide convincing evidence that they are truly a major class of geologic structures. But *how* did these structures form? In Chapter 14 we described the extensive fold terminology. In this chapter we will explore the mechanisms of folding, taking into consideration the environment of the rock body being deformed. The mechanism is strongly influenced by the factors that govern other kinds of deformation: temperature, pressure, fluid, and overall properties of the rock body as determined by the composition and character of individual layers. The relative thickness and contrasting properties within and across layers—***anisotropy***—greatly influence the kind of fold mechanism affecting a rock body. Anisotropy is in turn affected by changes in temperature and pressure.

Attempts by Bailey Willis (1857–1949) to model folds and fold mechanisms date back nearly a century. His goal was to discover the relationships between folds and thrusts in the Appalachian Valley and Ridge Province by duplicating them in a pressure box (Willis, 1893; Figure 10–1). H. M. Cadell, a British geologist of the early twentieth century, used a pressure box to produce folds in sediment with layers having contrasting properties. The studies by Cadell and Willis, though primitive by present standards, laid the foundation for our present-day understanding of fold mechanics. Willis's break thrusts and shear thrusts of 1893 are the same as John Suppe's fault-propagation folds and fault-bend folds of 1985.

As we know, folds bear a remarkable similarity to waves, and some mathematical studies of folding are based on wave mechanics. In recent years, modeling of folds has been done both experimentally

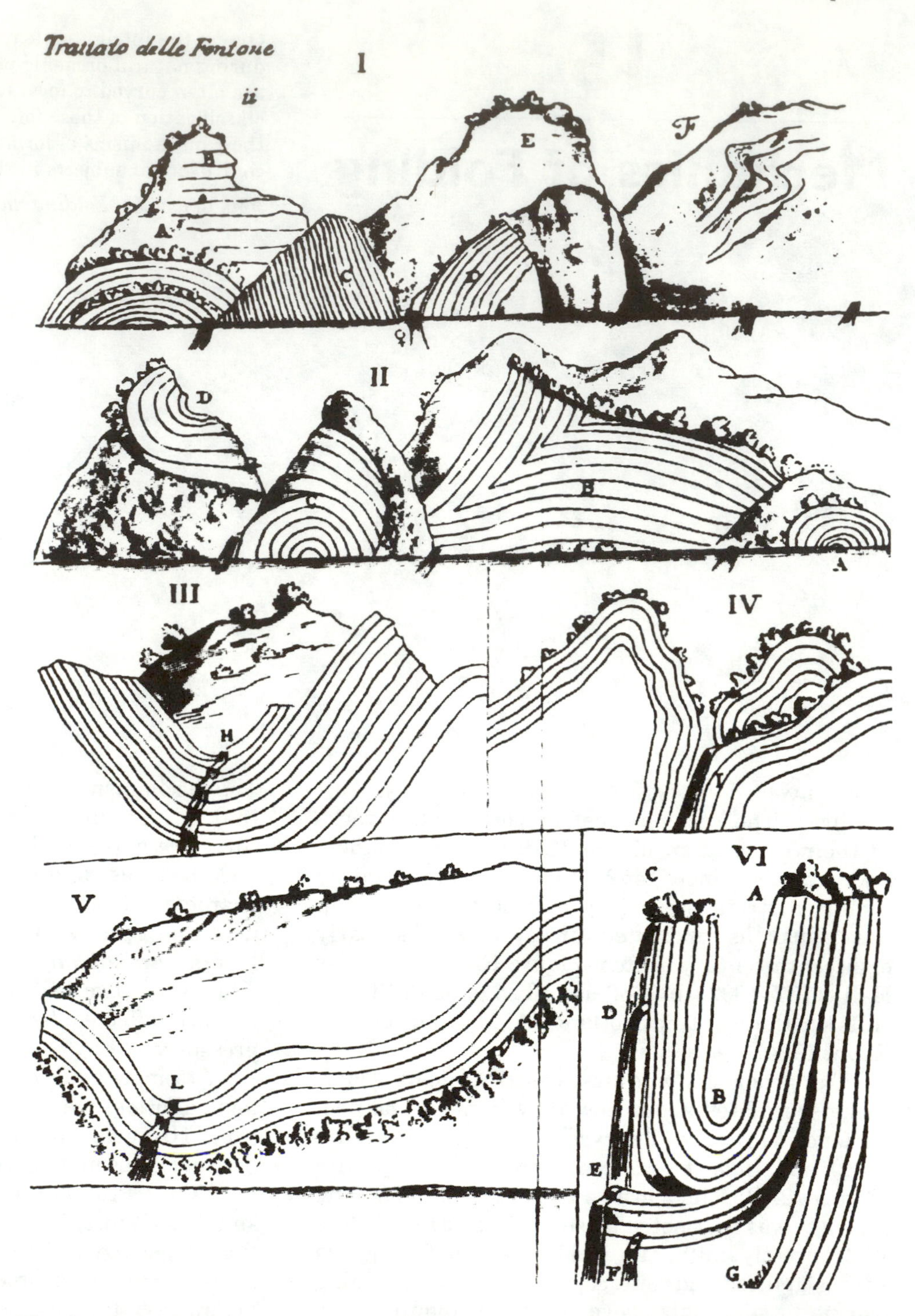

FIGURE 15–1
Folded rocks from sketches by Antonio Vallisnieri of folds on mountainsides in the Alps of Switzerland and Germany. (From *The Birth and Development of the Geological Sciences* by Frank Dawson Adams, © 1938, 1966, Dover Publications, Inc.)

and by computer simulation. The latter provides a way to model strain distribution (using the finite-element method) and to otherwise simulate fold mechanisms.

FOLD MECHANISMS AND ACCOMPANYING PHENOMENA

Several fold mechanisms have been identified (Figure 15–3a), including *buckling, bending,* and *passive (ductile) flow;* the last is sometimes called *passive strain amplification.* These mechanisms may be accompanied by *flexural slip* and *flexural flow.* Groshong (1975a) has discussed each and evaluated its relationships to deformation mechanisms involved in a rock mass during folding in unmetamorphosed sedimentary rocks in the Appalachians of Pennsylvania. There Groshong recognized that pressure solution was an important deformation mechanism and that a model including twin gliding, translation gliding, and adjustments along grain boundaries accounted for some of the deformation accompanying folding; he also concluded that much of the deformation during folding is accomplished by displacement along fractures. As we saw in Chapter 11, fault-bend and fault-propagation folds involve bending and flexural slip associated with thrust faulting.

Parallel-concentric folds and similar-type folds may be created by more than one mechanism. Some folds are the product of only one mechanism; others, of more than one. Thus the geometric form of a particular fold may be the end product of one or more than one fold mechanism—whether it is parallel (Ramsay's Class 1B), concentric, parallel-concentric, modified concentric (Class 1C), similar (Class 2), or similar-like (Class 1C or Class 3). Each process will operate according to the nature of the deforming rock mass and the physical conditions imposed on it.

One fold mechanism, such as buckling accompanied by flexural slip, may operate during the early history of a fold, and then buckling accompanied by flexural flow may dominate as the fold tightens and pressure increases during progressive deformation (Figure 15–3b). Under high temperature and pressure, layers no longer control the shapes of the folds and serve only as strain markers. Folds form by ductile flow across layers and displace them. Only seldom can we prove that a sequence of multiple fold mechanisms (like that just described) affected a rock mass undergoing progressive deformation. Relict structures formed by several mechanisms have been observed in rocks, indicating that a sequence of mechanisms operated during folding.

Flexural Slip

Layers play a dominant role in folding of most rocks at the low temperature and pressure found at shallow depths in the Earth. For layers to maintain constant thickness during folding of a mass of uniformly layered strong rocks—such as thickly bedded carbonate or sandstone—they must slip past one another (Figures 15–3a and 15–4). (For an instant demonstration of flexural slip, simply flex the pages of this book.) Where the mechanical properties of successive layers differ, as with interlayered limestone and shale, flexural slip still occurs but the shale may become crumpled into the hinges of folds; that produces thickening in the shale layers without actual flow of material (Figure 15–5a). The phenomenon of flexural slip commonly accompanies the bending and buckling mechanisms and is recognized by slickensides or fibers on bedding surfaces (Figure 15–5b). If flexural slip does occur, the slickensides will be oriented perpendicular to fold hinges.

FIGURE 15–2
Folds produced by ductile flow several kilometers in amplitude in ice and moraine in the Malaspina Glacier, southeastern Alaska. The St. Elias Mountains are in the background. (Austin Post, U.S. Geological Survey.)

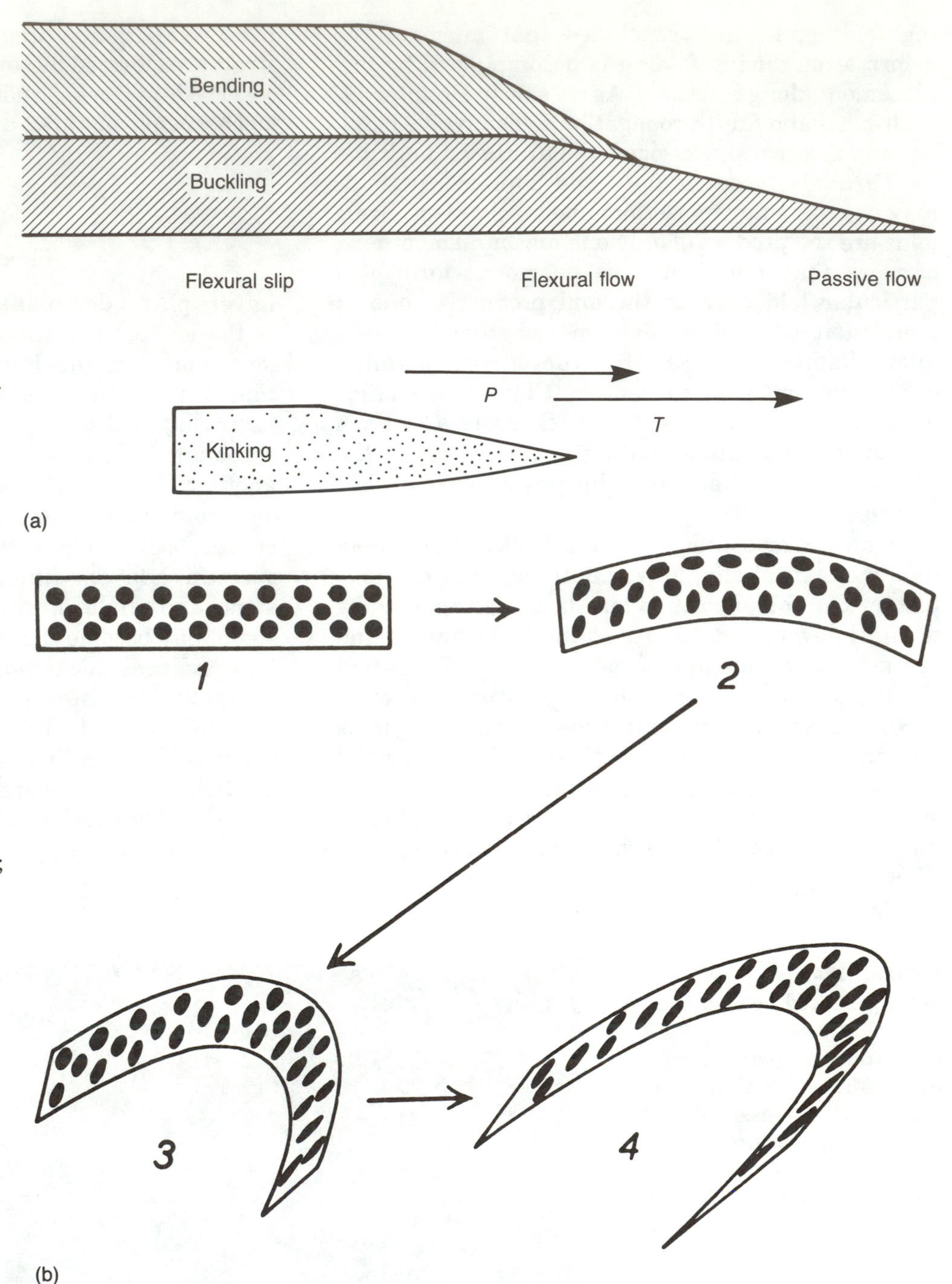

FIGURE 15–3
(a) Possible relationships of fold mechanisms. The horizontal axis relates increasing temperature and pressure to fold mechanism. The vertical axis reflects contrasting mechanisms under different conditions. Bending and buckling occur wherever layering influences fold shape. Kinking occurs similarly but independently of buckling and bending. Buckling and bending probably occur at early and later stages over the same temperature range. (b) Stages in the development of a folded layer showing several different processes occur at early and later stages during the progressive deformation of the layer by folding. Stage 1 is the undeformed condition. Stage 2 involves slight buckling (or bending) of the layer accompanied by internal heterogeneous strain involving stretching of the outer arc and compression of the inner arc (tangential-longitudinal strain; see Figure 15–14). Stage 3 involves progressive deformation of the layer by heterogeneous simple-shear strain, which is later superimposed (Stage 4) by homogeneous strain and pure shear (b is from J. G. Ramsay, *Folding and Fracturing of Rocks,* © 1967, McGraw-Hill Book Company. Reproduced with permission.)

Bending

Rock layers may be subjected to ***bending*** (Figure 15–6), a mechanism that involves application of force across layers. It produces folds that for the most part are very gentle. The term has also been used by Hans Ramberg (1963) and John Ramsay (1967) to describe a different mechanism, one that involves passive flow across layers. Folds produced by bending, in the sense intended here, are common in continental interiors—cratons—where vertical stress may be directed at high angles to originally horizontal bedding producing the broad domes, basins, swells, and arches so common in cratons. Flexural bending of lithospheric plates also occurs at subduction zones and adjacent to oceanic ridges, at smaller accumulations of volcanic rock in the oceans (such as guyots, seamount chains, and aseismic ridges), and where loading by ice or sediment causes the lithosphere to undergo flexural readjustment (Turcotte and Schubert, 1982). Another example of bending is arching of cover rocks above an intruding pluton.

In rocks subjected to bending, layers are bent like an elastic beam that has been supported at the

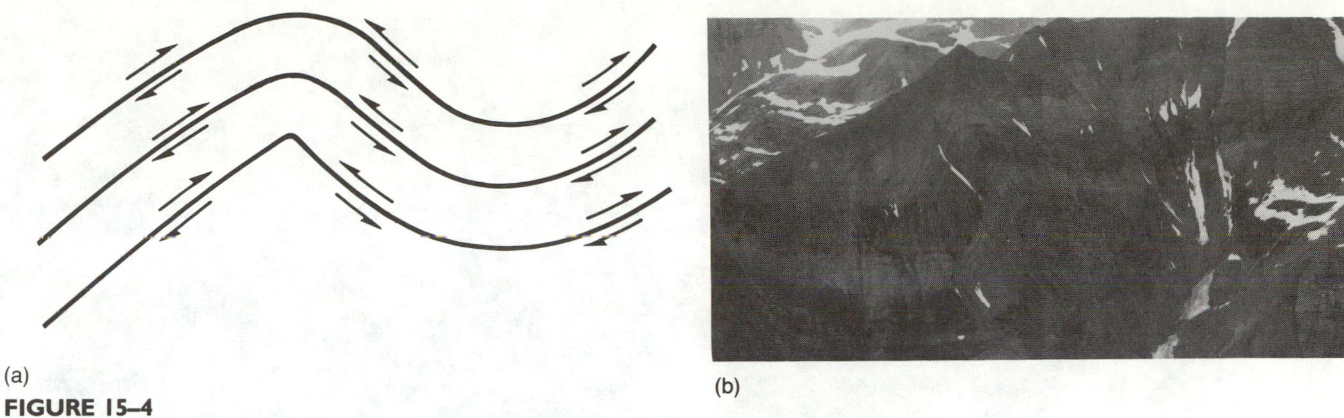

(a) (b)

FIGURE 15–4
(a) Flexural-slip folds. Arrows indicate relative motion on each layer surface during folding. (b) Flexural-slip buckle folds in Late Proterozoic Hamill Group sandstone and shale, Selkirk Mountains, British Columbia. (RDH photo.)

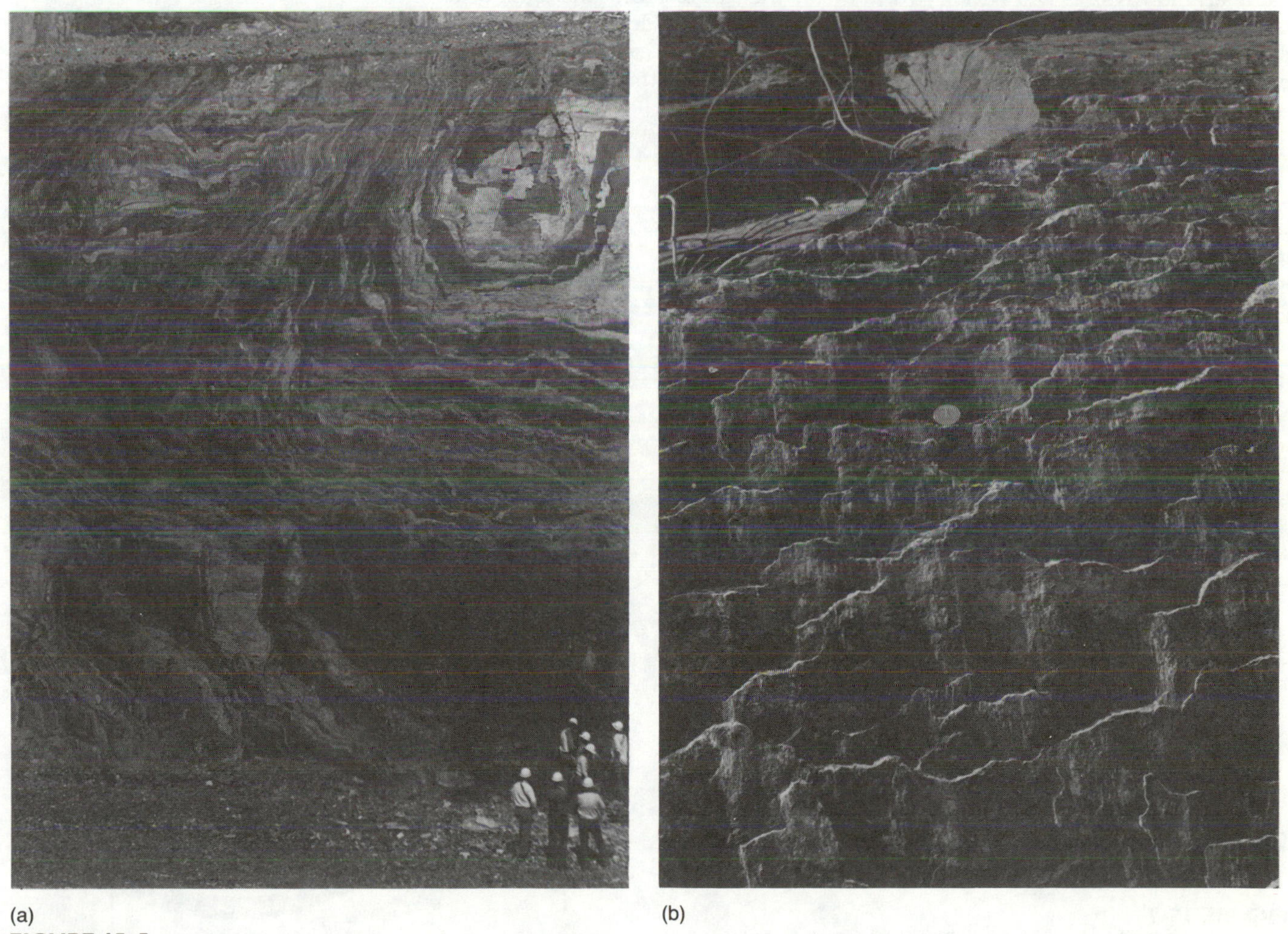

(a) (b)

FIGURE 15–5
(a) Abandoned barite mine west of Glenwood, Arkansas, exposing interlayered sandstone and shale near the base of the upper Mississippian Stanley Shale. Shale layers were thickened by crumpling into hinges of folds rather than by ductile flow. (b) Slickenfibers on folded bedding surfaces in calcareous sandstone produced by flexural slip, Middle Ordovician Tellico Sandstone near Tallassee, Tennessee. Note that dip of bedding decreases toward the top background of the photo into the hinge of the fold. (RDH photos.)

FIGURE 15–6
Candidate for a bending flexural-slip fold in Middle Ordovician limestone on the Cumberland River at Old Hickory (near Nashville), Tennessee. This could be a buckle fold, but it is located in the "stable interior" where large domes and basins, and possibly smaller folds, are thought to be related to vertical tectonic movement caused by loading of the crust during the late Paleozoic by thrust faulting farther east in the Appalachians. (RDH photo.)

ends and loaded in the middle (Figure 15–7). The layers undergo flexural slip, sliding past each other just as if they were buckled, and the layers are deflected into gentle folds. Mathematically modeling the bending of layers as elastic beams works well for small folds, but becomes inexact for large segments of the lithosphere because of inelastic behavior.

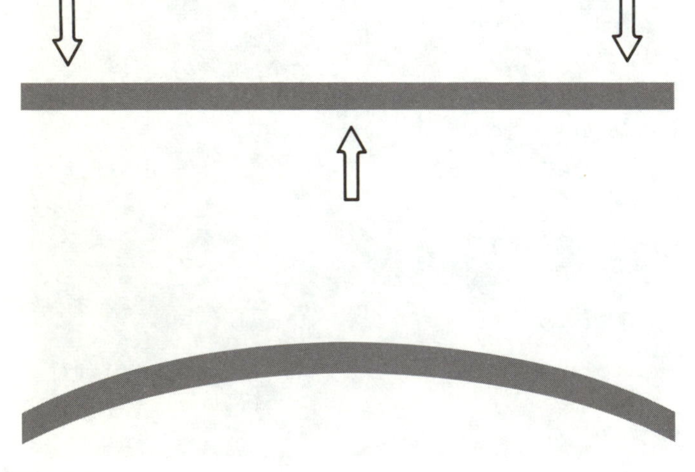

FIGURE 15–7
Formation of folds by bending. The layer is folded by deflection in the same directions as the applied stresses. This produces the same shape as is produced by buckling, but the actual mechanism involves bending of the ends of a beam that is buttressed in the middle. Compare with Figure 15–8.

Fault-bend folds (Suppe, 1983; Figure 11–11a) form by the bending of a thrust sheet as it passes over a ramp. Such folds may be the only kind of bending folds formed in orogenic belts by tectonic forces.

Buckling

Folds form by ***buckling*** where stress is applied parallel to layering in rocks (as at the ends of an elastic beam) and the easiest direction of strain is normal to the direction of stress application (Figure 15–8). The layers are said to buckle, and the structures formed are called *buckle folds*. Flexural slip commonly accompanies buckling at low tempera-

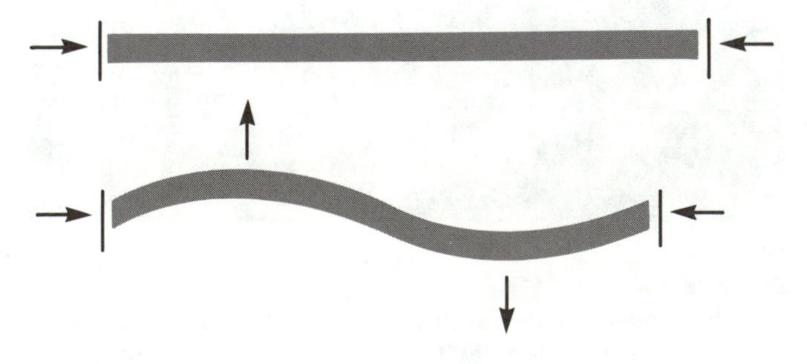

FIGURE 15–8
Formation of buckle folds. The layer is folded by deflection normal to the compression direction.

ture and pressure (Figure 15–9), whereas buckling of a rigid suspended layer (with high viscosity, strength, and competence) in a less-rigid medium may occur at higher temperature and pressure without flexural slip (Figure 15–10). The result at low temperature is sinusoidal parallel and parallel-concentric folds. At high temperature, other mechanisms, such as flexural flow (to be discussed later), may accompany buckling and produce similar-like folds. Buckling followed by formation of a thrust fault between an anticline and syncline is the mechanism that produces fault-propagation folds (Chapter 11).

Buckling is commonly preceded by layer-parallel shortening. The layers are shortened laterally and thickened perpendicular to layering and to the direction of applied stress. Buckling accompanied by inelastic strain within strong layers may modify the shape of the fold and may accelerate strain rate. For example, folds formed by a combination of buckling and pressure-solution strain (Figure 15–11) maintain the shapes of buckle folds but may develop a strong cleavage because of the associated flattening. This process may be thought of as strain-softening because the strain within the strong layers in the deforming mass would increase at constant or decreasing stress during progressive deformation.

Numerous theoretical models of buckle folds were proposed in the 1960s and 1970s. Some of them (Biot, 1961; Biot, Odé, and Roever, 1961; Chapple, 1968, 1969; Hudleston and Stephansson, 1973; Ramberg, 1960, 1961; Parrish and others, 1976) involved buckling of a single elastic layer suspended in a medium of lower viscosity. Mathematical models have also been devised by M. A. Biot (1963, 1965a) and William M. Chapple (1970) for buckling of multilayer sequences. Viscosity ratios are assumed to be 100:1, or higher, to simulate the contrast between strong rocks, like dolomite or sandstone, and weak rocks, like shale or coal. The ratio is actually geologically indeterminable for most situations where such contrasts occur, focusing some criticism on modeling techniques.

These theoretical studies have clarified the relationship between the thickness of the dominant (strong) member in the sequence being folded and the wavelength of buckle folds produced. Biot (1961) and Ramberg (1961) showed independently that an isolated strong (high-viscosity) layer embedded in a weaker (lower-viscosity) material will form buckle folds wherein wavelength is given by

$$\lambda_d = 2\pi t \sqrt[3]{\frac{\mu_1}{6\mu_2}} \qquad \textbf{(15–1)}$$

where λ_d is the dominant wavelength (of folds in the strong layer), t the thickness of the strong layer, μ_1 the viscosity of the strong layer, and μ_2 the viscosity of the weak supporting medium or matrix. Biot (1961) recognized that where viscosity ratios are 100:1 or slightly less, layer-parallel shortening becomes important and produces folds. Using Biot's observation, J. A. Sherwin and Chapple (1968), and then Peter J. Hudleston (1973a), modified equation 15–1 to

$$\lambda_d = 2\pi t \sqrt[3]{\frac{\mu_1}{6\mu_2}\frac{1}{2}\frac{S+1}{S^2}} \qquad \textbf{(15–2)}$$

where S is the ratio of quadratic elongations $(\lambda_1/\lambda_2)^{1/2}$ of the homogeneous deformation on which

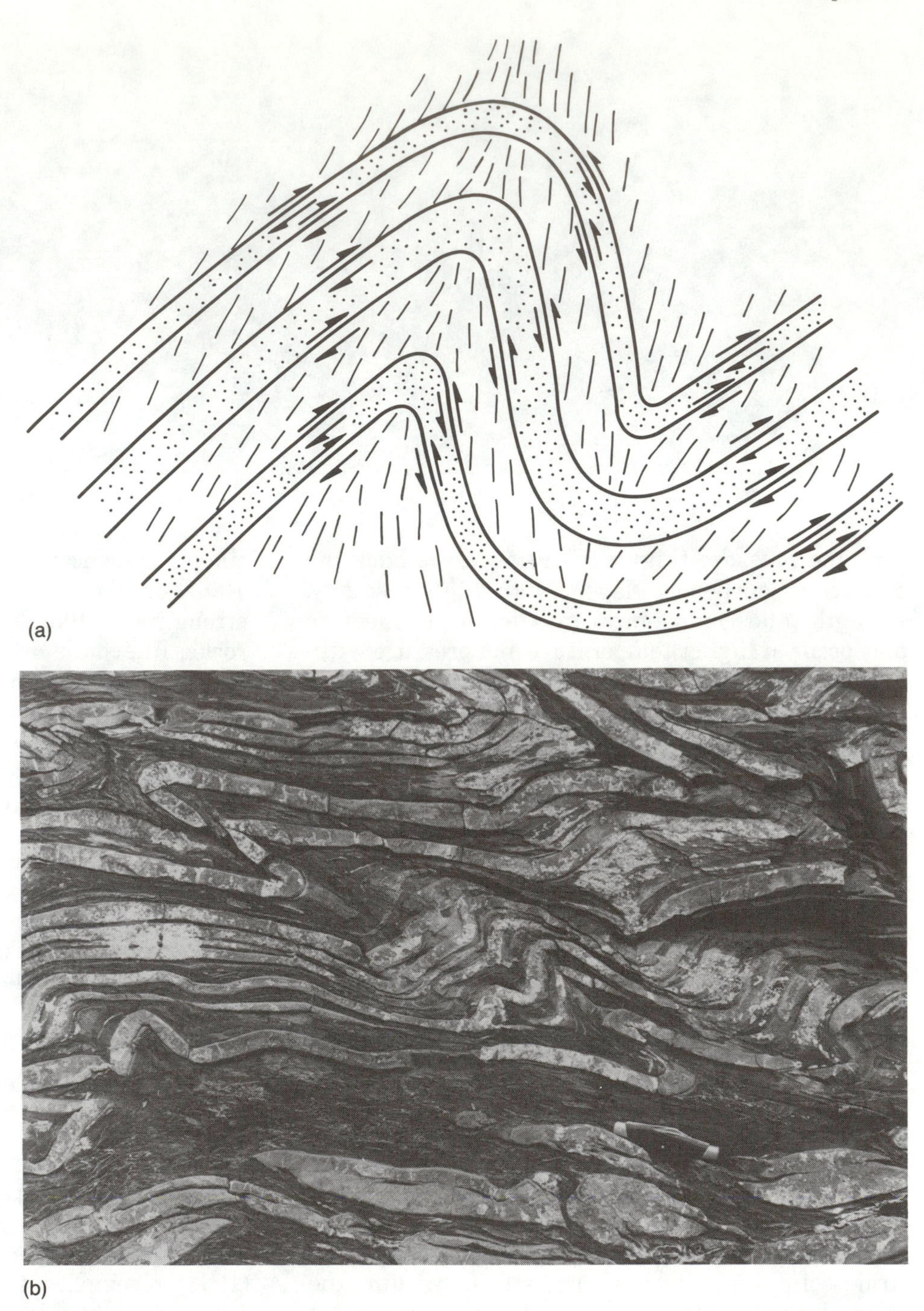

FIGURE 15–9
(a) Buckling and flexural slip at low temperature and pressure. The strong layer is stippled. The weak layer contains a cleavage parallel to fold axial surfaces. (b) Strongly disharmonic buckle folds of limestone (light) in shale (dark) near Whitehall, New York. The differences in thickness between strong limestone layers interlayered with weak shale prevent the folds from attaining the same shape throughout. (RDH photo.)

buckling is superimposed. Equations like these accurately depict the geometry and associated strains of ptygmatic folds (Chapter 14; Figure 15–12). Equations 15–1 and 15–2 show that thick layers produce folds with long wavelengths, and thin layers produce folds with short wavelengths.

Buckling theory enables us to predict the geometry of simple folds and demonstrates the utility of the relationship between the thickness of the dominant (strong) member and the wavelength. J. B. Currie, H. W. Patnode, and R. P. Trump (1962) used theory and photoelastic and other laboratory experiments to predict fold shapes and related fold wavelength to the thickness of the dominant member in buckle folds. The equation for the line in Figure 15–13 reveals that the wavelength of real buckle folds is about 27 times the thickness of the strong unit. That most folds measured fall on or near the line suggests that the rocks have been folded to the maximum curvature possible for the strong unit. Remarkably, the relationship holds without regard to rock type.

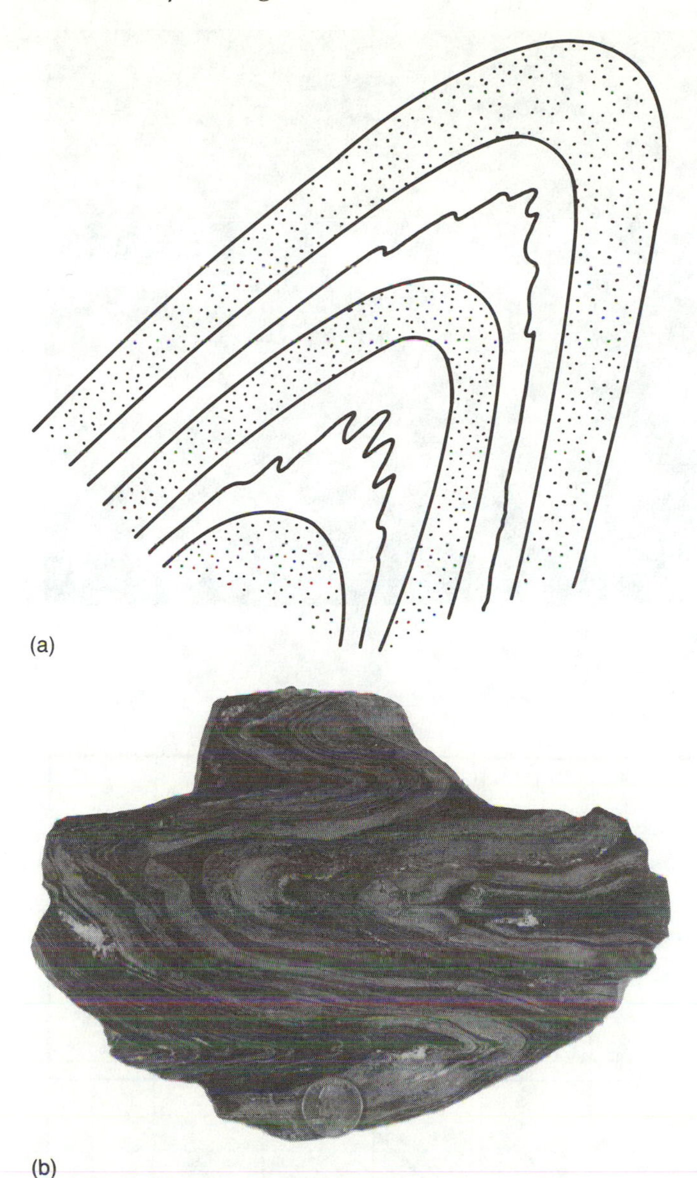

FIGURE 15–10
(a) Buckling and flexural flow at high temperatures and pressures, producing similar-type folds without flexural slip. Strong layers are stippled. (b) Sawed specimen of Poor Mountain amphibolite from near Hendersonville, North Carolina. Light layers are quartz- and feldspar-rich; dark layers are amphibole-rich and flowed more during folding than the light layers. These similar-type flexural-flow buckle folds were produced by folding under lower amphibolite facies conditions. (Specimen courtesy of Timothy L. Davis, University of Tennessee.)

FIGURE 15–11
Pressure solution on the inner arcs of a fold may weaken the layer and produce strain-softening during buckling.

A characteristic strain geometry related to bending and buckling was termed ***concentric-longitudinal strain*** by Ramberg (1961) and ***tangential-longitudinal strain*** by Ramsay (1967). It was first suggested by L. U. de Sitter (1959) as a fold mechanism involving internal deformation of layers of finite thickness. The inner arcs of these layers are subjected to layer-parallel compression, whereas the outer arcs of each layer are subjected to layer-parallel extension (Figure 15–14a). This results in a *neutral surface* of no finite strain that in each layer separates the zone of extension from the zone of compression. Folds exhibiting tangential-longitudinal strain may be identified by small structures that indicate compression in the inner arcs of folds (Figure 15–14b); examples are crenulations and pressure-solution cleavage. In the outer arcs, the layers may be broken by extension fractures or contain other structures (such as boudinage) that indicate extension.

Ramberg (1961) has shown that a simple relationship exists between strain and dimensions of a fold: the concentric shear strain (γ_{max}) and maximum concentric strains (ϵ_{max})—shear strains parallel to layering—in the crests of folded layers of this type with small ratios of amplitude to wavelength may be related to the thickness of the dominant layer (t) and length of arc of one fold (L) as

$$\frac{\gamma_{max}}{\epsilon_{max}} = 2\pi\frac{t}{L} \qquad (15\text{–}3)$$

The equation may be useful in determining the ratio of shearing to longitudinal strains in layers that have undergone tangential-longitudinal strain, if some strain marker is present. In derivation, Ramberg assumed that the layers composing the folds behave as viscous materials.

Edward C. Beutner and Frederick A. Diegel (1985) studied fibers that developed in pressure shadows and veins during formation of buckle folds in the Martinsburg Slate in New Jersey. They found that the hinges in anticlines and synclines actually migrated away from each other by lengthening the connecting common limb (Figure 15–15). The folds were then flattened during the later stages of folding, bringing the fold axial surfaces—but not the hinges—closer again.

Passive Slip—A Problematical Mechanism

Passive slip has been described (Knopf and Ingerson, 1938; Donath and Parker, 1964; Ramsay, 1967)

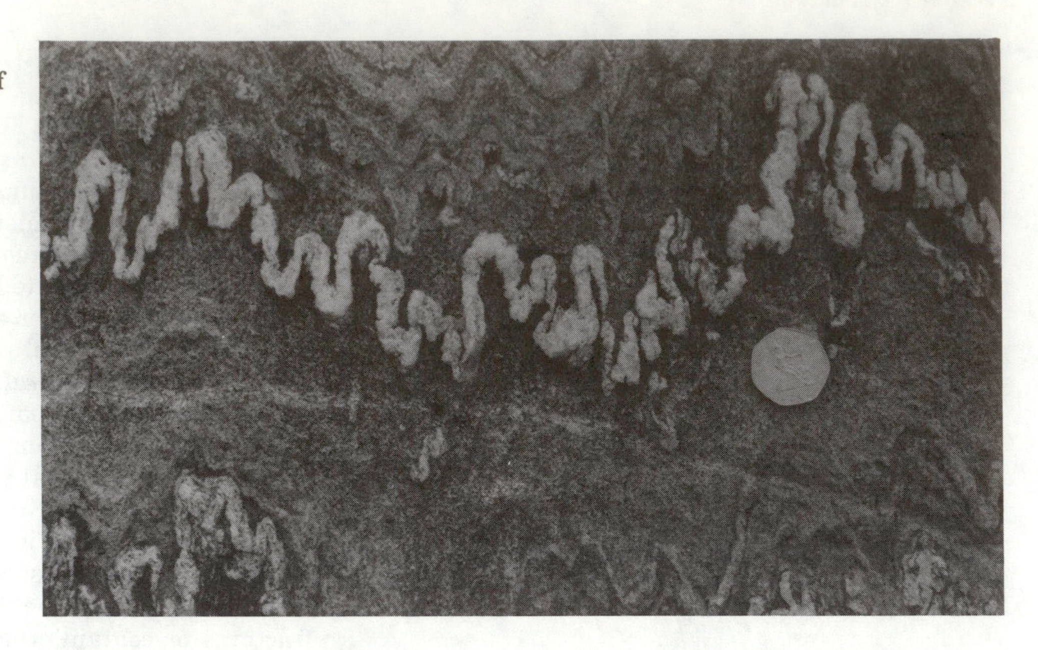

FIGURE 15–12
Ptygmatic folding (buckling) of a pegmatite layer in Middle Proterozoic Moine Series clastic rocks, Scottish Highlands. (Anthony L. Harris, University of Liverpool.)

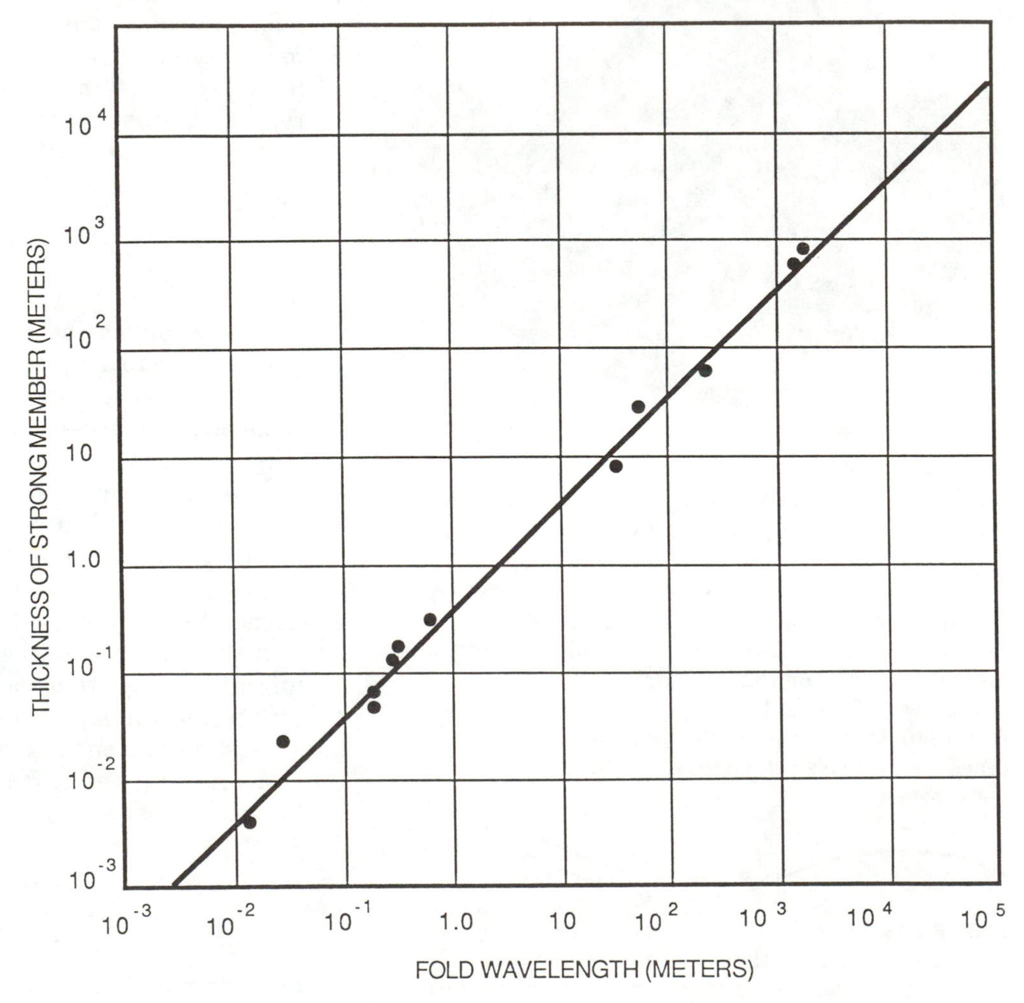

FIGURE 15–13
Log-log plot of thickness of strong layer versus wavelength (ranging from a few cm to over 1 km) of buckle folds from West Virginia, New York, and Alberta. The slope of the curve is about 27 times the thickness of the strong unit. (From J. B. Currie, H. W. Patnode, and R. P. Trump, 1962, Geological Society of America *Bulletin*, v. 73.)

as slip at an angle to layering; it generally results in a cleavage or schistosity that accommodates movement parallel to the new surface (Figure 15–16). Associated folds have been called *passive-slip folds* (Donath and Parker, 1964) or *shear folds* (Ramsay, 1967). Bedding or compositional layering serves only as a strain marker, recording displacement parallel to the cleavage or schistosity. Even so, passive slip is considered by many geologists to not be a viable fold mechanism because cleavage and foliation planes have been shown in many places to form parallel to the *XY* plane of the strain ellipsoid (Chapter 17). Paul Williams (1976) reviewed the evidence for shear parallel to cleavage, indicating cleavage may not always form parallel to the *XY* plane and movement may occur parallel to cleavage if the strain is not coaxial as during folding. It is hard to understand how a structure formed parallel to the *XY* plane of the strain ellipsoid and to the axial surfaces of folds can be immediately transformed into shear planes and take part in the folding process. Consequently, passive slip *should not be considered a fold mechanism* under conditions where cleavage forms parallel to the *XY* plane, but may be possible under conditions of noncoaxial shear deformation. Associated folds probably form by buckling, with or without flexural slip, and the cleavage forms as buckling produces a strong component of pure shear in the rock mass. Yet displacement parallel to axial-plane foliations has been observed in rocks of higher metamorphic grade (Figure 15–16), as well as in low-grade rocks. A different orientation of the strain ellipsoid that permits the foliation, once formed by flattening, to rotate into parallelism with a shear plane in the strain ellipsoid may require a two-stage mechanism (Figure 15–17). Ramsay (1962) has pointed out that a variable amount of flattening in a rock mass can lead to shear that parallels the axial surfaces of the folds.

The passive-slip mechanism was originally intended to explain the apparent displacement of a marker layer parallel to a cleavage or foliation surface. The apparent displacement can also be produced by removing part of the layer by pressure solution (Figure 15–18), but displacement has been observed in some low-grade slates unaffected by pressure solution (Onasch, 1983). Actual passive slip may exist in shear zones (Chapter 10) if foliations develop parallel to shear directions, and folds related to the foliations could be passive-slip folds. Folds of this kind are rarely observed, however, because shear zones are dynamic, producing penetrative structures that shear out (or *transpose*) the original marker layers until they parallel the shear planes.

Kink Folding

As defined earlier, kink and chevron folds have long straight limbs and narrow angular hinges. They form in minerals and rocks and occur on any scale from crystal lattices to map-scale folds (Figure 15–19). Kink bands and folds form in regular geometric shapes, either as isolated structures or as *conjugate* (paired) sets. Paterson and Weiss (1966) showed that thinly layered rocks compressed by 10 percent (or less) parallel to the layers will develop conjugate kink bands at 55° to 65° to the compression direction. As strain increases from 10 to 30 percent, kinks propagate through the entire mass at the same angle, forming two dominant crossing sets. At strain of 45 percent, the two sets have almost joined, with one set becoming dominant, and, at strains of 50 percent or more, we find a single set of crenulation folds (Figure 15–20). Ramsay (1967) has suggested two other models for kink formation involving simple shear and rotation of constant-length segments of kinked layers. A third model by Ramsay, like one proposed by Paterson and Weiss (1966), involves migration of axial surfaces.

Why do kinks form in some rocks and buckle folds form in others? Each is a product of an instability in the layering as compressive stress is applied parallel to the layering. Arvid Johnson (1977) has formulated two ways to predict whether kinks or sinusoidal folds will form. For his first criterion (K_1), he assumed that initiation of kink folds requires no slippage between layers (flexural slip); if slippage can occur, buckle (sinusoidal) folds will form. Johnson also assumed that the rock mass has the axial load needed to induce slippage between layers. This would occur at critical inflection points on layers whose contact strength approximates the critical load required to nucleate kinks. As a result, his first criterion may be stated as

$$b \cong \tau / G_a \qquad (15\text{–}4)$$

where b is the dip of layering where folding is initiated, τ is shearing stress (or contact strength between layers), and G_a is average shear modulus of the entire layer sequence. If the shear stress is the same in all layers (both strong and weak), and the contact strength decreases to zero, equation 15–4 becomes

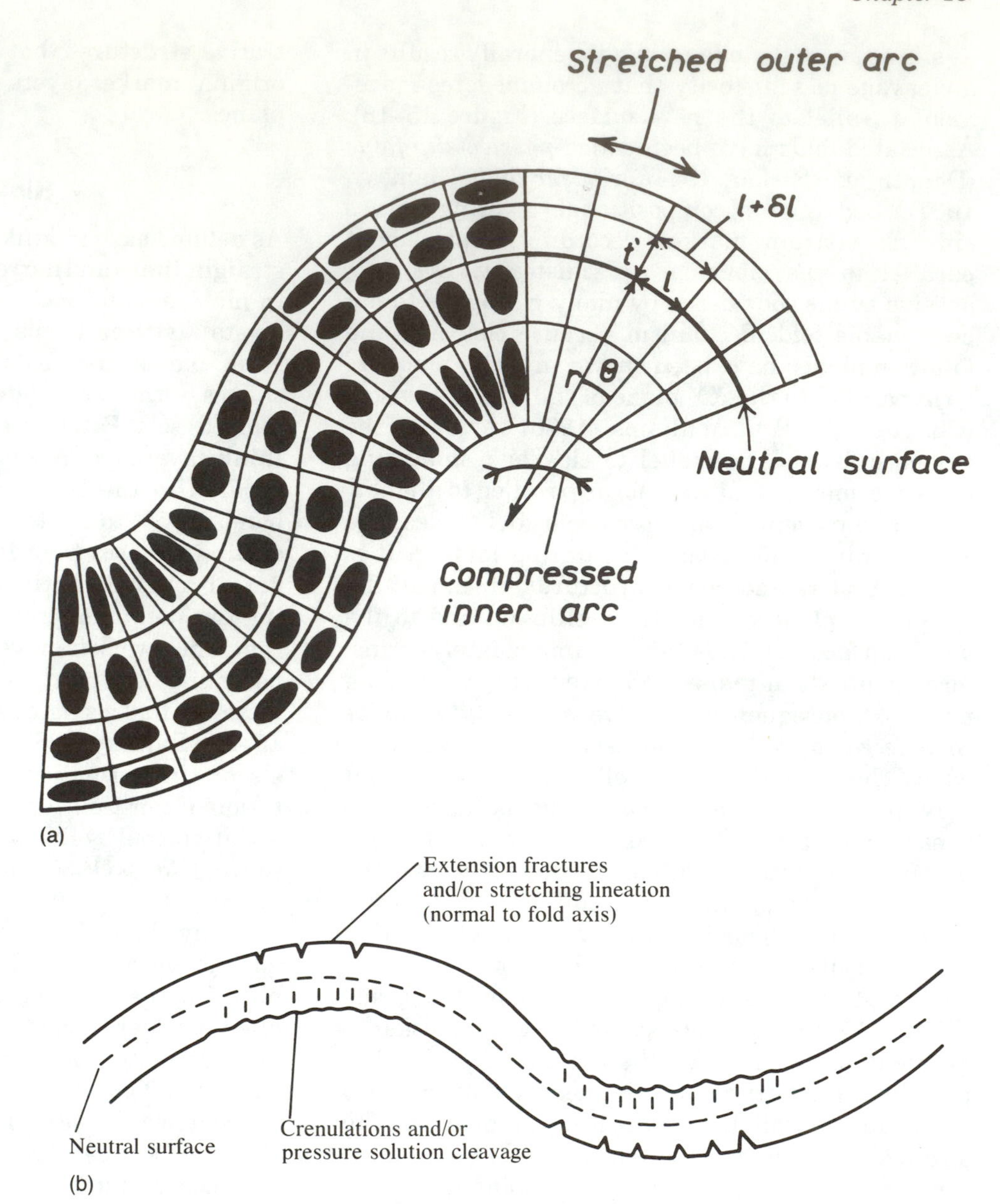

FIGURE 15–14
(a) Strain within a layer folded by buckling and tangential-longitudinal strain. Note that layer-parallel extension and shortening are separated by a neutral surface of no finite strain. (From J. G. Ramsay, *Folding and Fracturing of Rocks,* © 1967, McGraw-Hill Book Company. Reproduced with permission.) (b) Small structures produced by compression of inner arcs and extension of outer arcs of folds during tangential-longitudinal strain. (c) (opposite page) Folds formed by buckling and tangential-longitudinal strain from Ordovician Walloomsac phyllite near Hoosick Falls, New York. Crenulations in the synform and the stretching lineation oriented normal to the fold hinges suggest extensional motion normal to the antiformal axis and compression associated with the synformal axis.

$$G_a = 0 \text{ at } \tau = 0 \qquad (15\text{–}5)$$

This situation may occur regardless of the thicknesses or elastic properties of the layers. Therefore kinking is favored if the contact strength between layers is much less than the shear modulus of the strong layers.

If the ratio τ/G_a is small, the slopes of the layers are small when slippage occurs, and the buckle folds may not be clearly distinguishable from kinks formed at the same time. If the ratio is zero, slippage between layers occurs everywhere and only buckle folds are formed. If the ratio is large, no slippage may occur between layers and the buckle folds may be transformed into concentric and then chevron shape, rather than into tight kinks. So, if $G_a >> \tau$ ($\tau \neq 0$), kinks should be prominent. If $\tau = 0 = G_a$, or, if $G_a \leq t$, sinusoidal buckle folds should form.

Johnson based his second criterion (K_2) on the assumption that an estimate for the critical load for kink folding is the load required to overcome the contact strength between layers along the waveform at the midpoint of the sequence (Johnson, 1977).

(c)

FIGURE 15–14
(continued)

Calculations of the critical load, P_B, indicate that, for kink folds to be favored over sinusoidal buckle folds,

$$P_0 < P < P_B$$

or

$$1 < (P/P_0) < (P_B/P_0) = K_2 \qquad \textbf{(15–6)}$$

where P is axial load (parallel to layering), P_0 is minimum axial load. Using this criterion, conditions favor kink folding where $K_2 >> 1$, but sinusoidal buckle folds form where $K_2 \sim 1$.

Most real folds represent some intermediate shape between ideal kink and ideal sinusoidal buckle folds, but many folds approach ideal end-member behavior. So this model serves a practical purpose and is not merely a theoretically and experimentally derived model for imaginary structures.

Before Johnson's work, Biot (1965a) and Cobbold, Cosgrove, and Summers (1971) formulated criteria based on the anisotropy of a deforming multilayer rock body to determine whether kink or buckle folds would form. But Johnson's work indicates that anisotropy is *not* a factor. His criteria are no doubt applicable at lower temperature and pressure where the contrasts between kink and buckle folds are commonly observed in crustal rocks. At higher temperature, flexural slip need not accompany buckling because ductile flow occurs in the weaker layers; thus, Johnson's first criterion would not hold under these conditions. The exception seems intended.

Flexural Flow

A layered sequence is said to deform by ***flexural flow*** if some layers flow ductilely while others remain brittle and buckle. Flexural flow requires moderate to high ductility contrast between layers. The entire rock mass may be in a state of ductile flow, but some rocks, such as amphibolite and calc-silicate, have greater strengths under moderate

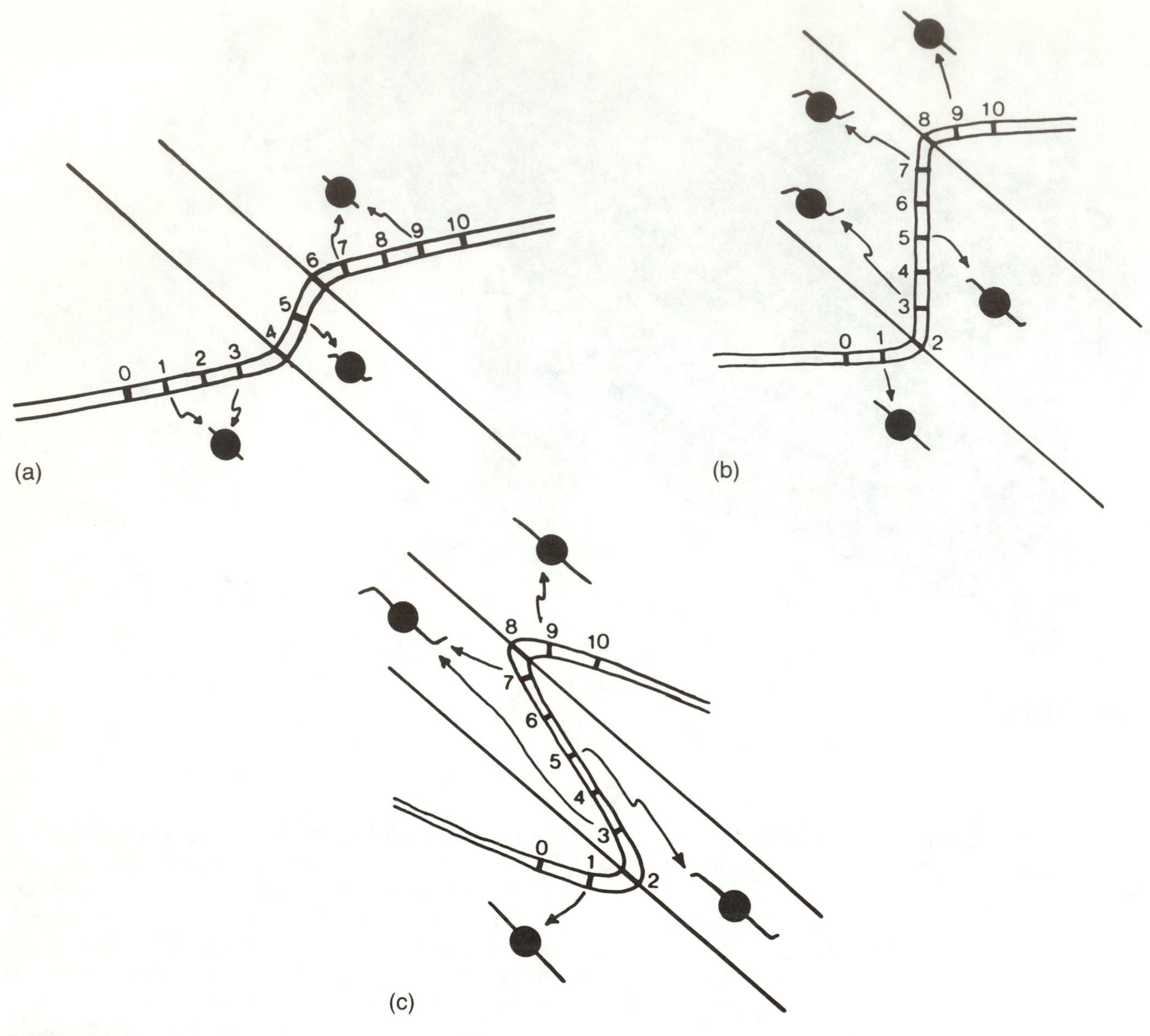

FIGURE 15–15
Beutner and Diegel migrating-hinges model. Fold is initiated (a); the common limb lengthened (b); then flattened and overturned (c). Fibrous pressure shadows on pyrite grains are the strain and shear-sense indicators. (From E. C. Beutner and F. A. Diegel, Determination of fold binematics from syntectonic fibers in pressure shadows, Martinsburg Slate, New Jersey, Jan. 1985, *American Journal of Science,* v. 285, p. 16–50. Reprinted by permission of American Journal of Science.)

temperature (300° to 500°C) and high pressure than interlayered weaker rocks, such as schist, carbonate rocks, impure sandstone, and felsic volcanic rocks. As a result, some strong layers may not undergo appreciable thickness changes, but weak layers may undergo extreme thickness changes (Figure 15–21). Because of the ductility contrast, the stronger layers control buckling and the fold geometry, although not so much that their control occurs in flexural-slip folds, where the shape of the folds changes. The products of flexural flow are mostly similar-type folds (Class 1C and 3) and rarely ideal similar folds (Class 2), which preserve identical curvature in all layers.

FIGURE 15–16
Possible passive-slip fold mechanism in Precambrian Moine Series metasandstone at Loch Monar, Scottish Highlands. Note that layering is displaced parallel to fold-axial surfaces, but buckling probably also occurred here. (RDH photo.)

In flexural flow, fold amplitude and wavelength may be controlled by the thickness, spacing, and strength of the strong layers, but will also be partly influenced by the amount of ductile strain parallel to the axial surfaces. But the relationships between thickness of the strong layer and the wavelength of the buckle folds still hold.

Passive Flow (Passive Amplification)

Passive flow involves uniform ductile flow of the entire rock mass, with layering (bedding, foliation, or gneissic banding) serving only as a strain marker (Figures 15–22 and 15–23; see also Figure 2–31b). If passive-flow folds are to form, there must be little or no ductility contrast between layers, even if the composition differs markedly, and there must be flow across the layering. Ramsay's suggestion (1962) of a variable amount of flattening producing simple shear across layers may prove important here. This mechanism will grade into flexural flow as ductility contrast increases and the layers begin to influence the shape of the folds. The most likely mechanism for ideal similar folds (Class 2), preserving identical shapes throughout the layered sequence, is passive

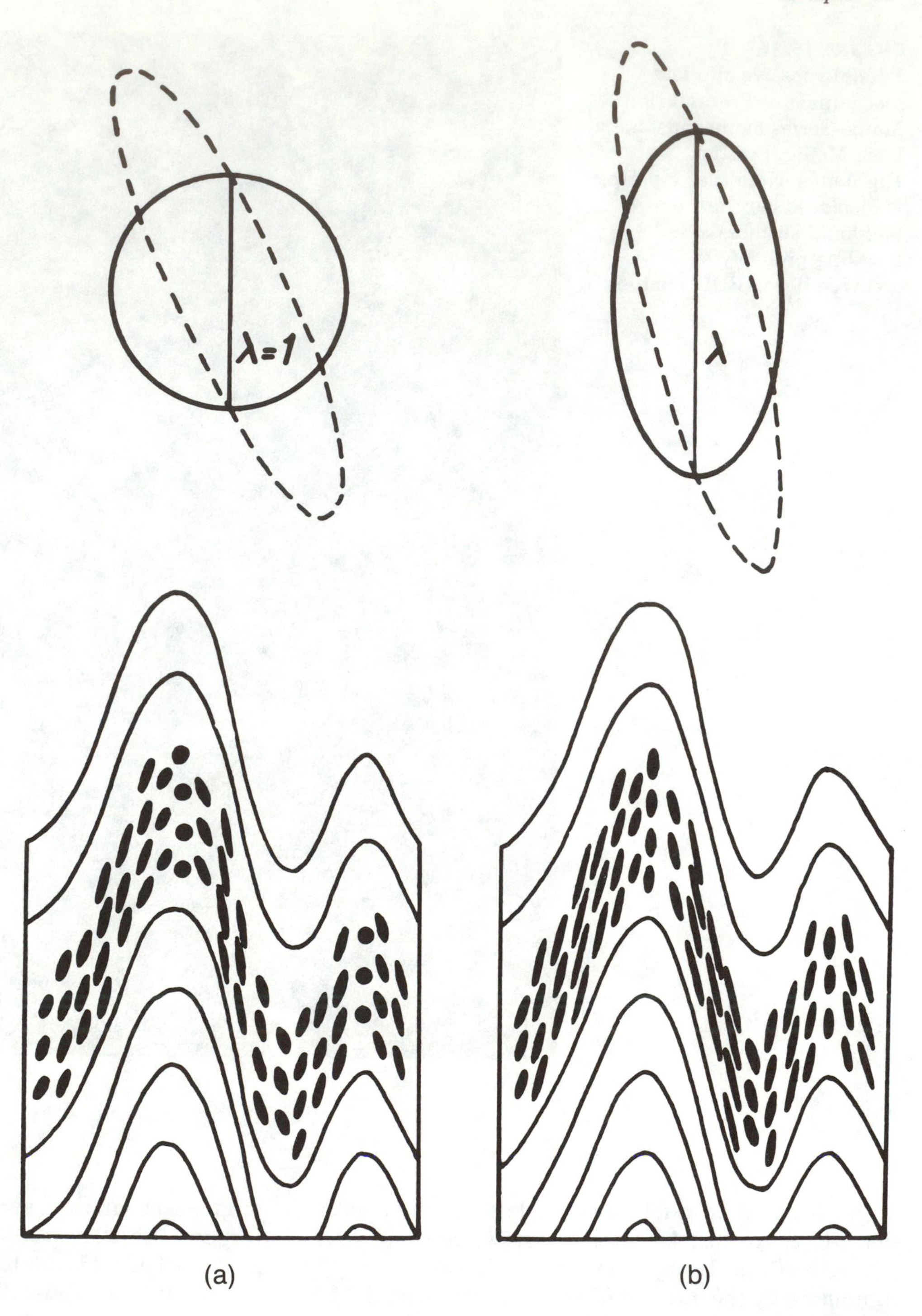

FIGURE 15–17
Reorientation of the strain ellipse as a possible explanation for passive slip. Alternative explanations of folding by passive slip include (a) simple shear and heterogeneous strain or (b) heterogeneous simple-shear strain followed by homogeneous deformation. (From J. G. Ramsay, *Folding and Fracturing of Rocks*, © 1967, McGraw-Hill Book Company. Reproduced with permission.)

(a)

(b)

FIGURE 15–18
(a) Folding of layering by axial-planar pressure-solution cleavage in Middle Ordovician Womble Shale, near Little Rock, Arkansas, suggesting passive slip; in reality, folds were formed by buckling accompanying cleavage formation. (b) Crenulation folds formed in talc schist in a shear zone at the base of the Karmøy ophiolite, near Haugesund, Island of Karmøy, southern Norway. (RDH photos.)

(a) (b) (c)

FIGURE 15–19
(a) Kinked kyanite crystal in graphitic schist near Black Mountain, North Carolina. Note the kinks are concentrated in the light area; unkinked kyanite is the dark portion to the right. Width of field is approximately 1 mm. Crossed polars. (b) Kink folds in phyllite from Tête Juene Cache, British Columbia. Knife is approximately 7 cm long. (c) Kink folds in thin-bedded late Precambrian quartzite near Oppdal, southern Norway. (RDH photos.)

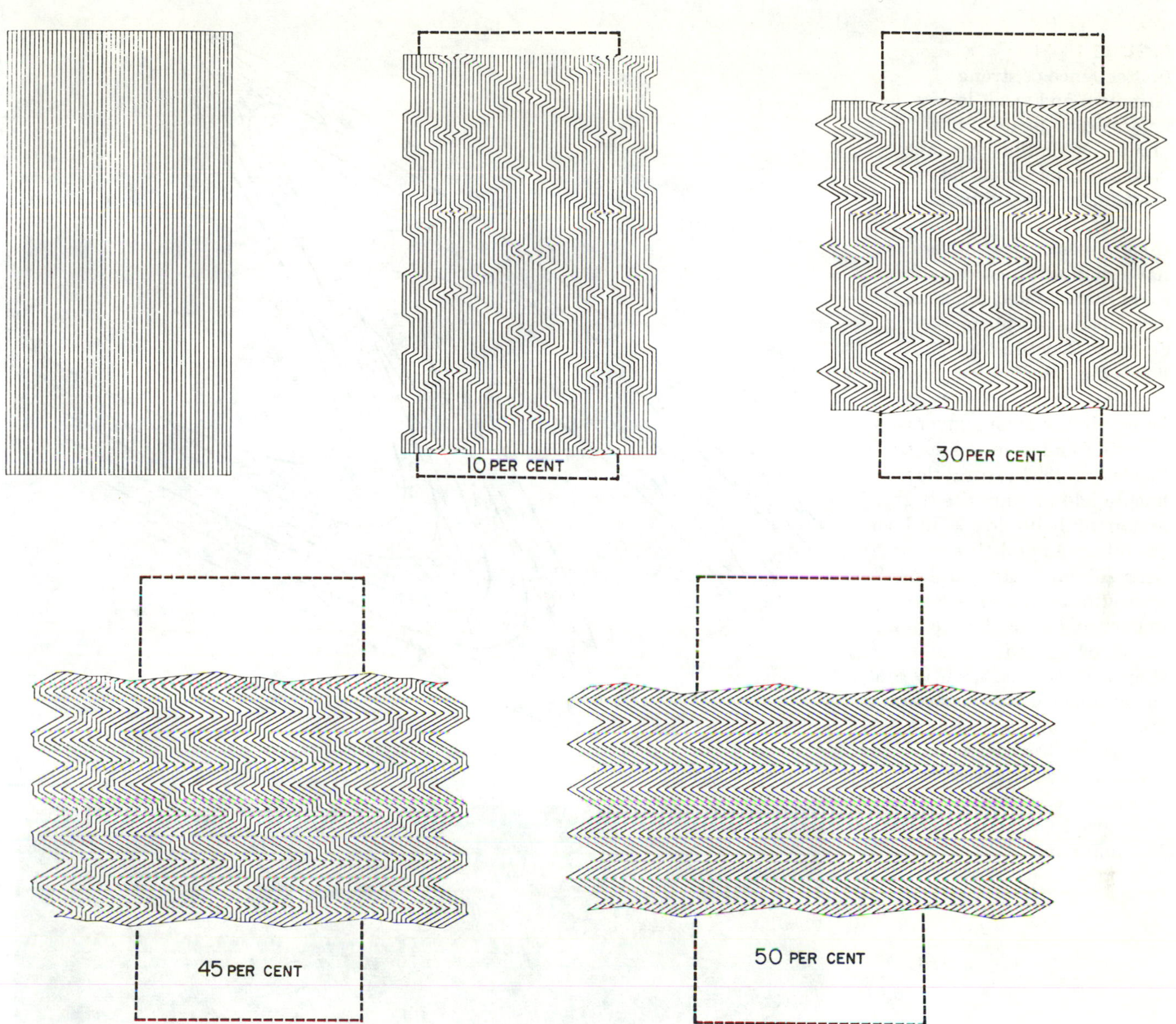

FIGURE 15–20
Kink folds produced at 0, 10, 30, 45, and 50 percent shortening strain. (From M. S. Paterson and L. E. Weiss, 1966, Geological Society of America *Bulletin,* v. 77.)

FIGURE 15–21
(a) Sequence of strong (stippled) and weak layers deformed by flexural flow, producing similar-type folds. Note that the strong layers have undergone minimal thickness changes but the thicknesses of weak layers have changed drastically from axial zones to fold limbs and developed axial-plane cleavage. Ductile flow—indicated by arrows—occurs in the weak layers; in the strong layers, buckling and minimal ductile flow occur. (b) Flexural-flow buckle fold in quartzite (light) and amphibolite (dark) in Poor Mountain amphibolite from near Salem, South Carolina. Quartzite layers were undergoing some ductile flow but layering still controlled shapes of folds sufficiently that the stronger layers separated during folding, providing a void where the coarse crystalline quartz was deposited (center of photograph). (Photo by D. G. McClanahan.)

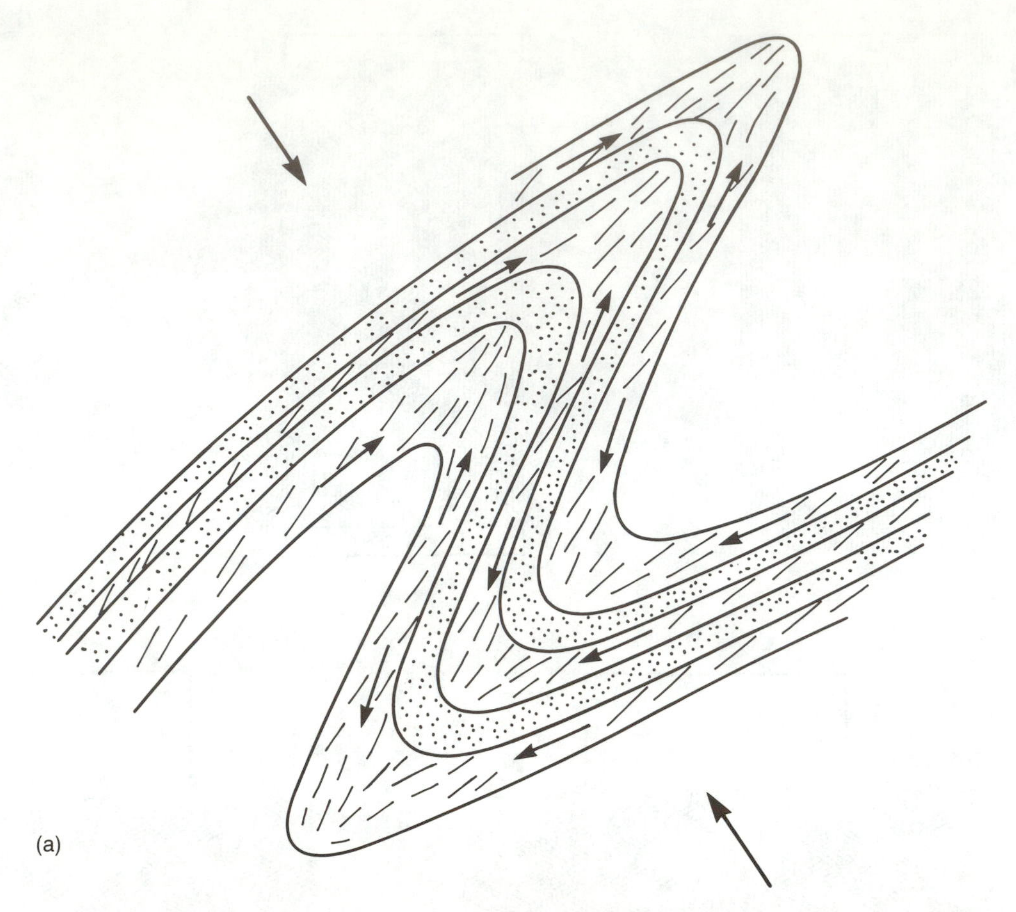

(a)

(b)

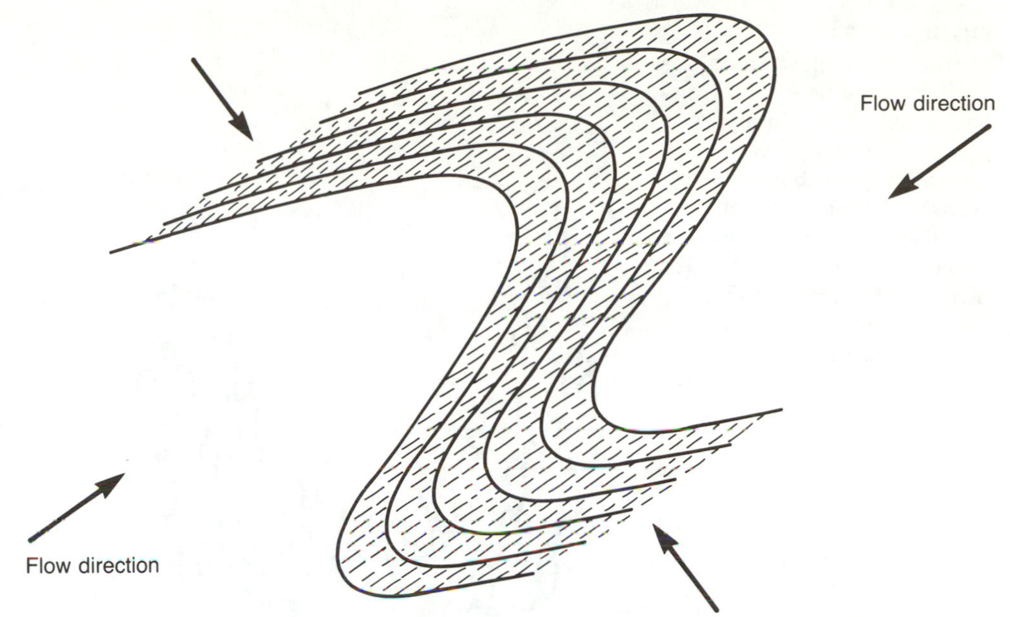

FIGURE 15–22
Passive-flow mechanism. Note that the thickness of each layer changes by the same amount and produces ideal (Class 2, or nearly ideal) similar folds by ductile flow; all hinges have the same curvature. Flow can occur in any direction with respect to layering, but must cross layering to produce folding. Arrows show compression direction. Axial-plane foliation suggests a strong component of flattening.

FIGURE 15–23
Passive-flow folds in migmatitic biotite gneiss, Thor-Odin dome, Shuswap metamorphic complex, southern British Columbia. Folding here is ductile and all layers have changed thickness to the same degree. (RDH photo.)

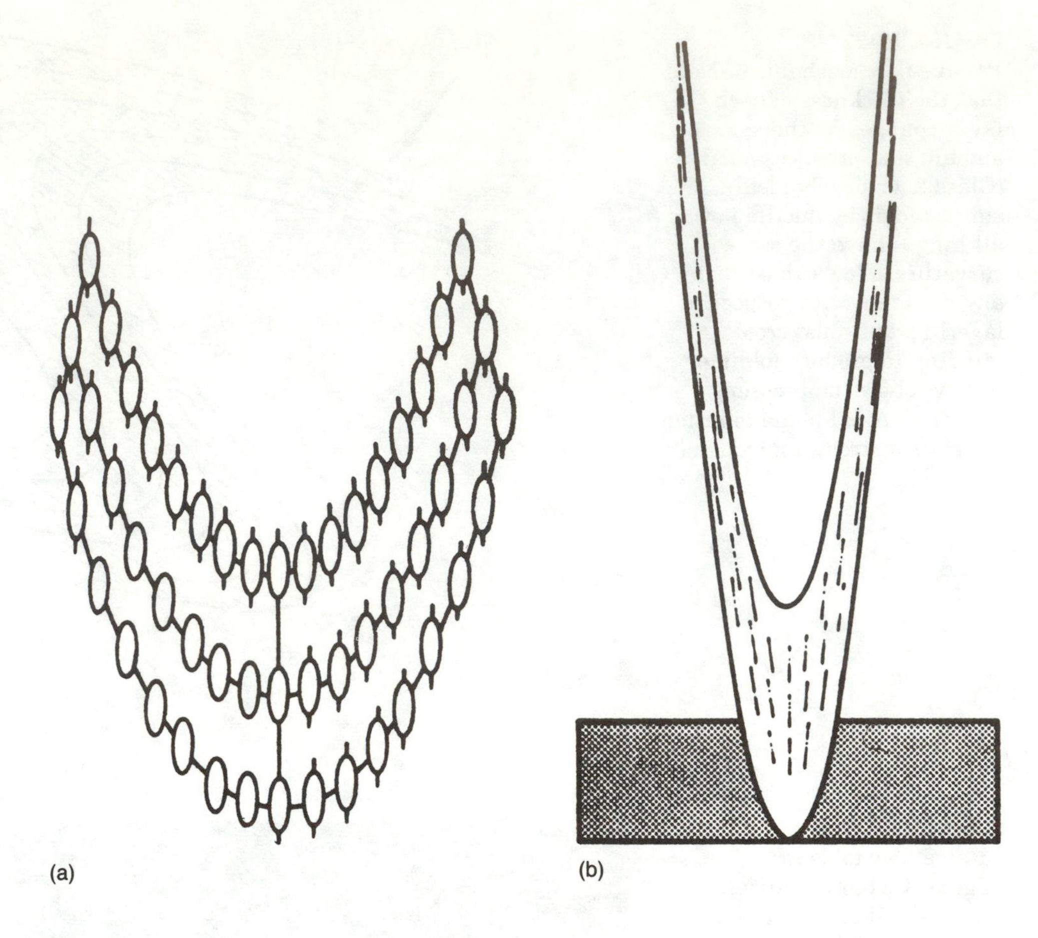

FIGURE 15–24
Strain (a) and flow directions (b) in passive flow involving homogeneous strain and greater than 50 percent shortening; (a) shows strain ellipses; (b) long axes of the strain ellipse, suggesting maximum flattening and flow. (From B. E. Hobbs, 1971, *Tectonophysics*, v. 11, p. 329–375, Elsevier Science Publishers.)

flow because of the lack of contrast of properties from layer to layer. In addition, similar-type (Class 3 and 1C) folds are produced by passive flow, frequently in combination with buckling.

The actual mechanism by which passive strain markers are folded in passive flow is nonuniform laminar ductile flow produced by heterogeneous pure shear or simple shear (Wynne-Edwards, 1963). Heterogeneous simple shear directed parallel to fold axial surfaces and homogeneous pure shear—whether parallel, oblique, or normal to layering—produces folding of passive layers (Figure 15–24).

Combined Mechanisms

Fold mechanisms, like deformation mechanisms, compete with one another at various temperatures and pressures and in various rock types. The mechanisms of passive flow and flexural flow, which require more or less ductility contrast, may compete at higher temperature in normal silicate-dominant rocks. In a rock mass being folded under brittle conditions, flexural-slip phenomena may be initiated with buckling; however, as the folds tighten and friction reduces layer slip, cleavage-forming mechanisms may take over and produce layer-parallel shortening and deformation parallel to the axial surfaces of folds (Figure 15–25). This occurs most frequently in the outer parts of the metamorphic cores of orogens and in the innermost parts of foreland fold-and-thrust belts. Attempts to pigeonhole folds by geometric or even generic criteria may fail because competing mechanisms produce folds that do not reveal recognizable patterns (Figure 15–26). Strain analysis sometimes helps resolve problems with fold mechanisms.

Buckling plus flexural slip is probably the most usual combination of folding processes (Figure 15–27). Here brittle rocks yield by forming open parallel or parallel-concentric folds, but, as the fold tightens, room problems arise in the core of the fold. Then disharmonic folding of weaker layers takes place, with or without fracturing, to solve the room problem. In ductile rocks, buckling in some layers may combine with ductile flow in others and produce flexural-flow folds. Disharmonic folds result from strong anisotropic contrasts in properties from layer to layer.

FIGURE 15–25
(a) Buckle fold in Ocoee Series (late Precambrian) rocks near Walland, Tennessee, containing evidence of both buckling and flexural slip. (b) Bedding surface on the southeast limb of the fold in (a) showing slickenlines truncating cleavage. Knife is parallel to cleavage. Note slickenlines on bedding surfaces and axial-plane cleavage in (b). Cleavage formed during homogeneous strain of the rock mass as the layers became unable to slip past each other, but some flexural slip occurred after the cleavage formed. (RDH photos.)

(a)

(b)

FIGURE 15–26
Isogon analysis of several mesoscopic folds from the Milton area of the Appalachian Piedmont in Virginia and North Carolina. Note that the folds show variations in convergent and divergent characteristics of isogons from layer to layer (of different composition); they would fit more than one of Ramsay's fold classes. Roman numerals identify particular fold sets. (From O. T. Tobisch and Lynn Glover, III, 1971, Geological Society of America *Bulletin*, v. 82, p. 2209–2230.)

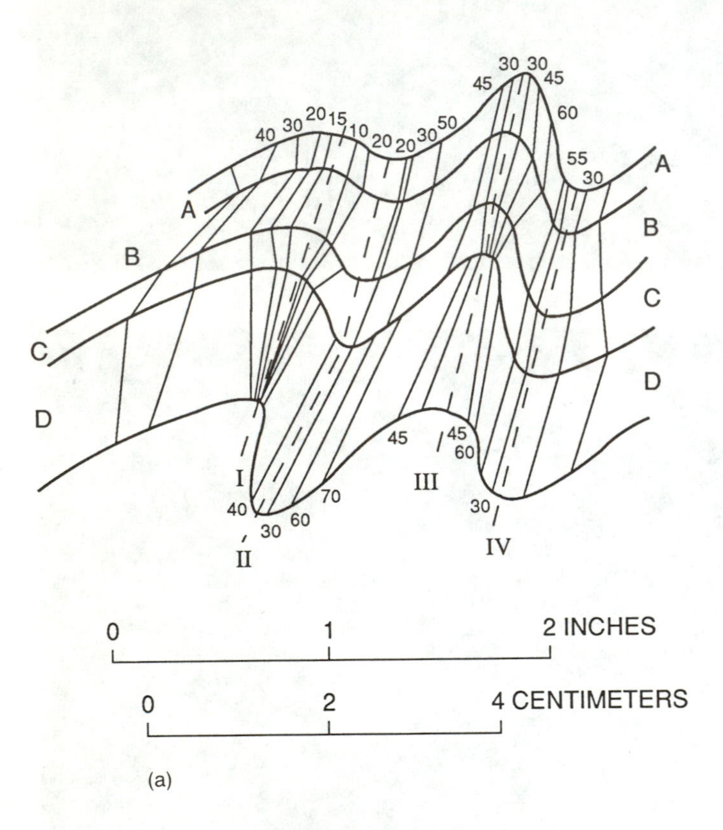

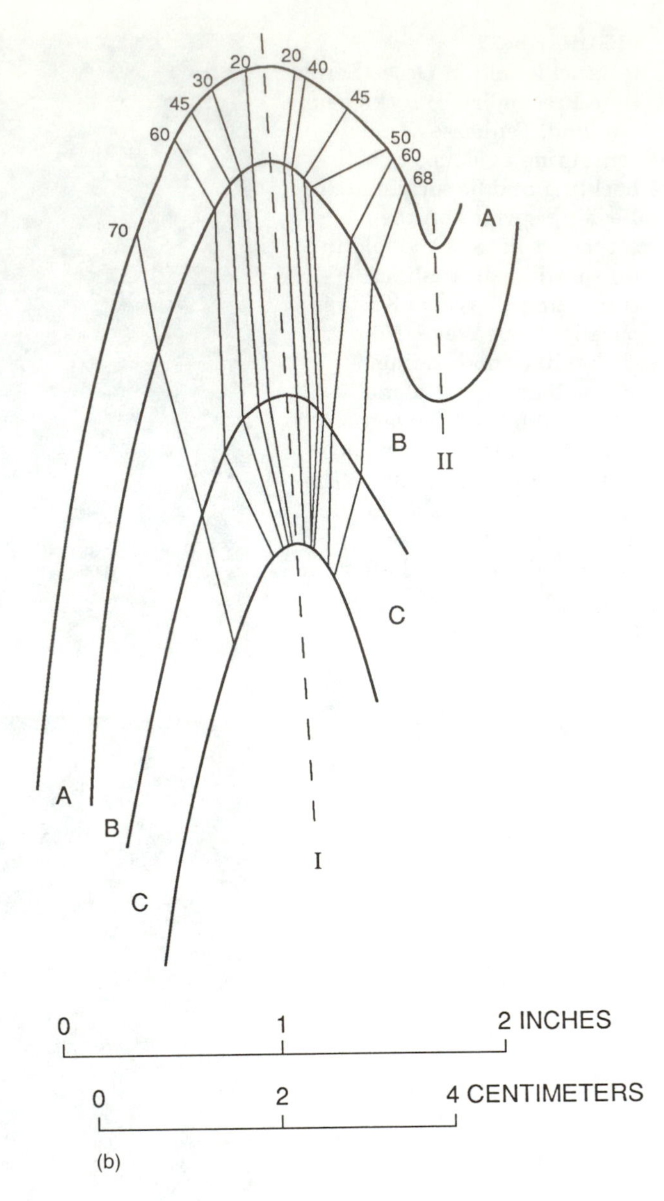

DEFORMATION MECHANISMS AND STRAIN

Fold mechanisms in rocks are determined largely by temperature and pressure during deformation. Consequently, the behavior modes and deformation mechanisms we discussed in Chapters 5 and 6 also play an important role in determining the mechanisms and ultimate shapes of folds. At the low-temperature end, layers that remain elastic bend or buckle and yield brittlely; at the high-temperature end, flow processes dominate and layers undergo ductile flow and recrystallize. At intermediate temperatures, some layers reach the thresholds for pressure solution and diffusion flow, and others remain strong and undergo little ductile flow. In an interlayered sequence of schist (weak) and amphibolite (strong), the schist may undergo ductile flow with recrystallization and the amphibolite may still undergo brittle deformation. At still higher temperature, the amphibolite may flow, but less than the schist.

Attempts have been made to determine the amount and distribution of strain in folds by using structures observed within layers or at layer boundaries. Such features include natural strain markers, like oöids (Cloos, 1971) and reduction spots (Wood, 1973); theoretical studies (Ramberg, 1961; Chapple, 1969); experimental rubber and clay models and photoelastic studies (Roberts and Strömgård, 1972); and computer modeling (Dieterich, 1970).

David Roberts and K.-E. Strömgård (1971) have studied the relationships of cleavage to weak and strong layers in buckle folds, then attempted to model observed strain patterns using rubber, Plasticene™, and putty and photoelastic models. The rubber layers exhibited considerable rigidity and

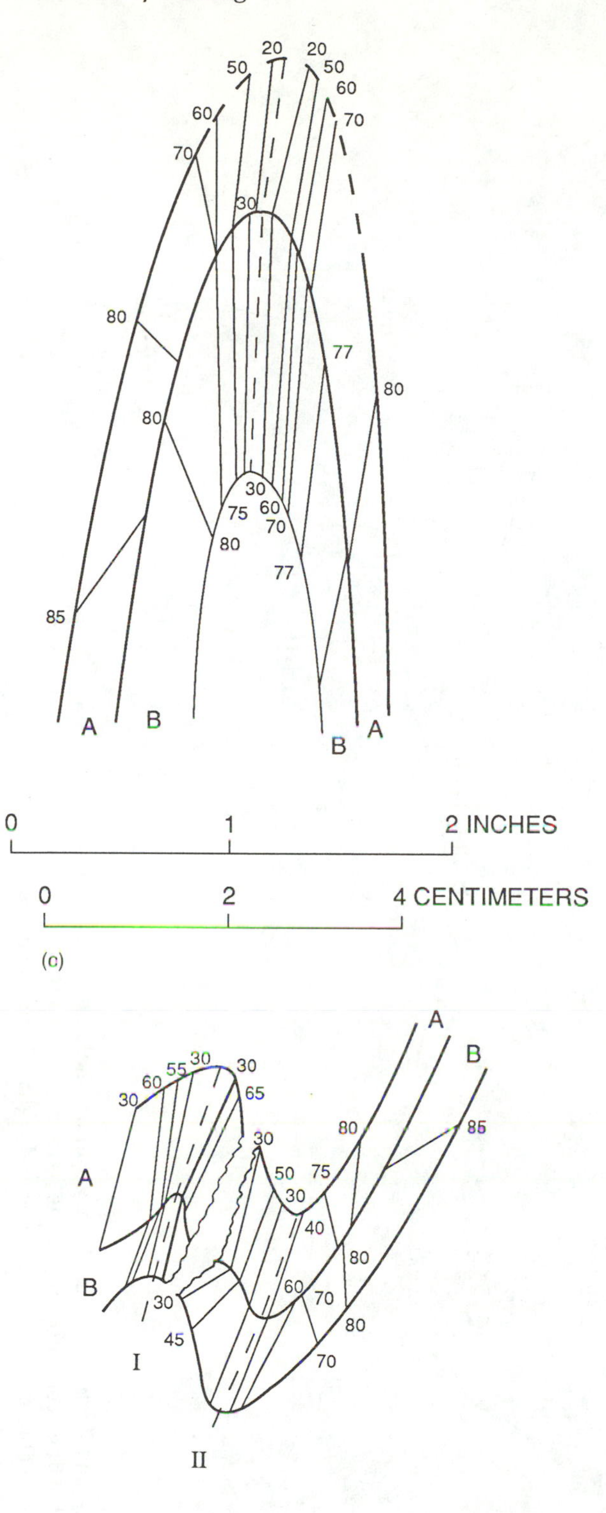

FIGURE 15–26 *(continued)*

clearly displayed tangential-longitudinal strain within a buckled layer (Figure 15–28). Multilayer putty and Plasticene models provided somewhat different results in different layers (Figure 15–29). The Plasticene layers, being stronger, exhibited tangential-longitudinal strain; the weaker putty layers exhibited more-nearly uniform flattening in the axial zones of folds (parallel to the axial surfaces) and thinning of limbs—a flexural-flow mechanism. But strain was refracted in the putty at layer boundaries, particularly on the outer arcs of the strong layers. Photoelastic models yielded similar results (Figure 15–30).

Chapple (1970) has studied strain variability in theoretical folds involving a single layer of high viscosity adjacent to a layer of much lower viscosity. He found considerable variation in the nature of strain at layer boundaries as the dip on the fold limb increases from about 23° to about 89°.

Among the first computer models of strain distribution in folds are those of James Dieterich (1970), who based them on the finite-element method. His models of strain distribution (Figure 15–31) reveal patterns like some of the experimental results obtained by Roberts and Strömgård (1972) and even show some of the variations they reported at layer boundaries.

Most of the models just described clearly illustrate the effects of various combinations of homogeneous and inhomogeneous strain on fold shape and contrasts in strain at layer boundaries. Homogeneous strain before, during, or after folding by heterogeneous strain (such as by buckling) profoundly affects the shapes of folds and determines whether the folds end up as modified concentric (Class 1C) or similar (Class 2) folds, or some other type. Likewise, layer-parallel shortening before flattening, whether during folding or afterward, also modifies the shapes of folds. Thus, the sequence of homogeneous and heterogeneous strain is also an important factor in fold mechanisms.

NONCYLINDRICAL AND SHEATH FOLDS

The importance of ***noncylindrical*** and ***sheath folds*** as products of inhomogeneous strain has been recognized only recently. In a strict sense, all folds are noncylindrical if traced far enough because their axes become curved as they terminate, unless truncated by faulting; however, the axes of some folds are strongly curved over short distances and so they are easily recognized as noncylindrical.

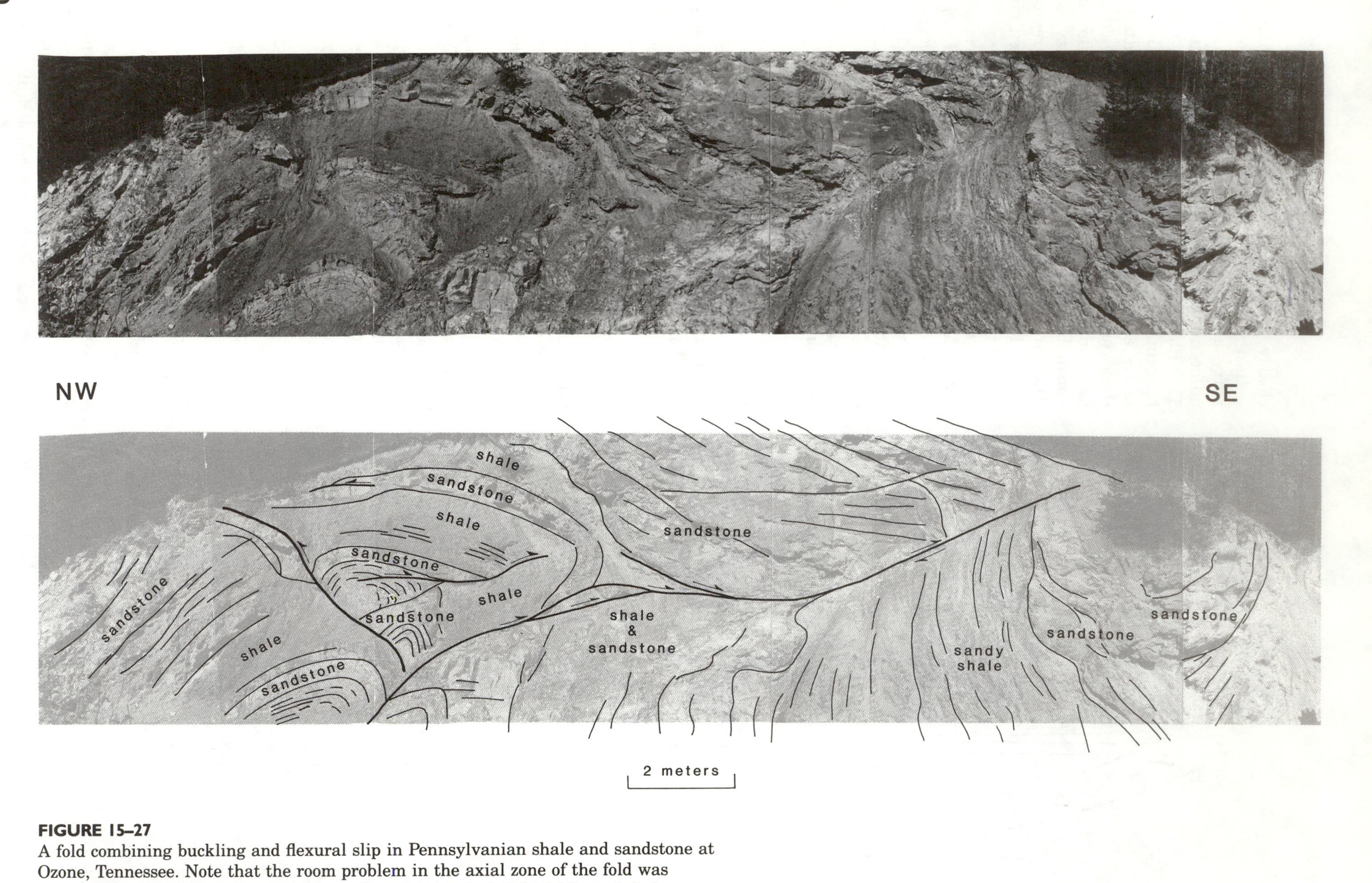

FIGURE 15–27
A fold combining buckling and flexural slip in Pennsylvanian shale and sandstone at Ozone, Tennessee. Note that the room problem in the axial zone of the fold was solved by brittle deformation (faulting) of sandstones and tight folding of shales. (Photograph by Alice L. Stieve, University of South Carolina.)

FIGURE 15–28
(a) Sketch of bedding and cleavage traces (light lines) on a buckle fold from the Varanger Peninsula of Norway. The models in Figures 15–28, 15–29, and 15–30 attempt to duplicate the strain, cleavage, and geometry associated with this fold. (b) Rubber-layer model produced by buckling of a layer of soft rubber between layers of stiff rubber (each 13 mm thick). Top, undeformed. Bottom, deformed, showing distinct differences in strain from soft layer to stiff. (From David Roberts and K.-E. Strömgård, 1971, *Tectonophysics,* v. 14, p. 105–120, Elsevier Science Publishers.)

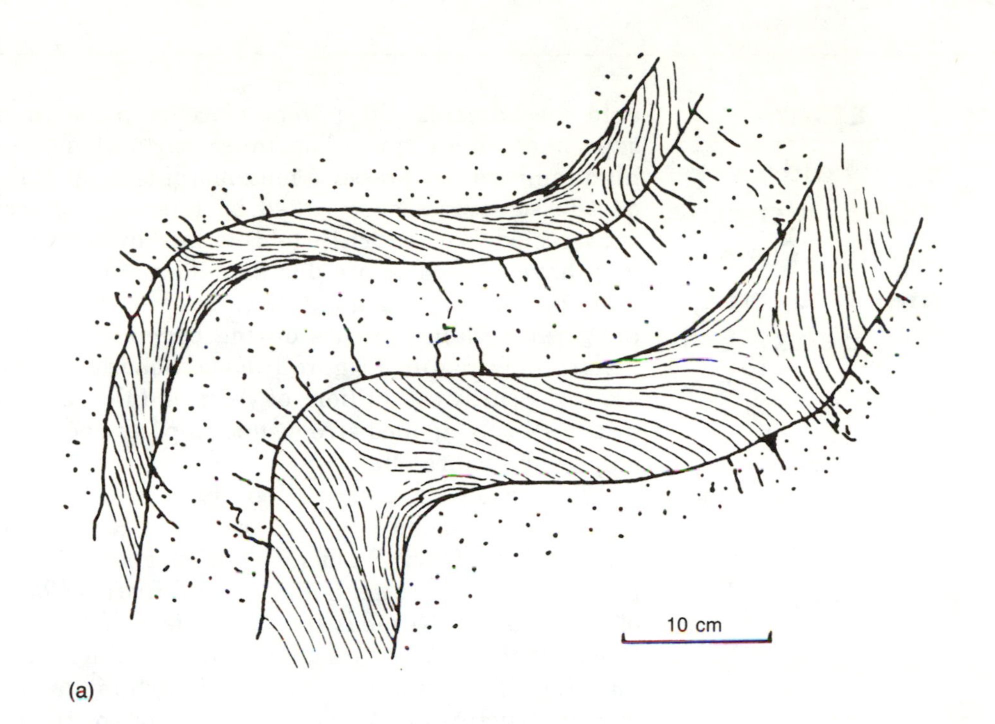

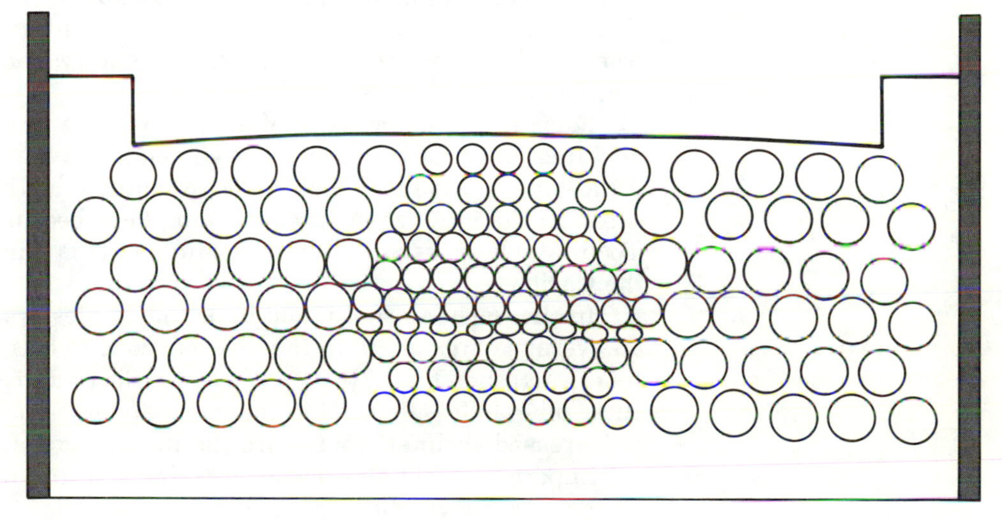

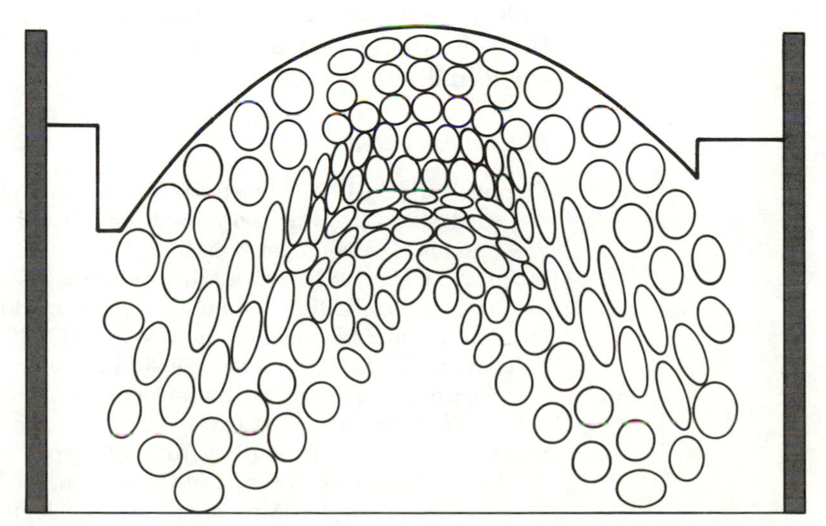

ESSAY

Fold Mechanisms, Space, Time, and Orogenic Belts

Fold mechanisms differ from place to place in mountain chains and are superimposed in time. The inner parts of a mountain chain subjected to high-temperature/pressure metamorphism generally preserve an early set of ductile structures overprinted by later brittle structures, formed after the mass cooled. Earlier brittle folds formed by flexural-slip or buckle mechanisms may have formed before ductile deformation (that is, before reaching maximum pressure and temperature), but would be modified or destroyed at high temperature and pressure during penetrative ductile strain and passive or flexural-flow folding (Figure 15E–1). Generally, once the thermal peak is past, cooling occurs over millions of years and the rocks become progressively more brittle. Newly applied stress superimposes open buckle and kink folds accompanied by flexural slip.

Transitions in deformation style from the outer to the inner zones of mountain chains have been described by Richard B. Campbell (1973), Arthur Snoke (1980), David Sanderson (1979), and D. C. Murphy (1987)—specifically, the Cariboo Mountains of east-central British Columbia, the Ruby Mountains of northeastern Nevada, and the Variscan fold belt in southwestern England. Remarkable parallels are present, despite differences in plate-tectonic settings and time of deformation. Each chain exhibits a similar transition from more ductile structures in its inner parts to more brittle structures in the outer parts (Figure 15E–2). The Cariboo and Ruby mountains underwent deformation from middle to late Mesozoic and early Tertiary, and the Variscan chain in the late Paleozoic.

Folds in the outer zones of all chains were produced by flexural slip and buckling. Folds are generally open, and cleavage is weak or nonexistent. Axial-plane cleavage becomes more prominent as the folds tighten toward the cores. In the inner zones of each chain, folds become similar-like and tight to isoclinal; passive flow becomes dominant. This transition is most obvious in the Cariboos.

In the Variscan chain, fold interlimb angles systematically decrease from an average of about 80° in the external zone to less than 45° in the innermost zone. Correspondingly, the dip of axial surfaces decreases from about 80° in the outer zones to less than 15° in the core. Sanderson has attributed these changes to increased shear strain toward the inner parts of the Variscan chain.

Superposed folding becomes prominent in the inner zone of the Cariboos. We do not know whether superposition is related to deformational events widely separated in time or to superposition of ductile folds during one event. It seems likely that superposition occurred both during the major thermal event and afterward.

References Cited

Campbell, R. B., 1970, Structural and metamorphic transitions from infrastructure to suprastructure, Cariboo Mountains, British Columbia: Geological Association of Canada Special Paper 6, p. 67–72.

Murphy, D. C., 1987, Suprastructure/infrastructure transition, east-central Cariboo Mountains, British Columbia: Geometry, kinematics and tectonic implications: Journal of Structural Geology, v. 9, p. 13–29.

Sanderson, D. J., 1979, The transition from upright to recumbent folding in the Variscan fold belt of southwest England: A model based on the kinematics of simple shear: Journal of Structural Geology, v. 1, p. 171–180.

Snoke, A. W., 1980, Transition from infrastructure to suprastructure in the northern Ruby Mountains, Nevada, *in* Crittenden, M. D., Jr., Coney, P. J., and Davis, G. H., eds., Metamorphic core complexes: Geological Society of America Memoir 153, p. 287–333.

(a)

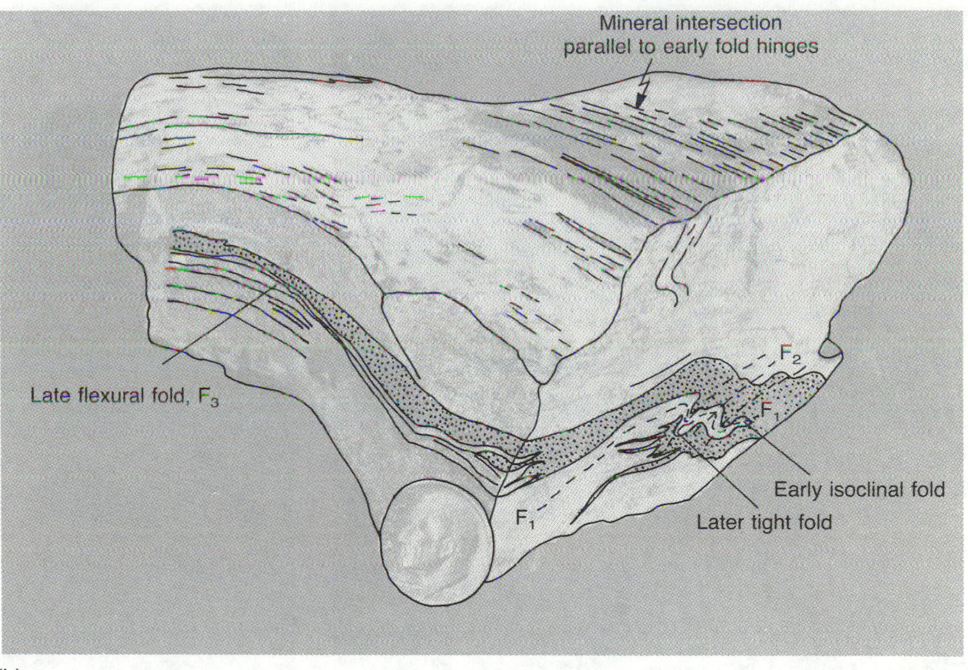

(b)

FIGURE 15E–1
(a) Overprinting of early ductile isoclinal folds by later brittle flexural folds in Poor Mountain amphibolite and feldspathic quartzite from near Toccoa, Georgia. (b) Early (F_1) ductile isoclinal recumbent passive-flow folds on the right side of the specimen were overprinted by more upright (F_2) ductile flexural-flow folds with similar axial trend to the earlier folds, then by brittle flexural buckle(?) folds visible on the left side of the specimen that trend perpendicular to the trend of the earlier folds. Note the strong mineral lineation on top of the specimen parallel to the earlier folds, probably an intersection lineation (Chapter 18) formed by intersection of axial-planar foliations (Chapter 17) of F_1 and F_2 folds.

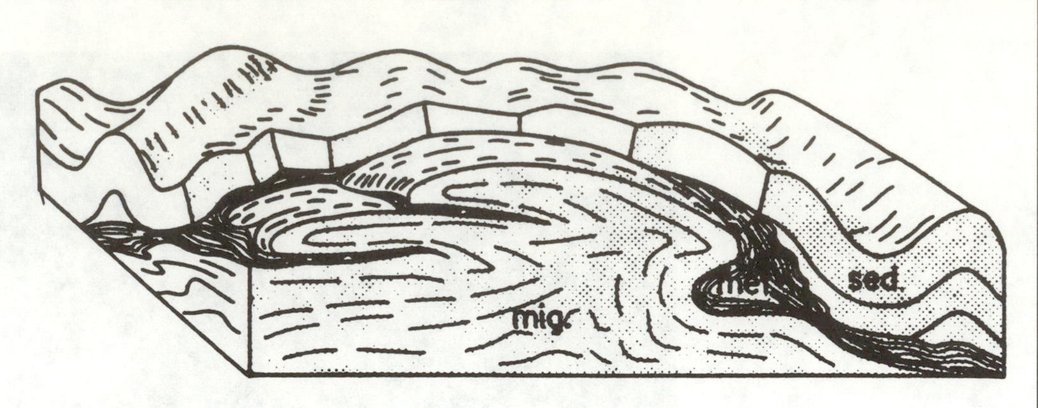

FIGURE 15E–2
Contrasting and progressive changes occurring in fold style from the outer zone (top of section) to inner zone (bottom); sed.—sedimentary rocks; mig.—migmatite. (From John Haller, 1956, *Geologische Rundschau,* v. 45.)

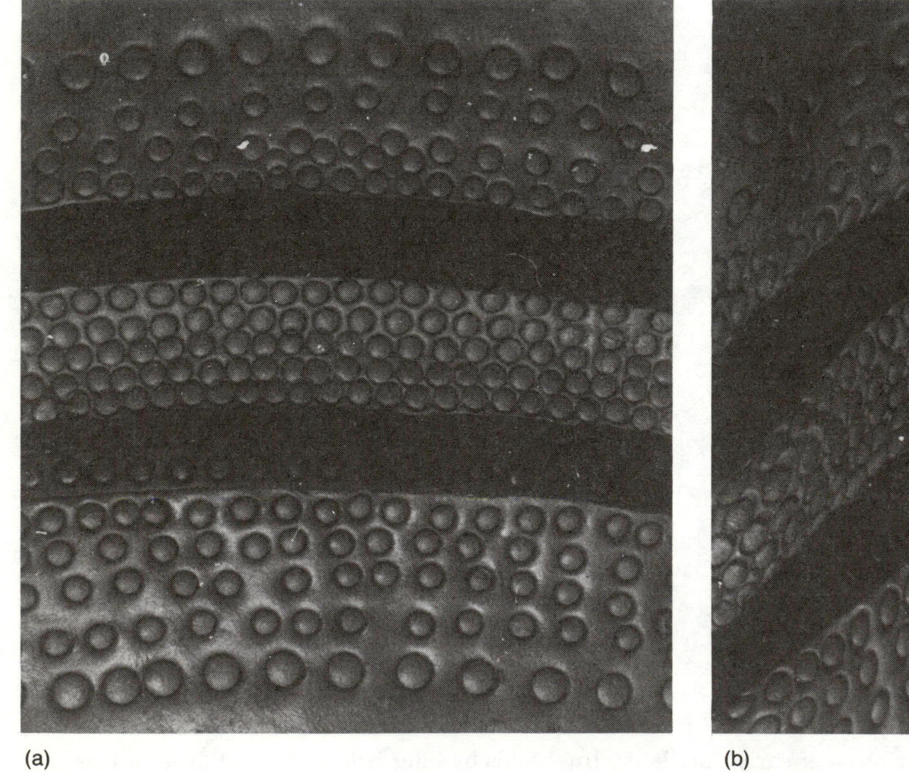

(a)

(b)

FIGURE 15–29
Model made of putty enclosing two stiff layers of Plasticene. (a) Undeformed. (b) Deformed, showing strain variations produced by heterogeneous strain and variations near layer boundaries. (From David Roberts and K.-E. Strömgård, 1971, *Tectonophysics,* v. 14, p. 105–120, Elsevier Science Publishers.)

FIGURE 15–30
(a) Photoelastic model of 18 percent shortening of a multilayer. Isochromatic bands indicate strain pattern in the weaker material. (b) Finite-strain trajectories constructed using the model in (a). They suggest that cleavage may undergo refraction according to these patterns. The dark area near the center of the model is a gash that developed as the layers separated. (Parts a and b from David Roberts and K.-E. Strömgård, 1971, *Tectonophysics,* v. 14, p. 105–120, Elsevier Science Publishers.) Note similarities of strain trajectories and cleavage in the fold in Figure 15–28a. (c) (page 346) Folded quartzite from Poor Mountain formation near Salem, northwestern South Carolina; layers have separated, and gash zones filled with new quartz. (d) (page 346) Sketch of fold in (c). New (vein) quartz is stippled.

(a)

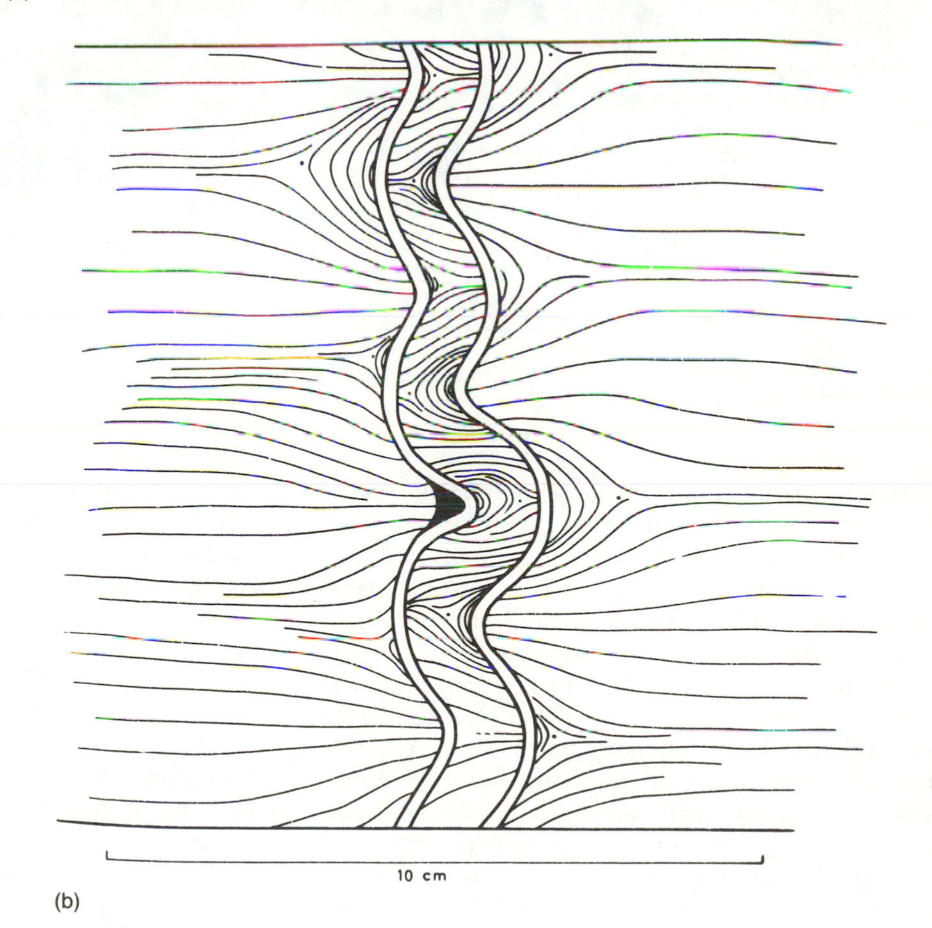

(b)

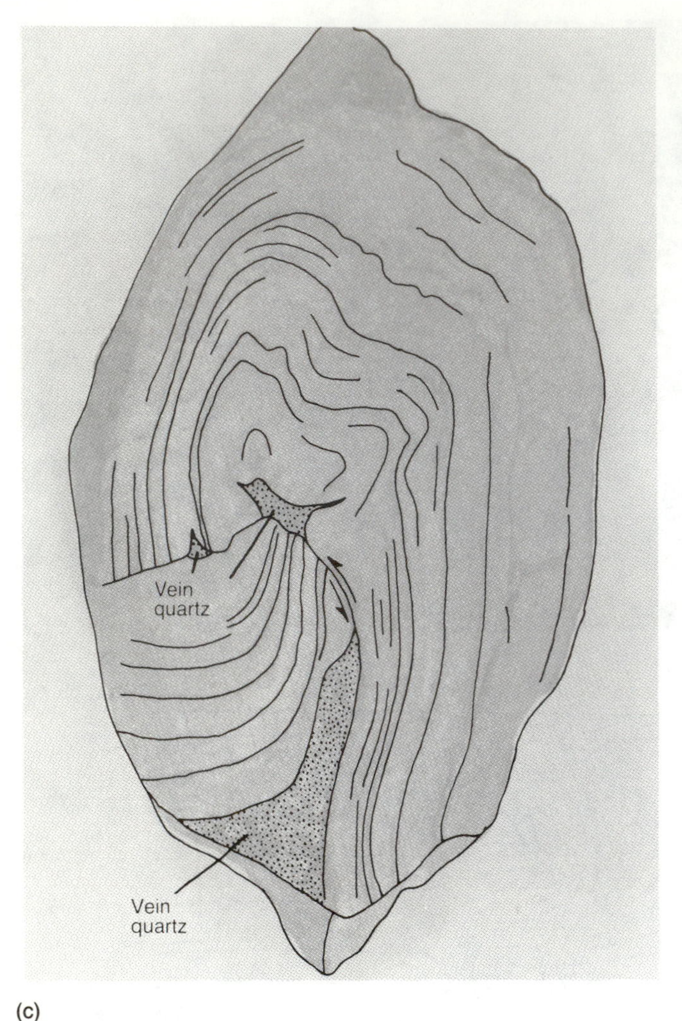

(c)

(d)

FIGURE 15–30
(continued)

Studies by Donald Ramsay and Brian Sturt (1973a) of the mechanisms and the occurrences of noncylindrical and aberrant folds, and experiments by Peter Cobbold and H. Quinquis (1980) on the formation of sheath folds, have led to explanations of how they form. In most cases, mildly to strongly noncylindrical folds are produced if the fold axis forms initially as part of an ordinary cylindrical fold; but the strain that produces the fold either diminishes or becomes inhomogeneous and a strong component of simple shear develops. Inhomogeneous simple shear deforms the preexisting fold axes so they become curved within the fold axial surfaces (Figure 15–32). Parts of the same fold axis are transported differentially, greatly distorting the originally cylindrical folds. If the process continues long enough, fold axes end up subparallel to the transport direction (Figure 15–33) and the folds are sheath folds. If the process ends before the axes are rotated into parallelism with the transport direction, the folds are noncylindrical. Sheath folds commonly form in ductile shear zones, but some layers may remain more brittle than the surrounding material.

LINEATIONS AND FOLD MECHANISMS

Lineations (Chapter 18) may indicate the nature of fold mechanisms. Mineral lineations commonly form as elongations down the dip of fold axial surfaces. They may also form parallel to the fold axis but become discordant to the fold axis as noncylindrical folding begins (Figure 15–34). As a result, the lineation may be parallel along the fold hinge but diverge toward a limb as it is traced to another part of the fold (Ramsay and Sturt, 1973b). Similarly, in sheath folds mineral lineations commonly parallel

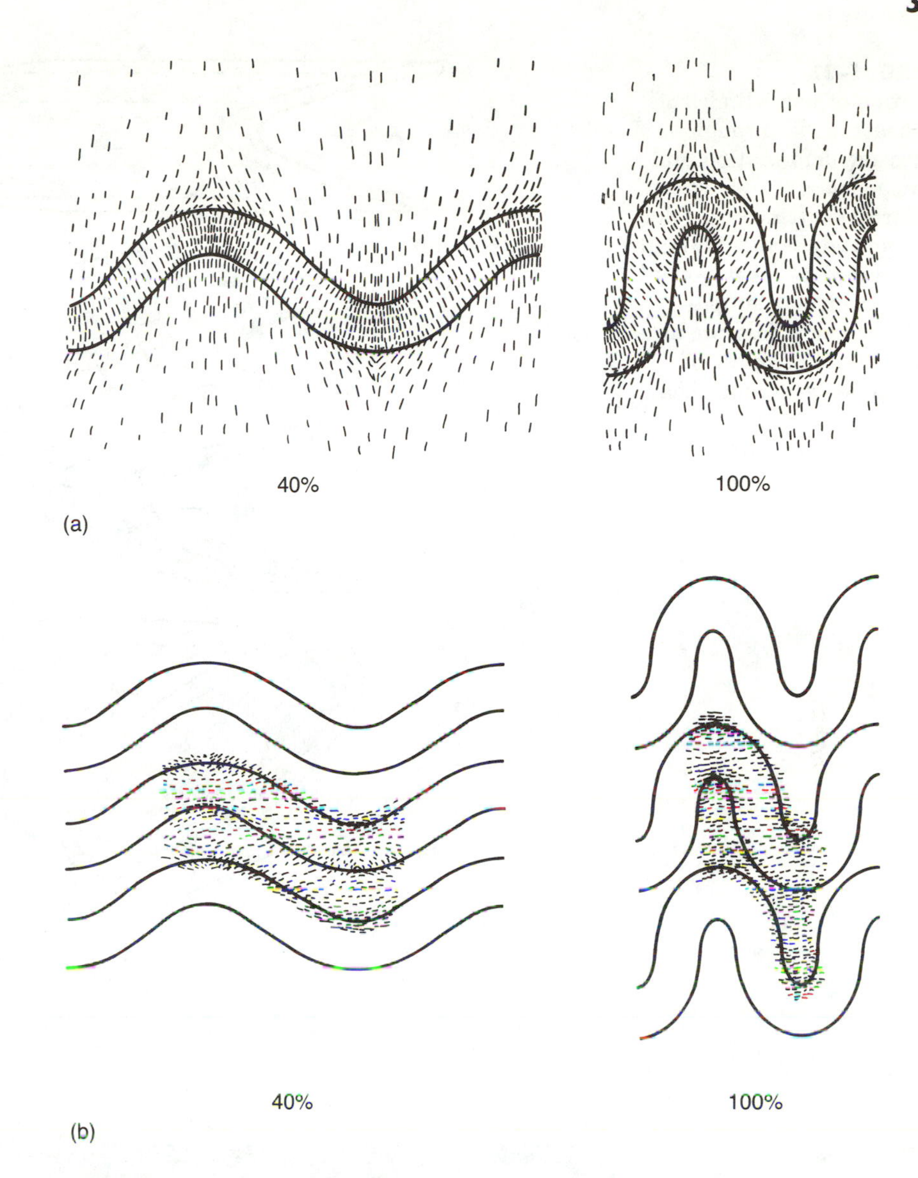

FIGURE 15–31
(a) Computer models of buckle folds. Short lines indicate orientation of the X-axis of the strain ellipsoid for a single layer shortened by 40 percent and 100 percent. (b) Model buckled multilayer. Short lines are parallel to σ_1. (From J. H. Dieterich, 1970, *Canadian Journal of Earth Science,* v. 7.)

the sheath-fold axes (with orientations that vary from place to place within the shear zone), and lineations are oriented down the dip of fold axial surfaces.

PARALLEL FOLDS AND SIMILAR FOLDS

Up to this point, our discussion of folds has traversed fold classifications and mechanics. Before considering complex folds in the next chapter, we will now relate fold mechanics of the two fundamental fold geometries—parallel folds and similar folds.

Layering plays a pivotal role in parallel folding, so buckling or bending is the dominant mechanism. Parallel folds formed by buckling or bending of massive sandstone or limestone layers represent an ideal example. Interlayering of strong sandstone or carbonate rocks with shale or other weak material that changes thickness during folding produces buckle folds that are not ideal parallel folds and have similar-like characteristics. Ductile layers having identical mechanical properties and folded by passive flow produce ideal similar folds. As layering begins to participate in folding, ideal similar folds are no longer produced and buckling becomes dominant.

Our examination of fold mechanics should provide the background for a discussion of complex folds in the next chapter.

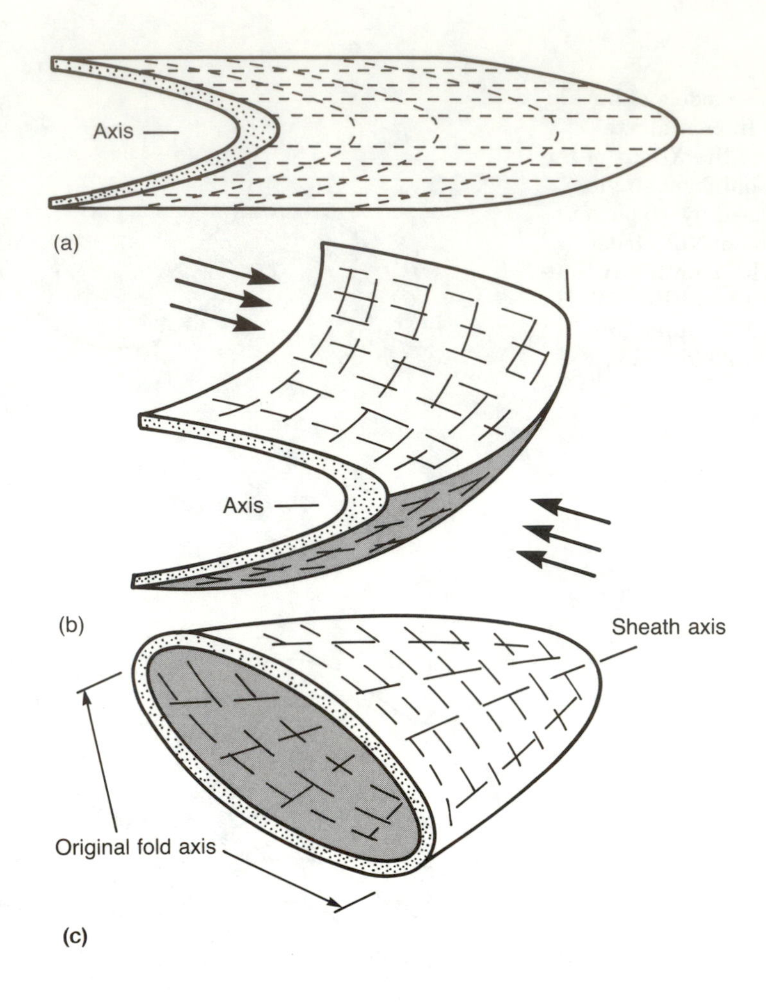

FIGURE 15–32
Formation of noncylindrical and sheath folds through progressive deformation and inhomogeneous simple shear. (a) "Normal" cylindrical fold. (b) Moderately noncylindrical fold produced by a component of inhomogeneous simple shear. (c) Sheath fold produced as an extreme example of inhomogeneous simple shear.

FIGURE 15–33
Map view of a sheath fold in metasandstone in the Middle Proterozoic Moine Series, Scottish Highlands. (Anthony L. Harris, University of Liverpool.)

FIGURE 15–34
(a, b) Folded mineral lineation (dark lines on a and b) showing changes in orientation of long axes of minerals in a noncylindrical fold. Some remain parallel to the fold axes (FA); others are rotated in the transport (simple-shear) direction. Fabric diagram in (b) shows the changes in orientation of lineations in different parts of the fold. (c) Small folds and plots of loci of lineations on a fabric diagram that are both parallel to and divergent from the axis of a noncylindrical fold. (From D. M. Ramsay and B. A. Sturt, 1973, *Tectonophysics,* v. 18, Elsevier Science Publishers.)

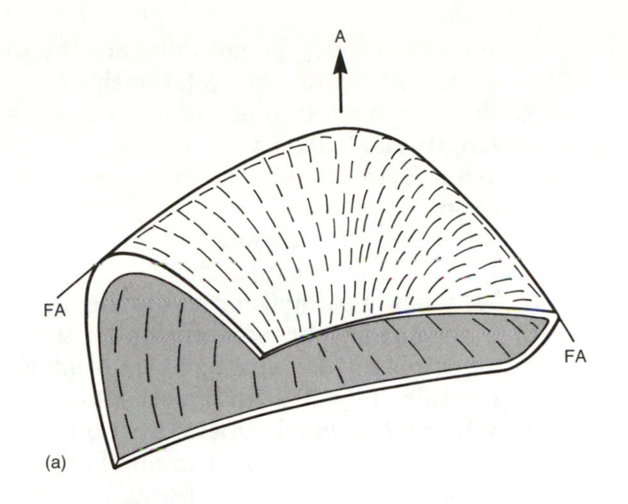

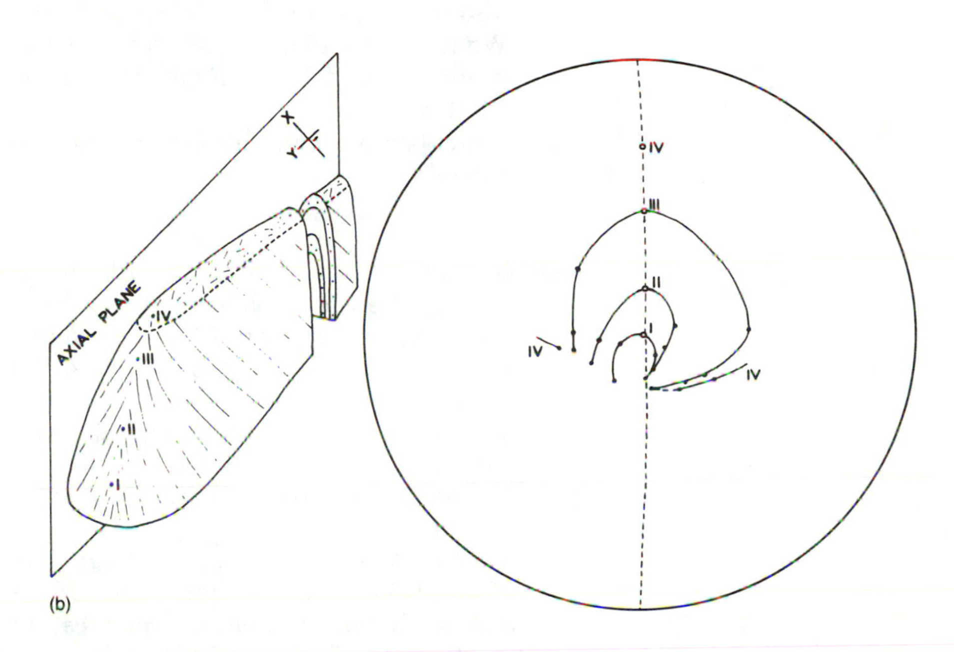

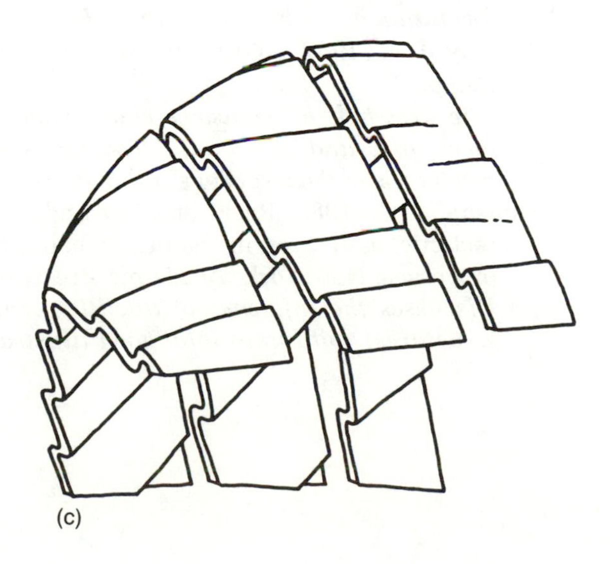

Questions

1. Why does rock type not influence the thickness of the dominant member/dominant wavelength relationship?
2. What is the origin of the neutral surface in folds formed by tangential-longitudinal strain?
3. What factors determine which mechanism dominates in folding a mass of rocks?
4. What are the physical requirements for flexural-slip buckle folding? Passive flow? Flexural flow?
5. Why is the passive-slip mechanism problematical?
6. Considering Johnson's criteria for the formation of kink and buckle folds, explain why both kinks and sinusoidal folds occur in some deformed single crystals of kyanite and in others only kinks?
7. Why do occasional mineral lineations parallel a fold axis in one part of the fold and diverge from it in another?
8. Where in an orogenic belt would you expect passive-flow folds? Why do we also see them in soft sediment and salt?
9. What was the sequence of application of homogeneous and heterogeneous strain in the folds in Figure 15–17? The rubber-layer model in Figure 15–28?
10. Could passive-flow folds form by the mechanism proposed by Beutner and Diegel?

Further Reading

Hobbs, B. E., Means, W. D., and Williams, P. F., 1976, An outline of structural geology: New York, John Wiley & Sons, 571 p.
The chapter on folds deals with fold mechanics from a point of view somewhat different (but equally useful) from that just presented and in the other references in this list.

Hudleston, P. J., 1986, Extracting information from folds in rocks: Journal of Geological Education, v. 34, p. 237–245.
Discusses the three principal fold mechanisms and the environments of formation.

Johnson, A. M., 1977, Styles of folding: Mechanics and mechanisms of folding of natural elastic materials: Amsterdam, Elsevier, 406 p.
Johnson's introduction summarizes the state of knowledge about fold mechanics in the late 1970s and describes rigorously developed ideas on the formation of buckle and kink folds.

Ramsay, J. G., 1967, Folding and fracturing of rocks: New York, McGraw-Hill, 568 p.
The first half of Ramsay's chapter on folds and folding deals with fold mechanics and discusses buckling, shear, and compressive strain and mechanisms that lead to similar folds.

Ramsay, J. G., 1983, Rock ductility and its influence on the development of tectonic structures in mountain belts, *in* Hsü, K. J., Mountain building processes: New York, Academic Press, p. 111–128.
Discusses the influence of ductility contrast on folding style (and other structures), with unusually good illustrations.

16

Complex Folds

Thus the Loch Monar synform has folds which are superimposed upon its limbs, which cut across the major structure and which appear to be unrelated to it. Individual second folds can be traced across the first synformal structure from the north-east, where they are found on the gently dipping limb of the synform to the south-west, where they contort the steeply dipping limb of the synform. . . . A remarkable feature created by the second-fold structures crossing the first folds is that any one stratigraphical horizon is folded at least twice by each second fold.

JOHN G. RAMSAY, Quarterly Journal of the Geological Society of London

SO FAR IN THIS BOOK, WE HAVE CONCENTRATED MORE ON discussions of single folds than on fold systems. Yet, isolated folds are less common than groups. Most form in sets with a common orientation of axes and axial surfaces. Folds formed under the same conditions also have the same sense of vergence. The strain field can change from homogeneous to heterogeneous (or vice versa), or the orientation of principal stresses can change, and either can produce a new set of folds in the same area; the new set may overprint existing folds or extensively modify the geometry. Systems of ***superposed,*** or ***interference, folds*** result where one set of folds overprints another. The shapes of interference folds are related to the orientation of the two fold sets and to the physical conditions of the rock body being deformed.

FIGURE 16–1
Interference pattern produced by water waves crossing in the lee of a headland on the coast of Brazil. The waves radiating from the headland are interfering with linear waves to produce domes and basins, particularly well-developed just south of the headland. (U.S. Geological Survey.)

Fold interference can result in patterns like those produced in water when waves cross and interfere with one another (Figure 16–1).

Noncylindrical and sheath folds, too, form geometrically complex systems. Like other fold systems, they usually involve either continued heterogeneous deformation or changes in the strain field through time.

OCCURRENCE AND RECOGNITION

Complex folds occur in a variety of settings. Two folding events may be separated by minutes or by millions of years, or a continuum may exist; in any case, two events are likely to yield fold-interference patterns and thus complex folds. Ductile shear zones

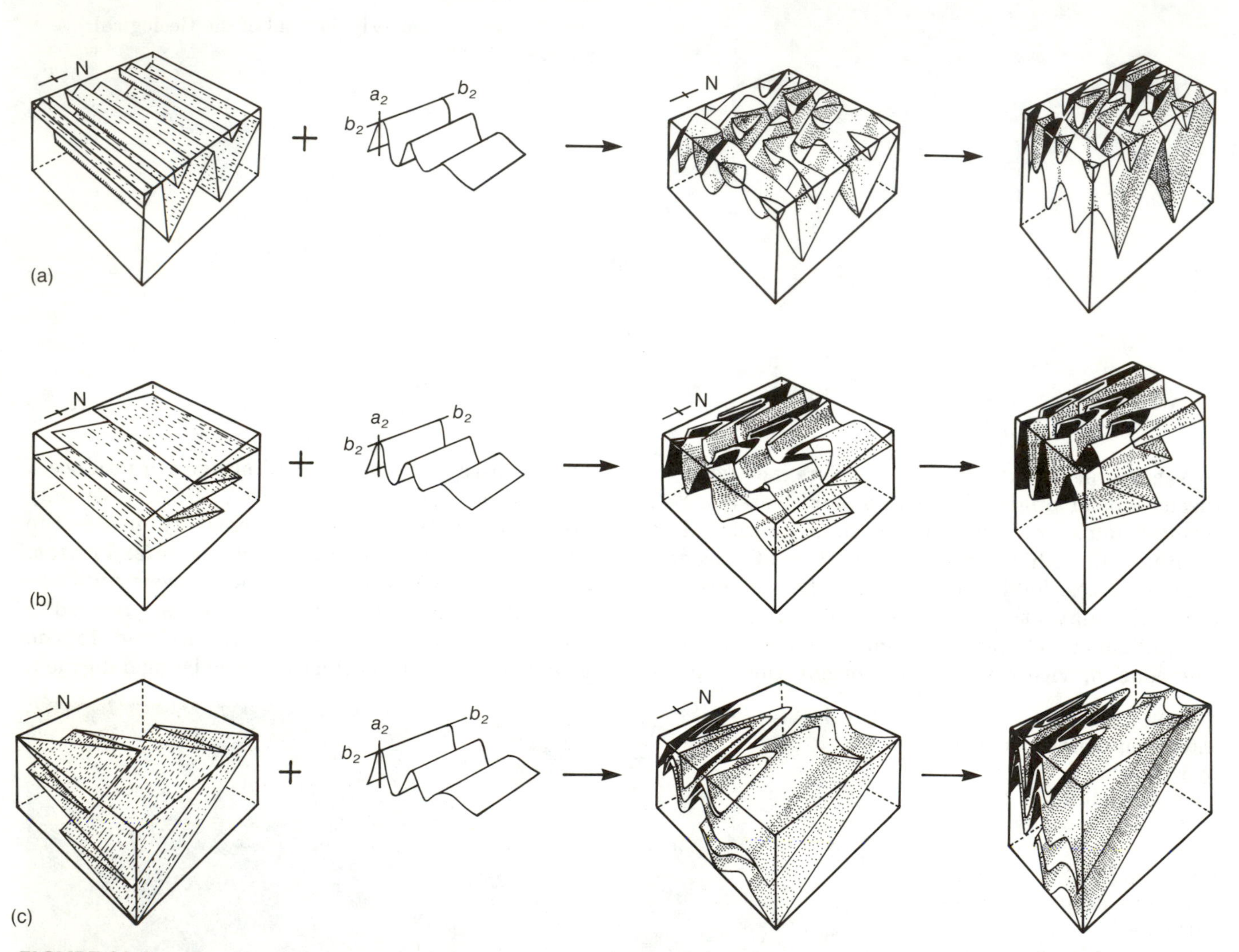

FIGURE 16–2
Ramsay's three fundamental types of fold-interference patterns, showing basic geometry, outcrop pattern, and fabric diagrams with axes and axial surface orientations. (a) Type 1. Dome-and-basin pattern: early folds with steeply dipping axial surfaces are overprinted by a second set, also having steeply dipping axial surfaces, but with an orientation perpendicular to the first set. (b) Type 2. Early tight folds with gently dipping axial surfaces are overprinted by a set of folds with near-vertical axial surfaces oriented perpendicular to the first set. (c) Type 3. Early tight reclined folds are overprinted by a later set with the same axial orientation, but with vertical axial surfaces. (Modified from J. G. Ramsay, *Folding and Fracturing of Rocks,* © 1967, McGraw-Hill Book Company. Reproduced with permission.)

and strain variation within ductile rock masses make possible the formation of complex folds. Such conditions arise most commonly in the cores of mountain chains where a continuum of deformation and metamorphism occurs. The same conditions may also occur at plate boundaries—in subduction zones, spreading ridges, or transform fault zones—where varying amounts of thermal energy combine with a strong component of simple or pure shear.

FOLD-INTERFERENCE PATTERNS

Near Lock Monar in the northern Scottish Highlands, John Ramsay (1958, 1962) studied the rocks and saw that multiple fold episodes produced three fundamental types of fold-interference patterns (Figure 16–2). In Australia and elsewhere, S. Warren Carey (1962) also recognized the importance of fold interference and interference patterns. Here we will consider these fundamental types, along with intermediate types that often form rather than ideal end members.

Type 1

If two sets of folds have upright axial surfaces that intersect at high angles, they interfere and produce a structure called ***egg-carton structure*** or ***dome-and-basin pattern*** (Figures 16–1, 16–2a, and 16–3). Ideally, both fold generations have the same style—the same fold mechanism. Axes of the first set of folds are oriented at high angles (70° to 90°) to those of the later fold set, and the axial surfaces dip steeply. Tight folds interfere and produce high-amplitude dome-and-basin structures, and open folds produce gentle, low-amplitude dome-and-basin structures. They may occur on all scales and in any part of an orogen. In the Glen Cannich area of the Scottish Highlands (Figure 16–4), Othmar Tobisch (1966) reported a large dome-and-basin pattern. Tobisch and others (1970) have suggested that similar patterns are widespread in that region.

Type 2

If folds having inclined or reclined axial surfaces are superposed by folds with steeply dipping axial surfaces and axes at high angles to those of the first set, a ***boomerang pattern*** results (Figures 16–2b, 16–5). Generally, the first folds are inclined isoclinal folds; the second, upright similar-type folds. Their shape depends on the angle between the first and second folds, the dip of the axial surfaces of the early folds, and the plunge of axes of early and late folds. The boomerang may become highly asymmetric if the angle between the two axial surfaces is *not* 90°. A boomerang may be greatly extended and opened to tongue shape if the axial surfaces of the earlier folds dip gently. As the dip of the axial surfaces of the early folds steepens relative to the later folds, boomerangs also grade into

FIGURE 16–3
Dome-and-basin (Type 1) interference pattern in Archean grunerite chert banded iron formation, Opapimiskan Lake area, northwestern Ontario (North Caribou Lake greenstone belt, Superior province). (Fred W. Breaks, Ontario Geological Survey.)

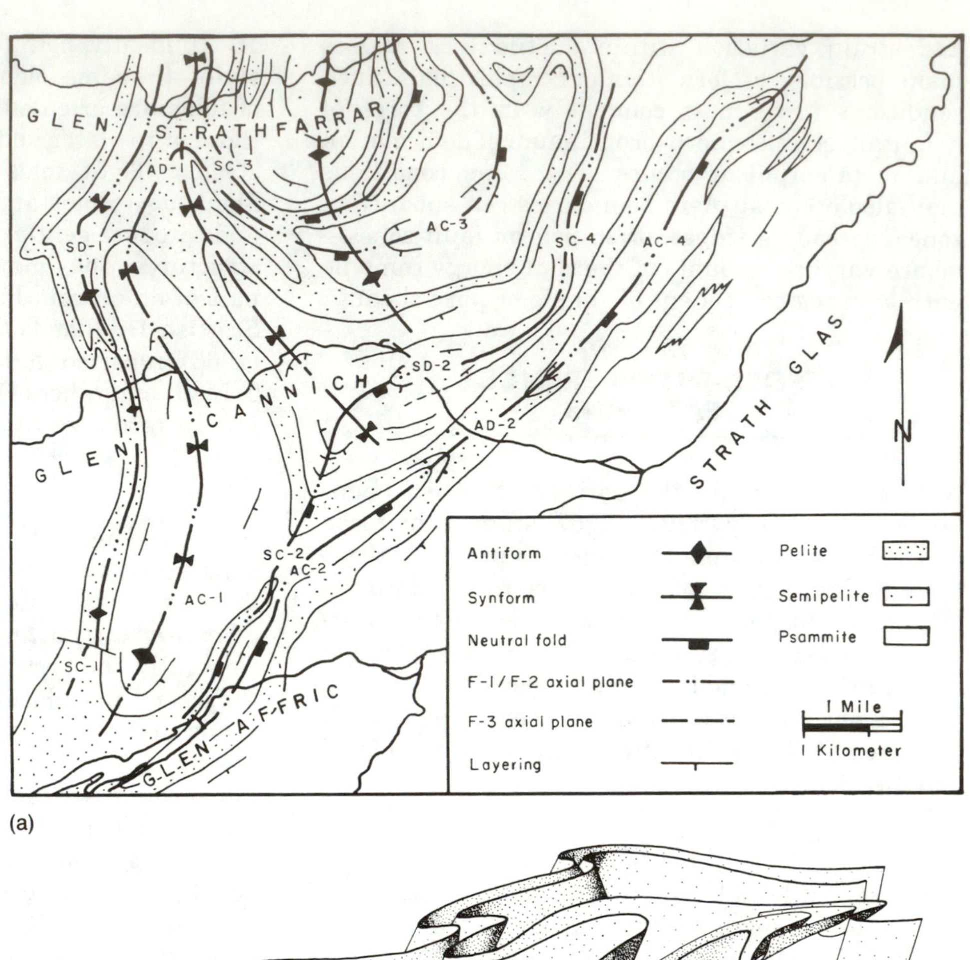

(a)

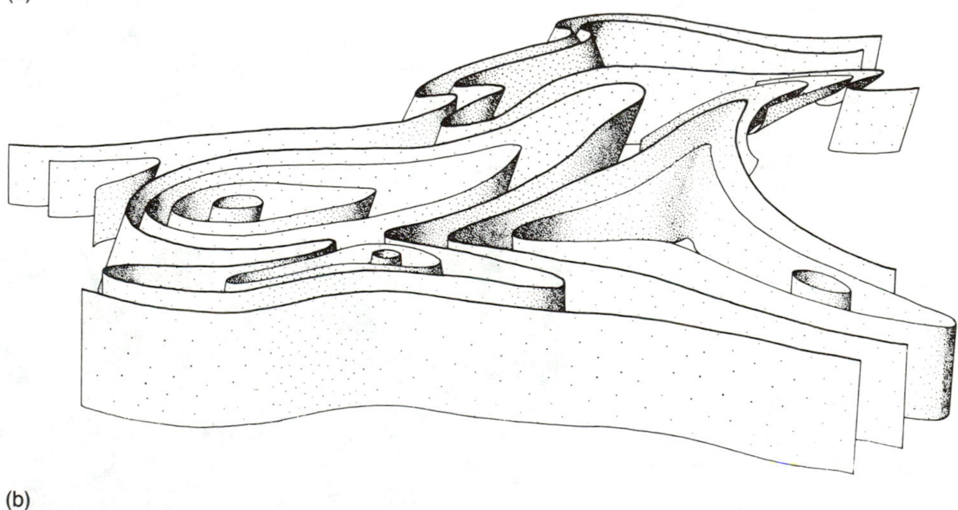

(b)

FIGURE 16–4
(a) Tectonic sketch map of the Glen Cannich and adjacent areas in the Scottish Highlands showing the outcrop pattern of the dome-and-basin interference pattern.
(b) Simplified three-dimensional diagram of Type 1 folds in the Glen Cannich area. (From O. T. Tobisch, 1966, Geological Society of America *Bulletin*, v. 77.)

dome-and-basin (Type 1) interference folds. A large-scale boomerang pattern exists in the core of the Toxaway dome in South Carolina (Hatcher, 1977; Figure 16–6), where several episodes of folds on all scales are superposed.

Type 3

If early-formed tight to isoclinal folds are refolded about the same axis, they produce an interference pattern called a ***hook*** (Figure 16–7) and may form on small scales as well as large (Figure 16–8). A hook pattern may result from continued folding during the same event (Figure 16–9) or from overprinting by a second generation of folds with the same axial orientation as the first set. To form Type 3 interference patterns, the two fold sets must not be coplanar. It may be difficult to resolve the conditions that form Type 3 folds because we need to know whether folding ended at one time and then resumed later under the same stress conditions—or whether it was continuous within a single event and produced coaxial refolding (Wynne-Edwards, 1963).

FIGURE 16–5
Boomerang (Type 2) interference pattern in Upper Proterozoic Tallulah Falls Formation metasedimentary rocks near Rainy Mountain, northeastern Georgia. (RDH photo.)

Modifications

Combinations of all three types of interference patterns have been found in nature. As noted earlier, they grade into one another where fold axial surfaces steepen, or become gentler, or approach parallelism. Type 2 interference patterns may grade into Type 3 where the axial trends of early- and late-formed folds become more nearly parallel. Outcrop patterns reflect these gradational fold patterns. Boomerangs appear as tongue-shaped folds in map view or appear more elongate, with one end open in a hooked outcrop pattern (Figure 16–10). Various possible combinations of outcrop patterns of interference fold shapes have been worked out by Richard Thiessen and Winthrop Means (1980), using cuts at different angles through a hypothetical cube of repeatedly deformed crust. Their work also helps elucidate three-dimensional aspects of superposed folding.

RECOGNITION OF MULTIPLE FOLD PHASES

Several characteristics help identify multiple fold phases. The geometry of outcrop patterns on geologic maps or on the mesoscopic scale is the best key to superposed folds (Figure 16–10), but in some areas exposure of superposed folds are rare and other clues are needed. Multiple fold orientations and styles of folds in the same area can be used, with caution, for determining whether superposed folds exist, but alone they do not suffice; exposures of superposed folds must be found, too. Multiple foliations, folded foliations, and folded lineations also indicate multiple deformation and superposed folds.

NONCYLINDRICAL AND SHEATH FOLDS

All folds become noncylindrical if traced far enough and if they are not truncated by faults. Folds that are noncylindrical over short distances are products of localized heterogeneous deformation. The transition from aberrant folds through noncylindrical to sheath folds (see Figure 15–32) is related to increase in inhomogeneous strain.

Sheath folds resemble dome-and-basin interference folds (Figure 16–11), but they are generally more isolated. They may be strongly asymmetric and elongate rather than symmetrical, but many sheath folds yield highly symmetric *eyed* or *bulls-eye* sections in plan view. Antiformal and synformal sheath folds may be strongly aligned in a single zone of strong simple shear, such as a shear zone, whereas Type 1 interference folds may be developed over a broad area, but the distinction is not always clear.

An area where polyphase folding occurs must be studied carefully to ascertain that fold-interference patterns are not really sheath or strongly noncylindrical folds. Dome-and-basin interference patterns

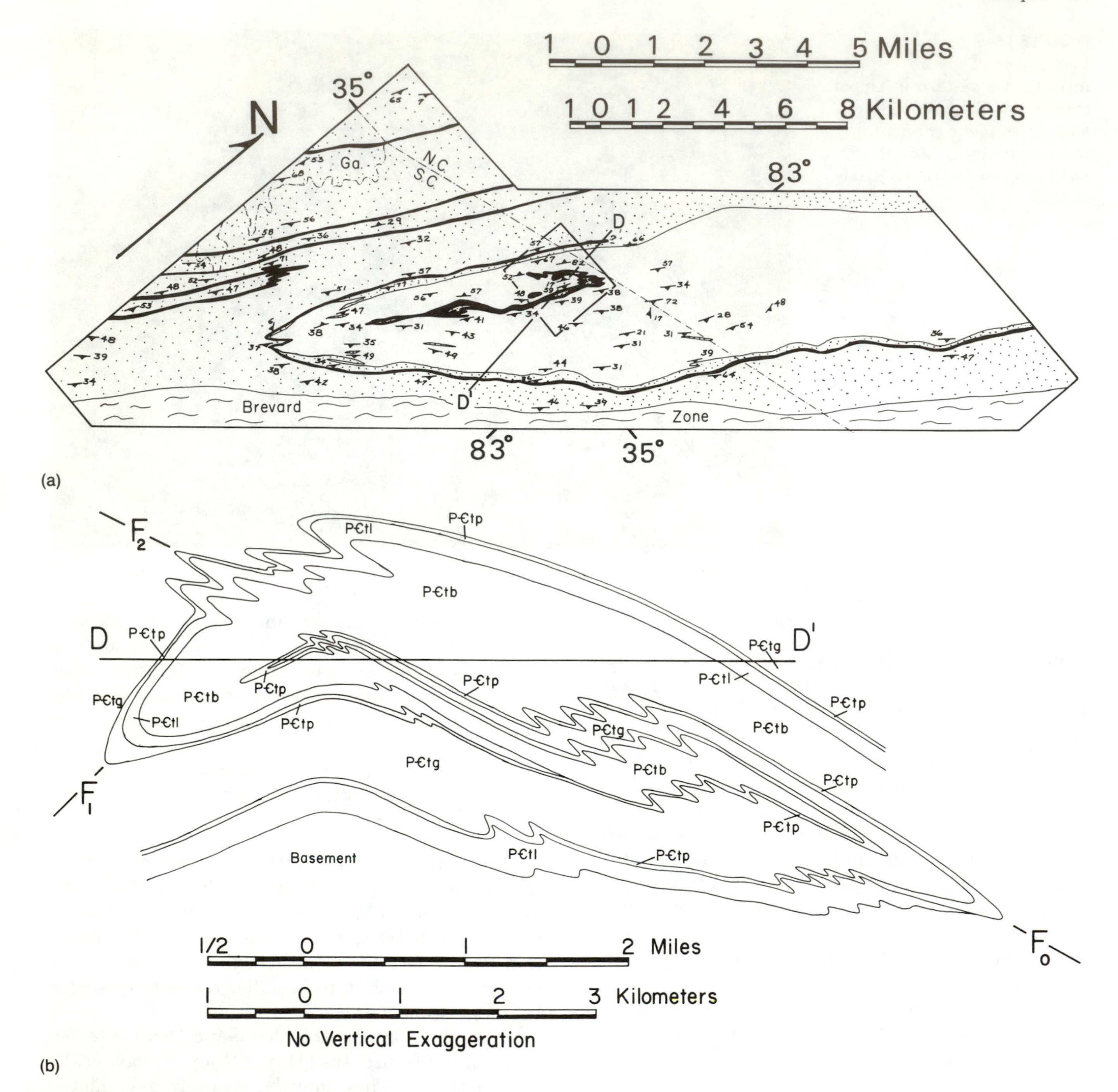

FIGURE 16–6
(a) Geologic map of part of the Toxaway dome, South Carolina–North Carolina, showing the map-scale boomerang Type 2 interference pattern developed by superposed folding of basement (unpatterned) and cover rocks (stippled and dark pattern). (b) North-south geologic section (D-D′) constructed normal to the orientations of the early folds. Note that an earlier ("pre-F_1") set of folds was resolved here after recognizing that the oldest rocks (basement) must have been beneath the cover before deformation. Basement rocks are indicated as pЄtb, cover rocks as pЄtl, pЄtp, and pЄtg. (From R. D. Hatcher, Jr., 1977, Geological Society of America *Bulletin*, v. 88.)

FIGURE 16–7
Hook (Type 3) interference pattern to left of hammer in Upper Proterozoic Tallulah Falls Formation metasandstone, Holcombe Creek Falls, near Clayton, Georgia. (RDH photo.)

may be mistaken for sheath folds and must be distinguished as one or the other because of the implications concerning fold mechanisms and deformation history. It is therefore necessary to work out the fold mechanisms as well as the timing of the different fold phases. Sheath folds may be confined to a linear zone of ductile shear or to a large fault zone; fold-interference patterns may be distributed regionally and not confined to particular deformation zones.

FORMATION OF COMPLEX FOLDS

If superposed folds are to form, strain states must vary in both space and time. Analysis of the geometry of superposed folding may begin with consideration of *coaxial* and *noncoaxial deformation*. Noncoaxial deformation is required to form Type 1 and Type 2 interference patterns. Where Type 1 folds develop, the strain axes are reoriented after the formation of the early folds so that the second folds trend normal to the first—and the rocks remain in much the same rheological state. Likewise, with Type 2 and Type 3 interference patterns there must be some reorientation of strain axes as well. But formation of Type 1 and Type 2 patterns requires reorientation of the strain axes rapidly enough that the mechanical properties of the rocks are not changed by cooling or by diminishing pressure. This is a more difficult set of circumstances to achieve, especially considering the amount of reorientation required, but it could occur during plate collision if a promontory or other irregularity were encountered on either plate boundary—or possibly by migration of a triple junction. In contrast, Type 3 interference patterns form by coaxial deformation (but not coplanar) as a continuation of the same folding process that affected the rocks earlier. Wynne-Edwards (1963; Figure 16–9) has described the circumstances under which Type 3 superposed folds can form during the same event by deformation more or less uninterrupted; the process forms the first folds, then superposes the second set coaxially but not along the original axial surfaces.

MECHANICAL IMPLICATIONS OF COMPLEX FOLDING

The geometry of superposed and noncylindrical folding implies certain things about the mechanical processes that form them. That they are readily observed as interfering water waves demonstrates that they form in all ductile media. Interference

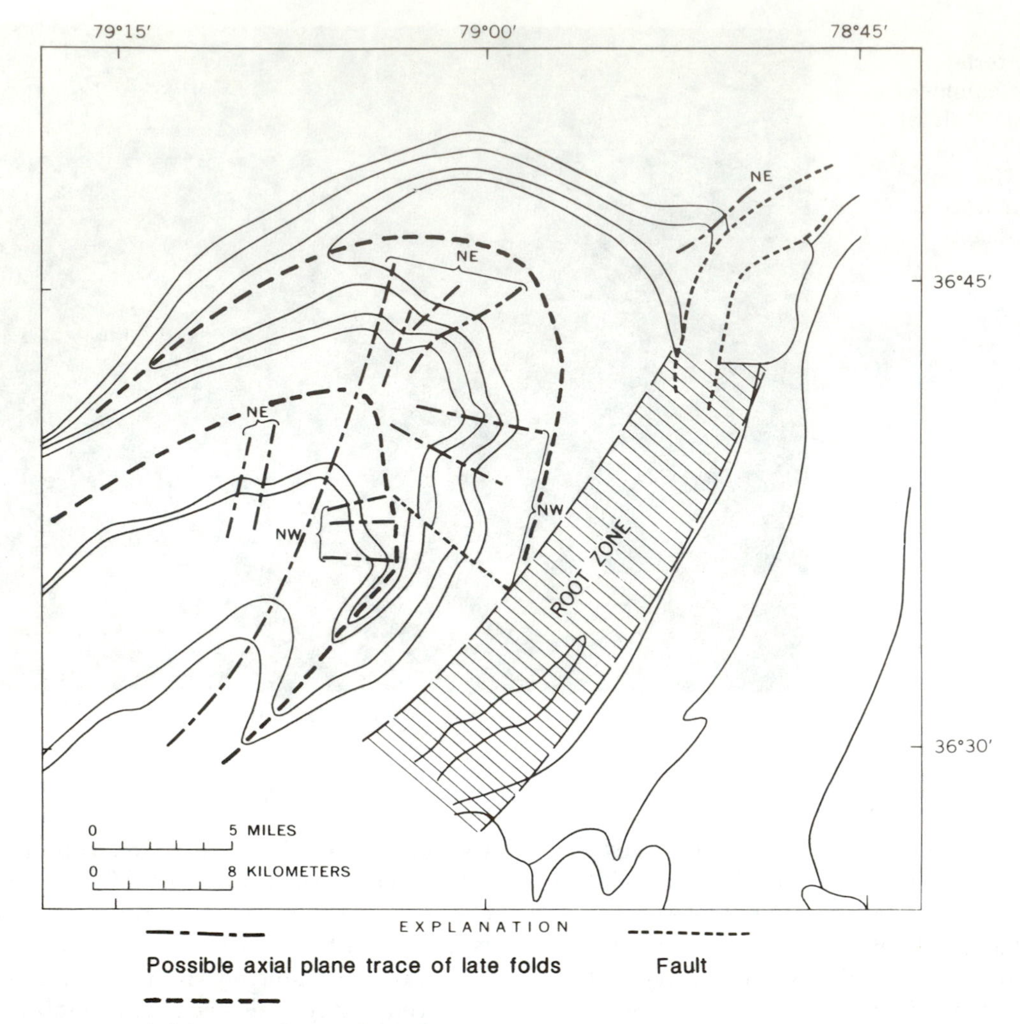

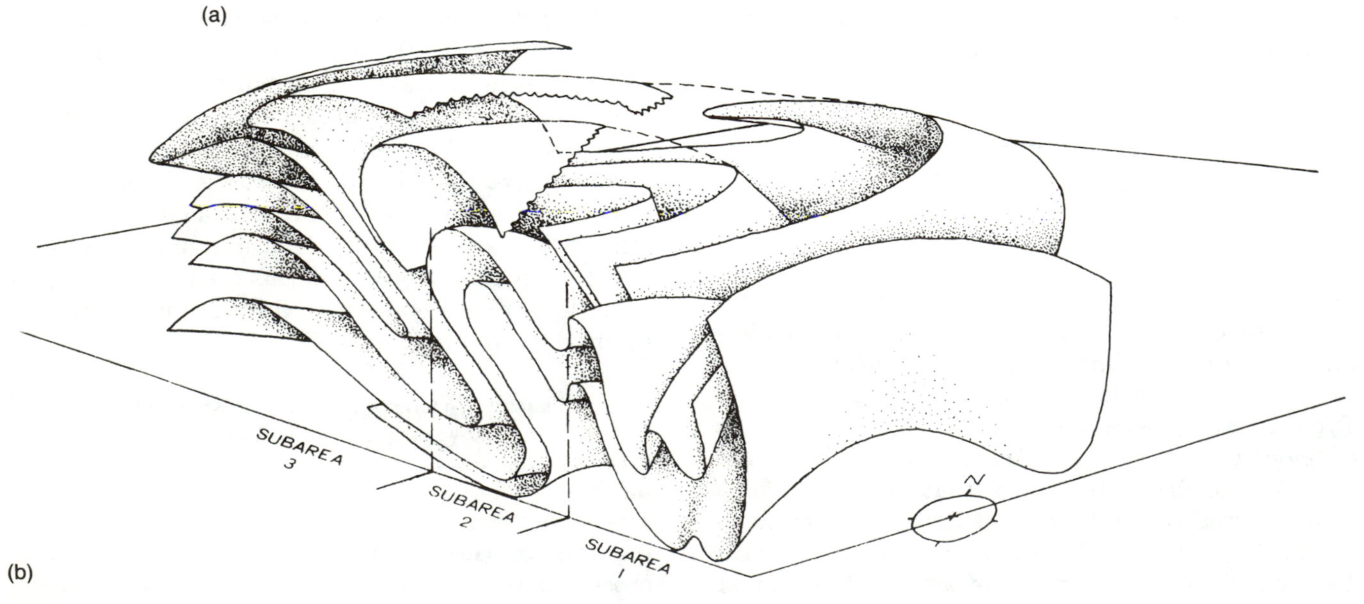

FIGURE 16–8
(a) Map-scale Type 3 interference pattern in the Milton area north of Greensboro, North Carolina. (b) Sketch of the three-dimensional relationships of folds in the Milton area. (From O. T. Tobisch and Lynn Glover, III, 1971, Geological Society of America *Bulletin,* v. 82.)

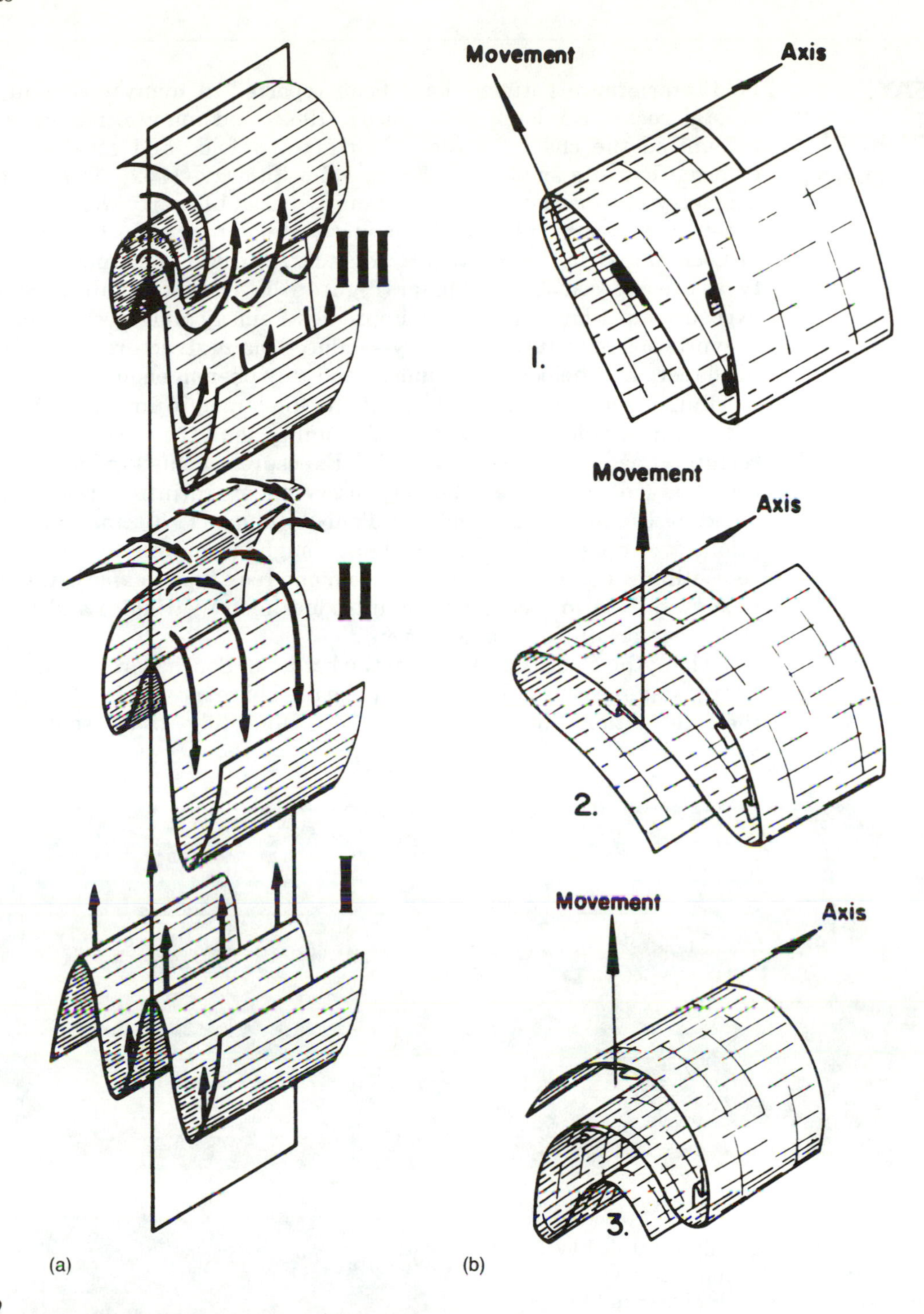

FIGURE 16–9
Formation of Type 3 interference folds by progressive deformation of an already-formed fold without varying axial orientation. (a) Formation by migration of the hinge of an advancing antiform next to a stationary or more slowly advancing synform. (b) Formation by change in direction of transport. (From H. R. Wynne-Edwards, Flow folding, Nov. 1963, *American Journal of Science,* v. 263, p. 793–814. Reprinted by permission of the American Journal of Science.)

ESSAY

The Value of Rosetta Stones

Fold-interference patterns have been reported in many places in deformed crustal rocks. No doubt Sam Carey (1962) and many other geologists had recognized the characteristics of superposed folds and considered multiple deformation in a single area before John Ramsay (1962, 1967) described the geometry of superimposed folds. Ramsay based his early work largely on one excellent exposure of Precambrian metasedimentary rocks of the Moine Series at Loch Monar in the Scottish Highlands. All three fundamental interference types are exposed at Loch Monar (Figure 16E–1). Finding all three in a single exposure must have profoundly impressed Ramsay and—because of his ability to synthesize structural geometry—induced many structural geologists to read his papers and helped them understand this phenomenon.

Ramsay might have worked out the fundamental interference types even without a single exposure of such quality, but the Loch Monar exposure certainly made his task much easier. Exposures of this kind are useful in any complexly deformed terrane for observing overprinting relationships and avoiding the pitfalls described by Paul Williams (1970), in which fold generations are recognized and correlated throughout an area by their style alone. As Williams pointed out, changes from one rock type to another, in metamorphic grade and in fluid pressure, introduce many variables and thus produce style changes during a single event.

Along the Chattooga River on the border of South Carolina and Georgia is a large flat exposure called Woodall Shoals. This exposure is a Rosetta stone for the deformational history of the eastern Blue Ridge of the southern Appala-

(a)

(b)

(c)

FIGURE 16E–1
Interference-fold patterns originally identified by John Ramsay in Middle Proterozoic Moine Series metasandstone and schist at Loch Monar in the Scottish Highlands. (a) Type 1. Dome-and-basin pattern in center of photo. (b) Type 2. Boomerang pattern in schistose (dark) layer beneath coin. (c) Type 3. Hook pattern to left and below knife. Type 1 pattern is present above knife. (RDH photos.)

chians: it preserves several generations of folds and enables us to reconstruct the map-scale deformational history of much of the eastern Blue Ridge. The rocks are predominantly sillimanite-grade migmatitic biotite gneiss, amphibolite, anatectic granitoids, and pegmatite belonging to the Late Proterozoic Tallulah Falls Formation. Six or seven separate fold generations are preserved here, along with Type 1 and 2 fold interference patterns, and possibly Type 3, as well as several folds that may be sheath folds. This does not mean that six or seven deformational events separated in time affected these rocks, but it does mean that favorable conditions existed to superimpose several sets of folds during perhaps two events (or at most, three) separated by thousands of years.

Before Woodall Shoals was studied carefully, geologists were uncertain which fold generation here was the earliest—the same uncertainty that exists in other complexly deformed areas. It had been assumed that the earliest folds had sheared out limbs (*intrafolial folds*) and lay within the dominant foliation—commonly designated F_1 folds; those that overprint the F_1s were designated F_2, etc. The map-scale (macroscopic) outcrop pattern on the Toxaway dome several kilometers to the northeast (Figure 16–6) strongly suggested that a pre-F_1 fold set existed in the eastern Blue Ridge (Hatcher, 1977). At Woodall Shoals, similar-like folds were preserved in pods (boudins) of less-ductile amphibolite that were wrapped by the dominant regional foliation in the biotite gneiss (Figure 16E–2). This occurrence confirmed that one earlier set of passive or flexural flow folds was formed—possibly two—before the intrafolial folds lying within the dominant foliation.

(a)

(b)

(c)

FIGURE 16E–2
Early structures at Woodall Shoals in South Carolina and Georgia. (a) Amphibolite boudin showing foliation in dark amphibolite truncated by enclosing dominant foliation in biotite gneiss. Note small intrafolial fold to the left. (b) Fold in amphibolite boudin is earlier than the earliest folds in the enclosing biotite gneiss. (c) Deformed amphibolite boudin at Woodall Shoals showing structure in enclosing rocks. (RDH photos.)

This example demonstrates the usefulness of places like Loch Monar and Woodall Shoals in deciphering both deformational processes and the sequence of overprinting of deformations.

References Cited

Carey, S. W., 1962, Folding: Journal of the Alberta Society of Petroleum Geologists, v. 10, p. 95–144.

Hatcher, R. D., Jr., 1977, Macroscopic polyphase folding illustrated by the Toxaway dome, South Carolina–North Carolina: Geological Society of America Bulletin, v. 88, p. 1678–1688.

Ramsay, J. G., 1962, Interference patterns produced by the superposition of folds of "similar" type: Journal of Geology, v. 60, p. 466–481.

_____ 1967, Folding and fracturing of rocks: New York, McGraw-Hill, 568 p.

Williams, P. F., 1970, A criticism of the use of style in the study of deformed rocks: Geological Society of America Bulletin, v. 81, p. 3283–3296.

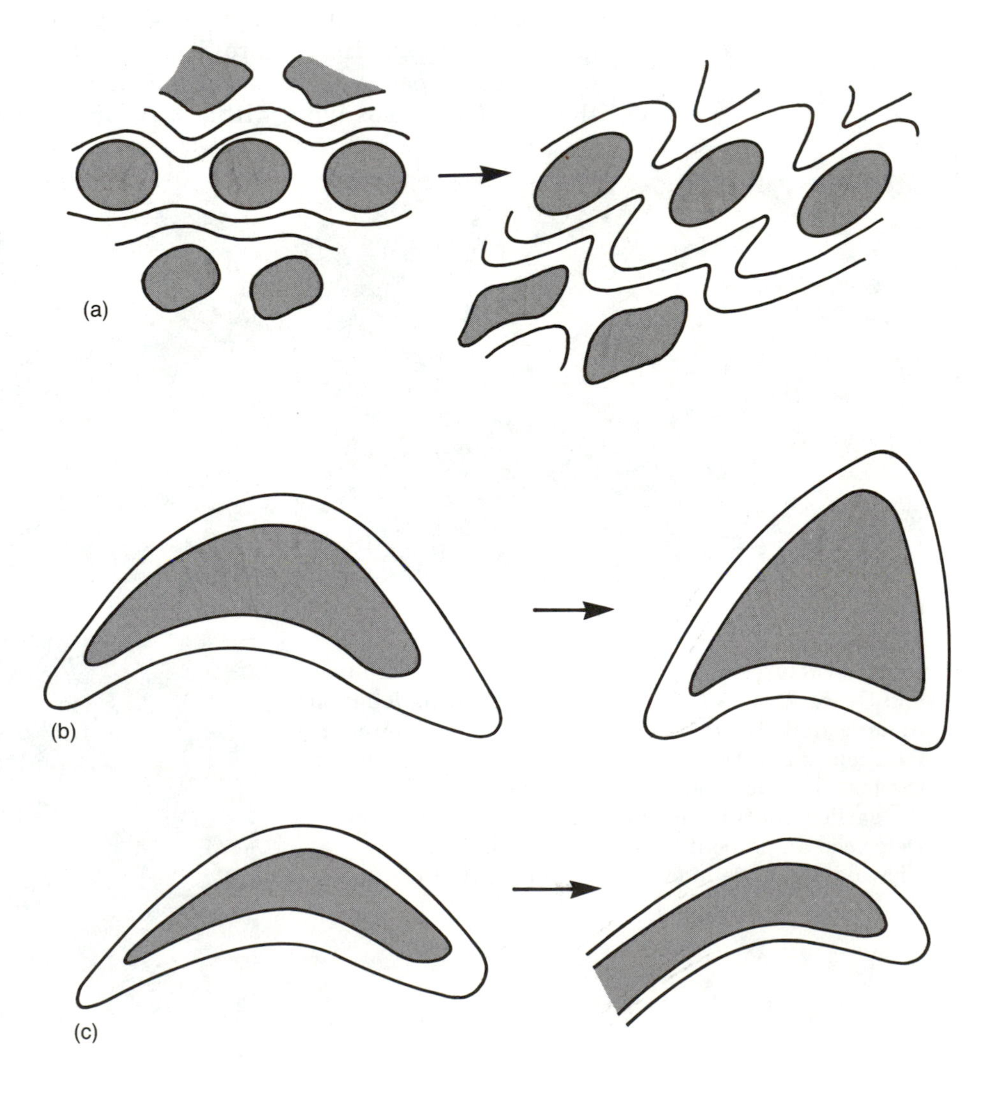

FIGURE 16–10
Changing outcrop pattern with modification of Type 1 domes to elongate domes (a), ideal Type 2 boomerang folds to tongue-shaped outcrop patterns (b), and opening up of one end of a Type 2 fold to become a Type 3 (c).

FIGURE 16–11
Sheath folds in Archean gneiss near Cape Jermain, Foxe fold belt, Melville Peninsula, Canada. (Reprinted with permission from *Journal of Structural Geology,* v. 3, J. R. Henderson, Structural analyses of sheath folds with horizontal X–axes, Northeast Canada, copyright 1981, Pergamon Press PLC.)

patterns and strongly asymmetric shears resembling folds also appear in clouds, particularly with the passage of strong frontal systems. Their appearance in ductilely deformed rocks further indicates the similarity of flow processes in all media.

Fold-interference patterns can form by flexural slip or by ductile flow. The standard Type 1, 2, and 3 interference patterns are usually associated with a single rheological state, either brittle or ductile, but the same states may also combine and produce interference patterns. Time is required to cool originally ductile rock to a brittle condition. Early buckle folds formed by flexural flow may be superposed by later buckle folds that were formed by flexural slip. Other combinations of passive flow, flexural flow, and flexural slip with buckling are also possible. Deformation mechanisms within individual layers again control fold geometry, as with single folds.

If folds are nucleated by symmetrical buckling (homogeneous strain) and later are deformed by heterogeneous strain, they are transformed to noncylindrical folds and then to sheath folds. The process may occur in either brittle or ductile rocks. Ductile shear zones are probably the best places for formation of noncylindrical and sheath folds, but they may form in a purely ductile mass of rocks being deformed by heterogeneous strain, which produces a different rate of ductile flow.

Here we end our discussion of folds and folding, but will use this knowledge as we now turn to cleavage and foliations, lineations, structural analysis, and geophysics.

Questions

1. How do Type 2 interference folds form?
2. How would you distinguish Type 1 interference patterns from sheath folds?
3. Why do sheath folds commonly occur in ductile shear zones and fault zones?
4. Why is it difficult to say that the early and late folds making up Type 3 interference folds formed at different times?
5. The most common interference fold types are Type 3; next are Type 2, and Type 1 folds are least common. Why?
6. How could you distinguish between complex tectonic folds and those formed by soft-sediment deformation?
7. Describe conditions in which late folds in a Type 3 pattern could be separated from earlier folds.

8. Here is a geologic map of a boomerang (Type 2) interference fold. Photocopy the fold and sketch in the traces of early and late fold axes with arrows indicating direction of plunge, and show which fold axial trace is older or younger. The cross-hatch pattern is Precambrian gneiss; the stipple pattern, Ordovician quartzite; outside the patterned area is Silurian marble.

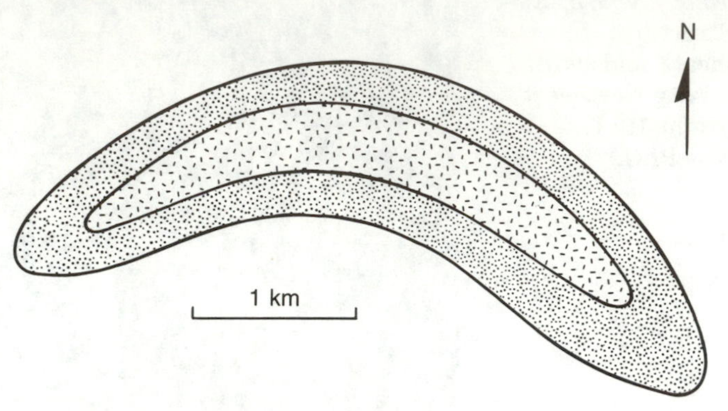

Further Reading

Cobbold, P. R., and Quinquis, H., 1980, Development of sheath folds in shear regimes: Journal of Structural Geology, v. 2, p. 119–126.
Discusses the mechanics of formation of sheath folds and establishes that they form in zones of inhomogeneous simple shear.

Ramsay, J. G., 1962, Interference patterns produced by the superposition of folds of "similar type": Journal of Geology, v. 60, p. 466–481.

Ramsay, J. G., 1967, Folding and fracturing of rocks: New York, McGraw-Hill, 568 p.
The 1962 paper and Chapter 10 of the 1967 book first systematized interference fold types.

Thiessen, R. L., and Means, W. D., 1980, Classification of fold interference patterns: A reexamination: Journal of Structural Geology, v. 2, p. 311–316.
Redefines Ramsay's interference fold types, using the angular relationships between the axes of the first folds and the pole to the axial planes of the second folds. Thiessen and Means realized that a greater variety of interference pattern was possible within Ramsay's basic types.

PART FIVE

Fabrics, Structural Analysis, and Geophysics

OUTLINE

17 Cleavage and Foliations
18 Linear Structures
19 Structural Analysis
20 Geophysical Techniques

THIS SECTION CONSISTS OF CHAPTERS DEALING WITH DEFORMAtion fabrics in rocks—cleavage and foliations and linear structures—and a chapter on geophysical techniques commonly used to solve structural problems. These chapters discuss two additional groups of individual structures and the methods for deciphering the structure of simple or complex areas. The amount of high-quality crustal seismic reflection, refraction, gravity, magnetic, and other geophysical data available to the structural geologist increases every year. These data provide information on the generally inaccessible third dimension of the crust and help with the deciphering of large-scale geologic structures. Knowledge of the techniques outlined in Chapter 20 is very useful in working out the structure of areas of poor exposure and areas of complex structure. Chapters 19 and 20 are intended to provide a springboard from the context of thinking about individual structures on the microscopic to map scales to structures on the map to continent scales—the subject of tectonics.

17

Cleavage and Foliations

Rocks affected by slaty cleavage have suffered a compression of their mass in a direction everywhere perpendicular to the plane of cleavage and an expansion of their mass in the direction of cleavage dip.

DANIEL SHARPE, Quarterly Journal of the Geological Society of London

A TENDENCY TO SPLIT ALONG PLANES OTHER THAN BEDDING was recognized more than a century ago as a fundamental property of many deformed rocks. The significance was probably not known at first. Early geologists did observe bedding cut at a high angle by a prominent planar structure and soon realized that it was a major structure—cleavage—related to both deformation and metamorphism. In this chapter, we will trace the early ideas on the origin of slaty cleavage and foliation and again see the fundamental relationship between small and large structures first noted by Raphael Pumpelly in the early twentieth century.

We have several reasons to understand cleavage in rock. Cleavage is directly linked to other deformation processes—especially folding—and to metamorphism. It can help determine the fold geometry in an area, and, if we know the cleavage-forming mechanism in a rock mass, it will lead to better understanding of the deformation processes and also help reconstruct the physical conditions during deformation.

DEFINITIONS

The term ***fabric*** is used in describing the relationships of planar and linear structures—including bedding, cleavage, and the orientation of minerals—to texture in rocks. Fabric is an attribute of both fine- and coarse-grained rocks.

Structures that pervade the rock mass at the scale of observation (Figures 17–1, 17–2) are ***penetrative.*** Ideally, penetrative means present at all scales of observation, but some structures, such as joints, are not present at all scales. We can cut thin sections or even break large specimens from a jointed mass and not intersect a joint; but on the map scale, in the same body of rock, joints may be so closely spaced that they must be considered penetrative. *Slaty cleavage* is generally regarded as penetrative; it consists of parallel grains of thin-layer silicates (clay minerals or micas) or thin anastomosing subparallel zones of insoluble residues. Slaty cleavage is observable on the scale of a hand specimen and in thin sections; it also may be represented on maps and so is a good example of penetrative structure.

On the other hand, ***nonpenetrative*** structures (Figure 17–2) include unique nonrecurring structures like fault planes. An isolated fault obviously cannot be penetrative. Likewise, an isolated fold hinge or axial surface is nonpenetrative, but folds visible on all scales are penetrative structures.

In an early work on this subject, Bruno Sander (1930), an Austrian geologist, suggested that planar structures (including curved planar structures) be called ***S-surfaces*** in rocks generally having a tec-

FIGURE 17–1
Folded slate in the Ocoee Series (Upper Proterozoic), near Walland, Tennessee, showing bedding and penetrative axial-plane slaty cleavage. Note relationships of dip of cleavage and bedding on the upright and overturned limbs of the fold—cleavage dips steeper than bedding on the upright limb, less steeply than bedding on the overturned limb. (Arthur Keith, U.S. Geological Survey.)

tonic origin (Figure 17–3). S-surfaces include all cleavage and foliation surfaces commonly thought of as penetrative structures. They also include one nontectonic structure, bedding. To aid description of S-surfaces, a shorthand system has been devised. In areas of multiple S-surfaces, a series of subscripts is assigned: bedding, being oldest, is designated S_0; S_1 is the first cleavage (or foliation), and any later structures are given numerically higher subscripts in chronologic order (Figure 17–3). Corresponding multiple-fold sets are designated F_1, F_2, and so on; other linear structures, L_1, L_2, and so on. Multiple deformations may be designated D_1, D_2, and so on.

Cleavage (or foliation) in fine-grained rocks may be ***continuous*** or ***spaced*** (Figure 17–4). Continuous cleavage pervades the rock mass on all scales, but spaced cleavage can be resolved into domains of uncleaved rock separated by cleavage planes with a spacing ranging from less than a millimeter to several centimeters (Figure 17–5). Uncleaved zones between cleavage surfaces are ***microlithons*** (Figure 17–4).

Most cleavage in rocks, according to C. McA. Powell (1979), is *domainal,* or spaced, if the rock is examined closely enough, but continuous cleavage may be more widespread than Powell realized. He used these characteristics to distinguish types of cleavage (Figure 17–6): spacing of cleavage domains, shape of cleavage domains, microlithon fabric, and proportion of the rock occupied by cleavage domains. Thus he created two basic subdivisions, spaced and continuous cleavage, and divided spaced cleavage into ***disjunctive*** (cross-cutting, and not related to original layering) and ***crenulation cleavage*** (which crenulates preexisting layers). Crenulation cleavage may be further divided into *discrete* and *zonal* types; disjunctive cleavage may be divided into *stylolitic, anastomosing, rough,* and

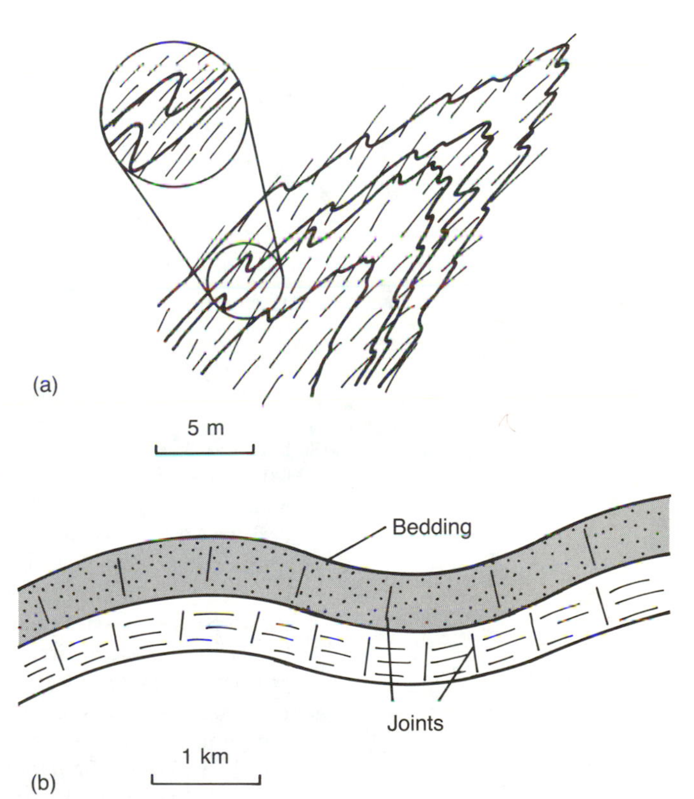

FIGURE 17–2
Penetrative and nonpenetrative structures. (a) Mesoscopic fold with small parasitic folds and an axial-plane foliation. Small folds and foliation are penetrative structures, for they occur throughout the larger fold on all scales. (b) Unique fold hinges and widely spaced joints that occur on only one scale and are not repeated on other smaller or larger scales are nonpenetrative.

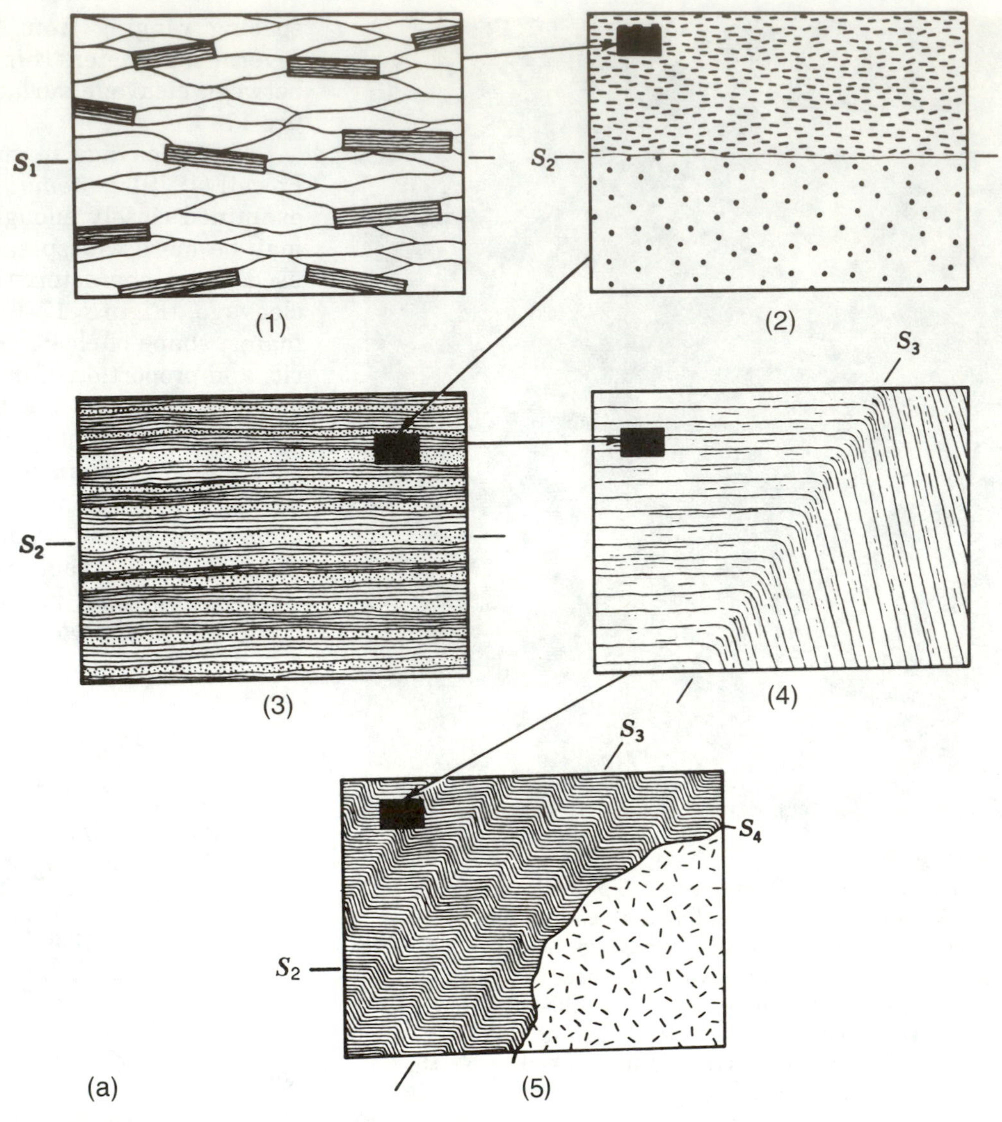

FIGURE 17–3
(a) Planar discontinuities in the same body of rocks at several different scales. Microscopic scale (1) showing preferred orientation of grain boundaries and minerals defining a weakly penetrative planar structure S_1. Grains (2) define a penetrative planar structure S_1 in the upper layer on the mesoscopic scale. The compositional boundary S_2 is nonpenetrative on this scale. Alternating layers (3) parallel to S_2 make S_2 a penetrative structure at this scale (larger mesoscopic). Map (macroscopic)-scale structures (4) in which a kink S_3 divides the body into two sectors or domains having different strikes. Larger map scale (5) in which S_3 becomes a series of closely spaced kink surfaces and is penetrative. Another nonpenetrative compositional boundary (S_4) appears on this scale. (From F. J. Turner and L. E. Weiss, *Structural Analysis of Metamorphic Tectonites,* © 1963 McGraw-Hill Book Company. Reproduced with permission.) (b) (page 369) Metasandstone-schist at Sill Vinson's Rock near Otto, North Carolina. Original bedding (now transposed) comprises the earliest foliation. A later foliation dips toward the left (west) parallel to the axial surfaces of the folds. A later foliation—a crenulation cleavage—dips steeply to the right (east), where it is visible above the small ledge and parallels the axial surfaces of the folds at the bottom of the photo. Crenulations occur here only in the schist because the sandstone layers are too quartz-rich. White layers are quartz-feldspar veins. Note that thin quartz-feldspar veins in schist form ptygmatic folds. Sandstone layers are light gray; schist layers are dark gray. (RDH photo.)

(b)

smooth, depending on the shape of the cleavage domains. Spacing of true slaty cleavage may range from less than 0.01 mm (continuous cleavage) to less than 0.1 mm, that of crenulation cleavage from 0.1 mm to 3 cm, and that of stylolitic and anastomosing cleavage from a few millimeters to several centimeters. Different rock types in the same body, such as interlayered sandstone (or pure limestone) and shale, may display cleavage in contrasting ways. Shale may display closely spaced slaty cleavage produced by pressure solution (or recrystallization); the sandstone (or limestone) may contain more widely spaced pressure-solution cleavage (Figure 17–7).

Pressure solution produces spaced cleavage by dissolving the most soluble parts of a rock mass, leaving behind discrete residues in irregular planar zones that define the cleavage (Figure 17–8). Here the terminology is a bit confusing because (as we saw in Chapter 6) pressure solution is a deformation mechanism, but pressure-solution cleavage has become a common term for a cleavage-forming process as well. Spacing of pressure-solution surfaces may range from less than a millimeter to more than a centimeter. They may be somewhat irregular (stylolitic to anastomosing to rough) where poorly developed, to smooth, where the rock mass is more severely deformed.

Slaty cleavage is a planar tectonic structure resulting from parallel orientation of clays or other layer silicates (such as chlorite and muscovite), or seams of insoluble residues (or both), in a fine-grained rock. It is penetrative and thus visible at all scales in the rock mass (Figure 17–1). At microscopic to submicroscopic scale (Figure 17–9), slaty cleavage may be defined by layer-silicate minerals that are aligned and separated by *microlithons* containing layer silicates that are not aligned. Spacing of slaty-cleavage planes may range from a fraction of a millimeter up to one or two millimeters. Rocks

(a)

(b)

FIGURE 17–4
(a) Continuous cleavage in slate in the Upper Proterozoic Wilhite Formation on U.S. 129 along Chilhowee Lake, southeastern Tennessee. (b) Spaced cleavage in Ordovician Cumberland Head Limestone, Lessor's Quarry, South Hero Island, Vermont. (c) Microlithons of earlier foliation between cleavage planes in phyllite near Enosburg Falls, Vermont. (RDH photos.)

that commonly develop slaty cleavage are fine-grained sedimentary and volcanic rocks such as shales, mudstones, siltstones, and tuffs and their equivalents at low metamorphic grade (anchizone—to subchlorite—to chlorite zone).

Cleavage marked by small-scale crinkling or crenulations is called ***crenulation cleavage*** (Figure 17–10). Most crinkles are spaced and asymmetric, and the short limb is usually a plane of breakage. It usually forms by deformation of an earlier cleavage or (much less frequently) bedding. The axial zones or limbs of the crenulations may be aligned with or transected by cleavage planes, but the limbs of the crenulations generally remain in-

FIGURE 17–4
(*continued*)

(c)

tact. Microlithons between the crenulation cleavage planes preserve the earlier crenulated cleavage or foliation (Figure 17–11).

The various cleavage types, including slaty and crenulation cleavage (some geologists add bedding), are collectively termed ***foliations.*** The term is also used for the planar structure occurring in coarser-grained metamorphic rocks, such as phyllite, schist, and gneiss, where the parallel orientation of at least one mineral dominates the overall fabric (Figure 17–12). Foliation in coarser-grained rocks (such as schist and gneiss) may be produced by parallel orientation of micas (mostly biotite and muscovite), amphiboles, and flattened quartz grains and even by

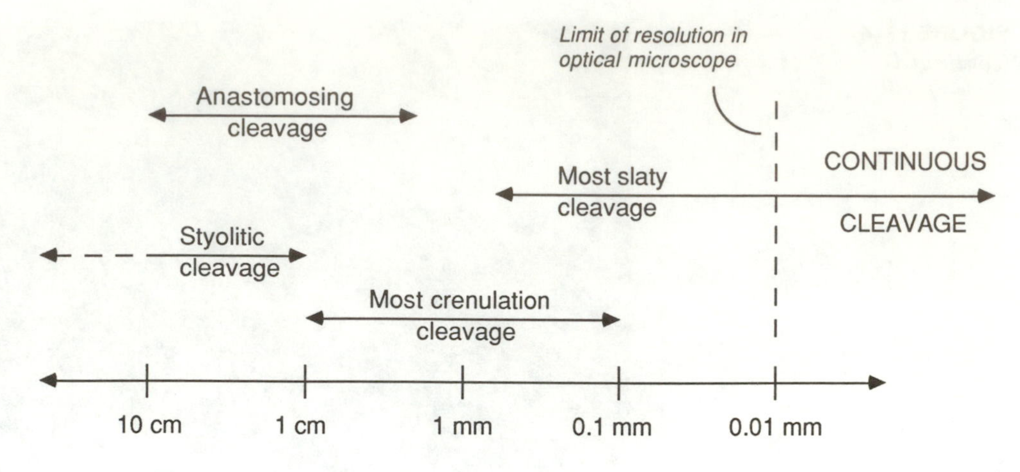

FIGURE 17–5
Type and spacing of cleavages. (From C. McA. Powell, 1979, *Tectonophysics,* v. 58, p. 21–34, Elsevier Science Publishers.)

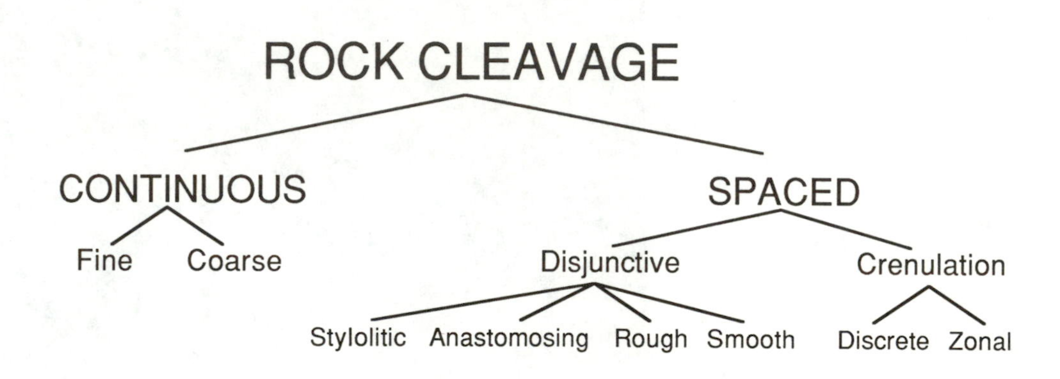

FIGURE 17–6
Powell's classification of cleavages based upon morphology. (From C. McA. Powell, 1979, *Tectonophysics,* v. 58, p. 21–34, Elsevier Science Publishers.)

FIGURE 17–7
Contact between Middle Ordovician slate and sandstone near Smiths Basin, south of Whitehall, New York, showing closely spaced cleavage in the slate and widely spaced pressure-solution cleavage in the sandstone. (RDH photo.)

the interlayering of quartz, feldspar, and mica layers. ***Schistosity*** refers to foliation in schistose and—occasionally—gneissose rocks (Figure 17–12b). Banded foliation in gneissic rocks results partly from original differences in composition and partly from the orientation of platy or elongate minerals. Secondarily developed layering is called ***gneissic banding*** (Figure 17–12c). This foliation is easily recognized and may be found where layers dominated by quartz and feldspar alternate with layers dominated by micas or amphiboles.

Metamorphic differentiation involves formation of new layering by pressure solution or recrystallization. It reorganizes the chemical components of a rock and produces new minerals with new orientation. Foliation produced during progressive deformation and metamorphism, largely through recrystallization, is termed ***differentiated layering*** (Figure 17–13). Layering in the original rock mass may be enhanced by this process at high temperature and pressure and produce gneissic banding; formation of spaced slaty and crenulation cleavages may produce differentiated layering at low temperature and pressure.

Fracture cleavage consists of parallel to subparallel fractures, generally spaced 1 to 3 cm apart (Figure 17–14). As the name implies, this structure is formed by a brittle mechanism, but rocks may also fracture along pressure-solution seams. The latter was originally called fracture cleavage because the rock fractures along planes of weakness that appear to not be of the same kind as in the slaty cleavage just described. Fracture cleavage (as used here) tends to develop in slightly coarser-grained rocks than rocks with slaty cleavage, but it has also been observed in fine-grained rocks. If its origin is traceable to fracturing, many structural geologists call it jointing. But it cannot be true cleavage because most joints are extensional, and most cleavage results from compression and pure shear. In any case, the term fracture cleavage is entrenched in the literature of structural geology. Probably it should be used only as a last resort to describe cleavage that appears to have formed by "fracturing."

CLEAVAGE-BEDDING RELATIONSHIPS

Geologists have long known that folding is related to slaty cleavage within folds and also that these structures are formed synchronously. The angular relationship between cleavage and bedding can be used to determine whether one is observing an upright or an overturned limb of a fold that is incompletely exposed: *If bedding dips less steeply (at a lower angle) than cleavage in rocks dipping in the same direction, the rocks are on upright limbs of folds. Conversely, if bedding dips more steeply than the cleavage (that is to say, the cleavage dips less than bedding) the rocks are on the overturned limb of a fold* (see Figure 17–1). The rule may be used in areas where the geologist cannot observe complete folds and must project outcrop data to map scale. Conversely, cleavage-bedding relationships can be used to locate the observer's place on a known structure.

Care must be taken to determine whether the cleavage being used to determine position within a structure is the only cleavage present and whether the cleavage formed at the same time as the structure of interest. If more than one cleavage is found, the cleavage-bedding relationship in each generation of cleavage and folds must be separated before drawing any conclusion about position in the structure (Figure 17–15). In complexly deformed areas with several generations of cleavages, independent information—such as the stratigraphic order based on primary sedimentary structures—should be obtained routinely as a check.

CLEAVAGE REFRACTION

A phenomenon frequently observed in cleaved rocks is refraction of cleavage from layer to layer and occasionally (as in graded beds) within layers. ***Cleavage refraction*** occurs when strain varies from layer to layer in rocks of different ductility or competence. The angle between cleavage and bedding changes, or refracts, as the cleavage passes from one layer to another (Figure 17–16). Apparent refraction occasionally results from flexural slip on bedding, producing drag of cleavage surfaces.

Cleavage refraction produces contrasting effects during folding. Most slaty cleavage forms parallel to axial surfaces in folds, but may be displaced or fanned with respect to the hinge as folding proceeds; thus it is no longer parallel to the axial surface (Figure 17–17a). In a different approach, Susan Treagus (1983) considered the cleavage-refraction phenomenon as a product of inhomogeneous strain between layers, rotating cleavage—after it had formed by pure shear—by a component of differential simple shear parallel to bedding but by an

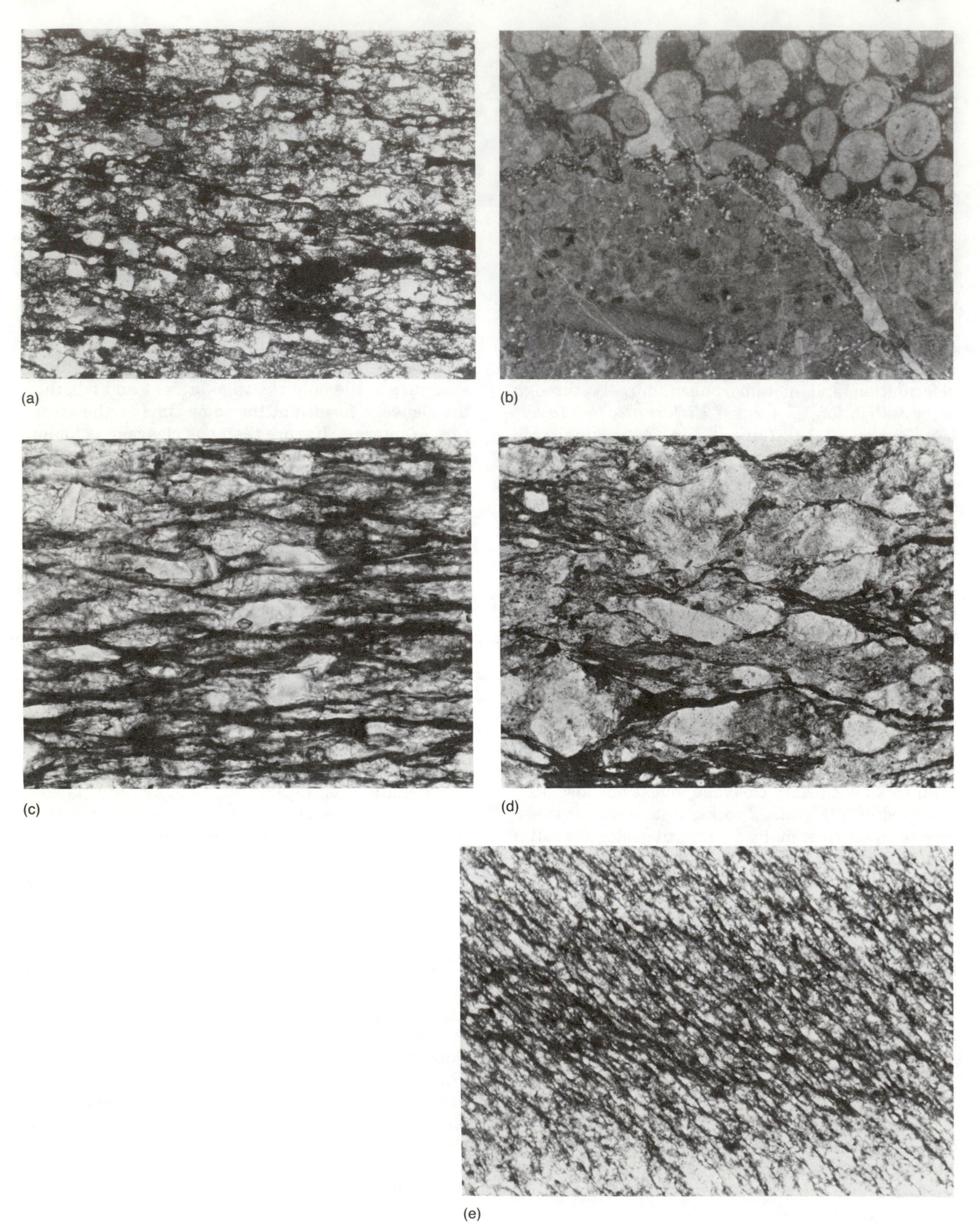
(a)
(b)
(c)
(d)
(e)

amount that varied from bed to bed. Hence the phenomenon of apparent refraction (Figure 17–18).

Refracting cleavage has been confused with cross bedding where the cleavage is traced as a continuous curve from coarser- to finer-grained parts of the same graded bed: the steeper-dipping cleavage (relative to dip of bedding) occurs in the coarser (basal) part of the bed; the gentler-dipping cleavage is in the finer-grained part at the top of the bed (Ramsay, 1967). This indicates a facing direction opposite the direction indicated by graded bedding. In such cases, careful observation of relationships between primary sedimentary and tectonic structures should readily solve the problem.

EARLY IDEAS OF THE ORIGIN OF SLATY CLEAVAGE

Now we turn to the history of ideas and arguments related to the origin of slaty cleavage. Development of the concepts is presented in greater detail by Dennis Wood (1974) and also by Andrew Siddans (1972). The historical perspective bears on present-day concepts of foliation in rocks as well as on related structures.

FIGURE 17–8
Pressure-solution cleavage and residues. (a) Residues of insoluble minerals truncating quartz grains along cleavage surfaces in Upper Ordovician Martinsburg Slate, Delaware Water Gap, New Jersey. Plane light. Field is approximately 2 mm wide. (Thin section courtesy of Timothy L. Davis, University of Tennessee.) (b) Stylolitic seam truncating oöids, producing apparent displacement of an earlier calcite vein in limestone in Upper Cambrian Nolichucky Shale, Oak Ridge, Tennessee. Plane light. Field is approximately 7 mm wide. (Thin section courtesy of Peter. J. Lemiszki, University of Tennessee.) (c) Anastomosing slaty cleavage in Devonian Marcellus Shale near Cumberland, Maryland. Plane light. Width of field is approximately 2 mm. (Charles M. Onasch, Bowling Green State University.) (d) Rough cleavage in graywacke in Upper Ordovician Martinsburg Slate near Front Royal, Virginia. Pressure-solution seams have a very irregular shape because of the different amounts of insoluble and soluble constituents present. Plane light. Width of field is approximately 2 mm. (Charles M. Onasch, Bowling Green State University). (e) Penetrative pressure-solution "smooth" cleavage in Upper Ordovician Martinsburg Slate, Delaware Water Gap, New Jersey. Plane light. Width of field is approximately 7 mm. (Thin section courtesy of Timothy L. Davis, University of Tennessee.)

The origin of slaty cleavage has been debated for a century and a half. Some concepts we now consider fundamental evolved very early but by the turn of the century had been almost forgotten—only to be revived in the 1970s.

Among the first and fundamental deductions about the nature of slaty cleavage was that cleavage is related to folding and that cleavage often parallels the axial planes of folds. This was recognized independently in the mid-nineteenth century by Adam Sedgwick (1835), Charles Darwin (1846), and Henry D. Rogers (1856).

In the 1830s and '40s, Sedgwick (1835), J. Phillips (1844), and D. Sharpe (1849) thought slaty cleavage developed after folding. Sedgwick understood the axial-plane relationship of cleavage and folding, but not that they must be contemporaneous. Apparently J. Tyndall (1856) was the first to see that slaty cleavage and folding develop at the same time.

Several early geologists recognized the relationship between formation of slaty cleavage and the process of metamorphism. Darwin (1846) considered slaty cleavage the end point of the metamorphism, again probably including the idea that slaty cleavage was related to a late stage of folding.

Another fundamental observation was that strain markers, such as fossils, are distorted when cleavage forms. Phillips (1844) saw that fossils in Welsh slates might be distorted parallel to cleavage planes. Sharpe (1849) noted that shells are most distorted in layers that are most slaty and concluded that the maximum principal shortening direction is normal to cleavage planes.

Among the most important contributions made during the mid-nineteenth century was that by a British geologist, Henry C. Sorby. He used reduction spots in Cambrian slates of Wales to show that the XY plane of the strain ellipsoid parallels cleavage planes and estimated that there was as much as 75 percent shortening normal to cleavage planes. (Sorby pioneered the use of petrographic techniques: he is credited with making the first thin section, and he advanced the idea that rotation of inequant grains tends to align their long axes in the plane of cleavage.) Sorby also observed that preferred orientation could be enhanced by recrystallization in the plane of cleavage.

Sorby was probably the first to recognize pressure solution (1863; Figure 17–7) and suggest it as an important mechanism of material transfer in rocks. But he felt that all other processes in formation of slaty cleavage were secondary to rotation produced by compression.

FIGURE 17–9
Scanning electron micrograph of domainal slaty cleavage from Upper Ordovician Martinsburg Slate near the Delaware River, Pennsylvania. Spacing between cleavage planes (strongly oriented narrow zones) is about 20 μm. (From B. G. Woodland, 1982, *Tectonophysics,* v. 82, p. 89–124, Elsevier Science Publishers.)

FIGURE 17–10
Crenulation cleavage formed by crenulating an earlier slaty cleavage near Wells, Rutland County, Vermont. (Charles D. Walcott, U.S. Geological Survey.)

(a)

(b)

FIGURE 17–11
Microlithons of earlier deformed material between crenulations at both meso- (a) and micro- (b) scales: (a) is crenulated siltstone in the Upper Proterozoic Hamill Group near Golden, southern British Columbia. (RDH photo.) (b) is chlorite schist from the Wissahickon Group near Westminster, Maryland. Plane light. Width of field is approximately 7 mm. (Charles M. Onasch, Bowling Green State University.)

Around the end of the nineteenth century Charles R. Van Hise suggested—like Sorby—that slaty cleavage developed normal to direction of principal shortening (the minimum principal elongation). Unlike Sorby, Van Hise believed that new mineral growth in the plane of flattening was the principal factor.

So, by the beginning of the twentieth century, all the alternative hypotheses about the origin of slaty cleavage that we recognize today had been outlined. The dispute about rotation versus recrystallization has continued throughout much of the twentieth century. Oddly, the idea that pressure solution is important in the formation of slaty cleavage was almost forgotten until recently and was not mentioned as a viable cleavage-forming mechanism in elementary textbooks of structural geology until the 1980s. As a potential solution to the problem of rotation versus recrystallization, Ramsay (1967) had suggested that rotation dominates at low temperature, but, as temperature increases, recrystallization becomes dominant. His notion is no doubt partly correct, but the important role of pressure solution remained in the shadows until recently.

Another controversy focused on whether slaty cleavage formed by simple shear or pure shear. The question arose in the nineteenth century, being argued by Phillips (1844), A. Laugel (1855), O. Fisher (1884), and G. F. Becker (1896, 1904). Becker advocated development by simple shear parallel to one of the shear planes of the strain ellipsoid. In this century, several geologists, including Bruno Sander (1930), F. J. Turner (1948), R. Hoeppener (1956), and G. Voll (1960), also thought this was the case. Early advocates of the pure shear or flattening mechanism include Sharpe (1847, 1849), Sorby (1853, 1856), Tyndall (1856), S. Haughton (1856), Alfred Harker (1885, 1886), and Van Hise (1896). Still later, structural geologists such as Ramsay (1967), James Dieterich (1969), Andrew Siddans (1972), and Dennis Wood (1973) concluded that pure shear is more likely the dominant cleavage-forming mechanism.

Pure shear as the mechanism is supported by studies of relationships between finite-strain markers and the orientation of cleavage planes and the relationships to fold axes or axial surfaces. Even so, it has been shown that in some places slaty cleavage started forming at a high angle to bedding as the rocks were folded and was later sheared to lower angles by flexural slip along bedding. This may imply at first glance that cleavage did not form parallel to the *XY* plane of the strain ellipsoid, but in fact the changing orientation is related to more than one fold mechanism operating at the same time, rather than to the cleavage not being parallel to the *XY* plane.

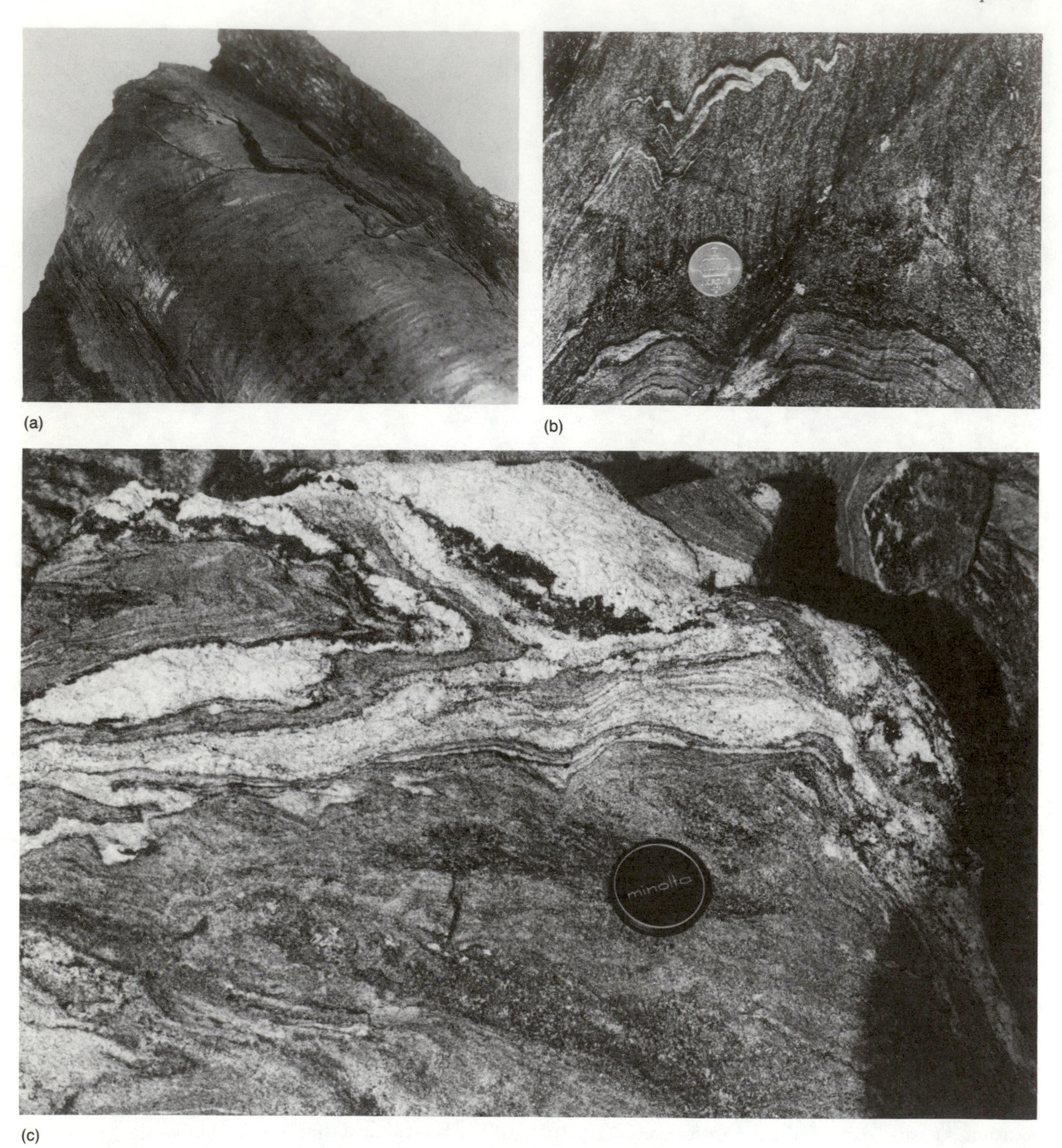

FIGURE 17–12
Foliations in three rocks: (a) Phyllite from the Ordovician Walloomsac phyllite near Hoosick Falls, New York. Note the fine grain size and lustrous foliation. (b) Schistosity in crenulated amphibole-biotite schist containing several quartz-feldspar layers on Blåhø, Trollheimen region, south-central Norway. (c) Gneiss containing gneissic banding composed of quartz-feldspar–rich layers alternating with biotite-rich layers, Thor-Odin dome, Shuswap metamorphic complex, south-central British Columbia. (RDH photos.)

FIGURE 17–13
Differentiated layering in gneiss (a), Toxaway Gneiss near Whitewater Falls on the North Carolina–South Carolina line. (RDH photo.) (b) Pressure-solution deformed fine-grained sandstone of the Lower Ordovician Halifax Formation, Little Harbor, Tor Bay, Nova Scotia. Earlier cleavage is steeply dipping, but the later cleavage is anastomosing and nearly horizontal. (J. Duncan Keppie, Nova Scotia Department of Mines.)

(a)

(b)

MECHANICS OF SLATY-CLEAVAGE FORMATION

Thus the origin of all types of slaty cleavage has been controversial. The concepts of rotation versus recrystallization, pure shear versus simple shear, and, more recently, pressure-solution mechanisms versus those without pressure solution have been debated throughout much of two centuries. Today most structural geologists accept Daniel Sharpe's conclusion (1847) that slaty cleavage forms by compression normal to the *XY* plane, but debate persists about whether cleavage refraction actually occurs or

FIGURE 17–14
Weak fracture cleavage (most visible lower left) that fans around folds in Cambrian limestone and dolomite near Danby, Vermont. Is this a true cleavage? (Arthur Keith, U.S. Geological Survey.)

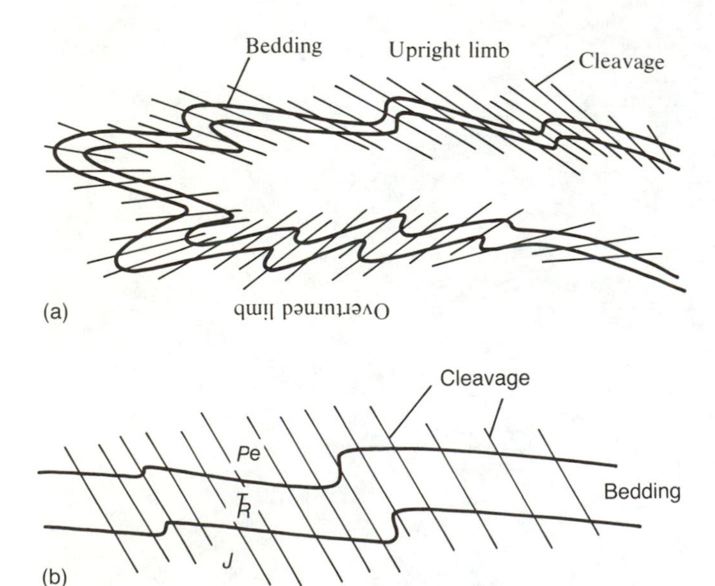

FIGURE 17–15
Cleavage-bedding relationship and its use in determining relative position in a fold. (a) Relationships between orientation of cleavage surfaces and the position on an upright or overturned limb. (b) Cleavage-bedding relationships indicating that the rocks are upright in a sequence that is really overturned, thus showing the cleavage was emplaced after overturning. J—Jurassic; Ћ—Triassic; Pe—Permian.

is a product of later inhomogeneous simple shear between layers (Figure 17–18; Williams, 1976; Henderson and others, 1986; Treagus, 1983, 1988). Both appear possible in light of available data. Paul Williams (1976) considered the questions of whether cleavage develops parallel to principal planes and remains so throughout the history of deformation and whether cleavage develops in some other (shear) orientation and rotates toward parallelism during progressive deformation to be more fundamental. He concluded that, if the strain is coaxial (parallel to principal planes), the cleavage will develop parallel to the XY plane and remain so. If the strain is noncoaxial (simple shear strain), cleavage will not develop parallel to the XY plane. Formation of cleavage during folding generally involves noncoaxial strain.

An important piece of evidence relates to the role of pressure solution in formation of slaty cleavage. In calcareous rocks, the relative solubility of carbonate and noncarbonate minerals favors carbonate pressure dissolution in the presence of acidic fluid; this has long been known, but the concept of a large volume of material being removed from fine-grained clay or silty rocks did not gain acceptance until recently. (One reason is that the solubility of silicate minerals under alkaline conditions was not

FIGURE 17–16
Refracting cleavage in layers of slate (dark), fine sandstone, and coarse sandstone in the Upper Proterozoic Walden Creek Group, Ocoee Gorge, southeastern Tennessee. The closely spaced, gently dipping penetrative cleavage in the slate at the bottom of the photo refracted into the fine sandstone at a steeper dip and coarser spacing, then into the thickly bedded sandstone as a spaced cleavage. (RDH photo.)

fully appreciated.) We also know today that during orogenesis a huge volume of water is fluxed out of the deforming rock mass (Figure 11–2; Oliver, 1986). Several factors—the density of shale and slate, the potential for rearranging minerals during compaction, and the loss of soluble minerals (like quartz in fine-grained rocks)—all suggest an upper limit for volume loss through rearrangement and dissolution. The limit is about 20 percent. But, as we will see, greater loss is possible. Studies by Ramsay and Wood (1973) revealed that a 10 percent loss in volume may occur without actual loss of material. Pressure solution along cleavage planes does remove much material and reduces volume with little change in density, but material dissolved during pressure solution need not leave the rock body. As chemical conditions change, the dissolved minerals may be precipitated again nearby.

Loss of 20 to 40 percent volume has been estimated by Ramsay and Wood (1973) for slates in Wales and by Richard A. Groshong (1975b) in Pennsylvania. A maximum of 60 percent loss during formation of pressure-solution cleavage has been estimated by Thomas Wright and Lucian Platt (1982) and Charles Onasch (1983, 1984) in interbedded graywacke and shale in the Martinsburg Formation of Pennsylvania and by Edward Beutner and Emmanuel Charles (1985) in the Hamburg sequence in Pennsylvania and New York. In the Borrowdale volcanic rocks of England's Lake District, Tim Bell (1985) could not account for all strain accompanying formation of slaty cleavage by invoking pressure solution alone. He traced strain paths by using strain markers and concluded that a combination of plane strain—such as initial layer-parallel shortening—and volume loss were probably responsible for formation of slaty cleavage.

Determining the actual mechanism that formed slaty cleavage at a particular locality may be difficult because the evidence is often consistent with more than one mechanism. (If an easy solution existed, structural geologists would not have debated the question for more than a century.) Slaty cleavage is often accompanied by some recrystallization and reorientation of clays in shales and siltstones. Studies of the structure of slates by X-ray and optical and scanning electron microscopes have shown that reorientation of grains does occur and that reorientation is usually accompanied by recrystallization (Figures 17–9 and 17–19). A paper by J. H. Lee and others (1986) concluded that layer-silicate grains oriented parallel to cleavage in the Martinsburg Slate in Pennsylvania and New Jersey are virtually free of strain. They said pressure solution also occurred, dissolving phyllosilicate minerals originally oriented parallel to bedding. Edward Beutner (1978) suggested that dissolution of

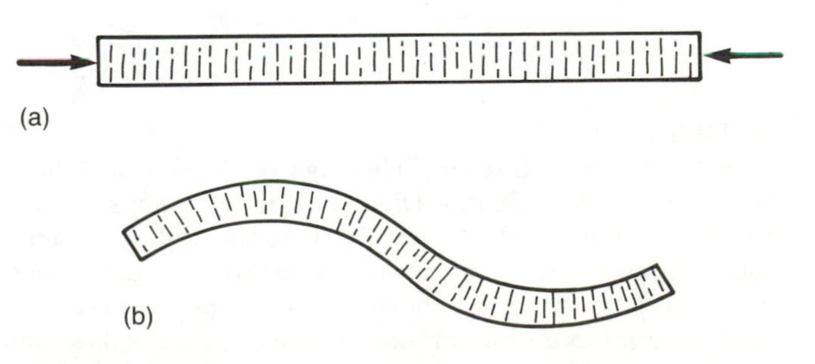

FIGURE 17–17
Formation of a fanned cleavage (b) by the cleavage forming by layer-parallel shortening before folding (a).

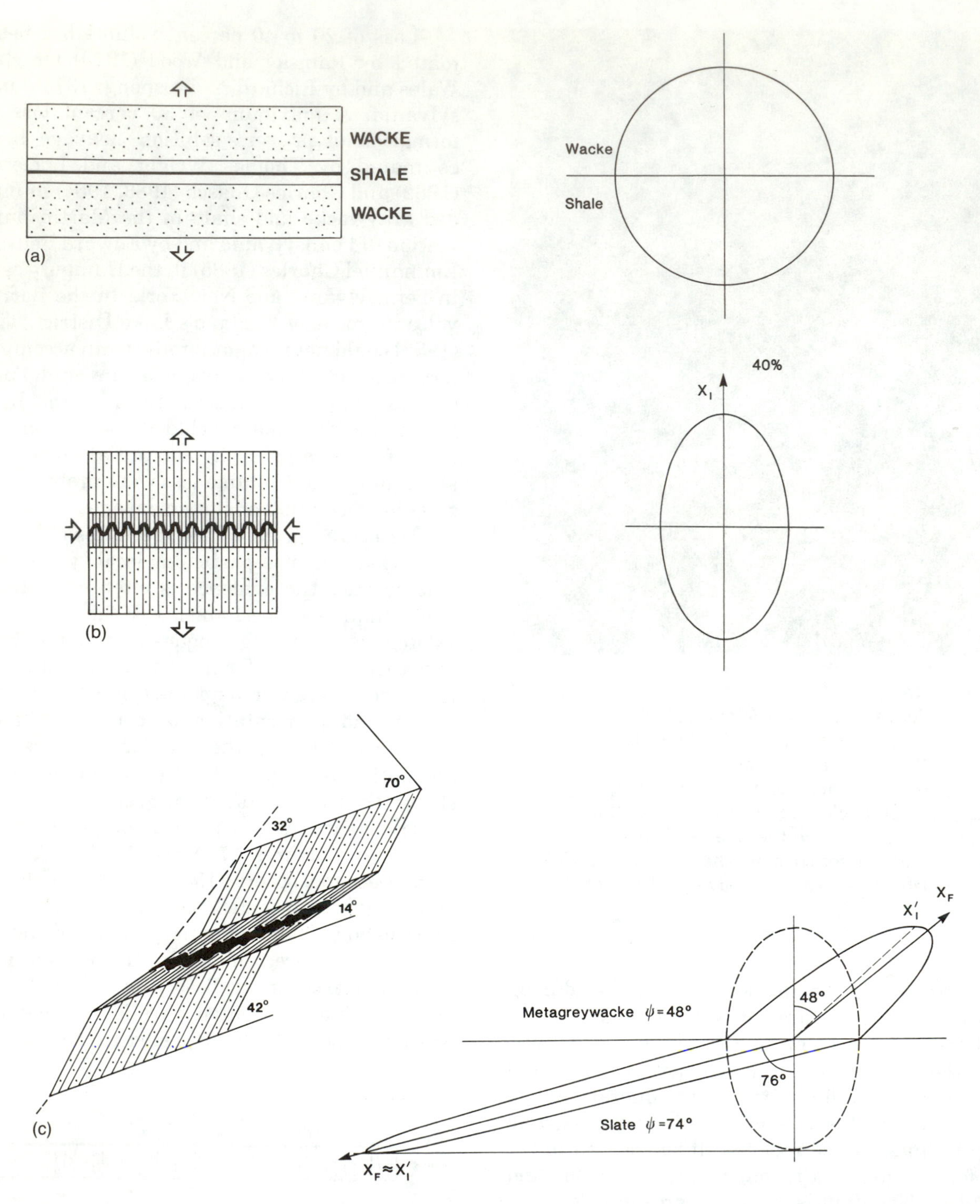

FIGURE 17–18
Formation of a "refracting" cleavage in Cambrian Goldenville Formation sandstone and shale in Nova Scotia. Diagrams on left represent the original rock mass and structures that developed within it; to the right is a strain ellipse for each stage of deformation. (a) Extension normal to layering, producing a vein (black layer) in the shale. (b) Layer-parallel shortening of 40 percent occurred, buckling the vein and producing an axial-planar slaty cleavage. (c) Displacement of cleavage by differential simple shear parallel to bedding. X_F—initial principal greatest-strain axis; X'_1—principal greatest-strain axis after simple shear; Ψ—simple shear strain. (Modified from J. R. Henderson, T. O. Wright, and M. N. Henderson, 1986, Geological Society of America *Bulletin,* v. 97, and S. H. Treagus, 1988, Geological Society of America *Bulletin,* v. 100.)

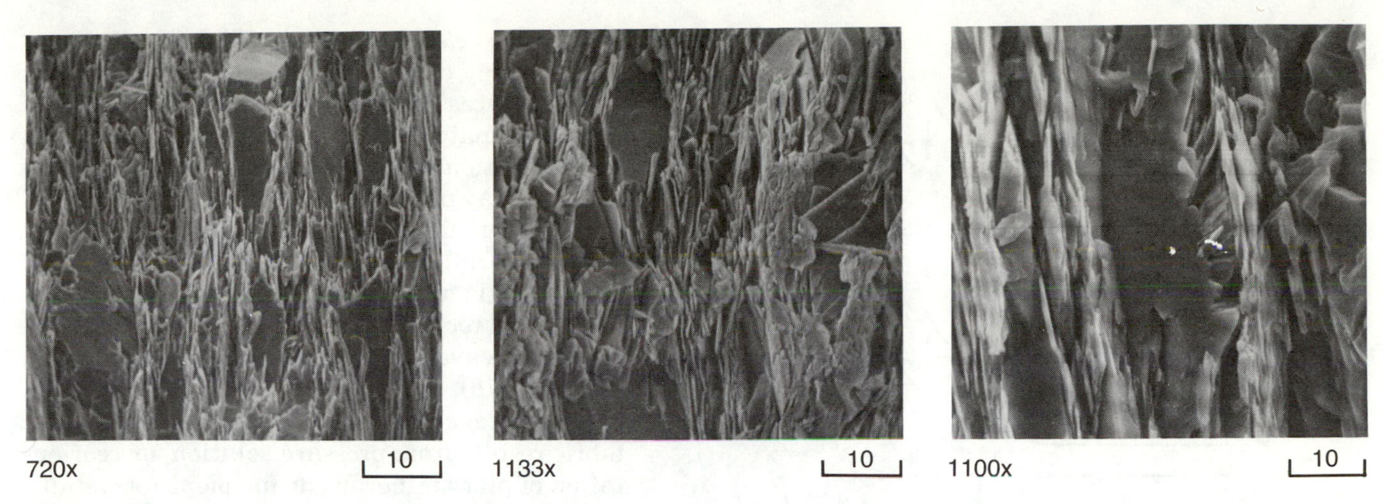

FIGURE 17–19
Scanning electron micrographs of reoriented and recrystallized micas in mica beards from Ordovician slate from the south shore of New World Island, Newfoundland. (From B. A. Van der Pluijm, 1984, *Geologische Rundschau,* v. 73.)

even the layer-silicate minerals occurs in proportion to the original orientation of grains relative to the orientation of cleavage surfaces (Figure 17–20). Therefore he concluded that little systematic reorientation of grains occurs during formation of slaty cleavage.

Microscopic study of thin sections cut perpendicular to cleavage reveals either that cleavage planes are spaced and contain residues of insoluble minerals (clays, micas, and iron oxides), and a relative lack of soluble minerals (quartz and calcite), or that they consist of a continuous oriented fabric of recrystallized grains (Figure 17–21). Relict partial grains of quartz or calcite truncated at cleavage planes are often found in pressure-dissolved rocks. Slaty cleavage formed by recrystallization (with reorientation or without) does not contain residues or partial grains along cleavage planes. All grains are thus oriented parallel to cleavage planes, and new unstrained minerals are evident. These mechanisms may compete and overlap (Lee and others, 1986). Pressure solution at low temperature gives way to recrystallization at higher temperature of the greenschist facies where chlorite, micas, and other minerals begin to form in fine-grained rocks.

Dissolution of calcite under stress in fine-grained carbonate rocks is directly related to the amount of impurities. Stephen Marshak and Terry Engelder (1985) determined that pressure-solution cleavage develops more readily in fine-grained carbonate rocks (such as marl) that contain more silt than pure limestone. The same relationship probably holds for pressure-solution effects in pure quartzite as compared with less-pure argillite and fine-grained sandstone. Such argillite and sandstone also contain more water. Impurities in large amounts may provide more points to nucleate dissolution, thereby making the process more efficient. Clays, too, may catalyze pressure solution because of their small grain size, and they may serve as both a source and a conduit for movement of water.

PROGRESSIVE DEVELOPMENT IN FINE-GRAINED SEDIMENT

Even as sediment is being deposited, it begins adjusting to the new environment. As the weight of the overlying sediment weighs down clay particles, the particles begin aligning normal to the vertical maximum principal lithostatic stress and form a *bedding fissility.* Much of the change comes during expulsion of water and initial compaction, which is the first stage in the transformation of sediment into sedimentary rock. Further development may occur during the early stages of folding. Later, tectonism may transform the rock into a strongly cleaved or foliated mass (Figure 17–22). The entire process has been divided by Ramsay and Huber (1983) into six stages of increasing tectonic strain. Those stages, slightly modified, are as follows:

A. *Undeformed condition*—Bedding fissility develops.
B. *Earliest deformation stage*—Loss of volume, from reorientation of grains and expulsion of water, enhances bedding fissility.

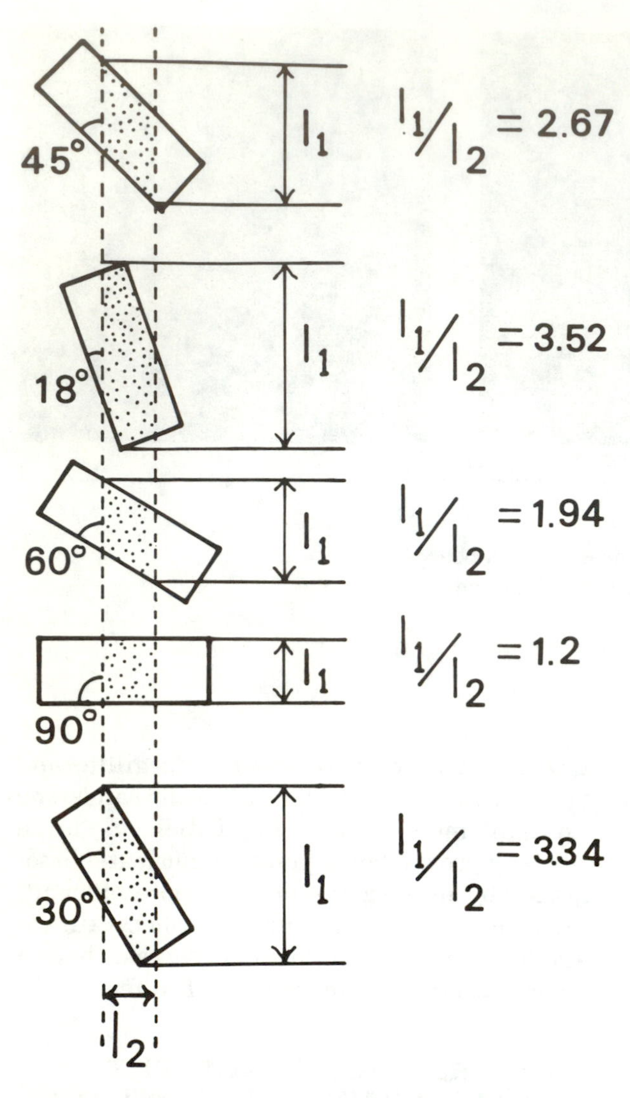

FIGURE 17–20
Corrosion of grains in different orientations to produce new shapes (stippled) proportional to grain orientation. Dashed lines indicate the initial locations of cleavage surfaces, and all material outside those lines is removed. New axial ratios are shown as l_1/l_2 from an initial ratio for all grains of 2.64. (From E. C. Beutner, Slaty cleavage and related strain in Martinsburg Slate, Delaware Water Gap, New Jersey, Jan. 1978, *American Journal of Science,* v. 278, p. 1–23. Reprinted by permission of American Journal of Science.)

C. *Pencil structure* (Figures 17–22b and 17–23)—Elongate pencil-like fragments are produced by intersection of bedding and cleavage, with no continuously organized planar cleavage. Such structure is best shown by homogeneous silty shales. Pencils generally parallel fold axes. This stage may be found in unmetamorphosed fine-grained rocks.

D. *Embryonic cleavage stage* (Figure 17–22c)—Weak, poorly developed cleavage parallel to fold axial surfaces results from pressure solution of silty carbonate or argillaceous rocks. Minor recrystallization of new minerals—illite, quartz, or calcite—may parallel the cleavage, and an intersection structure of pencil-bedding cleavage may develop parallel to the axes or hinges of folds. This stage is most likely to be found in unmetamorphosed rocks. Layer-parallel shortening may produce cleavage before folding (Engelder and Geiser, 1979).

E. *Cleavage stage* (Figure 17–22c)—A strong planar fabric results from pressure solution, or reorientation of platy minerals, or incipient recrystallization of clays. Generally, cleavage has an axial planar relationship to folds that form at this stage, accompanied by well-developed intersection lineations of cleavage and bedding. Rocks displaying this cleavage stage range in metamorphic grade from anchizone to lower greenschist facies (chlorite zone).

F. *Strong cleavage with mineral (stretching?) lineation* (Figure 17–22d)—The planar character of slaty cleavage is better developed than in earlier stages, and a faint mineral-elongation lineation (quarry workers' *grain*) appears on the cleavage. This lineation is generally oriented normal to fold axes, parallel to the *X* direction (down the dip of cleavage planes), and lies in the *XY* plane, but the orientation of the mineral lineation can vary widely in the cleavage plane. The bedding-cleavage intersection lineation is usually well developed. This is the highest stage of cleavage development and may occur throughout the chlorite zone of regional metamorphism; sometimes it persists into the biotite zone. At the upper extreme of this stage, the rock mass is totally recrystallized, with deformational/thermal structures becoming dominant. Recrystallization processes (dislocation and diffusion creep) and formation of other new minerals become dominant as the metamorphic temperature increases.

Pressure solution as a cleavage-forming mechanism dominates in the early stages, but the consolidation of cleavage planes in the later stages is probably related to increasing recrystallization. Many strongly cleaved rocks are phyllonites with well-developed S-C structure (Chapter 10).

Progressive development of tectonic cleavages in fine-grained rocks is a means by which planar fabrics are organized as deformation increases (Figure 17–24). Beyond the last stage in Ramsay and

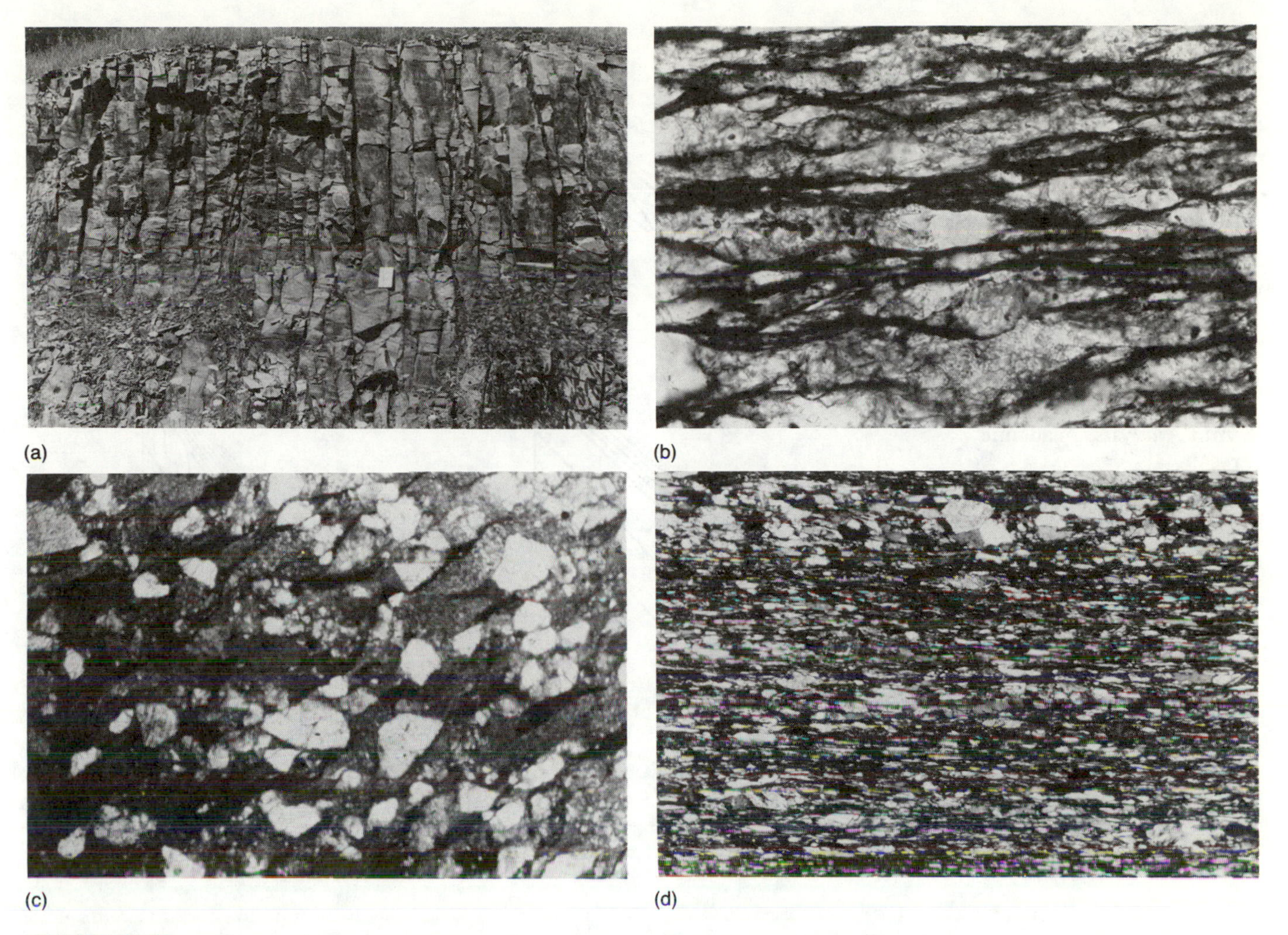

FIGURE 17–21
(a) Pressure-solution spaced cleavage in Middle Ordovician silty carbonate rocks, South Hero Island, Vermont. Cleavage is near vertical; bedding (faint color banding) is nearly horizontal. (RDH photo.) (b) Residues of insoluble minerals along cleavage planes in Devonian Marcellus Shale near Cumberland, Maryland, produced by pressure solution. Plane light. Width of field is approximately 3 mm. (Charles M. Onasch, Bowling Green State University.) (c) Partly dissolved quartz grains next to pressure-solution cleavage planes in Upper Ordovician Martinsburg Slate, Delaware Water Gap, New Jersey. Plane light. Width of field is approximately 2 mm. (Thin section courtesy of Timothy L. Davis, University of Tennessee.) (d) Slaty cleavage in Upper Ordovician Arvonia Slate, Arvonia, Virginia, produced by recrystallization or rotation of grains. Note the absence of residues parallel to cleavage. Plane light. Width of field is approximately 2 mm. (RDH photo.)

Huber's scheme of cleavage and foliation development is transition to coarser grain size. A similar scheme was devised by C. McA. Powell (1979) for slaty-cleavage formation.

Solution cleavage may be classified under a system devised by Walter Alvarez, Terry Engelder, and Peter Geiser (1978), who used the terms weak, moderate, strong, and very strong. Classification depends on the character of cleavage surfaces, spacing of cleavage, and percentage of shortening. Cleavage classified as weak is usually stylolitic, with no preferred orientation; spacing is more than 5 cm, and shortening is 0 to 4 percent. Moderate cleavage has discrete surfaces, orientation from parallel to 120°; spacing is 1 to 5 cm, and shortening is 4 to 25 percent. Strong cleavage is wispy to locally

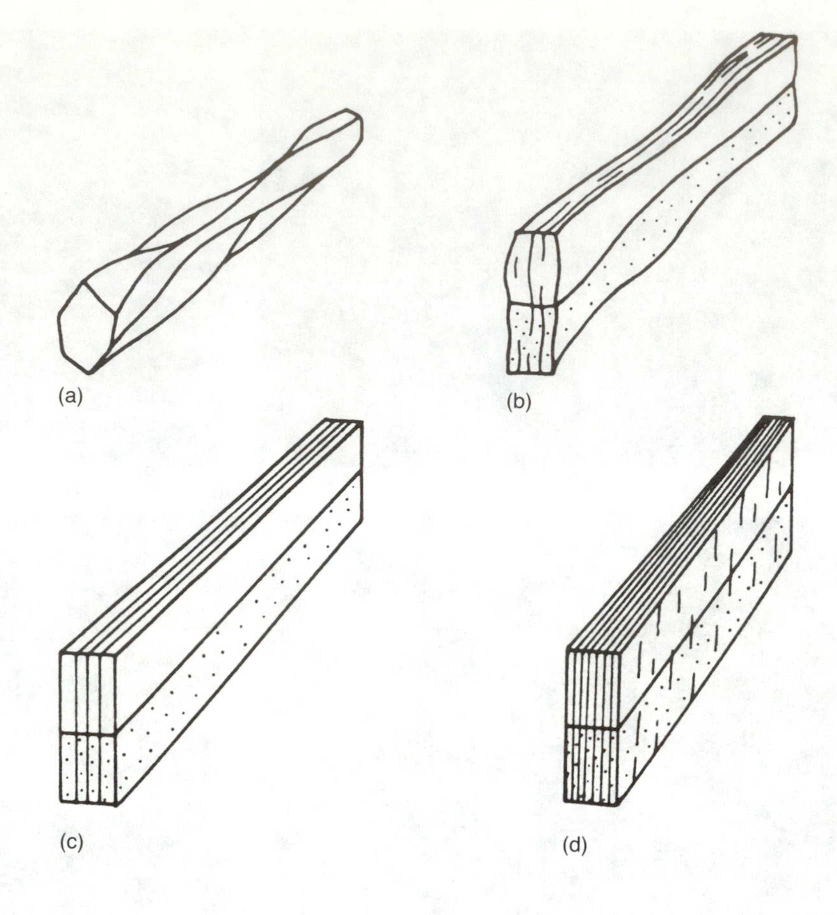

FIGURE 17–22
Stages of cleavage development: (a) Pencil structure. (b) Embryonic cleavage stage. (c) Cleavage stage with accompanying lineation formed by intersection of cleavage and bedding. (d) Well-developed cleavage with mineral lineation. (Photographs and diagrams.) (After J. G. Ramsay and Martin Huber, 1983, *The Techniques of Modern Structural Geology:* Volume 1: *Strain Analysis,* Academic Press.)

FIGURE 17–23
Large pencils developed in siltstone in the Upper Proterozoic Sandsuck Formation near Reliance, Tennessee. (RDH photo.)

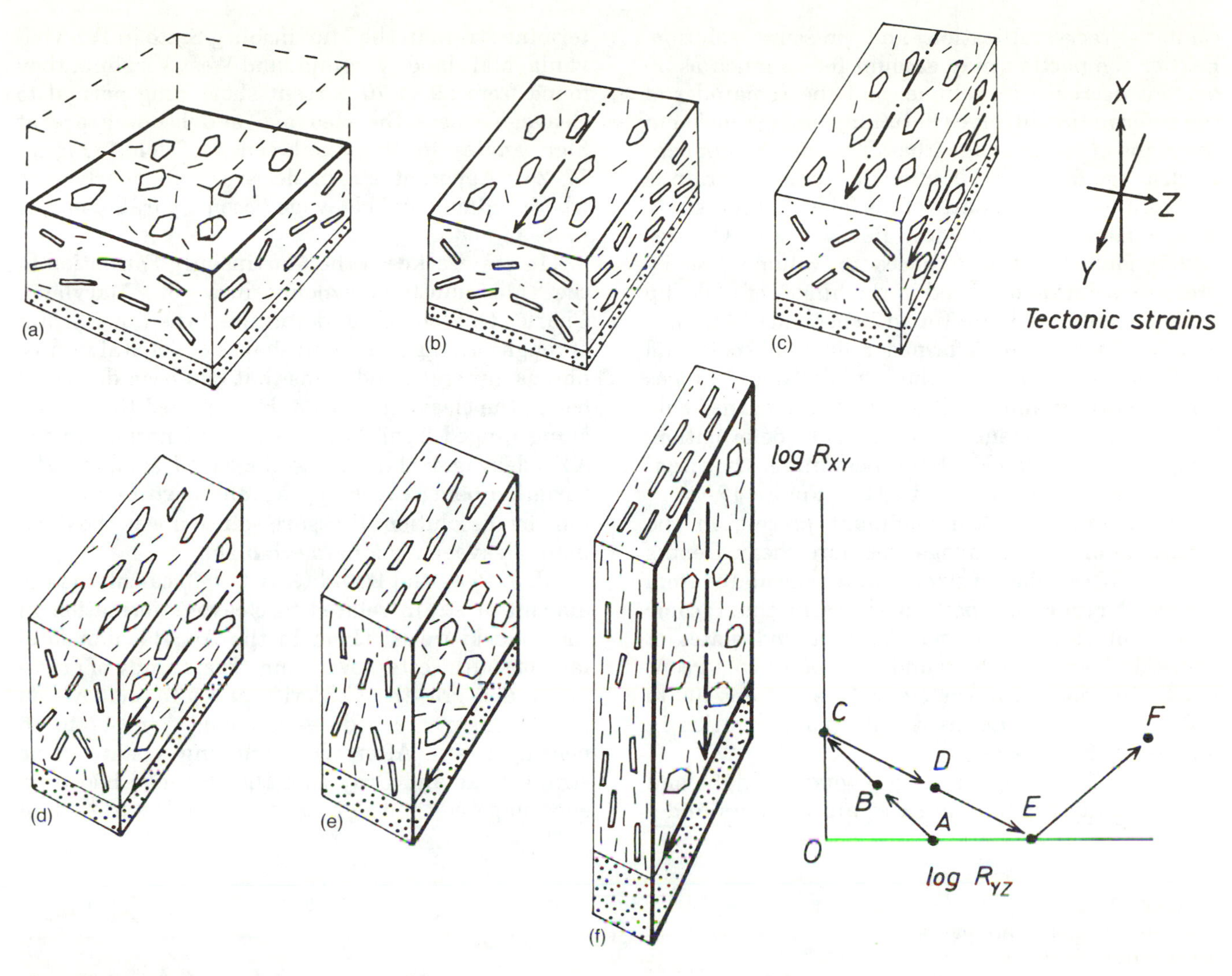

FIGURE 17–24
Relationships between two-dimensional strains and the organization of planar fabrics in rocks; (a) represents the compaction stage; (b) involves the first deformation and minor reorganization of grains; (c) involves development of pencils; (d) is the embryonic cleavage development stage; (e) is the stage of development of a well-developed cleavage; (f) is the stage of further cleavage development where a prominent mineral lineation forms. The Flinn diagram to the right of (f) shows the deformation path from one stage to another. (From J. G. Ramsay and Martin Huber, 1983, *The Techniques of Modern Structural Geology:* Volume 1: *Strain Analysis*, Academic Press.)

anastomosing, concentrating in major surfaces; spacing is 0.5 to 1 cm, and shortening is 24 to 35 percent. Very strong cleavage is sigmoidal with transposed bedding; spacing is less than 0.5 cm, and shortening is more than 35 percent. The rock mass contains abundant calcite veins perpendicular to bedding.

STRAIN AND SLATY CLEAVAGE

The amount of strain accompanying slaty cleavage directly indicates the degree of deformation that accompanies folding. Cleavage-forming mechanisms combine recrystallization and pressure solution. Earlier compaction may account for as much as 30 percent decrease in volume, and the remainder of the deformation involves flattening and extension in the plane of the cleavage (the *XY* plane). *Strain path* in cleavage formation is irregular because strain is different in different parts of the folds. In their study of the Hamburg sequence, Beutner and Charles (1985) found that differences in volume loss by pressure solution occurred in the hinges of folds (up to 59 percent) and limbs (up to 29 percent). Measurements of strain in deformed reduction spots and fossil conodonts indicate that the limbs and hinges were deformed along different strain paths. Fold limbs were flattened throughout deformation; hinges were flattened at the beginning, then constricted, and later flattened again (Figure 17–25).

Assuming that the dominant process in the formation of slaty cleavage was pure shear, Dennis Wood (1973) compared strain values calculated from deformed reduction spots in slates in the Taconic klippes of New York and Vermont and slates of Wales. In both cases he found from 50 to 75 percent shortening normal to the cleavage and from 100 to 400 percent extension as a principal elongation in the plane of cleavage.

Graptolites were used by Thomas Wright and Lucian Platt (1982) as finite-strain markers to determine strain in the Martinsburg Slate in Pennsylvania, Maryland, Virginia, and West Virginia; they found from 50 to 70 percent shortening normal to cleavage where the cleavage and bedding are at high angles in the axial zones of folds (Figure 17–26). Apparent strain decreases to nearly zero where bedding and cleavage become parallel on the limbs of folds.

In graywackes of the Martinsburg Formation in the Massanutten synclinorium of Maryland, Charles Onasch (1983) determined the finite strain of rough cleavage. He used shapes of detrital grains, fibrous minerals, and veins that had been deformed before the cleavage formed. He reported that shortening ranged from 29 to 55 percent normal to the *XY* (cleavage) plane in well-cleaved samples, the variation being caused by an unknown loss of volume by dissolution. Pressure solution was the dominant cleavage-forming mechanism.

J. J. Reks and David Gray (1983) calculated the amount of strain related to cleavage formation in the Pulaski thrust sheet in the Appalachian foreland of southwestern Virginia. Their study of orientation and growth of chlorite pressure shadows on framboidal pyrite nodules revealed from 17 to 35 percent strain. After reconstructing a state of no strain, Reks and Gray said the cleavage had been superimposed parallel to the axial surfaces of

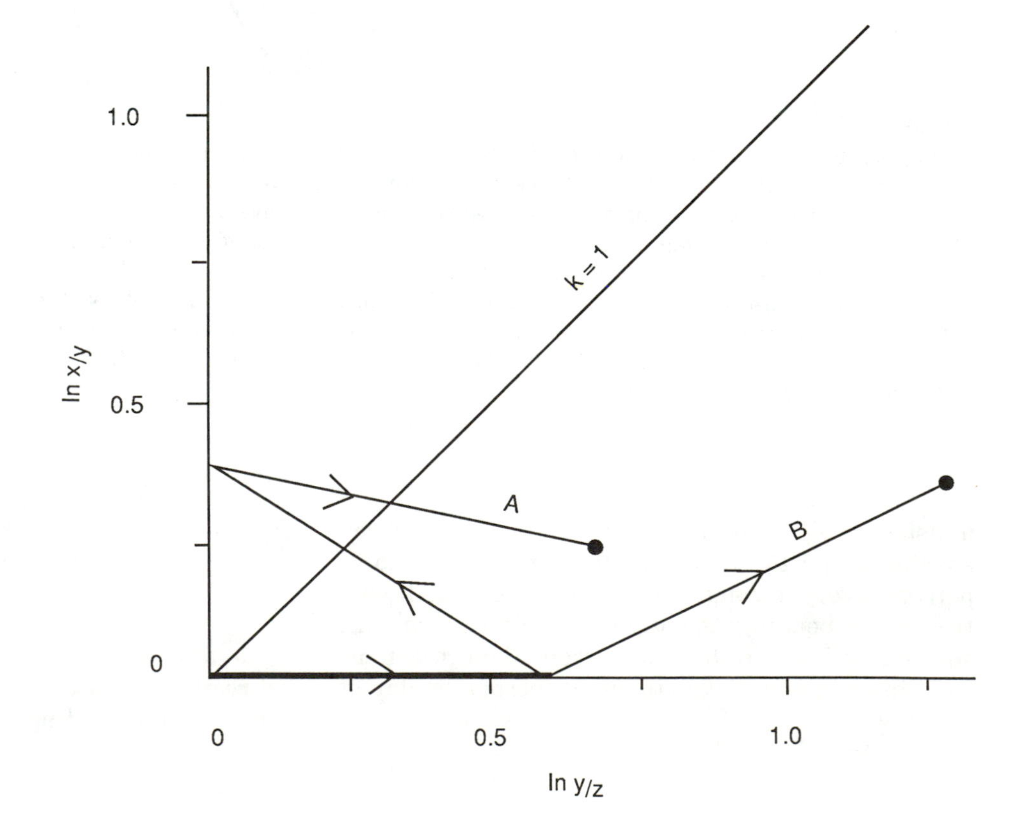

FIGURE 17–25
Flinn diagram of strain paths derived from study of strain indicators in hinges (curve A) and limbs (curve B) of near-isoclinal folds in the Hamburg sequence near Shartlesville, Pennsylvania. Note that the strain in fold hinges (A) was traced from the field of increasing dominant flattening strain to the dominant extension strain field, back into dominant flattening, but with a strong component of triaxial strain. Strain in the limbs of the fold remains in the dominant flattening field and traces toward a greater component of triaxial strain. (From E. C. Beutner and E. G. Charles, 1986, *Geology,* v. 13.)

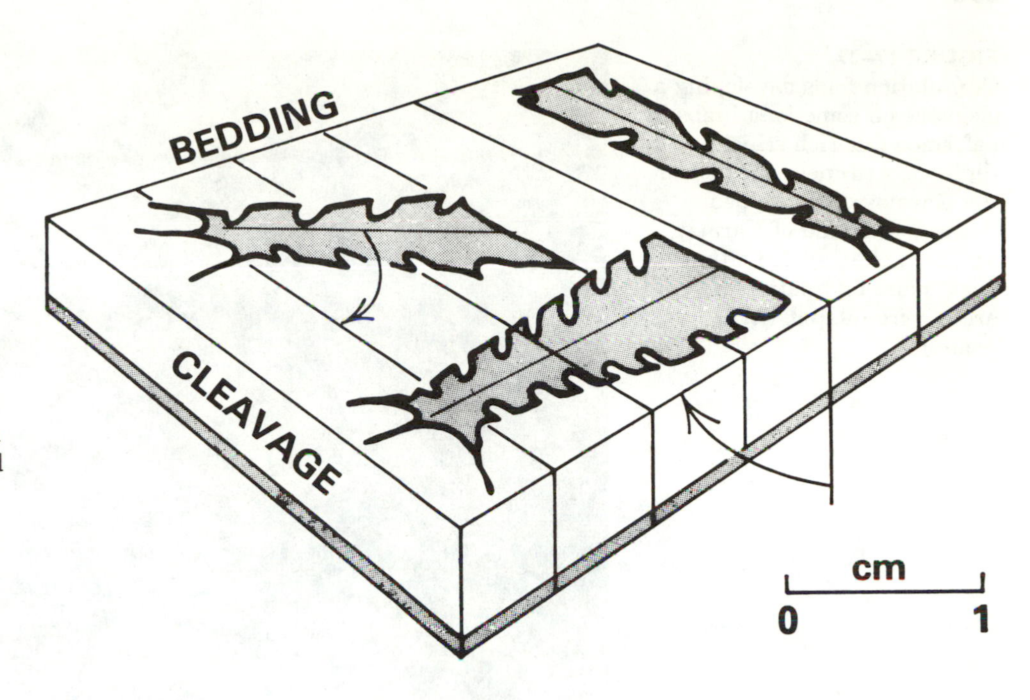

FIGURE 17–26
Relationship of graptolites on a bedding plane to shortening across cleavage. Graptolites parallel to the trace of cleavage on bedding are narrower than usual; those perpendicular to the cleavage trace are shorter. The original spacing of indentations (thecae) on each graptolite is constant for both adults and juveniles of the same species, so the amount of shortening in any direction can be attributed to tectonic strain. (From T. O. Wright and Lucian Platt, Pressure dissolution and cleavage in the Martinsburg shale, Feb. 1982, *American Journal of Science,* v. 282, p. 122–135. Reprinted by permission of American Journal of Science.)

already-formed Class 3 buckle folds. Deformation, they concluded, was largely by homogeneous strain, with folds and cleavage forming at different stages of the same event.

The amount of shortening during cleavage formation decreases from the high values just cited to much lower values toward the outer edges of orogenic belts. Under proper conditions, fabrics that are related to pressure solution and accompany layer-parallel shortening can extend to the outer edges of the deformed zone. In the central Appalachians, as Terry Engelder (1979a, 1979b), Engelder and Richard Engelder (1977), and Engelder and Peter Geiser (1979) have shown, the strain was accommodated by layer-parallel shortening above a detachment zone in the rocks beneath.

CRENULATION CLEAVAGE

Development of crenulation cleavage generally involves formation of a new structure that overprints an existing cleavage or foliation (Figure 17–27). Crenulations may also form on bedding; they have been observed even in glacial silts. *Crenulation cleavage,* as defined by David Gray (1977a), occurs in zones of mineral differentiation coincident with the limbs of microfolds in crenulated rock fabrics. Crenulations generally overprint an existing cleavage by crinkling and crenulating the earlier structure. Differentiated layering frequently begins with development of a crenulation structure and continues with pressure-solution removal of quartz from the limbs of crenulations and redeposition in the hinges.

The origin of crenulation cleavage has long been debated. Many structural geologists support its origin by pure shear, but others have found evidence for a simple-shear mechanism. In shear zones, simple shear best explains crenulations (Chapter 10), but the widespread regional crenulation cleavage is best explained by pure shear.

Gray (1977a) constructed a classification of crenulation cleavage in which two classes are recognized: discrete and zonal (as used in Powell's classification cited earlier). *Discrete crenulations* involve sharply defined cleavages that truncate the older fabric in the rock mass. His *zonal crenulations* occur as wide diffuse zones, and the original rock fabric continues uninterrupted (Figure 17–28).

Differentiated crenulation cleavages in Australia and Switzerland led Gray (1977b) and Gray and David Durney (1979) to suggest that the crenulations formed by buckling and fluid-induced mobility. The process was driven by stress and was related to the wavelengths of microfolds and the grain-boundary contacts, for grain-boundary diffusion (Chapter 6) was declared the deformation mechanism. The researchers also concluded the crenulations formed by pure shear.

By way of contrast, consider a study by Simon Hanmer (1979), who suggested that crenulation cleavage forms on microshears developed by heter-

FIGURE 17–27
Crenulation folds developing a cleavage on some long limbs in deformed talc-rich schist (light-colored) near the base of the Karmøy ophiolite near Hagesund, Island of Karmøy, southern Norway. Note that dark (more feldspathic) layers are not crenulated. (RDH photo.)

ogeneous strain parallel to the shear planes in the strain ellipsoid (Figure 17–29). He concluded that the character of the existing planar and linear fabrics and porphyroblasts that form during recrystallization all affect initiation and propagation of crenulations.

Evidence can thus be cited to show that crenulations can form by pure shear or by simple shear. Paul Williams (1972) concluded that crenulations and slaty cleavage grade into one another in the Bermagui region of Australia, implying a common origin by pure shear. We also know that crenulations form in shear zones (Chapter 10) as both *extensional crenulations* (Platt and Vissers, 1979) and also as *reverse-sense crenulations* (Dennis and Secor, 1987). So they must form by simple shear, too. Hanmer's study shows that crenulations also form in fine-grained rocks by simple shear, not as part of a shear zone.

CLEAVAGE FANS AND TRANSECTING CLEAVAGES

A fundamental property of most slaty cleavage is the parallelism (or a fanning relationship) between the orientations of fold axial surfaces and slaty cleavage, providing a genetic link between the two structures (Figure 17–1). Yet examples can be found that lack parallelism between fold axial surfaces and cleavage (Figure 17–30). Folds where the cleavage and fold axial surfaces are not coincident—or where the fold axis does not lie in the cleavage—are ***transected folds.***

Convergent and divergent cleavage fans may indicate discordant timing between folding and cleavage formation. Causes range from the obvious noncontemporaneity of formation to subtler relationships resulting from the sequencing of homogeneous and inhomogeneous strain, as well as discordance in strain rate and strain variation from bed to bed.

In the simplest relationship between transected folds and cleavage, folds and cleavage formed at different times with differently oriented principal strain axes (Figure 17–31). The differences in timing and orientation can be explained if the folds formed as soft-sediment structures, or if the folds formed tectonically (earlier than the cleavage), or if the folds formed later than cleavage. Powell (1974) has suggested that slaty cleavage forms as a principal-plane cleavage parallel to the XY plane of an incremental strain ellipsoid; but, because folding occurred later, Graham Borradaile (1978) suggested that cleavage does not parallel a plane in the finite-strain ellipsoid.

In Ramsay's view (1963), cleavage and folds form about the same time and cleavage forms parallel to the XY plane, but noncoaxial strain is superimposed during the same deformation so that the principal plane of the strain ellipsoid and the XY cleavage plane rotate to positions for cutting through both fold limbs. Later, Borradaile (1978) said it was unlikely that the strain ellipsoid can be rotated so that cleavage will transect folds.

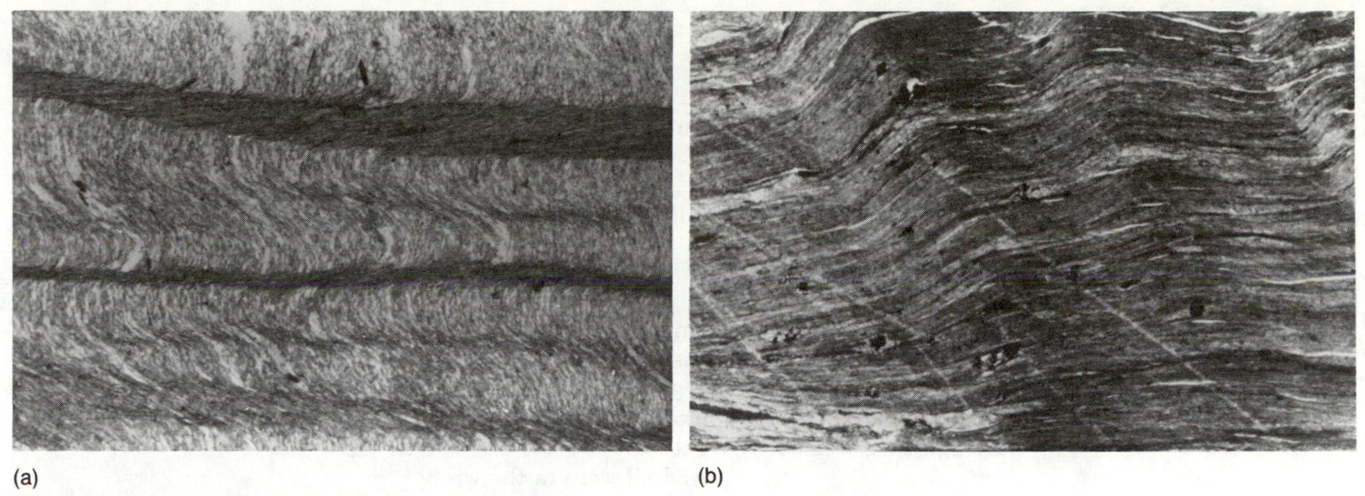

FIGURE 17–28
(a) Discrete crenulations in Lower Silurian Vakdal Formation phyllite near Ulven, southeastern Norway. Plane light. Width of field is approximately 16 mm. (b) Zonal crenulations in (Ordovician?) Mineral Bluff Formation phyllite near Murphy, North Carolina. Plane light. Width of field is approximately 16 mm. (RDH photos.)

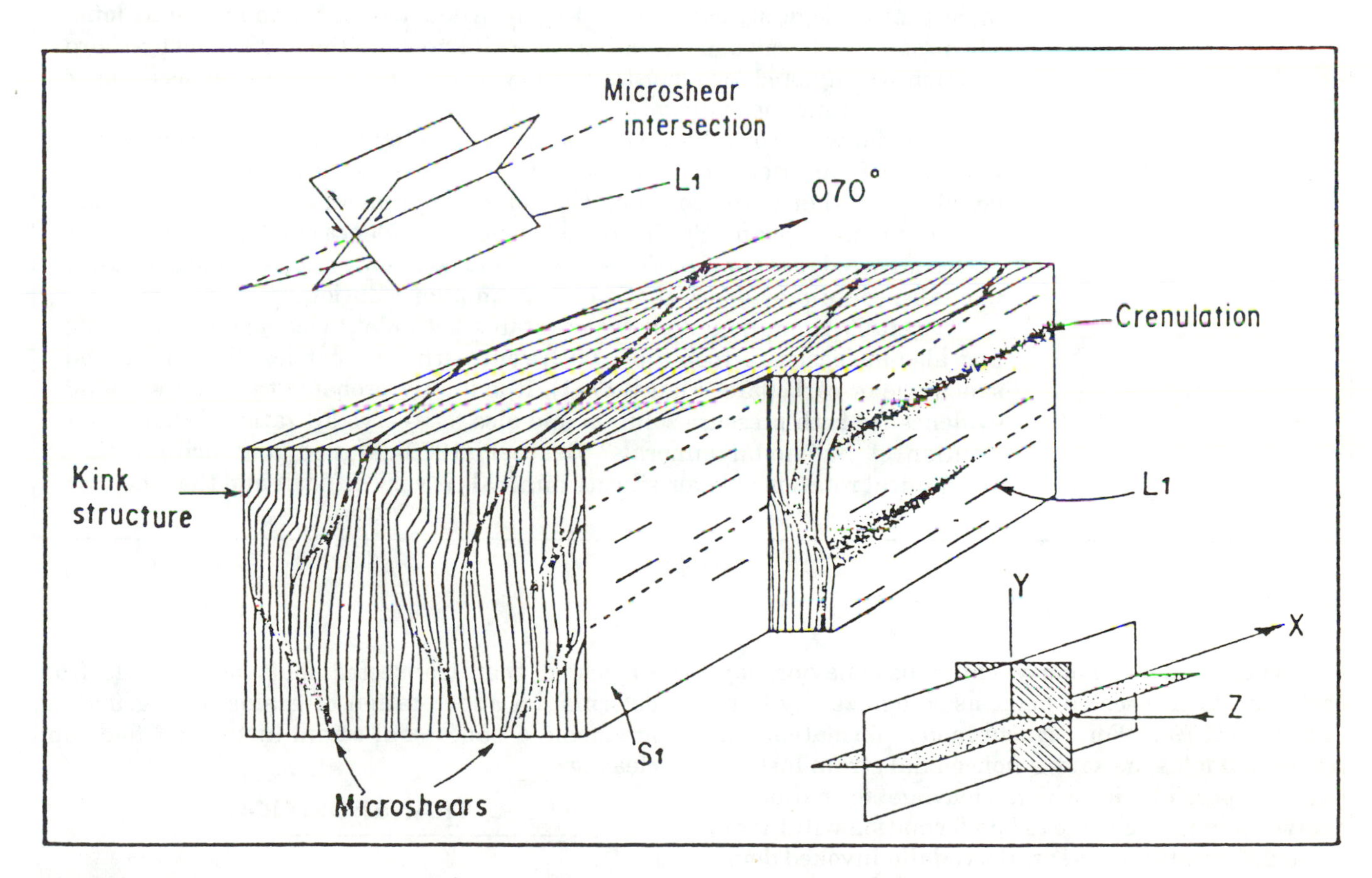

FIGURE 17–29
Vertical S_1 cleavage surfaces, arbitrarily oriented ENE, containing a subhorizontal mineral lineation (L_1). Sections cut normal to *XY* and *YZ* planes show relationships of microshears, crenulations, and kinks to principal planes of the strain ellipsoid. (Reprinted with permission from *Journal of Structural Geology,* v. 1, S. K. Hanmer, The role of discrete heterogeneities and linear fabrics in the formation of crenulations, © 1979, Pergamon Journals Ltd.)

ESSAY

Cleavage Formation and Identification of Elephants

It was six men of Indostan
To Learning much inclined,
Who went to see the Elephant
(though all of them were blind),
That each by observation
Might satisfy his mind.

. . .

And so these men of Indostan
Disputed loud and strong,
Each in his own opinion
Exceeding stiff and strong
Though each was partly right
And all were in the wrong!

from John Godfrey Saxe,
The Blind Men and the
Elephant—A Hindoo Fable

For a century and a half, geologists have recognized cleavage as an important geologic structure, and they have debated its origin almost as long. The processes of cleavage formation were all known in the 1800s, and pressure solution was ignored for almost a century before being resurrected as a major cleavage-forming mechanism.

Why did we ignore the obvious for so long? Why have we not been able to resolve the question of its origin? The answer is not simple, and so—possibly—neither is the solution. Consider the problem of the six blind men examining an elephant. Similarly, independent evidence can be found to favor each suggested mechanism of cleavage formation, but each cleavage examined may also contain evidence that supports another solution.

Carefully examine a rock that contains both slaty cleavage and bedding and also at least two rock types with contrasting properties—like slate and sandstone or carbonate (Figure 17–1). It is highly probable that you will find evidence for both pressure solution and also for recrystallization. What about rotation of individual minerals? Beutner (1978) argued convincingly that mechanical rotation of layer-silicate minerals is very difficult and that they too

Synchronous and nonsynchronous behavior may each lead to transected folds, as recognized by Borradaile (1979). For synchronous formation of transected folds, he said, a noncoaxial strain history may be possible in which cleavage formation is delayed slightly relative to fold formation within the same deformational event. Borradaile invoked dominance by a competing mechanism, such as dewatering, which prevents cleavage from forming initially during folding; when cleavage does begin to form, it transects the existing folds. Transected folds may also form in shear zones. Borradaile (1981) later suggested that transected folds may result from deformation and rotation of grains during folding, producing a noncoaxial strain history for folds and cleavage.

TRANSPOSITION

For many years, geologists have recognized the progressive dominance of cleavage and foliation with increasing deformation and metamorphic grade. They have also known layering in rocks of high metamorphic grade is probably not original

must be partly dissolved. But Borradaile (1981) has argued that grain-boundary sliding may be an important factor in reorientation of minerals during cleavage formation. So, after a century and a half, we may have to accept two mechanisms, with pressure solution dominating at low temperature. But this points again to the complexity of nature—and our attempts to simplify and pigeonhole it. The dispute over the relationships of cleavage planes to the strain ellipsoid is largely resolved; most structural geologists agree with Williams (1976) that cleavage forms parallel to the *XY* plane.

If your rock is also crenulated, you may find evidence favoring either a pure-shear or simple-shear mechanism. Crenulations and kinks may have a similar origin, but most geologists believe crenulations develop by pure shear parallel to the *XY* plane (Gray, 1980). Still, some crenulations may form by simple shear (Hanmer, 1979).

The origin of the different cleavages is now partly known. There probably is more evidence for continuous cleavage than Powell (1979) believed. There probably are some crenulations that originated by simple shear. There may be some rotation of layer-silicate grains during formation of slaty cleavage, although wholesale reorientation of most grains in a rock seems unlikely. More facts about the chemistry of pressure solution and the nature of recrystallization processes have helped enormously in answering some of the questions, but no doubt the problem of cleavage formation will be debated for many decades.

References Cited

Beutner, E. C., 1978, Slaty cleavage and related strain in Martinsburg slate, Delaware Water Gap, New Jersey: American Journal of Science, v. 278, p. 1–23.

Borradaile, G. J., 1981, Particulate flow of rock and the formation of cleavage: Tectonophysics, v. 72, p. 305–321.

Gray, D. R., 1980, Geometry of crenulation-folds and their relationship to crenulation cleavage: Journal of Structural Geology, v. 2, p. 187–204.

Hanmer, S. K., 1979, The role of discrete heterogeneities and linear fabrics in the formation of crenulations: Journal of Structural Geology, v. 1, p. 81–91.

Powell, C. McA., 1979, A morphological classification of rock cleavage: Tectonophysics, v. 58, p. 21–34.

Williams, P. F., 1976, Relationships between axial-plane foliations and strain: Tectonophysics, v. 30, p. 181–196.

bedding (Figure 17–32). Transformation of original bedding by folding, ductile shear, or other process into parallelism with cleavage or foliation through progressive deformation is called ***transposition.*** Sander (1911) may have been the first to consider transposition as a separate process. Its importance was recognized by Jonas (1927) during her field studies in the central Appalachians and by Knopf and Ingerson (1938) in their book on structural petrology.

Transposition of bedding, or of any layering, occurs progressively. It begins with the initial stages of slaty-cleavage formation described earlier. Cleavage thus formed cuts through the axial zones of folds at a high angle, but through the limbs at a low angle, thus making cleavage and bedding almost parallel (Figure 17–33). As deformation continues, the short limbs of the fold are progressively flattened and thinned, and layering in the axial zones that originally crossed foliation at a high angle is flattened into increased parallelism with the cleavage (or foliation). Recrystallization is favored as deformation continues and temperature rises. Flattening proceeds to an even greater degree, and

FIGURE 17–30
Transected fold in Devonian slate from near Zell, Mosel Valley, West Germany. The transecting cleavage (S_2) forms an intersection lineation that obliquely crosses the F_2 fold shown here. (From G. J. Borradaile, M. B. Bayly, and C. McA. Powell, eds., 1982, *Atlas of Metamorphic and Deformational Rock Fabrics,* Springer-Verlag.)

layering that initially was at a high angle to cleavage planes is flattened subparallel to cleavage, even in the axial zones of folds (Figure 17–34). Finally, all layering is parallel to the foliation and original bedding is no longer recognizable as a primary structure—unless primary structures like graded bedding or cross bedding survive within the layers. Transposition is complete.

Earlier cleavages or foliations may be transposed in much the same way. An initial foliation or cleavage may be overprinted by a new cleavage, possibly crenulation cleavage. With continued deformation, the older S-surface is reoriented till subparallel to the new foliation (Figure 17–32). S_0, the original bedding in the sequence of rocks, may be transposed till parallel with S_1 so that the designation $S_0 = S_1$ may be appropriate. Foliation in rocks with no original layering, such as massive plutonic rocks, is designated S_1. A second and younger foliation S_2 may overprint the older one, then transpose both S_0 and S_1 till subparallel to the younger structure. Ability to recognize this overprinting relationship helps unravel the history of complex areas and identify structures that indicate transposition.

Several indicators of transposition occur in rocks. Small to large isolated relict isoclinal folds lying within a foliation, ***intrafolial folds,*** are common indicators (Figure 17–34a). They commonly take the form of isolated hinges where the limbs have been attenuated and sheared off through deformation. Boudins of resistant rocks, such as amphibolite or calcsilicate, may preserve earlier folds and foliations.

Microlithons may preserve earlier structures on microscopic scale (Figure 17–34c). They may provide evidence of an earlier foliation in a deformed and metamorphosed rock mass. Additional indicators of earlier deformation are *inclusion trails in porphyroblasts* or other coarse crystals as seen in thin section (Figure 17–34d). Garnets are particularly useful in preserving inclusion trails that represent earlier foliations, but garnets may be rotated several times during the same deformation. So, the existence of inclusion trails at a high angle to the dominant foliation may or may not identify transposed layering. Other minerals such as staurolite, feldspar, and kyanite may contain inclusion trails that can be used similarly.

We will briefly return to a discussion of cleavage in Chapter 19 after discussing linear structures in Chapter 18. The role played by cleavage and foliations in the formation of certain types of linear structures and its importance in structural analysis should then become clear.

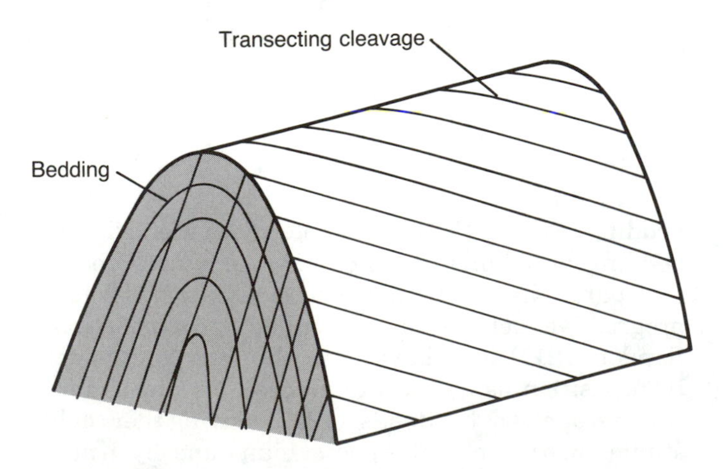

FIGURE 17–31
Transected fold showing relationships between the hinge and noncoplanar cleavage.

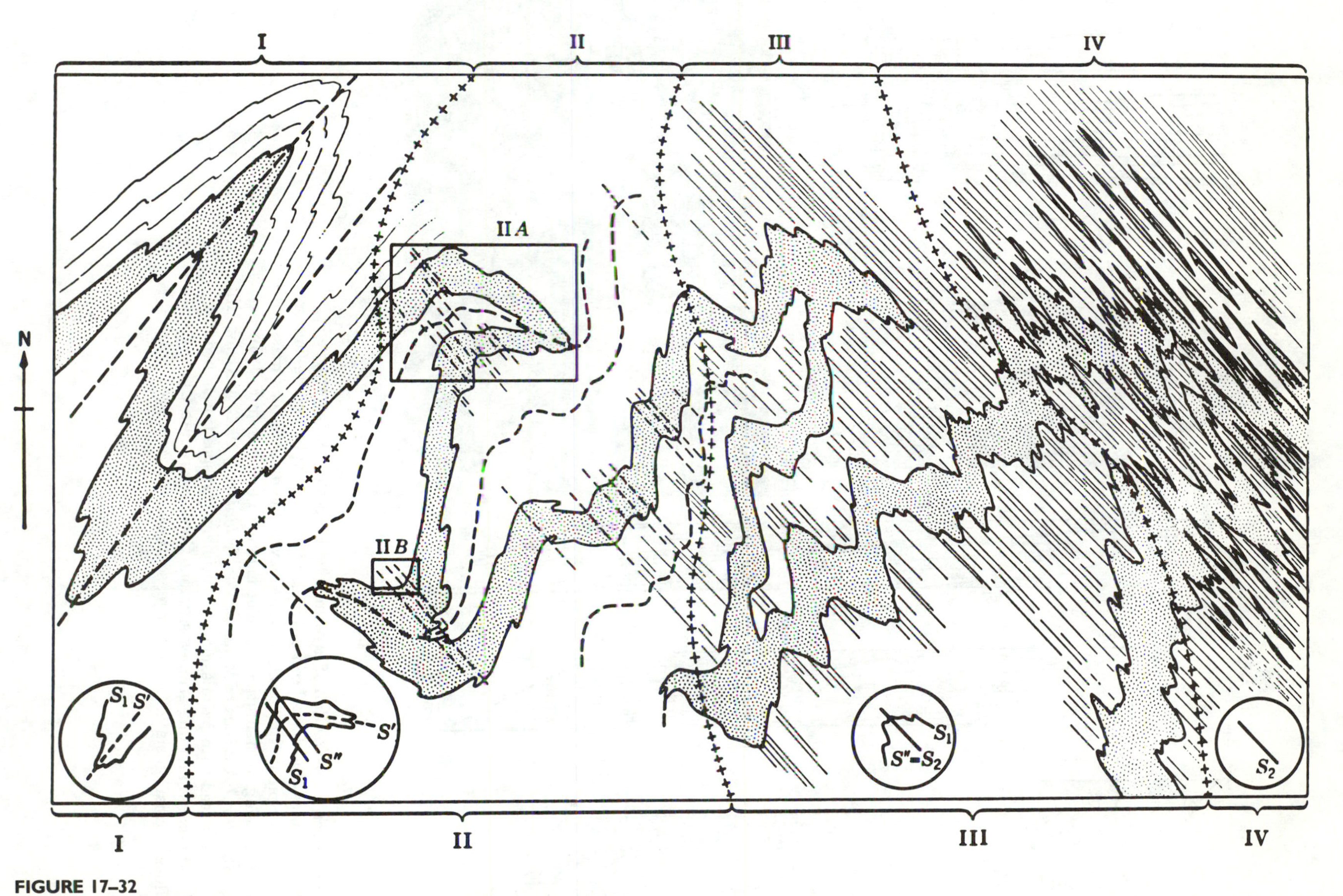

FIGURE 17–32
Progressive deformation and transposition of earlier bedding and cleavage, producing a rock mass containing foliations that are all parallel. (From F. J. Turner and L. E. Weiss, *Structural Analysis of Metamorphic Tectonites,* © 1963, McGraw-Hill Book Company. Reproduced with permission.)

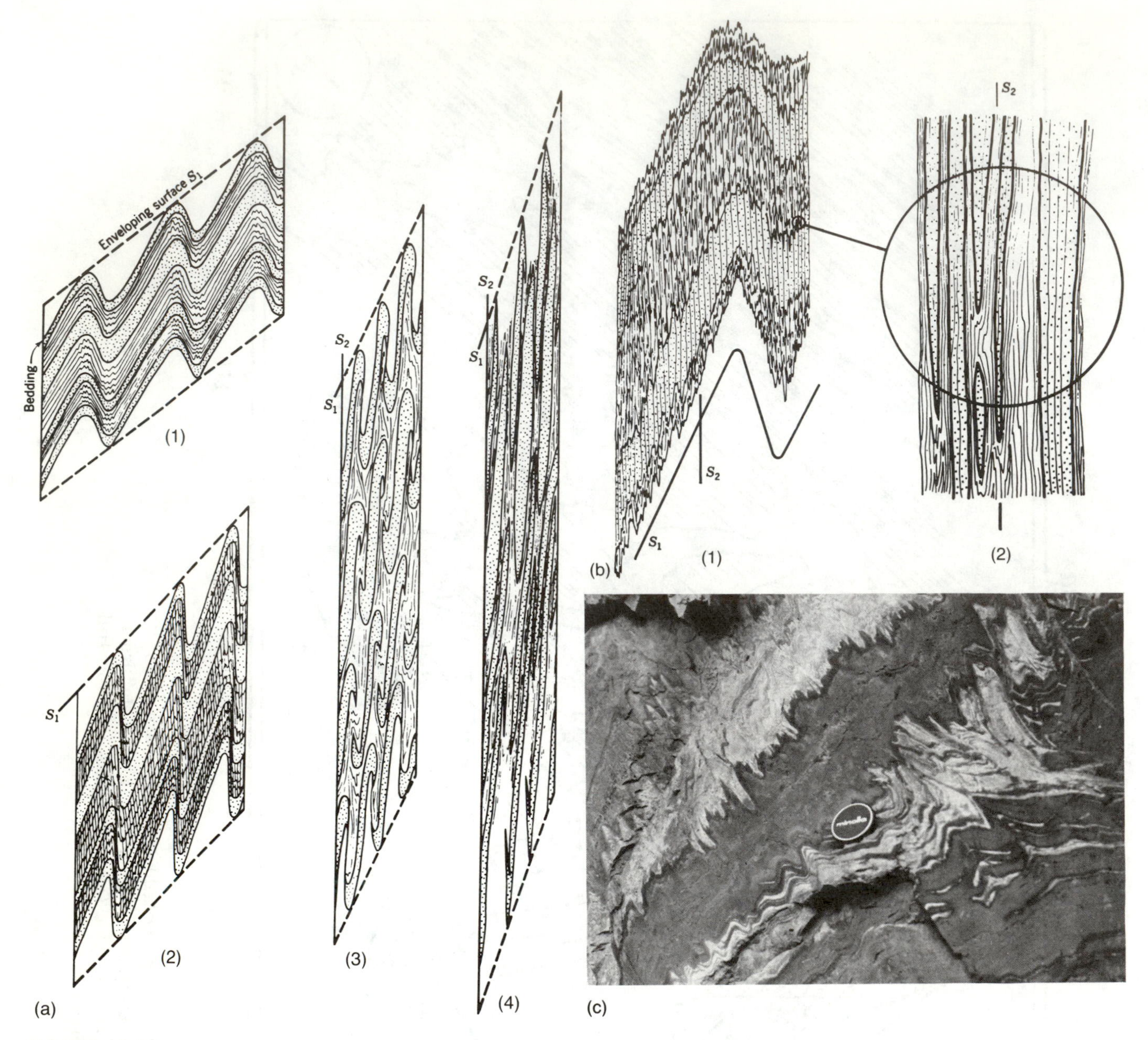

FIGURE 17–33
(a) Relationship between transposition of layering and hinges on the overturned (short) limbs of small folds that have been progressively removed by ductile flow or pressure solution. (b) Incipient transposition in a cleavage-dominated rock mass with closeup of bedding showing reorientation by deformation. (a and b from F. J. Turner and L. E. Weiss, *Structural Analysis of Metamorphic Tectonites*, © 1963, McGraw-Hill Book Company. Reproduced with permission.) (c) Incipient transposition of bedding in interlayered slate (dark) and fine-grained siltstone (light-colored) in Upper Proterozoic Wilhite Formation slate, Ocoee Gorge, southeastern Tennessee. (RDH photo.)

FIGURE 17–34
(a) Isoclinal intrafolial folds in Middle Proterozoic Toxaway Gneiss, near Whitewater Falls on the North Carolina–South Carolina border. (b) Boudins of amphibolite preserving an earlier foliation truncated and wrapped by the foliation in the enclosing rocks, Shuswap metamorphic complex near Revelstoke, British Columbia. (c) Microlithons of pressure-solution cleavage (dark) preserving bedding (light and dark layers) between the cleavage planes in Upper Proterozoic Wilhite Formation slate, Great Smoky Mountains, eastern Tennessee. Note that the layer above the striped bed is graded. Small circular objects are bubbles in the section. Plane light. Field is approximately 16 mm wide. (Thin section courtesy of Nicholas B. Woodward, Knoxville, Tennessee.) (d) Garnet porphyroblast from Upper Proterozoic metasedimentary gneisses near Franklin, North Carolina, containing inclusion trails truncated by the enclosing foliation. Plane light. Field is approximately 7 mm wide. (RDH photos.)

Questions

1. Several long-standing controversies concern the origin of slaty cleavage. Why?
2. How can you distinguish slaty cleavage from “fracture cleavage”?
3. What is cleavage refraction?
4. Explain how slaty cleavage that starts out as a pressure-solution cleavage could evolve into cleavage formed by recrystallization.
5. How can you estimate cleavage-related strain in a rock mass?

6. Purity or lack of purity of a rock affect its ability to exhibit strain when pressure solution forms slaty cleavage. How?
7. How has it been determined that the amount of strain in slate is more than 100 percent in some directions?
8. Why does pencil structure form as a precursor to cleavage?
9. How can you prove transposition has occurred in a rock mass?
10. Suggest two mechanisms for forming strongly discordant cleavage and fold axial orientation.
11. The drawing shows discontinuous exposures of cleaved rock. The heavy solid lines are bedding; light dashed lines are slaty cleavage. Photocopy this page and connect the bedding surfaces to show the larger structure.

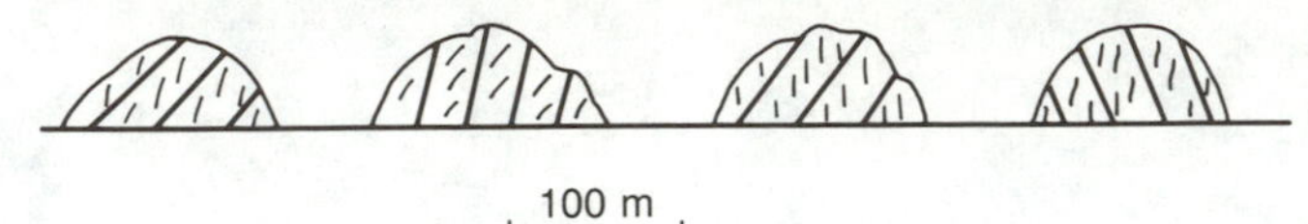

100 m

12. How can cleavage formed parallel to the *XY* plane contain rotated porphyroblasts?

Further Reading

Borradaile, G. J., Bayly, M. B., and Powell, C. McA., eds., 1982, Atlas of deformation and metamorphic rock fabrics: New York, Springer-Verlag, 551 p.
Contains excellent photographs of most kinds of cleavage and foliations.

Gray, D. R., 1977, Morphological classification of crenulation cleavage: Journal of Geology, v. 85, p. 229–235.
Classifies crenulation cleavage and pigeonholes the different types.

Platt, Lucian, 1976, A Penrose Conference on cleavage in rocks: Geotimes, v. 21, p. 19–20.
Summarizes a research conference that considered some of the latest ideas on the origin and nature of cleavages. Several papers, such as Powell's (in this list), were later published.

Powell, C. McA., 1979, A morphological classification of rock cleavage: Tectonophysics, v. 58, p. 21–34.
Clarifies the descriptive classification of cleavage types.

Ramsay, J. G., and Huber, M. I., 1983, The techniques of modern structural geology, Volume 1: Strain analysis: New York, Academic Press, 307 p.
Contains useful discussions of cleavage/foliation development and summarizes Sorby's contributions to our knowledge of the nature of slaty cleavage.

Siddans, A. W. B., 1972, Slaty cleavage, a review of research since 1818: Earth Science Reviews, v. 8, p. 205–232.

Williams, P. F., 1976, Relationships between axial-plane foliations and strain: Tectonophysics, v. 30, p. 181–196.
Clarifies the evidence and theoretical basis for formation of cleavage whether parallel or nearly parallel to a principal plane of the strain ellipsoid.

Wood, D. S., 1974, Current views of the development of slaty cleavage: Annual Reviews of Earth and Planetary Sciences, v. 2, p. 369–501.
Each reviews fundamental concepts and ideas that led to our present state of knowledge. Reference lists are quite useful.

18

Linear Structures

Linear structures are important traces of movement and of the direction of tectonic transport. They are more useful than planes because movements along lines are restricted to two directions, whereas movements in planes can occur in all directions.

ERNST CLOOS, *Lineation*

LINEAR STRUCTURES WERE FIRST DESCRIBED IN 1833 AND IN 1888 first appeared as symbols on a published geologic map. The map was made by a Norwegian geologist named Hans Reusch, who measured and plotted orientations of the long axes of deformed fossils in Karmøy, in southern Norway. As Reusch's map shows, linear structures are natural vectors: they indicate orientation of fold axes, sense and direction of movement along fault surfaces, and in a mass of rock the direction of penetrative ductile flow. Interpreting them may be easy and straightforward, or difficult and controversial, or anything in between—depending on the kind of linear structure and its setting.

Any structure that can be expressed as a real or imaginary line is a ***linear structure,*** or ***lineation*** (Figure 18–1). Hinge lines of folds (Chapters 14 through 16) are linear structures. Linear structures

FIGURE 18–1
Pulled-apart layer of calcsilicate quartzite more competent than the groundmass, producing boudinage in Upper Proterozoic Coleman River Formation micaceous metasandstone, Chunky Gal Mountain, North Carolina. Garnets formed in pressure shadows at ends of boudins, and the boudins have been rotated. (RDH photo.)

constitute one kind of ***fabric element.*** A *lineament* is a topographic feature consisting of aligned surficial features like valleys and ridges. Many lineaments in aerial photographs and satellite imagery have been interpreted as having a tectonic origin, but on further examination turn out to be nontectonic or even artificial. In this chapter we will discuss only tectonic lineations, from mesoscopic to microscopic scale.

DEFINITIONS

Nonpenetrative Linear Structures

Like foliations, linear structures are either penetrative or nonpenetrative. The latter kind are confined to isolated surfaces in a rock mass. Probably the most common nonpenetrative linear structure is ***slickenlines.*** Slickenlines are the direct result of frictional sliding and flexural slip, processes described in Chapters 8 and 15. The term ***slickensides*** refers to the entire movement surface. Slickensides may develop on the surfaces of faults, bedding, and foliations (Figure 18–2). Each involves relative movement parallel to the surface. As a result, the surface may be grooved and polished or may be covered with fibrous crystals of calcite, quartz, chlorite, iron oxides, or other minerals where long axes are oriented in the direction of movement. The slickenside surface is commonly stepped with the down sides of the steps indicating the direction of motion (Figure 18–3), but Win Means (1987) has described a very small-scale (less than a millimeter)

FIGURE 18–2
Slickenfibers of calcite on a movement surface in limestone, Romani Road near Rabat, Morocco. (RDH photo.)

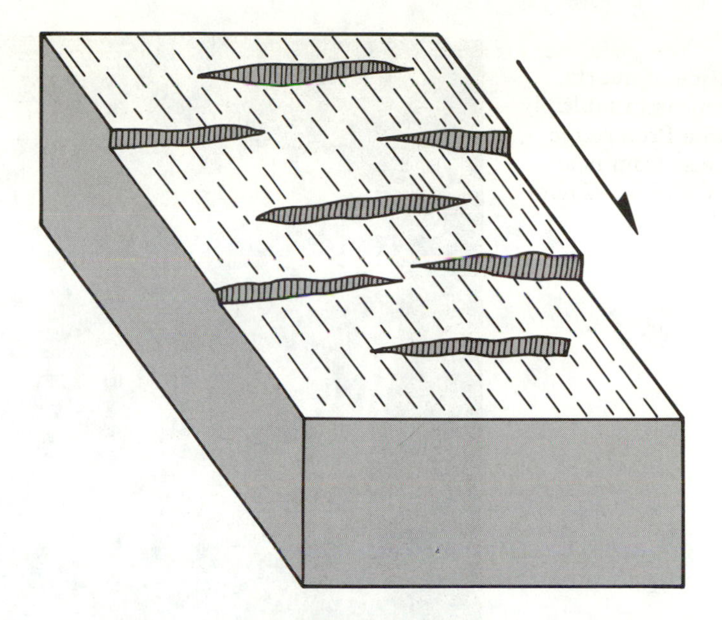

FIGURE 18–3
Formation of slickensides by relative movement. Steps are formed facing the direction of movement (arrow).

slickenside striation consisting of a series of valleys and ridges that nest exactly (ridges in valleys) in those of the opposing block. M. S. Paterson (1958) showed experimentally that for a very small displacement slickenlines may indicate motion in either direction, but for large displacements the slickenlines always step down in the direction of motion. If slickenlines occur on many layers in a flexural-slip folded sequence, it is penetrative on map scale and even on mesoscopic scale.

At the appropriate scale, an isolated fold hinge is a nonpenetrative linear structure. Large folds, like the Wills Mountain, Burning Springs, and Sequatchie anticlines, in the Appalachian Plateau from Pennsylvania to Alabama, are nonpenetrative folds. But if the fold hinge belongs to a family of large and small folds it is penetrative on map scale.

Penetrative Linear Structures

The most common linear structures are penetrative. They are usually penetrative because of their distribution on all scales throughout the rock mass; thus, they are also fabric elements. Most lineations are mesoscopic structures.

The line where two planes intersect is called an ***intersection lineation***—a penetrative linear structure (Figure 18–4). The most common intersection

(a)

(b)

FIGURE 18–4
Intersection lineations. (a) Cleavage-bedding intersection in Upper Proterozoic Kaza Group phyllite, Cariboo Mountains, British Columbia. (b) Intersection of two foliations in Ordovician(?) Mineral Bluff Formation phyllite at Murphy, North Carolina. View looks down on the earliest foliation that was cut by a younger domainal crenulation. Note that the prominent domainal crenulation transects the longer-wavelength folds. (RDH photos.)

FIGURE 18–5
Mineral lineation of quartz, feldspar, and micas in multiply deformed Middle Proterozoic Cranberry Gneiss from near Boone, North Carolina. (Photo by D. G. McClanahan.)

involves bedding (S_0) and cleavage (S_1), sometimes designated by the shorthand notation $L_{1\times0}$ ("L one cross zero") to show that it is an intersection lineation involving bedding and the earliest cleavage. The convention is to denote lineations associated with a particular foliation or deformation as L_1, L_2, etc., the subscript indicating relative age. The intersection of slaty cleavage and bedding is a structure often measured in low-grade metamorphic rocks. Generally, the intersection of cleavage and bedding parallels a fold axis, as a direct consequence of cleavage paralleling the fold axial surface.

Another intersection lineation is defined by two cleavages or other foliations. For example, $L_{2\times1}$ denotes the intersection of the earliest-formed foliation by the next earliest. The genetic relationship between the two may not be immediately obvious, but either cleavage may be related to folding, thereby producing one lineation (or both) parallel to a fold axis. In either case, the intersection of two cleavages or foliations produces a line that can be measured and perhaps related to larger structures.

Mineral lineations consist of aligned elongate mineral grains and grain aggregates. Amphiboles, micas, feldspars, and even quartz are among minerals that may be aligned (Figure 18–5). ***Pressure shadows*** of quartz, muscovite, chlorite, magnetite, or other minerals on either side of a single crystal of pyrite or garnet constitute a mineral lineation (Figure 18–6). Minerals grow in pressure shadows because the pressure near rigid grains is lower. The ductile matrix deforms around the grain.

The axes of rotation of ***rotated minerals*** may compose a mineral lineation, although not a true lineation in the same sense as the others described here. Garnets frequently contain a snowball structure or an S-shaped alignment of inclusion trails, which indicate that simple shear and rotation occurred during garnet growth (Figure 18–7).

Some mineral lineations are deformed into elongate ***rods***—the group noun is ***rodding***—or aggregates of one or more minerals, like quartz, feldspars, and micas (Figure 18–8). They also occur at intersections of two foliation planes, thereby obscuring their origin. Rodding of minerals occurs in many places, but it is common in ductile shear zones (Chapter 10).

Long axes of objects that are deformed into ***natural strain ellipsoids*** make up another category of lineations. The long axes of pebbles, boulders, reduction spots, oöids, and pisolites all become natural strain ellipsoids and form measurable lin-

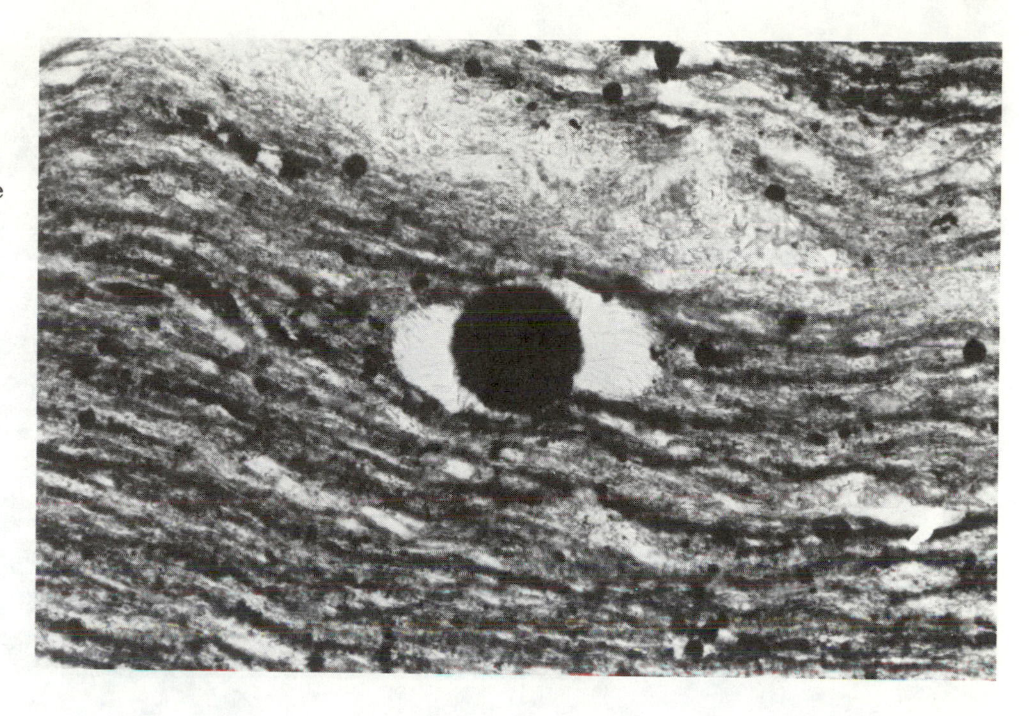

FIGURE 18–6
Quartz-fiber pressure shadows on a pyrite framboid in Lower Ordovician(?) rocks of the Hamburg sequence near Hamburg, Pennsylvania. Plane light. Width of field is approximately 1 mm. (Charles M. Onasch, Bowling Green State University.)

eations during deformation, depending on strain state (Figure 18–9). They were discussed in Chapter 7 as finite-strain markers, where we noted that most were not originally spherical and thus cannot be true strain ellipsoids. They constitute a linear structure, making it possible to measure the orientation of the principal axes of ellipsoids. The lineation to be measured parallels the X axis of the strain ellipsoid.

Mullions form at boundaries between differing rock types and consist of corrugated or scalloped surfaces (Figure 18–10). They result from contrast in competence or ductility from layer to layer. The smooth convex sides of a mullion are directed toward the layer with the higher viscosity, and the cusps toward the layer with the lower viscosity. A structure such as slaty cleavage may be well developed in one layer even though the one above it contains a poorly defined and widely spaced cleavage. Mullions form a corrugated boundary between layers. They may also form in uncleaved rocks by ductile deformation at the boundary of layers in which competence differs.

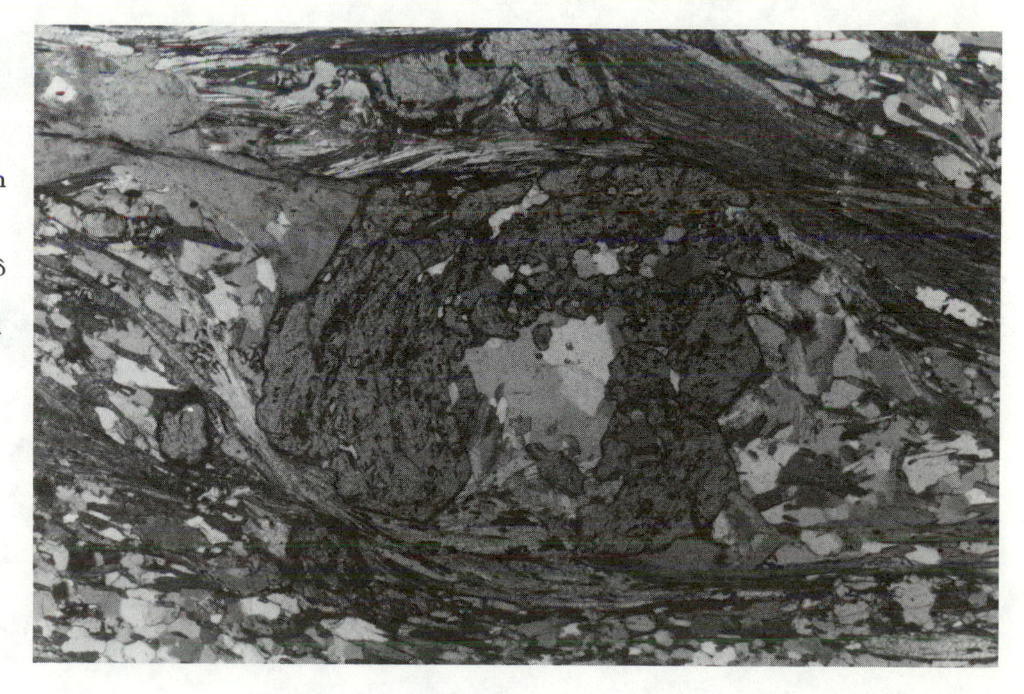

FIGURE 18–7
Garnet in metasiltstone of lower Paleozoic(?) Poor Mountain Formation near Salem, South Carolina, containing concentric inclusion trails in a porphyroblast that has undergone sinistral rotation after formation (as a δ porphyroclast)—indicated by pressure shadows of micas and quartz. Plane light. Garnet is approximately 2.5 mm in diameter. (RDH photo.)

FIGURE 18–8
Rodding of a quartz-feldspar layer in migmatitic Upper Proterozoic Tallulah Falls Formation metagraywacke at Woodall Shoals, South Carolina–Georgia. (RDH photo.)

Boudinage consists of lenticular segments of a layer that has been pulled apart and flattened in such a way that the layer is segmented (Figure 18–11). The layer being segmented is less ductile than the enclosing material and the degree of contrast in competence affects the shape of boudins: large contrast produces boudins with sharp edges; small contrast, boudins with rounded edges. Boudins can develop under conditions either ductile or brittle. Under brittle conditions, most boudins are angular and the space between them is filled with less-competent rock or fibrous minerals. ***Ordinary boudinage*** consists of segmented, sausage-shaped pieces of a single layer in which the lenticular segments parallel one another. It results from the layer being extended in a single direction. If layer-parallel extension has occurred in two directions, the resulting boudinage consists of a series of three-dimensional blocks called ***chocolate-block*** (or ***chocolate tablet***) ***boudinage.*** Boudins may be the

FIGURE 18–9
Natural strain ellipsoids lineation—deformed quartz pebbles—in Upper Proterozoic Bygdin conglomerate, Bygdin, Norway. Note that the length-to-maximum-width ratios of the pebbles are 10:1 or greater. (RDH photo.)

(a)

(b)

(c)

FIGURE 18–10
Mullions. (a) Cleavage-bedding-related mullions on a corrugated bedding surface in Cambrian shale near Bouznika, Morocco. Mullions are formed by flattening perpendicular to cleavage (perpendicular to mullions) and parallel to bedding. (RDH photo.) (b) Ductility (viscosity) contrast-related mullions without cleavage along a quartz-feldspar (light-colored)–mica-quartz (dark) boundary in mylonite from the Towaliga fault zone from near Indian Springs, Georgia. Specimen is approximately 10 cm long. (Collected by Robert J. Hooper, University of South Florida.) (c) Ductility contrast-related mullions along a boundary between amphibolite (dark) and marble (light) in lower Paleozoic(?) Gaffney Marble near Blacksburg, South Carolina. (Photo by D. G. McClanahan.)

most useful of all lineations because they are easiest to interpret: they yield information about strain, shear sense, and differences in competence.

If flattening is accompanied by simple shear after the blocks have formed, or if the extended layer is not perpendicular to the direction of shortening (Figure 18–12), boudins commonly become rotated. Boudinage may be related to the finite- and incremental-strain ellipsoids. It forms parallel to the *XY* plane, normal to *Z*. Boudinage is frequently seen in the limbs of folds, where most flattening and layer-parallel extension occurs. The "neck line" of the boudin is the lineation, commonly parallel to fold axes (Figure 18–13).

LINEATIONS AS SHEAR-SENSE INDICATORS

Several of the lineations just described may be used as shear-sense indicators. Slickensides directly indicate movement sense by the direction of their lines and steps. Boudins indicate the extension direction. Mineral lineations yield sense of shear if the linear mineral is segmented in the movement direction. Rotated minerals are shear-sense indicators: the direction of movement is normal to the lineation (rotation axis).

FOLDS AND LINEATIONS

Lineations are directly related to folds and can be used to help decipher a folded area (Figure 18–14). Flexural-slip folds frequently produce slickenlines on bedding surfaces (Figure 18–2). These lineations are commonly oriented normal to the fold axis.

Intersection lineations tend to parallel fold axes because of the nature of the lineation—the intersection of an axial planar cleavage (or foliation) with bedding or earlier compositional layering. Mullions

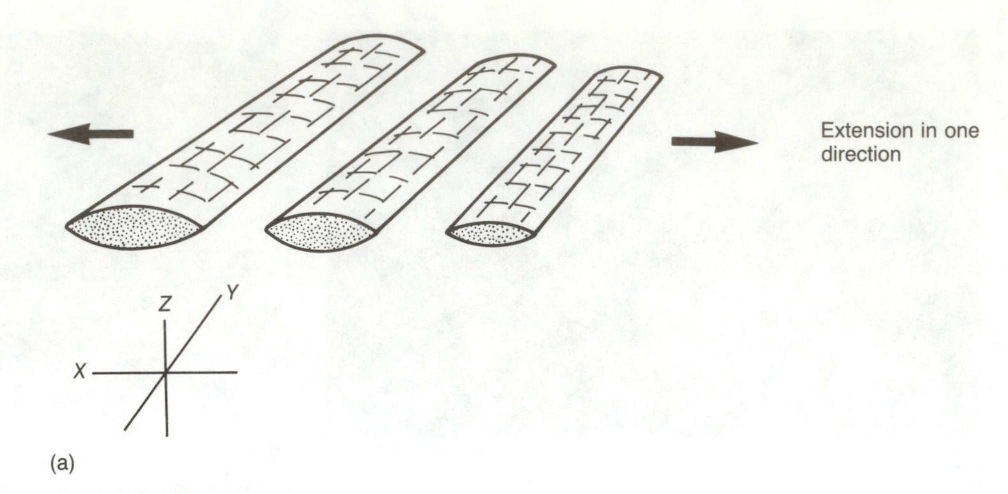

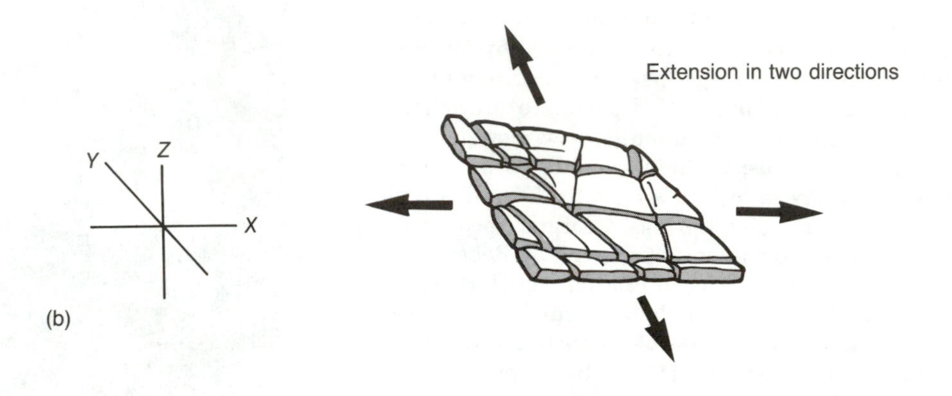

FIGURE 18–11
Boudinage. (a) Ordinary boudinage involving plane strain ($X > Y = 1 > Z$), and segmentation of a layer. (b) Chocolate-block boudinage involving $X > Y > Z$ or flattening with $X = Y > Z$.

FIGURE 18–12
Boudinage of competent sandy layers in less-competent limestone in the Berge limestone near Ange, Sweden. (RDH photo.)

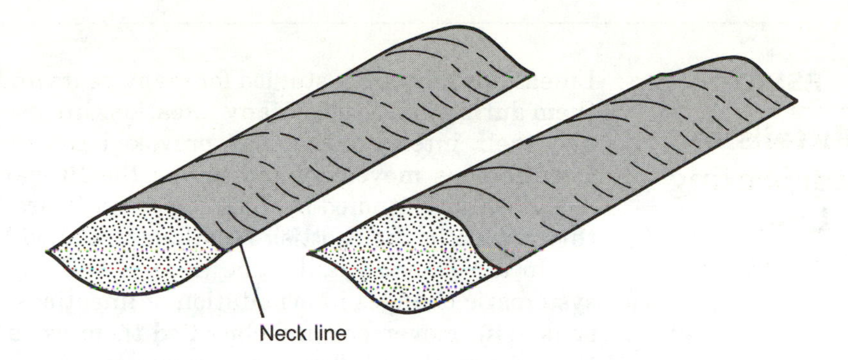

FIGURE 18–13
The actual lineation in boudins is the neck line.

and boudins generally parallel the fold axes. Intersections, mullions, and boudins all carry information about fold orientation. Mineral-elongation lineations sometimes parallel fold axes and are sometimes oriented oblique to normal to fold axes. Where such lineations parallel fold axes, they also indicate fold orientation in the area.

DEFORMED LINEATIONS

Once formed, lineations become strain markers that can be used to decipher later deformation. This deformation yields results both surprising and revealing. We might think that folding of linear structures would be easy to analyze and explain, but most plots of folded lineations in fabric diagrams yield complex patterns (Figure 18–15). If lineations are gently folded by flexural-slip buckling or bending so that the fold axial trend is normal to the lineations, they will plot on a great circle of the *equal-angle net.* If the fold axis is not perpendicular to the trend of the lineation, the deformed lines plot as small circles, spirals, or complex patterns difficult to interpret. A lineation *not* normal to the fold axis yields a small circle on the equal-area net for flexural-slip folding. Mineral lineations should be the only ones used for this kind of analysis because a cleavage superimposed on a surface already folded will produce an apparently folded intersection lineation—one actually *not* deformed. A cleavage superimposed on ripple marks would likewise produce a lineation deformed only apparently.

The reasons for this apparent complexity of folded lineations are many. One is that standard plotting nets cannot accommodate features that have been deformed into noncircular patterns. Most folded lineations are deformed into helixes, parabolas, or hyperbolas that may not be immediately evident in the field. Another is that different fold mechanisms deform lines differently. Small-circle patterns suggest a buckling or bending mechanism; great-circle patterns indicate passive flow. Whether

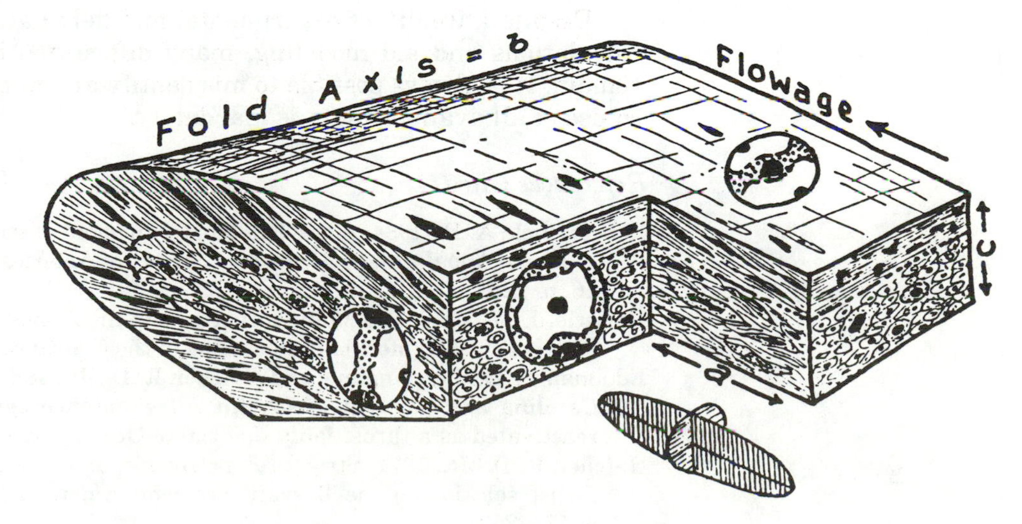

FIGURE 18–14
Relationships between a fold, intersection, oöid, and mineral lineations, and preferred orientations parallel to different planes in the strain ellipsoid. Fabric diagrams show orientation of dominant structures in different parts of the fold. (After Ernst Cloos, 1957, Geological Society of America Memoir 18.)

ESSAY

Pitfalls in Interpreting Linear Structures

Lineations have been studied for many years and geologists routinely measure them during field study. Many lineations are associated with large fault zones, and their interpretation has provoked controversy. Two examples involve lineations as movement indicators: the Brevard fault zone in the southern Appalachians, studied by Jack Reed and Bruce Bryant (1964), and the Moine thrust zone in the Scottish Highlands, studied by John Christie (1963).

In the Brevard fault zone (Figure 18E–1), Reed and Bryant interpreted systematic changes in orientation of lineations as indicating shear sense and strike-slip movement and classified them as "a"-lineations (indicating direction of movement). The shallow-plunging—northeast-to-southwest—mineral lineations (mostly quartz) were taken to indicate strike-slip. They also identified lineations trending to the northwest in the Blue Ridge northwest of the fault zone and interpreted them as a change of orientation of the same lineations into a northwest transport direction. But the northwest-trending lineations in the Blue Ridge parallel fold axes and are not obviously mineral-elongation lineations. They probably result from intersection of two foliations. Independent evidence for dip-slip along the fault zone in South Carolina, based on geologic map patterns, critical rock units, and parallelism of cylindrical folds with the strong northeast-trending mineral lineation (Hatcher, 1971, 1978), cast doubt on Reed and Bryant's interpretation of strike-slip. Thereafter the Brevard fault zone was presumed to be a large ductile dip-slip fault until Andy Bobyarchick (1984) used S-C structures to indicate a major component of dextral motion. Since then, Steven Edelman, Angang Liu, and I (1987) have suggested that both dip-slip and strike-slip occurred in the Brevard fault zone and that the strong quartz lineation is associated with the strike-slip events—as Reed and Bryant (1964) originally concluded. As we now understand it, the Brevard fault zone underwent early ductile dip-slip, then ductile strike-slip, and finally brittle dip-slip.

A similar problem existed in the Moine thrust zone. First, H. H. Read (1931) interpreted some of the linear structures as slickensides, indicating a component of strike-slip motion along the fault. Then John Christie (1963) concluded that the lineations were parallel to fold axes ("b"-lineations) and the transport direction 90° from that in Read's interpretation. Later still, Michael Johnson (1964) questioned Christie's interpretation of lineations and the basis for his conclusions. The debate about interpretation of linear structures is not finished.

Despite a wealth of experimental and field data, and computers to speed calculations and aid modeling, many difficulties in interpreting lineations remain. It is always possible to misidentify and misinterpret a structure, and so use of all available data is mandatory.

References Cited

Bobyarchick, A. R., 1984, A late Paleozoic component of strike-slip in the Brevard zone, southern Appalachians: Geological Society of America Abstracts with Programs, v. 16, p. 126.

Christie, J. M., 1963, The Moine thrust zone in the Assynt region, Northwest Scotland: University of California Publications in the Geological Sciences, v. 40, p. 335–440.

Edelman, S. H., Liu, Angang, and Hatcher, R. D., Jr., 1987, The Brevard zone in South Carolina and adjacent areas: An Alleghanian orogen-scale dextral shear zone reactivated as a thrust fault: Journal of Geology, v. 95, p. 793–806.

Hatcher, R. D., Jr., 1971, Structural, petrologic, and stratigraphic evidence favoring a thrust solution to the Brevard problem: American Journal of Science, v. 270, p. 177–202.

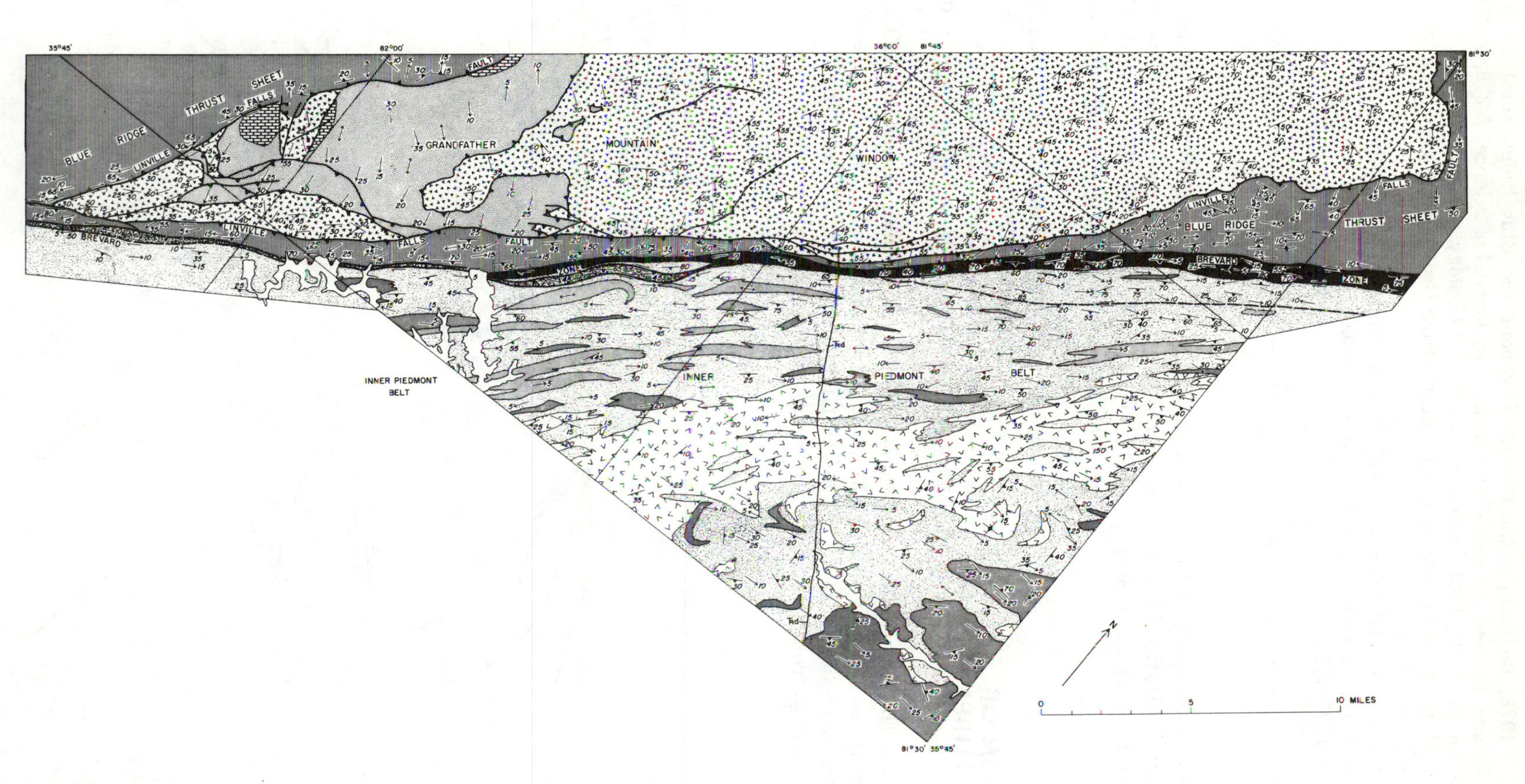

FIGURE 18E–1
Map showing trend and plunge of lineations (arrows) in the Inner Piedmont (irregular stipple, fine stipple, and large random rooftop patterns), Brevard fault zone (dark stipple), Blue Ridge (fine line pattern), and Grandfather Mountain window (small random rooftop, stipple, and limestone patterns), North Carolina. Faults are indicated by heavy toothed lines and heavy dashed lines. Heavy northwest-trending solid line in southeast part of map is a Jurassic diabase dike that cuts all earlier structures. (From Reed, J. C., Jr., and Bryant, Bruce, 1964, Geological Society of America *Bulletin,* v. 75, p. 1177–1195.)

_____ 1978, Tectonics of the western Piedmont and Blue Ridge, southern Appalachians: Review and speculation: American Journal of Science, v. 278, p. 276–304.
Johnson, M. R. W., 1964, Discussion: Journal of Geology, v. 72, p. 672–676.
Read, H. H., 1931, Geology of central Sutherland: Geological Survey of Great Britain Memoir, 238 p.
Reed, J. C., Jr., and Bryant, Bruce, 1964, Evidence for strike-slip faulting along the Brevard fault zone in North Carolina: Geological Society of America Bulletin, v. 75, p. 1177–1195.

the folds are tight or open is also significant. John Ramsay (1967) and Ramsay and Martin Huber (1987) have discussed the geometry of deformed lineations in greater detail.

INTERPRETATION OF LINEAR STRUCTURES

Linear structures may help reveal the deformation sequence of an area. Consequently, measuring and interpreting linear structures and associating them with other structures in the field is extremely important in deciphering their history. Generally, it is not hard to recognize mineral lineations, deformed pebbles, rods, boudins, mullions, and cleavage-bedding intersections. Interpreting most lineations is more difficult.

In the past some geologists have assumed that mineral alignment indicated direction of tectonic transport. It is true that quartz *c*-axes frequently parallel the quartz mineral lineations in deformed rocks (Figure 18–16), particularly in large fault zones (McIntyre, 1950; Christie, 1963, 1964; M. R. W. Johnson, 1960, 1964; Reed and Bryant, 1964). Thus many structural geologists assume that all mineral lineations yield transport direction, but

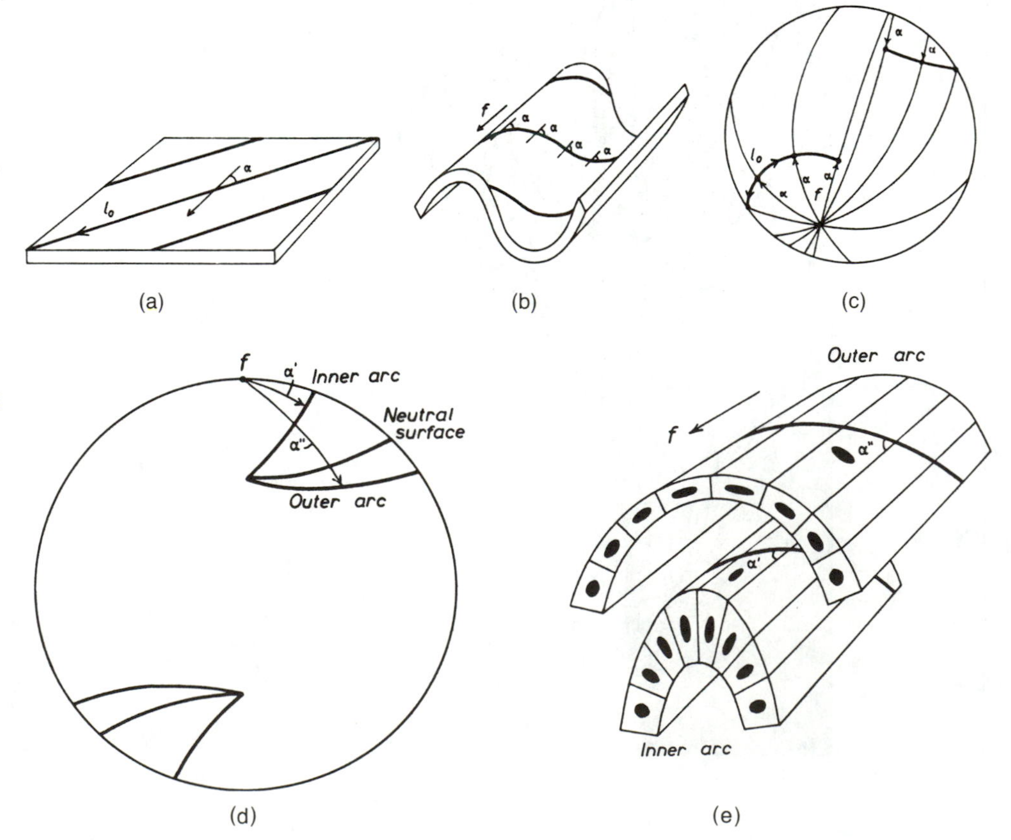

FIGURE 18–15
Deformed linear structures showing fabric diagrams and different potential orientations of lineations and fold axes. Folding layers (a) containing l_0 by flexural slip (b) results in l_0 plotting as the heavy line in the fabric diagram (c). α represents successive measurements of the orientation of layering in (b) plotted as a great circle in (c); (d) is a fabric diagram showing a plot of lineations folded by tangential longitudinal strain in (e). (From J. G. Ramsay, 1967, *Folding and Fracturing of Rocks:* New York: McGraw-Hill. © 1967. Reproduced with permission.)

FIGURE 18–16
Parallelism of quartz-feldspar mineral lineation and fold axes in feldspathic quartzite of the Sårv nappe along Bøfjorden at Staknes Point Lighthouse, Birlandet, south-central Norway. (RDH photo.)

field studies indicate a strong parallelism between quartz mineral elongation lineations and fold axes; thus a paradox. The quartz mineral lineation appears to be both an intersection involving an axial-plane foliation and an elongation lineation. In some cases, where independent evidence exists (such as boudins), quartz mineral lineations *do* indicate transport direction; in others they do not.

Many boudin and mullion axes parallel fold axes, showing direct relationship between folding and formation of boudinage and mullions. Rodding may result from stretching parallel to a major transport direction, especially in ductile shear zones, but rods also frequently parallel fold axes. The difficulty of interpreting mineral lineations, rods, and other linear structures may indicate that they developed by small-scale stretching parallel to fold axes during flattening or by simple shear directed normal to fold axes. Another possibility is that mineral lineations developed parallel both to fold axes and to transport directions in zones of inhomogeneous simple shear where sheath folds are forming, rotating fold axes into the transport direction. In many places, fold axes and mineral lineations (mostly quartz or micas) are parallel and no evidence has been found for reorientation of fold axes by any mechanism.

Many lineations parallel fold axes; others form high angles to fold axes. Interpreting those mineral lineations parallel to fold axes as elongation (stretching) lineations is difficult without independent evidence of movement sense.

Now that we have examined each major type of geologic structure, we can proceed to Chapter 19. There we will learn how to decipher combinations of structures in regions of varying complexity.

Questions

1. How does a lineation differ from a lineament?
2. What are intersection lineations? How do they originate?
3. Under what conditions can slickenlines be penetrative structures?
4. If all intersections, mineral lineations, and boudin axes are parallel to each other and to the dominant fold trend in an area, what can you say about their origin?
5. The data in the following table were collected from a set of folds that folded a mineral lineation (axial orientation 024, 3° NE, axial surface 034, 62° NW) in the Walloomsac phyllite just east of the Taconic allochthons near Hoosick Falls, New York (Figures 15–14c and 17–12a). Plot these data on an equal-area net and comment on the nature of the folding process, the effects of the orientation of the lineation and the fold axis, and the pattern observed in the equal-area plot.

119 29° SE	108 39° SE	144 54° SE	131 61° SE
106 36° SE	112 35° SE	136 63° SE	123 69° NW
102 36° SE	38 57° SE	138 44° SE	111 34° SE
108 54° NW	156 46° SE	146 36° SE	127 46° SE
112 52° NW	134 49° SE	123 54° SE	118 72° NW

6. How do boudins form? Mullions?
7. Why do pressure shadows form? Could they form in the same stress/strain environment as boudins? If so, how?
8. Identify the lineations in the fold below.

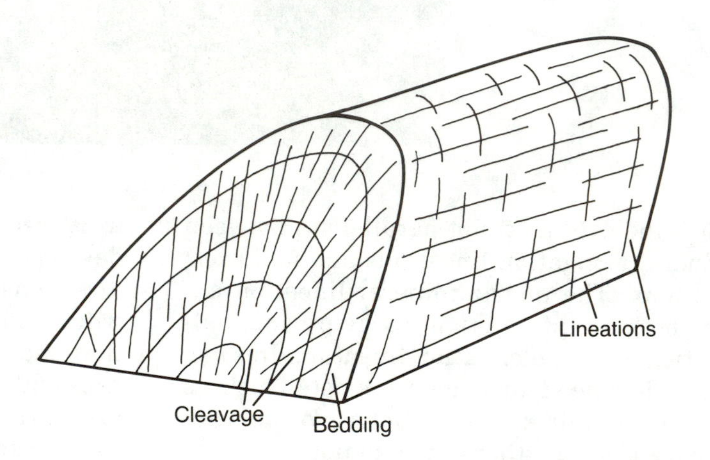

9. Sketch in the strain axes in the rectangular prisms, which represent boudinage. Was the contrast of competence during deformation large or small?

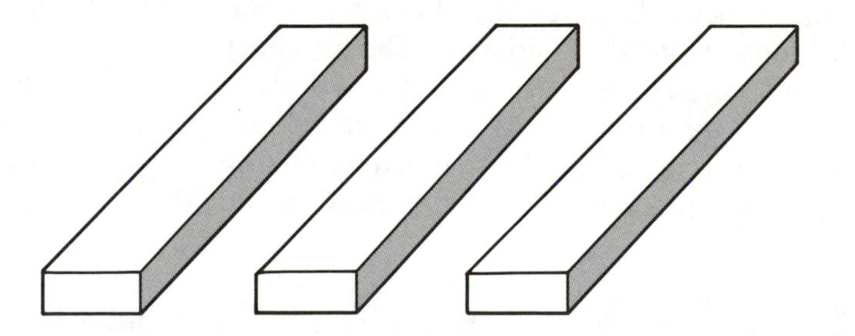

Further Reading

Cloos, Ernst, 1957, Lineation: A critical review and annotated bibliography: Geological Society of America Memoir 18, 136 p.
Comprehensively reviews lineations and discusses various kinds and interpretations of linear structures.

Wilson, Gilbert, and Cosgrove, J. W., 1982, Introduction to small-scale geological structures: London, Allen & Unwin, 128 p.
A readable compendium of mesoscopic structures, considered singly and in combination. Several sections concern linear structures.

19

Structural Analysis

On this profile I platted all the data of structure at every outcrop of rock. I realized then how valuable to me had been the close attention I had given to minute details of structure in my youthful studies in the mountains of Corsica, for this had led me to infer that the structure in a specimen might repeat in miniature that of the great rock masses from which the great rock masses had come.

RAPHAEL PUMPELLY, *My Reminiscences*

STRUCTURAL ANALYSIS CONSISTS OF DESCRIBING AND INTERPRETING all the structures in an area, on all scales (Figure 19–1). The technique complements the standard techniques of field geology—geologic mapping and stratigraphic analysis—and should be incorporated in any investigation of any deformed region. ***Structural analysis*** involves study of any of the structures described earlier in this book and provides a systematic framework for relating them to each other and to time. That is to say, it involves the observation, description, analysis, and interpretation of the kinds and orientations of folds and other linear structures, foliations and other planar structures, strain and displacement indicators, and faults.*

As first developed, structural analysis was a technique for geometric and kinematic analysis of the minerals in a penetratively deformed rock mass. The technique was developed for the microscopic scale, and then attempts were made to relate microscopic fabrics to the large-scale deformation plan of the rock mass.

Much structural analysis in the early twentieth century was developed by Bruno Sander, an Austrian structural geologist and metamorphic petrologist. His great work (1930) *Gefügekunde der Gesteine* (the English title is *An Introduction to the Study of the Fabrics of Geological Bodies*) was the basis for structural analysis and summarized advances that had been made in Germany. Later, Sander's concept of structural analysis was expanded to include all aspects of the deformational history of rock bodies. Further developments were reviewed by E. G. Knopf and E. Ingerson (1938) in *Structural Petrology;* their book enabled English-speaking geologists to apply the technique without extensive familiarity with the European literature. Then Francis Turner and Lionel Weiss (1963) published *Structural Analysis of Metamorphic Tectonites,* summarizing structural analysis on all scales. During the 1950s, knowledge of multiple deformation, and particularly of mesoscopic and macroscopic polyphase folding (Chapter 16) and their relationships to foliations, lineations, and other fabric elements, was enhanced by British geologists, including Basil King, Robert Shackleton, John Ramsay, Michael Fleuty, Lionel Weiss, Donald McIntyre, Nicholas Rast, Derek Flinn, and Brian Sturt. Their work, which dealt with specific areas in Scotland and pioneered modern techniques of geometric analysis, spawned many structural analyses in complexly deformed regions all over the world.

Microscopic structural studies require oriented thin sections, petrographic microscope, and universal stage, and so they are much slower than mesoscopic studies. But they are the principal means of

*Techniques for measurement and plotting structural data are outlined in Appendices 1 and 2.

FIGURE 19–1
Strongly deformed and multiply transposed biotite gneiss in the Thor-Odin dome, Shuswap metamorphic complex, British Columbia. (RDH photo.)

determining the orientations of crystallographic axes and other microscopic structures and provide a way to determine shear sense (Chapter 10) by measuring quartz *c*-axes and glide planes (Simpson and Schmid, 1983). Microtextural studies, preferably with oriented thin sections, provide useful information about vergence, chronology of deformation, and relative times of deformation and metamorphism.

Here in Chapter 19, we will emphasize mesoscopic analysis, where measurements are made in the field with a compass and clinometer (Appendix 2). The discussion focuses on a scheme for analyzing deformed areas, whether simple or complex. In any study, slightly different techniques may be used to obtain data for deciphering the history. Regardless of complexity, structural analysis should begin with the most fundamental of geologic data—the geologic map. If a detailed map does not exist, one must be made, and existing maps must be revised after gathering new data. Initial mapping and mesoscopic structural analysis of an area frequently lead to other and more specialized studies, but making all kinds of studies at one time may not be possible, or desirable.

DEFINITIONS

The term ***fabric*** encompasses the relationships between texture and structure in a rock. It relates the minerals, their orientations, and relative size to the structures in the rock.

The term ***tectonite*** is used in describing any rock containing a prominent fabric element that is penetrative and tectonic (Figure 19–2). The element may be either linear or planar. *S-tectonites* are dominated by foliation or cleavage; *L-tectonites,* by linear elements; *B-tectonites,* by fold axes. Combinations of linear and planar elements are possible—*LS-tectonites* if linear elements dominate and *SL-tectonites* if planar elements dominate. This terminology is falling into disuse, but it conveys useful information and appears in recent literature.

RESOLVING STRUCTURES IN MULTIPLY DEFORMED ROCKS

Most structural-analysis techniques were devised in complexly deformed areas—in Phanerozoic orogens or in the exposed roots of Precambrian mountain chains of the continental shields—where they may still be best applied. Here we will outline the general deformation plan for an orogen, explore difficulties of implementing mesoscopic analysis, and examine the problems of recognizing separate structures. In all cases the goal is to resolve the history of multiply transposed rocks, with ready application to less deformed rocks.

Deformation Plan in the Core of an Orogen

Mountain belts have many similarities in spatial arrangement of structures, sequence of development, and properties. All are repeated in orogenic belts whenever formed, from the Precambrian to the Tertiary (Figure 19–3). A ***foreland fold-and-thrust belt*** consists of a former trailing continental margin (formed at the beginning of a Wilson cycle; see Chapters 1 and 11) that was altered by thin-skinned deformation into a series of rootless folds and thrust faults directed toward the continent and away from the deforming orogen. Foreland fold-and-thrust belts vary greatly in width—even in the same orogenic belt. They have been found in most orogens, including the Alps, the Andes, the Appalachians, the British and Scandinavian Caledonides, the North American Cordillera, Taiwan, and the Proterozoic Wopmay orogen of Canada. Inside the foreland fold-and-thrust belt of most orogens is the ***metamorphic core,*** consisting of folded, faulted, cleaved, foliated, and intruded rocks ranging in metamorphic grade from low to high (Figure 11–3). The metamorphic core overlaps the foreland fold-and-thrust belt and may overlap any ***accreted terrane*** that has been sutured to the deforming mass.

Generally, the earliest recognizable structures in the metamorphic cores are produced by penetrative ductile deformation, which yields several generations of ductile faults (both dip-slip and strike-slip), folds, foliations, and linear structures. Later, crenulation cleavage may deform the early ductile structures, and then the crenulations may be overprinted by open flexural-slip buckle folds. Development of ductile faults, folds, tectonic slides (Chapter 11), and ductile-shear zones accompanies the early phase, which may then be overprinted by brittle structures. Deformation in the orogenic core is generally earlier than that in the foreland. Deformation that formed the early ductile structures in the inner parts of the orogen propagates outward to the outer parts and ultimately forms brittle structures. Later structures—open flexural-slip folds and brittle faults—primarily affect the foreland fold-and-thrust belt in its outer parts, but may develop throughout the entire orogen. Ductile deformation may occur in the core during the late stage and propagate to the outer zones as brittle structures.

Complex deformation in orogens requires an analytical scheme to resolve the succession of multiple events. One commonly used scheme begins with fundamentals: cross-cutting and overprinting relationships (Chapter 1). For example, if a foliation is folded, the foliation must have existed before the deformation that folded it; the radiometric age of a singly foliated pluton that cuts a sequence of multiply foliated metasedimentary rocks provides age constraints for both pre- and post-intrusion foliations.

Pitfalls in Using Style and Orientation in Polyphase-Deformed Rocks

Style and orientation of structures are frequently correlated to resolve the sequence of complex deformation. Differences in style and orientation generally show that the structures formed at different times, but they may not; similarities in style and orientation do not necessarily mean that the structures are contemporaneous. Obviously, the best means of deciphering the structure of an area is to find one place where the entire deformational sequence is preserved—a Rosetta stone like that discussed in Chapter 16. Such a place reveals overprinting and cross-cutting relationships of superimposed structures without forcing the observer to correlate structures from one small exposure to another. It permits resolution of representative styles and orientations of structures of different generations so that—where exposure is less complete—the observer can confidently separate individual structures and arrange them chronologically. Any area of 100 km^2 or more should contain at least one exposure revealing most mesocopic structures. Particular structural elements may be absent because of the rock type or other factors. At best, correlating particular events, and even resolving the questions of style, may still prove difficult.

In a critique of reasoning processes used in structural analysis, Paul Williams (1970) demonstrated that fold style or orientation may not be confined to one particular generation of folds. Williams's example was the well-exposed coastal section at Bermagui, New South Wales, where he concluded that fold generations in a complexly deformed area can be resolved only by use of overprinting relationships. Thus, an exposure where overprinting relationships may be observed is necessary to resolve the structural chronology and relationships between deformational episodes.

Generally, where contrasting behavior is observed, as in a ductile structure overprinted by a brittle structure, it can usually be assumed that enough time elapsed after development of the ductile structure that the rock mass cooled before overprinting by brittle deformation. An exception occurs where an abrupt increase in strain rate produces a brittle structure that otherwise would deform ductilely at a slower rate.

(a) (b) (c) (d) (e)

FIGURE 19–2
Different types of tectonites. (a) S-tectonite dominated by a single planar fabric; Middle Proterozoic Toxaway Gneiss near Whitewater Falls on the North Carolina–South Carolina border. (b) L-tectonite—an exposure dominated by mineral and crenulation lineations in the Middle Proterozoic Toxaway Gneiss at Whitewater Falls, North Carolina–South Carolina. The curved outline of the surface of the exposure is the enveloping surface of the smaller crenulation folds that define a larger fold. (c) B-tectonite—strongly crenulated amphibolite near Trondheim, west-central Norway. (d) LS-tectonite—quartzite layer dominated by linear fabrics but containing a strong foliation; near Burrells Ford, Oconee County, South Carolina. (e) SL-tectonite—Lower Cambrian Chilhowee quartzite at Linville Falls, North Carolina, dominated by a foliation but containing two lineations (an intersection and a mineral stretching lineation at 90°). (RDH photos.)

Problems of Multiply Transposed Rocks

In analyzing mesoscopic fabrics and working out the structural history of an area containing polydeformed rocks and rocks of moderate to high metamorphic grade, identification of the earliest structures may be difficult or impossible because of subsequent transposition. In such cases, the rock mass must remain sufficiently ductile that penetrative deformation can overprint and mask earlier structures. Such earlier structures are sometimes preserved in boudins formed from more competent layers—amphibolite, quartzite, or calcsilicate—in the mass of more ductile rock (Figure 19–4). Structures preserved in boudins may represent only an earlier pulse of a major deformation phase, and not a phase distinctly different. But formation during another phase of deformation is also possible. Multiple transposition may be deduced from intrafolial folds and microlithons containing earlier foliation.

MESOSCOPIC ANALYSIS

Now we will discuss terminology and fabric elements useful in conducting mesoscopic analysis, which involves resolution of the spatial-temporal relationships among mesoscopic structures. Fabric data may be plotted on maps, equal-area (Schmid) diagrams, and equal-angle (Wulff) diagrams using techniques described in Appendix 1. In any study area, all fabric elements should be examined for geometric style, orientation, and overprinting relationships.

Few areas are uniform. In one place some fabric elements—folds and other lineations, fold axial surfaces, and cleavage or foliation—may have a common orientation, but in another part nearby, the same elements may be oriented differently. An area where one fabric element is oriented the same throughout is called a ***homogeneous domain*** (Figure 19–5). Foliation or bedding may have the same dip and strike, the same kinds of lineations may have a similar trend and plunge, and all fold axes and axial surfaces of the same generation may have similar orientations. Homogeneous domains can be most easily defined by bedding or dominant foliation data because that is the structure most commonly measured (Figure 19–6). If enough measurements are available, folds and other linear fabrics also help distinguish homogeneous domains. It may be useful to define ***heterogeneous domains,*** which help recognize fold hinge zones, multiple folding events, ductile shears, and other structures; for a practical example, see the Essay accompanying this chapter.

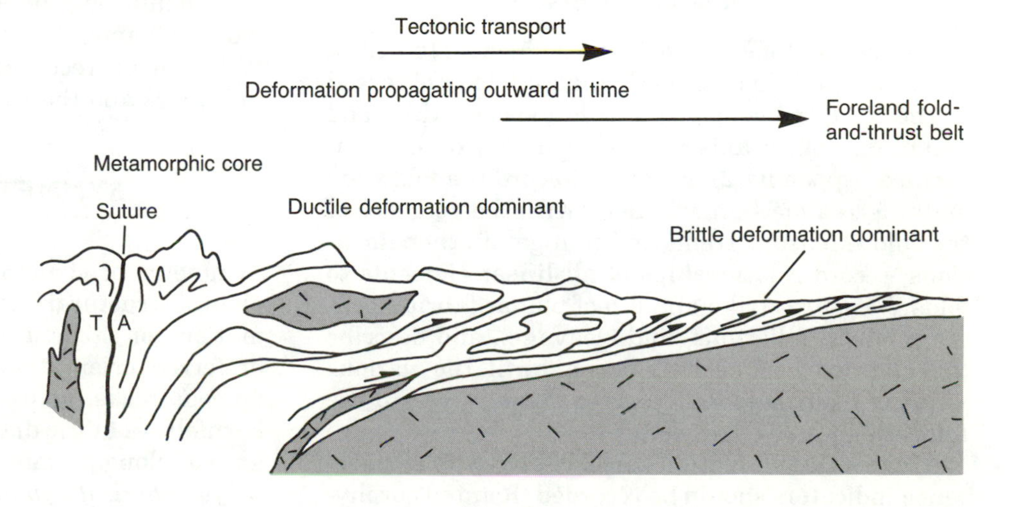

FIGURE 19–3
Sequence, style, and rheology of deformation in the inner and outer parts of an orogen.

(a)

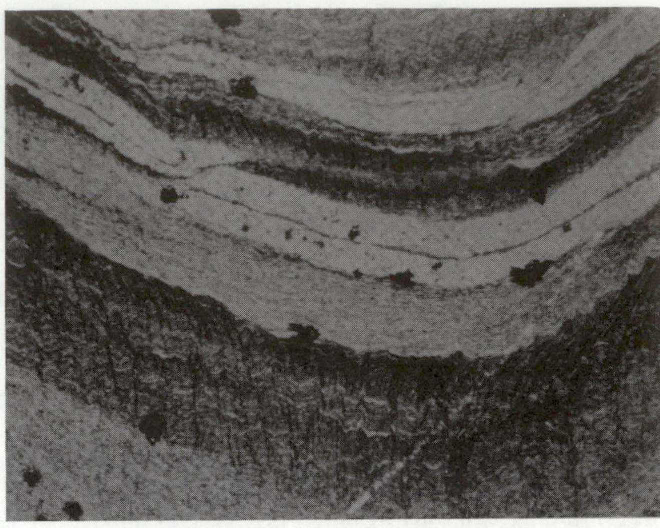

(b)

(c)

FIGURE 19–4
(a) Preservation of early folds and foliation in a boudin of amphibolite about one meter long, Trans-Canada Highway west of Revelstoke, British Columbia. The enclosing foliation truncates the foliation and folds in boudins. The structures could have formed at different times during the same event or during separate events. (RDH photo.) (b) Earlier foliation (transposed bedding) preserved as microlithons between later foliation developed as crenulation surfaces in the dark layers in Lower Devonian Seboomook Formation schist and metasandstone, Somerset County, Maine. (Arden L. Albee, U.S. Geological Survey.) (c) Earlier foliation (pseudo cross beds) preserved between later foliation in Middle Proterozoic Toxaway Gneiss near Whitewater Falls, North Carolina–South Carolina. (RDH photo.)

Structural Measurements

Folds are only one of the fabric elements studied in mesoscopic fabric analysis. When a fold is observed in the field, the geologist should measure the trend and plunge of the axis and strike and dip of the axial surface (Appendix 2). Further: Record the fold style and inferred mechanism, along with the vergence of the fold. Measure trend and plunge of other lineations; record relationships of all linear elements to folds. Measure strike and dip of planar fabrics (bedding, joints, foliations, and cleavage), and describe the character of each foliation until the various kinds of S-surfaces are resolved. Finally, record the inferred ages of all structures.

Presence and motion sense of observable shear-sense indicators should be recorded. Rotated porphyroclasts, shear bands (C-surfaces), and other indicators should help determine sense of movement on faults (Chapter 10). Rotation sense on the limbs of folds can be recorded after observing shear-sense indicators and the asymmetry of small folds.

SYMMETRY OF FABRICS

Symmetry of structures observed in the field and plots of structural data frequently yield useful information for structural analysis. In rocks with no preferred orientation of constituent grains—most igneous rocks, many sedimentary rocks, and some hornfels—a fabric diagram of orientation of the long axes of elongate minerals (such as hornblende) displays *spherical symmetry* (Figure 19–7a). A single preferred orientation—a mineral lineation—results in the symmetry of a cylinder—*axial symmetry* (Figure 19–7b). It may be produced by minerals in a strongly oriented L-tectonite.

Orthorhombic symmetry with two (or three) planes of symmetry is observed in plots of poles to bedding from symmetrical folds (Figure 19–7c). Plots of poles to two sets of joints oriented normal to each other also reveal orthorhombic symmetry. *Monoclinic symmetry* with one symmetry plane is found in asymmetric folds (Figure 19–7d). This lower symmetry is probably more common in nature than the other kinds described above. *Triclinic symmetry* may indicate multiple deformation—that is to say, more than one fabric element, such as two foliations (Figure 19–7e).

Fabric Axes

Early in the development of structural analysis, geologists found it possible to relate structure and fabrics to coordinate axes. The discussion here of fabric axes is included largely for its historical value, because today most structural geologists try to relate structures to the principal axes and planes of the strain ellipsoid, which, as we have seen, is another set of coordinate axes.

Geometric axes (Figure 19–8) describe the spatial relationships between particular structures and a set of axes *a*, *b*, and *c*. Here is one scheme, recommended by Turner and Weiss (1963) for orienting fabric axes:

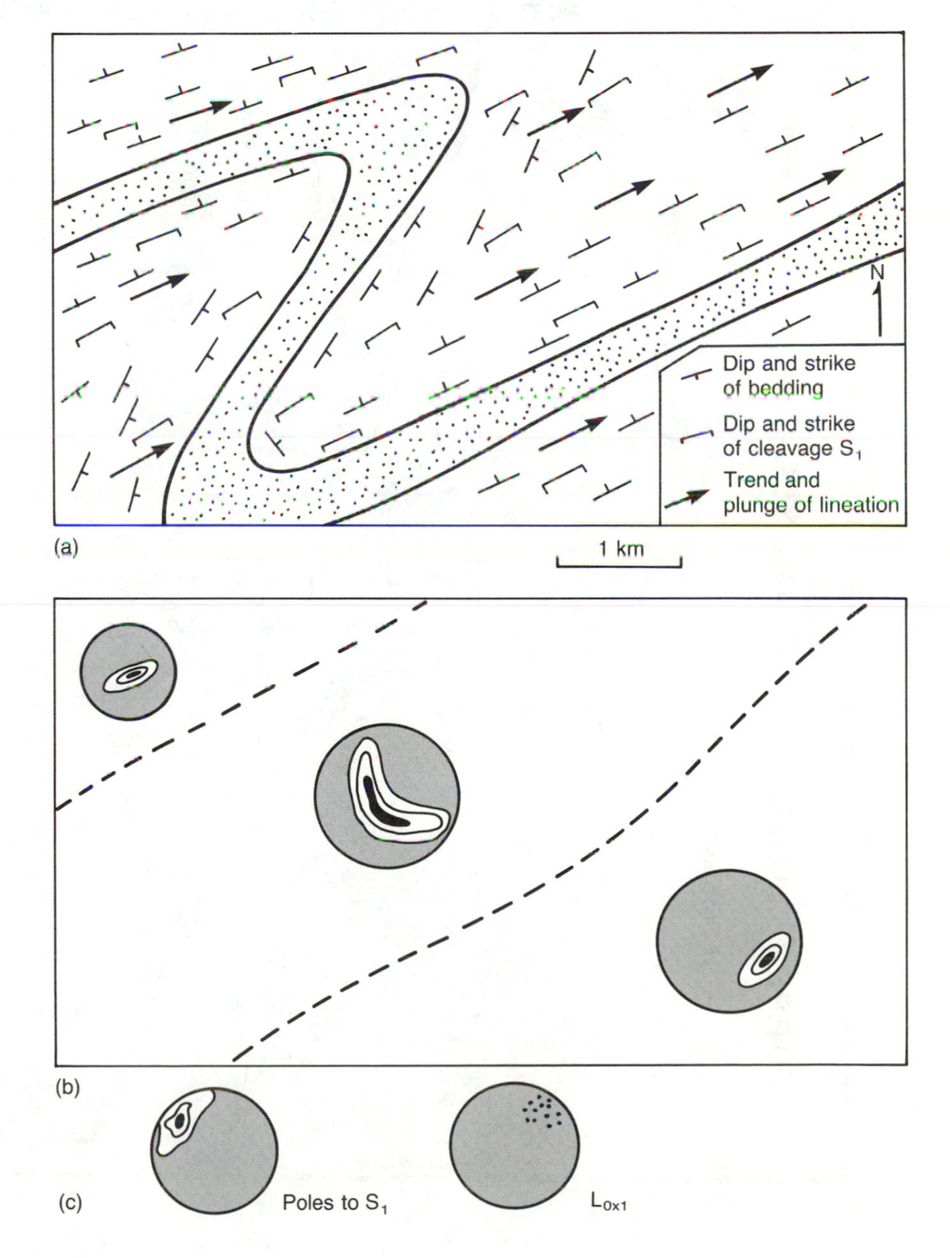

FIGURE 19–5
(a) Geologic map of a hypothetical area with structural data (bedding, S_1, $L_{0\times1}$). (b) Homogeneous domains of poles to bedding. (c) Fabric diagrams of $L_{0\times1}$ axes and poles to S_1.

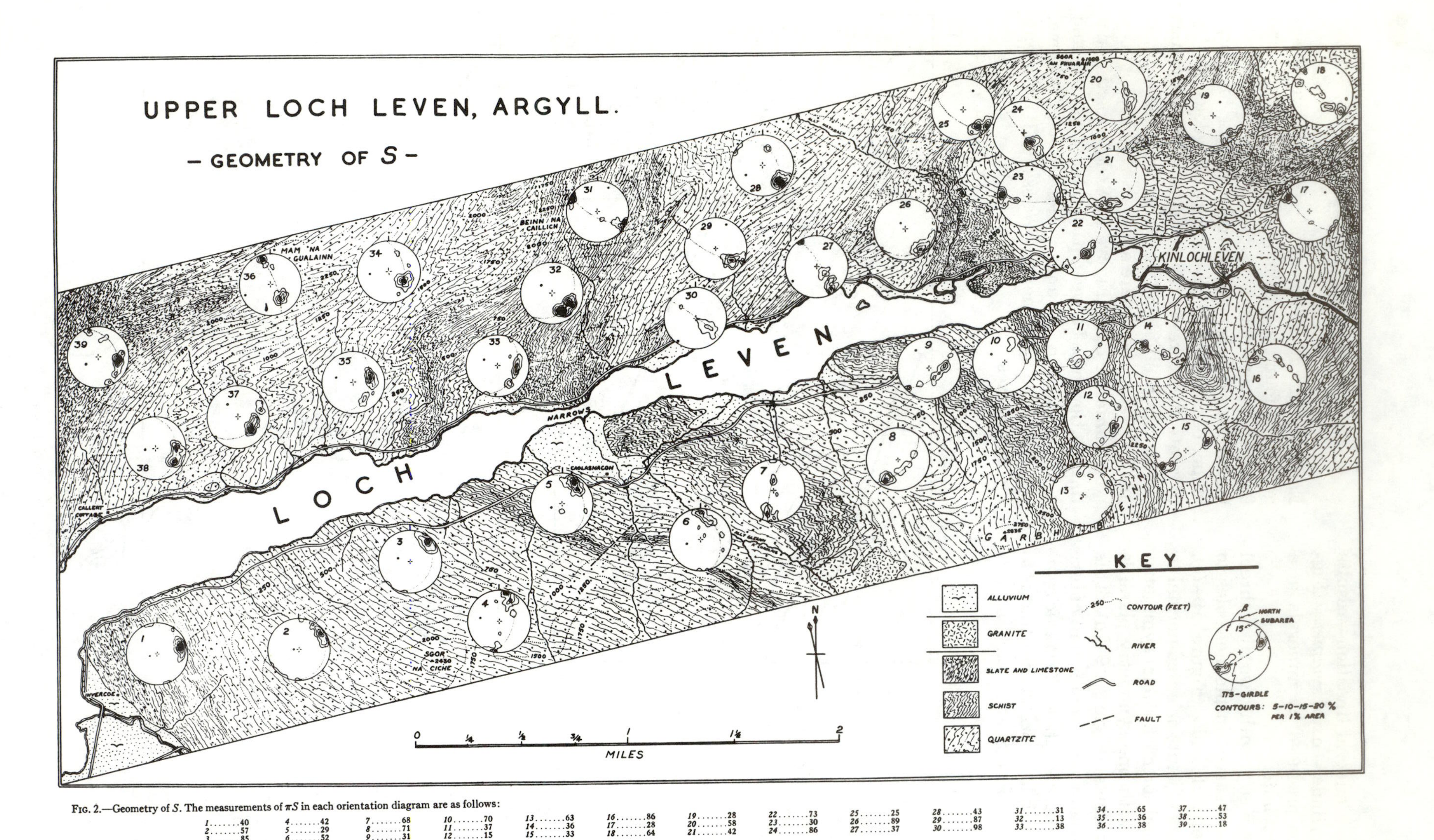

Fig. 2.—Geometry of S. The measurements of πS in each orientation diagram are as follows:

1.......40	4.......42	7.......68	10.......70	13.......63	16.......86	19.......28	22.......73	25.......25	28.......43	31.......31	34.......65	37.......47
2.......57	5.......29	8.......71	11.......37	14.......36	17.......28	20.......58	23.......30	26.......89	29.......87	32.......13	35.......36	38.......53
3.......85	6.......52	9.......31	12.......15	15.......33	18.......64	21.......42	24.......86	27.......37	30.......98	33.......38	36.......38	39.......18

FIGURE 19–6
Homogeneous domains of the dominant foliation in metasedimentary rocks from the Scottish Highlands. Numbers along bottom of diagram refer to the domains and number of measurements made in each. (From L. E. Weiss and D. B. McIntyre, 1957, Structural geometry of the Dalradian rocks at Loch Leven, Scottish Highlands, *Journal of Geology,* v. 65, p. 575–602. Published by permission of University of Chicago Press.)

1. The *ab* plane is oriented parallel to the most prominent planar structure in the rock mass, and a prominent lineation lying in *ab* is *b*. If no lineation is found, the orientation of *b* is fixed arbitrarily.
2. In a rock mass containing two or more planar structures that intersect in a common axis, the most prominent of the two planar structures is designated *ab* and the intersection *b*.
3. If more than two planar structures occur and do not intersect in a common axis, the most prominent structure is designated *ab*, and its intersection with the next most prominent structure is *b*.
4. In a rock body dominated by a strong lineation, the orientation of the lineation is *b*, and any direction normal to *b* is *a*. The *a*-axis is best oriented in a planar structure.

These rules permit relating fabric axes to fabric symmetry. In rocks dominated by linear fabrics (axial symmetry), *b* or *c* is the principal axis of symmetry. The *a*-, *b*-, and *c*-axes are normal to

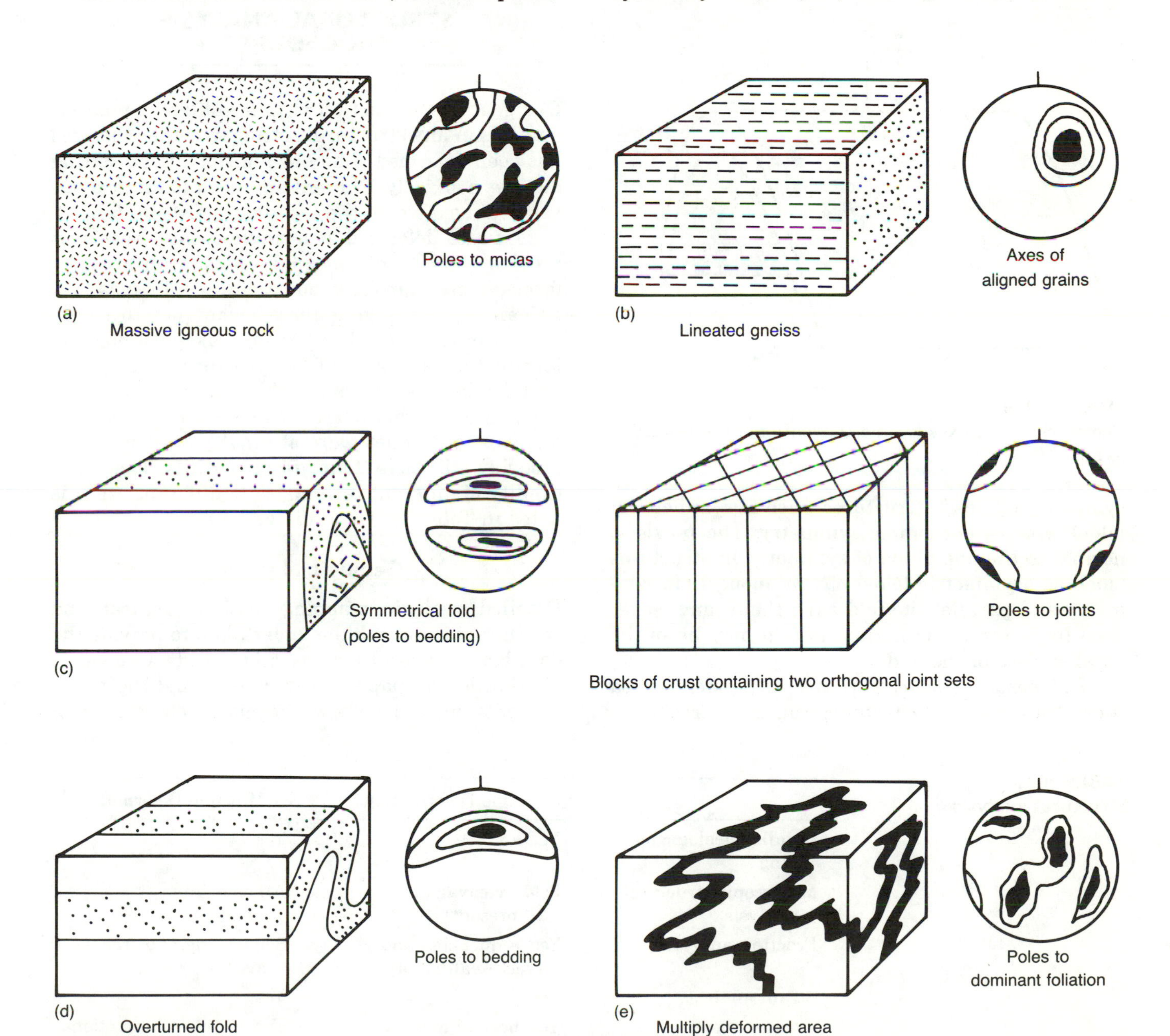

FIGURE 19–7
Fabric symmetries. (a) Spherical. (b) Axial. (c) Orthorhombic. (d) Monoclinic. (e) Triclinic. The side of each block nearest the reader faces south.

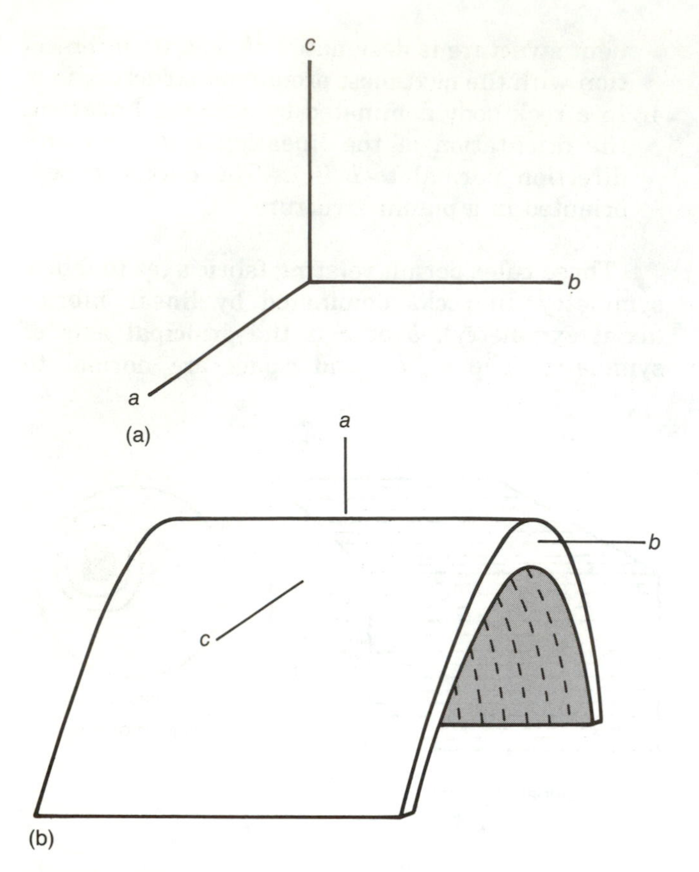

FIGURE 19–8
Geometric fabric axes (a) and conventional use related to a fold (b).

symmetry planes in structures (such as symmetrical folds) with orthorhombic symmetry. The *b*-axis is normal to the one plane of symmetry in structures (such as asymmetric folds) having monoclinic symmetry and parallel the fold axis; the *a*- and *c*-axes then lie in the plane normal to *b* (*a* may lie in the axial surface of the fold).

In *kinematic (movement) axes,* as Sander defined them, the *a*-axis is the tectonic transport direction, *b* is normal to *a* in the transport (fold axial) plane, and *c* is normal to the *ab* plane. Actually determining true movement directions in rocks with multiple overprinting is extremely difficult, and so is resolution of kinematic axes. Thus, nowadays, kinematic axes are seldom identified. Emphasis instead is on relating structures to the strain ellipsoid and determining orientations of strain axes—a more efficient way of describing structures and determining deformation paths.

STRUCTURAL-ANALYSIS PROCEDURES

Our goal in this section is to formulate a plan for analysis in any structural setting, whether the area has been deformed singly or multiply. Each setting dictates a slightly different approach to fabric analysis, but as a whole the techniques are similar.

An area deformed by a single event poses fewer problems for structural analysis than an area deformed more than once, but useful results may still be gained by applying these techniques to singly deformed areas. Applying them makes possible determination of aspects of the structural history that would otherwise be impossible. Areas affected by polyphase deformation are those where structural-analysis techniques were originally developed and probably still have their greatest utility. The applicability of different techniques to different areas is listed in Table 19–1 and discussed here.

Geologic Mapping

Detailed geologic mapping—both structural and stratigraphic—should be undertaken to provide the data base essential for any other study. Geometric distribution of map-scale structures and their constituents must be shown precisely on a map of

TABLE 19–1
Structural analysis

	Single Deformation	Multiple Deformation
Detailed geologic mapping	Yes	Yes
Mesoscopic structural analysis	Folds, cleavage (if present)	All elements
Fracture analysis	Yes, sequencing and crack-seal history	Sequencing and crack-seal history
Strain analysis	Yes	Yes
Cross sections (balanced)	Yes, normal and parallel to strike	Yes, in several directions
Microtextural studies	Yes, if several structures are present	Yes, including shear-sense indicators

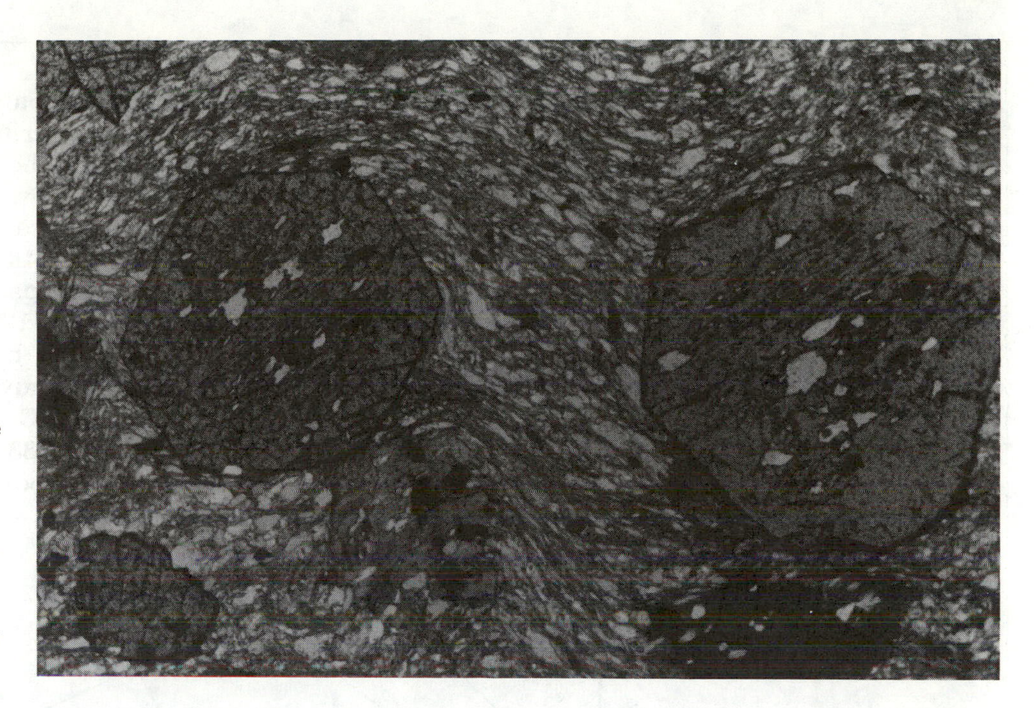

FIGURE 19–9
Earlier foliation preserved as aligned inclusions in garnet porphyroblasts, suggesting multiple deformation in a crenulated schist from the Upper Proterozoic Ammons Formation near Topton, North Carolina. The garnet has a younger rim devoid of inclusions, indicating additional growth after the event that formed the core of the crystal. Could this grain be detrital and the overgrowth developed in a new cycle of deformation and metamorphism? If so, is it still evidence of multiple deformation? Plane light. Width of field is approximately 7 mm. (Thin section courtesy of M. P. Ausburn. RDH photo.)

reasonably large scale (1/24,000, 1/12,000, 1/10,000, or larger). It may be useful to map particular exposures at scales of 1/120 or 1/60, using a grid laid out on the exposure itself and transferred to graph paper for a base map. Such mapping is the basis of all structural studies and is an absolute necessity for full understanding of the structural history, whatever the kind of analysis.

Mesoscopic Structural Analysis

Study of foliations, lineations, folds, and other mesoscopic structural elements is undertaken by first measuring them in the field, then plotting each element separately on a map overlay to display the spatial distribution. The data should also be plotted on fabric diagrams (Appendix 1) to show the relationships between structural elements of different kinds. Domain analysis should also be undertaken (Figures 19–5 and 19–6). Fold axes and other linear structures, plotted on geologic maps or overlays, will show spatial distribution, changes in plunge, and relationships to macroscopic folds.

Cleavage may or may not be uniformily present in singly deformed areas, but may occur locally in the hinges of mesoscopic folds as an incipient domainal pressure-solution cleavage, as penciling, or as a well-developed cleavage.

Microtextural Studies

Study of thin sections may reveal relationships between various fabric elements. Techniques involving cross-cutting and overprinting relationships may be applied in thin section more clearly than in the field. Early structures preserved in porphyroblasts of garnet, feldspar, staurolite, and other minerals (Figure 19–9) and in boudins may be a key to multiple deformation and thus the structural history. Many clues to shear sense (Chapter 10) are also easy to find in thin section.

Fracture Analysis

Formation of systematic joints and other fractures may be related to development of folds and faults within the area, so their measurement and analysis is worthwhile. Fracture analysis is commonly undertaken separately from standard fabric analysis: the goals of fracture analysis are frequently related to interpretation of brittle events affecting the rocks much later than those producing penetrative ductile deformation (Wise and students, 1979). Fracture patterns may be related to late folds and therefore may elucidate fold mechanisms.

Finite- and Incremental-Strain Studies

Rocks in an area deformed penetratively make possible study of finite and incremental strain (Chapter 5), using folds or other strain markers. The sequence of opening and filling of veins, and vein orientations, should be measured. Study of finite and incremental strain may begin with measurement of internal

ESSAY

Structural Analysis at Woodall Shoals

In this essay we present a problem in structural analysis and solve it. The area chosen is the Rosetta stone at Woodall Shoals, described earlier (Chapter 16) as a key to the structural history of a large area in the eastern Blue Ridge of the Carolinas and Georgia. (For background, review the Chapter 16 Essay.) Woodall Shoals is a large rapids and rock exposure of about 700 m^2 along the Chattooga River at the border of Georgia and South Carolina (Figure 19E–1). A geologic map of Woodall Shoals is published as part of another by the South Carolina Geological Survey (Hatcher and others, 1989, in press).

The Woodall Shoals exposure was mapped by geology students from Clemson University (Steven H. Poe) and the University of South Carolina (Jack P. Horkowitz, Daniel B. Wynne, Arlene C. Burns) and myself as a part-time project I coordinated from 1976 until 1983. First we marked a series of north-south and east-west grid lines on the exposure at a spacing of 5 feet.

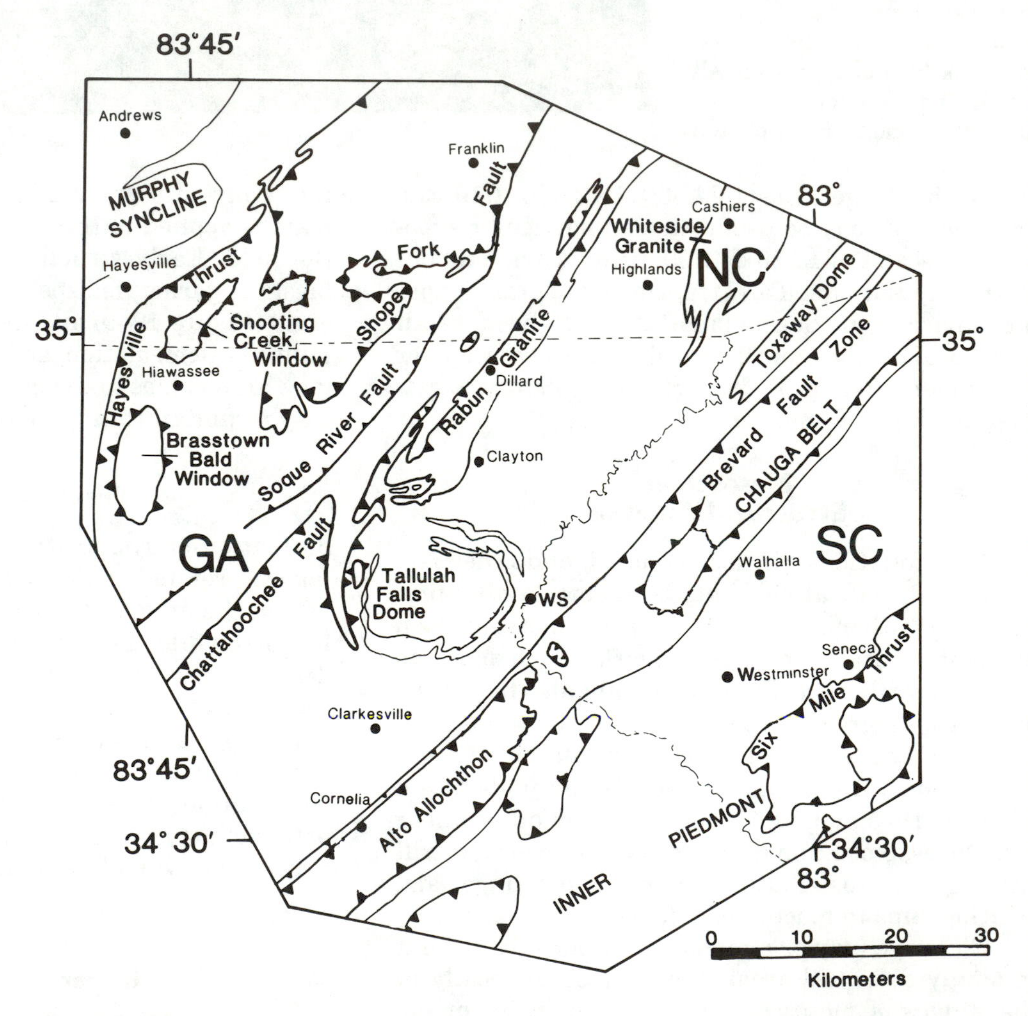

FIGURE 19E–1
Simplified tectonic map of northwestern South Carolina, northeastern Georgia, and adjacent North Carolina in the vicinity of Woodall Shoals (WS). Many of the outcrop patterns in this map are shown on much larger scale in the geologic map of Woodall Shoals (Figure 19E–2).

Then we used graph paper as a base to construct a geologic map showing the distributions of all rock units (Figure 19E–2) at a scale of 1 inch to 5 feet. We also mapped mesoscopic fabrics, joints (Figure 19E–3), and small late veins (Figure 19E–4).

We measured foliation and compositional layering in more than 850 places—enough for domain analysis of the exposure (Figure 19E–5). In addition, we identified several generations of folds and crenulations on the mesoscopic and map scales and measured the trends of numerous folds, crenulations, and other lineations (Figure 19E–6). All our data are compiled in Appendix 3.

Our tectonic and geologic maps (Figure 19E–2) indicate the complexity of this region, despite the small size of Woodall Shoals. The exposure is dominated by sillimanite-grade migmatitic metasedimentary rocks, mostly biotite gneiss, intruded by both early and late granites. Closer examination of the exposure reveals boudins of amphibolite, garnetiferous amphibolite, calcsilicate quartzite, and granitic gneiss—some preserving earlier foliation truncated by the dominant foliation in the biotite gneiss. Pressure shadows at the ends of some boudins contain quartz, feldspar, and garnet.

By constucting form lines from strike of dominant foliation in the biotite gneiss (Figure 19E–5), we were able to obtain an indication of the amount of folding. Foliation strikes dominantly northeast, but, as the map shows in many places, the strike may vary widely over a small area. We found this to be related to folding and fold interference.

We used the complete geologic map (with plotted structural data) to identify boundaries between areas of the same direction of strike and dip (homogeneous domains) and multiple orientations of strike and dip (heterogeneous domains). Thus we determined the locations of fold hinges and limbs, noting that homogeneous domains will locate limbs. (Most heterogeneous domains are probably fold hinges or zones of fold interference.) Once we plotted strike and dip of the dominant foliation (S_2) on fabric diagrams (Figure 19E–5), the distinction between homogeneous and heterogeneous domains became even more striking. Domains II, III, VII, and XII we identified as homogeneous domains because the diagrams contained clusters of points. Domains IV, V, VI, and XI we judged heterogeneous because points are scattered in the diagrams. Domains I, X, and XI contain enough diversity of orientations that they might be considered heterogeneous, but some clustering of points also occurs in the diagrams. It may have been possible to determine whether domains X and XI are homogeneous, but so far we did not have enough data. Inspection of the form line patterns in each of these domains indicated that domain I was homogeneous and that domains X and XI were heterogeneous.

The large number of folds present in this map area (Figure 19E–6) permitted us to identify several generations of superposed folds. The earliest folds occur in amphibolite boudins that survived strong transposition of the earliest foliation in the biotite gneiss and preserved a relict foliation in the boudins (see Figure 16E–2b). Later tight to isoclinal folds in the biotite gneiss may be responsible for dismemberment of the amphibolite layer (or layers) and formation of the boudins. Geologic mapping provided us with a perspective heretofore unavailable through studies I began here in the 1960s: it permitted mapping of a large recumbent fold, revealed by the distribution of large amphibolite boudins (Figure 19E–2), throughout the central part of the outcrop.

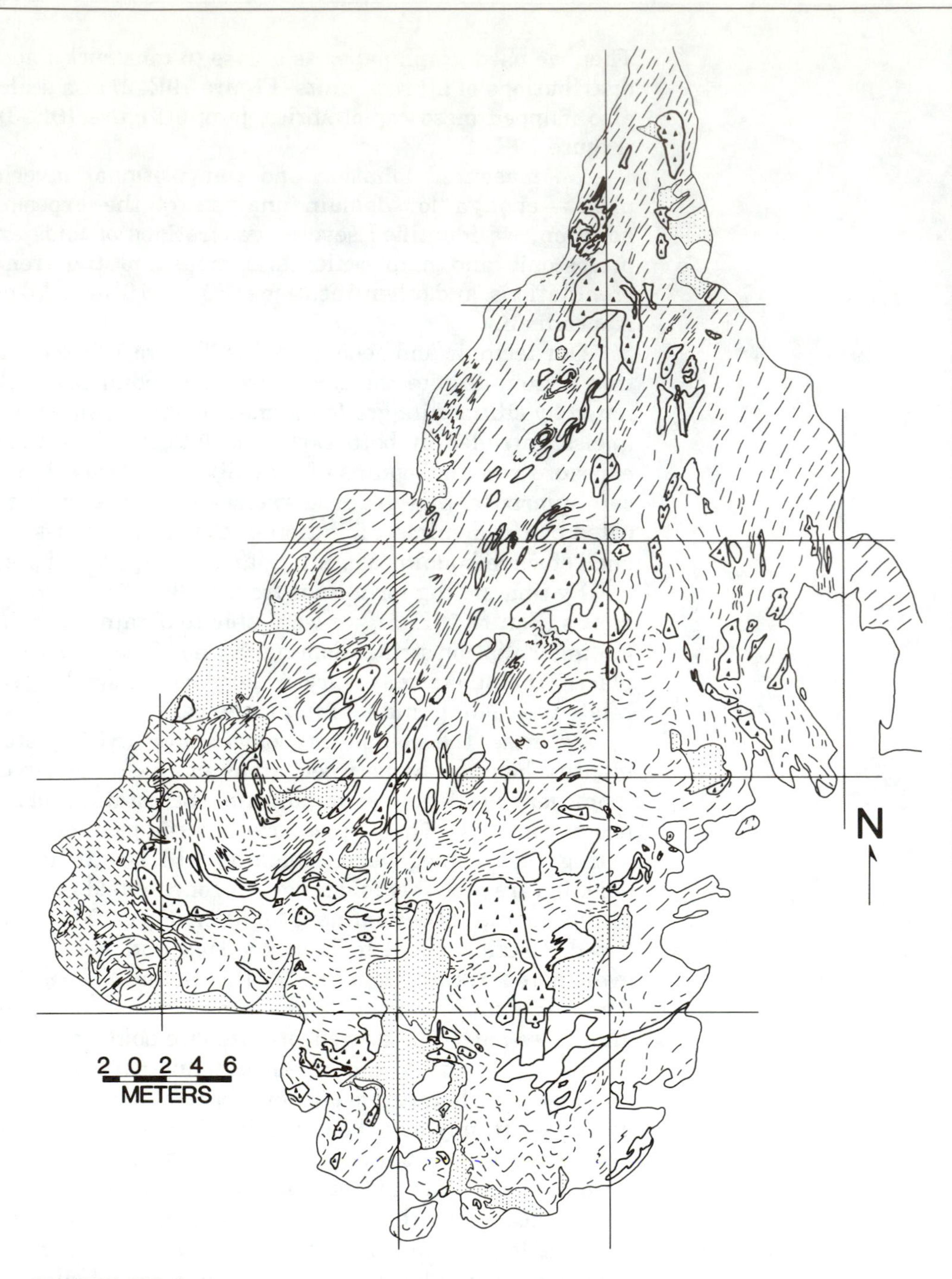

FIGURE 19E–2
Simplified geologic map of Woodall Shoals. The cross-hatch pattern is a late pegmatite intrusion; the short-line pattern is migmatitic biotite gneiss; small triangles mark amphibolite boudins; the unpatterned bodies are foliated granitic gneiss and pegmatite; the fine stipple pattern is recently deposited sand. (The stipple and cross-hatch patterns indicate the same units in all other maps in this Essay.)

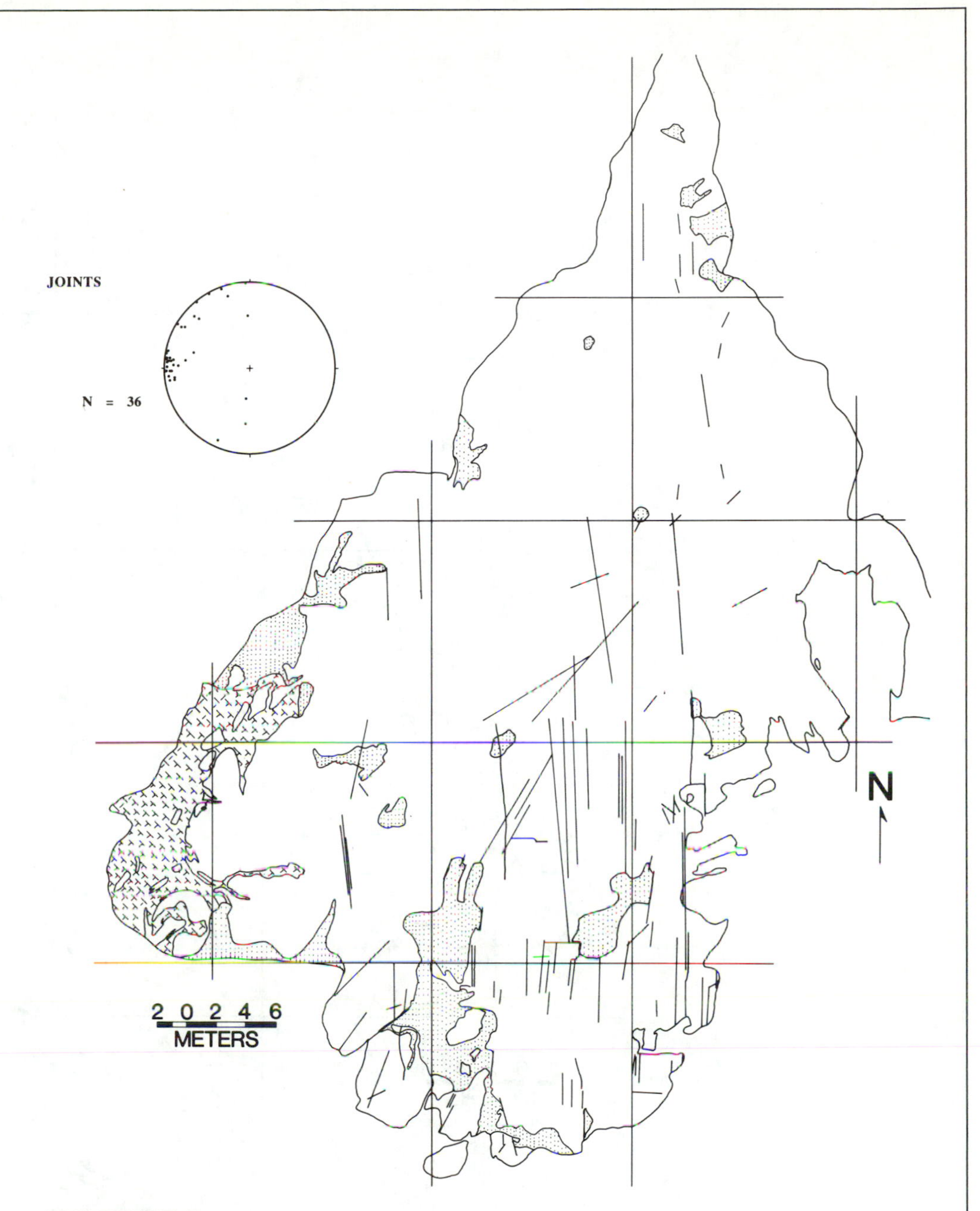

FIGURE 19E–3
Joint distribution at Woodall Shoals and fabric diagram of poles to joints. This and all other fabric diagrams in figures in this Essay were plotted using a Macintosh® computer and the Stereonet program written by R. W. Allmendinger, Cornell University.

The map of fold axial traces (Figure 19E–6) revealed several crossing fold sets. Most folds and crenulations here plunge gently northeast or southwest and verge northwest, as indicated in the fabric diagrams. A set of crossing folds in the southwestern part of the map consisted of earlier isoclinal folds (F_2?) crossed by a large (F_3?) northwest-verging tight fold (Figure 19E–6; see also Figure 16E–2). Limbs of small earlier folds produce an apparent plunge both

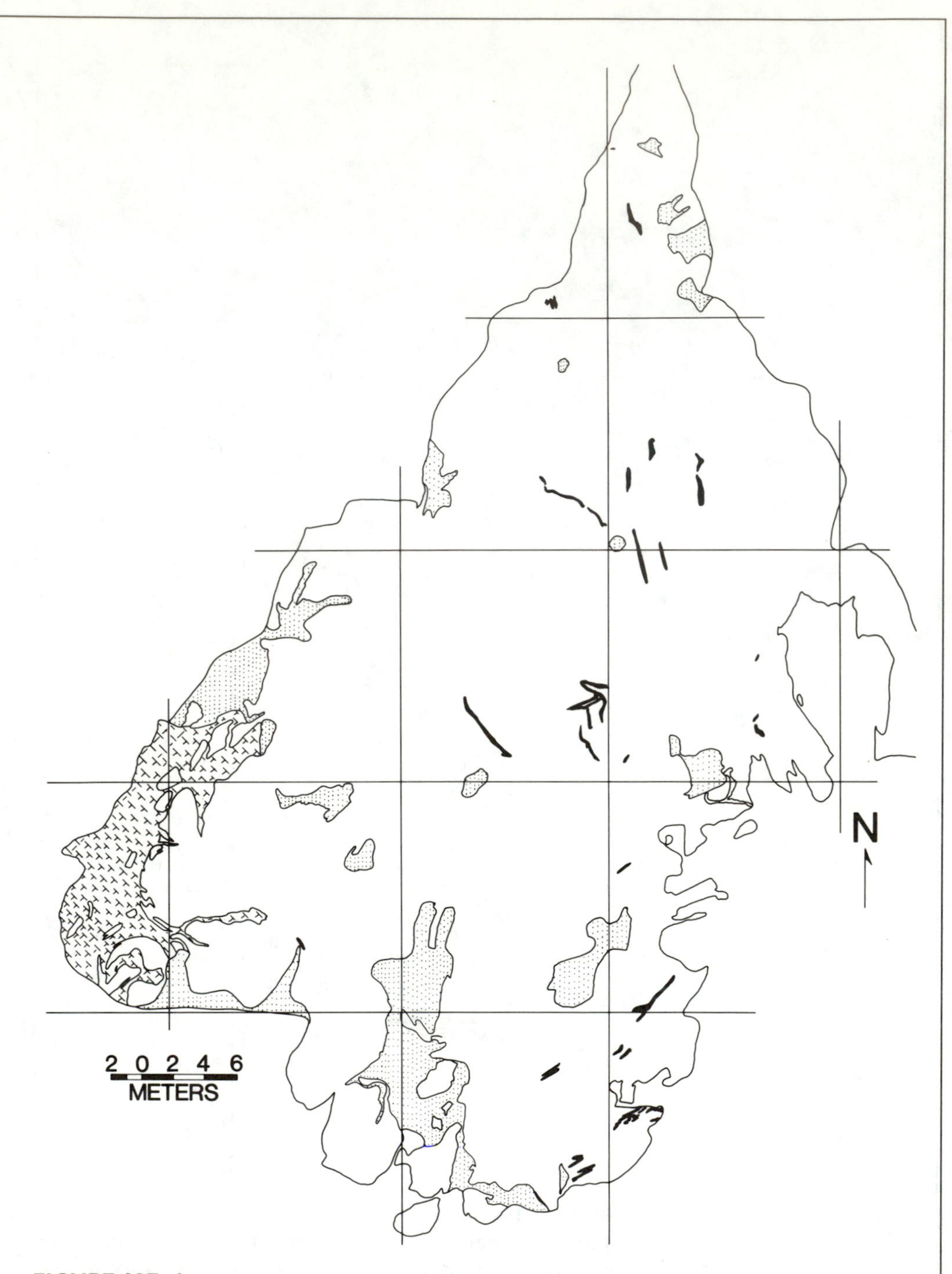

FIGURE 19E–4
Distribution of deformed granitic pegmatite veins (black) at Woodall Shoals.

northeast and southwest in the hinge of the later fold. Tracing the later fold a few meters northeast revealed that it really plunges northeastward.

The youngest structures found at Woodall Shoals were joints. Most of those we measured defined a set striking nearly north-south and dipping steeply southeast. A smaller number defined another set striking northeast and dipping southeast. We found few gentle dips. Although we measured only a small number, both sets recognized were regional sets (Acker and Hatcher,

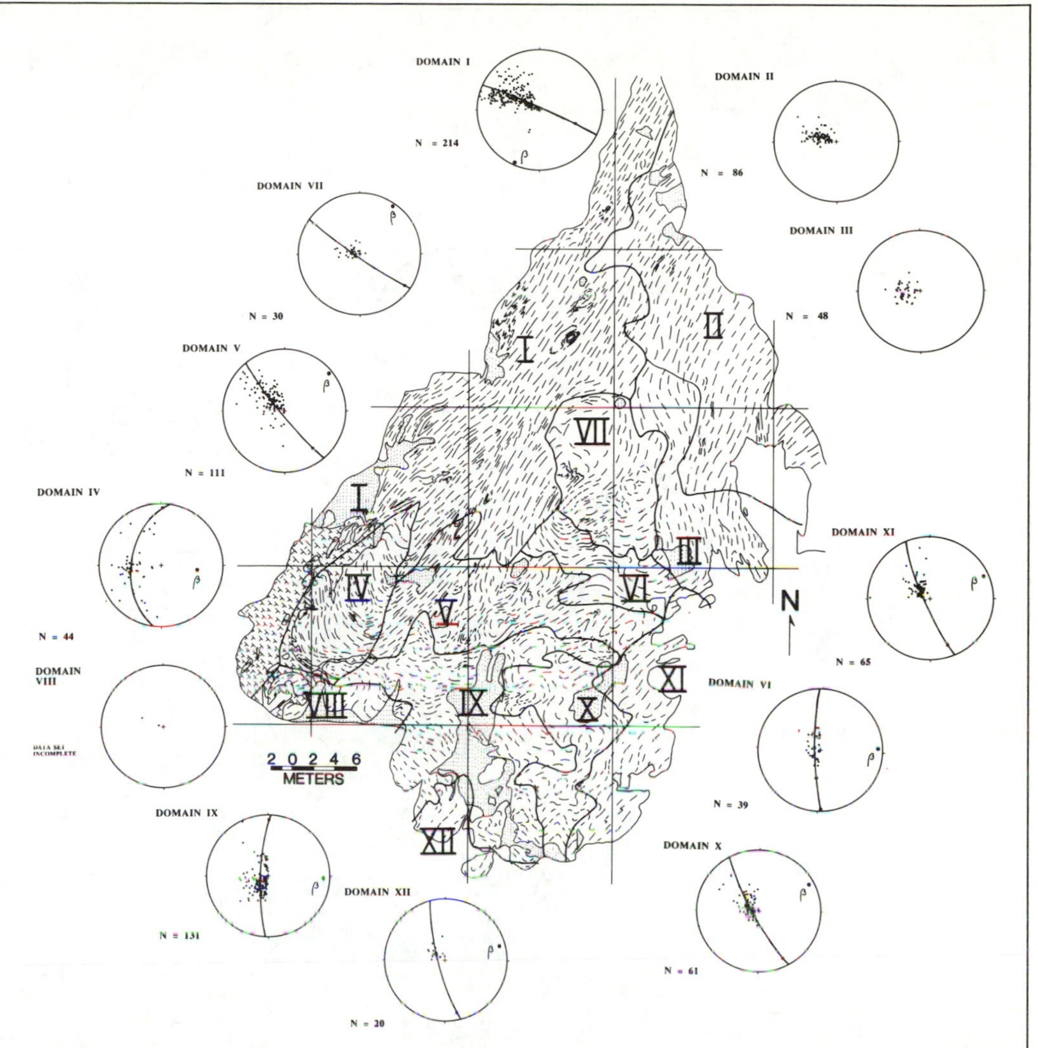

FIGURE 19E–5
Map showing form (strike) lines and domains of similar orientation and dip direction or multiple directions at Woodall Shoals. Fabric diagrams are equal-area plots of poles to the dominant foliation, S_2. Locations of the great circle and β were determined as a best fit by the computer program that plotted the data. All βs indicate a gentle plunge for the fold axis determined, but the trends of some may be biased by lack of scatter in the plot. Carefully examine each plot and position of the great circle and β relative to distribution of points.

1970; Schaeffer and others, 1979). Most joints were unfilled, but several contained quartz, epidote, and pink feldspar, indicating hydrothermal alteration after the fractures formed and while the rock mass was still under greenschist-facies pressure and temperature conditions. These filled joints were crossed by younger unfilled fractures.

A late unfoliated pegmatite body along the west side of the exposure (Figures 19E–2, 19E–4) contained xenoliths and roof pendants of biotite gneiss. We thought xenoliths and roof pendants could be distinguished by comparing the degree of parallelism of foliation in the gneiss to that in the enclosing rocks. Lack of parallelism led us to conclude that an inclusion was probably a xenolith.

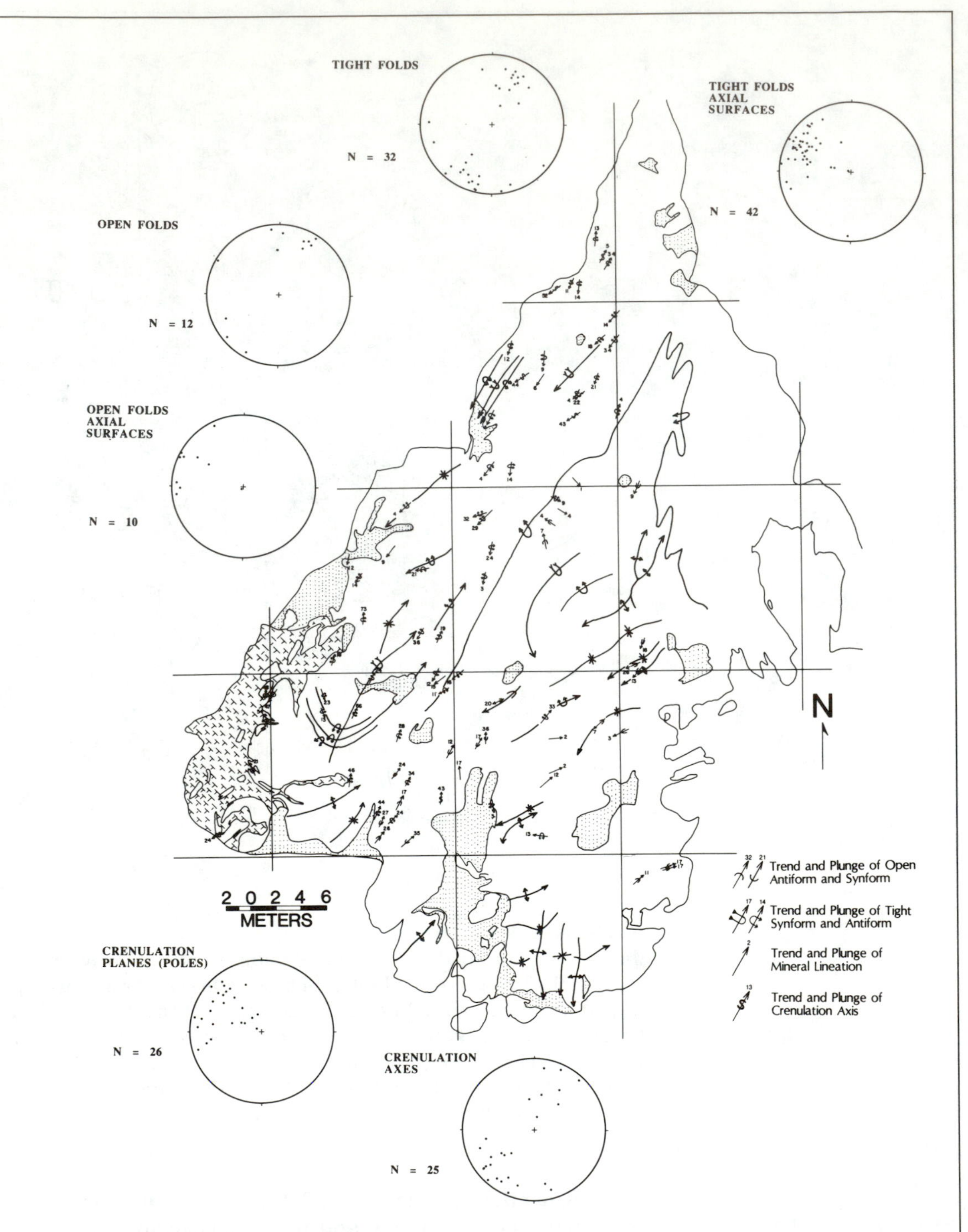

FIGURE 19E–6
Map showing mesoscopic folds and mineral lineations (small lines and symbols with numbers) measured at Woodall Shoals and map-scale folds determined from a combination of fabric data and direct observation. Equal-area plots of trend and plunge of fold axes and mineral lineations, and poles to axial surfaces of different classes of folds, are shown.

Analysis of a data set of this type compiled as maps and fabric data is a useful exercise; applying the technique will produce valid conclusions in areas of any size and provide immediate insight into the structural history of the area.

References Cited

Acker, L. L., and Hatcher, R. D., Jr., 1970, Relationships between structure and topography in northwest South Carolina: South Carolina Geologic Notes, v. 14, p. 35–48.

Hatcher, R. D., Jr., Acker, L. L., Liu, Angang, Zupan, Alan-Jon, and Mittwede, Steven, 1989, Geology and mineral resources of the Whetstone, Holly Springs, Rainy Mountain, and Tugaloo Lake quadrangles, South Carolina–Georgia: South Carolina Geological Survey, scale 1/24,000.

Schaeffer, M. F., Steffens, R. E., and Hatcher, R. D., Jr., 1979, *In situ* stress and its relationships to joint formation in the Toxaway Gneiss, northwestern South Carolina: Southeastern Geology, v. 20, 129–143.

strain in veins and any other strain markers; the goal is to determine the magnitude and orientation of strain as it affected the rocks at different times. Study of fibers in veins or pressure shadows (or both) is one of the best ways to measure noncoaxial progressive deformation.

Chronology of Development of Structures

Measurements of all structural elements, on all scales, and results of microtextural studies must be integrated into a complete structural analysis so that the chronology of development can be worked out. Once completed, it will provide a reasonably complete picture of the kinematics and sequence of deformation in the area. Such is the principal goal of structural analysis.

Cross-Section Analysis

If a single set of structures is dominant, cross sections normal and parallel to strike make a useful tool. Construction of a standard geologic cross section in a singly deformed area requires careful transfer of contacts from geologic maps to the section line (preferably showing topography, too) and systematic projection of the contacts between rock units to depth (Chapter 11). If this is done properly, thicknesses of rock units will remain constant. Vertical and horizontal scales should be the same to depict structural geometry without distortion. Balancing of cross sections involves retrodeforming the section to its undeformed state to see whether the section can be taken apart and then reconstructed (see Figures 11–36 and 11–37). It is desirable to balance the cross sections, an easy task if the rocks are not penetratively deformed. Balancing through retrodeforming sections should be carried out because it tests the interpretation: If a section will not balance, it cannot be correct; if it does balance, it *may* be correct (Hossack, 1979; Price, 1981; Elliott, 1983). Even in a complexly deformed area, constructing and balancing of cross sections should not be attempted without detailed knowledge of strain states or without proof that plane strain dominated the deformation. In practice, cross sections here are not often balanced because in areas of multiple deformation the structures have more than one dominant orientation. Probably the better course is to construct accurate cross sections along lines in different orientations—better than not constructing them at all.

In a classic example, John Ramsay (1981) resolved the differences in strain between three internally deformed thrust sheets in the Helvetic Alps (Figure 19–10). The lowest thrust sheet (the Morcles) is deformed more than the two above (the Diablerets and the Wildhorn), and the lowest part of the Morcles and the down-dip parts of the two higher are internally deformed more than other parts of the sheets.

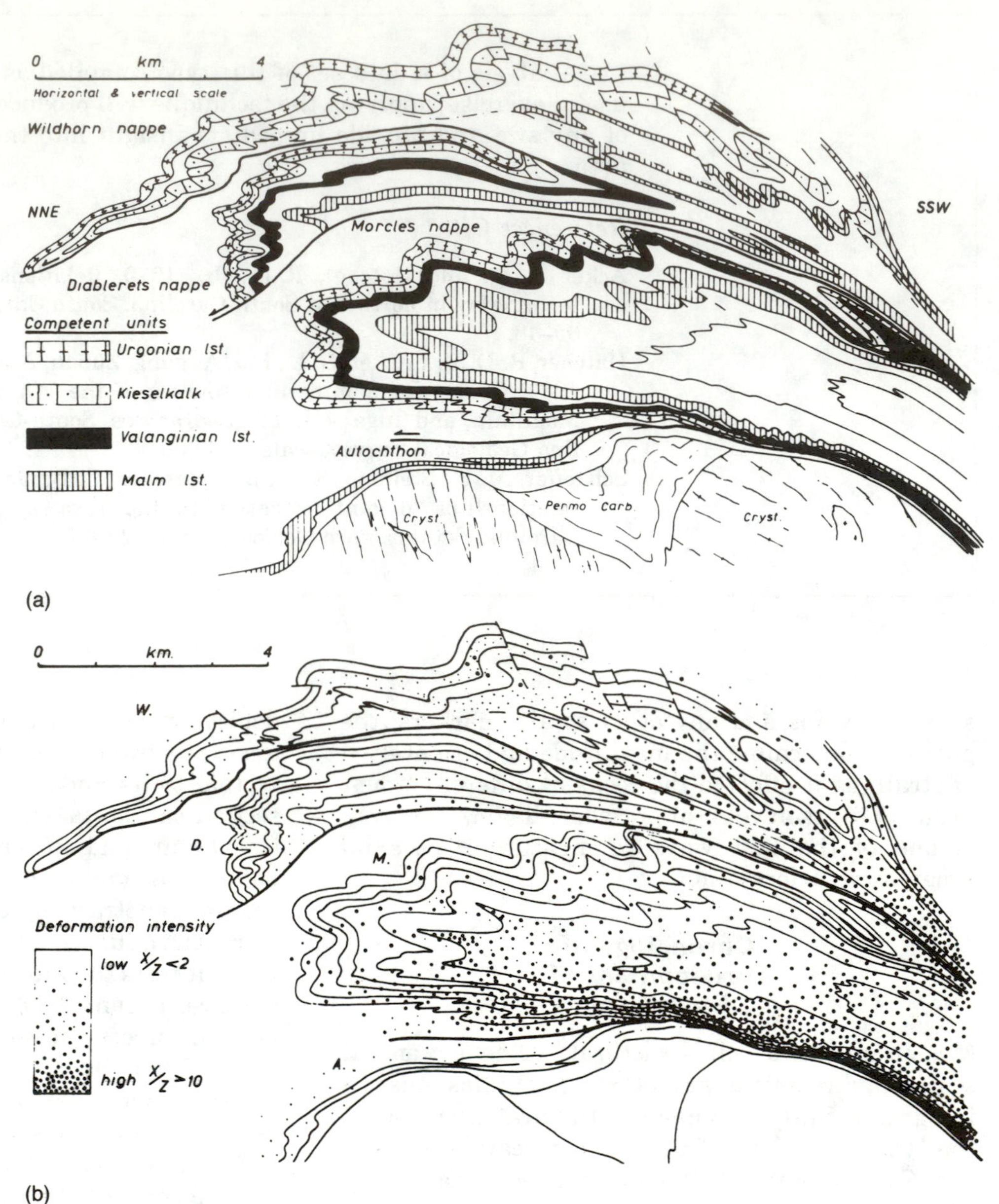

FIGURE 19–10
Cross sections through three internally deformed thrust sheets in the Helvetic Alps in Switzerland. (a) Geologic section showing relationships between rock and tectonic units and internal deformation in each thrust sheet. (b) Intensity of deformation in the cross section (indicated by density of stipple pattern), determined by study of strain markers within each sheet. (From J. G. Ramsay, 1981, *in* N. J. Price and K. R. McClay, eds., *Thrust and Nappe Tectonics:* Geological Society of London Special Publication 9, Blackwell Scientific Publications Limited.)

This completes our survey of geologic structures and structural analysis. The remaining chapter discusses geophysical techniques that provide useful information toward understanding of large-scale structures—on the map to continent scale, which is the subject of tectonics.

Questions

1. Why does the deformation plan of most orogenic belts follow that of early ductile structures overprinted by late brittle structures? Why not the reverse order?
2. How does use of similar styles and orientations in interpretation of multiple deformation sometimes lead to erroneous results?
3. What evidence is needed to resolve multiple transposition in rocks?
4. Here is a set of fabric diagrams of poles to foliation in several homogeneous domains in an area in the Blue Ridge. What kind of structure is represented? Photocopy this page and sketch in appropriate structural symbols (fold axes showing plunge, faults).

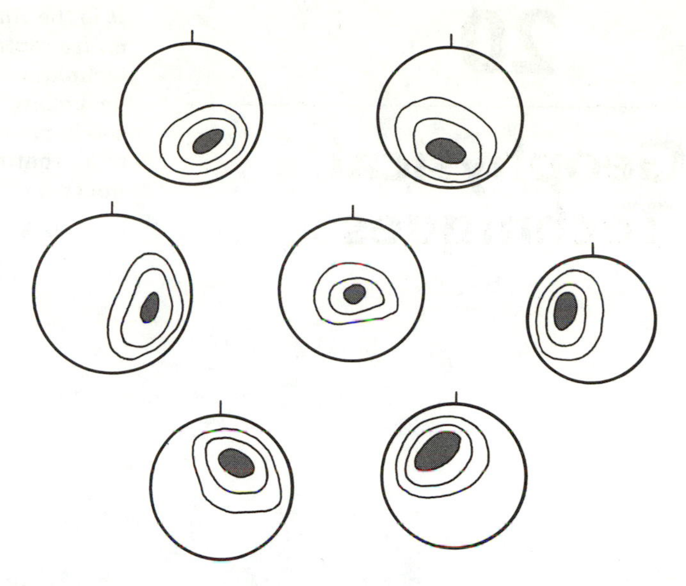

5. What are two interpretations of triclinic symmetry of a fabric diagram of poles to foliation?
6. Why is identification of kinematic axes in most areas considered useless or even impossible?
7. How would you undertake structural study of an area deformed by one event and containing penetratively deformed (cleaved) rocks, crack-seal veins, and mesoscopic- and map-scale folds and faults?
8. Sketch in major fold axes (showing plunge) on a photocopy of Figure 19–6 using only the outcrop patterns and form surfaces on the map. Then examine the orientation of *b* in each of the domain diagrams. Assuming β axes are parallel to folds, what do they tell you about the small and large structure in that area?

Further Reading

Hossack, J. R., 1979, The use of balanced cross-sections in the calculation of orogenic contraction: A review: Journal of the Geological Society of London, v. 136, p. 705–711.

Reviews principles of cross-section balancing and techniques of line and area balancing.

Ramsay, J. G., and Huber, M. I., 1983, The techniques of modern structural geology, Volume 1: Strain analysis: New York, Academic Press, 307 p.

Ramsay, J. G., and Huber, M. I., 1987, The techniques of modern structural geology, Volume 2: Folds and fractures: New York, Academic Press, 392 p.

This two-volume work covers the full spectrum of modern structural geology. Strongly recommended as a standard reference for anyone undertaking structural analysis.

Turner, F. J., and Weiss, L. E., 1963, Structural analysis of metamorphic tectonites: New York, McGraw-Hill, 545 p.

Contains a step-by-step approach to plotting fabric diagrams, analyzing rock fabrics on all scales, and interpreting complex structure.

Whitten, E. H. T., 1966, Structural geology of folded rocks: Chicago, Rand-McNally, 678 p.

A different approach to study of structure and fabrics in penetratively deformed rocks. Assumes familiarity with fabric diagrams.

Williams, P. F., 1985, Multiply deformed terrains—problems of correlation: Journal of Structural Geology, v. 7, p. 269–280.

Deals with the problem of correlating foliations in complexly deformed regions; discusses structures of microscopic- through map-scale.

20

Geophysical Techniques

It is the time in history for thorough exploration of the entire continental crust. The seismic reflection profiling technique . . . will almost certainly be the principal tool for probing the deep continental basement. . . . One can safely predict an era in which the deep crustal features of all continents are discovered, mapped, named, understood, and made familiar to all earth scientists.

JACK OLIVER, American Geophysical Union Geodynamics Series

GEOPHYSICAL TECHNIQUES AND CONCEPTS ARE WIDELY USED TO resolve individual geologic structures and to explore the structure of the Earth. They help resolve near-surface structure and extract information about the nature of the deeper crust and mantle. Techniques useful in delineating geologic structure include gravity, magnetism, paleomagnetism, electrical properties, seismic reflection, seismic refraction, and earthquake seismology. The resolution and imaging capability of each technique are quite variable. Seismic-reflection profiling is probably the most precise because it produces a vertical section that resembles a geologic cross section, followed by magnetics, gravity, and seismic refraction; the techniques will be discussed here with examples of applications to structural geology.

This chapter is not intended to provide in-depth background in geophysics, but to demonstrate the importance of geophysical techniques in modern structural geology and kindle interest in the subject (Figure 20–1). The techniques to be discussed here help most in interpreting large structures—faults and folds of map scale or larger, plutons, and crustal boundaries—and as such provide a springboard to tectonics.

POTENTIAL-FIELD METHODS

Gravity and magnetism—called *potential fields*—provide useful insights into structure of the crust and upper mantle. Each has its own limitations and degree of precision and resolution, but magnetism has the greater resolving power in the crust.

Terrestrial Magnetism

Earth magnetism was known to the Chinese at least as early as the eleventh century, and about 1295 Marco Polo brought magnetite (lodestone) from China to Europe. In the sixteenth century, William Gilbert (1540–1603), an English physicist, noted that a sliver of magnetite hung by a string would orient itself more or less north-south.

The Earth's magnetic field is believed to originate in the core. Yet temperatures in the core must

FIGURE 20–1
(a) (opposite page) Map of seismic reflectors at 2.8 seconds (approximately 3000 m below sea level) depth in Tracts 1 and 5 in the Green Canyon area in the Gulf of Mexico south of New Orleans, Louisiana, showing nearly circular salt diapirs that have intruded the sedimentary cover. These data were processed through three-dimensional migration. Note the similarity to map patterns in a region of multiple deformation in the core of a mountain chain (see Figures 2–22 and 19E–2). Arrows locate the vertical section in (b) (page 436). (b) North-south seismic-reflection profile showing a large salt dome near the north end and smaller structures to the south. Subhorizontal layering in sediments is folded upward by diapiric intrusion of salt. (Used by permission and courtesy of Geophysical Service Inc.)

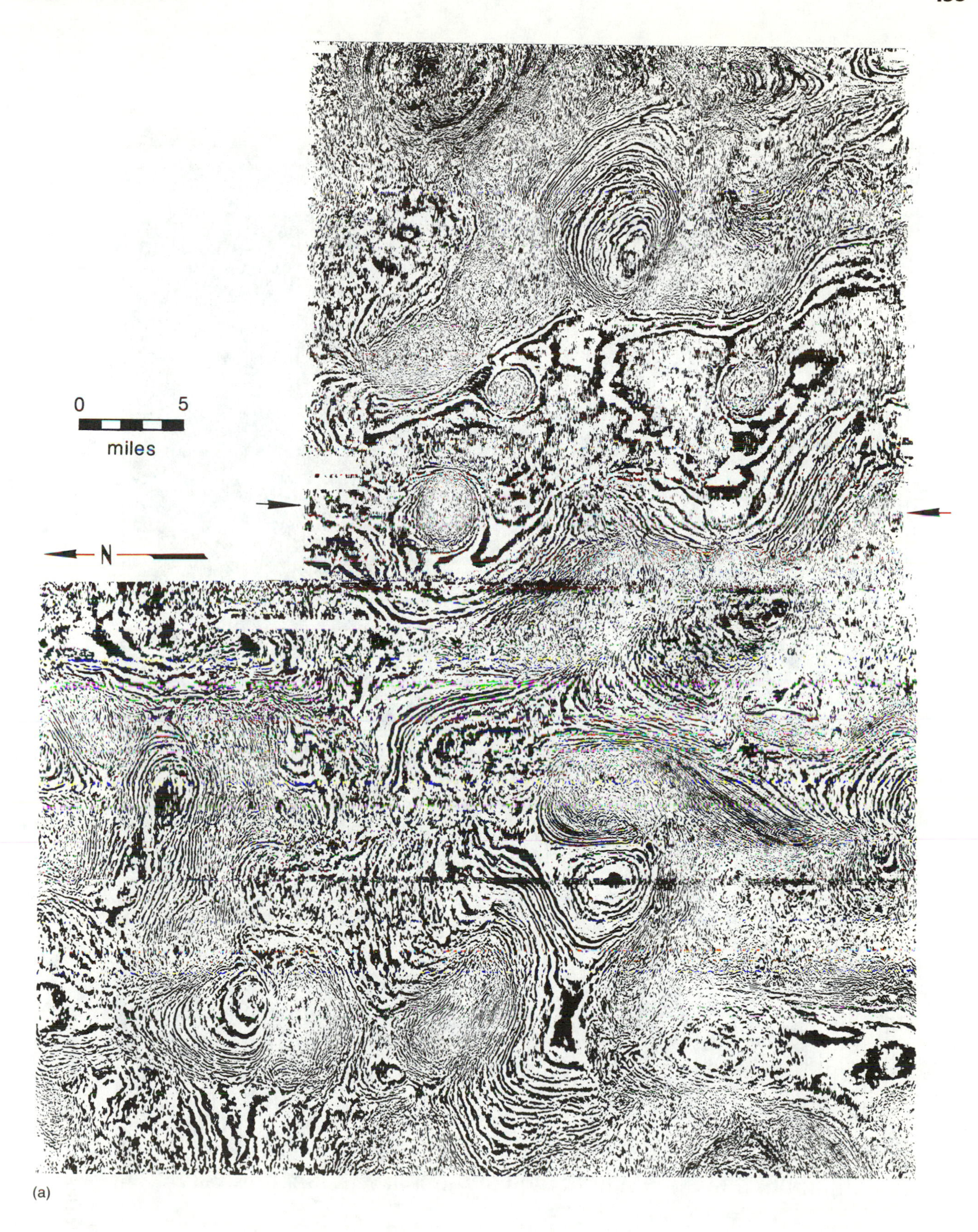

(a)

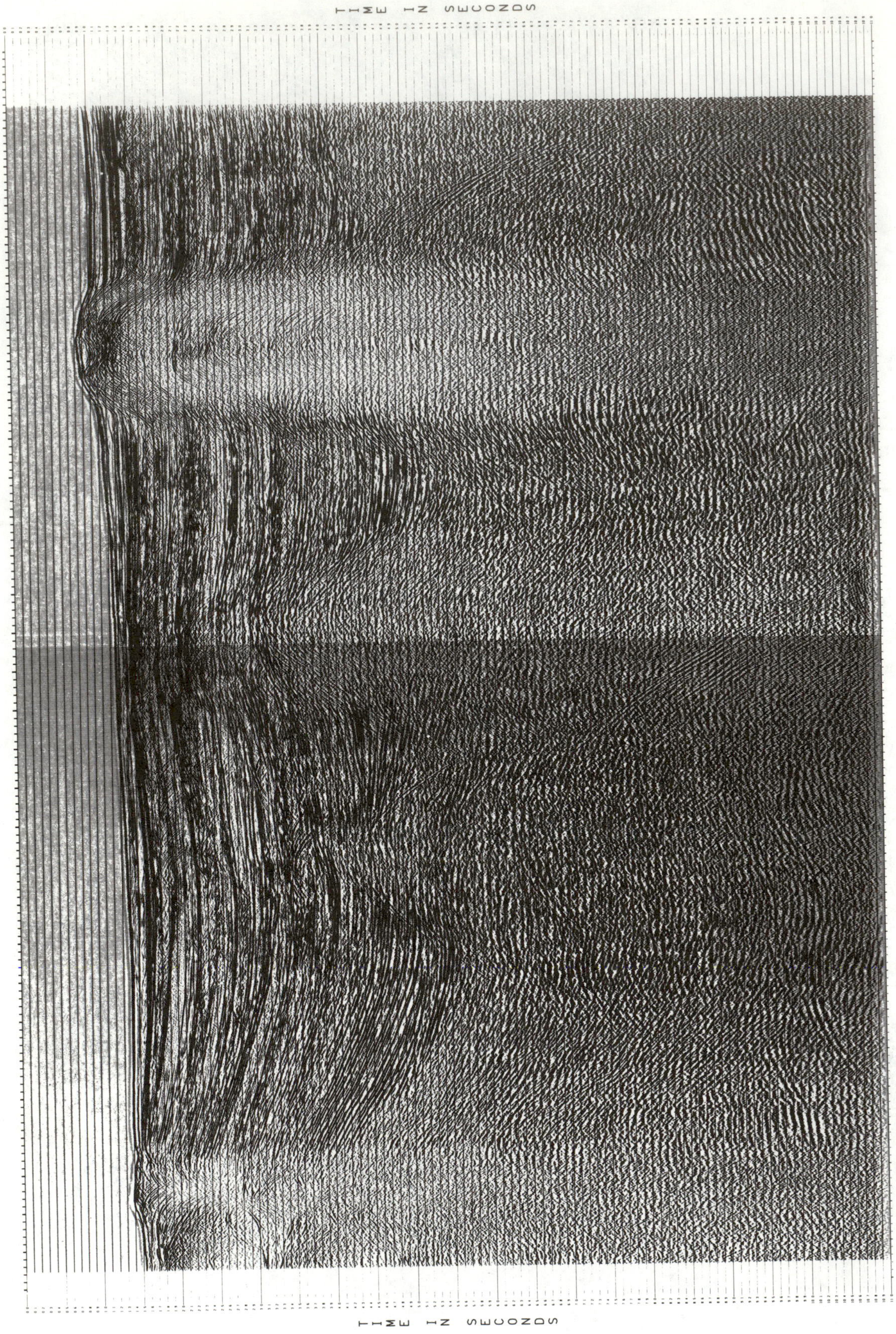

TIME IN SECONDS

TIME IN SECONDS

(b)

FIGURE 20–1 *(continued)*

TABLE 20–1
Curie temperatures for common magnetic minerals

Mineral	Curie temperature
Magnetite	578°C
Hematite	670°C
Iron	770°C
Pyrrhotite	316°C

be well above the Curie temperatures for most materials, including metallic iron, which probably is the dominant element in the core (90 percent iron, 10 percent nickel). The ***Curie temperature*** (after Pierre Curie, 1859–1906, a French physicist) is the temperature above which strongly magnetic (ferromagnetic) materials lose their ability to interact with a magnetic field (Table 20–1); a weaker (paramagnetic) property, like that of most materials, remains. Although it is not exactly certain why the Earth has a magnetic field, the best model is the ***self-exciting dynamo,*** first proposed by a British mathematician, Joseph Larmor (1857–1904), to explain the magnetic field of our sun. His idea was refined and applied by Walter M. Elsasser and Sir Edward Bullard to explain Earth magnetism (Rikitaki, 1966). The theory is based on the premise that convection currents exist in the liquid core, transport hotter material upward, and return cooler material to the interior. The convection cells are assumed to be oriented by the Earth's rotation. The convective transfer of liquid iron-nickel through a thermal gradient generates an electrical potential difference with an associated magnetic field. Once the magnetic field is generated, the convective motion enhances and perpetuates it, producing a self-exciting dynamo.

A major difficulty is that the self-exciting dynamo theory does not account for reversals: at intervals since the late Paleozoic the Earth's magnetic field has spontaneously reversed polarity many times (Figure 20–2). The intervals are not uniform, and the time it took to actually reverse polarity was probably on the order of a few thousand years—perhaps hundreds (Fuller and others, 1979). The dipole magnetic field may be lost during the reversal, leaving a weaker non-dipole field. During reversal, magnetic field strength decreases to about 15 percent of its normal strength, remaining there until the reversal process is complete (Watkins, 1969). K. A. Hoffman (1988) has summarized evidence showing that the strength of the present-day magnetic field has been decreasing for a century—that if continued at the same rate it would decrease to zero in 1500 years. Thus a polarity reversal may be imminent.

The reversal phenomenon raises many questions about the effect on life forms (among other things) because the magnetic field acts as a radiation shield. Removal of the shield will allow more radiation to reach the surface of the Earth, increasing the mutation rate.

Intensity of the magnetic field may be measured using a ***magnetometer*** on the ground, or at sea, or from an airplane. Vertical ***magnetic-intensity profiles*** may be plotted along the line of traverse or the

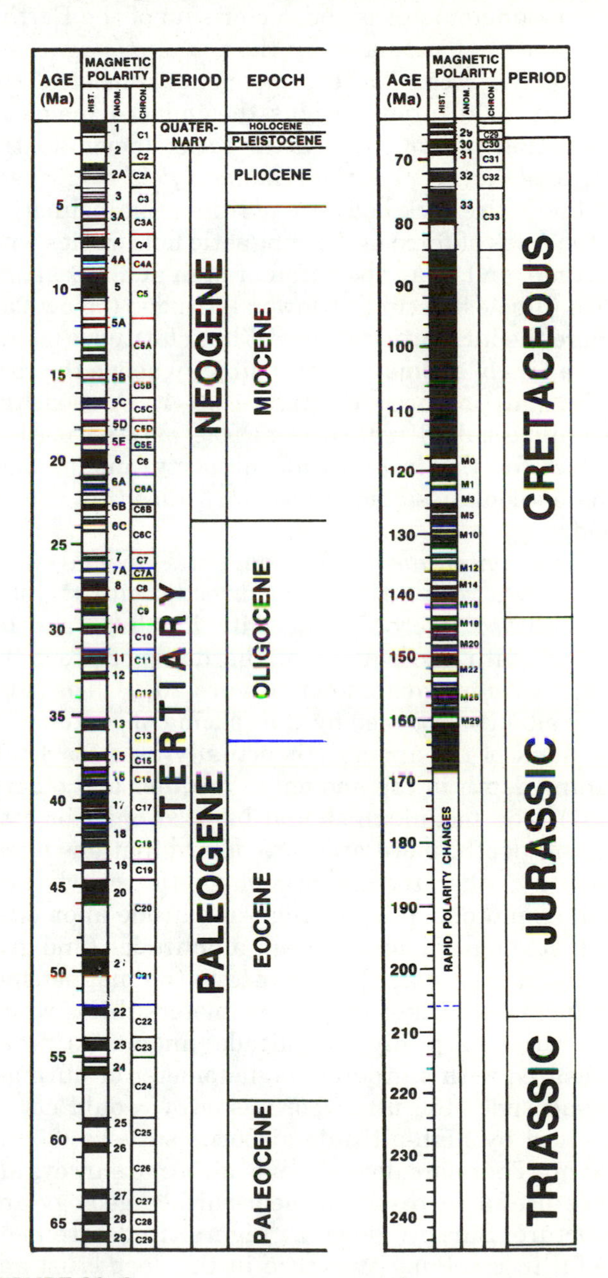

FIGURE 20–2
Magnetic polarity reversals from the Mesozoic to the present. Black intervals are normal polarity—as at the present-day—and white are reversed. (From A. R. Palmer, Geological Society of America DNAG Time Scale; used by permission.)

data contoured between successive traverse lines to produce a ***magnetic-anomaly map*** (Figure 20–3). An ***anomaly*** is a deviation from an assumed average value for magnetic-field intensity. Generally, rocks that contain a high percentage of magnetically susceptible minerals (such as magnetite) produce *positive anomalies,* and rocks that contain a low percentage of magnetically susceptible minerals produce *negative anomalies.*

The total intensity of a magnetic field measured by a magnetometer is the vector sum of the Earth's field and that produced by the *induced magnetization* of the rock. These components consist of the primary field intensity plus the induced magnetic susceptibilities of all crustal rocks (down to the Curie isotherm) plus the *remanent field* component for the entire rock body. In addition, the inclination of the lines of force of the magnetic field varies with latitude, and so a separation of high and low anomalies (dipoles) occurs at lower latitudes (below 25°) where the inclination is low; at high latitudes (above 70°) a single anomaly occurs directly above the rock body that produced it (Figure 20–4). A modeling technique called *reduction to pole* may be used to correct for the low-latitude property and produce single anomalies above the rock bodies that caused them.

The *amplitude* of the magnetic anomaly produced by a rock body varies directly with the susceptibility of the body and with its shape and inversely with the square of the distance from the body to the magnetometer (Figure 20–5). The slope (or gradient) produced by the spacing of contours on the flank of the anomaly is actually used for determining depth to the anomaly (Vacquier and others, 1951). The technique should be used only for estimating depth, more accurate determinations being made by other techniques. Magnetic rock bodies, such as plutons, produce high-amplitude anomalies with steep gradients wherein amplitudes (and gradients) decrease as the distance to the magnetometer increases. Dikes only a few meters thick, which generate sharp high-amplitude anomalies when measured with a ground magnetometer or during a low-altitude (150 m) airborne survey, would not be resolved by high-altitude airborne surveys (300 to 800 m). Therefore, low-altitude airborne surveys are more useful in resolving near-surface geology and structure. Surveys flown at high altitude are more useful in resolving structure in the deep crust and upper mantle.

Suppose contrasting sets of anomalies are surveyed from the same altitude (Figure 20–6): generally, broad, long-wavelength anomalies will be produced by deeper crustal features, and high-frequency, short-wavelength anomalies will be produced by near-surface features. Major boundaries and crustal blocks may be identified by studying magnetic-anomaly maps.

Problems arise when working with anomalies, even those reduced to the pole, because the total magnetic field is the vector sum of the induced and remanent fields. The remanent field component affects the amplitude and slope of contours on the flanks of anomalies by vectorially adding to (or subtracting from) the induced field. The ***Königsberger ratio, Q,*** is represented by

$$\mathbf{Q} = \frac{\mathbf{J_R}}{\kappa_b \mathbf{F}} \tag{20–1}$$

where $\mathbf{J_R}$ is remanent magnetism, κ_b is the magnetic susceptibility of the rock body producing an anomaly, and **F** is the induced magnetic-field intensity. Values of **Q** for older rocks are generally less than for younger rocks. Oceanic rocks generally have very high **Q** values relative to continental rocks, partly reflecting age but also a fundamental difference in properties. High **Q** is one reason the polarity-reversal pattern is strongly expressed in ocean crust. That the Königsberger ratios for continental rocks are low—about 0.1 to 1—indicates that a significant remanent component still exists and may strongly influence the total field anomaly, but reliable calculations of depth to magnetic anomalies using the slope technique of Victor Vacquier and others (1951) have been made where independent checks are available.

FIGURE 20–3
Magnetic-anomaly (a) (opposite page) and Bouguer gravity-anomaly maps (b) (page 440) of part of the north-central United States, showing distribution of anomalies in the Precambrian basement. The structure that produced the prominent northeast-southwest linear gravity and magnetic high in the center of (a) and (b) is the Mid-Continent rift, formed during the Late Proterozoic. The high is interpreted (c) (page 441) as having been produced by mafic igneous rocks that have been traced to the surface as the Keweenawan basalts of northern Michigan. The flanking gravity low is interpreted to have been produced by rift-related sedimentary rocks. (Magnetic data from Isidore Zietz, 1982, Composite magnetic-anomaly map of the conterminous United States: U.S. Geological Survey. Gravity data are from G. P. Wollard and H. R. Joesting, 1964, Bouguer gravity-anomaly map of the United States: U.S. Geological Survey.)

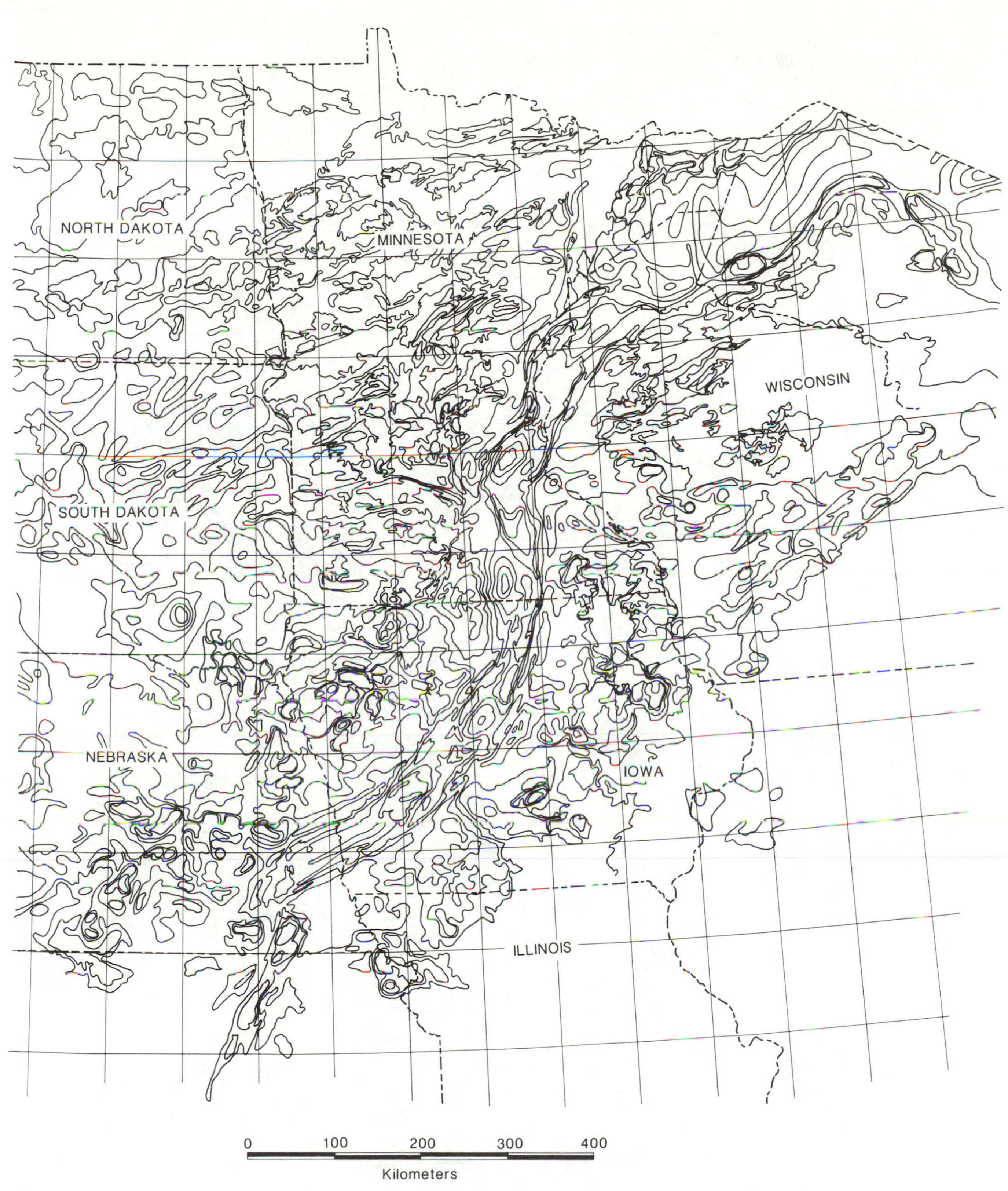
NORTH DAKOTA
MINNESOTA
WISCONSIN
SOUTH DAKOTA
NEBRASKA
IOWA
ILLINOIS
0
100
200
300
400
Kilometers

(a)

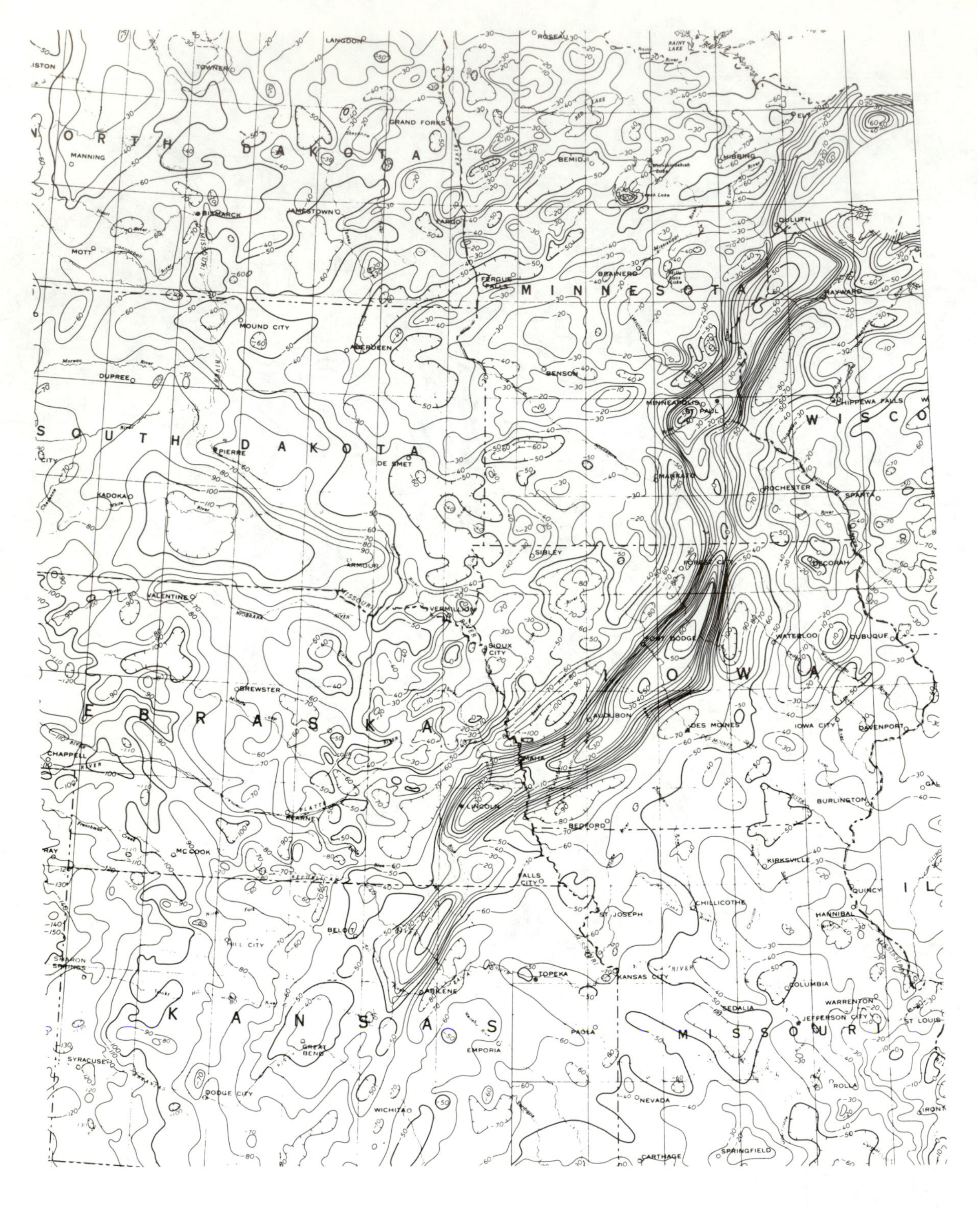

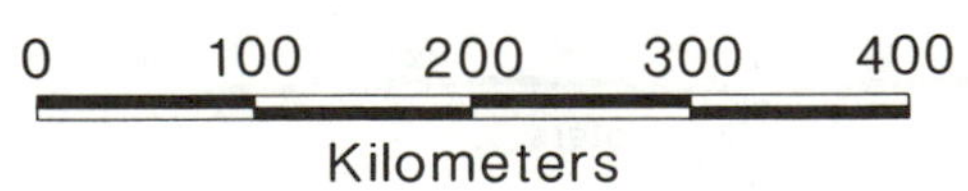

(b)

FIGURE 20–3
(continued)

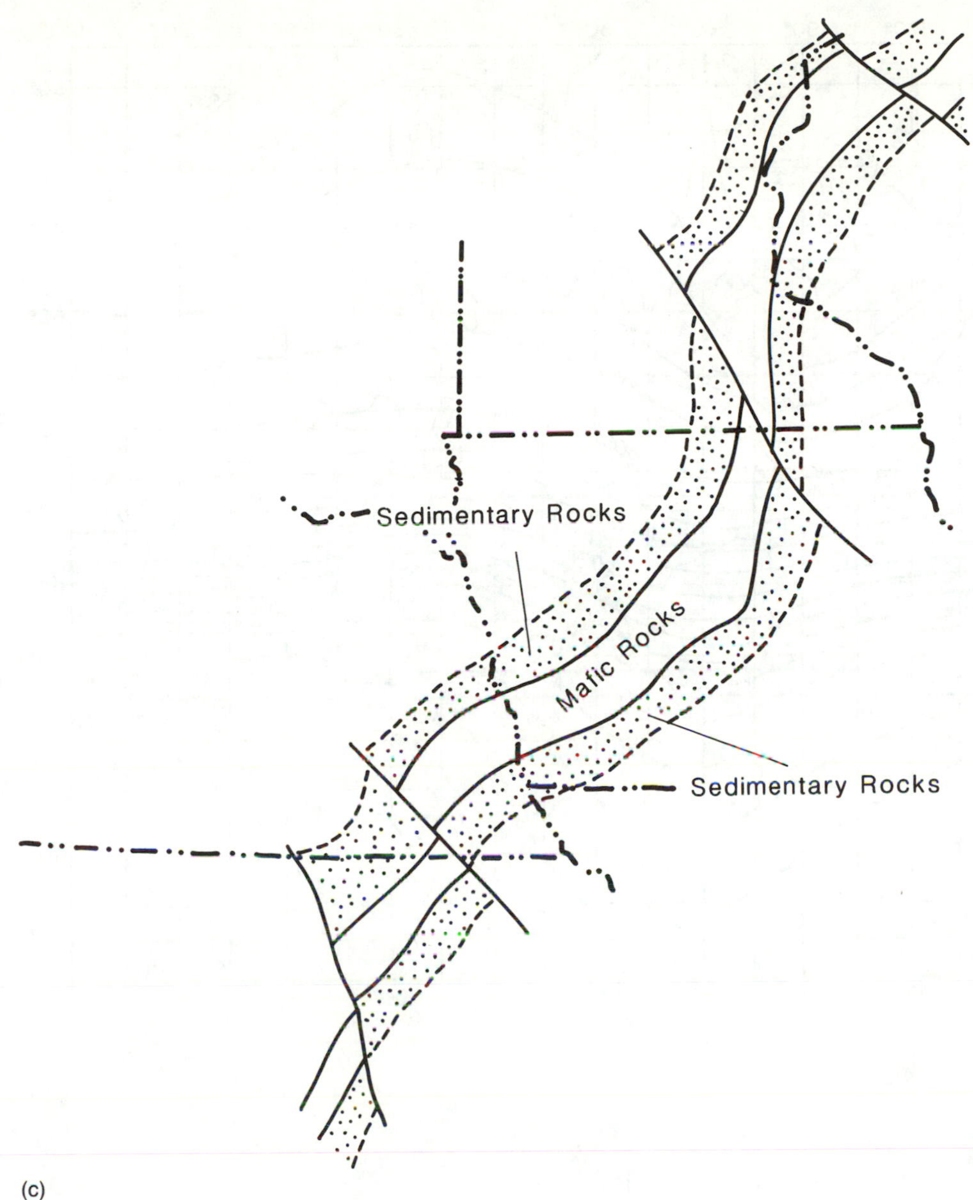

(c)
FIGURE 20–3
(continued)

Remanent Magnetism

Remanent magnetism develops in rocks when they crystallize or are deposited. The magnetic axes of magnetically susceptible minerals, such as magnetite, are aligned in the magma chamber or the depositional environment parallel to the magnetic field that existed during crystallization or deposition. In a magma, alignment does not occur until the temperature falls below the Curie temperature for that mineral (Table 20–1). Some minerals may crystallize completely and cool below their Curie temperatures before the entire magma congeals. Thus the orientation of the Earth's magnetic field at the time of crystallization or deposition is locked in. The study of ***paleomagnetism*** involves measurement of these remanent fields and attempts to reconstruct the positions of the ancient magnetic poles, continents, and oceans.

Remanent magnetism in igneous rocks is termed ***thermo-remanent magnetism (TRM)***. It is measured by collecting an oriented sample and shielding it from the present-day magnetic field so that the orientation of the ancient magnetic field can be measured. The measurement provides an azimuth toward the magnetic pole—the ancient pole position. Comparison with pole positions of other rock bodies of the same age may yield the amount of rotation experienced by the rock body.

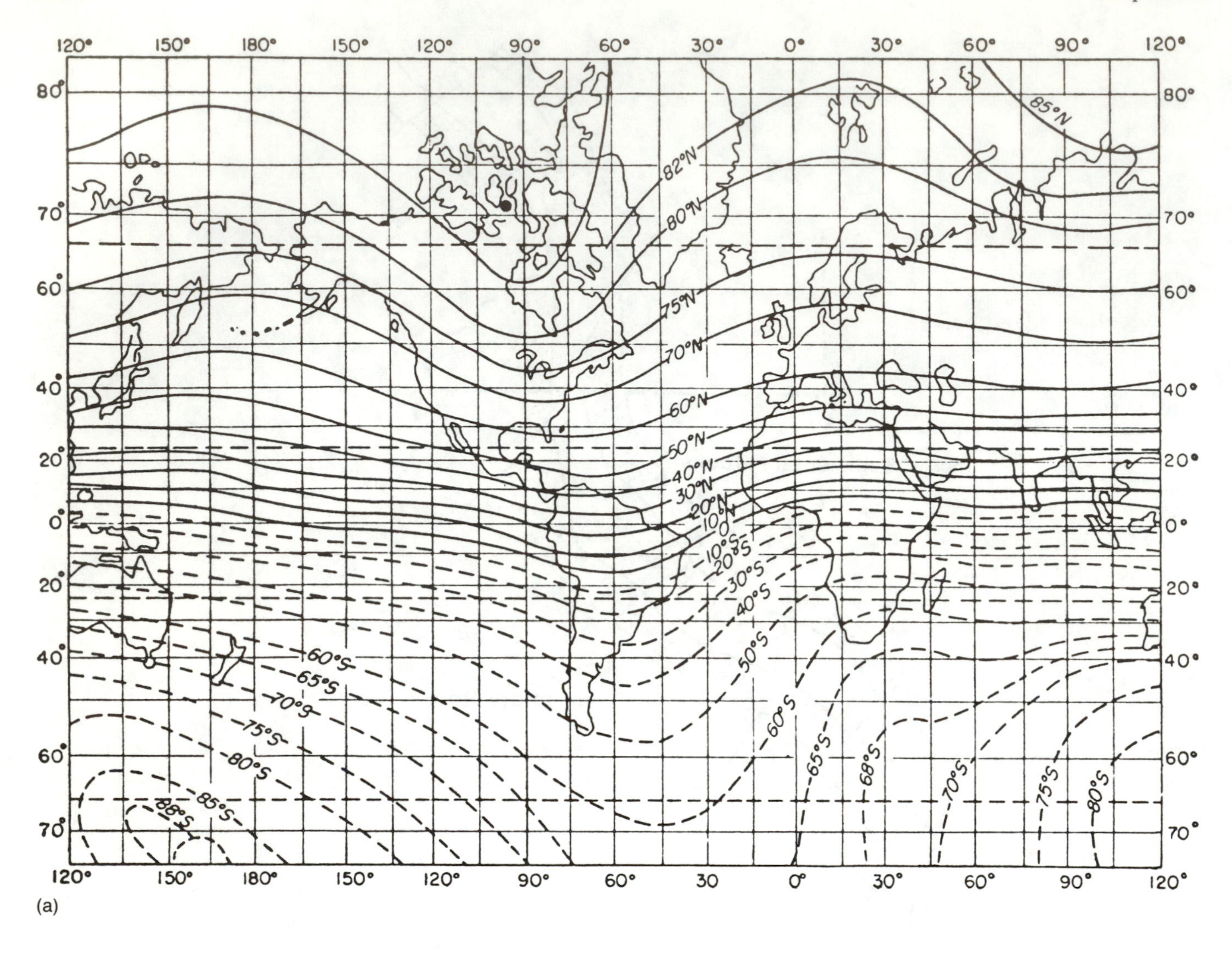

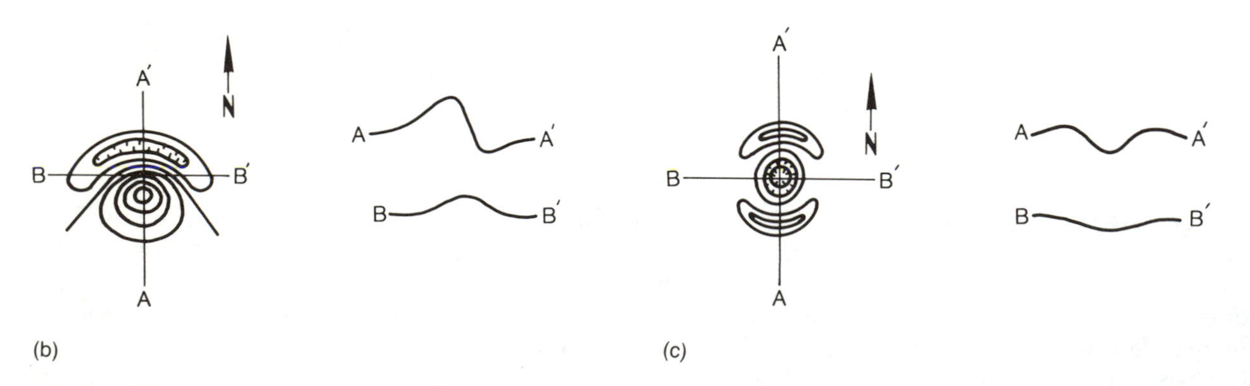

FIGURE 20–4
(a) Latitudinal variation in the inclination of the Earth's magnetic field. (From U.S. Naval Hydrographic Office.) (b) Magnetic-anomaly map showing a dipole anomaly over a pluton with a field inclination of 60° at 60° south. (c) Map from low latitude showing separation of anomalies from a single magnetic source into a dipolar pair. (b) and (c) from S. Breiner, 1973, Applications manual for portable magnetometers, E. G. & G. Geometrics, Inc.; used by permission.)

Similar measurements can be made of paleomagnetism in sedimentary rocks, where the property is called ***depositional remanent magnetism.*** Redbeds are most useful sedimentary rocks for paleomagnetic study because of the concentration of magnetic minerals. Measurements can also be made—with greater difficulty—on fine-grained limestone and other sedimentary rocks. Sedimentary and volcanic rocks are more desirable for study than most plutonic rocks because bedding provides an originally horizontal reference surface to which rotations and translations may be related.

Many rocks undergo changes after they form, changes that can alter and even erase paleomagnetic signatures. Chemical changes in composition or oxidation state may alter remanent properties. Such changes can sometimes be partly eradicated by treating rocks with acid or complexing agents to remove altered material; this may restore the original signature sufficiently to permit measurements. Thermal overprinting, particularly on some Paleozoic rocks in North America, hinders paleomagnetic measurement. Controlled reheating of samples may alleviate the problem. Paleomagnetic measurements made in a *cryogenic magnetometer,* which permits measurement inside a superconducting ring, may compensate for some difficulties, particularly where paleomagnetic signatures are weak.

A check called a ***fold test*** is performed on all samples to determine if paleomagnetic measurements can be used. The orientation of samples collected from opposite limbs of folds, or from otherwise tilted rocks, are plotted on an equal-area net or stereonet (see Appendix 1) and then rotated back to the horizontal. Paleomagnetic measurements for each sample are compared, and, if the rotation does not produce a consistent orientation, the locality may be abandoned or the results discarded as spurious. Further thermal or chemical treatment of samples may help.

Applications Using Magnetism

The concepts of terrestrial magnetism are widely applied in solving structural problems. Faults may be recognized by associated linear trends and by truncation of magnetic trends in rock bodies (Figure 20–7). Folds may be identified by noting the curvature of contrasting magnetically susceptible units. Plutons may be recognized by the shape of the magnetic anomaly. The bulk mafic or felsic composition of plutons may be estimated by noting the kind of associated magnetic anomaly (positive or negative) in conjunction with the corresponding gravity anomaly (positive or negative). Magnetic-anomaly maps may be used to identify boundaries and assess the crustal character of suspect terranes. Linear magnetic anomalies in the seafloor have been dated and correlated to provide a better understanding of the evolution of oceanic crust during the past 200 million years.

Data from paleomagnetic studies are useful in determining the paleolatitude of a rock body or a large crustal block, but cannot be used directly to determine longitude. Local rotation of a rock body due to faulting may be measured. ***Apparent polar wandering (APW) curves*** derived from paleomagnetic studies are useful in reproducing ancient positions of a continent. Such studies have been used to determine the original positions of cratonic blocks in Precambrian shields and also the positions of larger continents and suspect terranes during later geologic periods.

Gravity

Gravity is the force of mutual attraction among all bodies in the universe. It is exerted on and by all components of the Earth, including rock bodies. The more dense and voluminous the body, the stronger the force of gravity, and so denser rocks such as gabbro possess a stronger gravitational force. ***Newton's law of gravitation*** states that the force of gravity (**F**) is directly proportional to the masses (m) of the objects involved and inversely proportional to the square of the distance (r) between them, or

$$\mathbf{F} = G\frac{m_1 m_2}{r^2}\boldsymbol{\gamma}_1 \qquad (20\text{–}2)$$

where G is the universal gravitational constant, 6.67×10^{-8} dyne-cm^2/s^2, and $\boldsymbol{\gamma}_1$ is a unit vector directed from m_1 to m_2.

Every object in our solar system exerts a force of attraction on every other body. The magnitude of gravity on the Earth is related to the fact that the Earth is large in comparison to objects on it or any one rock body in it. Because we are relatively close to the source of gravity on the Earth, as compared to other objects in space, this force holds us on the Earth. Newton's law and the force of gravity are used to calculate the amount of fuel needed to power spacecraft from the Earth to our moon and other planets.

To measure variations in the force of gravity on the Earth, we use a *gravity meter.* First a value is assumed for average density of the crust where the mea-

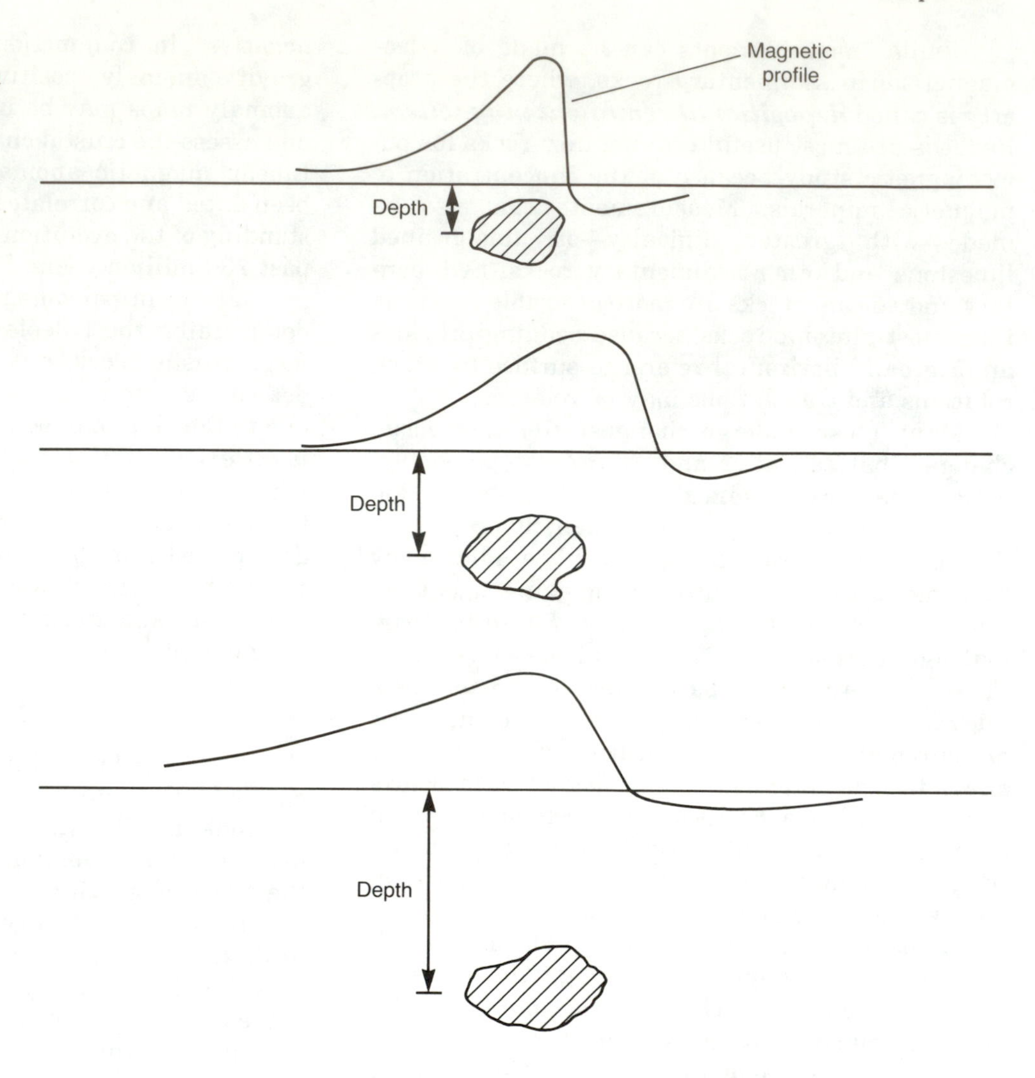

FIGURE 20–5
Depth dependence of the width of magnetic anomalies. As the depth to the source of the anomaly increases, the amplitude decreases and the anomaly widens.

surements are made. A density of 2.65 (the density of average granite) is commonly assumed for continental crust, and 3.0 (the density of average basalt) is assumed for oceanic crust. These densities should lead to an acceleration of gravity of 1 gal (1 cm/s^2)—a unit named after Galileo (1564–1642). The average value of gravity is 980 gals. Measured values of gravity that deviate from estimated values define ***gravity anomalies.*** Measurements of differences in the gravity field are made in units of milligals, and modern gravity meters (Figure 20–8) are capable of measurements within ±0.04 milligals. If the measured value of gravity at a station is greater than predicted by using the density assumed, a ***positive gravity anomaly*** exists; if it is less, a ***negative gravity anomaly.*** Gravity profiles may be constructed or, if the gravity measurements are distributed widely enough, the data may be contoured as a gravity anomaly map (Figures 20–3 and 20–9). If the map is corrected only for elevation above sea level, it is called a ***free-air gravity map;*** if data are corrected for differences in both elevation and density between sea level and the elevation where measurements were made, a ***Bouguer gravity anomaly map*** is produced (the name honors a French surveyor and hydrographer of the eighteenth century, Pierre Bouguer). Free-air measurements are most useful for determining the closeness

FIGURE 20–6
Magnetic-anomaly map showing contrasts in deep (broad anomalies) and shallow (high-frequency anomalies) magnetic sources in Georgia and South Carolina. (Data from Isidore Zietz, 1982, Composite magnetic anomaly map of the conterminous United States: U.S. Geological Survey.) The east-west linear anomalies crossing the entire northern part of the map area and passing through the middle of South Carolina record a large late-Paleozoic dextral fault zone. Circular anomalies are plutons at various depths.

NC
GA
SC
0 25 50 100
Kilometers

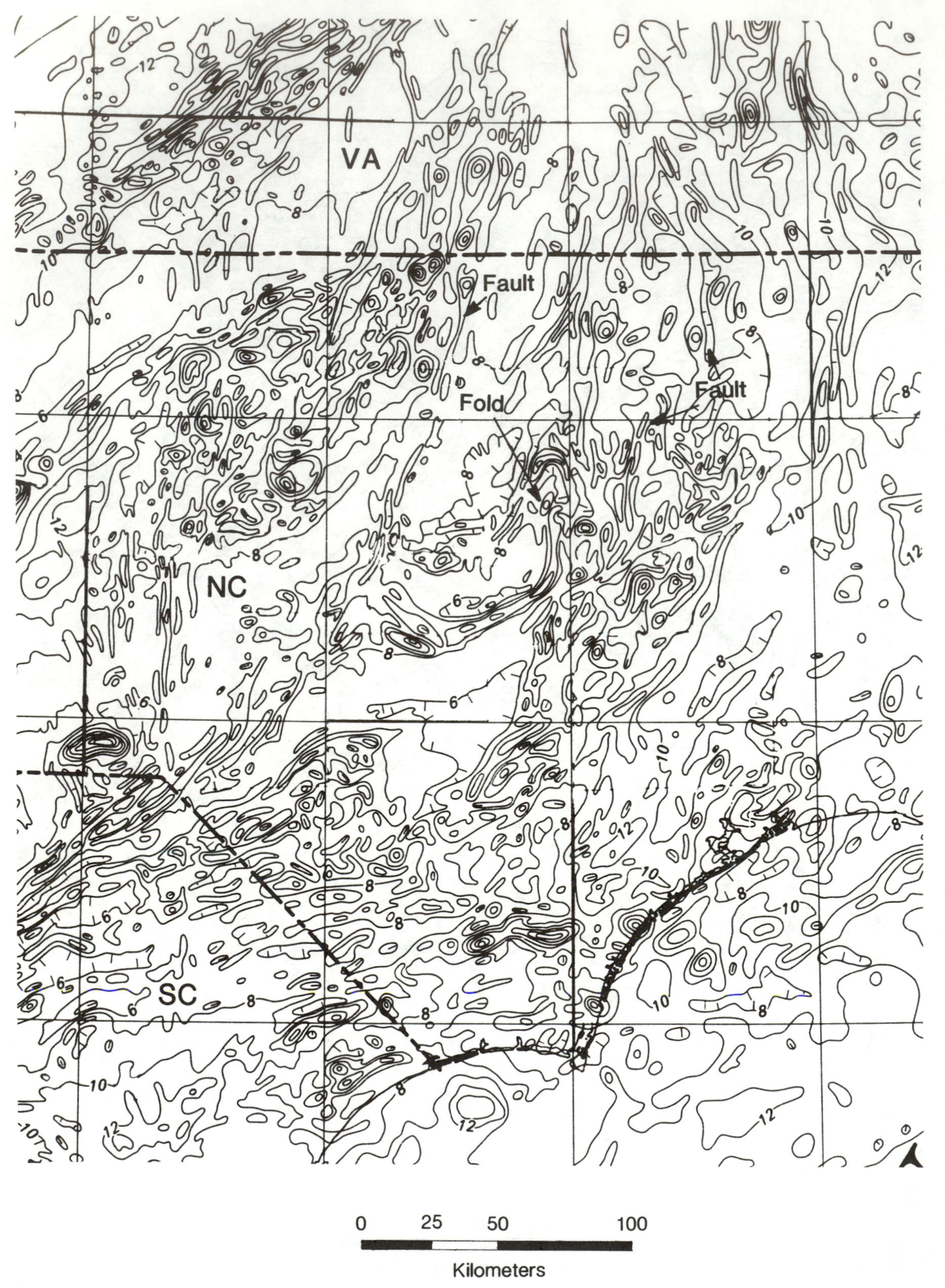

(a)

FIGURE 20–7
Magnetic-anomaly map of part of the North Carolina Piedmont (a) (opposite page) showing truncation of folds by faults (b) (below). The fold is defined by highly magnetic amphibolites; a swarm of Jurassic diabase dikes is suggested by subparallel linear anomalies. (From Isidore Zietz, F. E. Riggle, and F. P. Gilbert, 1984, U.S. Geological Survey Map GP-958.) (c) (page 448) Magnetic-anomaly map showing several plutons in the Piedmont of North Carolina and South Carolina. (From Isidore Zietz and F. P. Gilbert, 1980, U.S. Geological Survey Map GP-936.) (d) (page 449) Seafloor magnetic anomalies in the Pacific off the Pacific Northwest coast. (From W. J. Morgan, *Journal of Geophysical Research,* v. 73, p. 1959–1982, 1968, copyright by the American Geophysical Union.)

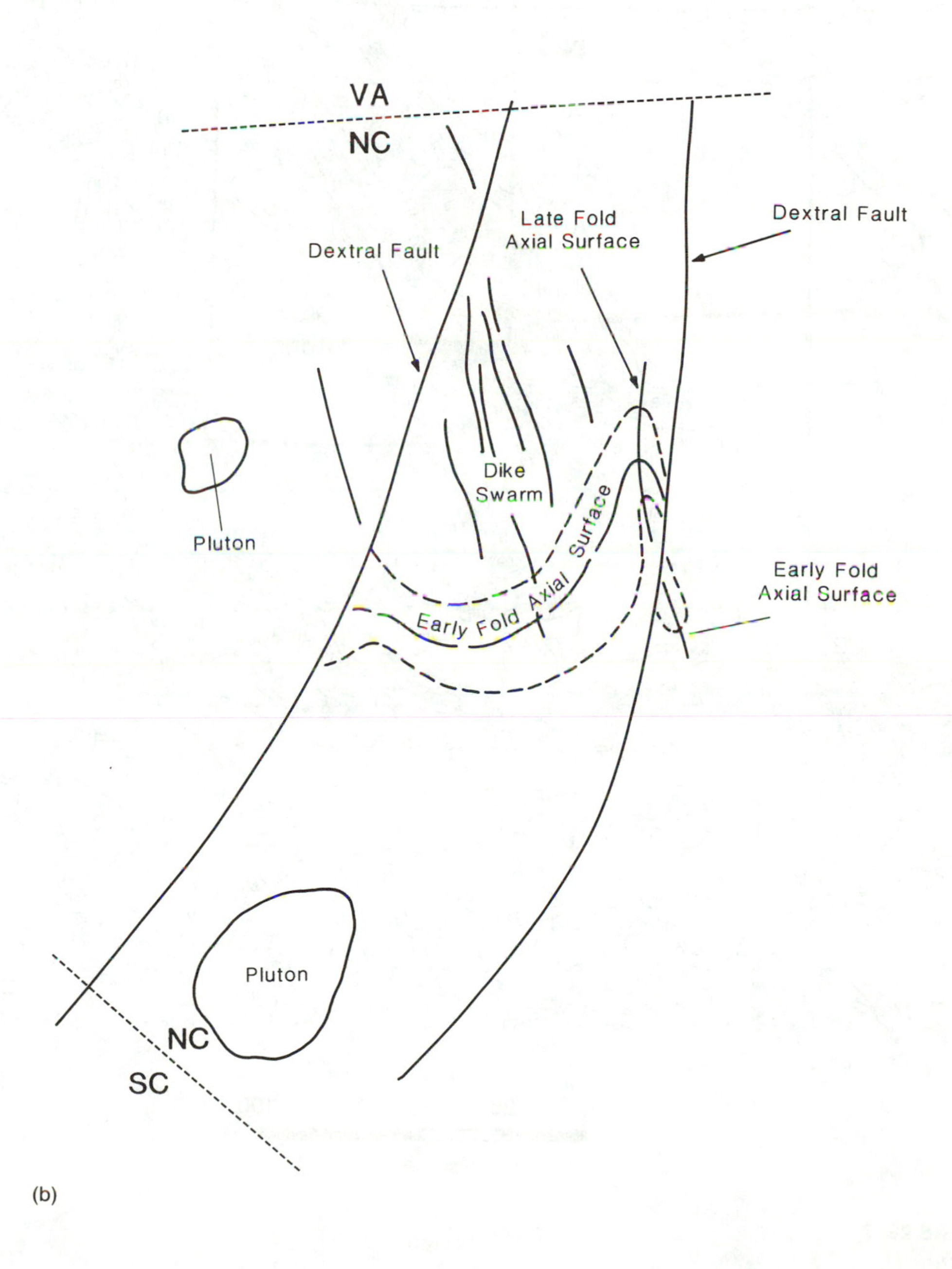

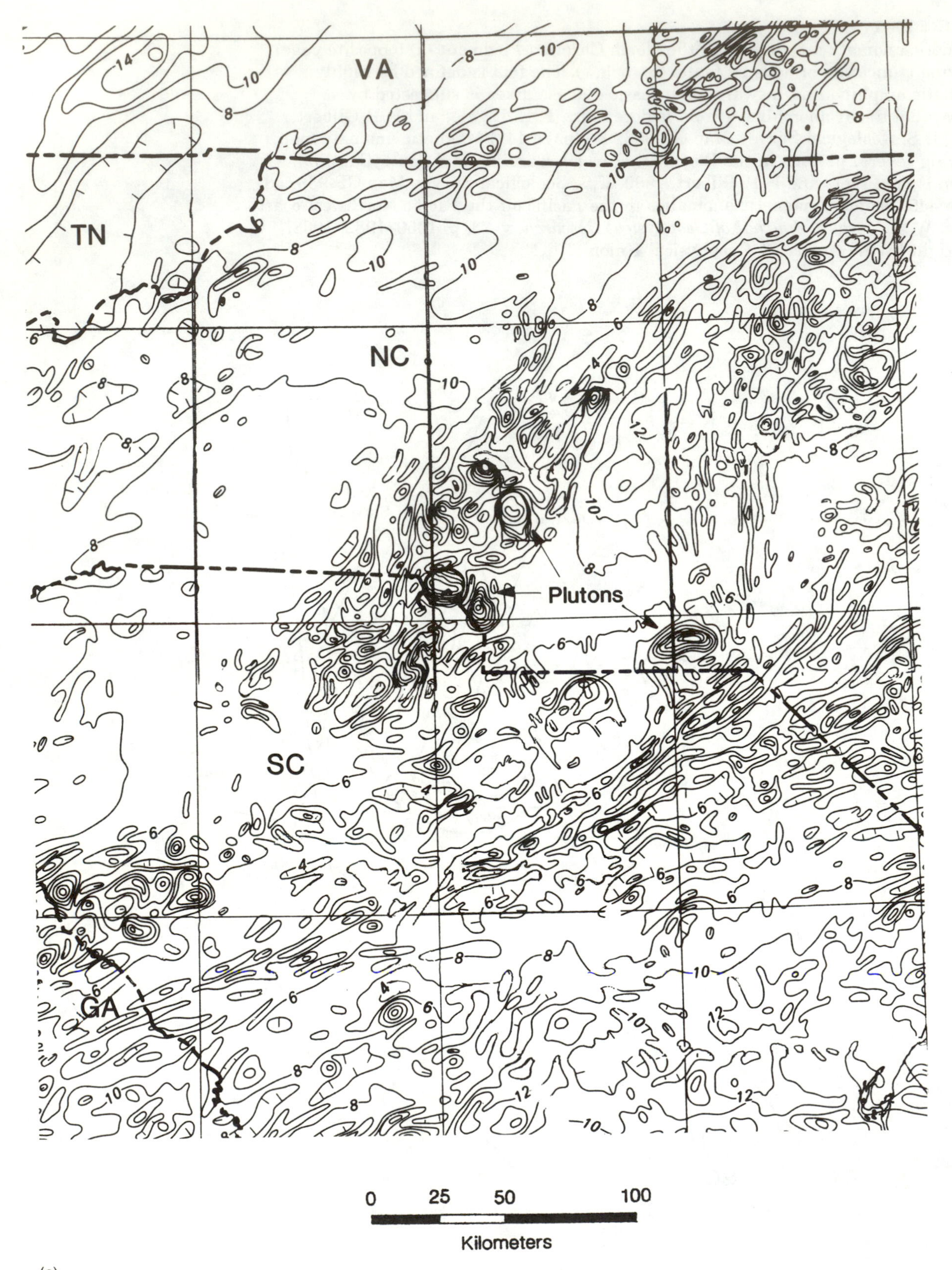

(c)
FIGURE 20–7
(continued)

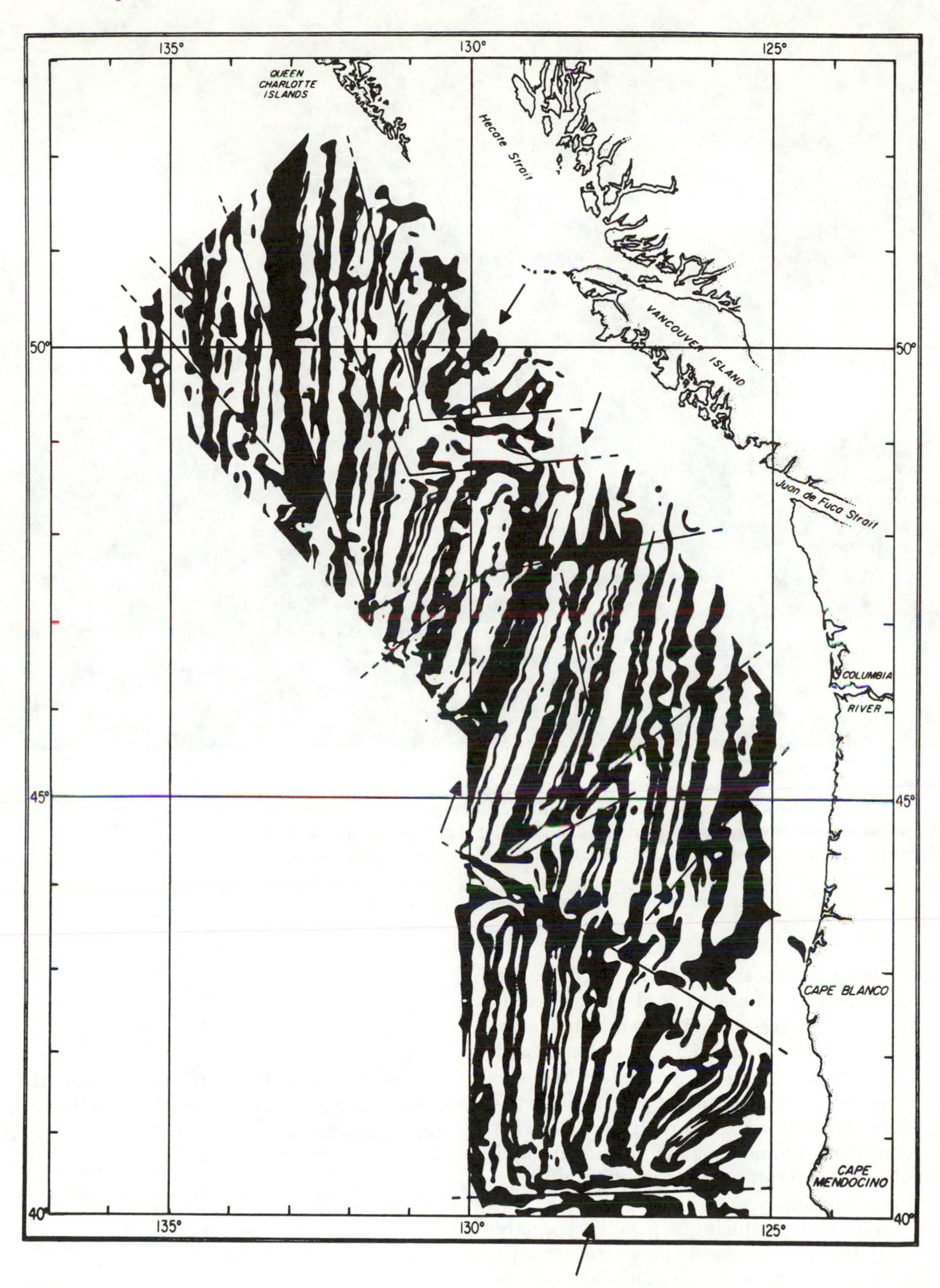

(d)

FIGURE 20–7
(continued)

FIGURE 20–8
Gravity measurement being made with a LaCoste and Romberg Model G gravity meter. This instrument combines the principles of the long-period seismograph and a zero-length spring—with tension proportional to the length of the spring. (RDH photo.)

to isostatic equilibrium of the crust. If a large free-air anomaly is present, the mass is out of isostatic equilibrium. A *terrain correction* is frequently made for Bouguer data, and if such corrections are made and plotted as a map the result is a *complete Bouguer anomaly map.* The magnitude of the gravity field varies with latitude, elevation, and Earth tides. Gravity measurements are commonly made from ships and may be made from aircraft but with greater difficulty and much less reliability.

Depth to a mass producing a gravity anomaly may be calculated after assuming a particular shape for the body causing the anomaly. The calculations remain simple if the shape is assumed to be a sphere or a cylinder. If the source of an anomaly of amplitude g_0 is assumed to be a spherical body, the depth to the anomaly may be calculated from the half-width of a profile through the anomaly (Figure 20–10a) as

$$g(x) = g_0 \frac{1}{\left(1 + \frac{x^2}{z^2}\right)^{3/2}} \tag{20–3}$$

where x is the width of the gravity anomaly and z is the depth to the anomaly. If the half-width of the anomaly is used,

$$x_{1/2} = \frac{1}{\left(1 + \frac{x_{1/2}^2}{z^2}\right)^{3/2}} \tag{20–4}$$

and

$$z = 0.77x_{1/2}$$

Calculation of the depth of a cylindrical body (Figure 20–10b) is even simpler:

FIGURE 20–9
Bouguer gravity-anomaly map of part of the South Carolina Piedmont shown in Figure 20–7b. Circular anomalies indicate plutons. Gravity lows are granitic plutons; highs are mostly gabbros. (After Pradeep Talwani, L. T. Long, and S. R. Bridges, 1975, Simple Bouguer Map of South Carolina, South Carolina Geological Survey.)

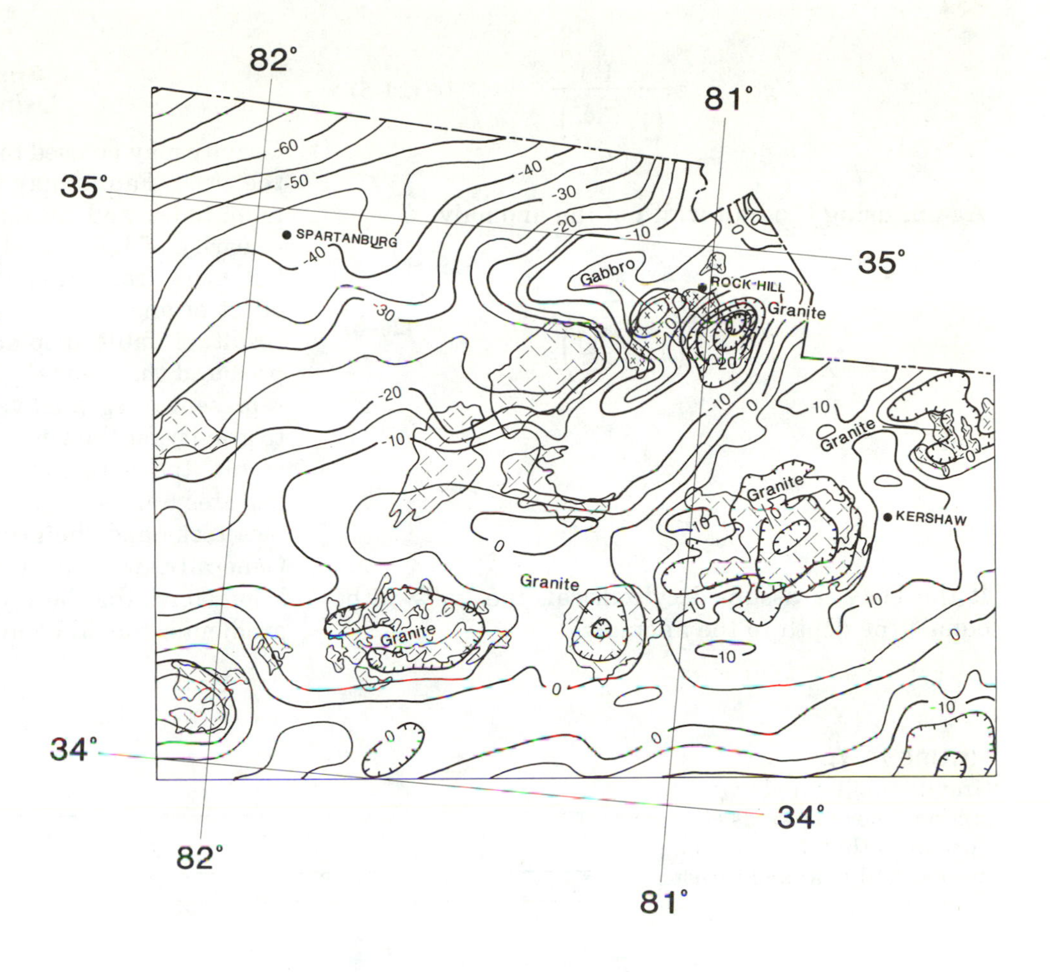

FIGURE 20–10
Calculation of the depth to spherical (a) and cylindrical (b) bodies from gravity data.

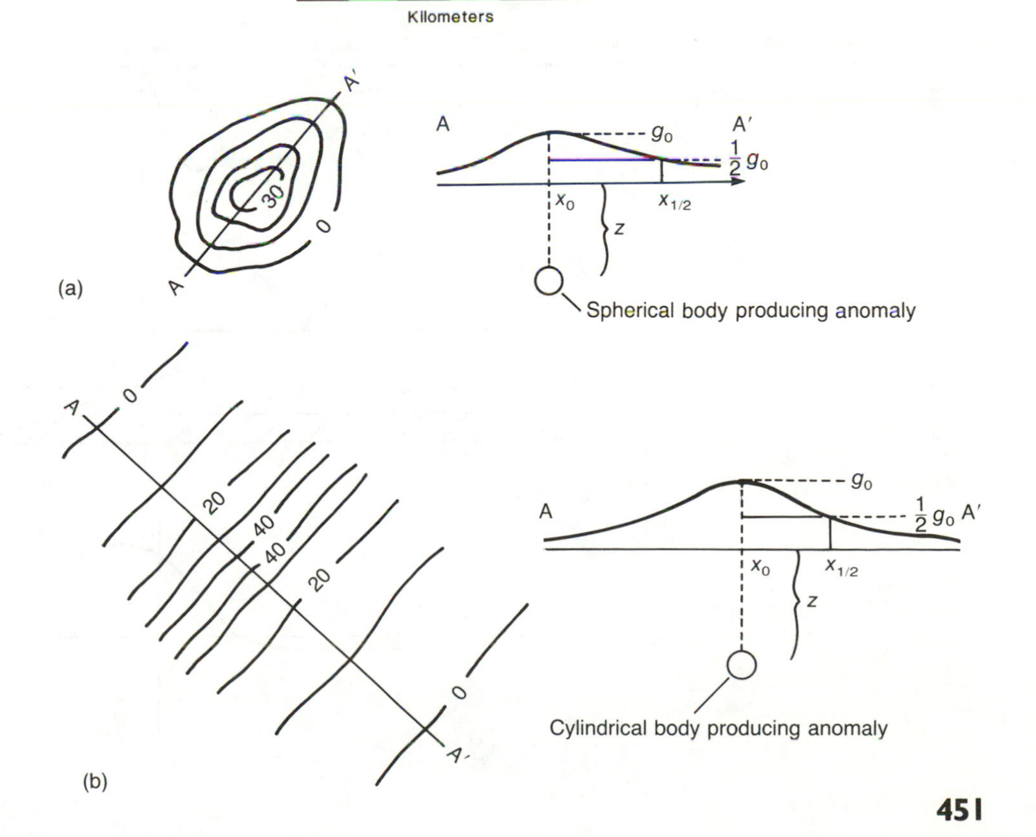

$$g(x) = g_0 \frac{1}{\left(1 + \frac{x^2}{z^2}\right)} \qquad \textbf{(20–5)}$$

Again, using the half-width of the anomaly,

$$x_{1/2} = \frac{1}{\left(1 + \frac{x_{1/2}^2}{z^2}\right)} \qquad \textbf{(20–6)}$$

and

$$x_{1/2} = z$$

If the body is assumed cylindrical, the half-width equals the depth to the anomaly.

Applications Using Gravity

Gravity may be used to characterize large regions of the crust. Faults may be recognized by contrasting amplitudes and patterns on adjacent blocks. The linearity of the actual fault sometimes is obvious, but more frequently the truncation, deflection, or offset of anomalies at faults is readily observed, a result of fault displacement of the masses that produced the anomaly (Figure 20–11). Gravity data represent integrated values for the acceleration due to gravity for the whole Earth at a single point. As a result, the data are less specific for near-surface features unless large density contrasts exist in surface rocks and the gravity survey is very detailed. Generally, with a station spacing of two to five kilometers, the data yield only regional patterns, gradients, crustal boundaries, very large faults, and

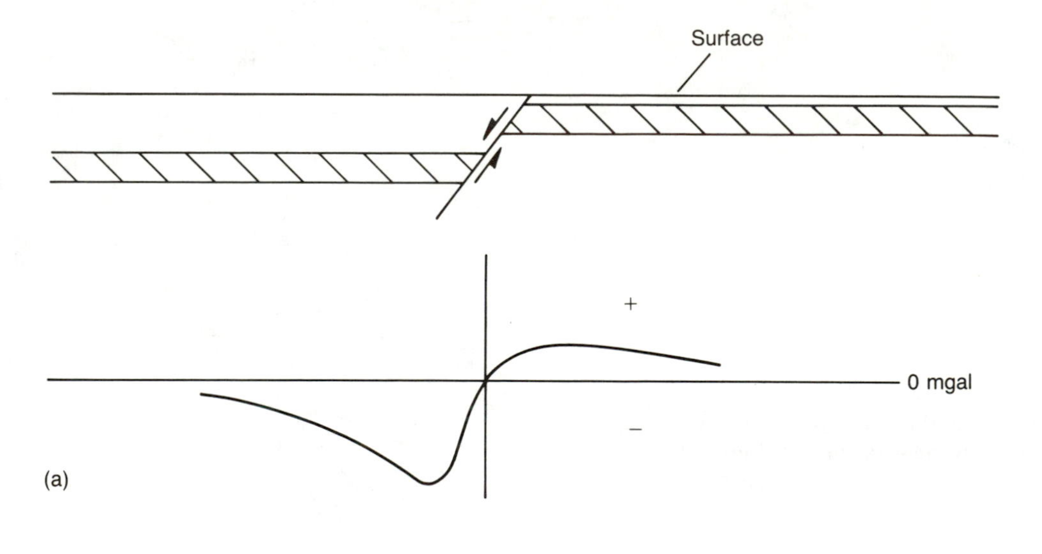

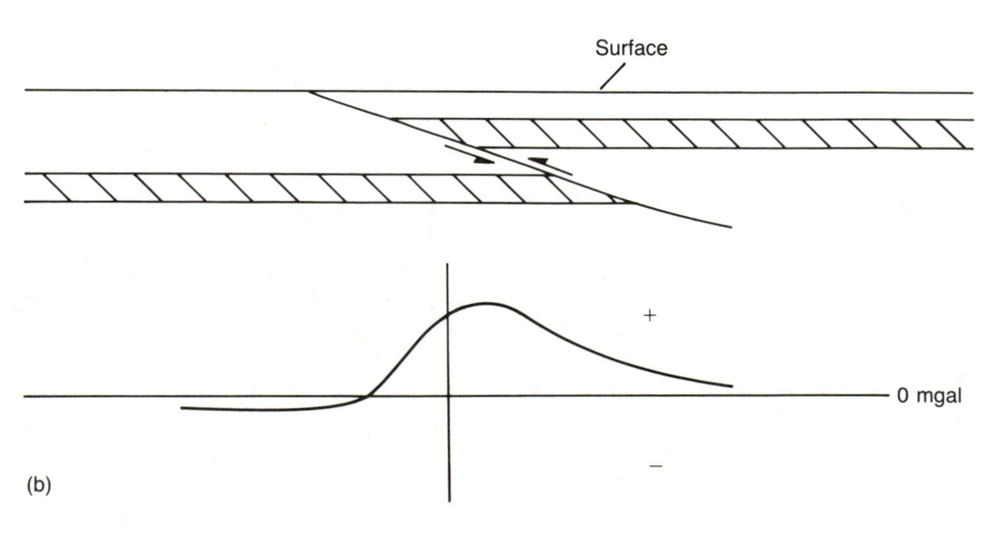

FIGURE 20–11
Gravity profiles and cross sections showing kinds of anomalies that develop along normal faults (a) and thrust faults (b).

plutons. Gravity data are useful for determining the shape and broad composition of plutons. Granitic plutons produce gravity lows because of their lower density, and mafic plutons yield gravity highs because of their higher density. Although neither gravity nor magnetic data alone yield precise identification of individual features, both used together are a powerful tool for delineating faults, plutons, and other structures (Figure 20–11). Generally, granitic rocks produce both gravity lows and magnetic lows, but if a granite contains enough magnetite it will yield a gravity low and a magnetic high. Mafic plutons tend to be both gravity highs and magnetic highs.

Gravity data can be used to calculate a *gravity model* for a part of the crust (Figure 20–12). Magnetic modeling is more difficult, for it requires full knowledge or accurate assumptions about susceptibility of the material being modeled. John McBride and Douglas Nelson (1988) have used a combination

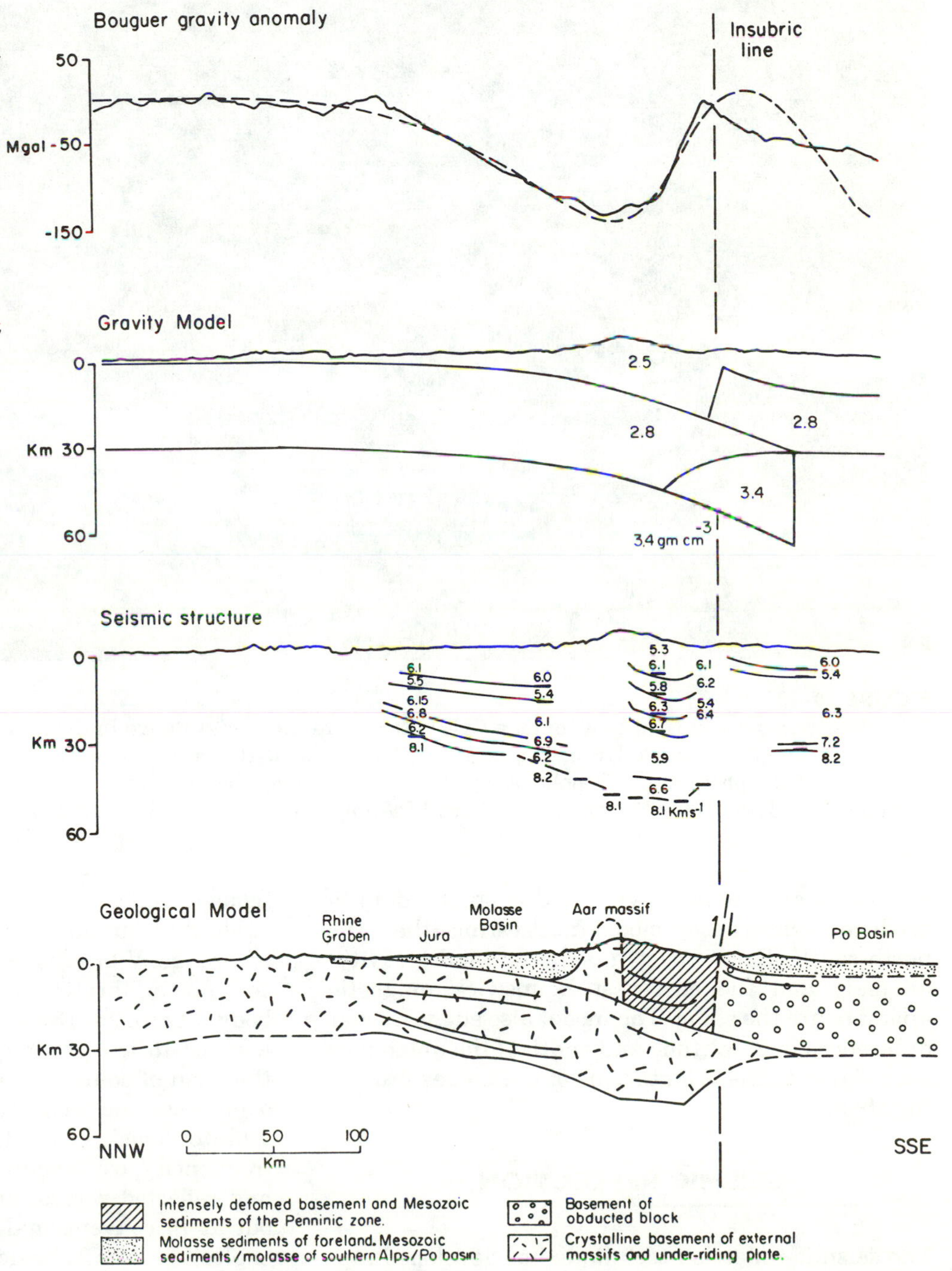

FIGURE 20–12
Gravity profiles observed (solid line in top figure) and calculated (dashed line) for the Alps shown as a geologic model and a gravity model. The seismic structure was determined from refraction studies. (From G. D. Karner and A. B. Watts, *Journal of Geophysical Research,* v. 88, p. 10,449–10,477, 1983, copyright by the American Geophysical Union.)

FIGURE 20–13
Vibroseis trucks working in tandem on a COCORP line in the House Range in the Basin and Range of western Utah. The recording truck (silhouette) is shown on the right edge of the photograph. Geophones would be buried along the roadside. (Douglas Von Tish, Larry Brown, and Richard W. Allmendinger, Cornell University.)

of seismic reflection, gravity, and magnetic data to produce a magnetic model suggesting that the north-south East Coast magnetic (and gravity) anomaly and the east-west Brunswick magnetic anomaly are the same. The model also suggests that differences in amplitude and width of the anomalies are related to the orientations of the bodies producing them.

SEISMIC REFLECTION

The seismic-reflection technique has developed rapidly during recent decades because computers and digital processing have speeded data processing. Seismic reflection was originally developed using explosives, but another development is the Vibroseis technique. Vibroseis uses a truck (or more than one) that has a vibrating steel plate mounted under the body (Figure 20–13); at regular intervals the plate is lowered to the ground, where vibration energy in the form of sound waves is passed into the ground over a specific frequency range. If more than one vibrator truck is used, the spacing and vibration frequencies are synchronized. To record transmitted and reflected waves, an array of *geophones* is laid out as in conventional seismic surveys (which use explosives as a seismic source) and the waves are recorded similarly, but with greater control over frequency and recording time. Multiple traces may be recorded and processed over the same structures, producing sharper records, which in turn provide greater latitude in reprocessing data to aid interpretation of crustal structure (Figure 20–14). An airgun may be used in much the same way, but it transmits less energy into the Earth from land than through water. Airguns are used for marine surveys. For maximum transmission of energy, and therefore a maximum return of reflected energy, explosives are still the best.

Seismic reflection is best suited to recognition of nearly horizontal structures (such as faults and

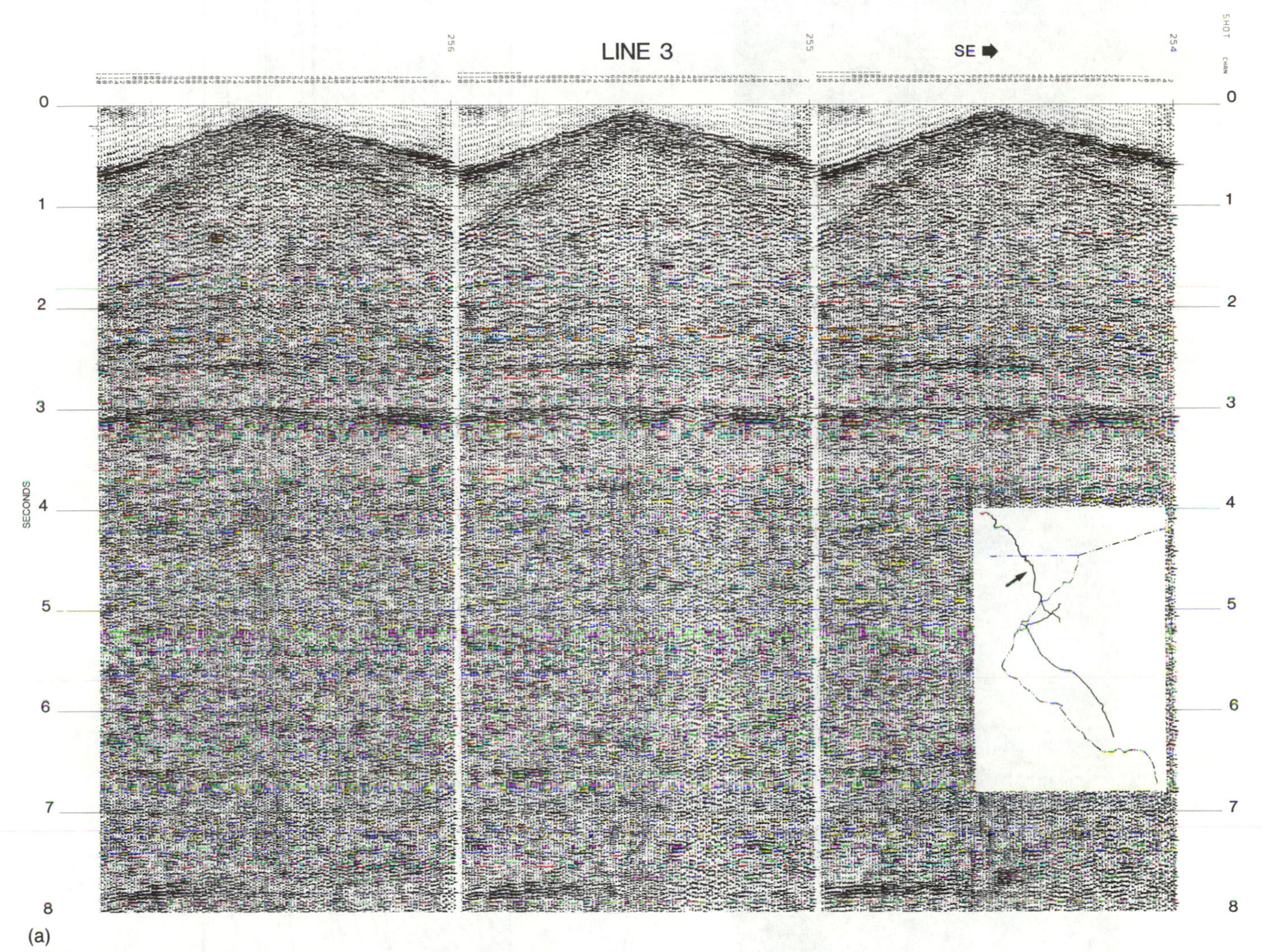

FIGURE 20–14
(a) Field record for three vibrator shot points collected as part of a Vibroseis line as part of the Appalachian Ultradeep Core Hole (ADCOH) site investigation in northeastern Georgia (arrow on inset map) showing major reflectors at 2.6, 3.1, and 8 seconds two-way travel time. Field records for each shot point are correlated and stacked using the computer to construct the initial section—called a "brute stack." (b) (page 456) Processed section constructed from field records like those in (a). Data collected as Line 1 (location shown on map in a) from northwestern South Carolina as part of the ADCOH Project site investigation. (Data processed at Virginia Tech by J. K. Costain and Cahit Çoruh.) (c) (page 456) Geologic interpretation of the line shown in (b). Striped area is the Blue Ridge–Piedmont crystalline thrust sheet. Reflectors in the right third of the seismic-reflection profile image recumbent folds that are mappable on the surface. Reflectors are produced by strong acoustic contrasts of the folded granitic rocks and amphibolite. Strong reflectors around 3-s two-way travel time are derived from flat-lying sedimentary rocks on top of crystalline basement. Inclined reflectors immediately beneath the 3-s reflectors are interpreted as small Cambrian rift basins. The Brevard fault zone is the distinct set of inclined reflectors that project to the surface near shot-point 700.

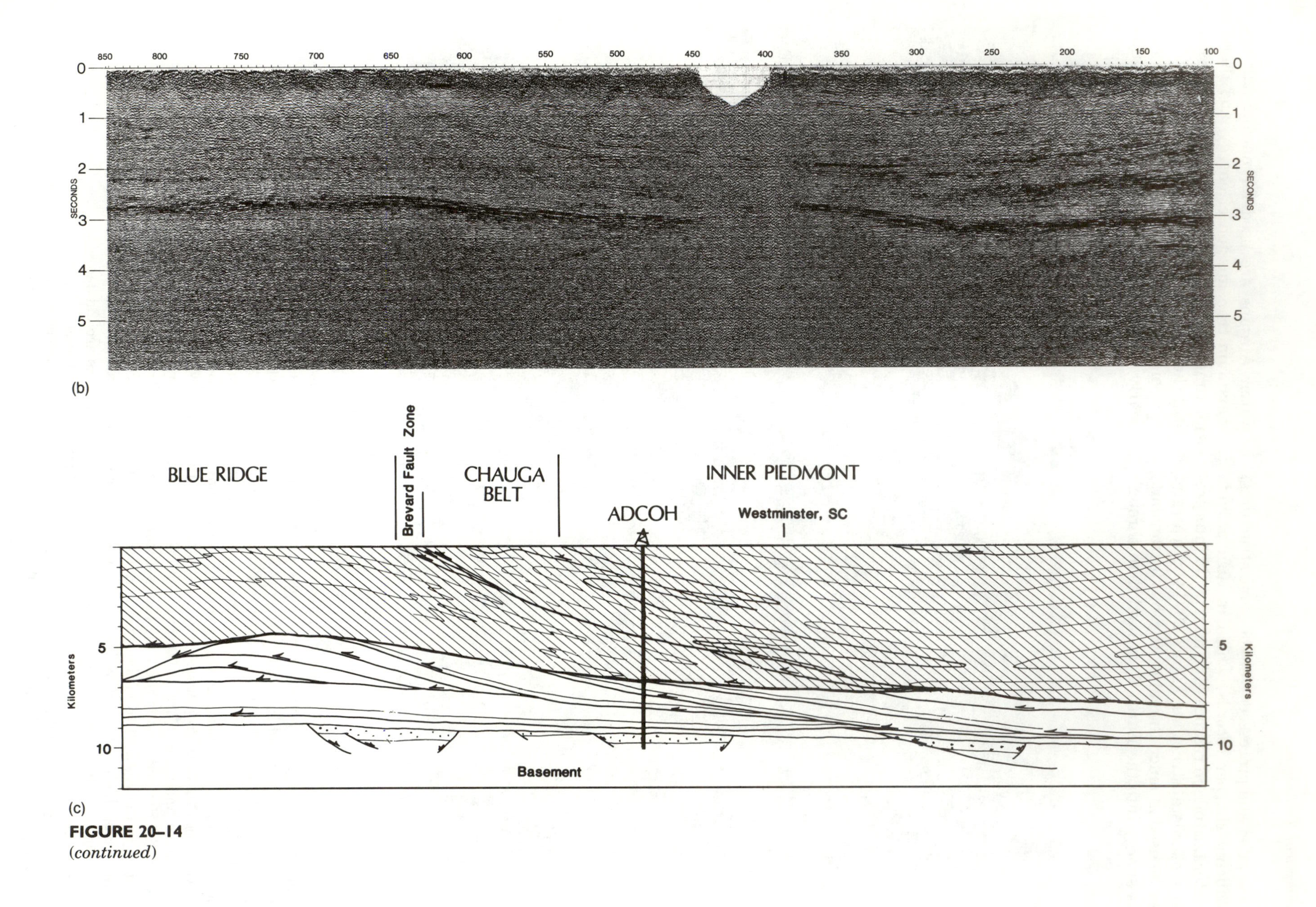

FIGURE 20–14
(continued)

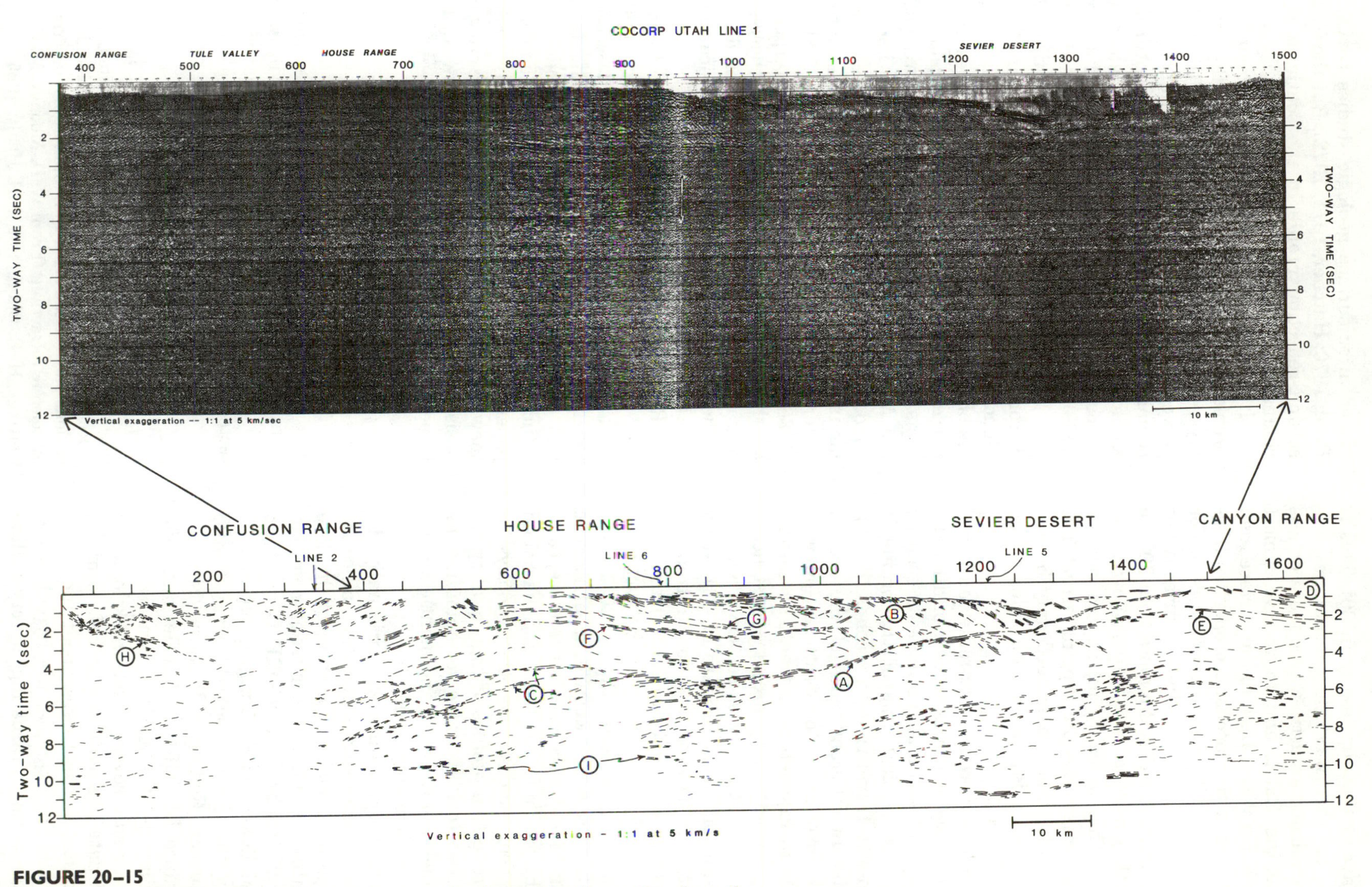

FIGURE 20–15
COCORP seismic-reflection profile and tracing of reflectors in the section from the Basin and Range province of Utah. (From Allmendinger and others, 1983, *Geology,* v. 11.) Structural interpretation of this section is shown in Figure 13–3.

packages of layers) that dip less than 30°. This is entirely due to the way the technique was designed—to transmit acoustic waves more or less vertically into the Earth and to receive almost vertical reflection of waves back to the surface from rock layers of contrasting acoustic properties (Figure 20–14). Theoretically it is possible to record or process reflections from steeply dipping surfaces, but in practice few reflectors that dip more than 30° are observed, and they may be selectively removed during processing. This probably introduces bias or even significant error. Steeply dipping structures, such as some faults, may be imaged only by displaced gently dipping layers on both sides of the structure (see Figure 11–34).

A processed seismic-reflection profile consists of a two-dimensional section through part or all of the crust (see Figure 11–34) and resembles a geologic cross section. The vertical axis of the profile is commonly two-way travel time in seconds. Knowledge of velocity in different parts of the section permits conversion of a time to a depth section, with a vertical scale in kilometers or feet. The seismic section may be used (with care) to construct a well-constrained structural cross section along the line of the seismic profile (see Figure 11–34). Many artifacts (unnatural flaws in the data introduced during data acquisition and processing) are only partly removed during processing. Construction of high-quality geologic sections (Figure 20–15; see also Figure 13–3) requires knowledge of seismic velocities and field and processing parameters, careful search of the profile for artifacts, and maximum use of surface geologic data and other geophysical data. Seismic-reflection profiles of high quality are probably the best tool available for imaging the structural geometry of the crust and mantle and for projecting surface structure to depth.

Seismic-reflection profiling was originally designed as a tool for petroleum exploration. Since the mid-1970s it has been adapted for study of the deep crust of the continents and the interiors of mountain chains (Figure 20–16) by the COCORP (Consortium for Continental Reflection Profiling) group at Cornell University, the U.S. Geological Survey, BIRPS (British Institutions Reflection Profiling Studies), ECORPS in France, DECORP in West Germany, and others, including the CALCRUST consortium in California. As a result, several thousand kilometers of new crustal data are available to help confirm, constrain, or disprove hypotheses about the structure of the crust. For example, the hypothesis of large-scale thin-skinned thrusting of crystalline rocks in the southern Appalachians had been suggested after surface geologic studies (Bryant and Reed, 1970; Hatcher, 1971, 1972). But the COCORP seismic-reflection profile in the southern Appalachians gave it greater credibility.

It has also raised new problems. Major unsolved problems remain: determination of the nature of layered reflectors deep in the crust, the nature of curved reflectors, deep-seated "bright spots" (exceptionally strong reflectors originally identified in the shallow crust as hydrocarbons, but which in the deeper crust may be magma; de Voogd and others, 1988), and the zones of no reflectors called transparent zones. The Mohorovičić discontinuity (M-discontinuity or Moho, named in honor of Andrija Mohorovičić, 1857–1936, a Croatian seismologist) is visible on reflection profiles as a series of nearly horizontal layered reflectors. Among related questions: What does the Moho actually represent? Why are Moho reflections often several rather than a single reflector? Does the Moho represent a zone of laminar flow at a major rheological discontinuity? When did the present Moho form? Jack Oliver (1988) has brought forth alternative interpretations of the Moho, including some of these—interpretations that we were incapable of discussing without the abundant crustal geophysical data we now have.

Reflection profiling has brought to light many aspects of crustal structure previously unknown, made possible the confirmation and description of several types of structure in the continental crust, and permitted formulation of new models for the structure of the continental crust. Compendia of papers (such as Barazangi and Brown, 1986; Matthews and Smith, 1987) interpreting the crust on the basis of these data and presenting new ideas continue to appear.

SEISMIC REFRACTION

The principle of refraction of waves as they pass from one layer to another is used in ***seismic-refraction*** studies. Sound or other waves are detected after they are produced by an explosion or an

FIGURE 20–16
(a) Ray traces from source to receiver array in a seismic-refraction experiment in Saudi Arabia, and a velocity model (lower right). (b) Seismic velocities for various parts of the crust in Saudi Arabia, determined by seismic refraction. (From W. D. Mooney, M. G. Gettings, H. R. Blank, and J. H. Healy, 1985, *Tectonophysics,* v. 111, Elsevier Science Publishers.)

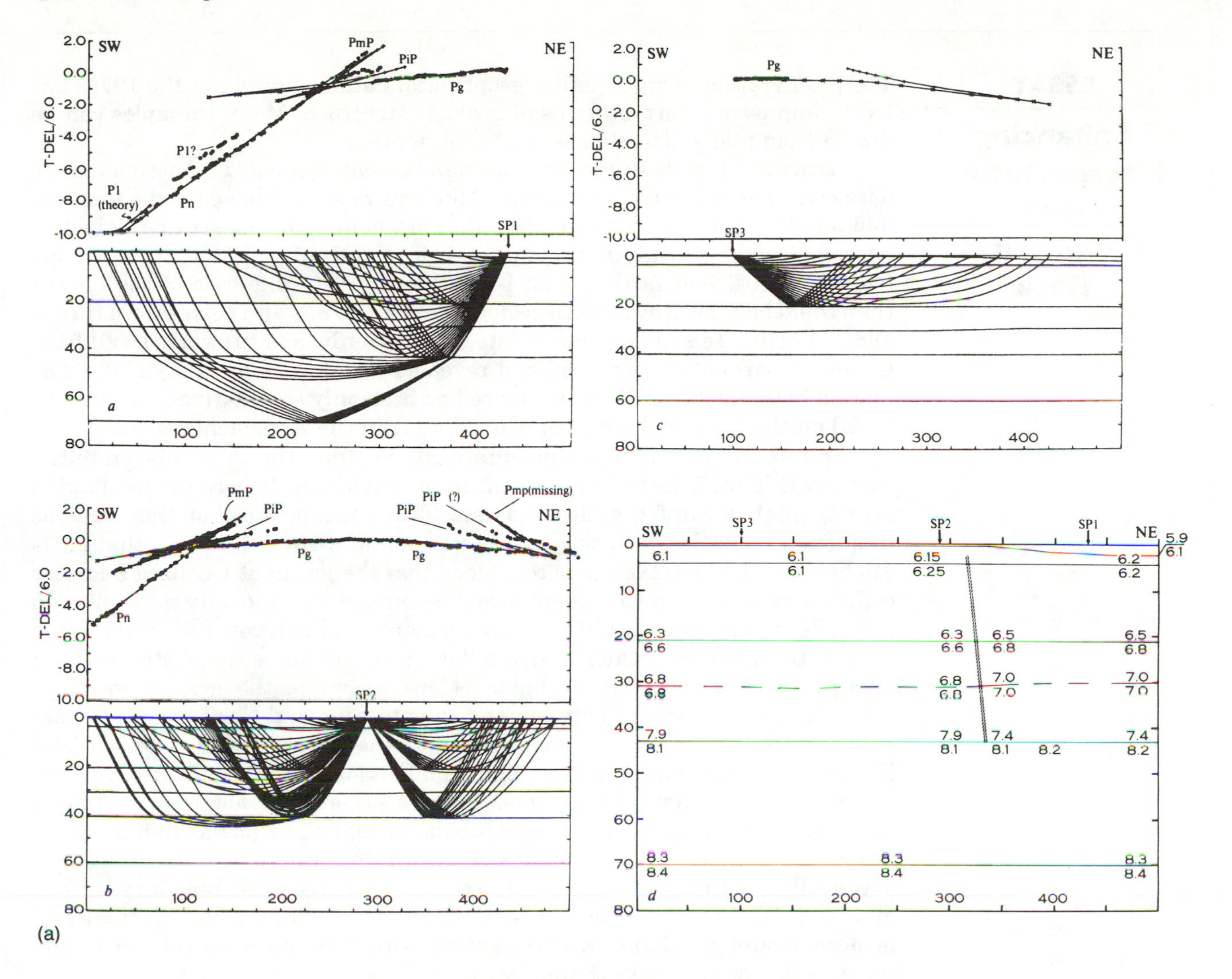
SW
NE
PmP
PiP
Pg
P1?
P1 (theory)
Pn
T-DEL/6.0
SP1
SP2
SP3
PiP (?)
Pmp(missing)
a
b
c
d

(a)

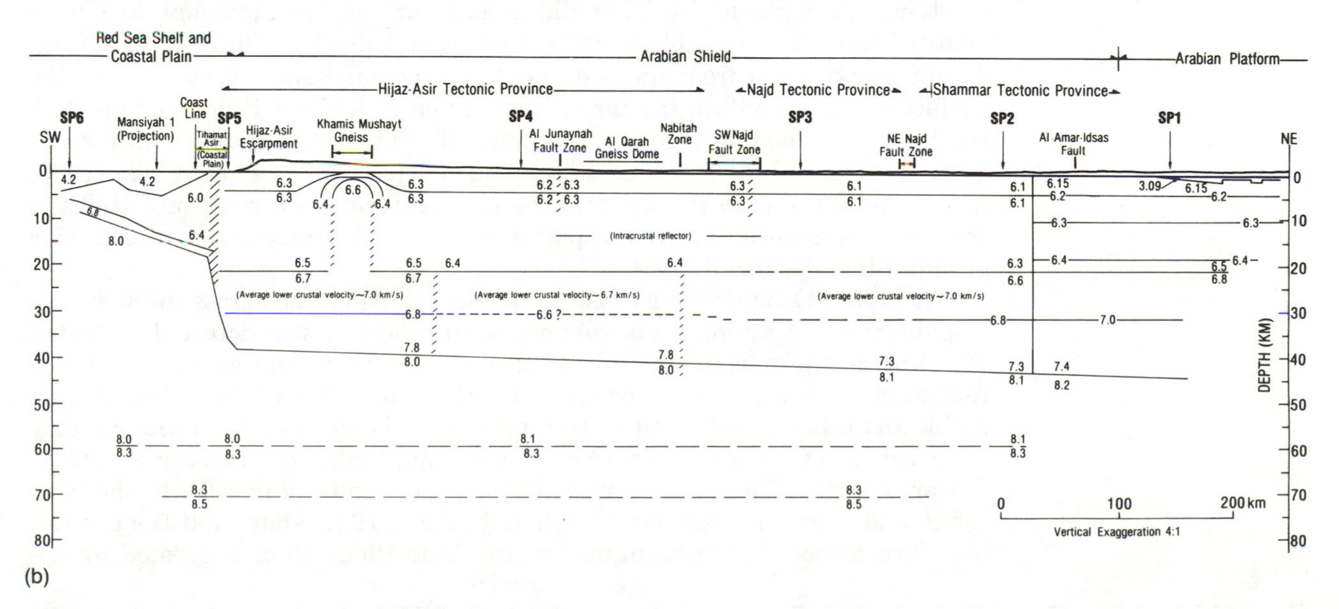
INTERPRETIVE CRUSTAL SECTION ALONG THE 1978 SEISMIC REFRACTION PROFILE
KINGDOM OF SAUDI ARABIA
Red Sea Shelf and Coastal Plain
Arabian Shield
Arabian Platform
Hijaz-Asir Tectonic Province
Najd Tectonic Province
Shammar Tectonic Province
SP6
Mansiyah 1 (Projection)
Coast Line
SP5
Tihamat Asir (Coastal Plain)
Hijaz-Asir Escarpment
Khamis Mushayt Gneiss
SP4
Al Junaynah Fault Zone
Al Qarah Gneiss Dome
Nabitah Zone
SW Najd Fault Zone
SP3
NE Najd Fault Zone
SP2
Al Amar-Idsas Fault
SP1
(Intracrustal reflector)
(Average lower crustal velocity ~7.0 km/s)
(Average lower crustal velocity ~ 6.7 km/s)
DEPTH (KM)
Vertical Exaggeration 4:1

(b)

ESSAY

Enhancing Interpretation with Geophysical Data

The proliferation of high-quality geophysical data beginning in the 1970s has led to improved interpretations of geologic structure. Many examples can be drawn from many places on several continents.

Structural trends in the Precambrian basement imaged by aeromagnetic data continue uninterrupted many kilometers westward beneath the foreland fold-and-thrust belt of the Canadian Rockies, helping demonstrate the lack of basement involvement in the thrust belt. Continuity of the thin-skinned fold-and-thrust belt farther west beneath the crystalline core of the orogen than could be proved by surface geologic data alone has also been shown (Price, 1981). Earlier, seismic-reflection data (Bally, Gordy, and Stewart, 1966) from the outer parts of the foreland had revealed the thin-skinned style of deformation here. Earlier still, its existence had been only speculative (but correct), based on the work of John Rich (Chapter 11) in the Appalachians.

An integral part of many mountain chains, from the Precambrian to the Tertiary (Figure 20E–1), is a gradient from gravity low toward the foreland to gravity high toward the closed ocean. There the position of the gradient frequently coincides with the surface position of a suture, and on that basis Michael D. Thomas (1983) has concluded that the gradient localized a former collision zone. Changes in the nature of aeromagnetic anomaly patterns also coincide in many places with gravity gradients and sutures. But in the same orogen the gravity gradient may follow a suture for some distance, then diverge from it. The Appalachians is one example: the gravity gradient coincides with the suture between an exotic terrane and North America from Alabama to Virginia, then diverges from the suture to the north. Thus Deborah Hutchinson, John Grow, and Kim Klitgord (1983) concluded that the Appalachian gravity gradient, and perhaps others, did not have a unique origin. Also, some geologists have speculated that the Appalachian gradient formed during Mesozoic extension and crustal thinning related to the opening of the Atlantic Ocean and not at all to contraction processes. But Gary Karner and Alan Watts (1983) declared that the gravity field over both ancient and modern mountain chains was consistent with the processes related to the construction of mountain chains by compression.

The extent of the huge Appalachian Blue Ridge–Piedmont thrust sheet has been better delineated by geophysical data (Figures 20–14b and 20–14c). Its existence throughout the Blue Ridge and part of the Piedmont had been deduced from surface geologic data (Bryant and Reed, 1970; Hatcher, 1972). Aeromagnetic data from the southern Appalachians show very little of the surface structure within the thrust sheet in both the Blue Ridge and much of the Piedmont, leading Hatcher and Zietz (1978) to report that the thin sheet extends across all of the Blue Ridge and much of the Piedmont. We (1980) incorporated the gravity data into the interpretation and confirmed that the changes in magnetic and gravity patterns occur at the same place—the gravity gradient just discussed.

In 1978, COCORP ran a crustal seismic-reflection line across the southern Appalachians. A strong set of subhorizontal reflectors was detected at depths of 2 km at the northwest end to 12 km or more to the southeast. These reflectors were initially interpreted as the Appalachian–Blue Ridge thrust (Cook and others, 1979), but were later said to be derived from sedimentary rocks beneath the thrust sheet (Ando and others, 1983; Cook and others, 1983). Better resolution in new seismic-reflection data acquired in 1984 in the Blue Ridge and western Piedmont (Çoruh and others, 1987) show that the crystalline thrust sheet is much thinner in the Blue Ridge than suggested by the

COCORP data and that the platform sedimentary sequence beneath is deformed into a series of duplex structures that, in turn, have arched the crystalline sheet above the duplexes. Large recumbent folds have also been imaged within the crystalline sheet in the Piedmont (Figure 20–14b), but little resolution of near-surface structure is visible in the Blue Ridge, probably because the dip there is moderate to steep.

Geophysical data promise to provide more exact solutions to crustal structure in the future—and imaging of structures heretofore unknown will lead to new controversies. The greatest strides in our knowledge of crustal structure will continue to be made by structural geologists and geophysicists working together to interpret geologic and geophysical data.

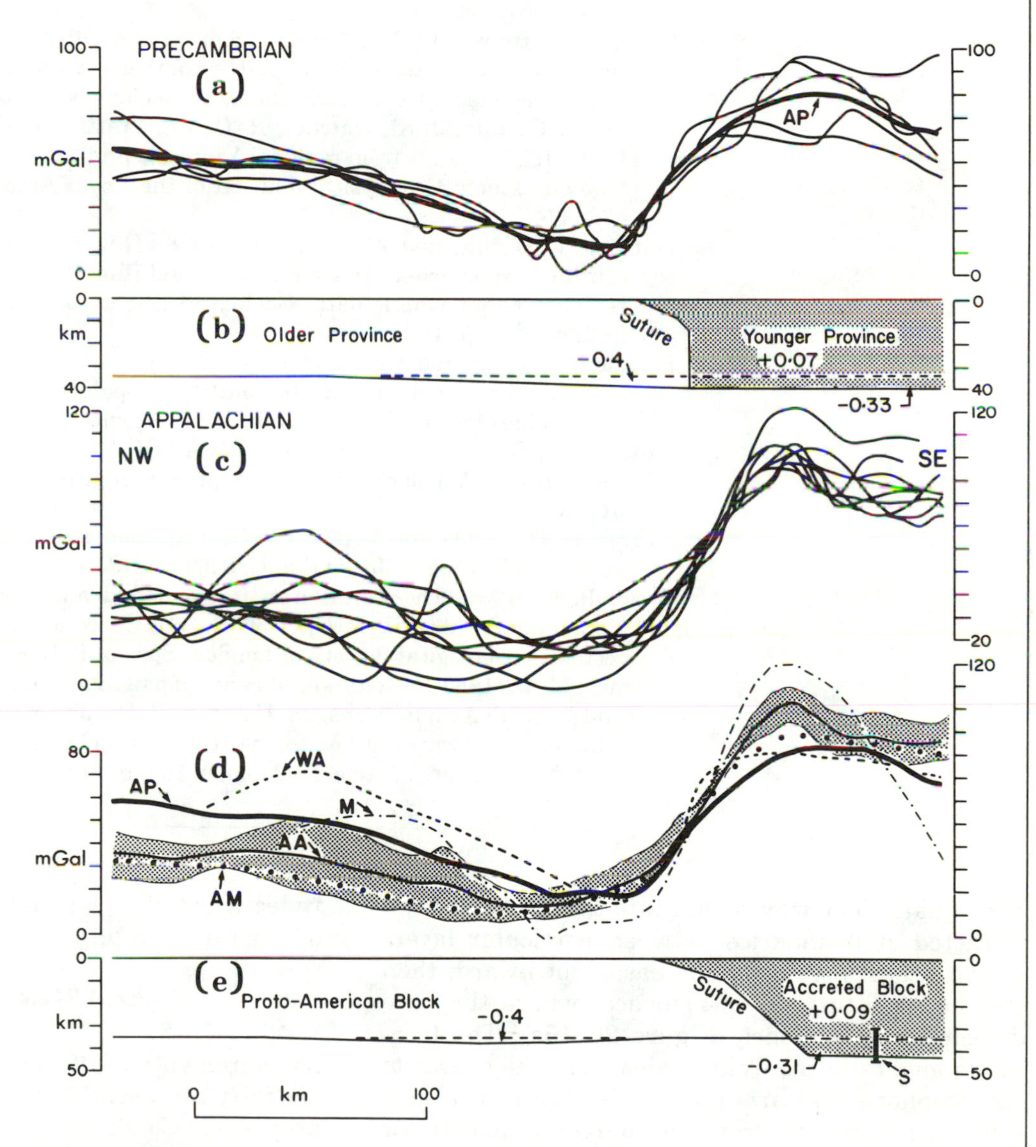

FIGURE 20E–1
Gravity profiles across several Precambrian orogens in the Canadian shield (a) and the southern and central Appalachians (c). The gravity low is always found closer to the older province and the high closer to the internal parts of the mountain chain. (After M. D. Thomas, 1983, Geological Society of America Memoir 158.)

References Cited

Ando, C. J., Cook, F. A., Oliver, J. E., Brown, L. D., and Kaufman, Sidney, 1983, Crustal geometry of the Appalachian orogen from seismic reflection studies, *in* Hatcher, R. D., Jr., Williams, Harold, and Zietz, Isidore, eds., Contributions to the tectonics and geophysics of mountain chains: Geological Society of America Memoir 158, p. 83–102.

Bally, A. W., Gordy, P. L., and Stewart, G. A., 1966, Structure, seismic data, and orogenic evolution of the southern Canadian Rockies: Canadian Petroleum Geology Bulletin, v. 14, p. 337–381.

Cook, F. A., Albaugh, D. S., Brown, L. D., Kaufman, Sidney, Oliver, J. E., and Hatcher, R. D., Jr., 1979, Thin-skinned tectonics in the crystalline southern Appalachians; COCORP seismic-reflection profiling of the Blue Ridge and Piedmont: Geology, v. 7, p. 563–567.

Cook, F. A., Brown, L. D., Kaufman, Sidney, and Oliver, J. E., 1983, The COCORP seismic reflection traverse across the southern Appalachians: American Association of Petroleum Geologists Studies in Geology, v. 14, 61 p.

Çoruh, C., Costain, J. K., Hatcher, R. D., Jr., Pratt, T. L., Williams, R. T., Phinney, R. A., 1987, Results from regional Vibroseis profiling: Appalachian ultradeep core hole site study: Geophysical Journal of the Royal Astronomical Society, v. 89, p. 473–474.

Hatcher, R. D., Jr., and Zietz, Isidore, 1978, Thin crystalline thrust sheets in the southern Appalachian Inner Piedmont and Blue Ridge: Interpretation based upon regional aeromagnetic data: Geological Society of America Abstracts with Programs, v. 10, p. 417.

Hatcher, R. D., Jr., and Zietz, Isidore, 1980, Tectonic implications of regional aeromagnetic and gravity data from the southern Appalachians, *in* Wones, D. R., ed., The Caledonides in the USA: Virginia Tech Geological Sciences Memoir 2, p. 235–244.

Hutchinson, D. R., Grow, J. A., and Klitgord, K. D., 1983, Crustal structure beneath the southern Appalachians: Nonuniqueness of gravity modeling: Geology, v. 11, p. 611–615.

Karner, G. D., and Watts, A. B., 1983, Gravity anomalies and flexure of the lithosphere at mountain ranges: Journal of Geophysical Research, v. 88, p. 10,449–10,477.

Price, R. A., 1981, The Cordilleran foreland fold and thrust belt in the southern Canadian Rockies, *in* McClay, K. R., and Price, N. J., eds., Thrust and nappe tectonics: Geological Society of London, Special Publication 9, p. 427–448.

Thomas, M. D., 1983, Tectonic significance of paired gravity anomalies in the southern and central Appalachians, *in* Hatcher, R. D., Jr., Williams, Harold, and Zietz, Isidore, eds., Contributions to the tectonics and geophysics of mountain chains: Geological Society of America Memoir 158, p. 113–124.

earthquake. The waves pass into the Earth and are refracted at boundaries between particular layers where acoustic velocity increases downward; then the waves return to the surface, where they are detected by geophones (Figure 20–16a). The technique generally involves angles from the source to the geophone that are much wider than in seismic reflection. Distances from the source to the detector are known, and so they may be calibrated with the travel times of acoustic waves of known frequency. Refraction is one of the best techniques for calculating seismic velocity and exploring the layered structure of the crust and mantle (Figure 20–16b). It is particularly useful in conjunction with seismic-reflection studies, which require detailed knowledge of crustal velocities.

ELECTRICAL METHODS

Measurements of ***electrical conductivity*** and ***resistivity*** are useful for determining certain crustal properties. Conductivity studies have been used for many years in the search for sulfide ore bodies because metallic sulfides are better electrical conductors than most other rock minerals. Resistivity and conductivity are usually measured by setting electrodes in the ground and recording for various lengths of time. Airborne techniques also exist.

Rock layers of greater conductivity than neighboring layers are detectable, but depth to the conducting layer may be determined only approximately. Also conductive, and detectable, are graphite-rich layers and water containing small quantities of dissolved solids. Electrical-resistivity anomalies may be exhibited by bodies composed of silicate rocks in a terrane otherwise dominated by carbonate or by evaporites in a sedimentary terrane otherwise dominated by clastic rocks.

Questions

1. What is the basis of the self-exciting dynamo theory?
2. We assume that the Earth's core consists of 90 percent iron and 10 percent nickel and that the mantle is peridotite. What is our basis?
3. Why is gravity less precise as an imaging technique than seismic reflection?
4. Calculate the depth at which most common magnetically susceptible minerals will no longer interact with the Earth's magnetic field.
5. Why are magnetic anomalies over a single source separated into high and low pairs at low latitudes but at high latitudes consist of a single anomaly?
6. How can paleomagnetic measurements that do not meet a fold test be rendered usable?
7. What are the differences between free-air and Bouguer gravity maps?
8. What does a gravity model for part of the crust or a body represent?
9. Why does seismic reflection not detect steeply dipping features?
10. Why are paleomagnetic measurements unable to determine longitude?
11. Calculate the depth to a 7-km-wide gravity anomaly thought to be produced by a nearly spherical pluton.

Further Reading

Bally, A. W., 1983, Seismic expression of structural styles: American Association of Petroleum Geologists, Studies in Geology 15, 3 volumes.
A compendium of seismic-reflection profiles showing various kinds of structures and providing examples of artifacts.

Barazangi, Muawia, and Brown, L. D., eds., 1986a, Reflection seismology: A global perspective: American Geophysical Union Geodynamics Series, v. 13, 311 p.

Barazangi, M., and Brown, L. D., eds., 1986b, Reflection seismology: The continental crust: American Geophysical Union Geodynamics Series, v. 14, 339 p.
Each volume applies reflection seismology to solving structural and tectonics problems and recognizes new unsolved problems. Some papers present new data not previously published; others interpret existing data.

Breiner, S., 1973, Applications manual for portable magnetometers: E. G. & G. Geometrics, Sunnyvale, California, 58 p.
Primarily for use of Geometrics magnetometers, but summarizes interpretation of magnetic data and adds some elementary theory. May not be widely available.

Dobrin, M. B., 1976, Introduction to geophysical prospecting: New York, McGraw-Hill, 630 p.
Compiles seismic-reflection theory and applications; less emphasis on other techniques.

Hoffman, K. A., 1988, Ancient magnetic reversals: Clues to the geodynamo: Scientific American, v. 256, no. 5, p. 76–83.
Discusses the Earth's magnetic field, polarity reversals, and evidence that decrease in magnetic-field intensity during the past century may presage a reversal of polarity in the next 1500 years or so.

Sharma, P. V., 1976, Geophysical methods in geology: Amsterdam, Elsevier, 428 p.
Surveys the various geophysical methods, probably treating gravity and magnetism best.

Telford, W. M., Geldart, L. P., Sheriff, R. E., and Keys, D. A., 1976, Applied geophysics: Cambridge University Press, 860 p.
A balanced treatment of geophysical methods, giving the limitations of each.

APPENDIX ONE

Fabric Diagrams

Two plotting nets are commonly used in geology for plotting planes and lines: the stereographic (Wulff) net and the equal-area (Schmidt) net (Figures A1–1 and A1–2). Each consists of a circular protractor that has been divided, using circular arcs, into a number of smaller segments. The principal difference between the two projections is the means by which the projection is constructed. In the case of the stereographic projection, plotting a point is accomplished using the radius of the reference sphere, whereas in the equal-area projection, a point is plotted using twice the radius of the reference sphere (Figure A1–3). It should also be noted that the subdivisions of the equal-area projection all have the same area throughout the entire protractor—hence the name—whereas the area differs in different parts of the stereographic projection. With both protractors, the projection attempts to portray in two dimensions planes or lines that may otherwise be expressed on the surface of a sphere. A common use of the stereographic projection is in mineralogy to express in two dimensions the relative positions and symmetry of crystal faces. Plotting crystal faces, or other planar or linear features, may be accomplished using either stereographic or equal-area projections.

All of the discussion that follows involves the equal-area projection as the standard reference for plotting; however, it is possible to plot all lines and planes on a stereographic projection. When data are to be contoured, the equal-area projection should be used for the quasi-statistical principles of contouring to be meaningful.

HOW TO BEGIN

An equal-area net should be obtained. The standard net has a 10-cm radius. The net should be prepared by attaching a rigid backing of cardboard or plastic, placing a transparent cover over the front of the protractor, and inserting a thumb tack through the exact center of the protractor from the back. The tack is used so that overlays may be placed on the protractor and rotated to conveniently plot data without sliding the overlay out of alignment (Figure A1–4).

It is common practice to begin plotting by taking a transparent piece of paper, Mylar™, or acetate film, and placing it over the protractor. The first step in plotting any element on an equal-area (or stereographic) net is to mark off the primitive circle on the overlay, then mark a convenient reference point so that you can always return to the starting position on the overlay for additional plotting of points or planes. This common reference point in structural geology is the position of north on the primitive circle located at the top of the circle at zero degrees on the protractor.

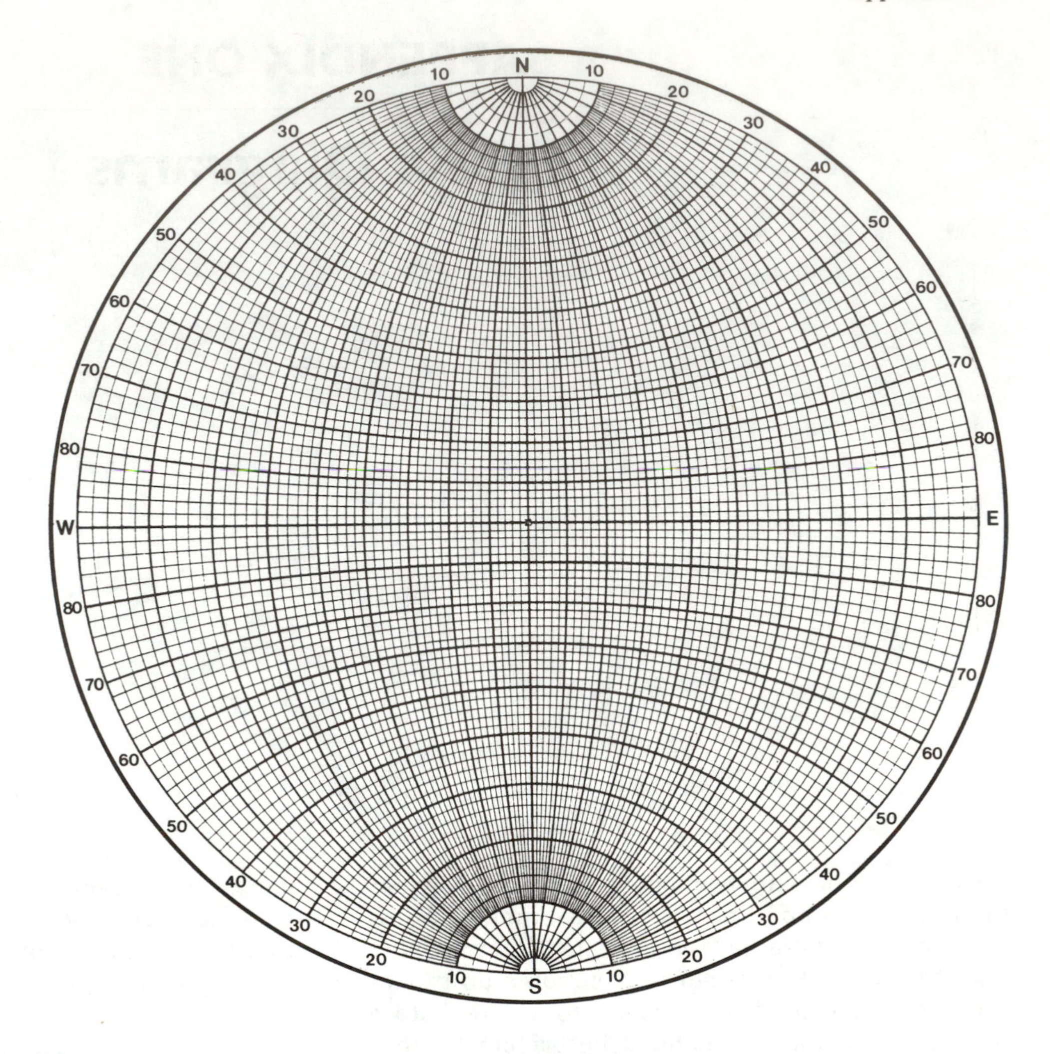

FIGURE A1–1
Stereographic (Wulff) net.

PLANAR STRUCTURES

Planar structures, including bedding, foliation, cleavage, fault surfaces, joints, and other planar structures, can be accomplished by two means on an equal-area or a stereographic projection. Planes may be plotted either as poles to planes or as the trace of the plane within the circumference of the reference circle. Traces of the plane are great circles that intercept the reference, or *primitive,* circle of the projection. *Poles* are points that are normal to the plane of interest and would therefore plot on the opposite side of the diagram from the great circle representing the corresponding plane.

Let's plot a dip-strike measurement on the equal-area net. It will be plotted both as a plane and as a pole to the plane. The measurement has a strike of N 37° W and a dip of 28° SW (alternatively 323, 28° SW). The first step is to count 37° counterclockwise (left) along the edge of the net—to the west of north—along the margin of the protractor and mark that point with a pencil on the overlay (Figure A1–5). This represents the strike of the plane. Next, the overlay should be rotated 37° clockwise (right) so that the point that was just marked on the overlay lines up with the north pole of the net. Then, count in along the equatorial line 28° from the west (left) edge of the protractor. The great circle connecting the north and south poles of the protractor should be traced out on the overlay. The circle corresponds to the trace of the plane in the projection, so the plane has been plotted as a trace having a strike of N 37° W and a dip of 28° SW. Finally, the pole to that plane may be plotted without moving the overlay by counting 28° from the center outward on the opposite (east) side along the east-west (equatorial) line across the protractor. A point representing the pole of the plane N 37° W, 28° SW is then placed on the overlay.

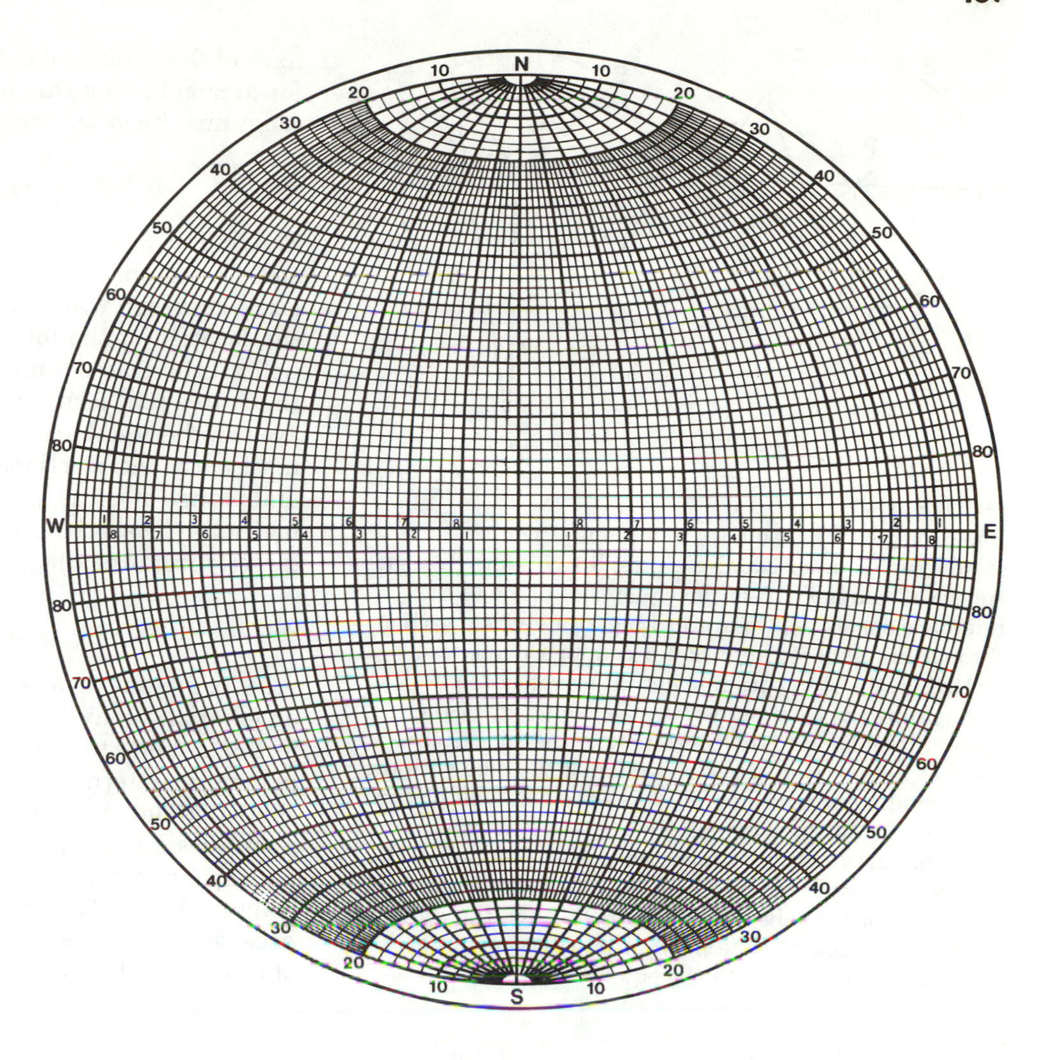

FIGURE A1–2
Equal-area (Schmidt) net.

LINES

An overlay is prepared for the equal-area net in exactly the same way as a plane is plotted. The primitive circle is sketched on the overlay and a mark is made to identify north.

Let's plot a linear structure on the equal-area net. The trend and plunge of a fold axis with a trend of N 63° E plunging 39° NE (alternatively expressed 063, 39°) could be plotted. The first step in plotting the line is to mark the strike along the primitive circle (Figure A1–6), as was done for plotting the trace of a plane. In this case, because the strike is N 63° E, you must count 63° to the east (right) of north and place a pencil mark at that point on the primitive circle. Next, the mark on the primitive circle is rotated until it coincides with the east-west (equatorial) diameter of the protractor. Now it is necessary to count in 39° from the east edge of the primitive circle and place a point there. This point corresponds to the point of intersection of the line with the plane of the projection.

Using this method of plotting lines, the orientations of many structural features can be expressed on equal-area diagrams. Fold axes, boudin axes, intersections,and mineral lineations may be plotted, as well as the orientations of crystallographic axes of minerals in thin sections of deformed rocks measured under the microscope using a universal stage.

LOCATING FOLD AXES USING EQUAL-AREA PLOTS: β AND π DIAGRAMS

It is often desirable to determine the locations of fold axes using dip-strike data where measurements of small fold axes cannot be made in sufficient numbers to accurately determine the position of major

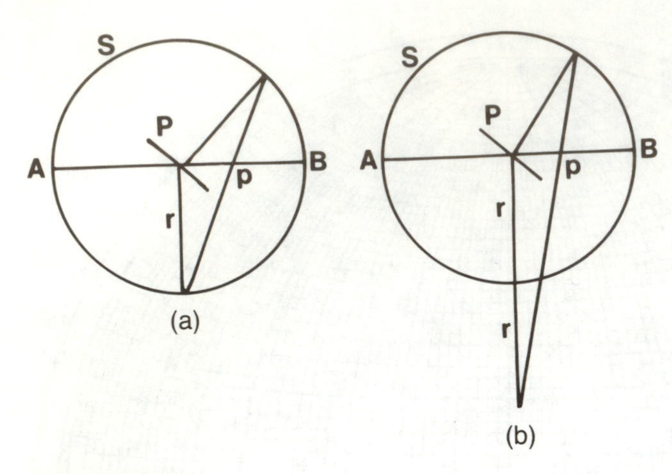

FIGURE A1–3
Basis for plotting a pole (p) to a plane (P) on a stereographic (a) or equal-area net (b). S is the reference sphere of radius r. AB is the equatorial plane of the reference sphere, so we are looking at a vertical section through the sphere. Note that a pole to a plane is plotted on the stereographic net (a) by reflecting a normal upward from plane P to the boundary of the sphere, then back to the south pole of the sphere. The point of intersection of the reflection of the pole with the equatorial plane is the point where the pole will plot in the diagram. In contrast, a pole to plane P plotted on the equal-area net (b) is reflected to the boundary of the sphere, then reflected to a point at a distance $2r$ from the center of the sphere that passes through the south pole. Point p again is the point of intersection of the reflection of the pole with the equatorial plane.

axes of first-order folds in an area. Two techniques for accomplishing this are by constructing β and π diagrams. Each will be explained below.

β Diagrams

Bedding or foliation planes plotted as traces on an equal-area projection will yield a family of great circle curves in which the intersections approximate the location of the fold axis (Figure A1–7). Rarely will all the traces intersect at a common point, but the traces will form many intersections in a small area on the equal area diagram so that the center of mass of the intersections will approximate the trend and plunge of a line, β, commonly assumed to be the fold axis. The orientation of β may then be read directly from the diagram as a trend and plunge.

π Diagrams

A set of dip-strike measurements—of either bedding or foliation—may be plotted on an equal-area projection as poles to bedding (or foliation). If the points plot in a pattern that may be fitted to a great (or small) circle, the pole to the great circle yields a point designated β on the diagram that should correspond to the trend and plunge of the fold axis (Figure A1–8). This diagram is called a π diagram because poles to planes are used instead of the traces of a plane to determine the position of β.

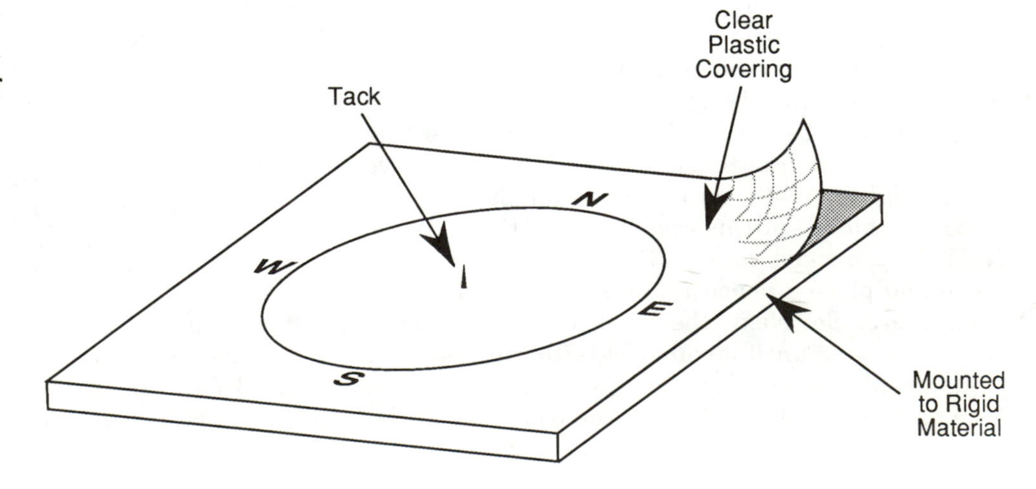

FIGURE A1–4
Mounting a plotting net. After mounting the net on a piece of cardboard or other rigid material, the net should be covered with a piece of clear, durable plastic and a tack pushed through the center of the protractor from the back.

FIGURE A1–5
Plotting a plane with a strike of N 37° W and a dip of 28° NE on the equal-area net. The plane plots as a great-circle arc and the pole (*P*) plots as a point.

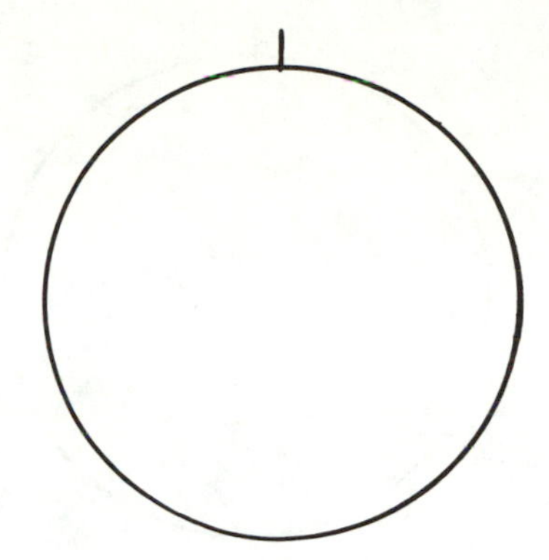

Draw Primitive
Circle on Overlay
Mark North

Mark
Strike
N37°W

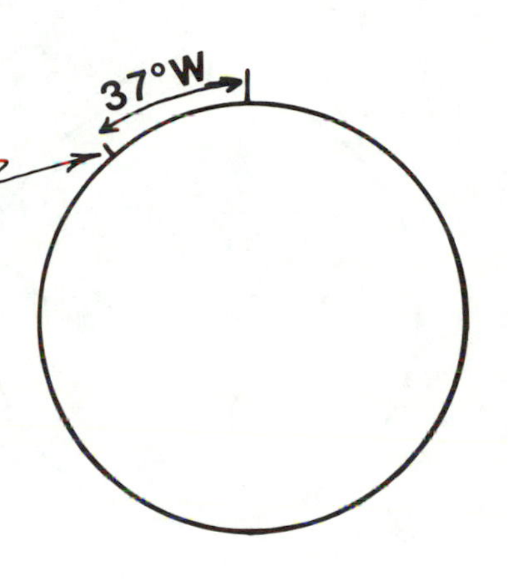

28°
28° P

Rotate Strike Mark
to North Position

Trace off 28°
Great Circle Arc

Mark Pole to Plane
28° to Right of
Center

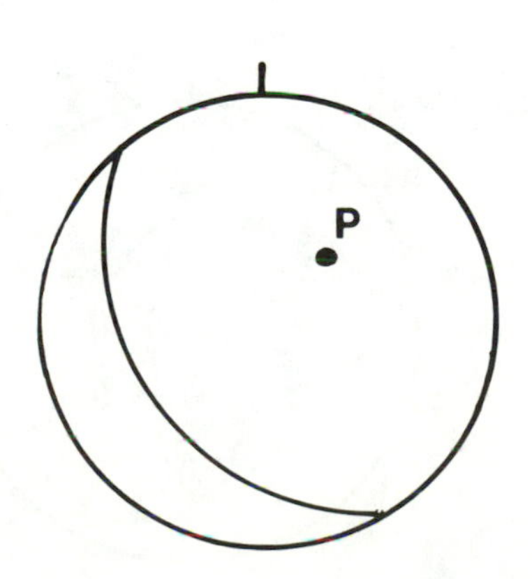

Finished Plot of
Plane and Pole
for N37°W, 28°SW

FIGURE A1–6
Plotting a line that has a trend of N 63° E and a plunge of 39° NE on the equal-area net. The line plots as a point.

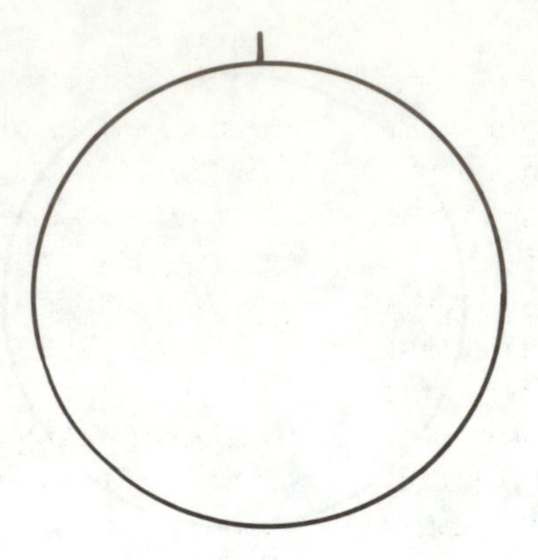

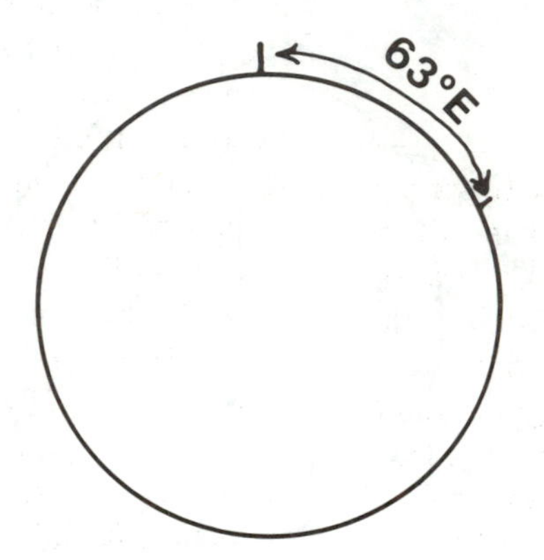

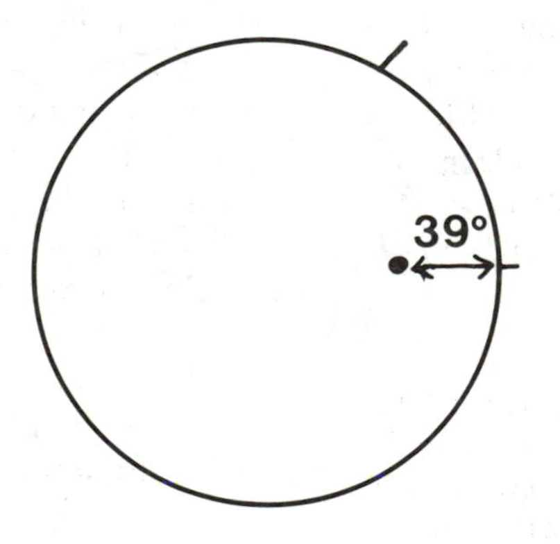

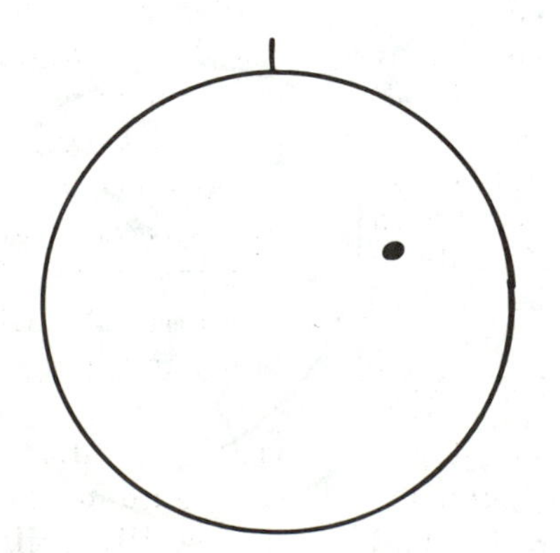

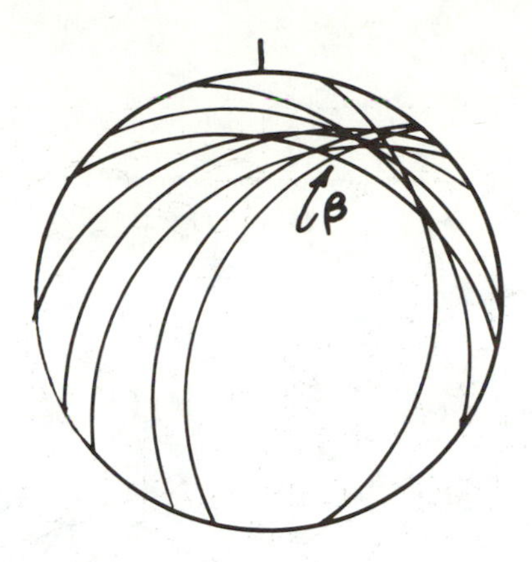

FIGURE A1–7
A β diagram consisting of a set of dip-strike data for a group of planar structures that plot as a series of great-circle arcs that intersect in one particular area of the fabric diagram. The center of this area of intersection is designated β and is assumed to approximate the trend and plunge of a fold axis.

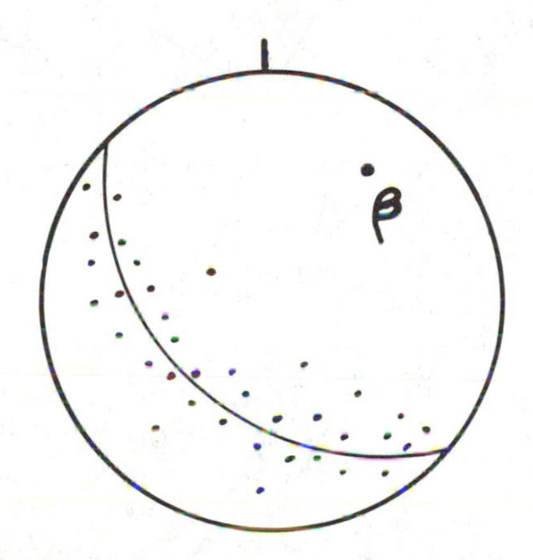

FIGURE A1–8
A π diagram consisting of poles to a set of dip-strike data for a group of planar structures. The fabric diagram that results consists of a series of points that scatter about a great-circle arc constructed as the best-fit line for the points. The pole to the great circle is designated β, assumed to approximate the trend and plunge of a fold axis—here the same fold axis as in Figure A1–7.

CONTOURING DATA

A set of points, poles to planes—or lines (axes, lineations, etc.)—plotted on an equal-area projection is a *point diagram.* It is useful to show the relative distribution of points throughout the diagram. Although it is possible by inspection to readily observe concentrations of points, a more efficient means of locating concentrations is by contouring the data. The standard technique of contouring fabric data involves the use of an equal-area plot of the data set and counting devices called *center* and *peripheral counters* (Figure A1–9). These counting devices are employed in conjunction with a grid to determine the number of points per one percent area within and along the edge of the diagram. The center counter is used to determine the number of points throughout the interior parts of the diagram.

The standard equal-area protractor has a diameter of 10 cm. The dimensions of the counters shown in Figure A1–9 are designed for a 10-cm net. If nets of other dimensions are used, these counting devices cannot be used. In addition, a 1-cm grid must be prepared to cover the entire area of the 10-cm equal-area projection. The process of preparing a contoured diagram will be described stepwise. If the grid is to be used over and over, it should be constructed on a piece of translucent Mylar or acetate film and the lines of the grid inked in a fine line pattern.

1. The grid is placed over the point diagram and a transparent paper or Mylar overlay placed over both of these. The outline of the primitive circle is drawn on the overlay and the position of north marked on it. The center counter is placed over each intersection of the 1-cm grid so that two lines of the grid intersect at the center of the circle. The number of points lying within the circle is counted and the number written in the center of that circle (Figures A1–10a and A1–10b). The center counter is then moved to the next position of intersecting lines on the 1-cm grid and the process repeated. It is moved again and again until a number appears in every position inside the primitive circle. Once all of the interior grid intersections have a number on them, the peripheral counter is used for positions along the edge of the circle, where the intersection on the grid causes part of the center counter to fall outside the primitive circle. Here the number of points is counted and summed in each partial circle on both ends of the peripheral counter, and the same number is written on both sides of the diagram in the center of each circle of the peripheral counter (Figure A1–10b).
2. The total number of points in the point diagram must be counted. A second overlay is then prepared on which the numbers previously written at the grid points are converted to percentages (by dividing each number by the total number of

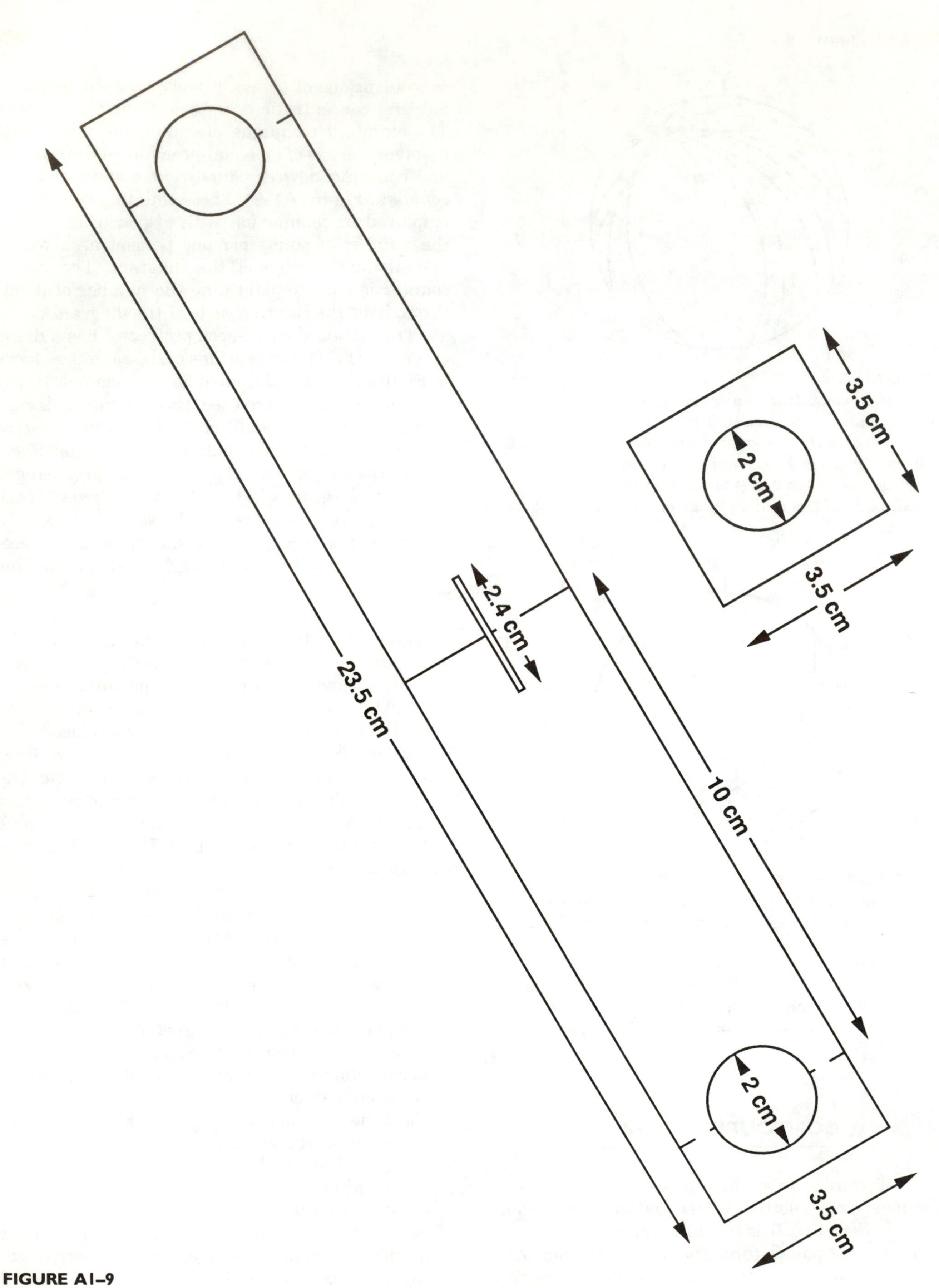

FIGURE A1–9
Dimensions of the center and peripheral counters for counting points prior to contouring an equal-area diagram.

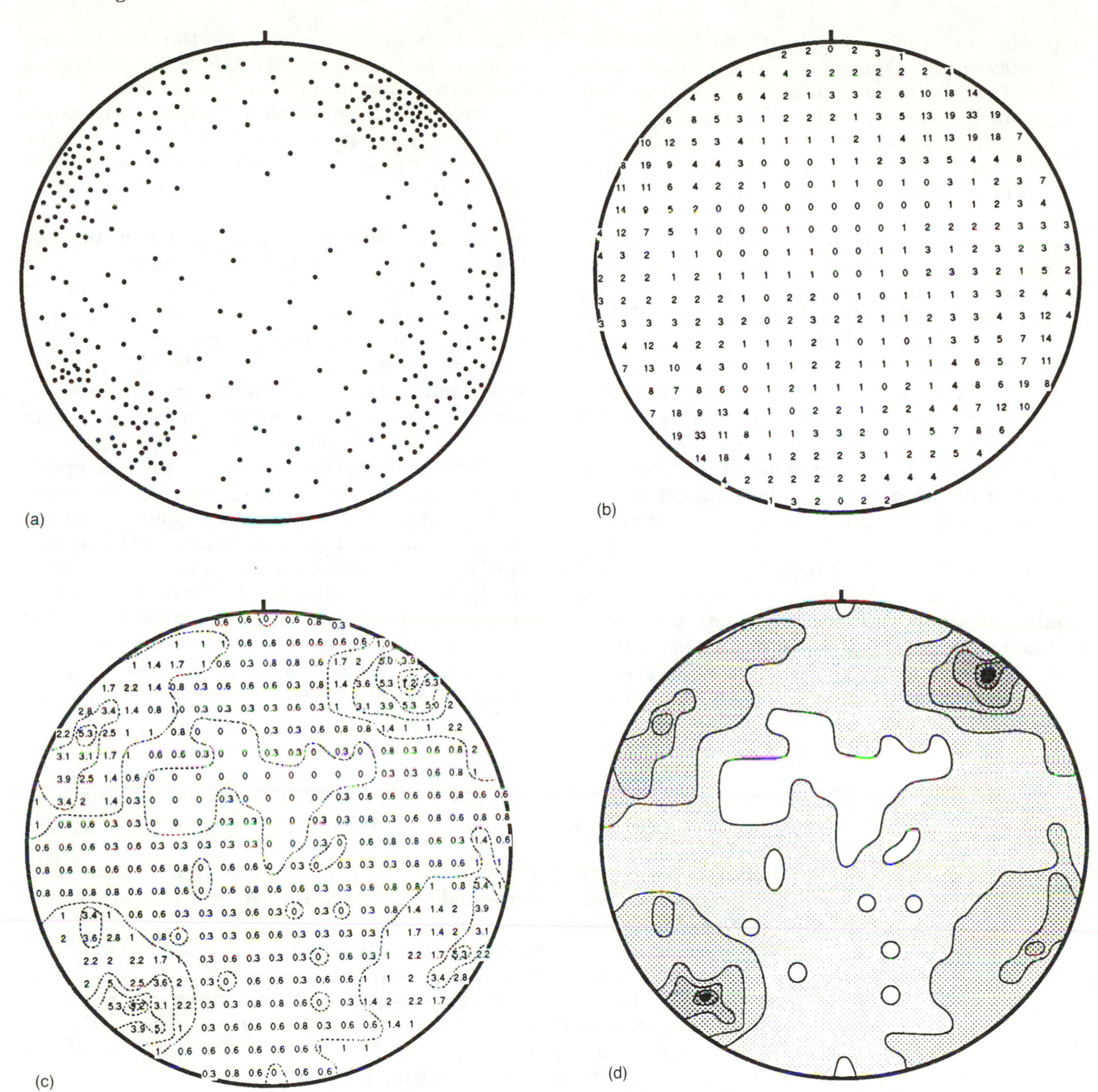

FIGURE A1–10
Contouring a set of points. (a) Point diagram consisting of 357 poles to joints measured in basement rocks in central Colorado. (b) Results of counting points in (a) using the center and peripheral counters and a 1-cm grid. The numbers on this diagram represent the number of points inside the 1-cm circle with a grid point as the center of the circle for areas inside the diagram. Numbers along the edge of the diagram represent the sum of points counted in both circles of the peripheral counter for grid points that cause the circle to overlap the edge of the diagram. (c) Percent diagram. Numbers here represent points counted in (b) converted to percent per one percent area by dividing each by the total number of points in the diagram. Contours are sketched here for transfer to the final contour diagram. (d) Contoured diagram with contours 0.3, 1, 3, 5, 7, and 9 percent per one percent area. Note that the noninterval contour, one percent, was drawn to show the boundary between no data and the smallest number of data.

points) and each percent written on the new overlay at the appropriate grid point (Figure A1–10c). These percentages represent percent per one percent of the area of the 10-cm circle.

3. The values of percent per percent area are contoured using an interval chosen by inspection that best fits the percentages represented on the diagram. Rather than always choosing an equal increment from the lowest to the highest, it might be useful to draw a contour that separates the lowest percentage—such as 0.3—from zero, thus showing the absolute distribution of data over the diagram (Figure A1–10d). Then an equal increment is chosen for the remainder of the contours—2, 4, 6, 8, or 5, 10, 15, 20, or whatever. The final contouring then is carried out on the percent overlay and the contour interval indicated. If it is desirable to reproduce the contour diagram, this may be accomplished by transferring the contours to another overlay.

The method described here is slightly modified from that described by Turner and Weiss (1963). A number of other contouring methods have been devised to accomplish the same result. One of these is the use of a counting net divided into small triangles, six of which form a hexagon equal to one percent of the area of the equal-area net (Kalsbeek, 1963). Some have suggested that the standard free-counter method (described above) for contouring introduces a bias in the shape of contours.

Mellis (1942) invented a method involving the use of overlapping circles that does work well for diagrams containing fewer than 150 points, and particularly for drawing the contour of lowest point density. The center counter is placed above each plotted point and a circle drawn around the inside of the counting circle on an overlay, producing a mesh of overlapping circles. The contour representing two percent per one percent area outlines the area where two circles overlap, four percent per one percent area is outlined by drawing a line through the places where four circles overlap, and so on. This contouring technique is less subjective; it is described in greater detail in Turner and Weiss (1963).

The mechanical process of conversion of a point diagram to a contour diagram may be easily accomplished today using a number of computer programs available in the literature (such as Jeran and Mashey, 1970; Cooper and Nutall, 1981) or easily obtained commercially. Several have been written specifically for IBM™ and Macintosh™ personal computers. The diagrams in the Chapter 19 Essay were plotted using a program written by R. W. Allmendinger at Cornell University for the Macintosh computer.

References Cited

Cooper, M. A., and Nutall, D. J. H., 1981, GODPP: Programs for presentation and analysis of structural data: Computers and Geosciences, v. 7, p. 267–285.

Jeran, P. W., and Mashey, J. R., 1970, A computer program for stereographic analysis of coal fractures and cleats: U.S. Bureau of Mines Information Circular 8454, 34 p.

Kalsbeek, F., 1963, A hexagonal net for the counting out and testing of fabric diagrams: Neues Jahrbuch für Mineralogie, Monatshefte, v. 7, p. 173–176.

Mellis, O., 1942, Gefügediagramme in stereographischer Projecktien: Zeitschrift Mineralogie und Petrographie Mittellungen, v. 53, p. 330–353.

Turner, F. J., and Weiss, L. E., 1963, Structural analysis of metamorphic tectonites: New York, McGraw-Hill, 545 p.

APPENDIX TWO

Structural Measurements and Observations

A number of structural measurements and observations are commonly made in the field. These range from the orientation of planar and linear structures, requiring two measurements to be made, to the orientation of folds, requiring four to five measurements for a complete description of orientation. Determination and recording of shear sense (vergence) of structures is also important. Accurate sketches of structures and the relationships between overprinted structures in your field book will prove useful. Photographs of representative and critical mesoscopic structures should also be made to augment—not to replace—sketches. Accurate sketches require a greater amount of observation than taking a photograph; details of a structure will become better known if you take the time (sometimes an hour or more) to accurately sketch it in your notebook.

Accompanying the requirement that measurements be made correctly is another that the location of the site where measurements are being made be accurately known, so that they can be related spatially to other measurements. Accurate location of each measurement is needed if you are constructing a geologic map, making measurements on a particular map-scale structure to investigate the mechanics by which it formed, or making detailed measurements of different kinds of structures in a roadcut, stream, or other small exposure.

ORIENTATION OF PLANES: STRIKE AND DIP

Two measurements are used to describe the orientation of a plane in the Earth: strike and dip. The ***strike*** is the trend (compass direction, azimuth) of a horizontal line lying in an inclined plane (Figure A2–1). The ***dip*** of a plane is the largest angle made by the plane with the horizontal and is measured perpendicular to strike in the vertical plane. Angles measured within the vertical plane that are less than the maximum are not perpendicular to strike and are measurements of *apparent dip*. Strike is measured using a compass and dip is measured with a clinometer, commonly built into compasses designed for making these measurements in the field (Figure A2–1).

Measurement of strike and dip can be made in several ways by using a compass. A common way is to place a smooth board (or thin sheet of aluminum) on the inclined surface to be measured to even out any irregularities, then place the edge of the compass against the inclined surface. Once the compass is leveled, the strike is read directly. The direction perpendicular to the strike is the dip direction; the angle of dip is then measured in this direction, down the plane, using the built-in clinometer in the compass. This method is very good for learning how to

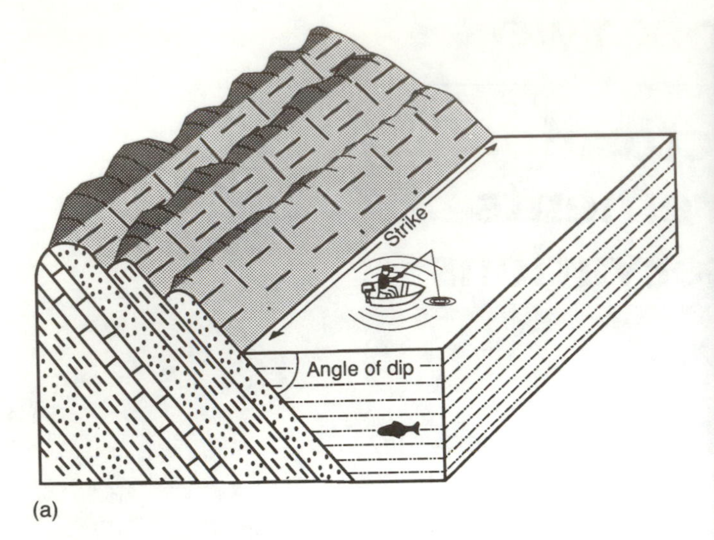

(a)

(b)

(c)

FIGURE A2–1
(a) Strike and dip of inclined bedding. Note that the strike is the line of intersection of the inclined bedding surface with the water surface; the dip is the acute angle between the inclined bedding and the water surface. (b) Measurement of the strike of compositional layering in 1.2-billion-year-old Toxaway Gneiss at Whitewater Falls in western North Carolina. (c) Measurement of dip on an inclined crenulated foliation surface at the same locality as (b). Compositional layering is visible on the end of the exposure. (RDH photos.)

measure the strike of planar surfaces, but can easily yield erroneous measurements if irregularities on the rock surface are allowed to tilt the board out of parallelism with the true dip of the rock surface. By making hundreds of measurements and developing an ability to visualize the correct orientation of an inclined plane to be measured, more accurate measurements of strike may be obtained by sighting along the strike line of the surface to be measured, then aligning the clinometer parallel to the dip of the inclined surface to obtain the dip. Recording the strike and dip in proper geographic orientation in your field book provides a good means of checking your measurement.

Direct measurement of strike can be made by aligning the compass parallel to the line formed by the intersection of an inclined surface with a water surface (Figure A2–1a). The dip is the acute angle the inclined planar rock surface makes with the water surface—a natural horizontal reference surface. Inclined surfaces also produce mesoscopic *cuestas* and *hogbacks,* the crests of which form natural strike ridges. So, to accurately measure the strike of an inclined layer producing such a feature, the compass is aligned parallel to the crest of the hogback and the strike read directly. Dip is measured by one of the methods described above.

Strike may be recorded in both azimuthal and quadrant format. Quadrant format is commonly measured and recorded relative to north. Azimuthal format records strike direction clockwise from 0° to 360°. For example, a plane with a northwest strike is recorded as N 45° W in quadrant format, or 315 in azimuthal format, not as S 45° E. Dip is usually expressed as an angle and a direction, here 37° NE. Less commonly, it may also be expressed as 37° N 45° E (or 37° 045), thereby also providing the strike orientation.

Measurement of strike and dip may be made on any inclined surface, including bedding, foliation, joints, and faults, as well as surfaces not of primary interest in structural studies, such as roads, topographic slopes, artificial cuts in rocks, and erosion surfaces. Measurement accuracy decreases where the surface being measured has a dip of less than 15° because of small surface irregularities and the difficulty of making measurements on gently dipping surfaces. Measurements of dip should be accurate to within ±2°–3°. At angles less than 15°, even if this level of accuracy can be maintained, ±2°–3° error is a significantly greater proportion of the angle being measured than at steeper dip, so the effective accuracy of measurements decreases where the dip is gentle.

ORIENTATION OF LINES: TREND AND PLUNGE; RAKE

The orientation of linear structures, such as mineral lineations, intersections, boudin axes, and fold axes, is expressed by the trend (azimuth) and plunge (Figure A2–2). The trend is recorded as the compass direction measured in the direction of plunge of the structure. For example, if a lineation plunges southwest, the trend is recorded as S 45° W, or 255. The

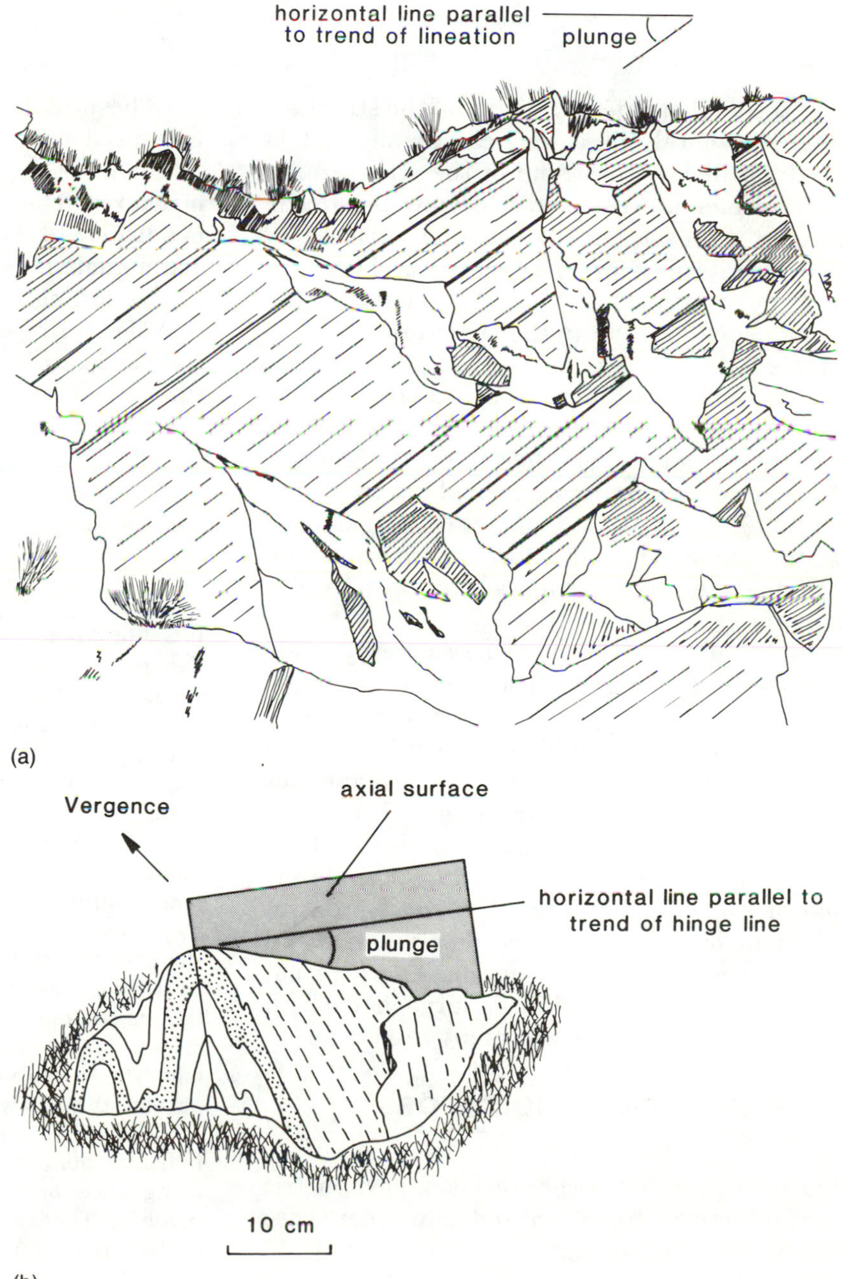

FIGURE A2–2
(a) Sketch of intersection lineation in Figure 18–4a showing the trend and plunge of the lineation. (b) Determination of the orientation of a mesoscopic fold. Note that the strike of the axial surface need not necessarily be coincident with the fold hinge line, or axis. Measurements to be made are the trend and plunge of the axis, strike and dip of the axial surface, and the direction of vergence.

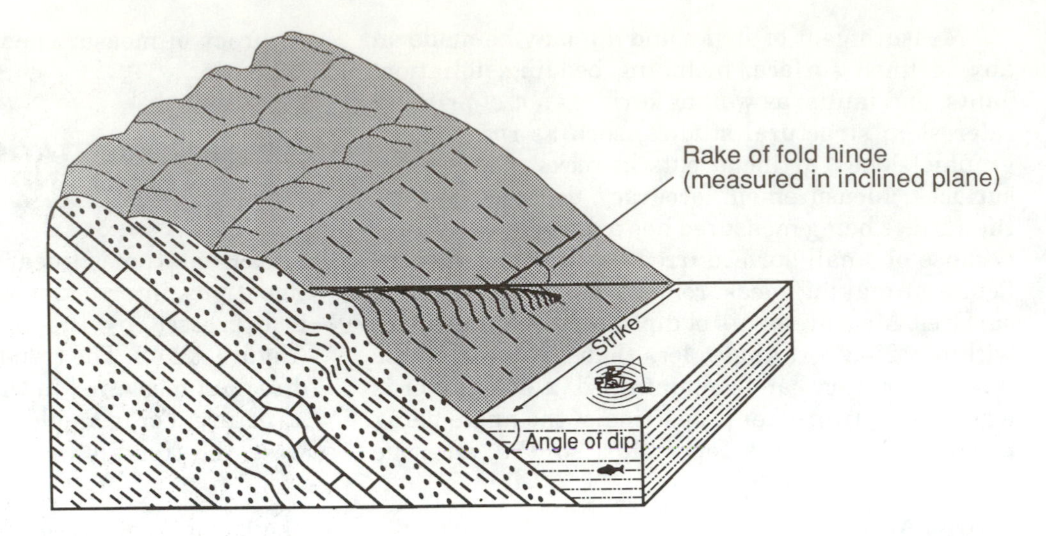

FIGURE A2–3
Rake of a fold hinge line is the angle made by the hinge line measured in the inclined bedding surface.

plunge is the angle made by the axis of the structure with the horizontal, in a vertical plane, and is measured parallel to the trend of the structure by using a clinometer. Always try to view the linear structure in at least two nonparallel surfaces so apparent orientations are not recorded. Measurement of linear structures may be facilitated by aligning a pencil parallel to the structure.

Complete description of the orientation of folds requires two additional measurements: the dip and strike of the axial surface and the direction of overturning (vergence). The fold axial surface is a hypothetical surface, unless an axial-planar cleavage or foliation is present. It may also be useful to record the symmetry and approximate interlimb angle of folds (either a number or open, tight, isoclinal, etc.).

An additional measurement occasionally made is the ***rake*** (Figure A2–3). The rake is an angle between a line lying in a plane with a horizontal line which also lies in the plane. Therefore, it is an angle not measured in a horizontal or vertical plane, but in the plane containing both the line and the horizontal line. Measurement of the rake of the displacement vector on a fault plane (see next section) is one example of the use of this kind of measurement. Measurement of the rake of the displacement implies that all motion is translational and does not involve rotation, but rotation of large and small blocks has been documented in many areas.

SHEAR-SENSE INDICATORS

S-C structures, rotated porphyroclasts, asymmetric folds, and other structures that indicate shear sense may be observed in the field.

The orientations of S- and C-surfaces can be measured as any other planes, and the intersections of S- and C-surfaces are lines that can also be measured. The shear direction can be recorded perpendicular to the intersection in the direction of the acute angle between S- and C-surfaces.

Recording the shear direction and sense of rotation of porphyroclasts and asymmetric folds provides the needed information to characterize these structures.

FIGURE A2–4
Two facing pages from a field book for part of the Satolah quadrangle, in the Blue Ridge of northeastern Georgia, showing one way structural measurements may be recorded and a sketch of a passive-flow fold. At the top of the page is the name of the quadrangle, the date, and the location of the traverse. Note that rock types are abbreviated. Rose is the pencil color used for biotite gneiss; ggn is granitic gneiss; amph. is amphibolite; peg. is pegmatite. A measurement followed by a rock type is a planar structure; the symbol beside the measurement indicates the kind of planar structure. PF(I)F indicates the orientation (trend and plunge) of the axis of a passive-flow isoclinal fold; AS is the orientation (strike and dip) of the axial surface of the fold. The sketch was carefully drawn to make sure all parts of the fold were shown correctly and in the right proportions; rock types were labeled, and the shading indicates the layers of different composition were colored using the colors chosen for the specific rock types involved. The scale and approximate orientation of the sketch are also indicated.

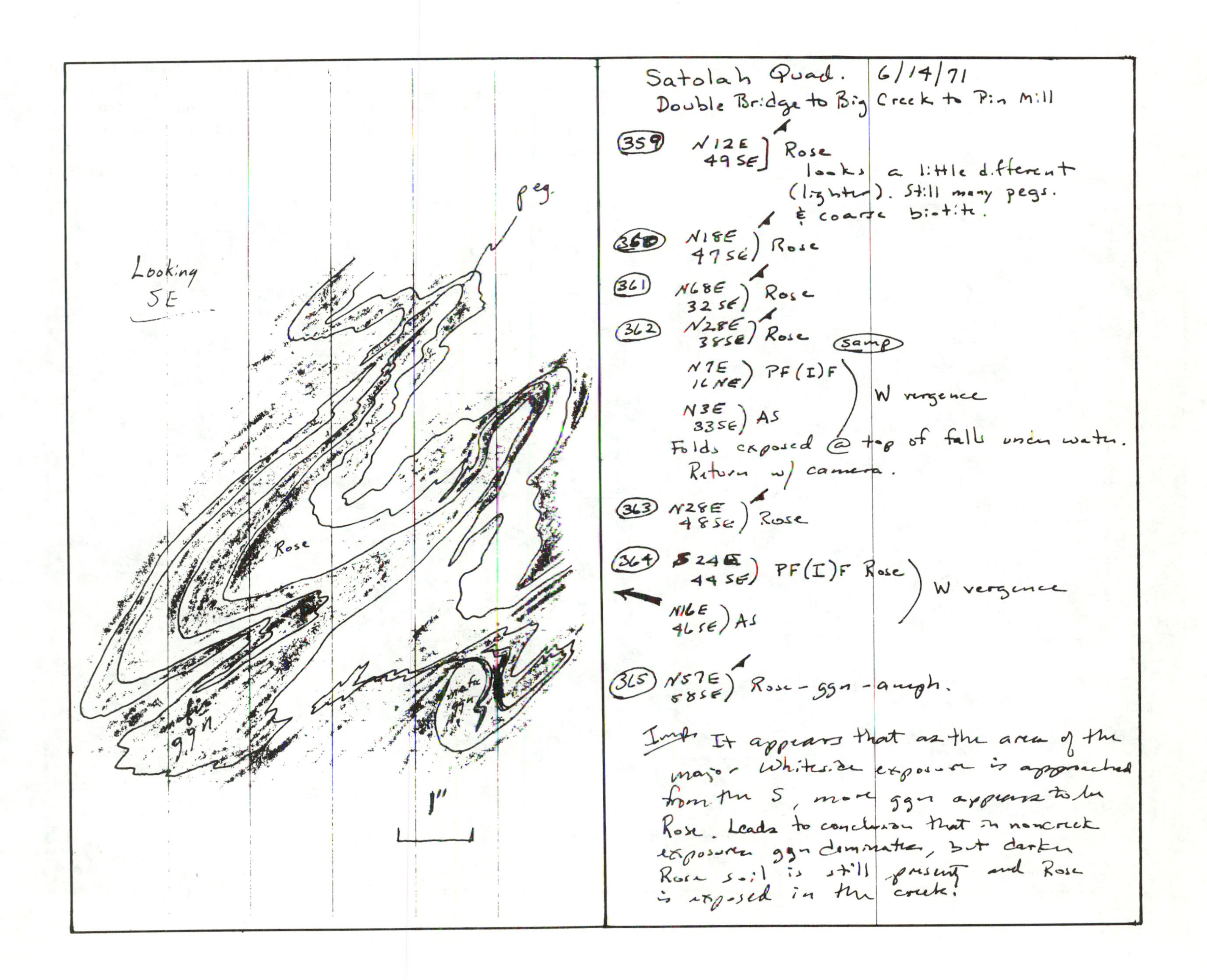
Looking SE
peg.
Rose
mafic ggn
mafic ggn
1"
Satolah Quad. 6/14/71
Double Bridge to Big Creek to Pin Mill
359 N12E 49SE] Rose
looks a little different (lighter). Still many pegs. & coarse biotite.
360 N18E 47SE) Rose
361 N68E 32SE) Rose
362 N28E 38SE) Rose
samp
N7E 16NE) PF(I)F
N3E 33SE) AS
W vergence
Folds exposed @ top of falls under water.
Return w/ camera.
363 N28E 48SE) Rose
364 S24E 44SE) PF(I)F Rose
N16E 46SE) AS
W vergence
365 N57E 58SE) Rose-ggn-amph.
Imp. It appears that as the area of the major Whiteside exposure is approached from the S, more ggn appears to be Rose. Leads to conclusion that in noncreek exposures ggn dominates, but darker Rose soil is still present and Rose is exposed in the creek!

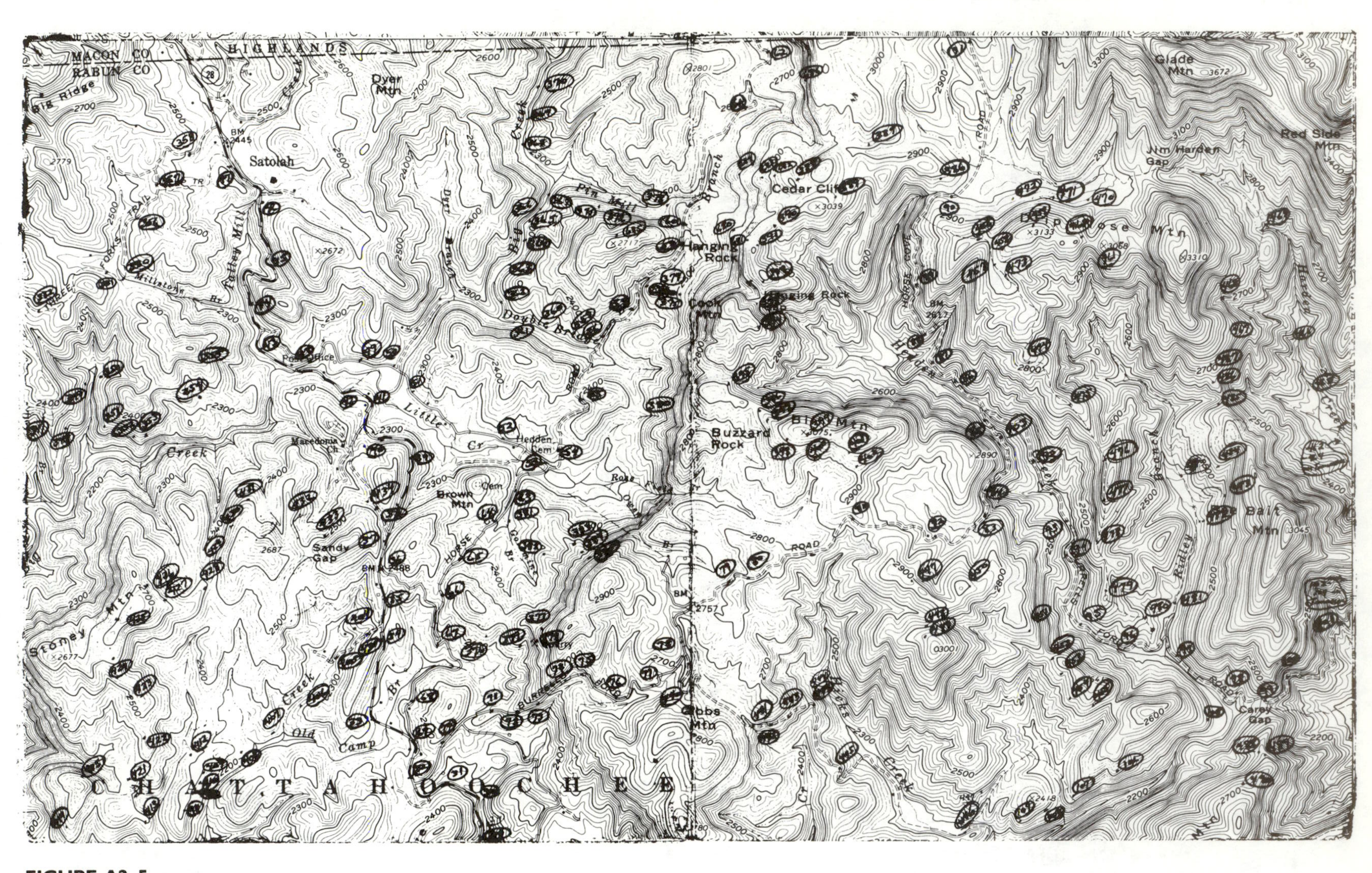

FIGURE A2–5
Part of a station map for the Satolah, Georgia, quadrangle. Circled numbers are keyed to numbers in the field notebook. The stations for the page shown in Figure A2–4 may be found, as indicated at the top of the page in the field notebook, beginning on Double Branch, traversing upstream into Big Creek to Pin Mill Branch, in the upper left center of the figure.

RECORDING DATA

Several ways exist to accurately record field data. The simplest is to write out the descriptions of all stratigraphic and other lithologic descriptions (as needed), write out structural measurements, and formulate a means to relate your data to a location on a map or photograph. Alternatively, field data may be tabulated as it is collected in the field. Data may also be preserved, using a tape recorder, to be transcribed later, but it is more difficult to search back through a tape in the field to review a note or measurement than to flip back several pages in a notebook.

It may be useful to invent a shorthand system for recording structural and other data. The system should be efficient and understandable to someone other than yourself. I have used a system for many years that involves the abbreviation of rock type and a detailed record of structural data for a particular station. Each page in the field book (Figure A2–4) has the date and location of the data set—by specific locality, name of a quadrangle, a locality in a quadrangle, or some other means of immediately determining the site where the data set on that page is located. Each place data are collected is assigned a separate station number—keyed to a station map (Figure A2–5). The first entry after the station number is the orientation of the dominant planar fabric element at the station—bedding, compositional layering, cleavage, foliation, schistosity—the most likely structure to be plotted on a geologic map at that station. The map symbol for the structure may be written beside the measurement: shorthand for which of the planar elements is represented by the measurement. Following that are the other structural measurements made at that station. A small Roman numeral (or other symbol; e.g., D_2) may be written next to any measurement to indicate the deformational event that might have produced it.

Extensive notes may be written about rock units in an area that has never been described before or one that is unfamiliar to the structural geologist. Once the rock units become familiar and recognition becomes routine, a shorthand system may be devised for describing particular rock units or rock types. Abbreviations like qzt. for quartzite, ss. for sandstone, sh. for shale, amph. for amphibolite, ls. for limestone; standard map unit abbreviations, like Dm for Marcellus Shale, ℙf for Fountain Sandstone, Mk for Kaibab Limestone, may be used; alternatively, the number code for the colored pencils used for particular rock types or units is a shorthand I have used; e.g., 745 is the number of the carmine red color of the Berol Verithin™ I have used for metagraywacke, and 742½ is the lavender Berol Verithin color I use for pelitic (muscovite-biotite) schist. Rather than writing metagraywacke or pelitic schist in my field book for the dominant lithology at a station, I will write 745 or 742½ opposite the dominant planar structure, or, if both are present, 745–742½. Following this scheme, only notes about geologic oddities, variations in rock type, structural relationships, or other nonroutine observations need be extensively described. Sometimes on a sunny day in the field—sometimes even on a cloudy day—a magnificant idea will reveal itself, demanding space in the field notebook.

Some of the best detailed notes I can recall describing a nonroutine observation were written one summer by a senior undergraduate student working in the thickly forested southern Blue Ridge in North Carolina. Unfortunately, these notes did not describe a new idea about the geology, but the appearance and proximity of a large black bear that walked to within <5 meters of the student as he was making structural measurements!

APPENDIX THREE

Woodall Shoals Fabric Data

Woodall Shoals Domain I—Measurements of S_2

16 54NW	39 33NW	51 55SE	67 38SE	58 38SE	14 64NW
24 49SE	42 64SE	68 16NW	29 57NW	22 39NW	24 39NW
28 56NW	39 12NW	01 18NW	17 67SE	12 52NW	38 39NW
52 34NW	03 53NW	11 62NW	22 69NW	34 22NW	16 69NW
16 13NW	52 48NW	56 41NW	26 68NW	21 33NW	58 23NW
62 55SE	22 63NW	16 40NW	79 51NW	63 27NW	24 19NW
14 52NW	23 33SE	39 44NW	14 68NW	19 51NW	24 26SE
52 39SE	07 67NW	44 63NW	43 28SE	63 23SE	38 24NW
29 41NW	29 54NW	48 14SE	24 61NW	43 03NW	22 42NW
11 49NW	58 54NW	353 32SW	17 67NW	24 54NW	12 59NW
54 73NW	12 46NW	27 39NW	11 66NW	24 46NW	41 75SE
27 33NW	23 29NW	16 38NW	26 41NW	28 47SE	13 47NW
37 36NW	30 14SE	23 06SE	52 36SE	52 27NW	27 48NW
28 33NW	65 19SE	18 27NW	32 52SE	42 72NW	22 58SE
28 18NW	30 24SE	19 67NW	27 41SE	62 28NW	23 47SE
20 17SE	25 17SE	23 62NW	11 68SE	293 39SW	19 55NW
08 07NW	35 28SE	16 73NW	27 52SE	294 34SW	27 63NW
25 34NW	32 12SE	21 56NW	68 56SE	41 29NW	30 23NW
21 34SE	38 62NW	47 23NW	09 48NW	23 25NW	295 30SW
12 28SE	43 62NW	26 48SE	19 51NW	22 25NW	20 10NW
24 7NW	24 70NW	66 22SE	21 68NW	27 34NW	20 03NW
29 25NW	07 47NW	47 22SE	13 73NW	20 15NW	22 15NW
32 46NW	27 21NW	47 16NW	23 44NW	23 21NW	00 10E
34 47NW	32 64NW	48 42NW	19 63NW	23 19NW	00 03W
22 37NW	52 34SE	51 21NW	23 72NW	25 11NW	15 10NW
12 62NW	21 43SE	26 39NW	38 24NW	22 16NW	25 01NW
19 57NW	22 68NW	61 24NW	30 52NW	320 01SW	16 32SE
19 71NW	51 61NW	27 26NW	02 29NW	16 11NW	19 02NW
07 51SE	49 73NW	28 24NW	17 68NW	25 14NW	30 36NW
29 36NW	43 82NW	42 46NW	12 79NW	25 18NW	42 54NW
36 32NW	74 51NW	08 69NW	33 11SE	10 15NW	39 11NW
14 43NW	54 53NW	17 74NW	22 41NW	18 19NW	47 36NW
10 84NW	20 17SE	12 51NW	22 43NW	49 64NW	43 42NW
31 27SE	16 44NW	18 89NW	37 24NW	13 88NW	56 20SE
24 16NW	14 44SE	48 54NW	27 37NW	22 19SE	18 52NW
64 69NW	48 24NW	38 06NW	25 12SE		

Woodall Shoals Domain II—Measurements of S_2

34 64SE	06 11SE	00 46E	26 49SE	15 18SE	55 27SE	50 30SE
04 14SE	79 12SE	70 18SE	17 20SE	78 12SE	14 16SE	10 7SE
30 25SE	31 18NW	25 16SE	07 10SE	30 20SE	16 12SE	20 42SE
15 22SE	11 23NW	15 06SE	15 09NW	16 17SE	00 29E	16 16SE
21 34SE	40 15SE	352 25SW	25 10SE	05 31SE	59 26SE	05 39SE
24 15NW	340 22NE	19 17SE	350 10SW	05 38SE	350 6NE	35 31SE
354 14NE	19 19SE	359 21NE	12 34SE	07 22SE	23 21SE	15 24SE
56 41SE	14 28SE	23 29SE	35 30SE	25 21SE	20 25SE	12 25SE
15 10SE	19 21SE	11 12SE	14 25SE	20 25SE	24 27SE	11 09SE
24 28SE	16 30SE	20 29SE	30 15SE	12 31SE	35 12SE	07 40SE
28 14SE	16 47SE	25 28SE	16 42SE	05 24SE	12 37SE	10 24SE
05 23SE	15 25SE	09 35SE	16 11SE	356 24NE	04 10SE	10 36SE
05 09SE	01 42SE					

Woodall Shoals Domain III—Measurements of S_2

39 18SE	10 18SE	345 34SW	00 15E	301 21NE	343 26NE	02 47SE
341 27NE	06 30SE	301 20NE	333 26NE	320 23NE	26 18SE	60 28SE
02 32SE	17 29SE	08 29SE	340 27NE	340 24NE	346 22NE	26 30SE
23 26SE	21 29SE	16 16SE	02 26SE	350 24NE	12 04NW	336 27NE
330 18NE	349 24NE	00 00	339 08SW	05 20SE	12 17SE	75 12SE
00 00	02 07SE	07 07SE	18 30SE	330 35NE	340 16NE	317 23NE
331 36NE	334 15NE	00 30NE	305 12NE	350 25NE	348 22NE	

Woodall Shoals Domain IV—Measurements of S_2

73 52NW	02 47SE	314 41NE	02 37SE	326 28NE	326 27NE	339 47NE
346 32NE	354 51SE	08 29SE	347 34NE	347 57NE	333 79SW	298 56NE
78 43SE	333 56NE	352 48SE	30 33SE	319 78SW	04 43E	348 12NE
348 40NE	348 43NE	00 69E	357 53NE	14 47SE	356 46NE	293 57NE
08 47SE	357 58NE	64 52SE	82 59SE	344 44NE	274 74NE	354 44NE
03 28SE	01 32SE	54 73NW	341 12NE	347 54NE	348 45NE	39 73SE
77 67NW	287 67NE	38 14SE*	04 37SE*	49 73NW*	319 57NE*	273 42NE*
354 45NE*	03 28SE*	353 42NE*				

*Not plotted on fabric diagrams.

Woodall Shoals Domain V—Measurements of S_2

52 22SE	33 05SE	27 07NW	303 27SW	24 10SE	61 14NW	62 34SE
62 28NW	40 39SE	75 37SE	46 36NW	68 24SE	13 28SE	47 21NW
00 00	52 17SE	31 47SE	23 24NW	42 19SE	24 52NW	41 37NW
358 27NE	06 37NW	43 22NW	13 24SE	24 12SE	00 00	57 35SE
67 36NW	36 78SE	41 21SE	24 21NW	44 53SE	42 36NW	346 16NE
08 68SE	44 28SE	322 39NE	36 67NW	12 20SE	86 14NW	29 69SE
32 47SE	09 24NW	05 07SE	42 71NW	317 27SW	53 07SE	52 48NW
272 18NE	63 08SE	44 32SE	46 29NW	35 25SE	56 24NW	27 38NW
10 13SE	58 67SE	84 03SE	03 19SE	51 53SE	82 23NW	65 14NW
49 51SE	69 34NW	61 14NW	48 64SE	38 18NW	57 07NW	46 31SE
282 03NE	317 16NE	43 82NW	82 08SE	14 19NW	72 43SE	37 52NW
48 17NW	48 43NW	31 32NW	281 21NE	33 54SE	351 28SW	43 78NW
44 21NW	57 57SE	29 19NW	32 34SE	18 40NW	39 48SE	81 19NW
53 32SE	11 14SE	46 18NW	69 27NW	33 11SE	32 16NW	49 21NW
23 14NW	36 12NW	63 29NW	358 13NE	42 23NW	56 17NW	56 44NW
24 10NW	73 21NW	22 19NW	69 22NW			

Woodall Shoals Domain VI—Measurements of S_2

319 09SW	283 12NE	303 03SW	84 24SE	84 30SE	40 39SE	288 22SW
273 38SW	277 39SW	78 7NW	79 30SE	357 18SW	344 16SW	295 14SW
75 19SE	32 20NW	275 21SW	301 12SW	321 13SW	87 20SE	65 25SE
80 16SE	329 15SW	313 14SW	330 08SW	75 17SE	319 12SW	85 21SE
66 20SE	59 23SE	287 13SW	279 13SW	50 26SE	85 20SE	16 16SE
86 16SE	275 16SW	05 11SE				

Woodall Shoals Domain VII—Measurements of S_2

271 39NE	46 16NW	351 28NE	11 34SE	57 04SE	343 16NE	317 07NE
09 17SE	329 11NE	03 26SE	14 14SE	48 11NW	20 18NW	20 10SE
14 31SE	22 07SE	24 21NW	357 05SW	39 10SE	27 12SE	00 12E
358 10NE	00 00	89 04SE	345 07NE	74 18SE	89 15NW	330 08NE
353 18NE						

Woodall Shoals Domain VIII—Measurements of S_2*

14 6SE	23 30SE

*Incomplete data set.

Woodall Shoals Domain IX—Measurements of S_2

83 24SE	352 11NE	333 37NE	334 14NE	72 16SE	308 43NE	09 29SE
88 22NW	277 24SW	355 39NE	303 23NE	65 89SE	317 07SW	356 31NE
351 38NE	333 06NE	88 06SE	325 05NE	05 33SE	72 08SE	63 31NW
321 05NE	335 33NE	302 23NE	277 34SW	282 04NE	344 27NE	82 13NW
300 15NE	329 26NE	317 25NE	305 20NE	283 31NE	290 07NE	327 30NE
315 8NE	82 27NW	337 03NE	313 11NE	330 12NE	334 17NE	337 12NE
320 23NE	279 11NE	303 09NE	344 30NE	357 01NE	83 12SE	307 09NE
346 18NE	336 16SW	280 15NE	350 06NE	318 52SW	87 10NW	318 15NE
83 21NW	85 11NW	311 41NE	277 18NE	23 23SE	323 04NE	277 12NE
317 34NE	312 19NE	60 38NW	284 17NE	282 25NE	68 22NW	307 16NE
326 12NE	55 22NW	300 15NE	281 23NE	72 13NW	325 20NE	321 19NE
82 24NW	307 04SW	311 18NE	308 22NE	310 18NE	311 24NE	02 18NE
283 13NE	272 11NE	293 19NE	275 28SW	281 22NE	342 17NE	295 17NE
305 27NE	292 26NE	294 16NE	00 09E	09 29SE	329 13NE	355 06NE
283 22NE	03 10SE	303 10SW	326 18NE	285 12NE	291 10SW	309 06NE
87 26NW	288 08SW	293 29NE	282 27NE	326 07SW	289 28NE	78 30NW
276 25SW	318 19NE	346 11NE	309 34SW	353 21NE	280 52NE	297 19SW
293 23NE	346 20NE	69 29SE	291 18NE	305 25NE	293 37SW	53 58NW
349 11NE	79 15SE	288 32NE	311 29NE	10 12SE		

Woodall Shoals Domain X—Measurements of S_2

353 14NE	25 11SE	353 20NE	77 27NW	358 08NE	15 17SE	68 16SE
15 18SE	44 25NW	09 22NW	07 09NW	315 18NE	62 12NW	28 62SE
19 31SE	75 17NW	63 23SE	87 29SE	00 05W	357 25NE	06 14NW
51 62SE	16 12SE	36 61SE	11 27SE	323 14NE	343 18SW	282 15NE
38 33NW	275 20NE	72 21SE	303 08NE	57 31SE	35 20SE	51 32NW
20 36SE	07 11SE	341 11NE	340 11NE	26 08NW	342 15NE	45 21SE
44 11SE	61 20SE	63 25NW	55 27SE	310 11NE	46 16SE	300 15NE
01 15SE	336 06NE	19 19SE	07 14SE	24 19SE	75 02SE	38 19SE
61 20SE	275 14NE	70 38NW	50 28NW	08 12NW		

Woodall Shoals Domain XI—Measurements of S_2

54 23SE	51 21SE	49 22SE	55 05SE	56 18SE	85 55SE	25 15SE
37 44SE	60 15SE	62 36SE	43 27SE	43 22SE	28 27SE	43 16SE
01 86NW	79 73SE	38 35SE	10 16SE	36 39SE	04 28SE	26 20SE
55 33SE	15 24SE	62 25SE	40 20SE	40 11SE	69 20SE	57 34SE
27 30SE	24 24SE	77 24SE	24 13SE	69 24SE	41 25SE	53 16SE
48 13SE	55 13SE	12 08SE	42 22SE	34 15SE	77 33SE	39 13SE
48 22SE	336 06NE	48 23SE	65 39SE	70 08SE	51 19SE	64 15SE
49 21SE	54 19SE	54 19SE	48 24SE	64 25SE	21 17SE	50 22SE
38 14SE	48 15SE	14 14SE	75 20SE	46 13SE	10 11SE	38 12SE
26 20SE	53 15SE					

Woodall Shoals Domain XII—Measurements of S_2

84 32NW	72 09NW	30 27SE	38 20SE	25 13SE	34 15SE	28 16SE
52 28SE	01 18SE	17 18SE	89 05SE	28 25SE	52 21SE	33 12SE
27 18SE	30 08SE	35 07SE	09 11NE	30 13SE		

Woodall Shoals—Orientations of Joints

12 88NW	358 81NE	70 88NW	32 79SE	353 78NE	02 73	276 30NE
88 51NW	05 85SE	01 87SE	04 87NW	07 88SE	06 79SE	07 89SE
02 89SE	274 55NE	05 82SE	06 88SE	87 89SE	01 87SE	09 87SE
15 58NW	351 78SW	43 78NW	73 77SE	02 79NW	07 65NW	51 89SE
358 89NE	358 88SW	44 72SE	294 82NE	351 84SW	31 89SE	31 83SE
354 85NE	358 83NE	07 84SE	61 89SE			

Woodall Shoals—Orientations of Crenulation Axial Surfaces*

351 61NE (IV)	344 82NE (IV)	348 68NE (IV)	04 79SE (IV)	12 85SE (IV)
52 78SE (IV)	28 21NW (IV)	42 70NW (IV)	16 35NW (IV)	52 35NW (IV)
52 51NW (IV)	24 25SE (IV)	54 64SE (I)	22 66SE (I)	51 72SE (I)
37 56NW (I)	49 64SE (I)	44 63SE (I)	12 75SE (I)	9 58SE (I)
32 6SE (I)	32 69NW (I)	42 84NW (I)	72 59SE (I)	79 20NW (I)
57 19NW (I)				

*Domain number is indicated in parentheses.

Woodall Shoals—Trend and Plunge of Crenulation Axes

234 43	202 04	231 22	217 14	229 44	224 52	12 48
09 13	32 34	212 14	222 04	251 21	259 32	237 29
351 54	164 13	168 23	43 49	12 75	232 12	208 24
42 06	196 19	232 16	232 11	24 08		

Woodall Shoals—Orientations of Axial Surfaces of Tight Folds*

24 59NW (I)	14 59NW (I)	33 63NW (I)	31 72SE (I)	21 62SE (I)
18 58NW (I)	13 23SE (I)	13 49NW (I)	42 71SE (I)	09 52SE (I)
43 49SE (I)	12 69SE (I)	32 69SE (II)	18 4SE (VII)	272 82SW (X)
352 51NE (IX)	06 88SE (IX)	27 71SE (IX)	22 77SE (IX)	28 81SE (IX)
44 58SE (IX)	32 81SE (IX)	26 58SE (IX)	11 79SE (IX)	39 75SE (VIII)
62 74NW (V)	48 76SE (V)	28 88NW (V)	06 88SE (V)	47 79NW (V)
33 55NW (V)	16 54SE (V)	26 68SE (V)	350 81SW (IV)	51 77NW (IV)
32 48SE (IV)	23 74NW (IV)	23 77SE (IV)	22 71SE (IV)	31 58SE (IV)
72 37NW (IV)	13 19SE (IV)	343 77NE (IV)		

*Numbers in parentheses are domain where each measurement was made.

Woodall Shoals—Trend and Plunge of Tight Folds

204 21	212 19	350 21	194 04	129 07
51 50	213 34	272 13	212 47	211 24
172 03	23 18	201 11	06 43	23 66
198 14	29 34	22 24	193 03	22 28
31 31	193 24	28 24	252 16	222 04
44 27	13 19	198 14	32 19	163 10
223 18	26 44	192 12	11 46	

Woodall Shoals—Trend and Plunge of Mineral Lineations

192 12	311 15	139 04	03 07	68 02	83 2	226 15

Woodall Shoals—Orientations of Axial Surfaces of Open Folds

28 89SE	28 89SE	32 65SE	358 78NE	64 84SE	03 83SE	24 88SE
26 83NW	299 89NE	32 48SE	353 84SW			

Woodall Shoals—Trend and Plunge of Open Folds

28 14	208 07	32 17	358 36	244 20	229 07	03 13
24 17	26 27	299 03	32 05	353 07		

Glossary

Aberrant folds. Folds that deviate slightly from ideal cylindrical folds.

Accident. Large, steeply dipping fault with uncertain motion sense.

Accreted terrane. Crustal-scale rock mass that has been tectonically joined to another mass; identified by stratigraphic, metamorphic, or deformational history that contrasts with that in the adjacent block.

Accretionary tectonics. Process whereby suspect and exotic terranes ranging in size from less than continental proportions to masses the size of continents are moved by plate motion to collision with each other or with other continents.

Accretionary wedge. Sediments in a subduction zone along an active margin that are scraped off a descending slab; wedge consists of an imbricate stack of thrusts with transport out of the subduction zone.

Active-roof duplex. Roof thrust in a duplex (thrust or strike-slip fault system) having considerable displacement, in contrast to a passive-roof duplex.

Allochthon. Large remnant of a far-traveled rock mass, generally a sheet moved by thrust faulting. Commonly rocks of an allochthon contrast in stratigraphy, metamorphic history, and structural history with rocks on which they presently rest.

Amontons' laws. Amontons' first law: Tangential force parallel to a movement surface necessary to initiate slip is directly proportional to the force normal to the movement surface. Amontons' second law: The frictional resistance to motion is independent of the contact area on a movement surface.

Amplitude. Half the crest-to-trough height of any wave. Half the distance from the crest of an anticline to the trough of an adjacent syncline, measured parallel to the axial plane.

Amygdules. Vesicles filled with secondary minerals.

Angular shear. Angle of rotation (ψ) of reference lines in a rock mass subjected to simple shear. The tangent of this angle is the shear strain.

Angular unconformity. Angular relationship that exists in rocks where a sequence has been deposited and later tilted, followed by erosion (or nondeposition), then renewed deposition.

Anisotropy. Characteristic of material in which properties vary with direction.

Anomaly. Deviation from an assumed value of magnetic-field intensity or gravity.

Anticlinal theory. Theory stating that hydrocarbons tend to accumulate in anticlines.

Anticline. Fold in which the layering is concave toward older rocks in the structure. An anticline contains older rocks in the center.

Antiform. Fold that is concave downward and has the shape of an anticline.

Antiformal stack duplex. Duplex (in a thrust-faulted terrane) in which the imbricate thrust sheets have moved forward to overlap each other, producing an antiformal arch of the imbricates and the roof thrust.

Antiformal syncline. Concave-downward fold wherein layering dips away from the axis, but the rocks in the center are younger. Also called downward-facing syncline.

Anti-Riedel shears. Secondary brittle fractures that form at a high angle (75° to 80°) to the primary fractures in a brittle shear zone and have a motion sense opposite to the dominant motion sense of the fracture zone.

Antithetic normal fault. Normal fault that dips oppositely to join a larger normal fault.

Apparent polar wandering (APW) curve. Curve that plots the positions through time of ancient magnetic poles for a continent; provides a record of "polar wandering" when the continent is actually moving.

Arrest line. Line marking direction of propagation of joint planes.

Asperities. Microscopic irregularities and imperfections on a fault surface.

Astrobleme. Structure produced by extraterrestrial impact, frequently consisting of concentrically and radially arranged horsts and grabens about a central chaotic zone. Impact structure.

Asymmetric fold. Fold having one limb that dips more steeply than the other.

Aulocogen. Tectonic trough, bounded by normal faults and formed within continental crust at a high angle to a nearby continental margin.

Axial elongation. Strain resulting from elongation parallel to a unique axis (hexagonal, tetragonal, or cylindrical symmetry); produces deformed objects (e.g., pebbles or oöids) resembling hot dogs (prolate spheroids).

Axial flattening. Strain resulting from shortening parallel to a unique axis, producing deformed objects resembling hamburgers (oblate spheroids).

Axial line. Line on a fold that separates dip in one direction from dip in the opposite direction; the fold axis.

Axial plane or **axial surface.** Plane or surface that results by connecting fold axes on successive folded surfaces of the same fold.

Axial symmetry. Symmetry of a cylinder; generally refers to deformation dominated by linear structures, such as folds or a mineral lineation.

Balanced cross section. Cross section demonstrating that by measuring lengths or areas of deformed layers, the section can be restored to an undeformed condition.

Basement. Crystalline rock that is the product of an earlier orogenic cycle; underlies a less deformed and less metamorphosed cover. The cover records only the younger orogenic cycle(s).

Basin. Unique bowl-shaped synform in which layering dips inward toward a central point; generally a concave-up fold.

Bedding. Primary layering in sediments and sedimentary and some volcanic rocks.

Bedding (bedding-plane) fault. Fault that follows bedding or occurs parallel to the orientation of bedding planes.

Bedding fissility. Paper-thin layering in fine-grained sediment wherein platy minerals (e.g., clays) are aligned horizontally, perpendicular to the vertical maximum principal lithostatic stress.

Bedding plane. Zone of mechanical weakness that forms in sediment because of slight compositional or textural differences at the interface between adjacent beds.

Bedding thrust. Thrust faulting localized parallel to bedding; commonly occurs in a weak rock unit, such as shale, coal, or evaporite, but may also occur parallel to bedding in strong units.

Bending. Fold mechanism that involves application of force across rock layers.

Blind thrust. Thrust in which displacement decreases upward within the sedimentary section; blind thrusts never reach the surface.

Blocking temperature. Temperature at which radiogenic isotopes are locked into a crystal lattice; commonly applied to K-Ar dating.

Body force. Force that acts equally on all parts of a body.

Bolide. Body from outer space (e.g., asteroid or meteor) that explodes on impact with the Earth or other planet.

Boomerang pattern. Fold pattern produced by interference of folds with gently dipping axial surfaces superposed by folds trending perpendicular to the first set that have near-vertical axial surfaces. Ramsay Type 2 fold interference pattern.

Boudinage. Sausage-shaped segments of a layer that has been pulled apart. The layer being segmented is much less ductile than the enclosing material and the degree of contrast in competence affects the shape of boudins. Shapes of boudins range from angular (brittle layer segmented) to rounded (ductile layer segmented).

Bouguer gravity-anomaly map. Gravity-anomaly map (generally contoured) that incorporates in the gravity data differences in both elevation and density between sea level and the elevation where measurements were made.

Bouma sequence. Sequence first recognized in sedimentary rocks composed of five intervals that make up a complete turbidite succession. It develops as a graded interval (A), a lower interval of parallel laminations (B), an interval of current ripple laminations (C), an upper interval of parallel laminations (D), and a fine-grained shaly (pelitic) interval (E). Forms a sequence that grades upward from one part into another (A to B). Partial sequences, particularly A and B, are most commonly preserved.

Branch line. Line formed by the intersection of two fault surfaces.

Break thrust. Thrust formed during folding of the connecting limb of an anticline-syncline pair—synonymous with fault-propagation fold.

Brittle behavior. Failure in the elastic range.

Brittle deformation. Discontinuous deformation.

Broken formation. Zone of broken rock that develops beneath a detachment fault, perhaps by hydrofracturing under high pore pressure.

B-tectonite. Foliated rock dominated by fold axes.

Buckle fold. Fold formed by buckling.

Buckling. Fold mechanism involving application of stress parallel to layering in rocks.

Bulls-eye fold. See *Eyed fold*.

Burgers circuit. Loop traverse that will not close at the starting point, as in a perfect crystal, because the traverse crosses a dislocation.

Burgers vector. Magnitude of the failure of a loop traverse to close in a crystal containing a dislocation.

Burrow. Indirect indicator of previous organism activity.

Byerlee's law of rock friction. The coefficient of friction in Amontons' first law is independent of rock type and depends solely on values of shear and normal stress.

Cataclasis. Brittle granulation of rock at low temperature and low to moderate confining pressure; may be accelerated by high fluid pressure.

Cataclasite. Internally undeformed rock fragments formed by brittle deformation; composed of meter- to micrometer-size original rock or minerals. Breccia and gouge are cataclastic rocks.

Cataclastic flow. Ductilelike flow of fine-grained material; produced by brittle deformation.

Center-to-center method. Method of determination of the shape and orientation of the strain ellipsoid in a rock mass by using the distance and angular relationships of strain-oriented, closely packed objects in the deformed aggregate. Involves measurement of the distances and angles between a reference grain and the nearest neighbors.

Chevron folds. Folds with straight limbs and sharp angular hinges. Kink folds.

Chocolate-block (tablet) boudinage. Square to rectangular boudinage resulting from layer-parallel extension in two directions.

Class 1 folds. Folds where dip isogons converge toward the concave part of the fold in Ramsay's standard fold classification.

Class 1A folds. Class 1 folds that have strongly convergent isogons.

Class 1B folds. Class 1 folds with convergent isogons corresponding to parallel-concentric folds.

Class 1C folds. Class 1 folds having weakly convergent isogons; may correspond to flexural-flow folds.

Class 2 folds. Folds with parallel isogons; ideal-similar folds.

Class 3 folds. Similar-like folds with isogons that diverge from the concave part of the fold.

Cleavage refraction. Change in the angle between cleavage and bedding as the cleavage passes from one layer to another.

Closed fold. Fold in which the limbs make an interlimb angle of 70° to 30°.

Coaxial deformation. Deformation parallel to the principal axes of the incremental strain ellipsoid.

Coaxial strain. Strain parallel to principal axes (and planes) of the strain ellipsoid.

Coefficient of internal friction. Material property expressed as the ratio of the shear strength to the yield stress of the material.

Competent. Strong.

Complete Bouguer anomaly map. Gravity-anomaly map (generally contoured) that includes corrections in the gravity data for elevation, density, and terrain.

Compositional banding. Layering in an igneous body that results from crystal settling, differentiation, fractional crystallization, multiple parallel intrusions, or flow processes that flatten xenoliths.

Compressional stress. Stress that decreases the volume of or shortens a body.

Concentric faults. Arcuate faults that form concentric about a point.

Concentric folds. Parallel folds in which folded surfaces define circular arcs and maintain the same center of curvature.

Concentric-longitudinal strain. See *Tangential-longitudinal strain.*

Concordant mineral. Zircon or other mineral that yields ages that plot on a concordia curve.

Concordia curve. Curve that shows variation in abundances of lead isotope ratios through time.

Conical folds. Noncylindrical folds with convergent axes.

Conjugate. Paired.

Conjugate shear angle (θ). Angle between maximum principal stress (σ_1) and a fracture that forms as a result of the application of stress.

Contact metamorphic zone (aureole). Zone of thermally induced recrystallization around a pluton.

Continuous cleavage. Foliation in fine-grained rocks that pervades the rock mass on all scales.

Continuous creep. Stable sliding along a fault.

Contractional fault. Faults along which bedding is shortened. Also called a wedge.

Coulomb criterion of failure. The absolute value of shear strength (τ_s) is the sum of the inherent shear strength (S_0) and the product of the coefficient of internal friction (μ) and normal stress (σ).

Coulomb wedge. Wedge of sedimentary rocks that is undergoing brittle deformation, generally by imbricate thrusting.

Creep process. Rate-dependent lattice-scale deformation mechanism that involves mass transport or diffusion of atoms or ions at grain boundaries, glide and climb of dislocations within a lattice, and diffusion of point defects through lattices.

Crenulation cleavage. Cleavage that overprints an existing cleavage or foliation by crinkling and crenulating the earlier structure.

Crest. Highest point of a fold (or wave).

Critical taper. In a thrust-faulted terrane, the angle between the surface slope and the basal detachment. Once attained, the angle is maintained by erosion, internal deformation of the wedge, and/or displacement of the wedge.

Cross bedding. Layering inclined to the principal bedding planes in sedimentary rocks composed of sand-size particles. Normal (planar) cross beds cannot be used to determine facing direction; trough (festoon) cross beds can be used to determine facing direction.

Cross-section analysis. Study of the structure of an area by construction of vertical cross sections from surface geologic data, down-plunge projection, and geophysical data.

Cryogenic magnetometer. Instrument that measures paleomagnetism inside a superconducting ring, allowing detection of weak signatures.

Crystal lattice. Systematic internal array of atoms, ions, or molecules in a compound.

Crystalline solid. Regular geometric arrangement of atoms, ions, or groups to form a repeat unit in a compound. Crystalline solids need not be single homogeneous crystals, but can be crystal aggregates.

Crystalline thrust. Thrust involving crystalline (metamorphic and igneous) rocks.

C-surface. Foliation in shear zones that develops parallel to shear-zone boundaries. As shearing continues, the C-surfaces may be rotated to an angle of 18°–25°.

Curie temperature. Temperature above which strongly magnetic (ferromagnetic) materials lose their ability to interact with a magnetic field.

Current ripple marks. Translational ripples that form where a prevailing direction of transport and deposition of sediment occur. They present the same shape whether upright or overturned.

Cylindrical (cylindroidal) fold. Fold that can be generated by moving the fold axis parallel to itself.

Décollement. Movement surface (detachment) produced by folding or faulting; zone of weakness (shale, coal, or other weak rock, or a strong unit susceptible to strain softening) through which a fault may propagate. Separate zones of contrasting deformation above and below the décollement. French for "ungluing."

Defect. Imperfection in a crystal lattice.

Deformation map. Plot that shows the experimentally determined range of conditions of several deformation mechanisms for particular rock types or minerals.

Deformation path. Stages of progressive deformation affecting a rock mass.

Deformed state. Strained state.

Delta structure. See *Triangle structure.*

Depositional remanent magnetism. Paleomagnetism in sedimentary rocks; recorded by orientation of grains of magnetite (or other magnetically susceptible minerals) in the Earth's magnetic field at the time of deposition.

Desiccation cracks. See *Mud cracks.*

Detachment. See *Décollement.*

Deviatoric stress. Nonhydrostatic component of stress.

Dextral strike-slip fault. See *Right-lateral strike-slip fault.*

Diamictite. Rock composed of angular, unsorted to poorly sorted generally noncalcareous material of a wide range of particle size.

Diapir. Salt, mud, magma, or other material of lower density than the enclosing rocks that moves upward gravitationally and intrudes the overlying sediments or the crust. Most are cylindrical domes.

Differential stress. Difference between the maximum and minimum principal normal stresses.

Differentiated layering. Foliation involving segregation of minerals into layers of different composition (such as quartz-feldspar and biotite-rich layers); produced largely through recrystallization during progressive deformation and metamorphism.

Dilation. Positive or negative change in volume.

Dip. Angle made by an inclined surface with the horizontal, measured perpendicular to strike.

Dip isogon. Line connecting points of equal slope, or dip, on successive layers in a fold.

Dip-slip movement. Movement down or up parallel to the dip direction of a fault.

Disconformity. Unconformity produced by deposition of a sequence of sedimentary rocks followed by uplift or a drop in sea level. Uplift and erosion produce a surface having topographic relief without tilting or other deformation of the sequence, followed by subsidence and renewed deposition. Bedding in the rocks above and below the unconformity remains parallel.

Discordant mineral. Zircon or other mineral having lead isotope ratios that do not plot on the concordia curve.

Discrete crenulations. Sharply defined domainal crenulation cleavage that truncates the older fabric and preserves the older fabric in microlithons.

Disharmonic folds. Folds that change shape from layer to layer. Involves moderate to high ductility contrast. Same as quasi-flexural folds.

Disjunctive cleavage. Cross-cutting spaced cleavage, not related to original layering.

Dislocation. Point or line defect.

Dislocation creep. Combination of glide-and-climb motion of dislocations through a crystal lattice; occurs at moderate to high temperature.

Dislocation density. Relative concentration or absence of dislocations in a crystal, indicating the degree of deformation or recovery/recrystallization in the lattice.

Dislocation glide and climb mechanism. Mechanism in which vacancies in a crystal lattice diffuse away from, or less commonly toward, a deformed area and out of the extra half plane of the dislocation from one slip plane to the next.

Dislocation map. Map showing distribution, density, and kinds of dislocations in a crystalline solid.

Displacement. Total amount of motion measured parallel to the movement direction. Amount of relative motion measured on opposite sides of a fault. Net slip.

Distortion. Strain involving change in shape, length, or volume.

Doctrine of uniformitarianism. Processes occurring today upon and within the Earth have probably gone on similarly in the past and will continue in the future.

Domainal cleavage. Cleavage localized in particular zones. A type of spaced cleavage.

Dome. Unique antiform wherein layering dips in all directions away from a central point.

Dome-and-basin pattern. Fold pattern consisting of systematically related domes and basins. May be produced by two generations of folds that overprint each other at right angles, producing a Type 1 fold interference pattern, or one event with shortening from two directions.

Down-to-basin fault. Listric normal fault in which the downthrown side is toward an adjacent basin. Commonly forms during crustal rifting or filling of a basin.

Downward-facing syncline. See *Antiformal syncline.*

Drag. Folding of layering in rocks next to a fault that occurs by friction during movement along a fault.

Drag fold. Fold that forms during movement because of friction near a thrust or normal fault surface. It produces evidence of motion sense on the fault through observation of the fold asymmetry along the fault.

Drape fold. Fold in basement and cover produced by high-angle normal or thrust faults in the basement that may only partly break the sedimentary cover.

Ductile-brittle transition. Transition from brittle behavior in the upper crust to ductile behavior in the lower crust, where brittle behavior is inhibited because strain softening occurs as a result of increases in temperature and pressure with depth.

Ductile deformation. Continuous deformation occurring by plastic or viscous flow.

Ductility. Permanent strain in the form of viscous or plastic deformation that reflects the capacity for rocks to accommodate large amounts of strain homogeneously (affecting all of a rock mass) or heterogeneously (localized in ductile shear zones).

Ductility contrast. Contrast in ability to flow between layers making up a layered rock sequence or between a dike or sill and the enclosing rocks.

Duplex. Most commonly, a system of thrust faults bounded by two parallel to subparallel subhorizontal thrusts of roughly equal displacement separated by a deformed interval dominated by smaller imbricate faults that connect the two master faults. The imbricates have the same sense of movement as the master faults. The upper fault is the roof thrust; the lower is the floor thrust. Duplexes may also form in normal and strike-slip fault zones.

Dynamic recovery. Processes during deformation that reduce strain, dislocation density, and dislocation interaction and increase the rate of glide and climb of dislocations.

Dynamic recrystallization. Processes occurring during deformation that reduce dislocation density and dislocation interaction; these processes involve strain softening and occur at low strain rate and high temperature.

Edge dislocation. Edge of an extra partial layer (extra half plane) in a crystal lattice inserted parallel to existing layers. Edge dislocations may propagate in a lattice subjected to simple shear.

Effective normal stress. Normal stress minus fluid pressure.

Egg-carton structure. Dome-and-basin fold pattern produced by Type 1 fold interference.

Elastic strain. Strain that is recovered instantaneously on removal of applied stress, so the object returns to the original undeformed shape once the strain is removed.

Elasticas. Fold in which the interlimb angle is negative.

Elastic limit. Point on the stress-strain curve beyond which the material begins to undergo permanent deformation or ruptures (or both), initiating plastic or brittle behavior.

Elasticoviscous behavior. Combination of elastic and viscous behavior.

Elastic-plastic behavior. Combination of elastic and plastic end-member behavior.

Elastic rebound. Rebound commonly accompanied by earthquakes following rapid movement on a fault after long accumulation of elastic strain.

Electrical conductivity. Ability of a material to conduct electric current.

Electrical resistivity. Resistance of a material to the flow of electricity.

Elongation. Type of strain measured by the ratio of the length of lines in a deformed mass minus the original length to the original length.

Emergent thrust. See *Erosion thrust.*

***En echelon* faults.** Faults that strike roughly parallel to one another; they occur in short, sometimes overlapping, segments.

Entropy. Measure of the amount of energy unavailable for useful work in a system.

Enveloping surface. Surface constructed by connecting the crests or troughs of small folds on a larger fold by a surface tangent to the small folds.

Episodic lead loss. Loss of lead during one or more thermal events subsequent to crystallization of zircons (or other minerals); produces discordant lead isotope ratios that plot on a chord rather than the concordia curve.

Equilibrium. State of rest or balance. Excess energy in one part of a system is compensated by work performed in another, or flow of energy to the lower-energy state, until a state of balance is achieved.

Erosional outlier. Remnant of a formerly more extensive rock mass isolated by erosive processes.

Erosion thrust. Thrust that ruptures the ground surface; an emergent thrust.

Exotic terrane. Block of crustal dimensions that bears no resemblance (structural or metamorphic history, faunal assemblage) to the mass to which it is presently attached; commonly derived from the opposite side of a major ocean. Boundaries of exotic terranes with the masses to which they are attached are always tectonic.

Extensional crenulation cleavage. Cleavage (foliation) that develops in a shear zone opposite to the direction of shear-zone movement—in the extension direction—and opposite to C-surfaces. May form as normal or reverse-slip crenulations.

Extensional fault. Fault in which bedding undergoes layer-parallel extension.

Extra half plane. Inserted layer of an edge dislocation.

Eyed fold. Sheath fold that yields a highly symmetric section viewed perpendicular to the hinge. Also known as a bulls-eye fold.

Eyelid window. Window formed where a thrust sheet is arched into an antiformal fold, then broken along the antiform; erosion produces a map pattern exposing younger rocks in the window, but faults occur in the thrust sheet connected to the fault framing the window. Instead of having a near-circular shape, an eyelid window has a more elongate eye shape.

Fabric. Relationships between planar (S) and linear (L) structures (including bedding, cleavage, and the orientation of minerals) to texture in rocks.

Fabric element. Any structure, such as foliations, lineations, or small folds, that contributes to the fabric (texture plus structure) in rocks.

Facing direction. Direction toward the top of a sequence determined using primary sedimentary structures, faunal succession, or, less commonly, radiometric age data.

Failed arm. Initial rift that does not continue to develop and spread apart. Also called a failed rift.

Failed rift. See *Failed arm.*

Fault. Fracture having appreciable movement parallel to the plane of the fracture.

Fault-bend fold. Fold formed above a thrust fault as the surface changes from steeper dip to shallower in an up-dip direction as it passes over a ramp, forming generally parallel bending folds. A shear thrust.

Fault-line scarp. Fault scarp produced by differential erosion along a fault, lowering the level of the topographic surface and removing a resistant layer in the hanging wall so that the original displacement may or may not be indicated by the present-day topography.

Fault mechanics. Study of the physical processes and physics that produce faults and faulting.

Fault plane. Actual movement surface along a fault.

Fault-propagation fold. Fold formed during folding as a fold tightens and a thrust fault begins to propagate through the sedimentary sequence along the common tightened limb between the anticline and syncline. A break thrust.

Fault scarp. Scarp formed where a topographic surface is offset by dip-slip motion along a fault. The present topographic displacement indicates the movement sense on the fault.

Fault zone. Series of interleaving anastomosing brittle faults near the surface, or a ductile shear zone produced by faulting at great depth. Yields a zone of measurable thickness and not a discrete plane.

Fenster. See *Simple window.*

Field relations. Structural measurements made at outcrop scale that permit working out outcrop-scale to map-scale structures.

Filled joint. Joint that has been filled with minerals during or after formation of the joint.

Finite-element method. Numerical method for dividing a continuum into a grid that can be used to understand how structures form by studying the changes that occur in the grid as it is deformed.

Finite strain. Comparison of the difference in the present state of strain with some previous less deformed state. Relates to the instantaneous shape of a rock mass at one time relative to the initial undeformed shape.

First-order folds. Largest folds in an area.

First-rank tensor. Tensor that has three components and consists entirely of vectors.

Flexural flow. Flow in which some layers in a sequence flow ductilely during folding whereas stronger layers remain brittle (or are less ductile) and buckle. Moderate to high ductility contrast between layers is required.

Flexural-flow folds. Mostly similar-like folds that form in low to moderate metamorphic-grade rocks, but may include some parallel folds. Some layers maintain constant thickness; others are thickened into the axial zones and thinned in the limbs as a result of moderate to high ductility contrast.

Flexural folds. Folds in which shape is controlled by the layering in rocks.

Flexural-slip folds. Folds that form by buckling, bending, and slip parallel to layering. They correspond in shape and mode of formation to parallel or parallel-concentric folds.

Flinn diagram. Plot of elongation (X/Y) vs. flattening (Y/Z) strain.

Floor thrust. Lower of the two master faults of a duplex.

Flower structure. Upward and outward branching faults (cross-section view) in a strike-slip fault system that have the shape of a bouquet of flowers. Same as palm (-tree) structure.

Flute casts. Spoon-shaped structures that form in the sedimentary environment when currents scour and erode a surface.

Fold axis. Axial line of a fold that separates dip in one direction from dip in the opposite direction.

Fold test. Test of paleomagnetic data whereby the measurements made on oriented samples collected from opposite limbs of a fold are plotted on a stereonet and then rotated back to the horizontal. Paleomagnetic measurements for samples from opposite limbs are then compared for consistent orientation of paleomagnetic poles. A valid fold test is necessary to evaluate the usefulness of paleomagnetic measurements.

Foliation. Layering in deformed and metamorphosed rocks imparted by the parallel orientation of platy or elongate minerals. Schistosity, gneissic banding, slaty cleavage, and crenulation cleavage are all foliations.

Footwall. Rock mass beneath a fault plane.

Force. Vector that produces a change in the velocity, direction, or acceleration of a body.

Foreland fold-and-thrust belt. Belt of thrust faults and related folds—generally thin-skinned—that occurs in sedimentary and metamorphic rocks between the undeformed craton and the metamorphic core of nearly every mountain chain.

Form lines. Lines—commonly strike lines—on a map that outline the trend and shape of a structure.

Fossil. Remains of an organism preserved in the geologic record.

Fracture cleavage. Cleavage formed by parallel to subparallel fractures, generally spaced 1–3 cm apart.

Free-air gravity map. Gravity-anomaly map corrected only for elevation above sea level.

Fry method. Simpler version of the center-to-center method for determining strain. It produces a diagram containing a set of points with a central circular to elliptical blank area in which relative shape and orientation are proportional to the shape and orientation of the strain ellipse. A circular area indicates there is no strain.

General strain. Strain that produces the shape of a triaxial ellipsoid. Plots in the vicinity of the $k = 1$ line in the Flinn diagram.

Gentle fold. Fold with an interlimb angle of 180° to 120°.

Geochemical cycle. See *Rock cycle*.

Geometric axes. Spatial relationships between structures and a set of mutually perpendicular axes.

Geophone. Detector used in seismic exploration that produces an electrical signal proportional to the energy (velocity, amplitude) of acoustic waves detected. Can be designed to detect waves over a narrow or wide frequency range.

Gneissic banding. Foliation in gneissic rocks; commonly alternating layers of different texture, composition, and color.

Graben. Block that has been dropped down between two subparallel normal faults that dip toward one another.

Graded bedding. Bedding formed where sediment of widely different particle size is deposited rapidly in the same depositional environment so that the largest particles settle to the bottom first.

Grain-boundary diffusion creep (Coble creep). Deformation mechanism involving mass transport through diffusion at grain boundaries, occurring at low temperature and overlapping the pressure-solution realm ($<300°C$) up to about 400°C.

Grain-boundary sliding. Deformation mechanism where grains slip past one another at moderate temperature in association with Coble creep, or at low temperature under a strong component of simple shear.

Gravitational spreading. Theory that an orogen compensates for uplift derived from horizontal compression by flowing laterally under the influence of gravity.

Gravity anomalies. Measured values of gravity that deviate from estimated or assumed values.

Gravity fault. Normal fault, so-called by implication that gravity is the primary motive force. Also called extensional fault.

Gravity model. Model used to determine the shape and broad composition of plutons, or other rock bodies, by plotting gravity data.

Griffith cracks. Elliptical microscopic cracks that exist in glass; they suggest that stress is concentrated and magnified by several orders of magnitude at crack tips.

Growth fault. Fault—generally a normal fault—that involves simultaneous deposition and fault motion.

Hackle marks. Corrugations on a joint surface indicating a zone where a joint propagated rapidly. Twist hackle forms in response to changing stress orientation or possibly a component of simple shear.

Half-graben. Block bounded by a normal fault on one side; the other side passes into a gentle bending fold.

Hanging wall. Rock mass resting on (above) a fault plane.

Hartman's rule. Maximum principal stress (σ_1) bisects the acute angle between the conjugate shear planes of the stress ellipsoid.

Heave. Horizontal component of fault displacement.

Heterogeneous domain. Domain of a single kind of fabric data (e.g., foliation); contains multiple orientations.

High-angle boundary. Boundary with microscopic texture involving new grains formed to relieve strain in a strained crystal. New grains contain lattices oriented at a high angle to the lattice in the parent crystal and to each other.

Hinge. Part of a fold that has the greatest curvature.

Hinge line. Line where layering changes orientation from dip in one direction to dip in the opposite direction across the crest or trough of a fold.

Homocline. Structure with uniform dip of bedding in one direction.

Homogeneous domain. Area where one fabric element is oriented the same throughout.

Homogeneous material. Material in which properties are the same throughout. Isotropic.

Homogeneous strain. Strain in which lines originally straight and parallel before deformation remain the same after deformation.

Hook fold pattern. Pattern produced by refolding of an isoclinal fold by folds having axes parallel to the first folds. Type 3 fold interference pattern.

Horizontal compression. Force oriented in the horizontal that tends to push a mass together.

Horse. Fragment of material transported within a fault zone beneath a thrust sheet. A slice.

Horse-tail structure. Branching, frequently anastomosing, array of faults.

Horst. Structurally high block between two oppositely dipping normal faults.

Hummocky laminations. See *Ripple cross laminations.*

Hydraulic joints. Joints produced by abnormally high pore pressure during burial and vertical compaction of sediment at depths greater than 5 km.

Hydrostatic state of stress. State in which normal stress is the same in all directions, and no shear stress exists.

Imbricate stack. Assemblage of imbricate faults, generally thrusts, arranged so they overlap one another and do not coalesce upward as in a duplex.

Imbricate thrust. Smaller thrust that converges down-dip into a master fault.

Imbricate zone. Zone of imbricate faults. May refer to the entire series of thrusts making up a foreland fold-and-thrust belt.

Impact structure. Kilometer- or larger-scale structure commonly made up of a complex of radial and concentric faults that have a circular or elliptical outline but are not obviously related to tectonic processes. Astrobleme.

Incompetent. Weak.

Incremental strain. Strain that occurs in events separable from other deformational events in the progressive deformation of a body.

Induced magnetization. Magnetization in a rock mass produced by interaction of the magnetically susceptible minerals and an external magnetic field.

Infinitesimal strain. Strain that occurs in infinitesimally small amounts. Restricted to very small strain relative to an initial condition.

Inherited zircons. Zircons assimilated by magma from another rock mass.

Inhomogeneous property. Property that varies spatially with location at the scale of reference, either microscopic, in a hand specimen, or in a region. Anisotropy.

Inhomogeneous strain. Strain in which lines that are straight or parallel before deformation do not remain so after deformation; angular relationships change nonuniformly.

Inlier. Erosionally exposed rock mass surrounded laterally by the unit above it and with a normal or unconformable contact with the unit below.

Interference folds. See *Superposed folds.*

Interlimb angle. Angle between the limbs of a fold.

Intersection lineation. Linear structure formed by the intersection of two planar structures; usually a penetrative structure. The most common intersection involves bedding (S_0) and cleavage (S_1), sometimes designated $L_{1 \times 0}$.

Interstitial atom. Atom present between lattice sites in a crystal.

Interstitial defect. Defect that disturbs the regular arrangement of a crystal lattice.

Intracontinental transform. Transform fault that separates two continental terranes.

Intrafolial folds. Small to large isolated folds (commonly relict isoclinal folds) lying within a foliation.

Inverted fault structure. Structure in which original movement sense has been reversed.

Inverted rift structure. Series of normal faults that originally formed by rifting (extension) but were later subjected to compression, reactivating the faults as thrust or strike-slip faults.

Isochron. Line produced by plotting analyses of $^{87}Sr/^{86}Sr$ versus $^{87}Rb/^{86}Sr$. The slope of the isochron is the Rb-Sr whole-rock or mineral age.

Isoclinal fold. Tight fold in which the axial surface and limbs are parallel.

Isostatic equilibrium. State of equilibrium between blocks within the continents or oceans and between continents and adjacent oceans.

Isotropy. Lack of contrast in physical properties with direction between and within individual layers of a rock mass. Materials have the same properties in all directions.

Joint. Fracture along which there has been no appreciable movement parallel to the fracture and only slight movement perpendicular to the fracture plane. A Mode I fracture.

Joint set. Joints that share similar orientation in the same area.

Joint system. Two or more joint sets in the same area.

Kinematic (movement) axes. Fabric axes in which the *a* axis is the tectonic transport direction, *b* is normal to *a* in the transport (fold axial) plane, and *c* is normal to the *a-b* plane.

Kink folds. See *Chevron folds.*

Klippe. A small erosional outlier (remnant) of a thrust sheet.

Königsberger ratio. Remanent magnetism divided by the product of magnetic susceptibility of the rock material that produced a magnetic anomaly and the induced magnetic-field intensity.

Lateral ramp. Along-strike change in detachment level where a thrust is forced by stratigraphic pinchout or other means to rise to a higher level.

Law of cross-cutting relationships. A body of rock is older than structures or igneous bodies that cut through it. See also *Law of igneous cross-cutting relationships* and *Law of structural relationships.*

Law of faunal succession. Fossil organisms in a sequence are more advanced toward the top of the sequence.

Law of igneous cross-cutting relationships. An igneous body must be younger than the rocks it intrudes.

Law of original horizontality. Bedding planes within sediments or sedimentary rocks form in a horizontal to nearly horizontal orientation at the time of deposition.

Law of structural relationships. A structure, such as a fold or fault, must be younger than the rocks it deforms or cuts through.

Law of superposition. Within a layered sequence, commonly sedimentary rocks, the oldest rocks will occur at the base of the sequence and successively younger rocks will occur toward the top, unless the sequence has been inverted through tectonic activity.

Layer-parallel shortening. Deformation resulting in shortening of a layered sequence parallel to the layering. Pressure solution is a common mechanism.

Left-handed screw dislocation. Screw dislocation that involves anticlockwise rotation.

Left-lateral strike-slip fault. Strike-slip fault where the left side has moved toward an observer looking along the fault; also called sinistral strike-slip fault.

Limbs. Straighter or least-curved segments of a fold.

Lineament. Topographic feature consisting of aligned surficial features like valleys and ridges.

Linear structure; lineation. Structure on any scale that can be expressed as a real or imaginary line.

Line defect. Type of dislocation produced by insertion of an extra half plane into a lattice. Edge and screw dislocations are both line dislocations.

Listric fault. Thrust or normal fault with concave-up geometry. These faults have steep dip near the surface but flatten with depth.

Lithosphere. The outer 100 km of the Earth; includes all of the crust and part of the mantle.

Lithostat. Weight of a column of rock per unit area above it.

Load casts. Depressions in sediment beneath a sand bed. Result from gravitational instability at the interface between a layer of water-saturated sand and underlying mud after deposition and dewatering. May be used to determine facing direction.

Low-angle boundary. Microscopic boundary of a subgrain within a strained grain, indicating a slight (low-angle) lattice misorientation across the boundary.

LS-tectonite. Deformed rock wherein linear structures dominate over foliation or cleavage.

L-tectonite. Rock dominated by a linear fabric element.

Macroscopic scale. Mountainside to map scale.

Macroscopic structure. Structure of mountainside or larger scale.

Magnetic-anomaly map. Map produced by plotting vertical magnetic-intensity profiles along the line of traverse or the data contoured between successive traverse lines.

Magnetic-intensity profile. Plot of vertical magnetic intensity along the line of a profile.

Magnetometer. Instrument used to measure intensity of the Earth's magnetic field.

McKenzie model of rift formation. Model depicting the idea that spreading is symmetrical about a rift axis, decreasing to zero along the axis at a pole of rotation.

Mean ductility. Ductility of an entire rock mass. Rocks of low mean ductility are strong; rocks of high mean ductility are weak.

Mean stress. The mean of the principal stresses: $(\sigma_1 + \sigma_2 + \sigma_3)/3$.

Mélange. Mixture of sheared tectonically mixed weak and strong rock materials. Fragments of strong material may range up to several kilometers.

Mesoscopic scale. Scale of structures in the size range from hand specimen to outcrop.

Metamorphic core. Internal part of most compressional mountain chains, consisting of folded, faulted, cleaved, foliated, and intruded rocks ranging from low to high metamorphic grade.

Metamorphic core complex. Domal uplift of deformed metamorphic and igneous rocks exposed beneath a tectonically detached unmetamorphosed cover of sedimentary and volcanic rocks, commonly associated with regions of crustal extension. The Basin and Range Province is a well-described region of large-scale crustal extension that contains many core complexes.

Metamorphic differentiation. Formation of new layering by pressure solution or recrystallization.

Mica beards. Microscopic texture of pressure shadows composed of micas (commonly muscovite and chlorite) on quartz or other strong grains in a clastic sedimentary rock undergoing deformation at lower greenschist-facies metamorphic conditions.

Microcontinent; microplate. Crustal mass of thrust sheet to subcontinental proportions.

Microlithon. Uncleaved zone between cleavage surfaces.

Microscopic scale. Scale of structures that require magnification to be observed.

Migmatite. Mixed igneous-metamorphic–looking rock produced by partial melting or metamorphic differentiation. New material (neosome) is commonly more feldspar-quartz rich; older material (paleosome) is commonly rich in mafic components (hornblende, biotite).

Mineral lineation. Lineation consisting of aligned elongate mineral grains and grain aggregates.

Mixing line. Best-fit line produced by plotting $^{87}Sr/^{86}Sr$ versus $^{87}Rb/^{86}Sr$, but in which the points do not plot close to the line to define an isochron and accurately determine rock or mineral age. Scatterchron.

Mode I fracture. Fracture formed by extension; a joint.

Mode II fracture. Fracture formed by sliding; a dip-slip fault.

Mode III fracture. Fracture formed by a tearing motion; a strike-slip fault.

Mohr circle for strain. Circle produced by plotting on a set of coordinate axes values of reciprocal quadratic elongation on the horizontal axis versus modified shear strain on the vertical axis.

Mohr construction. Circles produced by a plot of values of σ_1 and σ_3 on the σ axis of a plot of σ versus τ. Mohr circles for stress produced.

Mohr envelope. Tangent to Mohr circles.

Mohr's hypothesis. Shear strength is a function of normal stress.

Monocline. Structure in which an otherwise uniform regional dip is locally steepened.

Monoclinic symmetry. Symmetry in only one plane. Asymmetric folds have monoclinic symmetry.

Mud cracks. Polygonal cracks produced by drying in the surface of sediment that taper downward and terminate. Useful in determining facing direction. Also called desiccation cracks.

Mullions. Corrugated or scalloped surfaces resulting from folding, cleavage formation, or differential flow; form at boundaries between rock types of different relative ductility. Type of lineation.

Multiple working hypotheses. Principle enabling formulation of more than one possible explanation of the same data, evaluation of each, and selection of the most likely hypothesis.

Mylonite. Strongly foliated metamorphic rock that exhibits characteristics of high ductile strain. Decrease of grain size is characteristic of mylonite, along with ribbon quartz and rotated porphyroclasts. S-C fabrics are commonly present. Microfabric is commonly only partially recovered or recrystallized but may be totally recrystallized if deformation ceased during or before a thermal event.

Nabarro-Herring creep. See *Volume-diffusion creep.*

Natural strain ellipsoids. Features in rocks (oöids, pebbles, vesicles) that are nearly spherical before deformation and deform into ellipsoid shapes.

Negative anomaly. Gravity or magnetic feature that has a less than predicted intensity.

Net slip. See *Displacement.*

Neutral surface. Surface of no finite strain in each layer in tangential-longitudinal strain buckle folds; separates a zone of extension in the outer arcs from a zone of compression in the inner arcs.

Newtonian fluid. Fluid in which stress and shear strain rate are proportional.

Newton's law of gravitation. The force of gravity is directly proportional to the masses of the objects involved and inversely proportional to the square of the distance between them.

Node. Point where three or more dislocations meet.

Noncoaxial deformation. Deformation not parallel to the principal axes of the strain ellipsoid. Simple-shear deformation.

Nonconformity. Unconformity in which igneous or metamorphic rocks—or both—occur below the erosion surface and sedimentary rocks occur above.

Noncylindrical folds. Folds in which axes are not parallel on successive folds; fold axes converge toward a point.

Nonpenetrative structure. Isolated structure, such as a fault or isolated fold. Some structures, like joints, may be nonpenetrative at outcrop scale, but penetrative at map scale.

Nonsystematic joints. Irregular fractures that do not share a common orientation; fracture surfaces are usually highly curved and irregular.

Normal cross beds. Beds in which tabular (planar) layering is inclined to bedding. Cannot be used to determine facing direction.

Normal fault. Dip-slip fault in which the hanging wall has moved down relative to the footwall.

Normal-sequence thrust. Thrust fault in a foreland fold-and-thrust belt; produced by propagation outward from the internal core of a mountain chain.

Normal stress. Stress that acts perpendicular to a surface, parallel to the axes of the strain ellipsoid.

Oblique-slip motion. Combined dip-slip and strike-slip motion.

Obsequent fault-line scarp. Fault-line scarp that through erosion of a resistant layer faces opposite to the direction of the original fault scarp. An apparent opposite motion sense is inferred from the topography without subsequent movement.

Olistostrome. Deposit without internal order; made up of rock fragments of diverse size and composition that accumulated in finer sediment by submarine slumping or gravity sliding from an unstable slope. Blocks range from a few millimeters to several kilometers.

Oöid. Millimeter-size concentrically layered spheroidal to ellipsoidal concretionary body composed most frequently of $CaCO_3$. Commonly formed by precipitation around a nucleus (shell fragment, sand grain) in a high-energy (wave) zone. Good finite-strain indicator.

Open folds. Folds in which the limbs dip gently away from or toward one another, producing a large interlimb angle (>70°).

Ordinary boudinage. Boudinage formed by extension in one direction; consists of parallel, segmented, sausage-shaped remnants of a former continuous layer in a more ductile matrix.

Orogeny. Process of mountain building, accompanied by metamorphism, plutonism, and associated deformation, resulting from subduction, terrane accretion, and continent-continent collision.

Orthorhombic symmetry. Symmetry along two or three planes. Symmetrical folds and conjugate joints have orthorhombic symmetry.

Oscillatory ripple marks. Symmetrical ripple marks consisting of alternating sharp linear high and low crests and broad rounded troughs formed by back-and-forth motion of water. They may be used for determining facing direction.

Out-of-sequence thrust. Younger, higher thrust fault that breaks an older thrust sheet.

Outrageous hypothesis. Initially unacceptable hypothesis formulated to explain a phenomenon; the value of the hypothesis is not in being right or wrong but in focusing attention on an unsolved problem.

Overburden pressure. Pressure of overlying material.

Overturned folds. Folds that have one inverted limb.

Paleomagnetism. Remanent ancient magnetic fields preserved in rocks. Measurement allows reconstruction of the positions of the ancient magnetic poles, continents, and oceans.

Palm (-tree) structure. See *Flower structure*.

Paraconformity. Unconformity involving only sedimentary rocks with little relief on the erosion surface; bedding remains parallel on both sides.

Parallel folds. Folds in which the layering maintains constant thickness and layers remain parallel.

Passive flow. Uniform ductile flow of the entire rock mass, with layering serving only as a strain marker; little ductility contrast exists between layers.

Passive-flow folds. Ideal similar folds that involve plastic deformation. The layering acts only as a strain marker to record the deformation.

Passive folds. Folds in which layering serves only as a strain marker during folding.

Passive-roof duplex. Duplex in which the roof thrust has only minor displacement.

Passive slip. Slip (simple shear) at an angle to layering along a cleavage or schistosity. Associated folds are called passive-slip or shear folds.

Passive-slip folds. Similar-like folds that are thought to form by shearing along planes (cleavage, foliation) inclined to the layering.

Pencil structure. Elongate pencil-shaped structure in rock; produced by the intersection of bedding and cleavage.

Penetrative structure. Structure such as cleavage, foliations, and some folds that occurs on any scale chosen for observation. Some structures, such as joints, may not be penetrative at outcrop scale, but are penetrative at map scale.

Perfect crystal. Crystal lattice where all the sites are filled with the correct atoms, ions, or groups for a particular mineral; contains no dislocations or interstitial atoms between lattice sites.

Perfect fluid. Stationary fluid that will not transmit shearing stress.

Pileup. Accumulation of dislocations in a crystal lattice.

Pillow structure. Clustered volcanic structures with rounded tops and sharply pointed bases resembling pillows; forms where lava is erupted beneath or flows into water. Exteriors of pillows are fine-grained and scoriaceous to glassy, with grain size increasing toward the interior.

Pisolite. Concentrically layered spheroidal to ellipsoidal concretionary body that forms under conditions similar to those for formation of oöids. Similar in shape to an oöid but larger and more irregular. Good finite-strain indicator.

Planar (tabular) cross bed. Layering inclined to normal bedding planes that is planar and does not become tangent to the bottom of a bed; it is truncated at both the top and bottom of the bed. Facing direction cannot be determined using planar cross beds because they provide the same perspective upright or overturned.

Plane strain. Strain that occurs following deformation when a series of parallel planes remains undistorted and parallel to the same set of planes in the undeformed body. Strain parallel to the $k = 1$ line in the Flinn diagram.

Plastic strain. Permanent, nonrecoverable strain occurring without loss of cohesion. Results from rearranging chemical bonds in crystal lattices and generally affects an entire rock mass. Ductile shear zones are narrow zones of plastic strain.

Plate tectonics. Concept of global tectonics in which the Earth's surface is divisible into seven major plates, and several smaller ones, that contain all the continents and oceans. Plates are formed at the oceanic ridges as new oceanic crust forms and are consumed by subduction in the trenches. Involves plate generation, motion and destruction, and interaction.

Plumose joint. Joint that has a feathered surface texture.

Plunge. Bearing and amount of inclination of a fold axis or other linear structure from the horizontal.

Plunging fold. Fold with nonhorizontal axis.

Point defect. Substitution or interstitial defect involving atoms, ions, or molecules.

Poisson's ratio (υ). Measure of compressibility defined by the ratio of the lateral compression of a reference body to the longitudinal extension in the same body.

Polygon. Strain-free crystal formed by recovery or recrystallization. Generally has high-angle boundary but forms initially along low-angle boundaries.

Porphyroblasts. Large grains that have grown in a rock mass during deformation or metamorphism.

Porphyroclasts. Relict earlier large grains of one or more minerals that remain as a rock mass is deformed.

Positive anomaly. Gravity or magnetic feature that has a greater than predicted value.

Potential field. Gravity, magnetic, or electric field.

Pressure shadow. Mineral lineation composed of quartz, muscovite, chlorite, magnetite, or other mineral that forms in the extensional strain field on either side of a larger crystal or detrital grain of magnetite, pyrite, garnet, feldspar, or other mineral. Pressure shadows are commonly oriented parallel to a prominent foliation or lineation.

Pressure solution. Deformation mechanism that involves dissolution at grain boundaries under stress of soluble constituents such as calcite or quartz. Generally active at low to moderate temperature in the presence of water.

Principal axes of the strain ellipsoid. Three mutually perpendicular axes of unequal length that define the strain ellipsoid: X (greatest), Y (intermediate), and Z (least).

Principal strain. One of the axes of the strain ellipsoid.

Principal stresses. Three mutually perpendicular normal stresses σ_1, σ_2, and σ_3. σ_1 is the greatest principal normal stress, σ_3 the least.

Progressive deformation (strain). Strain that results when a series of deformational events produces an increase in the amount of strain in a rock body through time. Incremental finite strain.

Protolith. Original rock type from which a metamorphic rock is derived.

Ptygmatic folds. Near-parallel folds formed by folding of strong layers in a much more ductile matrix.

Pull-apart basin. Rhomb-graben (rhombochasm) basin, produced by *en echelon* dextral strike-slip faults with right step-overs or sinistral strike-slip faults with left step-overs.

Pumpelly's rule. Small structures are a key to and mimic the style and orientation of larger structures of the same generation.

Pure shear. Distortion—with or without translation but no rotation—involving homogeneous deformation. Principal strain axes are not rotated and, if a reference point traces a line during deformation that is parallel to principal axes of the strain ellipsoid, the line will remain parallel afterward.

Push-up range. Rhomb horst, produced by dextral strike-slip faults with left step-overs or sinistral faults with right step-overs.

Quadratic elongation. Square of the ratio of deformed length to original length.

Quasi-flexural folds. See *Disharmonic folds*.

Radial faults. Faults that converge toward a single point.

Rain imprints. Small crater-like structures produced where rain falls on fine sediment that is intermittently exposed to the atmosphere. They can be used to determine facing direction.

Rake of net slip. Angle from the horizontal of a line in an inclined plane measured in the plane. If the dip of the plane is vertical, the rake and plunge of the line are the same; otherwise they are not the same.

Ramp. High-angle segment along a thrust or normal fault.

Reclined folds. Folds that have moderately plunging axes and gently dipping axial surfaces.

Recovery. Process that reduces dislocation density in a crystalline material. Thermal softening of a dislocation-hardened, plastically deformed crystal resulting from energy release via migration of point defects and local rearrangement of tangles of dislocations, followed by formation of low-angle boundaries and initiation of polygonization.

Recumbent folds. Folds that have horizontal axes and axial surfaces.

Reduction spot. Spherical color alteration in sediment produced by a small grain or fragment that is chemically different from the surrounding mass of sediment. The chemical difference may produce a nearly spherical area of reduction expressed as a color change in the immediate vicinity of the grain in the otherwise oxidized sediment. Reduction spots cannot be used to determine the facing direction in a sequence, but they serve as an important indicator of strain if the time of formation can be resolved.

Reduction to pole. Modeling technique for aeromagnetic data used to correct for the low-latitude property of magnetic data where a high and low pair of anomalies for the same body are separated. Reduction to pole produces a single anomaly above the rock body that caused the anomaly.

Regional structural geology. Study of parts of mountain ranges, small parts of continents, trenches, and island arcs by geologic mapping, mesoscopic and microscopic structural studies, and the relationships of the resolved structural history to stresses and tectonic plates.

Release joints. Joints that form near the surface as erosion removes overburden and thermal-elastic contraction occurs.

Release spectrum. Release of argon from biotite, muscovite, or hornblende as a sample is heated through a range (spectrum) of temperature, commonly from room temperature to the melting point of the mineral. Used to identify excess argon that introduces an error into K-Ar ($^{40}Ar/^{39}Ar$) age determinations.

Remanent field. Magnetic field remaining in a rock after the present-day field has been removed.

Resequent fault-line scarp. Fault-line scarp along which erosion preserves the original motion sense of the fault.

Reversed graded bedding. Metamorphosed normal graded bedding where the fine-grained top of a bed is recrystallized to coarse micas (or other minerals) that are larger than the granular material at the original bottom of the bed. Also a primary structure formed during deposition of pumice fragments in water, where the smallest fragments are deposited first, followed by successively larger fragments.

Reverse drag. Drag that occurs where layers appear to have been dragged down-dip parallel to the movement direction, usually along a normal fault.

Reverse drag folds. Folds that form along growth (normal) faults where the part of the downthrown block close to the fault is displaced downward more than the parts farther away.

Reverse fault. Fault with moderate to steep dip (45° or more) in which the hanging wall has moved up relative to the footwall. Mechanically the same as a thrust fault.

Reverse-sense crenulations. Crenulations in a shear zone having opposite movement sense to that of the shear zone and C-surfaces.

R_f/ϕ method. Method of determination of homogeneous strain in a rock mass containing deformed initially elliptical objects; results in objects that remain elliptical. The shape of the final ellipse is determined by the initial shape and orientation (ϕ) of the starting ellipse relative to the shape and orientation of the strain ellipse.

Rhomb-graben basin. Basin formed along *en echelon* dextral strike-slip faults with right step-overs.

Rhomb horst. Push-up range produced by dextral strike-slip faults with left step-overs or sinistral faults with right step-overs.

Rhombochasm basin. Rhomb-graben basin or pull-apart basin.

Riedel shears. *En echelon* shear fractures formed at 10° to 15° to the principal shear (fracture orientation). These have the same displacement sense as the primary shear fractures.

Right-handed screw dislocation. Clockwise rotation along a screw dislocation.

Right-lateral strike-slip fault. Steeply dipping fault where the right side moves toward an observer looking parallel to the trace of the fault. Also called dextral strike-slip fault.

Rigid indenter. Promontory on a continental margin that is transported by plate motion into collision with another continent; the promontory produces localized deformation in the (more plastic?) collided continent.

Rigidity modulus. Elastic constant determined from the ratio of shear stress to shear strain.

Ripple cross laminations. Small-scale trough cross beds in fine-grained sediment; form under low-velocity conditions. Can be used to determine facing direction. Also known as hummocky laminations.

Ripple marks. Ripple marks formed where sediment finer than 0.6 mm is moved by a current or where the bottom sediment surface is otherwise disturbed by water moving above a threshold velocity. May be symmetrical or asymmetrical.

Rock cycle. Cyclic changes of energy fluxes ranging from the crystallization of magma to conversion of sedimentary or igneous rocks into metamorphic rocks. Also known as the geochemical cycle.

Rock mechanics. Application of principles of mechanics to the study of rocks.

Rodding. Lineation produced by linear alignment of coarse mineral aggregates, pebbles, and other objects to give the rock an appearance of being composed of rods.

Rods. Aggregates of one or more minerals (quartz, feldspars, micas), pebbles, or other features that have been deformed into elongate rodlike form. Rods form at intersections of two foliation planes, or by elongation, and are common in ductile shear zones.

Rollover anticline. Anticline that forms along growth faults where the part of the downthrown block close to the fault is displaced downward more than the parts farther away.

Roof thrust. Upper of the two master faults that bound a duplex.

Room problem. Unfilled voids resulting from improper construction of a cross section that suggest too much room (or volume) in the core of a fold or beneath a fault block.

Rotated mineral. Mineral that has been rotated during deformation.

Rotation. Angular displacement of a mass without distortion.

Rupture. Point at which nonelastic brittle behavior begins and permanent strain occurs by the material losing cohesiveness.

Saddle-reef deposit. Mineral deposit formed in a void along a fold hinge where strong layers have separated. May be found in successive layers in the hinge of the same fold.

Saint Venant behavior. Ideal plastic behavior.

Scalar. Number having only magnitude.

Scale. Relationship of the dimensions of a feature in the field to the same feature in a photograph or on a map. Map scale commonly expressed as a representative fraction (e.g., 1/50,000) or a bar scale. Also, scale is used to describe the relative size of features as microscopic, mesoscopic, and macroscopic.

Scatterchron. See *Mixing line.*

Schistosity. Foliation in schist and some gneisses.

Scour marks. Marks formed in sediment as currents scour a bedding surface.

Screw dislocation. Dislocation involving helical rotation about a dislocation axis that upon 360° rotation results in movement of the dislocation down one lattice spacing to the next layer.

Second law of motion (Newton). Force equals mass times acceleration.

Second-order folds. Smaller folds on the flanks of the largest (first-order) folds in an area.

Second-rank tensor. Tensor consisting of nine components that are all vectors.

Sedimentary facies. Lateral changes in the kind of sediment as the environment of deposition changes.

Seismic reflection. Acoustic (sound) waves, generated by explosives or a mechanical vibrating truck (vibroseis), transmitted through the Earth, reflected where velocity changes, and recorded by an array of geophones.

Seismic refraction. Change of velocity of acoustic waves, as they pass from one layer to another, recorded a large distance (tens to hundreds of kilometers) from the source of the waves.

Self-exciting dynamo. Premise that convection currents oriented by the Earth's rotation exist in the liquid iron-nickel core, transport hotter material upward, then return cooler material to the interior through a thermal gradient, thus generating an electric current and a magnetic field as a self-exciting dynamo.

Separation. Amount of apparent offset of a faulted surface measured in a specified direction. May be described as strike-separation, dip-separation, or net separation.

Shatter cone. Centimeter- to meter-size striated cone-shaped rock formed as a shock wave propagated down and away from an explosion or high-velocity impact. Common in many meteorite impact craters.

Shear-band foliation. C-surface foliation developed parallel or at 18° to 25° to the boundaries of a ductile shear zone.

Shear fold. Passive-slip fold.

Shear fractures. Modes II and III fractures.

Shear modulus. Rigidity modulus; an elastic constant determined from the ratio of shear stress to shear strain.

Shear strain (γ). Strain that results when parts of a rock body move past one another so that angles between parts of the rock mass are rotated from the original orientation.

Shear strength. Maximum shear stress a material is able to withstand and not undergo permanent strain.

Shear stress. Stress that acts parallel to a surface.

Shear thrust. Thrust fault formed by shearing parallel to and across layering independent of folding. Associated fault-bend folds form as dip flattens over a ramp.

Shear zone. Zone of closely spaced, interleaving, anastomosing brittle faults and crushed rocks near the surface or of anastomosing ductile faults and associated mylonitic rocks at great depth.

Sheath folds. Noncylindrical tubular folds that are closed at one end and with fold axes tightly curved within the axial surfaces.

Sheeting. Jointing that forms by unloading more or less parallel to surface topography, generally in massive rocks.

Similar folds. Folds that maintain the same shape throughout a section normal to the hinge, so they do not die out upward or downward, but maintain the same curvature in the hinge. Layer thickness (measured perpendicular to layering) changes uniformly at the same position in all layers.

Simple shear. Rotational constant-volume homogeneous plane strain in which some elements of a body are rotated but others that were straight and parallel remain straight and parallel after deformation. Also known as noncoaxial deformation.

Simple window. Erosional hole in a thrust sheet that results in footwall rocks being completely surrounded by hanging-wall rocks in map view. Also called a fenster.

Sinistral strike-slip fault. See *Left-lateral strike-slip fault.*

Slaty cleavage. Penetrative planar tectonic structure consisting of parallel grains of thin layer silicates (clay minerals or micas) or thin anastomosing subparallel zones of insoluble residues in a fine-grained rock, generally of low metamorphic grade.

Slice. See *Horse.*

Slickenfibers. Lineation (mostly nonpenetrative) formed by fibrous minerals that have grown in the direction of movement on a fault or bedding surface. Frequently form stepped surfaces.

Slickenlines. Striations on movement surfaces. The direct result of frictional sliding and flexural slip. A nonpenetrative linear structure if planes are isolated.

Slickensides. Striated stepped surfaces along a fault or bedding surface that have undergone movement (flexural slip). Refers to entire movement surface.

Slide. See *Tectonic slide.*

Slip. Movement parallel to a fault plane.

Slip lines. Lines produced by relative motion of reference points in successive layers during folding or other deformation. They may produce slickensides, fibers, or other visible lines indicating motion or may be projected lines based on motion sense.

Slip systems. Specific planes in crystals along which slip may occur.

SL-tectonite. Deformed rock in which planar elements dominate over linear elements.

Snake-head structure. Asymmetric folds visible in seismic sections or in vertical outcrops that form as a thrust sheet passes over a ramp and the hanging-wall anticline has not moved very far past the ramp, or result from propagation along the common limb of an anticline-syncline and the hanging-wall anticline is not displaced very far.

Sole marks. Marks formed during deposition on the undersides of sandstone beds that are deposited on shale or siltstone; produced by currents, readjustments of the sandstone-shale bedding interface by the weight of deposited sand, dewatering of sediment following deposition, or a combination of these processes. Flute casts, drag marks, channels, load casts, and others are sole marks.

Spaced cleavage. Foliation in fine-grained rocks that can be resolved into domains of uncleaved rock separated by cleavage planes with a spacing ranging from less than a millimeter to several centimeters.

Spherical symmetry. The symmetry of a sphere—a center and an infinite number of planes and axes. A fabric diagram that has an infinite number of symmetry axes.

Splay. Smaller fault that forms by branching of a larger fault.

S-surface. Penetrative planar tectonic structure (including curved surfaces) in rocks. Bedding is commonly included and designated S_0 despite having a nontectonic origin.

Stacking faults. Planar lattice defects consisting of irregularities in the repeat order in a series of layers in a close-packed lattice.

Static recrystallization. Recrystallization that occurs in a rock mass no longer undergoing deformation.

Steady-state deformation mechanism. Rate-dependent deformation mechanism; generally includes the creep mechanisms and pressure solution.

S-tectonite. Deformed rock in which foliation or cleavage is the dominant structure.

Step-over. Mechanism in faulting by which motion is transferred from one segment of a strike-slip fault to another through zones of extension or compression.

Stick-slip movement. Intermittent movement along a fault; produced by unstable frictional sliding.

Strain. Permanent deformation in the form of distortion, translation, and rotation.

Strain ellipsoid. Triaxial ellipsoid with three mutually perpendicular axes. The principal directions are $X > Y > Z$; X is the axis of greatest principal strain, Y is the axis of intermediate principal strain, and Z is the axis of least principal strain.

Strain hardening. Increased stress resistance to strain with increasing amount of plastic deformation.

Strain marker. Any deformed object in rocks where the original shape can be quantitatively inferred from the present deformed shape.

Strain partitioning. Separation of strain into different mechanisms in a rock mass or single crystal.

Strain path. Path described on a Flinn diagram during progressive deformation. May involve initial flattening or elongation, followed by transformation into the other field and possibly later return to the original field of the Flinn diagram.

Strain softening. Decreased resistance to strain with increased stress as amount of plastic deformation increases.

Strength. Stress required to cause permanent deformation.

Stress. Force applied per unit area ($\mathbf{F}/A$).

Stress ellipsoid. Triaxial ellipsoid with three mutually perpendicular axes—the three principal stresses σ_1, σ_2, and σ_3 ($\sigma_1 > \sigma_2 > \sigma_3$).

Stretch. The ratio of the deformed line length to the original length of a reference line.

Strike. Compass direction (azimuth) of a horizontal line in an inclined surface.

Strike-slip fault. Fault in which movement is parallel to the strike of the fault plane.

$^{87}Sr/^{86}Sr$ (strontium 87-strontium 86) initial ratio. $^{87}Sr/^{86}Sr$ at the time the rock formed. An initial ratio <0.706 indicates formation in the lower crust or upper mantle; initial ratio >0.706 indicates formation in the middle to upper crust.

Structural analysis. Description and interpretation of structures on all scales in an area. This involves the observation, description, analysis, and interpretation of the kinds and orientations of folds and other linear structures, foliations and other planar structures, strain and displacement indicators, and faults.

Structural terrace. Local flattening of a uniform regional dip.

Strut. Strong unit, such as massive limestone, dolomite, or sandstone, in a sequence of sedimentary rocks; serves as the dominant unit in determining the shape of folds or the size of fault blocks.

Stylolites. Irregular surfaces coated with insoluble minerals or organic matter in limestone, sandstone, or other rock type; results from partial dissolution by pressure solution.

Subgrains. Small parts of grains with lattice orientations that differ by small angles in adjacent parts of the same grain.

Substitution defect. Defect in which a foreign atom or ion is present in a crystal lattice.

Superplastic flow. Gliding along grain boundaries, involving both Coble creep and grain-boundary sliding at temperatures greater than half the melting temperature of the rock. Thought to occur also at low temperature in fine-grained rocks being deformed in fault zones by inhomogeneous simple shear.

Superposed folds. One set of folds that overprints another. Also called polyphase or interference folds.

Supratenuous folds. Folds commonly produced during deposition by differential thickness accumulation and compaction in which the troughs of synclines are thickened and the crests of anticlines are thinned.

Surface force. Force that acts on a surface.

Suspect terrane. Mass of fault block to crustal dimensions, in which the original position is questionable with respect to the adjacent terrane or stable continental land mass to which it is presently attached. Boundaries of suspect terranes are always tectonic.

Symmetrical folds. Folds in which the limbs dip at the same angle on either side of the hinge. Isoclinal folds and elasticas, like other folds, may also be symmetrical.

Syncline. Concave-upward fold in which layering is concave toward the younger rocks, and, as a result, younger rocks are found in the central part of the structure.

Synform. Fold in which layering is concave-up and dips toward the center.

Synformal anticline. Fold in which layering dips inward as in a syncline, but the rocks in the center of the structure are older rather than younger. Also called upward-facing anticline.

Syntectonic recrystallization. Recrystallization that occurs while the rock mass is undergoing deformation.

Synthetic normal fault. Normal fault that dips in the same direction and joins a larger normal fault.

Systematic joints. Planar Mode I fractures that have a roughly parallel orientation and regular spacing.

Tangential-longitudinal strain. Buckle-fold mechanism in which the inner arcs of folds are subjected to layer-parallel compression, and the outer arcs of each layer are subjected to layer-parallel extension. Zones of extension and compression are separated by a neutral surface of no strain. Same as concentric-longitudinal strain.

Tangle. Accumulation of dislocations in a crystal lattice.

Tear fault. Strike-slip fault that terminates a thrust sheet along strike. Tear faults merge with the thrust at depth and define the ends of the thrust sheet.

Tectogene concept. Concept that the subsiding crumpling crust forming mountain chains is driven by mantle convection.

Tectonic cycle. See *Wilson cycle.*

Tectonic joints. Joints that form at depths of less than 3 km in response to abnormal fluid pressure and involve hydrofracturing.

Tectonics. Study of large crustal or mantle features, such as mountain ranges, parts of continents, trenches and island arcs, oceanic ridges, mantle plumes, and entire continents and ocean basins, and the relationships to stresses and tectonic plates.

Tectonic slide. Thrust produced by ductile attenuation of the overturned limb between an antiform and synform.

Tectonic structure. Structure produced in rocks in response to stresses generated, for the most part, by plate motion within the Earth; includes faults, folds, cleavage, and other geologic structures.

Tectonite. Rock containing and generally dominated by a penetrative tectonic fabric element.

Tensional stress. Stress that pulls an object or rock mass apart.

Tensor. Mathematical way of representing a physical quantity by referring it to an appropriate coordinate system. Tensors have the same value in any coordinate system, but the magnitude of the components depends on the choice of coordinate system.

Terrain correction. Correction of gravity data for differences in elevation and proximity to areas of different relief.

Thermo-remanent magnetism (TRM). Remanent magnetism that results from orientation of magnetic minerals in the Earth's magnetic field as igneous rocks cool below the Curie temperature for each mineral.

Thick-skinned structure. Structure that incorporates basement. Refers primarily to thrust faults, but can refer to folds and other fault types.

Thin-skinned structure. Structure that does not involve basement.

Third law of motion (Newton). All forces acting to move an object in one direction are countered by equal and opposite forces acting to move the object in the opposite direction.

Throw. Vertical component of fault displacement.

Thrust fault. Fault with a low angle of dip (30° or less) in which the hanging wall has moved up relative to the footwall. Thrust faults are produced by compression.

Tight fold. Fold in which the limbs dip steeply toward or away from one another.

Torrential cross bed. Cross bed inclined to primary bedding planes. Provides the same perspective in overturned or upright position, so cannot be used for determination of facing direction.

Tracks and trails. Fossil impressions left by organisms as they walked across soft sediment.

Transcurrent fault. Fault (commonly strike-slip) that cuts across the dominant regional trend of the structures in an area.

Transected fold. Fold where cleavage and fold axial surfaces are not coincident, or where the fold axis does not lie in the cleavage.

Transform fault. Strike-slip fault produced by differences in motion between plates and within the same plate along spreading ridges. These faults compensate for relative motion between lithospheric plates; kinds include ridge-ridge, ridge-arc, arc-arc, and intracontinental transforms.

Translation gliding. Process by which deformation occurs in a crystal lattice by slip along an existing crystallographic plane.

Transposition. Transformation of original bedding through progressive deformation—folding, cleavage formation, ductile shearing, or other process—into parallelism with cleavage or foliation.

Transpression. Resolution of strike-slip into components of significant compressional dip-slip motion that commonly results from oblique plate convergence.

Transtension. Opposite of transpression.

Trend. Compass direction (azimuth) of the orientation of a linear structure.

Tresca's criterion. Plastic yield will begin under extensional conditions when the maximum shear stress reaches a critical value.

Triangle structure. Structure formed by two thrusts that dip in the same direction as the limbs of a fold and opposite to each other; commonly formed to solve the room problem. Also called delta structure.

Triboplastic behavior. Apparent megascopic ductile behavior shown to be brittle on the microscopic scale.

Triclinic symmetry. Lack of symmetry; in a fabric diagram usually the result of multiple deformation, such as two superimposed foliations.

Triple junction. Plate boundary where three tectonic plates meet.

Trough. Lowest point in a synclinal (synformal) fold.

Trough (festoon) cross beds. Layering that is truncated at the top by a primary bedding plane, but curves (concave-up) to tangency with the bedding plane at the bottom of the bed. Can be used for determination of facing direction.

Turbidite. Deposit produced by rapid (turbulent) flow of a sediment-laden turbidity current down a slope on the seafloor or in a lake. These deposits may develop as graded, channeled, massive, and cross-bedded sequences that grade upward into one another.

Twin gliding. Deformation of a crystal lattice by slip of a segment of the lattice along crystallographic planes, producing strain-induced twinning of the lattice.

Type I S-C mylonite. S-C mylonite in quartzofeldspathic rocks.

Type II S-C mylonite. S-C mylonite resulting from mylonitization of quartz-mica–rich rocks.

Type 1 (Ramsay) fold interference pattern. Pattern of superposed folds that have upright axial surfaces that intersect at high angles and interfere to produce an egg-carton or dome-and-basin pattern.

Type 2 (Ramsay) fold interference pattern. Pattern of folds having inclined or reclined axial surfaces that are superposed by folds with steeply dipping axial surfaces and axes at high angles to those of the first set, resulting in a boomerang outcrop pattern.

Type 3 (Ramsay) fold interference pattern. Fold pattern in which early-formed tight to isoclinal folds are folded about the same axes, but with different axial surfaces, producing a hook pattern.

Ultimate strength. Highest point on a stress-strain curve.

Unconformity. Break in the stratigraphic sequence where some portion of geologic history is missing. Produced by nondeposition, erosion, or both, resulting in a loss of strata for part of the geologic record.

Undeformed state. Unstrained state.

Undulatory extinction. Variable extinction related to slightly misaligned parts of microscopic crystals, produced by unrecovered strain. Most commonly observed in quartz.

Unfilled joint. Joint in which fractures have not been filled with minerals.

Unit cell. Building block in a crystal lattice.

Unloading joints. Joints that occur near the surface as erosion removes overburden and thermal-elastic contraction occurs.

Upright folds. Folds with vertical axial surfaces.

Upward-facing anticline. See *Synformal anticline.*

Vacancy. Unoccupied site in a crystal lattice.

Vector. Quantity that has both magnitude and direction.

Vergence. Direction of leaning or overturning of the axial surface, or sense of shear of a fold. Also used to indicate transport direction of thrust faults.

Vesicles. Cavities in volcanic rocks left by gas bubbles formed in magma as pressure decreases.

Viscoelastic behavior. Combination of viscous and elastic behavior.

Viscoplastic behavior. Combination of viscous and plastic behavior.

Viscous. Pertaining to fluid behavior. Viscous deformation is permanent and involves dependence of stress on strain rate.

Viscous behavior. Fluid behavior. Stress is proportional to strain rate.

Visual harmonic analysis. Visual comparison of the profile shape of a fold with 30 ideal fold forms based on combinations of six possible shapes and five possible amplitudes.

Volume-diffusion creep. Diffusion of point defects through crystals, occurring at high temperature and low stress. Also called Nabarro-Herring creep.

Von Mises criterion. Model for plastic deformation that assumes the strain energy of distortion is a constant determined from the sums of the squares of the principal stress differences (maximum deviatoric stress).

Wavelength. Distance from crest to crest of adjacent anticlines, or trough to trough of adjacent synclines.

Wedge. See *Contractional fault.*

Whole-rock age. Radiometric age determined by measuring the Rb and Sr isotopes in an entire rock sample rather than analyzing individual minerals (which yields a mineral age).

Wildflysch. Heterogeneous deep-water accumulation of blocks and finer sediment resulting from tectonic or nontectonic processes.

Wilson cycle. Tectonic cycle of opening and closing of ocean basins.

Window. See *Simple window* or *Eyelid window.*

Wrench fault. See *Strike-slip fault.*

Yield point. Point where Hooke's law no longer holds (nonelastic behavior begins).

Younging direction. Facing direction of the top of a sequence.

Young's modulus. Elastic constant determined from the ratio of stress to strain.

Zero-rank tensor. Scalar, with only one component.

Zonal crenulations. Crenulations occurring as wide diffuse zones, whereby the original rock fabric continues uninterrupted outside the crenulated zone.

References

Abbate, E., Bortolotti, V., and Passerini, P., 1970, Olistostromes and olistoliths: Sedimentary Geology, v. 4, p. 521–557.

Aggarwal, Y. P., and Sykes, L. R., 1978, Earthquakes, faults and nuclear power plants in southern New York and northern New Jersey: Science, v. 200, p. 425–429.

Allard, G. O., 1976, Doré Lake complex and its importance to Chibougamau geology and metallogeny: Québec Ministry of Natural Resources DP-368, 446 p.

Allmendinger, R. W., Sharp, J. W., Von Tish, D., Serpa, L., Brown, L., Kaufman, S., and Oliver, J., 1983, Cenozoic and Mesozoic structure of the eastern Basin and Range Province, Utah, from COCORP seismic-reflection data: Geology, v. 11, p. 532–536.

Alsop, L. E., and Talwani, M., 1984, The East Coast magnetic anomaly: Science, v. 226, p. 1189–1191.

Alvarez, W., Engelder, T., and Geiser, P. A., 1978, Classification of solution cleavage in pelagic limestones: Geology, v. 6, p. 263–266.

Alvarez, W., Engelder, T., and Lowrie, W., 1976, Formation of spaced cleavage and folds in brittle limestone by dissolution: Geology, v. 4, p. 698–701.

Anderson, E. M., 1942, 1951, The dynamics of faulting and dyke formation, with applications to Britain: Edinburgh, Oliver and Boyd, 191 p.

Argand, E., 1925 (Carozzi, A. V., trans., 1977), Tectonics of Asia: New York, Hafner Press, 218 p.

Armstrong, R. L., and Dick, H. J. B., 1974, A model for the development of thin overthrust sheets of crystalline rock: Geology, v. 1, p. 35–40.

Atwater, T., 1970, Implications of plate tectonics for the Cenozoic tectonic evolution of western North America: Geological Society of America Bulletin, v. 81, p. 3513–3536.

Aydin, A., and Nur, A., 1982, Evolution of pull-apart basins and their scale independence: Tectonics, v. 1, p. 91–105.

Aydin, A., and Page, B. M., 1984, Diverse Pliocene-Quaternary tectonics in a transform environment, San Francisco Bay region, California: Geological Society of America Bulletin, v. 95, p. 1303–1317.

Bailey, E. B., 1910, Recumbent folds in the schists of the Scottish Highlands: Quarterly Journal of the Geological Society of London, v. 90, p. 462–522.

Bailey, E. B., 1934, West Highland tectonics: Loch Leven to Glen Roy: Quarterly Journal of the Geological Society of London, v. 90, p. 462–522.

Bailey, E. B., 1935, Tectonic essays, mainly Alpine: Oxford, Clarendon Press, 191 p.

Bak, J., Sørensen, K., Grocott, J., Korstgård, Nash D., and Waterson, J., 1975, Tectonic implications of Precambrian shear belts in western Greenland: Nature, v. 254, p. 566–569.

Balk, R., 1937, Structural behavior of igneous rocks: Geological Society of America Memoir 5, 177 p.

Bally, A. W., Gordy, P. L., and Stewart, G. A., 1966, Structure, seismic data, and orogenic evolution of southern Canadian Rocky Mountains: Bulletin of Canadian Petroleum Geology, v. 14, p. 337–381.

Banks, C. J., and Warburton, J., 1986, 'Passive-roof' duplex geometry in the frontal structures of the Kirthar and Sulaiman mountain belts, Pakistan: Journal of Structural Geology, v. 8, p. 229–238.

Barazangi, M., and Brown, L., eds., 1986, Reflection seismology: A global perspective: Geodynamics Series, v. 13, American Geophysical Union, 311 p.

Beach, A., 1980, Numerical models of hydraulic fracturing and the interpretations of syntectonic veins: Journal of Structural Geology, v. 2, p. 425–438.

Becker, G. F., 1896, Schistosity and slaty cleavage: Journal of Geology, v. 4, p. 429–448.

Becker, G. F., 1904, Experiments on schistosity and slaty cleavage: U.S. Geological Survey Bulletin 241, p. 1–34.

Bell, T. H., 1985, Deformation partitioning and porphyroblast rotation in metamorphic rocks: A radical reinterpretation: Journal of Metamorphic Geology, v. 3, p. 109–118.

Berthé, D., Choukroune, P., and Jegouzo, P., 1979, Orthogneiss, mylonite, and noncoaxial deformation of granites: The example of the South Armorican shear zone: Journal of Structural Geology, v. 2, p. 31–42.

Beutner, E. C., 1978, Slaty cleavage and related strain in Martinsburg Slate, Delaware Water Gap, New Jersey: American Journal of Science, v. 278, p. 1–23.

Beutner, E. C., and Charles, E. G., 1985, Large volume loss during cleavage formation, Hamburg sequence, Pennsylvania: Geology, v. 13, p. 803–805.

Beutner, E. C., and Diegel, F. A., 1985, Determination of fold kinematics from syntectonic fibers in pressure shadows, Martinsburg Slate, New Jersey: American Journal of Science, v. 285, p. 16–50.

Biot, M. A., 1961, Theory of folding of stratified viscoelastic media and its implications in tectonics and orogenesis: Geological Society of America Bulletin, v. 72, p. 1595–1620.

Biot, M. A., 1963, Internal buckling under initial stress in finite elasticity: Royal Society of London Proceedings, v. 273, p. 306–328.

Biot, M. A., 1965a, Theory of viscous buckling and gravity instability of multilayers with large deformation: Geological Society of America Bulletin, v. 76, p. 371–378.

Biot, M. A., 1965b, Further development of the theory of internal buckling of multilayers: Geological Society of America Bulletin, v. 76, p. 833–840.

Biot, M. A., Odé, H., and Roever, W. L., 1961, Experimental verification of the theory of folding of stratified viscoelastic media: Geological Society of America Bulletin, v. 72, p. 1621–1632.

Bjornerud, M., 1989, Mathematical model for folding of layering near rigid objects in shear deformation: Journal of Structural Geology, v. 11, p. 245–254.

Borg, I. Y., and Handin, J., 1966, Experimental deformation of crystalline rocks: Tectonophysics, v. 3, p. 249–368.

Borg, I. Y., and Heard, H. C., 1970, Experimental deformation of plagioclases, *in* Paulitsch, P., ed., Experimental and natural rock deformation: Berlin, Springer-Verlag, p. 375–403.

Borradaile, G. J., 1978, Transected folds: A study illustrated with examples from Canada and Scotland: Geological Society of America Bulletin, v. 89, p. 481–493.

Borradaile, G. J., 1979, Strain study of the Caledonides in the Islay region, S.W. Scotland: Implications for strain histories and deformation mechanisms in greenschist: Journal of the Geological Society of London, v. 136, p. 77–88.

Borradaile, G. J., 1981, Particulate flow of rock and the formation of cleavage: Tectonophysics, v. 72, p. 305–321.

Borradaile, G. J., and Poulsen, K. H., 1981, Tectonic deformation of pillow lava: Tectonophysics, v. 79, p. T17–T26.

Bosworth, W., Lambiase, J., and Keisler, R., 1986, A new look at Gregory's Rift: The structural style of continental rifting: Eos, v. 67, p. 577–583.

Boullier, A. M., and Gueguen, Y., 1975, SP-mylonites: Origin of some mylonites by superplastic flow: Contributions to Mineralogy and Petrology, v. 50, p. 93–104.

Bouma, A. H., 1962, Sedimentology of some flysch deposits: Amsterdam, Elsevier Publishing, 169 p.

Boyer, S. E., and Elliott, D., 1982, Thrust systems: American Association of Petroleum Geologists Bulletin, v. 66, p. 1196–1230.

Brock, W. G., and Engelder, T., 1977, Deformation associated with the movement of the Muddy Mountains overthrust in the Buffington window, southeastern Nevada: Geological Society of America Bulletin, v. 88, p. 1667–1677.

Brun, J.-P., and Choukroune, P., 1983, Normal faulting, block tilting, and décollement in a stretched crust: Tectonics, v. 2, p. 345–356.

Bryant, B., and Reed, J. C., Jr., 1970, Geology of the Grandfather Mountain window and vicinity, North Carolina and Tennessee: U.S. Geological Survey Professional Paper 615, 190 p.

Bucher, W. H., 1936, Cryptovolcanic structures in the United States: 16th International Geological Congress, v. 2, p. 1055–1084.

Burchfiel, B. C., Fleck, R. J., Secor, D. T., Vincelette, R. R., and Davis, G. A., 1974, Geology of the Spring Mountains, Nevada: Geological Society of America Bulletin, v. 85, p. 1013–1022.

Burke, K., and Dewey, J. F., 1973, Plume generated triple junctions: Key indicators in applying plate tectonics to old rocks: Journal of Geology, v. 81, p. 406–433.

Busk, H. G., 1929, Earth flexures: Cambridge, England, Cambridge University Press, 106 p.

Butler, R. W. H., and Coward, M. P., 1984, Geological constraints, structural evolution, and deep geology of the northwest Scottish Caledonides: Tectonics, v. 3, p. 347–365.

Buxtorf, A., 1916, Prognosen and Befunde beim Hauensteinbasis- und Grenchenburg-tunnel und die

Bedeutung der letzteren für die Geologie des Juragebirges: Naturforschende Gesellschaft Basel Verhandlungen, v. 27, p. 184–254.

Byerlee, J. D., 1977, Friction of rocks, *in* Evernden, J. F., ed., Experimental studies of rock friction with application to earthquake prediction: Menlo Park, California, U.S. Geological Survey, p. 55–77.

Byerlee, J. D., 1978, Friction of rocks: Pure and Applied Geophysics, v. 116, p. 615–626.

Cadell, H. M., 1890, Experimental researches in mountain-building: Transactions of the Royal Society of Edinburgh, v. 35, p. 337–357.

Campbell, R. B., 1973, Structural cross section and tectonic model of the southeastern Canadian Cordillera: Canadian Journal of Earth Sciences, v. 10, p. 1607–1620.

Carey, S. W., 1953, The rheid concept in geotectonics: Journal of the Geological Society of Australia, v. 1, p. 67–117.

Carey, S. W., 1962, Folding: Bulletin of Canadian Petroleum Geology, v. 10, p. 95–144.

Carreras, J., and White, S. H., 1980, Preface *to* "Shear zones in rocks": Journal of Structural Geology, v. 2, p. 1.

Carter, N. L., 1971, Static deformation of silica and silicates: Journal of Geophysical Research, v. 76, p. 5514–5540.

Carter, N. L., and Avé Lallemant, H. G., 1970, High temperature flow of dunite and peridotite: Geological Society of America Bulletin, v. 81, p. 2181–2202.

Chapple, W. M., 1968, A mathematical theory of finite-amplitude rock-folding: Geological Society of America Bulletin, v. 79, p. 47–68.

Chapple, W. M., 1969, Fold shape and rheology: The folding of an isolated viscous-plastic layer: Tectonophysics, v. 7, p. 97–116.

Chapple, W. M., 1970, The finite-amplitude instability in the folding of layered rocks: Canadian Journal of Earth Sciences, v. 7, p. 457–466.

Chapple, W. M., 1978, Mechanics of thin-skinned fold-and-thrust belts: Geological Society of America Bulletin, v. 89, p. 1189–1198.

Christie, J. M., 1963, The Moine thrust zone in the Assynt region, northwest Scotland: University of California, Department of Geological Sciences Publication 40, p. 345–419.

Christie, J. M., 1964, Moine thrust: A reply: Journal of Geology, v. 72, p. 677–681.

Christie, J. M., Griggs, D. T., and Carter, N. L., 1964, Experimental evidence for basal slip in quartz: Journal of Geology, v. 72, p. 734–756.

Cloos, E., 1964, Wedging, bedding-plane slips, and gravity tectonics in the Appalachians, *in* Lowry, W. D., ed., Tectonics of the southern Appalachians: Virginia Polytechnic Institute and State University Department of Geological Sciences, Memoir 1, p. 63–70.

Cloos, E., 1968, Experimental analysis of Gulf Coast fracture patterns: American Association of Petroleum Geologists Bulletin, v. 52, p. 420–444.

Cloos, E., 1971, Microtectonics along the western edge of the Blue Ridge, Maryland and Virginia: Baltimore, The Johns Hopkins Press, 234 p.

Cobbold, P. R., and Quinquis, H., 1980, Development of sheath folds in shear regimes: Journal of Structural Geology, v. 2, p. 119–126.

Cobbold, P. R., Cosgrove, J. W., and Summers, J. M., 1971, Development of internal structures in deformed anisotropic rocks: Tectonophysics: v. 12, p. 23–53.

Coney, P. J., 1980, Cordilleran metamorphic core complexes: An overview, *in* Crittenden, M. D., Jr., Coney, P. J., and Davis, G. H., eds., Cordilleran metamorphic core complexes: Geological Society of America Memoir 153, p. 7–34.

Coney, P. J., Jones, D. L., and Monger, J. W. H., 1980, Cordilleran suspect terranes: Nature, v. 288, p. 329–333.

Cook, F. A., Albaugh, D. S., Brown, L. D., Oliver, J. E., Kaufman, Sidney, and Hatcher, R. D., Jr., 1979, Thin-skinned tectonics in the crystalline southern Appalachians: Geology, v. 7, p. 563–567.

Cook, F. A., Brown, L. D., Kaufman, S., and Oliver, J. E., 1983, The COCORP seismic reflection traverse across the southern Appalachians: American Association of Petroleum Geologists Studies in Geology 14, 61 p.

Cooper, B. N., 1964, Relations of stratigraphy to structure in the southern Appalachians, *in* Lowry, W. D., ed., Tectonics of the southern Appalachians: Department of Geological Sciences, Virginia Tech, Memoir 1, p. 81–114.

Costello, J. O., and Hatcher, R. D., Jr., 1985, Southern Appalachian foreland-hinterland transition and the boundary conditions of duplex nucleation: Geological Society of America Abstracts with Programs, v. 17, p. 554.

Coward, M. P., and Kim, J. H., 1981, Strain within thrust sheets, *in* McClay, K. R., and Price, N. J., eds., Thrust and nappe tectonics: Geological Society of London Special Publication 9, p. 275–292.

Crowell, J. C., 1962, Displacement along the San Andreas fault, California: Geological Society of America Special Paper 71, 61 p.

Crowell, J. C., 1974, Origin of Late Cenozoic basins in southern California, *in* Dickinson, W. R., ed., Tectonics and sedimentation: Society of Economic Paleontologists and Mineralogists Special Publication No. 22, p. 190–204.

Currie, J. B., Patnode, H. W., and Trump, R. P., 1962, Development of folds in sedimentary strata: Geological Society of America Bulletin, v. 73, p. 655–674.

Dahlen, F. A., Suppe, J., and Davis, D., 1984, Mechanics of fold-and-thrust belts and accretionary wedges: Cohesive Coulomb theory: Journal of Geophysical Research, v. 89, p. 10087–10101.

Dahlstrom, C. D. A., 1969, Balanced cross sections: Canadian Journal of Earth Science, v. 6, p. 743–757.

Darwin, C., 1846, Geological observations in South America: London, Smith-Elder.

Davis, D., Suppe, J., and Dahlen, F. A., 1983, Mechanics of fold-and-thrust belts and accretionary wedges: Journal of Geophysical Research, v. 88, p. 1153–1172.

Davis, G. A., and Burchfiel, B. C., 1973, Garlock fault: An intracontinental transform structure, southern California: Geological Society of America Bulletin, v. 84, p. 1407–1422.

Davis, G. A., and Lister, G. S., 1988, Detachment faulting in continental extension; Perspectives from the southwestern U.S. Cordillera: Geological Society of America Special Paper 218, p. 133–159.

Davis, G. H., 1980, Structural characteristics of metamorphic core complexes, southern Arizona, *in* Crittenden, M. D., Jr., Coney, P. J., and Davis, G. H., eds., Cordilleran metamorphic core complexes: Geological Society of America Memoir 153, p. 35–78.

Davis, G. H., Anderson, J. L., Frost, E. G., and Shackelford, T. J., 1980, Mylonitization and detachment faulting in the Whipple-Buckskin-Rawhide Mountains terrane, southeastern California and western Arizona, *in* Crittenden, M. D., Jr., Coney, P. J., and Davis, G. H., eds., Cordilleran metamorphic core complexes: Geological Society of America Memoir 153, p. 79–130.

Davis, W. M., 1926, The value of the outrageous geological hypothesis: Science, v. 63, p. 463–468.

de Charpal, O., Montadert, L., Gucunnoc, P., and Roberts, D. G., 1978, Rifting, crustal attenuation and subsidence in the Bay of Biscay: Nature, v. 275, p. 706–711.

Dennis, A. J., and Secor, D. T., 1987, A model for the development of crenulations in shear zones with applications from the southern Appalachian Piedmont: Journal of Structural Geology, v. 9, p. 809–817.

DePaor, D. G., 1986, Orthographic analysis of geological structures—II. Practical applications: Journal of Structural Geology, v. 8, p. 87–100.

DePaor, D. G., and Anastasio, D. J., 1987, The Spanish external Sierra: A case history in the advance and retreat of mountains: National Geographic Research, v. 3, p. 199–209.

de Sitter, L. U., 1959, Structural geology, 2nd ed.: New York, McGraw-Hill Book Company, 552 p.

de Voogd, B., Serpa, L., and Brown, L., 1988, Crustal extension and magmatic processes: COCORP profiles from Death Valley and the Rio Grande rift: Geological Society of America Bulletin, v. 100, p. 1550–1567.

Dewey, J. F., and Bird, J. M., 1970, Mountain belts and the new global tectonics: Journal of Geophysical Research, v. 75, p. 2615–2647.

Dewey, J. F., Pittman, W., III, Ryan, W. B. F., and Bonnin, J., 1973, Plate tectonics and the evolution of the Alpine system: Geological Society of America Bulletin, v. 84, p. 3137–3180.

Dibblee, T. W., Jr., 1977, Strike-slip tectonics of the San Andreas fault and its role in Cenozoic basin evolvement, *in* Nilsen, T. H., ed., Late Mesozoic and Cenozoic sedimentation and tectonics in California: Bakersfield, California, San Joaquin Geological Society, p. 26–38.

Diegel, F. A., 1986, Topological constraints on imbricate thrust networks, examples from the Mountain City window, Tennessee, U.S.A.: Journal of Structural Geology, v. 8, p. 269–280.

Dieterich, J. H., 1969, Origin of cleavage in folded rocks: American Journal of Science, v. 267, p. 155–165.

Dieterich, J. H., 1970, Computer experiments on mechanics of finite amplitude folds: Canadian Journal of Earth Sciences, v. 7, p. 467–476.

Dietz, R. S., 1960, Shatter cones in cryptoexplosion structures (meteorite impact?): Journal of Geology, v. 67, p. 496–505.

Dillon, J. T., Haxel, G. B., and Tosdal, R. M., 1989, in press, Structural evidence for northeastward movement on the Chocolate Mountains thrust, southeasternmost California: Journal of Geophysical Research, v. 94.

Donath, F. A., and Parker, R. B., 1964, Folds and folding: Geological Society of America Bulletin, v. 75, p. 45–62.

Eaton, G. P., 1979, Regional geophysics, Cenozoic tectonics, and geologic resources of the Basin and Range Province and adjoining regions, *in* Newman, G. W., and Goode, H. D., eds., Basin and Range symposium: Denver, Rocky Mountain Association of Geologists, p. 11–40.

Elliott, D., 1976a, The motion of thrust sheets: Journal of Geophysical Research, v. 81, p. 949—963.

Elliott, D., 1976b, The energy balance and deformation mechanisms of thrust sheets: Philosophical Transactions of the Royal Society of London, v. 283, p. 289–312.

Elliott, D., 1983, The construction of balanced cross sections: Journal of Structural Geology, v. 5, p. 101.

Elliott, D., and Johnson, M. R. W., 1980, Structural evolution in the northern part of the Moine thrust belt, N.W. Scotland: Transactions of the Royal Society of Edinburgh, v. 71, p. 69–96.

Engelder, T., 1979a, Mechanisms for strain within the Upper Devonian clastic sequence of the Appalachian Plateau, western New York: American Journal of Science, v. 279, p. 527–542.

Engelder, T., 1979b, The nature of deformation within the outer limits of the central Appalachian foreland fold and thrust belt in New York State: Tectonophysics, v. 55, p. 289–310.

Engelder, J. T., 1982, Is there a genetic relationship between selected regional joints and contemporary stress within the lithosphere of North America?: Tectonics, v. 1, p. 161–178.

Engelder, T., 1985, Loading paths to joint propagation during a tectonic cycle: An example from the Appalachian Plateau, USA: Journal of Structural Geology, v. 7, p. 459–476.

Engelder, T., and Engelder, R., 1977, Fossil distortion and décollement tectonics on the Appalachian Plateau: Geology, v. 5, p. 457–460.

Engelder, T., and Geiser, P., 1979, The relationship between pencil cleavage and lateral shortening within the Devonian section of the Appalachian Plateau, New York: Geology, v. 7, p. 460–464.

Etheridge, M. A., Hobbs, B. E., and Paterson, M. S., 1973, Experimental deformation of single crystals of biotite: Contributions to Mineralogy and Petrology, v. 38, p. 21–36.

Evans, D. M., 1966, Man-made earthquakes in Denver: Geotimes, v. 10, no. 9, p. 11–18.

Faill, R. T., 1973, Kink-band folding, Valley and Ridge Province, Pennsylvania: Geological Society of America Bulletin, v. 84, p. 1289–1314.

Fischer, M. W., and Coward, M. P., 1982, Strains and folds within thrust sheets: An analysis of the Heilan sheet, Northwest Scotland: Tectonophysics, v. 88, p. 291–312.

Fisher, O., 1884, On cleavage and distortion: Geological Magazine, v. 1, p. 268–276 and 396–406.

Fleuty, M. J., 1964, The description of folds: Proceedings of the Geologists' Association, v. 75, p. 461–492.

Flinn, D., 1962, On folding during three-dimensional progressive deformation: Geological Society of London Quarterly Journal, v. 118, p. 385–433.

Freund, R., 1965, A model of the structural development of Israel and adjacent areas since Upper Cretaceous times: Geological Magazine, v. 102, p. 188–204.

Freund, R., Zak, I., and Garfunkel, Z., 1968, Age and rate of the sinistral movement along the Dead Sea Rift: Nature, v. 220, p. 253–255.

Fry, N., 1979, Density distribution techniques and strained length methods for determination of finite strains: Journal of Structural Geology, v. 1, p. 221–229.

Fuller, M., Williams, I., and Hoffman, K. A., 1979, Paleomagnetic records of geomagnetic field records and the morphology of the transitional fields: Reviews of Geophysics and Space Physics, v. 17, p. 179–203.

Gee, D. G., 1978, Nappe displacement in the Scandinavian Caledonides: Tectonophysics, v. 47, p. 393–419.

Geiser, P., and Engelder, T., 1983, The distribution of layer parallel shortening fabrics in the Appalachian foreland of New York and Pennsylvania: Evidence for two noncoaxial phases of the Alleghanian orogeny: Geological Society of America Memoir 158, p. 161–175.

Goguel, J., 1965, Traité de Tectonique, 2nd ed.: Paris, Masson, 457 p.

Gray, D. R., 1977a, Morphologic classification of crenulation cleavage: Journal of Geology, v. 85, p. 229–235.

Gray, D. R., 1977b, Differentiation associated with discrete crenulation cleavages: Lithos, v. 10, p. 89–101.

Gray, D. R., and Durney, D. W., 1979, Crenulation cleavage differentiation: Implications of solution-deposition processes: Journal of Structural Geology, v. 1, p. 75–80.

Greenly, E., 1919, The geology of Anglesey: Great Britain Geological Survey Memoir, 980 p.

Groshong, R. H., Jr., 1975a, Strain, fractures, and pressure solution in natural single-layer folds: Geological Society of America Bulletin, v. 86, p. 1363–1376.

Groshong, R. H., Jr., 1975b, "Slip" cleavage caused by pressure solution in a buckle fold: Geology, v. 3, p. 411–413.

Groshong, R. H., Jr., Teufel, W., and Gasteiger, C., 1984, Precision and accuracy of the calcite strain-gage technique: Geological Society of America Bulletin, v. 95, p. 357–363.

Hafner, W., 1951, Stress distributions and faulting: Geological Society of America Bulletin, v. 62, p. 373–398.

Hamilton, W. B., 1979, Tectonics of the Indonesian region: U.S. Geological Survey Professional Paper 1078, 345 p.

Hamilton, W., 1982, Structural evolution of the Big Maria Mountains, northeastern Riverside County, southeastern California, *in* Frost, E. G., and Martin, D. L., eds., Mesozoic-Cenozoic tectonic evolution of the Colorado River region, California, Arizona, and Nevada: San Diego, Cordilleran Publishers, p. 1–27.

Hancock, P. L., 1985, Brittle microtectonics: Principles and practice: Journal of Structural Geology, v. 7, p. 437–457.

Hancock, P., and Engelder, T., 1989, in press, Neotectonic joints: Geological Society of America Bulletin, v. 101.

Hanmer, S. K., 1979, The role of discrete heterogeneities and linear fabrics in the formation of crenulations: Journal of Structural Geology, v. 1, p. 81–91.

Hansen, E., 1971, Strain facies: New York, Springer-Verlag, 207 p.

Harding, T. P., 1974, Petroleum traps associated with wrench faults: American Association of Petroleum Geologists Bulletin, v. 58, p. 1290–1304.

Harker, A., 1885, The cause of slaty cleavage: Geological Magazine, v. 2, p. 15–17.

Harker, A., 1886, On slaty cleavage and allied rock structures: Report to the British Association for the Advancement of Science 1885, p. 813–852.

Harland, W. B., 1971, Tectonic transpression in Caledonian Spitsbergen: Geological Magazine, v. 108, p. 27–42.

Harms, J. C., Southard, J. B., Spearing, D. R., and Walker, R. G., 1975, Depositional environments as interpreted from primary sedimentary structures and stratification sequences: Society of Economic Paleontologists and Mineralogists Short Course Lecture Notes 2.

Hatcher, R. D., Jr., 1971, Stratigraphic structural and petrologic evidence favoring a thrust solution of the Brevard problem: American Journal of Science, v. 270, p. 177–202.

Hatcher, R. D., Jr., 1972, Developmental model for the southern Appalachians: Geological Society of America Bulletin, v. 83, p. 2735–2760.

Hatcher, R. D., Jr., 1978, Eastern Piedmont fault system: Reply: Geology, v. 6, p. 580–582.

Hatcher, R. D., Jr., 1981, Thrusts and nappes in the North American Appalachian orogen, *in* McClay, K. R., and Price, N. J., eds., Thrust and nappe tectonics: Geological Society of London Special Publication 9, p. 491–499.

Hatcher, R. D., Jr., and Hooper, R. J., 1981, Controls of mylonitic processes: Relationships to orogenic thermal/metamorphic peaks: Geological Society of America Abstracts with Programs, v. 13, p. 469.

Hatcher, R. D., Jr., and Odom, A. L., 1980, Timing of thrust in the southern Appalachians, USA: Model for orogeny?: Quarterly Journal of the Geological Society of London, v. 137, p. 321–327.

Hatcher, R. D., Jr., and Williams, R. T., 1986, Mechanical model for single thrust sheets Part 1: Crystalline thrust sheets and their relationships to the mechanical/thermal behavior of orogenic belts: Geological Society of America Bulletin, v. 97, p. 975–985.

Hatcher, R. D., Jr., and Zietz, Isidore, 1979, Interpretation of regional aeromagnetic and gravity data from the southeastern United States, Part II—Tectonic implications for the southern Appalachians: Geological Society of America Abstracts with Programs, v. 11, p. 181–182.

Hatcher, R. D., Jr., and Zietz, Isidore, 1980, Tectonic implications of regional aeromagnetic and gravity data from the southern Appalachians, *in* Wones, D. R., ed., The Caledonides in the USA: Virginia Polytechnic Institute Department of Geological Sciences, Memoir 2, p. 235–244.

Haughton, S., 1856, On slaty cleavage and the distortion of fossils: Philosophical Magazine, v. 12, p. 409–421.

Henderson, J. R., Wright, T. O., and Henderson, M. N., 1986, A history of cleavage and folding: An example from the Goldenville Formation, Nova Scotia: Geological Society of America Bulletin, v. 97, p. 1354–1366.

Hobbs, B. E., Means, W. D., and Williams, P. E., 1976, An outline of structural geology: New York, John Wiley, 571 p.

Hodgson, R. A., 1961, Regional study of jointing in Comb Ridge–Navajo mountain area, Arizona and Utah: American Association of Petroleum Geologists Bulletin, v. 45, p. 1–38.

Hoeppener, R., 1956, Zum problem der Bruchbildung, Schieferung und Faltung: Geologische Rundschau, v. 45, p. 247–283.

Hoffman, K. A., 1988, Ancient magnetic reversals: Clues to the geodynamo: Scientific American, v. 258, no. 5, p. 76–83.

Hoffman, P., 1973, Evolution of an Early Proterozoic continental margin: The Coronation geosyncline and associated aulocogens of the northwestern Canadian shield: Philosophical Transactions of the Royal Society of London, v. 273, p. 547–581.

Hossack, J. R., 1968, Pebble deformation in the Bygdin area (southern Norway): Tectonophysics, v. 5, p. 315–339.

Hossack, J. R., 1979, The use of balanced cross sections in the calculation of orogenic contraction, A review: Journal of the Geological Society of London, v. 136, p. 705–711.

Howell, D. G., ed., 1985, Tectonostratigraphic terranes of the circum-Pacific region: Circum-Pacific Council for Energy and Mineral Resources, Houston, 581 p.

Hsü, K. J., 1968, Principles of mélanges and their bearing on the Franciscan-Knoxville paradox: Geological Society of America Bulletin, v. 79, p. 1063–1074.

Hsü, K. J., 1969, Role of cohesive strength in the mechanics of overthrust faulting and of landsliding: Geological Society of America Bulletin, v. 80, p. 927–952.

Hubbert, M. K., 1945, Strength of the Earth: American Association of Petroleum Geologists Bulletin, v. 29, p. 1630–1653.

Hubbert, M. K., 1951, Mechanical basis for certain familiar geologic structures: Geological Society of America Bulletin, v. 62, p. 355–372.

Hubbert, M. K., and Rubey, W. W., 1959, Role of fluid pressure in mechanics of overthrust faulting: Part 1. Mechanics of fluid-filled porous solids and its application to overthrust faulting: Geological Society of America Bulletin, v. 70, p. 115–166.

Hudleston, P. J., 1973a, Fold morphology and some geometrical implications of theories of fold development: Tectonophysics: v. 16, p. 1–46.

Hudleston, P. J., 1973b, The analysis and interpretation of minor folds developed in the Moine rocks of Monar, Scotland: Tectonophysics, v. 17, p. 89–132.

Hudleston, P. J., and Stephansson, O., 1973, Layer shortening and fold-shape development in the buckling of single layers: Tectonophysics, v. 17, p. 299–321.

Hull, D., 1975, Introduction to dislocations: Oxford, England, Pergamon Press, 268 p.

Isacks, B., Oliver, J., and Sykes, L. R., 1968, Seismology and the new global tectonics: Journal of Geophysical Research, v. 72, p. 5855–5900.

Jackson, M. P. A., and Talbot, C. J., 1986, External shapes, strain rates, and dynamics of salt structures: Geological Society of America Bulletin, v. 97, p. 305–323.

Jaeger, J. C., and Cook, N. G. W., 1976, Fundamentals of rock mechanics: London, Chapman and Hall, 585 p.

Johnson, A. M., 1970, Physical processes in geology: San Francisco, Freeman, Cooper, and Company, 577 p.

Johnson, A. M., 1977, Styles of folding: New York, Elsevier, 406 p.

Johnson, A. M., and Ellen, S. D., 1974, A theory of concentric, kink, and sinusoidal folding and of monoclinal flexuring of compressible, elastic multilayers. I. Introduction: Tectonophysics, v. 21, p. 301–339.

Johnson, M. R. W., 1960, The structural history of the Moine thrust zone at Lochcarron, Wester Ross: Royal Society of Edinburgh Transactions, v. 64, p. 139–168.

Johnson, M. R. W., 1964, The Moine thrust: A discussion: Journal of Geology, v. 72, p. 672–676.

Jonas, A. I., 1927, Geologic reconnaissance in the Piedmont of Virginia: Geological Society of America Bulletin, v. 38, p. 837–846.

Karner, G. D., and Watts, A. B., 1983, Gravity anomalies and flexure of the lithosphere at mountain ranges: Journal of Geophysical Research, v. 88, p. 10,449–10,477.

Kennedy, W. Q., 1946, The Great Glen fault: Quarterly Journal of the Geological Society of London, v. 102, p. 41–76.

Kerrich, R., 1978, An historical review and synthesis of research on pressure solution: Zentralblatt für Geologie und Palaeontologie, v. 5/6, p. 512–550.

Kerrich, R., and Allison, I., 1978, Flow mechanisms in rocks: Microscopic and mesoscopic structures, and their relation to physical conditions of deformation in the crust: Geoscience Canada, v. 5, p. 109–118.

Knopf, E. G., and Ingerson, E., 1938, Structural petrology: Geological Society of America Memoir 6, 270 p.

Krogh, T. E., 1973, A low-contamination method for hydrothermal decomposition and extraction of U and Pb for isotopic age determination: Geochimica et Cosmochimica ACTA, v. 37, p. 485–494.

Kulander, B. R., Barton, C. C., and Dean, S. L., 1979, The application of fractography to core and outcrop fracture investigations: Morgantown, West Virginia, DOE METC/SP-79/3, 174 p.

Lakes, R., 1987, Foam structures with a negative Poisson's ratio: Science, v. 235, p. 1038–1040.

Latham, J.-P., 1985, A numerical investigation and geological discussion of the relationship between folding, kinking and faulting: Journal of Structural Geology, v. 7, p. 237–249.

Laubscher, H. P., 1983, Detachment, shear, and compression in the central Alps: Geological Society of America Memoir 158, p. 191–211.

Laugel, A., 1855, Du clivage de roches: Compte Rendu Academie Science, v. 40, p. 182, 185, and 978–980.

Lee, J. H., Peacor, D. R., Lewis, D. D., and Wintsch, R. P., 1986, Evidence for syntectonic crystallization for the mudstone to slate transition at Lehigh Gap, Pennsylvania, U.S.A.: Journal of Structural Geology, v. 8, p. 767–780.

Le Pichon, X., 1968, Seafloor spreading and continental drift: Journal of Geophysical Research, v. 73, p. 3661–3697.

Lister, G. S., Etheridge, M. A., and Symonds, P. A., 1986, Detachment faulting and the evolution of passive continental margins: Geology, v. 14, p. 246–250.

Lister, G. S., and Snoke, A. W., 1984, S-C mylonites: Journal of Structural Geology, v. 6, p. 617–638.

Lister, G. S., and Williams, P. F., 1983, The partitioning of deformation in flowing rock masses: Tectonophysics, v. 92, p. 1–33.

Marshak, S., and Engelder, T., 1985, Development of cleavage in limestones of a fold-thrust belt in eastern New York: Journal of Structural Geology, v. 7, p. 345–359.

Matthews, D., and Smith, C., eds., 1987, Deep seismic reflection profiling of the continental lithosphere: Geophysical Journal of the Royal Astronomical Society Special Issue, v. 89, p. 1–494.

McBride, J. H., and Nelson, K. D., 1988, Integration of COCORP deep reflection and magnetic anomaly analysis in the southeastern United States: Implications for origin of the Brunswick and East Coast magnetic anomalies: Geological Society of America Bulletin, v. 100, p. 436–445.

McEachran, D. B., and Marshak, S., 1986, Teaching strain theory in structural geology using graphics programs for the Apple Macintosh computer: Journal of Geological Education, v. 34, p. 191–195.

McIntyre, D. B., 1950, Note on two lineated tectonites from Strathavon, Banfshire: Geological Magazine, v. 87, p. 331–336.

McKenzie, D. P., 1970, Plate tectonics of the Mediterranean region: Nature, v. 226, p. 239–243.

McKenzie, D. P., Davies, D., and Molnar, P., 1970, Plate tectonics of the Red Sea and east Africa: Nature, v. 226, p. 243–248.

McKenzie, D. P., and Morgan, W. J., 1969, Evolution of triple junctions: Nature, v. 224, p. 125–133.

McKenzie, D. P., and Parker, R. L., 1967, The North Pacific: An example of tectonics on a sphere: Nature, v. 226, p. 1276–1280.

McLaren, A. C., Retchford, J. A., Griggs, D. T., and Christie, J. M., 1967, Transmission electron microscope study of Brazil twins and dislocations experimentally produced in natural quartz: Solid State Physics, v. 19, p. 631–644.

Means, W. D., 1986, Three microstructural exercises for students: Journal of Geological Education, v. 34, p. 224–230.

Means, W. D., 1987, A newly recognized type of slickenside striation: Journal of Structural Geology, v. 9, p. 585–590.

Milici, R. C., 1975, Structural patterns in the southern Appalachians: Evidence for a gravity slide mechanism for Alleghanian deformation: Geological Society of America Bulletin, v. 86, p. 1316–1320.

Mitra, S., 1988, Three-dimensional geometry and kinematic evolution of the Pine Mountain thrust system, southern Appalachians: Geological Society of America Bulletin, v. 100, p. 72–95.

Monger, J. W. H., Price, R. A., and Templeman-Kluitt, D. J., 1982, Tectonic accretion and the origin of the two major metamorphic and plutonic welts in the Canadian Cordillera: Geology, v. 10, p. 70–75.

Morgan, W. J., 1968, Rises, trenches, great faults, and crustal blocks: Journal of Geophysical Research, v. 73, p. 1959–1982.

Morley, C. K., 1986, A classification of thrust fronts: American Association of Petroleum Geologists, v. 70, p. 12–25.

Muehlberger, W. R., 1968, Internal structures and mode of uplift of Texas and Louisiana salt domes: Geological Society of America Special Paper 88, p. 359–364.

Nickelsen, R. P., 1979, Sequence of structural stages of the Allegheny orogeny at the Bear Valley Strip Mine, Shamokin, Pennsylvania: American Journal of Science, v. 279, p. 225–271.

Nicol, J., 1861, On the structure of the North-Western Highlands and the relations of the gneiss, red sandstone, and quartzite of Sutherland and Ross-Shire: Geological Society of London Quarterly Journal, v. 17, p. 85–113.

Nicolas, A., 1987, Principles of rock deformation: Dordrecht, Holland, D. Reidel Publishing, 201 p.

Nicolas, A., and Poirier, J. P., 1976, Crystalline plasticity and solid state flow in metamorphic rocks: London, Wiley-Interscience, 444 p.

Norris, D. K., 1958, Structural conditions in Canadian coal mines: Geological Survey of Canada Bulletin 44, 54 p.

Odé, H., 1957, Mechanical analysis of the dike pattern of the Spanish Peaks area, Colorado: Geological Society of America Bulletin, v. 68, p. 567–576.

Oliver, J., 1986, Fluids expelled tectonically from orogenic belts: Their role in hydrocarbon migration and other geologic phenomena: Geology, v. 14, p. 99–102.

Oliver, J., 1988, Opinion: Geology, v. 16, p. 291.

Onasch, C. M., 1983, Dynamic analysis of rough cleavage in the Martinsburg Formation, Maryland: Journal of Structural Geology, v. 5, p. 73–81.

Onasch, C. M., 1984, Application of the R_f/ϕ technique to elliptical markers deformed by pressure-solution: Tectonophysics, v. 110, p. 157–165.

Oriel, S. S., 1950, Geology and mineral resources of the Hot Springs window, Madison County, North Carolina: North Carolina Department of Conservation and Development Bulletin 60, 70 p.

Oxburgh, E. R., 1972, Flake tectonics and continental collision: Nature, v. 239, p. 202–209.

Panza, G. F., and Müller, St., 1979, The plate boundary between Eurasia and Africa in the Alpine area: Memorie di Scienze Geologiche, v. 33, p. 43–50.

Parrish, D. K., Krivz, A. L., and Carter, N. L., 1976, Finite-element folds of similar geometry: Tectonophysics, v. 32, p. 183–207.

Passchier, C. W., and Simpson, C., 1986, Porphyroclast systems as kinematic indicators: Journal of Structural Geology, v. 8, p. 831–843.

Paterson, M. S., 1958, Experimental deformation and faulting in Wombeyan marble: Geological Society of America Bulletin, v. 69, p. 465–476.

Paterson, M. S., and Weiss, L. E., 1966, Experimental deformation and folding in phyllite: Geological Society of America Bulletin, v. 77, p. 343–374.

Peach, B. N., Horne, J., Clough, C. T., Cadell, H. M., and Dinhaus, C. H., 1888 (reprinted 1923), Assynt District: Geological Survey of Great Britain, scale 1:63,360.

Phillips, J., 1844, On certain movements in the parts of stratified rocks: Report of the British Association for the Advancement of Science, v. 1843, p. 60–61.

Platt, J.-P., and Vissers, R. L. M., 1979, Extensional structures in anisotropic rocks: Journal of Structural Geology, v. 2, p. 397–410.

Pohn, H. A., 1981, Joint spacing as a method of locating faults: Geology, v. 9, p. 258–261.

Poldervaart, A., and Walker, K. R., 1962, The Palisade sill: International Mineralogical Association, 3rd General Congress, Washington, D.C., Northern Field Excursion Guidebook, p. 5–7.

Pollard, D. D., and Holzhausen, G., 1979, On the mechanical interaction between a fluid-filled fracture and the Earth's surface: Tectonophysics, v. 53, p. 27–57.

Pollard, D. D., and Müller, O. H., 1976, The effects of gradients in regional stress and magma pressure on the form of sheet intrusions in cross section: Journal of Geophysical Research, v. 81, p. 975–984.

Powell, C. McA., 1974, Timing of slaty cleavage during folding of Precambrian rocks, northwest Tasmania: Geological Society of America Bulletin, v. 85, p. 1043–1060.

Powell, C. McA., 1979, A morphological classification of rock cleavage: Tectonophysics, v. 58, p. 21–34.

Price, N. J., 1977, Aspects of gravity tectonics, and the development of listric faults: Journal of the Geological Society of London, v. 133, p. 311–325.

Price, R. A., 1973, Large-scale gravitational flow of supracrustal rocks, southern Canadian Rockies, *in* DeJong, K. A., and Scholten, R., eds., Gravity and Tectonics: New York, John Wiley & Sons, p. 491–502.

Price, R. A., 1981, The Cordilleran fold and thrust belt in the southern Canadian Rocky Mountains, *in* McClay, K. R., and Price, N. J., Thrust and nappe tectonics: Geological Society of London Special Publication 9, p. 427–448.

Price, R. A., and Mountjoy, E. W., 1970, Geologic structure of the Canadian Rocky Mountains between Bow and Athabaska Rivers—A progress report: Geological Association of Canada Special Paper 6, p. 7–25.

Proffett, J. M., Jr., 1977, Cenozoic geology of the Yerington district, Nevada, and implications for the nature and origin of Basin and Range faulting: Geological Society of America Bulletin, v. 88, p. 247–266.

Raleigh, C. B., 1965, Glide mechanisms in experimentally deformed minerals: Science, v. 150, p. 739–741.

Ramberg, H., 1960, Relationships between length of arc and thickness of ptygmatically folded veins: American Journal of Science, v. 258, p. 36–46.

Ramberg, H., 1961, Relationship between concentric longitudinal strain and concentric shearing strain during folding of homogeneous sheets of rocks: American Journal of Science, v. 259, p. 382–390.

Ramberg, H., 1963, Evolution of drag folds: Geological Magazine, v. 100, p. 97–106.

Ramberg, H., 1967, Gravity, deformation and the Earth's crust as studied by centrifuged models: Academic Press, New York, 241 p.

Ramsay, D. M., and Sturt, B. A., 1973a, An analysis of noncylindrical and incongruous fold pattern from the Eo-Cambrian rocks of Söröy, northern Norway, Part I.

Noncylindrical, incongruous and aberrant folding: Tectonophysics, v. 18, p. 81–107.

Ramsay, D. M., and Sturt, B. A., 1973b, An analysis of noncyclindrical and incongruous fold pattern from the Eo-Cambrian rocks of Söröy, northern Norway, Part II. The significance of synfold stretching lineation in the evolution of noncylindrical folds: Tectonophysics, v. 18, p. 109–121.

Ramsay, J. G., 1958, Superimposed folding at Loch Monar, Inverness-Shire and Ross-Shire: Geological Society of London Quarterly Journal, v. 113, p. 271–308.

Ramsay, J. G., 1962, The geometry and mechanics of formation of "similar" type folds: Journal of Geology, v. 70, p. 309–327.

Ramsay, J. G., 1963, Structure and metamorphism of the Moine and Lewisian rocks of the northwest Caledonides, *in* Johnson, M. R. W., and Steward, F. H., eds., The British Caledonides: Edinburgh, Oliver and Boyd, Ltd., p. 1434–1475.

Ramsay, J. G., 1967, Folding and fracturing of rocks: New York, McGraw-Hill, 568 p.

Ramsay, J. G., 1980, Shear zone geometry, a review: Journal of Structural Geology, v. 2, p. 83–99.

Ramsay, J. G., 1981, Tectonics of the Helvetic nappes, *in* McClay, K. R., and Price, N. J., eds., Thrust and nappe tectonics, Geological Society Special Publication No. 9: Oxford, Blackwell Scientific Publications, p. 293–309.

Ramsay, J. G., and Graham, R. H., 1970, Strain variation in shear belts: Canadian Journal of Earth Sciences, v. 7, p. 786–813.

Ramsay, J. G., and Huber, M. I., 1983, The techniques of modern structural geology, Vol. 1: Strain analysis: London, Academic Press, 306 p.

Ramsay, J. G., and Huber, M. I., 1987, The techniques of modern structural geology, Vol. 2: Folds and fractures: London, Academic Press, 392 p.

Ramsay, J. G., and Wood, D. S., 1973, The geometric effects of volume change during deformation processes: Tectonophysics, v. 13, p. 263–277.

Ratcliffe, N. M., 1971, Ramapo fault system in New York and adjacent northern New Jersey: A case of tectonic heredity: Geological Society of America Bulletin, v. 82, p. 125–142.

Raymond, L. A., 1975, Tectonite and mélange—A distinction: Geology, v. 3, p. 7–9.

Reed, J. C., Jr., and Bryant, B., 1964, Evidence for strike-slip faulting along the Brevard zone in North Carolina: Geological Society of America Bulletin, v. 75, p. 1177–1196.

Reed, J. J., 1964, Mylonites, cataclasites, and associated rocks along the Alpine Fault, South Island, New Zealand: New Zealand Journal of Geology and Geophysics, v. 7, p. 645–684.

Reineck, H.-E., and Singh, I. B., 1975, Depositional sedimentary environments: New York, Springer-Verlag.

Reks, I. J., and Gray, D. R., 1983, Strain patterns and shortening in a folded thrust sheet: An example from the southern Appalachians: Tectonophysics, v. 93, p. 99–128.

Rich, J. L., 1934, Mechanics of low-angle overthrust faulting as illustrated by Cumberland thrust block, Virginia, Kentucky and Tennessee: American Association of Petroleum Geologists Bulletin, v. 18, p. 1584–1596.

Rickard, M. J., 1971, A classification diagram for fold orientations: Geological Magazine, v. 108, p. 23–26.

Riedel, W., 1929, Zur Mechanik geologischer Brucherscheinungen: Zentralblat fur Mineralogie, Geologie und Paleontologie, v. B, p. 354–368.

Rikitaki, T., 1966, Electromagnetism and the earth's interior: New York, Elsevier Publishing Company, 308 p.

Roberts, D., and Strömgård, K.-E., 1972, A comparison of natural and experimental strain patterns around fold hinge zones: Tectonophysics, v. 14, p. 105–120.

Rodgers, J., 1949, Evolution of thought on structure of middle and southern Appalachians: American Association of Petroleum Geologists, v. 33, p. 1643–1654.

Rodgers, J., 1964, Basement and no-basement hypotheses in the Jura and the Appalachian Valley and Ridge, *in* Lowry, W. D., ed., Tectonics of the southern Appalachians: Virginia Polytechnic Institute Department of Geological Sciences Memoir 1, p. 71–80.

Rogers, H. D., 1856, On the laws of structure of the more disturbed zones of the Earth's crust: Transactions of the Royal Society of Edinburgh, v. 21, p. 431–472.

Rubey, W. W., and Hubbert, M. K, 1959, Role of fluid pressure in mechanics of overthrust faulting, Part II: Overthrust belt in geosynclinal area of western Wyoming in light of fluid-pressure hypothesis: Geological Society of America Bulletin, v. 70, p. 167–206.

Rutter, E. H., 1976, The kinetics of rock deformation by pressure solution: Philosophical Transactions of the Royal Society, v. A283, p. 203–219.

Rutter, E. H., 1983, Pressure solution in nature, theory and experiment: Journal of Geological Society of London, v. 140, p. 725–740.

Rutter, E. H., 1986, On the nomenclature of mode of failure transitions in rocks: Tectonophysics, v. 122, p. 381–387.

Safford, J. M., 1856, A geological reconnaissance of the state of Tennessee: Nashville, Mercer, 164 p.

Sander, B., 1911, Über Zusammenhange Zwischen Teilbewegung und Gefüge in Gesteinen: Tschermaks Mineralogie und Petrographie Mittelungen, v. 30, p. 381–384.

Sander, B., 1930, Gefügekunde der Gesteine: Vienna, Springer (English translation: An introduction to the study of fabrics of geological bodies: Oxford, Pergamon Press, 641 p.).

Secor, D. T., 1965, Role of fluid pressure in jointing: American Journal of Science, v. 263, p. 633–646.

Sedgwick, A., 1835, Remarks on the structure of large mineral masses, and especially on the chemical changes produced in the aggregation of stratified rocks during different periods after their deposition: Transactions of the Geological Society of London, Series 2, v. 3, p. 461–486.

Segall, P., and Pollard, D. D., 1983, Joint formation in granitic rock of the Sierra Nevada: Geological Society of America Bulletin, v. 94, p. 563–575.

Seifert, K. W., 1965, Deformation bands in albite: American Mineralogist, v. 50, p. 1469–1472.

Sharpe, D., 1847, On slaty cleavage: Quarterly Journal of the Geological Society of London, v. 3, p. 74–105.

Sharpe, D., 1849, On slaty cleavage: Quarterly Journal of the Geological Society of London, v. 5, p. 111–129.

Sherwin, J. A., and Chapple, W. M., 1968, Wavelengths of single layer folds: A comparison between theory and observation: American Journal of Science, v. 266, p. 167–179.

Sibson, R. H., 1977, Fault rocks and fault mechanisms: Journal of the Geological Society of London, v. 133, p. 191–213.

Sibson, R. H., 1983, Continental fault structure and the shallow earthquake source: Journal of the Geological Society of London, v. 140, p. 741–767.

Siddans, A. W. B., 1972, Slaty cleavage, a review of research since 1815: Earth Science Reviews, v. 8, p. 205–232.

Sieh, K. E., and Jahns, R. H., 1984, Holocene activity of the San Andreas fault at Wallace Creek, California: Geological Society of America Bulletin, v. 95, p. 883–896.

Silver, L. T., and Schultz, P. H., eds., 1982, Geological implications of impacts of large asteroids and comets on the Earth: Geological Society of America Special Paper 190, 528 p.

Simpson, C., 1986, Determination of movement sense in mylonites: Journal of Geological Education, v. 34, p. 246–261.

Simpson, C., and Schmid, S. M., 1983, An evaluation of criteria to deduce the sense of movement in sheared rocks: Geological Society of America Bulletin, v. 94, p. 1281–1288.

Sorby, H. C., 1853, On the origin of slaty cleavage: New Philosophical Journal of Edinburgh, v. 55, p. 137–148.

Sorby, H. C., 1856, On slaty cleavage, as exhibited in the Devonian limestones of Devonshire: Philosophical Magazine, v. 11, p. 20–37.

Sorby, H. C., 1863, On the direct correlation of mechanical and chemical forces: Proceedings of the Royal Society, v. 12, p. 538–550.

Stockmal, G. S., 1983, Modeling of large-scale accretionary wedge deformation: Journal of Geophysical Research, v. 88, p. 8271–8288.

Stone, B. D., 1976, Analysis of slump slip lines and deformation fabrics in slumped Pleistocene lake beds: Journal of Sedimentary Petrology, v. 46, p. 313–325.

Stone, B. D., and Koteff, C., 1979, A late Wisconsin ice readvance near Manchester, New Hampshire: American Journal of Science, v. 279, p. 590–601.

Suppe, J., 1981, Mechanics of mountain building and metamorphism in Taiwan: Geological Society of China Memoir 4, p. 67–89.

Suppe, J., 1983, Geometry and kinematics of fault-bend folding: American Journal of Science, v. 283, p. 648–721.

Suppe, J., 1985, Principles of structural geology: Englewood Cliffs, New Jersey, Prentice-Hall, 537 p.

Swanson, M. T., 1982, Preliminary model for an early transform history in central Atlantic rifting: Geology, v. 10, p. 317–320.

Sykes, L. R., 1967, Mechanisms of earthquakes and nature of faulting on the mid-oceanic ridges: Journal of Geophysical Research, v. 72, p. 2131–2153.

Sylvester, A. G., 1988, Strike-slip faults: Geological Society of America Bulletin, v. 100, p. 1666–1703.

Sylvester, A. G., and Smith, R. R., 1976, Tectonic transpression and basement-controlled deformation in San Andreas fault zone, Salton Trough, California: American Association of Petroleum Geologists Bulletin, v. 60, p. 2081–2102.

Tapponnier, P., and Molnar, P., 1977, Active faulting and Cenozoic tectonics of China: Journal of Geophysical Research, v. 82, p. 2905–2930.

Tchalenko, J. S., 1970, Similarities between shear zones of different magnitudes: Geological Society of America Bulletin, v. 81, p. 1625–1640.

Templeman-Kluitt, D. J., 1979, Transported cataclasite, ophiolite, and granodiorite in the Yukon: Evidence of arc-continent collision: Canada Geological Survey Paper 79-14, 27 p.

Terzaghi, K. V., 1923, Die Berechnung der Durchlassigkeitsziffer des Zones aus dem Verlauf der hydrodynamischen Spannungserscheinungen: Sitzungisbericht der Akademie der Wissenschoftes Wien, v. 132, p. 105.

Teufel, L. W., and Logan, J. M., 1978, Effect of shortening rate on the real area of contact and temperatures generated during frictional sliding: Pure and Applied Geophysics, v. 116, p. 840–865.

Thiessen, R. L., and Means, W. D., 1980, Classification of fold interference patterns: A reexamination: Journal of Structural Geology, v. 2, p. 311–316.

Tobisch, O. T., 1966, Large-scale basin-and-dome pattern resulting from the interference of major folds: Geological Society of America Bulletin, v. 77, p. 393–408.

Tobisch, O. T., Fleuty, M. J., Merh, S. S., Mukhopadhyay, D., and Ramsay, J. G., 1970, Deformational and metamorphic history of Moinian and Lewisian rocks between Strathconon and Glen Affric: Scottish Journal of Geology, v. 6, p. 243–265.

Torneböhm, A. E., 1872, En geognostisk profil över den skandinaviska fjällryggen mellan Östersund och Levanger: Sveriges Geologiska Undersökning, v. 6, 24 p.

Treagus, S. H., 1983, A theory of finite strain variation through contrasting layers, and its bearing on cleavage refraction: Journal of Structural Geology, v. 5, p. 351–368.

Treagus, S. H., 1988, A history of cleavage and folding: An example from the Goldenville Formation, Nova Scotia: Discussion: Geological Society of America Bulletin, v. 100, p. 152–154.

Tullis, J., Christie, J. M., and Griggs, D. T., 1973, Microstructures and preferred orientations of experimentally deformed quartzites: Geological Society of America Bulletin, v. 84, p. 297–314.

Tullis, J., and Yund, R. A., 1985, Dynamic recrystallization of feldspar: A mechanism for ductile shear zone formation: Geology, v. 13, p. 238–241.

Tullis, T. E., 1971, Experimental development of preferred orientation of mica during recrystallization [Ph.D. thesis]: Los Angeles, University of California, 262 p.

Turcotte, D. L., and Schubert, G., 1982, Geodynamics: Applications of continuum physics to geological problems: New York, John Wiley & Sons, Inc., 450 p.

Turner, F. J., 1948, Mineralogical and structural evolution of the metamorphic rocks: Geological Society of America Memoir 30, 342 p.

Turner, F. J., Griggs, D. T., and Heard, H. C., 1954, Experimental deformation of calcite crystals: Geological Society of America Bulletin, v. 65, p. 883–934.

Turner, F. J., and Weiss, L. E., 1963, Structural analysis of metamorphic tectonites: New York, McGraw-Hill, 545 p.

Tyndall, J., 1856, Observations on "The theory of the origin of slaty cleavage" by H. C. Sorby: Philosophical Magazine, v. 12, p. 129–135.

Vacquier, V., Nelson, C. S., Henderson, R. G., and Zietz, I., 1951, Interpretation of aeromagnetic maps: Geological Society of America Memoir 47, 151 p.

Van Hise, C. R., 1896, Principles of North American Precambrian geology: U.S. Geological Survey Annual Report 16th (1894–1895), Part 1, p. 571–843.

Vernon, R. H., 1988, Sequential growth of cordierite and andalusite porphyroblasts, Cooma Complex, Australia: Microstructural evidence of a prograde reaction: Journal of Metamorphic Geology, v. 6, p. 255–269.

Voll, G., 1960, New work on petrofabrics: Geological Journal, v. 2, p. 503–597.

Wager, L. R., and Deer, W. A., 1939, Geological investigations in East Greenland, Part III: The petrology of the Skaergaard intrusion, Kangerdlugsuag, East Greenland: Meddlungen Grønland, v. 105, p. 1–352.

Walker, F., 1940, Different ratios of the Palisade Diabase, New Jersey: Geological Society of America Bulletin, v. 51, p. 1059–1106.

Watkins, N. D., 1969, Non-dipole behaviour during an Upper Miocene geomagnetic polarity transition in Oregon: Geophysical Journal of the Royal Astronomical Society, v. 17, p. 121–149.

Weijermars, R., and Rondeel, H. E., 1984, Shear band foliation as an indicator of sense of shear: Field observations in central Spain: Geology, v. 12, p. 603–606.

Weller, J. M., 1947, Relations of the invertebrate paleontologist to geology: Journal of Paleontology, v. 21, p. 570–575.

Wellman, H. G., 1962, A graphic method for analysing fossil distortion caused by tectonic deformation: Geological Magazine, v. 99, p. 348–352.

Wernicke, B., 1985, Uniform-sense normal simple shear of the continental lithosphere: Canadian Journal of Earth Sciences, v. 22, p. 108–125.

Wernicke, B., and Burchfiel, B. C., 1982, Modes of extensional tectonics: Journal of Structural Geology, v. 4, p. 105–115.

White, J. C., and White, S. H., 1980, High-voltage transmission electron microscopy of naturally deformed polycrystalline dolomite: Tectonophysics, v. 66, p. 35–54.

White, S. H., Burrows, S. E., Carreras, J., Shaw, N. D., and Humphreys, F. J., 1980, On mylonites in ductile shear zones: Journal of Structural Geology, v. 2, p. 175–187.

Williams, G. D., and Chapman, T. J., 1979, The geometrical classification of noncylindrical folds: Journal of Structural Geology, v. 1, p. 181–185.

Williams, H., and Hatcher, R. D., Jr., 1983, Appalachian suspect terranes, *in* Hatcher, R. D., Jr., Williams, Harold, and Zietz, Isidore, Contributions to the tectonics and geophysics of mountain chains: Geological Society of America Memoir 158, p. 33–53.

Williams, P. F., 1970, A criticism of the use of style in the study of deformed rocks: Geological Society of America Bulletin, v. 81, p. 3283–3296.

Williams, P. F., 1972, Development of metamorphic layering and cleavage in low grade metamorphic rocks at Bermagui, Australia: American Journal of Science, v. 272, p. 1–47.

Williams, P. F., 1976, Relationships between axial-plane foliations and strain: Tectonophysics, v. 30, p. 181–196.

Willis, B., 1893, The mechanics of Appalachian structure: U.S. Geological Survey 13th Annual Report 1891–1892, Part 2, p. 212–281.

Wilson, C. W., Jr., and Stearns, R. G., 1968, Geology of the Wells Creek structure, Tennessee: Tennessee Division of Geology, Bulletin 68, 236 p.

Wilson, J. T., 1965, A new class of faults and their bearing on continental drift: Nature, v. 207, p. 343–347.

Wilson, J. T., 1966, Did the Atlantic close and then re-open?: Nature, v. 211, p. 676–681.

Wilson, J. T., 1968, Static or mobile earth: The current scientific revolution: Proceedings of the American Philosophical Society, v. 112, p. 309–320.

Wilson, M. E., 1941, Noranda District, Québec: Geological Survey of Canada Memoir 229, 162 p.

Wilson, M. E., 1962, Rouyn-Beauchastel map area, Québec: Geological Survey of Canada Memoir 315, 118p.

Wise, D. U., 1963, An outrageous hypothesis for the tectonic pattern of the North American Cordillera: Geological Society of America Bulletin, v. 74, p. 357–362.

Wise, D. U., Dunn, D. E., Engelder, J. T., Geiser, P. A., Hatcher, R. D., Jr., Kish, S. A., Odom, A. L., and Schamel, S., 1984, Fault-related rocks: Suggestions for terminology: Geology, v. 12, p. 391–394.

Wise, D. U., Funiciello, R., Maurizio, P., and Salvini, F., 1985, Topographic lineament swarms: Clues to their origin from domain analysis of Italy: Geological Society of America Bulletin, v. 96, p. 952–967.

Wise, D. U., and students, 1979, Fault, fracture and lineament data for western Massachusetts and western Connecticut: Amherst, University of Massachusetts, Department of Geology and Geography, 253 p.

Wojtal, S., 1986, Deformation within foreland thrust sheets by populations of minor faults: Journal of Structural Geology, v. 8, p. 341–360.

Wood, D. S., 1973, Patterns and magnitudes of natural strain in rocks: Philosophical Transactions of the Royal Society, Series A, v. 274, p. 373–382.

Wood, D. S., 1974, Current views of the development of slaty cleavage: Annual Review of Earth and Planetary Sciences, v. 2, p. 369–401.

Woodward, N. B., 1987, Geological applicability of critical-wedge thrust-belt models: Geological Society of America Bulletin, v. 99, p. 827–832.

Woodward, N. B., Boyer, S. E., and Suppe, J., 1985, An outline of balanced cross-sections: Knoxville, University of Tennessee, Department of Geological Sciences, Studies in Geology 11, 170 p.

Woodward, N. B., Gray, D. R., and Spears, D. B., 1986, Including strain data in balanced cross-sections: Journal of Structural Geology, v. 8, p. 313–324.

Wright, T. O., and Platt, L. B., 1982, Pressure dissolution and cleavage in the Martinsburg Shale: American Journal of Science, v. 282, p. 122–135.

Wynne-Edwards, H. R., 1963, Flow folding: American Journal of Science, v. 261, p. 793–814.

Xiao, H.-B., and Suppe, J., 1986, Role of compaction in the listric shape of growth faults: Geological Society of America Abstracts with Programs, Annual Meeting, v. 18, p. 796.

AUTHOR INDEX

Abbate, E., 41
Acker, L. L., 428
Adams, F. D., 316
Aggarwal, Y. P., 247
Allard, G. O., 38
Allmendinger, R. W., 236, 258, 457
Alsop, L. E., 185
Alvarez, Walter, 213, 385
Amontons, Guillaume, 183
Anastasio, D. J., 219
Anderson, E. M., 161, 164, 176, 177, 178
Ando, C. J., 460
Argand, Emile, 12
Armijo, R., 253
Armstrong, R. L., 231
Atwater, Tonya, 246
Avé Lallemant, H. G., 125
Aydin, A., 142, 246, 247, 249

Bailey, E. B., 201, 224
Bak, J., 243
Balk, Robert, 157
Bally, A. W., 460
Banks, C. J., 213
Barazangi, M., 458
Barton, C., 143, 151, 153
Bayly, M. B., 394
Beach, A., 153
Becker, G. F., 377
Bell, T. H., 132, 194, 381
Benioff, Hugo, 111
Berthé, D., 194
Beutner, E. C., 323, 328, 381, 384, 388, 392
Bickel, L. D., 62
Billings, M. P., 24
Biot, M. A., 321, 327
Bird, J. M., 12
Birkhead, P. K., 154
Bjornerud, M., 194
Blank, H. R., 458
Bobyarchick, A. R., 172, 408
Bonnin, Jean, 12
Borg, I. Y., 125
Borradaile, G. J., 38, 390, 392, 393, 394
Bosworth, W., 265, 270
Bouguer, Pierre, 444
Boullier, A. M., 129
Bouma, A. H., 43
Bowden, F., 185
Boyer, S. E., 202, 218, 220, 235, 236
Breiner, S., 442
Bridges, S. R., 451
Brock, W. G., 227
Brown, L. D., 458
Brun, J.-P., 269, 274
Bryant, Bruce, 172, 205, 408, 409, 410, 458, 460
Bucher, W. H., 44, 46, 230
Bullard, Edward, 437
Burchfiel, B. C., 172, 219, 251, 265
Burke, Kevin, 264
Busk, H. G., 289, 302
Butler, R. W. H., 205
Buxtorf, A., 204
Byerlee, J. D., 184, 185, 186

Cadell, H. M., 176, 201, 224, 315
Campbell, R. B., 342
Carey, S. W., 112, 353, 360
Carreras, J., 190
Carter, N. L., 124, 125
Chapman, T. J., 287, 296
Chapple, W. M., 202, 229, 321, 338
Charles, E. G., 381, 388
Choukroune, Pierre, 194, 269, 274
Christensen, N. I., 104
Christie, J. M., 124, 135, 136, 408, 410
Clark, M. M., 249
Cloos, Ernst, 80, 83, 213, 217, 257, 260, 338, 399, 407
Clough, C. T., 201
Cobbold, P. R., 253, 327, 346
Coney, P. J., 12, 265, 272

Conley, J. F., 154
Cook, F. A., 205, 236, 460
Cook, N. G. W., 61, 150, 177
Cooper, M. A., 474
Coruh, C., 460
Cosgrove, J. W., 327
Costellow, J. O., 213
Coulomb, C. A., 61, 178
Coward, M. P., 85, 86, 205
Cox, John, 16, 32, 33
Crowell, J. C., 246, 247, 251
Curie, Pierre, 437
Currie, J. B., 322, 324

Dahlen, F. A., 228, 229, 230
Dahlstrom, C. D. A., 213, 214, 236
Dallmeyer, R. D., 18
Darwin, Charles, 375
Daubrée, Auguste, 76
Davies, Brian, 253
Davies, D., 264
da Vinci, Leonardo, 183, 315
Davis, Dan, 228, 229, 230
Davis, G. A., 219, 251, 265
Davis, G. H., 233, 265
Davis, W. M., 9
Dean, S. L., 143, 151, 153
de Boer, Jelle, 154
de Charpal, O., 269
Deer, W. A., 37
Dennis, A. J., 197, 198, 390
DePaor, D. G., 55, 219
de Saint-Venant, A.-J.-C. B., 105
de Sitter, L. U., 323
de Voogd, Beatrice, 458
Dewey, J. F., 12, 264
Dibblee, T. W., Jr., 247
Dick, H. J. B., 231
Diegel, F. A., 213, 214, 323, 328
Dieterich, J. H., 338, 339, 347, 377
Dietz, R. S., 10, 45
Dillon, J. T., 288
Dixon, H. R., 170
Donath, F. A., 100, 301, 307, 323, 325
Douglas, R. J. W., 214
Drake, A. A., 300
Drake, E. L., 280
Drummond, K. M., 154
Dunnet, D., 88
Durney, D. W., 389

Eaton, G. P., 190, 265
Edelman, S. H., 172, 408
Ellen, S. D., 278, 315
Elliott, David, 202, 205, 218, 219, 220, 221, 224, 227, 229, 230, 235, 431
Elsasser, W. M., 437
Engelder, Richard, 389
Engelder, Terry, 122, 147, 148, 151, 152, 156, 157, 213, 227, 383, 384, 385, 389
Escher von der Linth, Arnold, 201
Etheridge, M. A., 125, 132, 265, 272
Evans, D. M., 188, 189

Faill, R. T., 292
Fischer, M. W., 85
Fisher, O., 377
Fleck, R. J., 219
Fleuty, M. J., 224, 285, 292, 293, 302
Flinn, Derek, 85
Freund, R., 247
Fry, Norman, 92
Fullagar, P. D., 14
Fuller, M., 437
Funicello, Renato, 147, 149

Galileo, Galilei, 444
Gansser, Augusto, 44
Garfunkel, Z., 247
Gee, D. G., 226
Geiser, Peter, 213, 384, 385, 389
Genik, G. J., 270
Gettings, M. G., 458
Gibbs, A. D., 262, 264
Gilbert, F. P., 447
Gilbert, G. K., 256
Gilbert, William, 434
Glover, Lynn, III, 338, 358
Goguel, Jean, 225
Goldstein, Arthur, 170
Gordy, P. L., 460
Graham, Rod, 199
Gray, D. R., 237, 388, 389, 393
Greenly, E., 41
Griffith, A. A., 176
Griggs, D. T., 58, 135, 136
Grocott, J., 191
Groshong, R. H., Jr., 81, 315, 381
Grown, J. A., 460
Gueguen, Y., 129
Gwinn, V. E., 233

Hafner, W., 50, 178, 225
Haimson, B. C., 62
Hall, Sir James, 315
Haller, John, 344
Hamilton, W. B., 12, 265
Hancock, P. L., 151, 153
Handin, J., 125
Hanmer, Simon, 389, 391, 393
Hansen, Edward, 232, 288, 297
Harker, Alfred, 377
Harland, W. B., 249
Harms, J. C., 27
Hatcher, R. D., Jr., 12, 132, 172, 193, 194, 198, 203, 205, 208, 223, 224, 231, 354, 356, 361, 408, 424, 428, 458, 460
Haughton, S., 377
Haxel, Gordon, 288
Healy, J. H., 458
Heard, H. C., 58, 106, 125
Heim, Albert, 80
Hempton, M. R., 249
Henderson, J. R., 287, 363, 380, 382
Henderson, M. N., 382
Hess, Harry, 10
Hickman, S., 63, 65
Hietanen, Anna, 38
Hobbs, B. E., 125, 135, 336
Hodgson, R. A., 147, 151
Hoeppener, R., 377
Hoffman, K. A., 437
Hoffman, Paul, 264
Holzhausen, G., 150
Hooke, Robert, 102
Hooker, V. E., 62
Hooper, R. J., 132, 193, 194, 198
Horne, J., 201
Horton, J. W., 14
Hoskins, L. M., 176
Hossack, J. R., 431
Howell, D. G., 12
Howell, J. V., 280
Hsü, K. J., 41, 227
Hubbert, M. K., 56, 60, 63, 182, 183, 225, 226, 229, 230, 233
Huber, M. I., 73, 88, 89, 90, 91, 92, 383, 386, 387, 410
Hudleston, P. J., 293, 296, 299, 303, 306, 321
Hull, Derek, 116, 119, 121
Hutchinson, D. R., 460
Hutton, James, 6, 34

Ingerson, E., 301, 323, 393, 413
Isacks, Bryan, 10

Jackson, M. P. A., 44, 109
Jaeger, J. C., 61, 150, 177
Jahns, R. H., 246
Jegouzo, P., 194
Jeran, P. W., 474
Joesting, H. R., 438
Johnson, A. M., 158, 278, 315, 325, 326
Johnson, M. R. W., 205, 220, 224, 408, 410
Jonas, A. I., 172, 393
Jones, D. L., 12

Kalsbeek, F., 474
Karner, G. D., 453, 460
Kehle, R. O., 63
Keisler, R., 270
Keith, Arthur, 170

Kennedy, W. Q., 240
Kerrich, Rob, 123
Keyte, W. R., 280
Kim, J. H., 85
Klitgord, K. D., 460
Knopf, E. G., 301, 323, 393, 413
Koteff, C., 109
Krogh, T. E., 13
Kulander, B. R., 143, 144, 151, 153

Lakes, Roderic, 104, 115
Lambiase, J., 270
Lapworth, Charles, 201
Larmor, Joseph, 437
Latham, J-P., 113, 114
Laubscher, H. P., 205, 226
Laugel, A., 377
LeDain, A. Y., 253
Lee, J. H., 381, 383
LeFort, J.-P., 253
LePichon, Xavier, 10, 11
Levorsen, A. I., 280
Lister, G. S., 112, 191, 196, 265, 272
Liu, Angang, 172, 408
Livingston, J. L., 172
Logan, J. M., 185
Logan, William, 280
Long, L. T., 451
Lowell, J. D., 250, 270
Lowrie, W., 213
Lundgren, Lawrence, 170
Lyell, Charles, 176

Marshak, Stephen, 93, 122, 383
Mashey, J. R., 474
Matthews, Drummond, 10, 458
Maurizio, Parotto, 147, 149
Maxwell, J. C., 112
May, P. R., 154
McBride, J. H., 453
McCoy, A. W., 280
McEachran, D. B., 93
McGeary, Susan, 264
McIntyre, D. B., 410, 420
McKenzie, D. P., 10, 12, 247, 264
McKerrow, W. S., 84
McLaren, A. C., 125
Mead, W. J., 76
Means, W. D., 125, 135, 137, 355, 400
Mellis, O., 474
Meyers, W. B., 172
Milici, R. C., 230
Miller, D. J., 34
Milton, D. J., 14
Mitra, Shankar, 204, 207
Mohr, Otto, 61, 176
Molnar, Peter, 252, 264
Monger, J. W. H., 12
Mooney, W. D., 458
Morgan, W. J., 10, 11, 12, 245, 447
Morley, C. K., 213, 230
Mountjoy, E. W., 213
Muehlberger, W. R., 43, 45
Müller, O. H., 150, 205
Murchison, R. I., 201
Murphy, D. C., 342

Navier, M., 178
Nelson, K. D., 453
Nicol, J., 201
Nicolas, A., 107, 129, 130
Nickelsen, R. P., 147
Norris, D. K., 213
Nur, A., 246, 247, 249
Nutall, D. J. H., 474

Obermeier, S. F., 16
Odé, H., 150, 321
Odom, A. L., 205
Oliver, Jack, 10, 124, 203, 381, 434, 458
Onasch, C. M., 325, 381, 388
Oriel, S. S., 220
Oxburgh, E. R., 231

Page, B. M., 247, 249
Palmer, A. R., 437
Panza, G. F., 205
Parker, R. B., 301, 307, 323, 325
Parker, R. L., 247
Parrish, D. K., 321
Passchier, C. W., 193, 198
Paterson, M. S., 325, 333, 401
Patnode, H. W., 322, 324
Peach, B. N., 201, 202, 205, 212
Pegram, W. J., 14
Peltzer, G., 253
Phillips, J., 80, 375, 377
Pittman, Walter, III, 12
Platt, J. P., 197, 390
Platt, L. B., 124, 381, 388, 389
Playfair, John, 6, 176
Pohn, H. A., 149
Poirier, J. P., 107
Poisson, Simeon-Denis, 103
Poldervaart, Arie, 38
Pollard, D. D., 142, 150
Polo, Marco, 434
Poulsen, K. H., 38
Powell, C. McA., 367, 372, 385, 390, 393, 394
Price, Neville, 259
Price, R. A., 12, 213, 226, 230, 231, 431, 460
Proffett, J. M., Jr., 265
Pumpelly, Raphael, 9, 413

Quinquis, H., 346

Ragland, P. C., 154
Raleigh, C. B., 125
Ramberg, Hans, 5, 230, 318, 321, 323, 338
Ramsay, D. M., 283, 287, 346, 349
Ramsay, J. G., 67, 73, 83, 88, 89, 90, 91, 92, 93, 190, 198, 199, 296, 299, 304, 305, 318, 323, 325, 326, 330, 351, 352, 353, 360, 375, 377, 381, 383, 386, 387, 390, 410, 431, 432
Rankin, D. W., 172
Ratcliffe, N. M., 247, 263, 264
Raymond, J. C., Jr., 41
Read, H. H., 408
Reed, J. C., Jr., 172, 205, 408, 409, 410, 458, 460
Reed, J. J., 190
Reineck, H. E., 27
Reks, J. J., 388
Reusch, Hans, 399
Rich, J. L., 204, 242
Richardson, R. M., 65
Rickard, M. J., 285
Riedel, W., 198
Riggle, F. E., 447
Rikitaki, T., 437
Roberts, D., 338, 339, 341, 344, 345
Rodgers, John, 205, 206
Roeder, D. H., 235
Roever, W. L., 321
Rogers, H. D., 375
Rondeel, H. E., 196, 198
Rubey, W. W., 182, 183, 225, 229, 230, 233
Ruskin, John, 2
Russ, D. P., 16
Rutter, E. H., 104, 108, 123, 128
Ryan, W. B. F., 12

Safford, J. M., 201
Salvini, Francesco, 147, 149
Sander, Bruno, 366, 377, 393, 413
Sanderson, D. J., 342
Schaeffer, M. F., 429
Schmid, S. M., 191, 192, 194, 414
Schubert, G., 20, 184, 318
Schultz, P. H., 45
Secor, D. T., 150, 197, 198, 219, 390
Sedgwick, Adam, 375
Segall, P., 150
Seifert, K. W., 125
Shackleton, R. M., 24, 25, 36
Sharp, R. V., 249
Sharpe, Daniel, 80, 366, 375, 377, 379
Sherwin, J. A., 321
Sibson, R. H., 134, 243
Siddans, A. W. B., 375, 377
Sieh, K. E., 16, 17, 246
Silver, L. T., 45

Simpson, Carol, 191, 192, 193, 194, 196, 198, 414
Singh, J. B., 27
Sleep, N. H., 65
Smith, C., 458
Smith, M. J., 191
Smith, R. R., 249
Snider, F. G., 154
Snoke, A. W., 191, 196, 342
Soloman, S. C., 65
Sorby, H. C., 80, 122, 375, 377
Spears, D. B., 237
Speed, R. C., 15
Stearns, R. G., 45, 47
Steno, Nicholas, 8
Stensen, Niels, 8
Stephansson, O., 321
Stern, T. W., 14
Stewart, G. A., 460
Stille, Hans, 230
Stockmal, G. S., 230
Stone, B. D., 25, 109
Strömgård, K. E., 338, 339, 341, 344, 345
Sturt, B. A., 287, 346, 349
Summers, J. M., 327
Suppe, John, 120, 177, 183, 202, 208, 228, 229, 230, 236, 259, 292, 315, 320
Sutter, J. F., 14
Swanson, M. T., 264
Sykes, L. R., 10, 165, 247, 251
Sylvester, A. G., 240, 249
Symonds, P. A., 265, 272

Talbot, C. J., 44, 109
Talwani, Pradeep, 16, 32, 33, 451
Talwant, M., 185
Tapponnier, Paul, 252, 253
Tchalendko, J. S., 198
Templeman-Kluitt, D. J., 12, 226
Terzaghi, K. V., 150
Teufel, L. W., 185
Thiessen, R. L., 355
Thomas, M. D., 460, 461
Thomas, P. R., 37
Thomas, W. A., 206
Tobisch, O. T., 338, 353, 354, 358
Torneböhn, A. E., 205
Tosdal, Richard, 288
Treagus, S. H., 373, 380, 382
Trump, R. P., 322, 324
Trümpy, Rudolf, 233
Tsukahara, H., 63, 65
Tullis, J., 130, 135, 136, 137
Turcotte, D. L., 20, 184, 318
Turner, F. J., 58, 124, 301, 377, 395, 396, 413, 419, 474
Tyndall, J., 375, 377

Vacquier, Victor, 438
Vallisnieri, Antonio, 315, 316
Van Breeman, Otto, 18
Van der Pluijm, B. A., 383
Van Hise, C. R., 377
Verson, R. H., 194
Vincelette, R. R., 219
Vine, Fred, 10
Vissers, R. L. M., 197, 390
Voll, G., 377

Wager, L. R., 37
Walker, F., 38
Walker, K. R., 38
Waterson, John, 191
Warburton, J., 213
Watkins, N. D., 437
Watts, A. B., 453, 460
Weijermars, R., 196, 198
Weiss, L. E., 301, 325, 333, 395, 396, 413, 419, 420, 474
Weller, J. M., 24
Wellman, H. G., 87
Wernicke, Brian, 265
Westbrook, G. K., 191
Wettstein, A., 80
White, I. C., 280
White, J. C., 129
White, S. H., 129, 190, 196
Williams, G. D., 287, 296
Williams, Harold, 12
Williams, P. E., 125, 135
Williams, P. F., 112, 325, 380, 390, 393, 415
Williams, R. T., 203, 205, 223, 224, 231
Willis, Bailey, 176, 202, 208, 224, 315
Willis, D. G., 60, 63
Wilson, C. W., Jr., 45, 47
Wilson, Gilbert, 287
Wilson, J. T., 12, 164, 240, 247, 249
Wilson, M. E., 38
Wintsch, R. P., 170
Wise, D. U., 9, 133, 146, 147, 149, 198, 423
Witherspoon, W. R., 235
Wojtal, Steven, 213, 216
Wollard, G. P., 438
Wood, D. S., 80, 83, 338, 375, 377, 381, 388
Woodland, B. G., 376
Woodward, N. B., 202, 236, 237
Wright, T. O., 124, 381, 382, 388, 389
Wynne-Edwards, H. R., 336, 354, 357, 359

Xiao, Hong-Bin, 259

Yund, R. A., 130, 137

Zak, I., 247
Zietz, Isidore, 205, 438, 444, 447, 460
Zoback, M. D., 62, 63, 64, 65
Zoback, M. L., 63, 64

SUBJECT INDEX

Accident, 240
Accreted terrane, 415
Accretionary tectonics, 12
Accretionary wedge, 211, 212
Active-roof duplex, 213
Age dating methods
  Argon-40/argon-39 method, 18
    release spectrum, 18
  ^{14}C (Carbon-14) age, 16
  ^{14}C (Carbon-14) dating, 16
  Rubidium-strontium method, 14
  Samarium-neodymium method, 18
  Uranium-lead method, 12
Albany, New York, 214
Alpine fault, New Zealand, 185, 190, 240, 242, 251
Alps, 12, 226, 453
Ammons Formation, North Carolina, 423
Amontons' laws, 61, 183
Amygdule, 37, 84
Anatolian fault, Turkey, 185
Anchizone, 370
Anderson's fault classification, 165
Andes, 12
Anglesey, Wales, 41
Anisotropic material, 101
Anisotropy, 2, 116, 315
Anomaly (geophysical), 438
Anticline, 282
  upward-facing, 283
Antitaxial crystal growth, 93
Antithetic normal fault, 213
Appalachians, 154, 205, 236, 460
  Blue Ridge, 232, 233
  central, 389
  foreland, 230, 233
  New England, 170
  Piedmont, 94
  Southern Appalachians, 170
  Valley and Ridge, 235, 315
Appalachian ultradeep core hole, 455
Appalachian-Variscan Mountains, 12
Apparent dip, 475
Apparent polar wandering curve, 443
Arabian shield, 253
Arnabol thrust, 205
Arvonia Slate, Virginia, 385
Asperities, 184, 185
Assynt District, Scotland, 201, 212
Assynt window, 220
Astrobleme, 269
Augen
  asymmetric, 193
Aulocogen, 264, 270
Austin Glen Graywacke, New York, 34, 41
Axial elongation, 87
Axial flattening, 87
Axial line, 281
Axial plane, 281
Axial surface, 281

Balanced cross section, 236, 431
Bartletts Ferry fault, Georgia, Alabama, 133
Basement, 202
  involvement, 208
  uplift, 231
Basin, 283
Basin and Range Province, Utah, 21, 236, 256, 259, 457
Bay of Biscay, Spain, France, 259, 269
Bear Valley strip mine, Pennsylvania, 147
Bedding, 26
  fissility, 383
  plane fault, 165
  plane, 26
  thrust, 208
Beekmantown Group, 210
Beer-can experiment, 183
Behavior of materials
  brittle, 102
  elasticoviscous, 102, 112

Behavior of materials, *continued*
 elastic-plastic, 102
 plastic (Saint Venant), 101, 105
 viscoelastic, 112
 viscoplastic, 102
 viscous, 101, 102, 105
Belt Series, Montana, 202
Bending, 318
Beta (β) diagram, 468, 471
Bilateral symmetry, 87
Binnewater Sandstone, New York, 34
BIRPS, 458
Blåhø, Norway, 288, 297, 378
Blind thrust, 219
Blocking temperature, 13, 18
Blue Ridge-Piedmont thrust sheet, 226, 455, 460
Body force, 51
Bolide, 269
Boomerang pattern, 353, 355
Boone, North Carolina, 402
Borrega Mountain, California, 247
Boudinage, 394, 397, 404
 asymmetric, 193
 chocolate-block (tablet), 404, 406
 ordinary, 404, 406
Bouguer gravity-anomaly map, 444
Bouma sequence, 43
Bouznika, Morocco, 405
Bow-and-arrow rule, 219
Box Ankle fault, Georgia, 133
Branch line, 215, 218
Brazil twin, 125
Break thrust, 208, 315
Brevard fault zone, South Carolina, North Carolina, 132, 170, 195, 408, 455
Bright spot, 458
British Caledonides, 205
Brittle deformation, 2, 80
Broken formation zone, 213, 217
Brunswick magnetic anomaly, Georgia, 454
Buckling, 320
Bulls-eye fold, 355
Buoyant effect, 225
Burgers circuit, 119
Burgers vector, 119
Burning Springs anticline, Pennsylvania, West Virginia, 401
Burrells Ford, South Carolina, 417
Busk method, 302
Byerlee's law, 185
Bygdin, Norway, 83, 404
Bygdin conglomerate, Norway, 404

Calcite strain gage, 81
CALCRUST, 458
Caledonian deformation, 36
Caledonian orogeny, 264
Cambrian slates, Wales, 375
Canadian Cordillera, 205, 213
Canadian Rockies, 161, 204, 235, 460
Canadian shield, 20
Capability of tectonic structures, 16
Cape Jermain, Canada, 363
^{14}C (Carbon-14) age, 16
^{14}C (Carbon-14) dating, 16
Cariboo Mountains, British Columbia, 342, 401
Carthage-Colton mylonite zone, 133
Cataclasis, 117, 121
Cataclasite, 116, 121, 132, 162, 168
Cataclastic flow, 121, 129
Center counter, 471
Center-to-center method, 91
Central America, 242
Central Highlands, Scotland, 37
Chattooga River, 424
Chevron fold, 287, 308
Chibougamau, Québec, 38
Chief Mountain, Montana, 201, 202
Chocolate block (tablet) boudinage, 404, 406
Chocolate Mountains thrust, California, 288
Chunky Gal Mountain, North Carolina, 399
Cleavage
 continuous, 367
 crenulation, 367, 370, 389
  discrete, 389
  extensional, 196, 390
  normal-slip, 197
  reverse-sense (slip), 197, 390
  zonal, 389
 disjunctive, 367
 fan, 390
 fracture, 373
 refracting, 381
 refraction, 373
 slaty, 366, 369
 spaced, 367
 stages in progressive deformation, 383, 386
 transected fold, 390, 394
Closed fold, 293
Clouds Creek Granite, South Carolina, 195
Coble creep, 128
COCORP, 236, 457, 458, 461
Code of Federal Regulations, 16
Coefficient of internal friction, 178, 229
Coesite, 43
Collapse structure, 269
Columnar joint, 158, 159
Competent, 208
Complete Bouguer anomaly map, 450
Complex folds
 formation, 357
 mechanical implications, 357
Compositional banding, 36
Compressional stress, 52
Compressive strength, 59
Concentric fault, 165
Concentric fold, 285
Concentric (tangential)-longitudinal strain, 323
Concordant zircon, 13
Concordia curve, 13
Conical fold, 285
Conjugate fracture, 144
Conjugate shear angle, 59
Conococheague Limestone, Maryland, 83
Contact metamorphic zone (aureole), 37
Continuous creep, 184
Contouring point, 473
Coulomb criterion of failure, 61
Coulomb-Mohr failure envelope, 178
Coulomb wedge, 228, 229
Coyote Creek fault, California, 247
Crack-seal, 98
Cranberry Gneiss, North Carolina, 402
Craton, 318
Creep mechanism, 126
Creep process, 126
Critical taper, 229
Cross bed
 normal, 27
 planar (tabular), 27
 torrential, 27
 trough-festoon, 27
Crystal lattice, 116
Crystalline solid, 116
Crystalline thrust, 205, 221
C-surface, 194, 418
Cumberland, Maryland, 375, 385
Cumberland Head Limestone, Vermont, 370
Curie temperature, 437
Cylindrical fold, 283, 287

Dalradian Group, Scottish Highlands, 420
Dauphiné twin, 125
Davidson Formation, Newfoundland, 87
Dead Sea, 249
Dead Sea fault, 251
Décollement (detachment), 208
DECORP, 458
Deep-focus earthquakes, 111, 112

Defect, 117
  interstitial, 118
  point, 118
  substitution defect, 118
Deformation map, 126
Deformation mechanisms
  grain-boundary diffusion creep, 128
  grain-boundary sliding, 117, 129
  translation gliding, 124, 127
  twin gliding, 125, 127
  volume diffusion creep, 129
Deformation path, 68, 106
Delaware Water Gap, New Jersey, 98, 125, 126, 375, 385
Delta (δ) porphyroclast, 403
Delta structure, 231
Depocenter, 259
Deviatoric stress, 53
Dewatering structure, 31
Dextral strike-slip fault, 164, 166, 240
Diabase dike, 154
Diablerets nappe, 431
Diamictite, 41
Diapir, 43
Differential stress, 53
Differentiated layering, 373
Dilation, 70, 80
  negative, 72
  positive, 72
Dip, 475
Dip isogon, 299
  construction, 304
Dip separation, 163
Dip-slip, 162
Discordant zircon, 13
Discrete crenulation, 389
Disharmonic fold, 287
Dislocation, 117
  creep, 128
  density, 120
  edge, 118
  extra half plane, 118
  glide and climb, 120
  left-handed screw, 119
  map, 120
  negative edge, 118
  node, 119
  pileups, 121
  positive edge, 118
  right-handed screw, 119
  screw, 119
  stacking fault, 118
  tangles of, 121
Distortion, 68, 80
Doctrine of uniformitarianism, 6
Dome, 283
Dome-and-basin pattern, 353
Dominant layer, 323
Donath and Parker fold classification, 301, 307
Down-to-basin fault, 260
Downward-facing syncline, 283
Drag, 168
Drag fold, 169, 211, 258
Drape fold, 258
Draytonville metaconglomerate, South Carolina, 94
Ductile-brittle transition, 108, 203, 257
Ductile
  deformation, 2
  faulting, 162
  flow, 111
  shear zone, 192
  strain, 80
Ductility, 104
Ductility contrast, 301
Dunham Dolomite, Vermont, 187, 210
Duplex, 212
Dynamic recovery, 130

Earthquakes, 110, 252
  artificial, 188
East Coast magnetic anomaly, 454
East Pacific Ridge, 251
ECORPS, 458
Eidsvoll quarry, Norway, 28
Elastic, 101, 102
  limit, 103
  rebound, 111, 184
  shear strain, 112
Elasticas fold, 293
Elastic limit, 103
Elastic rebound, 111, 184
Elastic shear strain, 112
Electrical conductivity, 462
Electrical resistivity, 462
Elliptical marker, 88
Emergent thrust, 215, 219
En echelon faults, 165, 242, 247
Enosburg Falls, Vermont, 370
Entropy, 19
Episodic lead loss, 13
Equal-area (Schmidt) net, 465
Equilibrium, 18, 54
  Isostatic, 20
Erosional outlier, 221
Exotic terrane, 12
Extension
  asymmetric, 265
Extensional crenulation cleavage, 196, 390
Extensional fault, 213, 217, 256

Fabric, 366, 414
Fabric axes
  geometric, 419
  kinematic, 422
Fabric element, 400
Facing direction, 26
Failed arm, 264
Failed rift, 264, 270
Fairweather fault, Alaska, 242
Fault, 161, 249
  anatomy, 162
  Andersonian classification, 164
  antithetic, 257
  broken formation zone, 213, 217
  concentric, 165
  contractional, 213, 217
  dip-slip, 162
  displacement, 68, 163
  down-to-basin, 260
  en echelon, 165, 242, 247
  extensional, 213, 217, 256
  footwall, 162
  geometry, 247
  gravity, 256
  growth, 256, 259
  hanging wall, 162
  heave, 163
  horse-tail structure, 247
  listric, 256
  net separation, 163
  net slip, 163
  oblique-slip, 163
  plane, 162, 165
  radial, 165
  rake of net slip, 163
  reverse, 164
  separation
    dip, 163
    oblique, 163
    strike, 163
  slip, 162
  splay, 219, 257
  strike separation, 163
  synthetic, 257
  synthetic normal, 213
  throw, 163
Fault-bend fold, 210, 287, 315, 320
Faulting
  criteria for, 165
Fault-line scarp, 169
  obsequent, 173, 174
  resequent, 174
Fault mechanisms
  frictional sliding, 183
Fault-propagation fold, 208, 210, 287, 315
Fault scarp, 169, 173, 174
Fenster, 220
Finite-element method, 316
Finite-element studies, 237
Flexural-flow fold, 309
Flexural fold, 301
Flexural slip, 317
Flinn diagram, 85, 86, 97, 387, 388
Flower structure, 247, 250
Fluctuation, 89
Fluidity, 105
Fluid pressure, 151

Flynn Creek, Tennessee, 269
Fold
  aberrant, 283, 346
  amplitude, 281, 438
  anatomy, 279
  anticline, 282
  antiform, 282
  antiformal, 283
  asymmetric, 281
  axial line, 281
  axial plane, 281
  axial surface, 281
  axis, 281
  basin, 283
  bulls-eye, 355
  chevron, 287, 308
  classifications, 292
    Donath and Parker, 301
    Fleuty, 293
    Hudleston, 296
    Ramsay, 296, 301, 304, 305
  closed, 293
  concentric, 285
  conical, 285
  crest, 279
  cylindrical, 283, 287
  disharmonic, 287
  dome, 283
  downward-facing syncline, 283
  drag, 169, 211, 258
  drape, 258
  enveloping surface, 282
  eyed, 355
  first-order, 282
  flexural, 301
  flexural flow, 309
  flexural slip, 317
  footwall syncline, 208
  gentle, 293
  hanging-wall anticline, 208
  hinge, 279
  hinge line, 279
  interference, 351
  interlimb angle, 293
  intrafolial, 394
  inverted, 224
  isoclinal, 285, 293
  kink, 287, 292, 325, 333
  limb, 279
  neutral surface, 323
  noncylindrical, 285, 287, 339, 346, 355
  open, 285, 293
  overturned, 285
  parallel, 285, 298, 347
  parallel-concentric, 317
  parasitic, 288
  passive, 301
  passive-flow, 309
  passive-slip, 301, 325, 329
  plunge, 281, 477
  plunging, 281
  ptygmatic, 287, 322, 324, 368
  quasi-flexural, 301
  reclined, 285
  recumbent, 224, 285
  reverse drag, 258
  role of layers, 317
  second-order, 282
  sheath, 285, 287, 339, 355
  similar, 287, 300, 347
  similar-type, 317, 334
  superposed fold, 351
    recognition of, 355
  supratenuous fold, 287
  symmetrical, 281
  syncline, 282
  synform, 283
  synformal anticline, 283
  test (paleomagnetism), 443
  thrusts, 213
  tight, 285, 293
  transected, 390, 394
  trough, 279
  upright, 285
  upright limb, 224
  upward-facing anticline, 283
  vergence, 80, 281
  visual harmonic analysis, 296
Fold interference
  boomerang pattern, 353, 355
  dome-and-basin pattern, 353
  egg-carton pattern, 353
  fundamental types and patterns, 352
  hook fold pattern, 354, 357
Fold mechanisms
  bending, 318, 327
  buckling, 320, 327
  combined, 336
  concentric-longitudinal strain, 323
  flexural flow, 327
  flexural slip, 327
  kink, 327
  multiple, 317
  passive flow (amplification), 329
  passive-slip, 325, 329
  relationship to lineations, 346
Foliation, 367
  compositional banding, 36
  composite, 194
  differentiated layering, 373
  gneissic banding, 373
  schistosity, 373
Footwall, 162
Footwall syncline, 208
Force, 50, 51
Foreland fold-and-thrust belt, 201, 415
Foxe fold belt, Canada, 363
Fracture cleavage, 373
Fractured grains, 192
Fracture orientation, 149
Franklin, North Carolina, 397
Free-air gravity map, 444
Frictional-sliding mechanism, 183
Front Royal, Virginia, 375
Fry method, 92

Gaffney, South Carolina, 405
Garlock fault, California, 251
Gaspé Peninsula, 280
General strain, 87
Gentle fold, 293
Geochemical cycle, 20, 21
Geochronology, 12
Geophone, 454
Geopressured shale, 260
Georgian Bay, Ontario, 153
Glacier, 109
Glarus thrust, Switzerland, 201
Glen Cannich, Scotland, 353, 354
Glencoul thrust, Scotland, 212
Gneissic banding, 373
Goat Rock fault zone, Georgia, 133
Golden, British Columbia, 377
Goldenville Formation, Nova Scotia, 382
Graben, 164, 166, 257
Graded beds, 26
  reversed, 26
Grain-boundary diffusion creep, 128
Grain-boundary energy, 134
Grain-boundary sliding, 117, 129
Grandfather Mountain window, North Carolina, 220
Grand Tetons, Wyoming, 161
Gravitational spreading, 230
Gravity, 434
  foldbelt, 232
  gradient, 460
  meter, 443
  model, 453
Gravity anomaly, 444
  gabbro, 451
  granitic plutons, 451
  negative, 444
  positive, 444
Gravity fault, 256
Gravity versus compression, 230
Great Basin, 108
Greatest principal strain, 71
Great Glen fault, Scotland, 240, 251
Great Valley, Maryland, 83
Green Canyon, Louisiana, 434
Gregory rift, East Africa, 264, 270
Griffith cracks, 150, 176
Griffith material, 59
Growth fault, 256, 259

Gulf Coast, 227, 257, 259
Gulf of California, 251, 256
Gulf of Mexico, 434
Gulf Stream, 51

Half-graben, 257
Halifax Formation, Nova Scotia, 379
Hamburg sequence, Pennsylvania, New York, 381, 388, 403
Hamill Group, British Columbia, 310, 377
Hanging wall, 162
Hanging-wall anticline, 208
Hansen's method, 297
Hartman's rule, 178
Heave, 163
Helvetic Alps, Switzerland, 232, 431, 432
Heterogeneous domain, 417, 425
High-angle boundary, 134
High fluid pressure, 260
Highgate Springs thrust, Vermont, 187
High Shoals Granite, North Carolina, 14
Himalayas, 12, 20, 22
Hinge, 279
Hinge line, 279
Homestake shear zone, Colorado, 130, 132, 187
Homocline, 283
Homogeneous
  domain, 417, 425
  material, 100
  strain, 68, 70, 81
  stress state, 177
Hookean behavior, 102
Hooke's law, 102
Hoosick Falls, New York, 326, 378
Horse, 211, 240
Horse-tail structure, 247
Horst, 164, 166, 257
Hudson Valley, New York, 164
Hydraulic fracturing (hydrofracturing), 62, 259

Iberville Shale, Vermont, 187, 210
Illecillewaet River, British Columbia, 309
Imbricate thrust, 211
Impact structure, 43, 269
Inclusion trail, 394
Incompetent, 208
Indian Springs, Georgia, 405
Inherited zircon, 13
Inhomogeneous material, 100
Inlier, 220
Interference folds, 351
Intermediate principal strain, 72
Interstitial atom, 117
Intracontinental transform, 251
Intrafolial fold, 394
Inverted fault, 262
Inverted rift structure, 249
Isochron, 15
Isotropic material, 101
Isotropy, 2

Jacksboro fault, Tennessee, 242
Jephtha Knob, Kentucky, 43
Joint, 142
  arrest line, 151
  columnar, 158, 159
  filled, 144
  hackle mark, 151
  hydraulic joints, 152
  in plutons, 157
  nonsystematic, 143
  plumose, 144
  propagation, 151
  regional joint patterns, 145
  release, 152
  set, 143
  sheeting, 158
  system, 143
  systematic, 143
  tectonic, 152
  unfilled, 143
  unloading, 152
Jura Mountains, Switzerland, France, 204

Kapuskasing zone, Ontario, 127
Karmøy, Norway, 39, 390
Kaza Group, British Columbia, 401
Keweenawan basalt, Michigan, 438
Keystone thrust, Nevada, 225
Killarney Granite, Ontario, 153
Kinds of volume changes, 72
Kings Mountain belt, North Carolina, 94
Kink fold, 287, 292, 325, 333
Klippe, 221
Knockan Crag, Scotland, 201
Königsberger ratio, 438

Lake Champlain, New York, 210
Lake Char fault, Connecticut, 170
Lardeau Group, British Columbia, 309
Lateral ramp, 219
Law
  cross-cutting relationships, 8
  faunal succession, 8
  igneous cross-cutting relationships, 8
  original horizontality, 8
  structural relationships, 8
  superposition, 7
Least principal strain, 72
Lessor's Quarry, Vermont, 370
Lewisian Gneiss, Scotland, 35
Limestone Cove window, Tennessee, 213, 214
0.706 line, 15
Lineament, 400
  orientations, 149
Linear strain, 69, 71
Linear structure, 399
  interpretation, 410
  nonpenetrative, 401
  penetrative, 401
  pitfalls in interpretation, 408
Lineation, 399
  boudinage, 394, 397, 404
  deformed, 407
  elongation, 69
  intersection, 401, 477
  mineral, 402
  mullion, 403
  rodding (rod), 402
Linville Falls, North Carolina, 417
Listric geometry, 164
Lithosphere, 10, 111
Lithostat, 53
Little Harbor, Nova Scotia, 379
Loch Monar, Scotland, 329, 360
Low-angle boundary, 130
Low-melting-point organic compounds, 137
Low Rock Point, Lake Champlain, Vermont, 210

Magnetic anomaly map, 438
Magnetic intensity profile, 437
Magnetic model, 454
Magnetism, 434
  polarity reversal, 437
Magnetization
  induced, 438
Magnetometer, 437
Malaspina Glacier, Alaska, 317
Marcellus Shale, Maryland, 375, 385
Martinsburg Slate,
  Formation, Pennsylvania, New Jersey, 125, 126, 323, 375, 381, 385, 388
Massanutten Sandstone, Maryland, 126
Massanutten synclinorium, Maryland, 388
McKenzie model, 264
Mean ductility, 301
Mechanical twinning, 125, 128
Mélange, 41
Melville Peninsula, Canada, 363
Mesoscopic analysis, 414
Mesozoic extension, 460
Metamorphic core, 204, 415

Metamorphic differentiation, 373
Metawee Slate, New York, Vermont, 33
Meteor Crater, Arizona, 43, 269
Mica beards, 124
Microcontinent/microplate, 12
Microlithons, 369, 377, 394
Microshears, 389
Microtextural studies, 414
Mid-Atlantic Ridge, 65, 251
Milton area, North Carolina, 358
Mineral age, 15
Mineral Bluff Formation, North Carolina, 391, 401
Mississippi delta, 260
Mixing line, 15
Mode I fracture, 143
Mode II fracture, 143
Mode III fracture, 143
Modoc fault, South Carolina, 195
Models of extension, 272
Moho (Mohorovičić), 458
Mohr
  circle, 55
  circle for strain, 72
  circle for stress, 151
  construction, 55, 58
  diagram, 178
  effective normal stress, 182
  envelope, 59
  hypothesis, 61
Mohr-Coulomb criterion, 225
Moine Series, Scotland, 204, 329, 348, 360
Moine thrust, 204, 212, 224, 408
Monocline, 283
Montana Rockies, 161
Morcles nappe, 431
Mosel Valley, West Germany, 394
Motagua fault, Central America, 242
Mount Crandell, Alberta, 213, 214
Mud lump, 43
Mullion, 403
Multiple working hypotheses, 8, 10
Murphy, North Carolina, 401
Mylonite, 116, 121, 132, 162, 168

Nabarro-Herring creep, 129
Natural strain ellipsoid, 402
Negative magnetic anomaly, 438
Neutral surface, 323
New Orleans, Louisiana, 434
Newton
  law of gravitation, 443
  second law of motion, 51
  third law of motion, 54, 55
Newtonian fluid, 105
New World Island, Newfoundland, 383
Niger delta, West Africa, 259
Nolichucky Shale, Tennessee, 98, 375
Noncylindrical folds, 285, 287, 339, 346, 355
Nonpenetrative structure, 6
Nonsystematic joint, 143
Noranda, Québec, 38
Normal fault, 164, 181, 256
  antithetic, 213
  graben, 164, 166, 257
  half-graben, 257
  horst, 164, 166, 257
  Mohr-circle analysis, 182
  principal stress, 181
  strain ellipsoid, 181, 259
  synthetic, 213
Normal stress, 52
Normanskill Shale, New York, 41
North American Cordillera, 12
North Carolina Piedmont, 447
North Sea, 263

Oblate spheroid, 87
Oblique collision, 205
Ocoee Gorge, Tennessee, 381, 396
Ocoee Series, Tennessee, 337, 367
Ocotillo Badlands, California, 247
Oklahoma aulocogen, 264
Old Red Sandstone, Scotland, 264
Olistostrome, 41
Oppdal, Norway, 332
Orthonet, 55
Outrageous hypothesis, 9
Overburden pressure, 123
Overcoring, 62
Overpressured fluid, 259
Ozone, Tennessee, 340

Pacific Northwest, 447
Paleomagnetism, 441
Palisades sill, 38
Palm (-tree) structure, 247
Passive flow (amplification), 329
Passive-roof duplex, 213
Penetrative structure, 6, 366
Pennine Alps, Switzerland, 233
Perfect crystal, 117
Perfect fluid, 105
Peripheral counter, 471
Photoelastic studies, 237
Pi (π) diagram, 468, 471
Piedmont
  North Carolina, 338
  South Carolina, 451
  Virginia, 338
Pine Mountain thrust, Tennessee, Virginia, Kentucky, 204, 207, 242
Pipe Rock Sandstone, Scotland, 86, 204
Pitch (rake), 477, 478
Plastic deformation, 117
Plastic material
  ideal, 110
Plate tectonics, 5, 10
Plotting a pole, 468
Point diagram, 471
Poisson's ratio, 103
Pole, 466
Polygon, 134
Poor Mountain amphibolite, South Carolina, 334
Pore pressure, 151
Porphyroblast, 193
  asymmetric, 193
Porphyroclast, 193
  asymmetric, 193
Positive magnetic anomaly, 438
Potassium-argon method, 16
Potential field, 434
Prealps, Switzerland, 232
Precambrian shield, 243
Pressure box, 315
Pressure shadow, 192, 402
  asymmetric, 193, 194
Pressure solution, 72, 98, 117, 122, 369, 375, 379, 381, 383
  strain, 99
Primary sedimentary structures
  bedding, 26
  bedding fissility, 383
  Bouma sequence, 43
  burrow, 85
  cross stratification, 27
  current ripple marks, 29
  desiccation cracks, 28
  flute casts, 30
  fossil, 30, 83, 87
  graded bed, 26
  hummocky laminations, 27
  load cast, 30
  mud cracks, 28, 159
  normal cross bed, 27
  oöid, 83
  oscillatory ripple mark, 29
  pebble, 83
  pillow structure, 38, 84
  pisolite, 83
  planar (tabular) cross bed, 27
  rain imprints, 29
  reduction spot, 31, 82
  reversed graded bedding, 26
  ripple-cross laminations, 27
  ripple marks, 29
    current, 29
    oscillatory, 29
    translatory, 29
  scour marks, 30

Primary sedimentary structures, *continued*
sedimentary facies, 31
sole marks, 30
torrential cross bed, 27
tracks and trails, 30
trough-festoon cross bed, 27
turbidite, 41
Principal strain, 71, 72
greatest, 71
intermediate, 72
least, 72
Principal stress
intermediate, 58
maximum, 58
minimum, 58
Progressive deformation (strain), 68, 71, 81
Protolith, 5
Pulaski thrust sheet, Virginia, 388
Pull-apart basin, 247
Pumpelly's rule, 9, 279, 288
Push-up range, 247

Quadratic elongation, 70
Quartz beards, 124

Rabat, Morocco, 400
Rake (Pitch), 477, 478
Ramapo fault, New Jersey, 246, 264
Ramp, 208
Ramsay (standard) fold classification, 296, 304, 305
Rebound (glacial), 20
Recovery, 130
Recrystallization
dynamic, 130, 134, 136
static, 134, 136
Red Sea, 256, 264, 270
Reduction to pole, 438
Regional crustal extension, 264
Regional joint patterns, 145
Regional structural geology, 5
Remanent magnetism (field), 438
depositional, 443
thermal, 443
Revelstoke, British Columbia, 397, 418
Reverse drag, 169
Reverse-sense (-slip) crenulation, 197, 390
R_f/ϕ method, 88
Rheid, 112
Rhine graben, Switzerland, West Germany, 257
Rhomb graben, 269
basins, 247
fault system, 249
Rhomb horst, 247
Rhombochasm, 269
basin, 247
Ridge-push, 65
Rift basin, 455
Rift zone, 260
Rigid indenter, 252
Rigidity modulus, 103
Rincon Mountains, Arizona, 233
Rock cycle, 20, 21
Rock mechanics, 4
Rocky Mountain Arsenal, Colorado, 188
Rollover anticline, 258
Rome Formation, Tennessee, 216
Rose diagram, 146, 148
Rosetta stone, 360
Rouyn, Québec, 38
Ruby Mountains, Nevada, 342
Russell Fork fault, Virginia, 242

Saddle-reef deposit, 279
Saint Albans Bay, Vermont, 187
Saint Elias Mountains, Alaska, 317
Saint Venant (plastic) behavior, 105
Salton Sea, California, 249
Salt dome, 269
Salt structure, 43
Saltville thrust, Tennessee, Virginia, 124
Sams Creek Formation, Maryland, 84
San Andreas fault, California, 16, 17, 64, 108, 161, 185, 240, 242, 246, 251
Sandsuck Formation, Tennessee, 386
San Francisco Bay, California, 247, 249
Sårv nappe, Norway, 411
Scalar, 50
Scale, 6
macroscopic, 6, 279
mesoscopic, 6, 279
microscopic, 6, 279
Scandinavian Caledonides, 205, 226
Scatterchron, 15
Schistosity, 373
Schmidt (equal-area) net, 465
Scottish Highlands, 24, 30, 201, 353, 360, 420
Seafloor magnetic anomaly, 447
Seboomook Formation, Maine, 418
Seismic activity, 161
Seismic reflection, 204, 250, 454
data, 236
profile, 234, 258
Seismic refraction, 458
Self-exciting dynamo, 437
Selkirk Mountains, British Columbia, 310
Separation,
dip, 163
oblique, 163
strike, 163
Sequatchie anticline, Pennsylvania, Alabama, 401
Serpent Mound, Ohio, 43
Setters Quartzite, Maryland, 131
Seve thrust sheet, Sweden, 196
Shatter cone, 44, 46
Shear
angular, 70
anti-Riedel, 198, 247
displacement, 199
fractures, 143,150
modulus, 103, 112
pure, 73, 78
Riedel, 198, 247
simple, 73, 78
strain, 70, 71
strength, 61
stress, 52
thrust, 208, 315
zone, 161, 190, 243
ideal, 190
Shear-sense indicators, 191, 478
delta (δ) porphyroclast, 403
rotated crystal, 192
rotated mineral, 402
rotated porphyroclast, 192, 418
shear band, 418
shear-band foliation, 196
Shuswap complex, British Columbia, 25, 311, 335, 378, 414
Siccar Point, Scotland, 34
Sierra Nevada Mountains, California, 38
Siliceous cataclasite, 154
Silly Putty, 110
Silvretta nappe, Switzerland, 226
Skaergaard intrusion, Greenland, 37
Skolithos tube, 30, 85, 86
Slaty cleavage
early ideas, 375
mechanics, 379
Slice, 211, 240
Slickenlines, 163, 400
Slickensides, 163, 169, 400
Slide, 224
Slip line, 282
Slip system, 124
Snake-head structure, 236
Soft-sediment deformation, 315
Southern England, 187
Southern Norway, 28
South Hero Island, Vermont, 370, 385
South Mountain, Maryland, 80, 83, 131
Splay, 219, 257

S-surface, 194, 366
Stack of Glencoul, Scotland, 204
Standard state, 177
Stanley Shale, Arkansas, 319
Step-over, 247
Stereographic (Wulff) net, 465
Stick-slip mechanism, 184
Strain, 53, 68
  bulk, 99
  ellipsoid, 71, 73, 150, 259
  finite, 68, 81, 88
  hardening, 106
  incremental, 68, 81, 112
  infinitesimal, 68, 81
  inhomogeneous, 68, 70, 81
  irreversible, 101
  kinds, 80
    stretch, 69
  linear, 69, 71
  linear viscoelastic, 112
  marker, 80, 82
  partitioning, 112
  path, 388
  plane, 87
  principal axes, 71, 72
  quadratic elongation, 70
  rotation, 68
  rupture, 103
  shear, 70, 71
  softening, 106
  total, 112
Strain analysis methods (techniques), 87
  center-to-center, 91
  Fry method, 92
  R_f/ϕ, 88
  Wellman's, 87
Strength, 53
  compressive, 59
  tensile, 59
  ultimate, 103
  yield point, 103
Stress, 50
  at a point, 55
  compressional, 52
  differential, 53
  effective, 150
  effective normal, 149, 183, 226
  ellipsoid, 61
  hydrostatic stress, 52
  In situ, 62, 63
  mean, 53
  normal, 52
  on a plane, 53
  tensional, 52
  tensor, 57
Stretch, 69
Strike, 475
Strike separation, 163
Strike-slip fault, 162, 164, 181
  dextral, 164, 166, 240
  left-lateral strike-slip, 164, 166, 240
  mechanics, 246
  principal stress, 180
  properties and geometry, 240
  right-lateral strike-slip, 164, 166, 240
  sinistral, 164, 166, 240
  strain ellipsoid, 180
  tear fault, 219, 240, 242, 243
  termination of, 247, 250
  transcurrent, 240
  wrench, 240
$^{87}Sr/^{86}Sr$ (Strontium 87 - Strontium 86) initial ratio, 15
Structural analysis, 413
Structural terrace, 283
Strut, 208
Stylolite, 123, 126
Subgrain, 130
Sudbury irruptive, Ontario, 43
Superplastic flow, 129
Surface force, 52
Suspect terrane, 12
Swedish Caledonides, 192
Symmetrical extension, 265
Symmetry
  axial, 419
  fabric, 418, 421
  monoclinic, 419
  orthorhombic, 419
  spherical, 419
  triclinic, 419
Syntaxial crystal growth, 93
Syntectonic recrystallization, 134

Taconic klippes, New York, Vermont, 388
Taiwan thrust belt, 202
Tallulah Falls Formation, Georgia, 355, 361, 404
Tangential-longitudinal strain, 318, 323, 326
Tay nappe, Scotland, 24, 37
Tear fault, 219, 240, 242, 243
Tectogene, 10
Tectonic cycle, 22
Tectonic inheritance, 262
Tectonics, 5
Tectonic slide, 224, 231
Tectonite
  B-tectonite, 414
  L-tectonite, 414
  LS-tectonite, 414
  S-tectonite, 414
  SL-tectonite, 414
Tellico Sandstone, Tennessee, 319
Tensor, 50
  first-rank, 51
  second-rank, 51, 57
  zero-rank, 51
Terrain correction, 450
Tête Juene Cache, British Columbia, 332
Tethys, 8, 22
Texas Gulf Coast, 256
Thick-skinned structure, 203, 206
Thin-skinned
  fold-and-thrust belt, 460
  structure, 203, 206
  types, 231
Thor-Odin dome, British Columbia, 311, 335, 378, 414
Thrust
  antiformal stack duplex, 203, 212
    active-roof duplex, 213
    passive-roof duplex, 213
  blind, 219
  composite, 231
  crystalline, mechanics, 231
  erosion, 215
  fracture, 213
  horse, 211, 240
  imbricate, 211
  listric, 205
  ophiolite, 231
  ramp, 208
  roof, 212
  slip, 162
  strut, 208
  tear fault, 219, 240, 242, 243
  thin-skinned, 231
Thrust fault, 164, 179
  bedding, 208
  bow-and-arrow rule, 219
  branch line, 215, 218
  break thrust, 208, 315
  duplex, 212
  emergent, 215, 219
  fenster, 220
  floor, 212
  heave, 163
  hinterland-dipping duplex, 212
  in sequence thrust, 221
  klippe, 221
  lateral ramp, 219
  layer-shortening thrust, 213
  mechanics, 224
  Mohr-circle analysis, 180
  normal-sequence, 219
  out-of-sequence, 221
  paradox of, 226
  passive-roof duplex, 213
  principal stresses, 179
  propagation, 213
  room problem, 231
  slice, 211, 240
  strain ellipsoid, 179
  termination, 213, 220

Thrust fault, *continued*
throw, 163
wedge, 202
Tintina fault, British Columbia, 242
Toccoa, Georgia, 343
Torridonian Sandstone, Scotland, 35
Towaliga fault zone, Georgia, Alabama, 405
Toxaway dome, North Carolina, South Carolina, 354, 356, 361
Toxaway Gneiss, North Carolina, South Carolina, 379, 397, 417, 418
Trans-Canada Highway, British Columbia, 418
Transcurrent fault, 240
Transected fold, 390, 394
Transform fault, 12, 164, 240, 249
arc-arc, 165
ridge-arc, 165
ridge-ridge, 165, 246
Transposition, 393, 395, 396
Transpression, 249
Transtension, 249
Tresca's criterion, 107
Triangle structure, 231
Triassic-Jurassic basin, 154, 264
Triboplastic behavior, 121
Triple junction, 12
Trollheimen, Norway, 288, 378
Trondheim, Norway, 417
Tronfjell, Norway, 196
Tymochtee Dolomite, Ohio, 126
Type I S-C mylonite, 196
Type II S-C mylonite, 196

Ulven, Norway, 391
Unconformity, 32
angular, 33
disconformity, 32
nonconformity, 34
paraconformity, 33
Unconsolidated sediment, 260
Undeformed state, 67
Undulatory extinction, 130
Unit cell, 117
Upright limb, 224

Vacancy, 117
Varanger Peninsula, Norway, 341
Vector, 51
Vein, 168
Vergence, 80, 281
Vesicle, 37, 84
Vibroseis, 454
Viking graben, North Sea, 263
Viscoelastic behavior, 112
Viscoplastic behavior, 102
Viscous behavior, 101, 102, 105
Viscous material
ideal, 110
Visual harmonic analysis, 296
Volume change, 70, 72, 80
Von Mises criterion, 107
Vredefort structure, South Africa, 269

Walden Creek Group, Tennessee, 381
Walloomsac phyllite, New York, 326, 378
Watkins Glen, New York, 153
Wavelength, 281
Wave mechanics, 315
Weaverton Quartzite, West Virginia, 131
Wedge, 202
Wells Creek, Tennessee, 43, 269
Westminster, Maryland, 377
Whipple Mountains, California, 265, 273
Whitehall, New York, 278, 310, 322, 372
Whitewater Falls, North Carolina, South Carolina, 379, 397, 417, 418
Whole-rock age, 15
Wildflysch, 41
Wildhorn nappe, 431
Wilhite Formation, Tennessee, 370, 396, 397
Wills Mountain anticline, Pennsylvania, Alabama, 401
Wilson cycle, 12, 21, 22, 415
Window
eyelid, 220
simple, 220
Wissahickon Group, Maryland, 377
Womble Shale, Arkansas, 331
Woodall Shoals, South Carolina, Georgia, 360, 404, 424
fabric data, 482
Wopmay orogen, Northwest Territories, 264
Wrench fault, 240
Wulff (stereographic) net, 465
Wyoming-Montana thrust belt, 233

XY plane, 83, 85, 86, 325, 375, 379, 380, 384, 388, 390, 391, 393, 405

Yield point, 103
Younging direction, 26
Yukon-Tanana thrust sheet, Yukon, Alaska, 226
YZ plane, 85, 86, 147, 391

Conversion Factors

Convert	Symbol	To	Multiply By
ångstrom	Å	m	10^{-10}
bar	b	atm	0.98692
		dyne cm^{-2}	10^{6}
		lb in^{-2}	14.5038
		mm Hg	750.06
		MPa	10^{-1}
calorie	cal	joule	4.184
centimeter	cm	inch	0.39370
		m	10^{-2}
degree (latitude @ 40° N)		mi	69
		km	111.04
degrees	$x°$	radians	57.2958
degrees Celsius	°C	°F	9/5(°C) + 32
degrees Fahrenheit	°F	°C	5/9(°F − 32)
density (lb/in^3)	ρ	kg/m^3	2.768×10^{4}
dyne		g cm s^{-2}	1
erg		cal	2.39×10^{-8}
		dyne cm	1
		joule	10^{-7}
foot, feet	ft	m	0.3048
gal (acceleration of gravity)		cm s^{-2}	1
gallon (U.S.)	gal	liters	3.78541
gallon (Imperial)	gal	liters	4.54608
gamma	γ	gauss	10^{-5}
		tesla	10^{-9}
gauss		tesla	10^{-4}
gram	g	pound	0.0022046
inch	in.	cm	2.54
joule	J	erg	10^{7}
		cal	0.239006
kilogram	kg	g	10^{3}
		pound	2.20462
kilogram/square centimeter	kg/cm^2	Pa	9.807×10^{4}
kilometer	km	m	10^{3}
		ft	3280.84
		mile	0.621371
liter	l	cm^3	1000
		gal (U.S.)	0.26417
		in^3	61.0237
meter	m	ft	3.28084
micrometer	μ	m	10^{-6}
mile	mi	km	1.6093
newton	N	dyne	10^{5}
poise		g cm^{-1} s^{-1}	1
		kg m^{-1} s^{-1}	0.1
		Pa s	0.1
pound	lb	kg	0.453592
tesla	T	gauss	10^{4}
yard	yd	m	0.9144
year	y	s	3.1536×10^{7}